# CRC HANDBOOK OF BIOSOLAR RESOURCES

Editor-in-Chief

## Oskar R. Zaborsky, Ph.D.

## VOLUME I: BASIC PRINCIPLES

**Akira Mitsui, Ph.D.**
Professor
Division of Biology and Living Resources
School of Marine and Atmospheric Science
University of Miami
Miami, Florida

**Clanton C. Black, Jr., Ph.D.**
Professor
Boyd Graduate Studies Research Center
University of Georgia
Athens, Georgia

## VOLUME II: RESOURCE MATERIALS

**Thomas A. McClure, Ph.D.**
Senior Economist
Battelle Columbus Laboratories
Columbus, Ohio

**Edward S. Lipinsky**
Senior Research Leader
Battelle Columbus Laboratories
Columbus, Ohio

## VOLUME III: RESOURCE PRODUCTS/APPLICATIONS

**Fred Shafizadeh, Ph.D., D.Sc.**
Professor
Department of Chemistry
University of Montana
Missoula, Montana

# CRC Handbook of Biosolar Resources

### Editor-in-Chief

## Oskar R. Zaborsky
National Science Foundation
Washington, D.C.

# Volume I
# Part 2
# Basic Principles

### Editors

## Akira Mitsui
Professor
Rosenstiel School of Marine
and Atmospheric Science
University of Miami
Miami, Florida

## Clanton C. Black
Professor
Boyd Graduate Studies Research
Center
University of Georgia
Athens, Georgia

CRC Press, Inc.
Boca Raton, Florida

Library of Congress Cataloging in Publication Data
Main entry under title:

CRC handbook of biosolar resources.

Includes bibliographies and indexes.
CONTENTS: v. 1. Fundamental principles.--v. 2. Re-
source materials.
1. Biomass energy--Handbooks, manuals, etc.
2. Agriculture--Handbooks, manuals, etc. I. Zaborsky,
Oskar.
TP360.C18        333.95'3        79-27371
ISBN 0-8493-3470-5 (set)

Direct all inquiries to CRC Press, Inc., 2000 N.W. 24th Street, Boca Raton, Florida, 33431.

© 1982 by CRC Press, Inc.

International Standard Book Number 0-8493-3470-5 (set)
International Standard Book Number 0-8493-3471-3 (Volume I, Part I)
International Standard Book Number 0-8493-3472-1 (Volume I, Part II)

Library of Congress Card Number 79-27371
Printed in the United States

# HANDBOOK OF BIOSOLAR RESOURCES

## SERIES PREFACE

The purpose of this multivolume handbook is to serve as a reference source of the most pertinent information on biosolar resources ultimately intended for utilization. The term "biosolar" refers to biological systems (especially plant and microbial systems) which are directly dependent on solar energy for their life processes and growth. The series represents an integrated approach to critical information on the fundamental principles of photosynthetic organisms, the more significant biomass resources, and major products and applications. The series focuses on renewable resources for ultimate utility as energy, fuels, and other materials.

Since antiquity, biosolar resources have been essential to man's well-being and quality of life. Presently, this somewhat neglected and unappreciated resource (except for foods) is becoming a priority topic of study, discussion, and debate in many circles of society both at national and international levels. However, for the interested individual, there was not yet available a single source of information which presented the more relevant data in a convenient and understandable style. It is our hope that this three-volume series will fill this need. We also hope that the series will serve as a vehicle for organizing and disseminating additional information as it is generated.

From the outset of this endeavor, we recognized the immense diversity and complexity of biosolar resources as well as the delicate and subtle interdependency of the comprising elements. During the course of assembling the data, the generalization that the economic viability of biosolar resources is dictated by the fullest utility became even more appreciated. Regrettably we also recognized from the outset that certain difficult choices had to be made with regard to the subject matter to be included and the scope of the information presented to arrive at a manageable activity. The topics chosen were largely determined by their relevancy to biosolar resources, the availability of information, or the need to provide available data in an integrated manner suitable for use by individuals interested in biomass. A major objective was to provide data of present or potential commercial significance and not to be dominated by transitory research interests or subjective viewpoints. Yet, we equally recognized that new knowledge and directions must be found for the future and that biosolar resources represent an alternative option, but one which needs to be assessed fully and critically.

The initial series of three volumes seeks to present data at several levels for various intended users. The entire series is based on the natural flow of events from producing to using biosolar resources. At times, there may be some overlap between the volumes but we view this more as a benefit than as a detriment. The authors may have a different viewpoint or purpose for the particular resource described and individual volumes are somewhat self-consistent. We also recognized the existence of other CRC handbooks and other reference books and have avoided extensive duplication of readily available data.

The Editor-in-Chief, the Editors, and CRC Press, Inc. invite comments and criticisms from users on the selection of topics, accuracy and scope of the information presented. We particularly desire to hear about errors and superior data as they are generated.

The Editor-in-Chief expresses his gratitude to the hundreds of individuals who have contributed their time and expertise to this undertaking. In particular, I wish to thank my colleague Volume Editors and members of the Advisory Board for their outstanding assistance, constant objectivity, and critical review. I also thank the many authors for their able contributions prepared in a very short time. I give special thanks to Mrs.

Linda Pilkington for her able assistance. Lastly, I particularly wish to note the patience and understanding of my wife, Marcia Lee, during this endeavor and the moral stimulation from my children, Erik Christian and Claudia Lee.

Oskar R. Zaborsky
McLean, Virginia
August, 1980

## PREFACE: VOLUME I

Biosolar energy is a widely recognized renewable resource upon which all mankind is dependent. Volume I, Parts 1 and 2, presents the current state of knowledge on the basic biological processes for converting sunlight into chemical bond energy by photosynthetic organisms. This knowledge is collected with the hope that it will assist in more effectively utilizing photosynthetic organisms as renewable resources.

In preparing these volumes, skilled authors from around the world were asked to contribute concise articles containing up-to-date and pertinent information in several major areas. Section 1 presents the fundamentals of light absorption processes, photosynthetic pigments, and electron transport systems. Section 2 provides a description of assimilation and dissimilation of the major elements of photosynthetic organisms. Section 3 of Part 1 presents the major biosynthetic pathways essential for the production of biosolar resources. Section 4 presents a general classification of photosynthetic organisms and information on sources and collections.

Section 5 of Part 2 provides a general description of the characteristics of selected organisms. The response of photosynthetic organisms to major environmental factors is given in Section 6 along with information on photosynthetic organisms from under-exploited environments. Section 7 presents information on the primary productivity of biosolar organisms. Section 8 presents the physical resources and inputs needed for organisms to live, grow, and reproduce.

These categories were created with the intention of assisting users in finding information on related subjects. An attempt was made to deal completely with the broad subject of biosolar resources. However, restrictions on the time available for the formulation of the manuscripts and limitations on the realistic size of the volume have led us to omit some areas. The relative emphasis placed on different subjects could be debatable. For example, because of the Handbook's theme of solar energy and related resources, special emphasis has been placed on the physical, physiological, and metabolic aspects of light as an environmental resource.

Throughout these volumes, emphasis has been placed on providing information in a format which would provide understandable and useful data to persons from a wide variety of disciplines. For this reason, the contributors were asked to use tables and figures whenever possible and to provide concise statements of the current state of knowledge on their subject.

Even though an attempt was made to present information in as uniform a manner as possible, there is some variability among the presentations by each author. This is almost inevitable considering the number of authors involved. A case in point is the designation of photosynthetic bacteria and blue-green algae. Some contemporary systematists no longer use the term photosynthetic bacteria and refer to blue-green algae as cyanobacteria. However, since the former terms are still common in the literature their use has not been restricted. Another example of disparity in approach is the variability of units used in the measurement of light intensity. While current research favors the use of quantum units, a wide diversity of other units remains in use, including lux, foot-candles, and langley per minute. While the editors tried to make the units uniform, this was impossible in some cases.

In an effort such as this, editors encounter delinquent authors, and some subjects were covered by the editors and their laboratory staff. Though we planned contributions on subjects such as light and temperature responses of terrestrial photosynthetic organisms, nutrients in terrestrial environments, and other topics, the promised manuscripts were not prepared. Thus, in fairness to contributing authors and due to a time constraint, the editors hesitantly dropped these topics.

With all the inherent difficulties which go along with the formulation of a handbook aimed at broad goals, we believe the information will be a valuable aid to students, researchers, and administrators interested in the expanding field of biosolar resource utilization.

Sincerest thanks go to the skilled contributors who donated their valuable time and expertise in an effort to make this an authoritative and timely contribution to the field of biosolar resources. In addition, the volume editors would like to thank Dr. Oskar R. Zaborsky, the Editor-in-Chief, for his constant encouragement and enthusiasm during the innumerable hours spent putting these volumes together.

We also would like to express appreciation and thanks for the editorial assistance of Ms. Cecelia Langley, Ms. Gay Ingram, Ms. Tia Maria, Ms. Susana Barciela, Ms. Hedy Mattson, and Ms. Margaret Ahearn at the University of Miami.

Akira Mitsui
Clanton C. Black

## EDITOR-IN-CHIEF

**Oskar R. Zaborsky, Ph.D.**, is a Program Manager at the National Science Foundation in Washington, D.C.

Dr. Zaborsky attended the Philadelphia College of Pharmacy and Science (B.Sc. in Chemistry, 1964), the University of Chicago (Ph.D. in Chemistry, 1968), and Harvard University (postdoctoral fellow, 1968-69). Prior to his present position with the Foundation, Dr. Zaborsky was with the Corporate Research Laboratories at the Exxon Research and Engineering Company in Linden, New Jersey. At Exxon, his research dealt with the chemical modification of proteins and the immobilization of enzymes. At the Foundation, Dr. Zaborsky has been involved in programs dealing with enzyme technology, renewable resources, and substitute materials.

Dr. Zaborsky's present scientific and administrative interests include enzyme technology, biotechnology, catalysis, renewable resources, and biosolar resources. He has published numerous articles on these topics and is the author of the book, *Immobilized Enzymes,* which was the first comprehensive treatise on this subject. Dr. Zaborsky's research on immobilized enzymes at Exxon also led to patents, and he is a founding editor of the journal, *Enzyme and Microbial Technology.*

Dr. Zaborsky is a member of several professional societies, including the American Chemical Society and the American Association for the Advancement of Science.

# THE EDITORS

**Akira Mitsui, Ph.D.**, is Professor of Marine Biochemistry and Bioenergetics, Division of Biology and Living Resources, School of Marine and Atmospheric Science at the University of Miami.

Dr. Mitsui received his B.S. in Biology from the Faculty of Science at the University of Tokyo in 1953, his M.S. in 1955, and his Ph.D. in 1958 in plant physiology from the University of Tokyo. From 1958 to 1964 he was a teaching and research faculty member at the Institute of Applied Microbiology, University of Tokyo. From 1960 to 1963, on leave from the University of Tokyo, he was an exchange scientist between the U.S. and the Japanese governments at the Department of Cell Physiology, University of California, at Berkeley, California. From 1964 to 1970 he was an Associate Professor at the Department of Biochemistry, Yokahama Medical School, Yokahama, Japan. He was a visiting scientist at the Department of Plant Sciences, Indiana University at Bloomington, Indiana from 1970 to 1971, and at the Kettering Research Laboratory at Yellow Springs, Ohio, from 1971 to 1972. Since 1972 he has been a professor at the University of Miami.

Dr. Mitsui has published many papers and articles dealing with hydrogen photoproduction, nitrogen fixation, photophosphorylation, photosynthetic electron carriers, and utilization of marine photosynthetic organisms.

Throughout the years he has actively participated in many international biological solar energy conversion and hydrogen energy conferences as chairman, co-chairman, organizer, advisory member, and invited speaker. He is co-editor of several books related to the subject of solar energy bioconversion.

**Clanton C. Black, Jr., Ph.D.**, is a Professor of Biochemistry at the University of Georgia. He received a Ph.D. from the University of Florida in 1960. He then spent 2 years at Cornell University, Ithaca, as a National Institutes of Health Postdoctoral Fellow in the Biochemistry Department and 1 year as a Charles F. Kettering Foundation Fellow in Yellow Springs, Ohio. From 1963 to 1967 he was a Staff Scientist at the C. F. Kettering Research Laboratory and an Assistant Professor at Antioch College in Yellow Springs, Ohio.

Dr. Black is a Fellow of the American Association for the Advancement of Sciences. He is a member of the Research Advisory Committee of the Agency for International Development. He has served as President and in other positions in the American Society of Plant Physiologists. He is an Editor of *Plant Physiology*.

Dr. Black's research work centers on the biochemistry of photosynthetic organisms, with emphasis upon the variations and efficiencies at which higher plants assimilate their essential elements. He has lectured in many countries on these topics and published over 100 research papers on the biochemistry of photosynthesis.

CONTRIBUTORS
*Part 1*

Takashi Akazawa
Professor
School of Agriculture
Nagoya University
Chikusa, Nagoya, Japan

Louise E. Anderson
Professor
Department of Biological Sciences
University of Illinois at Chicago Circle
Chicago, Illinois

Sumio Asami
Postdoctoral Fellow
Research Institute for Biochemical
 Regulation
School of Agriculture
Nagoya University
Chikusa, Nagoya, Japan

Reinhard Bachofen
Professor
Institut für Pflanzenbiologie
Universität Zurich
Zürich, Switzerland

James A. Bassham
Senior Scientist
Lawrence Berkeley Laboratory
University of California
Berkeley, California

D. S. Bendall
Lecturer
Department of Biochemistry
University of Cambridge
Cambridge, England

Peter Böger
Professor Chair
Lehrstuhl fur Physiologie und
 Biochemie der Pflanzen
Universität Konstanz
Konstanz, West Germany

Herbert Böhme
Fakultät für Biologie
Fachbereich Biologie
Universität Konstanz
Konstanz, West Germany

Jeanette S. Brown
Research Biologist
Carnegie Institution of Washington
Stanford, California

Bob B. Buchanan
Professor
Section of Cell Physiology
Department of Plant and Soil Biology
University of California
Berkeley, California

David J. Chapman
Professor
Department of Biology
University of California
Los Angeles, California

Mitsuo Chihara
Professor
Institute of Biological Sciences
University of Tsukuba
Sakura-mura, Ibaraki-ken
Japan

Brian H. Davies
Senior Lecturer
Department of Biochemistry and
 Agricultural Biochemistry
The University College of Wales
Penglais, Aberystwyth
United Kingdom

Deborah Delmer
Associate Professor
MSU-DOE Plant Research Laboratory
Michigan State University
East Lansing, Michigan

Richard A. Dilley
Professor
Department of Biological Sciences
Purdue University
West Lafayette, Indiana

Eirik O. Duerr
Post Doctoral Associate
School of Marine and Atmospheric
  Science
University of Miami
Miami, Florida

L. N. M. Duysens
Professor
Department of Biophysics
Huygens Laboratory of the State
  University
Leiden, The Netherlands

Gerald E. Edwards
Professor
Department of Botany
Washington State University
Pullman, Washington

Waldemar Eichenberger
Associate Professor
Department of Biochemistry
University of Bern
Bern, Switzerland

William Fenical
Associate Research Chemist and
  Lecturer
Institute of Marine Resources
Scripps Institution of Oceanography
La Jolla, California

W. R. Finnerty
Professor and Chairman
Department of Microbiology
University of Georgia
Athens, Georgia

Howard Gest
Distinguished Professor
Photosynthetic Bacteria Group
Department of Biology
Indiana University
Bloomington, Indiana

Douglas Graham
Senior Principal Research Scientist
Leader, Plant Physiology Group
CSIRO Division of Food Research
North Ryde, N.S.W., Australia

Jürg R. Gysi
Postdoctoral Fellow
Department of Biology
University of California
Los Angeles, California

M. D. Hatch
Chief Research Scientist
Division of Plant Industry
Commonwealth Scientific and
  Industrial Research Organization
Canberra City, Australia

Takayoshi Higuchi
Professor and Director
Wood Research Institute
Kyoto University
Uji, Kyoto, Japan

Takekazu Horio
Professor
Division of Enzymology
Institute for Protein Research
Osaka University
Osaka, Japan

Johannes F. Imhoff
Department of Microbiology
University of Bonn
Bonn, West Germany

S. Izawa
Professor
Department of Biological Sciences
Wayne State University
Detroit, Michigan

K. W. Joy
Professor
Department of Biology
Carleton University
Ottawa, Canada

Tomisaburo Kakuno
Instructor
Division of Enzymology
Institute for Protein Research
Osaka University
Osaka, Japan

Sakae Katoh
Professor
Department of Pure and Applied
  Sciences
College of General Education
University of Tokyo
Tokyo, Japan

Bacon Ke
Senior Investigator
Charles F. Kettering Research
  Laboratory
Yellow Springs, Ohio

Donald L. Keister
Senior Investigator
Charles F. Kettering Research
  Laboratory
Yellow Springs, Ohio

S. B. Ku
Plant Physiologist
Department of Biochemistry
University of Georgia and
R. B. Russell Agricultural Research
  Center
USDA/SEA
Athens, Georgia

Shuzo Kumazawa
Post Doctoral Associate
School of Marine and Atmospheric
  Science
University of Miami
Miami, Florida

H. K. Lichtenthaler
Professor
Botanical Institute
University of Karlsruhe
Karlsruhe, West Germany

George H. Lorimer
Research Scientist
Central Research and Development
  Department
E. I. DuPont de Nemours and
  Company Experimental Station
Wilmington, Delaware

Michael Madigan
Assistant Professor
Department of Microbiology
Southern Illinois University
Carbondale, Illinois

Kazutosi Nisizawa
Professor
Department of Fisheries
College of Agriculture and Veterinary
  Medicine
Nihon University
Tokyo, Japan

Masayuki Ohmori
Assistant Professor
Ocean Research Institute
University of Tokyo
Tokyo, Japan

Glenn W. Patterson
Professor and Chairman
Department of Botany
University of Maryland
College Park, Maryland

Norbert Pfennig
Professor
Fakultät für Biologie
Universität Konstanz
Konstanz, West Germany

P. Pohl
Professor
Institut für Pharmazeutische Biologie
Universität Kiel
Kiel, West Germany

Jack Preiss
Department of Biochemistry and
  Biophysics
University of California
Davis, California

Karin Schmidt
Institut für Mikrobiologie der
  Universität
Göttingen, West Germany

Karel R. Schubert
Associate Professor
Department of Biochemistry
Michigan State University
East Lansing, Michigan

David S. Seigler
Professor
Department of Botany
University of Illinois
Urbana, Illinois

Horst Senger
Professor
Fachbereich Biologie der Philipps
  Universitatät
Marburg, West Germany

Masateru Shin
Associate Professor
Department of Biology, Faculty of
  Science
Kobe University
Kobe, Japan

Paul C. Silva
Research Botanist
Department of Botany
University of California
Berkeley, California

Mario Snozzi
Lecturer
Institut für Pflanzenbiologie
Universität Zürich
Zürich, Switzerland

Harry E. Sommer
Assistant Professor
School of Forest Resources
University of Georgia
Athens, Georgia

Tatsuo Sugiyama
Associate Professor
Department of Agricultural Chemistry
School of Agriculture
Nagoya University
Nagoya, Japan

Ian W. Sutherland
Reader
Department of Microbiology
University of Edinburgh
Edinburgh, Scotland

Tetsuko Takabe
Assistant Professor
Research Institute for Biochemical
  Regulation
School of Agriculture
Nagoya University
Chikusa, Nagoya, Japan

Manfred Tevini
Professor
Department of Botany
University of Karlsruhe
Karlsruhe, West Germany

John F. Thompson
Plant Physiologist
U.S. Plant, Soil and Nutrition
  Laboratory
U.S. Department of Agriculture
Ithaca, New York

Achim Trebst
Professor
Department of Biology
Ruhr-University
Bochum, West Germany

**Hans G. Trüper**
Professor
Department of Microbiology
Rheinische Friedrich-Wilhelms-
  Universitat
Bonn, West Germany

**Rienk van Grondelle**
Postdoctoral Fellow
Department of Biophysics
Huygens Laboratory of the State
  University
Leiden, The Netherlands

**George A. White**
Plant Introduction Officer
Germplasm Resources Laboratory
USDA Agricultural Research Center
Beltsville, Maryland

**David A. Young**
Assistant Professor
Department of Botany
University of Illinois
Urbana, Illinois

**Jinpei Yamashita**
Associate Professor
Division of Enzymology
Institute for Protein Research
Osaka University
Osaka, Japan

# CONTRIBUTORS
## *Part 2*

**Yusho Aruga**
Associate Professor
Laboratory of Phycology
Tokyo University of Fisheries
Tokyo, Japan

**Reinhard Bachofen**
Professor
Institut für Pflanzenbiologie
Universität Zürich
Zürich, Switzerland

**Melvin S. Brown**
Research Assistant
Division of Biology and Living
  Resources
Rosenstiel School of Marine and
  Atmospheric Science
University of Miami
Miami, Florida

**R. Harold Brown**
Professor
Department of Agronomy
University of Georgia
Athens, Georgia

**J. S. Bunt**
Director
Australian Institute of Marine Science
Townsville, Australia

**James H. Carpenter**
Professor and Chairman
Marine and Atmospheric Chemistry
Rosenstiel School of Marine and
  Atmospheric Science
University of Miami
Miami, Florida

**David Coon**
Associate Research Specialist
Marine Science Institute
University of California
Santa Barbara, California

**T. P. Croughan**
Professor
Rice Experiment Station
Louisiana State University
Crowley, Louisiana

**W. Marshall Darley**
Associate Professor
Department of Botany
University of Georgia
Athens, Georgia

**Clinton J. Dawes**
Professor
Department of Biology
University of South Florida
Tampa, Florida

**Eirik O. Duerr**
Post Doctoral Associate
School of Marine and Atmospheric
  Science
University of Miami
Miami, Florida

**Hubert Durand-Chastel**
General Manager
Texaco Company
Mexico City, Mexico

**Rana A. Fine**
Research Associate and Professor
Rosenstiel School of Marine and
  Atmospheric Science
University of Miami
Miami, Florida

**Leon A. Garrard**
Associate Professor
Department of Agronomy
University of Florida
Gainesville, Florida

**B. Clifford Gerwick**
Senior Research Biologist
Dow Chemical Company
Walnut Creek, California

Howard R. Gordon
Professor
Department of Physics
University of Miami
Coral Gables, Florida

Yoshio Hasegawa
Research Adviser
Marine Ecology Research Institute
Tokyo, Japan

Jean François Henry
Energy Planning and Design
  Corporation
Herndon, Virginia

Ulrich Horstmann
Research Scientist
Institut für Meereskunde
Universitat Kiel
Kiel, West Germany

Roland L. Hulstrom
Branch Chief
Renewable Resource Assessment
Solar Energy Research Institute
Golden, Colorado

Shun-ei Ichimura
Professor and Director
Institute of Biological Sciences
University of Tsukuba
Ibaraki, Japan

Johannes F. Imhoff
Department of Microbiology
University of Bonn
Bonn, West Germany

Akio Kamiya
Research Associate
Laboratory of Chemistry
Faculty of Pharmaceutical Sciences
Teikyo University
Kanagawa, Japan

Alfred R. Loeblich
Director
Marine Science Program
  and Associate Professor
Department of Biology
University of Houston
Houston, Texas

Jack R. Mauney
Plant Physiologist
Agricultural Research Service
U.S. Department of Agriculture
Phoenix, Arizona

Frank J. Millero
Professor
Rosenstiel School of Marine and
  Atmospheric Science
University of Miami
Miami, Florida

Akio Miura
Associate Professor
Laboratory of Algal Cultivation
Tokyo University of Fisheries
Tokyo, Japan

Shigetoh Miyachi
Professor
Institute of Applied Microbiology
University of Tokyo
Tokyo, Japan

Sigehiro Morita
Professor
Department of Environmental Science
  and Conservation
Tokyo University of Agriculture and
  Technology
Tokyo, Japan

John W. Morse
Associate Professor
Rosenstiel School of Marine and
  Atmospheric Science
University of Miami
Miami, Florida

Wheeler J. North
Professor
W. M. Keck Laboratory
California Institute of Technology
Corona del Mar, California

Edward J. Phlips
Post Doctoral Associate
Department of Biology and Living
  Resources
Rosenstiel School of Marine and
  Atmospheric Science
University of Miami
Miami, Florida

D. W. Rains
Professor
Department of Agronomy and Range
  Science and
Director
Plant Growth Laboratory
University ofCalifornia
Davis, California

Ferdinand Schanz
First Assistant
Hydrobiologcal-Limnological Station
University of Zurich
Kilchberg, Switzerland

Kurt Schneider
Institut für Pflanzenbiologie
Universität Zürich
Zürich, Switzerland

Horst Senger
Professor
Fachbereich Biologie der Universität
Lahnberge, West Germany

Ivan Show
IDS Associates
Encinitas, California

Bruce N. Smith
Professor
Department of Botany and Range
  Science
Brigham Young University
Provo, Utah

Samuel C. Snedaker
Associate Professor
Division of Biology and Living
  Resources
Rosenstiel School of Marine and
  Atmospheric Science
University of Miami
Miami, Florida

Mario Snozzi
Institut für Pflanzenbiologie
Universität Zürich
Zürich, Switzerland

Stan R. Szarek
Associate Professor
Department of Botany and
  Microbiology
Arizona State University
Tempe, Arizona

Masayuki Takahashi
Associate Professor
Institute of Biological Sciences
University of Tsukuba
Ibaraki, Japan

Howard J. Teas
Professor
Department of Biology
University of Miami
Coral Gables, Florida

Anitra Thorhaug
Professor
Department of Biology
Florida International University
Miami, Florida

Thai K. Van
Plant Physiologist
Science and Education
  Administration—Agricultural
  Research
United States Department of
  Agriculture
Fort Lauderdale, Florida

**Kurt Wälti**
Scientific Assistant
Hydrobiological Station
University of Zurich
Kilchberg, Switzerland


**William N. Wheeler**
Research Associate
Department of Biology
Simon Fraser University
Burnaby, British Columbia
Canada

**Bernt Zeitzschel**
Professor and Executive Director
Institut für Meereskunde
Universität Kiel
Kiel, West Germany

**Rod G. Zika**
Assistant Professor
Division of Marine and Atmospheric
  Chemistry
Rosenstiel School of Marine and
  Atmospheric Science
University of Miami
Miami, Florida

# TABLE OF CONTENTS

# SECTION 3: MAJOR BIOSYNTHETIC PATHWAYS

# SECTION 4: GENERAL CLASSIFICATION OF PHOTOSYNTHETIC ORGANISMS

*Part 2*

# SECTION 5: GENERAL CHARACTERISTICS OF PHOTOSYNTHETIC ORGANISMS

## SECTION 6: RESPONSE OF PHOTOSYNTHETIC ORGANISMS TO MAJOR ENVIRONMENTAL FACTORS

## SECTION 7: BIOLOGICAL RESOURCES: PRIMARY PRODUCTIVITY

# SECTION 8: PHYSICAL RESOURCES AND INPUTS

*Section 5*
*General Characteristics of Photosynthetic*
*Organisms*

# GENERAL CHARACTERISTICS OF PHOTOSYNTHETIC BACTERIA: *RHODOSPIRILLUM RUBRUM*

R. Bachofen and M. Snozzi

## MORPHOLOGY AND GROWTH*

*Rhodospirillum rubrum* is a photosynthetic bacterium of deep red color originally described as a purple bacterium.[9,10] Under optimal growth conditions, the cells are 7 to 10 $\mu$m long, have a diameter of 0.8 to 1.0 $\mu$m, and are vibriod or spiral shaped (see Figure 1). Depending on growth conditions, the size and shape may vary greatly. Multiplication occurs by binary fission. Cells are motile by means of polar flagella.[11-12] *R. rubrum* is Gram negative. The organism was originally isolated from river water and cultured with various additions of organic substrates. Its natural habitat is mud, soil, and water. *R. rubrum* can grow under various conditions: anaerobic in the light in the presence of an electron donor (phototroph), fermentative with fructose, or with an organotroph oxidative metabolism in the dark. The generation time for the wild-type strain S-1 under optimal conditions in the light is around 8 hr. A synthetic growth medium for anaerobic phototrophic growth, as well as for aerobic organotrophic growth, in the dark contains, in addition to mineral salts, ammonia, or glutamate as nitrogen source, an organic carbon source, e.g., malate, and the vitamin biotin. Stimulation of growth is often obtained by adding yeast extract. A medium for phototrophic growth is given by Ormerod,[13] and an alternative medium is available.[14] The optimal pH for growth is 6.8 to 7.0, with the optimal temperature being between 30 to 35°C. In light, growth is up to two times faster than under oxidative dark conditions. Stock cultures may be kept on slabs (growth medium plus 1.5% agar) in a refrigerator (6 to 10°C) under weak illumination. In addition to malate and glutamate, several other organic compounds may be assimilated, such as fatty acids, intermediates of the tricarboxylic acid cycle, ethanol, several amino acids, or fructose. No growth occurs with citrate, tartrate, gluconate, sugars other than fructose, and sugar alcohols.[15] *R. rubrum*, as well as the other members of the family Rhodospirillaceae (earlier named Athiorhodaceae or nonsulfur purple bacteria), shows a special type of bacterial photosynthesis which is described by the equation:

$$H_2A + CO_2 \rightarrow (CH_2O) + A$$

Simple organic compounds ($H_2A$) serve as the electron donor. Phototrophic growth of these organisms is strictly anaerobic. Instead of an organic electron donor, molecular hydrogen may serve as reducing power.

## INTERNAL STRUCTURE AND CHEMISTRY

Cells of *R. rubrum* contain an elaborate internal membrane system which is in continuity with the cytoplasmic membrane and is of vesicular type (see Figure 2). These internal membranes, also termed chromatophores, contain all the photoactive pigments.[16] The composition of these chromatophores is given in Table 1. Their formation, as well as the synthesis of photosynthetic pigments, is dependent on oxygen pres-

---

* More detailed information and data on photosynthetic bacteria are found in several books and review articles dealing with general aspects of bacterial photosynthesis.[1-8]

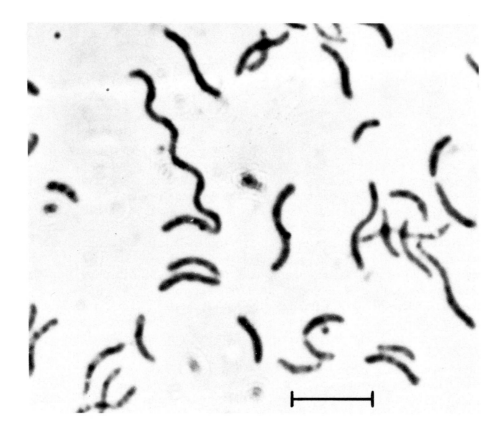

FIGURE 1. Microscopic view of cells of *R. rubrum* (bar equal to 5 μm). (Courtesy of Dr. W. Egger.)

sure and light intensity. The cells can be disrupted by sonication or by passing through a French press.

For measuring photosynthetic reactions in vitro, active membrane vesicles are isolated by standard procedures.[17,18]

The internal membrane system contains, in addition to the enzymatic system for light energy conversion, all the photosynthetic pigments, i.e., bacteriochlorophll *a*, bacteriopheophytin *a*, and several carotenoids. The chemical structure of bacteriochlorophyll *a* is given in Figure 3. The spectra of bacteriochlorophyll *a* in vivo (in chromatophores of *R. rubrum*) and after extraction in ether are shown in Figure 4. Two major absorption peaks in the near UV (350 to 400 nm) and near IR (865 nm in vivo, 780 nm after extraction) and some minor peaks at 590 nm and 800 nm (in vivo) are characteristic of the bacterial pigment.[19,20]

The synthesis of chlorophyll and carotenoids is controlled by the partial pressure of oxygen (p$O_2$), light intensity, and the reducing conditions in the medium.[15,21-27] At a p$O_2$ of more than about 10 mmHg $O_2$, practically no bacteriochlorophyll is formed. Maximum synthesis is observed at p$O_2$ of a few millimeters of mercury (equals semi-anaerobic conditions), as shown in Table 2. At low light, the bacteriochlorophyll content may reach 40 to 80 μg/mg protein; it decreases rapidly with increasing light intensity to 8 to 20 μg/mg protein at high light intensity (see Table 3). The ratio of chlorophyll to carotenoids is higher at low light conditions and with a more-reducing hydrogen donor.

The structural formula and the absorption maxima of the carotenoids occurring in *R. rubrum* are given in Table 4. The sequence of biosynthesis of the mentioned pigments is the following:[28,29]

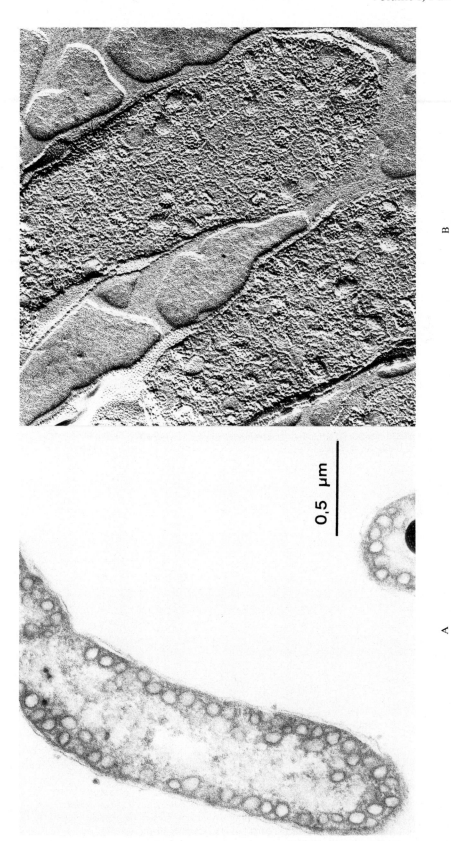

FIGURE 2.   Electron microscopic view of cells of *R. rubrum.* (A) thin section, Ba(MnO₄)₂ fixation; (B) freeze-fraction. (Bar equal to 0.5 μm in both figures). (Courtesy of Dr. R. Meyer.)

Table 1
CHEMICAL COMPOSITION OF INTERNAL MEMBRANES
( = CHROMATOPHORES)

| | Percent of dry weight | | |
| --- | --- | --- | --- |
| | *Rhodopseudomonas spheroides*[58] | *Rhodospirillum rubrum*[81] | *Rhodospirillum rubrum*[82] |
| Protein | 62 | 55 | 47 |
| Total lipids | 27 | — | 32 |
| Phospholipids | — | 16 | — |
| Bacteriochlorophyll | 7 | 3 | 5 |
| Carbohydrate | 4 | 6 | 14 |
| Nucleic acid | 0.2 | 0.4 | — |
| Carotenoids | 2 | — | — |

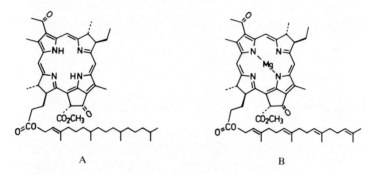

A                                                B

FIGURE 3.   Chemical structure of bacteriopheophytin *a* (A) and bacteriochlorophyll *a* (B) from *R. rubrum*.

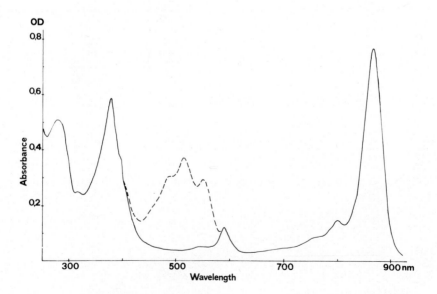

A

FIGURE 4.   Absorption spectra of pigments of *R. rubrum*. (A) absorption spectra of chromatophores from *R. rubrum*, carotenoidless strain G-9 (solid line) and wild-type strain S-1 with spirilloxanthin (dashed line). (B) absorption spectra of isolated bacteriochlorophyll *a* (dashed line) and the two possible forms of bacteriopheophytin *a* (solid line = phytyl ester, dashed line below 300 mm = geranyl-geranyl ester).[20]

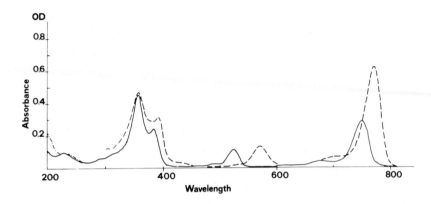

FIGURE 4B

Table 2
EFFECT OF PARTIAL
PRESSURE OF OXYGEN ON
SYNTHESIS OF
BACTERIOCHLOROPHYLL
IN THE DARK BY *R.
RUBRUM*[27]

| $pO_2$ (mmHg) | Bchl /mg protein after 6 hr of growth |
|---|---|
| 0 | 1.9 |
| 0—2 | 11.6 |
| 1—3 | 5.9 |
| 4 | 3.5 |
| 6 | 0.9 |
| 10 | 0.6 |
| 20 | 0.4 |
| Above 60 | 0 |

Table 3
EFFECT OF LIGHT
INTENSITY ON
CHLOROPHYLL CONTENT
OF ANAEROBICALLY
GROWN *R. RUBRUM*
CELLS[22]

| Light intensity (fc) | Bchl /mg protein in chromatophores ($\mu$g) |
|---|---|
| 2000 | 23.0 |
| 1000 | 43.0 |
| 200 | 58.1 |
| 100 | 77.6 |

lycopene→rhodopene→P-481→OH-P-481→monodemethylated spirilloxanthin→
spirilloxanthin

Cells of *R. rubrum* may accumulate polysaccharides, polyhydroxybutyrate, and polyphosphate as storage materials. The ratio of G and C in DNA is 63.8 to 65.8.

The protein of photosynthetic bacteria is rich in essential amino acids and vitamins,
comparable to proteins of animals or higher plants (see Table 5 and 6).[30-32] The cell
envelope consists of several layers, as in other Gram-negative bacteria, and contains
lipids, phospholipids, lipoproteins, and lipopolysaccharides in the outer membrane,
and murein in the sacculus (see Table 7).[33]

## METABOLISM RELATED TO LIGHT ENERGY TRANSDUCTION

### $CO_2$ Assimilation

For growth under phototrophic conditions, *R. rubrum*, as well as other photosynthetic bacteria, contains an efficient $CO_2$-fixation system. $CO_2$ is assimilated by the
cells through several pathways. Short-time incubation with [14]C-labeled $CO_2$ leads to
[14]C incorporation into phosphoglyceric acid, triose phosphates, sugars, intermediates

Table 4

## THE CHEMISTRY OF THE CAROTENOIDS IN ANAEROBIC-GROWN CELLS OF *R. RUBRUM*[28]

| | | Absorption maxima (nm) |
|---|---|---|
| Lycopene | | 445   470   502 |
| P-481 | | 454   481   514 |
| Rhodovibrin (hydroxy-P-481) | | 455   481   514 |
| Spirilloxanthin[a] | | 463   491   525 |
| Monomethylspirilloxanthin (hydroxyspirilloxanthin) | | 465   491   525 |
| Lycoxanthin (rhodopene) | | 445   470   501 |

[a] Accounts for 30% of the carotenoid pigments in young cells and between 90 and 95% in mature cells (after 96 to 158 hr of growth).

of the tricarboxylic acid cycle, amino acids, and other substances. These data point to the occurrence of several pathways for $CO_2$ incorporation, mainly through the reductive pentose phosphate cycle ($C_3$ cycle)[34] and the reductive tricarboxylic acid cycle.[35] Partial reactions of the latter cycle are especially important when $C_2$ compounds are present in the medium. The enzymes of the two cycles which have been detected in *R. rubrum*[34] are given in Tables 8 and 9. In addition to $CO_2$, many carbon compounds may be assimilated into cell material in a photoorganotrophic metabolism. The two cycles are driven by ATP and reduced NAD ($C_3$ cycle) or ATP, reduced NAD(P), and reduced ferredoxin. These latter compounds are formed on the bacteriochlorophyll-containing membranes in light in the presence of an appropriate H donor. With molecular hydrogen as the H donor, ferredoxin-dependent carboxylations, as well as the formation of $\beta$-hydroxybutyrate from acetate, are stimulated.

Table 5
## ESSENTIAL AMINO ACIDS IN PROTEINS FROM PHOTOSYNTHETIC BACTERIA

Percent of protein (chromatophores)

| Amino acid | *Rhodopseudomonas gelatinosa*[10] | *Rhodospirillum rubrum*[58] | *Rhodospirillum spheroides*[58] |
|---|---|---|---|
| Histidine | 3.4—3.9 | 2.0 | 2.0 |
| Isoleucine | 4.1—4.3 | 5.4 | 2.5 |
| Leucine | 7.4—7.9 | 9.6 | 10.0 |
| Lysine | 5.6—6.0 | 4.3 | 4.5 |
| Methionine | 3.0 | 2.0 | 2.7 |
| Phenylalanine | 4.3—4.6 | 6.1 | 3.7 |
| Threonine | 2.9—4.4 | 6.6 | 6.2 |
| Valine | 6.5—7.0 | 7.3 | 5.8 |

Table 6
## VITAMINS OF PHOTOSYNTHETIC BACTERIA[30]

| Vitamin | Photosynthetic bacteria ($\mu$g/100 mg dried cells) |
|---|---|
| Riboflavin ($B_2$) | 3,6 |
| Pyridoxine ($B_6$) | 3.0 |
| Folic acid | 2.0 |
| Cobalamin ($B_{12}$) | 0.2 |
| Ascorbic acid (C) | 10.0 |
| Cholecalciferol ($D_3$) | 10.0 |

Table 7
## CHEMICAL COMPOSITION OF CELL ENVELOPES[33]

| | mg/mg protein |
|---|---|
| Lipids | 0.18 |
| Phospholipids | 0.07 |
| Carbohydrates | |
| Rhamnose | 0.17 |
| Fucose | 0.16 |
| Glucose | 0.18 |
| Glucosamine | 0.049 |
| Diaminopimelic acid | 0.032 |

*Note:* Material analyzed of approximately 95% purity.

Table 8
## ACTIVITY OF C₃ CYCLE ENZYMES IN *R. RUBRUM* UNDER DIFFERENT GROWTH CONDITIONS[34]

| | $H_2$ $N_2$ phototroph | L-malate $(NH_4)_2SO_4$ phototroph | L-malate $(NH_4)_2SO_4$ dark heterotroph |
|---|---|---|---|
| Ribulose 1,5 diP carboxylase[a] | 240 | 7 | 4 |
| Ribulose 5-P kinase | 34 | 12[b] | 0[b] |
| Ribose 5-P isomerase | 40 | 51 | 66 |
| Glyceraldehyde 3-P dehydrogenase | 49 | 11 | 12 |
| Aldolase | 9.4 | 9.2 | 11 |
| Fructose 1,6-diPase (neutral) | 0 | 2 | 2 |
| Fructose 1,6-diPase (alkaline) | 27 | 1 | 1 |
| Enolase | 80 | 96 | 140 |

*Note:* Table gives $\mu$moles/min · g of protein.

[a]   Phosphoglycerate formed ($\mu$moles).
[b]   ATPase interferes.

Table 9
## ACTIVITY OF TRICARBOXYLIC ACID CYCLE ENZYMES IN *R. RUBRUM* UNDER DIFFERENT GROWTH CONDITIONS[34]

|  | $H_2$ $N_2$ phototroph | L-malate $(NH_4)_2SO_4$ phototroph | L-malate $(NH_4)_2SO_4$ dark heterotroph |
|---|---|---|---|
| Condensing enzyme | 140 | 35 | 140 |
| Aconitase | 110 | 110 | 170 |
| Isocitric dehydrogenase | 190 | 225 | 459 |
| α-Ketoglutaric dehydrogenase | 4.1 | 6.6 | 5.6 |
| Succinyl-CoA synthetase | 105 | 66 | 96 |
| Succinic dehydrogenase | 1560 | 1510 | 2220 |
| Fumarase | 580 | 510 | 800 |
| Malic dehydrogenase | 1900 | 1390 | 2350 |

*Note:* Table gives $\mu mol/min \cdot g$ of protein.

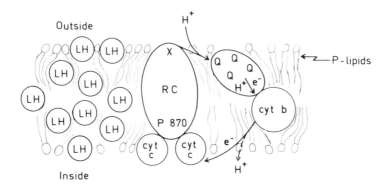

FIGURE 5.    Schematic view of the membrane of *R. rubrum* chromatophores and its components relevant to light energy conversion. RC = reaction center, P 870 = reaction center chlorophyll, LH = light-harvesting pigment protein complexes, X = primary acceptor, cyt c = cytochrome c, cyt b = cytochrome b, Q = ubiquinone.

## Synthesis of ATP and Reducing Power

While the enzymes of $CO_2$ assimilation are soluble in the cytosol, the production of ATP and reducing power occurs on the membranes of the chromatophores. A scheme for light-driven electron transport leading to a membrane potential and a proton gradient across the membrane and finally to the synthesis of ATP is shown in Figure 5.[36] Most of the individual components for the electron transport chain, cytochromes (*b* and *c* type), quinones, nonheme irons, and photo-active pigment-protein complexes have been isolated, purified, and characterized (see Table 10). ATP is synthesized on the coupling factor complex at the expense of a membrane potential and a proton gradient (= proton motive force) built up in light. Pyrophosphate is formed in the absence of ADP as the phosphate acceptor.[37,38]

The reduction of NAD by chromatophores also is a light-dependent reaction. Its mechanism in *R. rubrum* has not been elucidated fully.[39] It may be a light-driven non-cyclic electron transport from an electron donor of rather positive $E_o$ to NAD or an energy-dependent reversed electron flow (resulting from cyclic electron transport in the light) similar to the one observed in mitochondria (succinate + NAD → fumarate + $NADH_2$).

## Table 10
### PROTEINS RELEVANT TO LIGHT ENERGY CONVERSION ISOLATED FROM *R. RUBRUM*

| Name | Molecular weight (× 1000) | Characterization | Ref. |
|---|---|---|---|
| Hydrogenase | 65 | Membrane bound, thermostable, non-heme iron protein ($Fe_4S_4$), $O_2$ sensitive, synthesis only when grown in the light | 59 |
| Ferredoxin I | 9 | Soluble, nonheme iron protein, only synthesized in the light | 60,61 |
| Ferredoxin II | 14 | Soluble, nonheme iron protein, synthesized during growth in the light and in the dark, $E_m = -430$ mV | 60,61 |
| Ferredoxin III | 8.5 | Membrane-bound, nonheme iron protein, can be solubilized by detergents, very unstable, $E_m = -341$ mV | 62 |
| Ferredoxin IV | 14 | Membrane-bound, nonheme iron protein, soluble with detergents ($2 \times Fe_4S_4$), $E_m$ +355 mV and −380 mV | 62 |
| Reaction-center pigment-protein | 3 subunits, 21/24/29 | Membrane bound, soluble with detergents, contains 4 Bchl (2 as 800 nm, 2 as 865 nm; the latter bleach in the light reversibly), UQ-Fe, 2 Bpheophytin, and spirilloxanthin (missing in carotenoid-free mutants) | 63—65 |
| Light-harvesting pigment-protein complex | 11—14 | Membrane bound, soluble with detergents and in chloroform-methanol, contains bacteriochlorophyll and spirilloxanthin | 66,67 |
| Coupling factor ATPase BF$_1$ | 350 | Soluble from membrane at low ionic strength, 5 different subunits, ATP-hydrolyzing activity, after isolation cold labile, divalent cation necessary for activity | 68,69 |
| BF$_0$ | | Membrane bound, soluble with detergents, 8 subunits (?), 1 at least chloroform-soluble; oligomycin and DCCD binding, acts as proton channel | 70,71 |
| Ribulosebisphosphat-carboxylase | 120 | Activity controlled by metabolites (citrate, phosphate, 3-phosphoglyceric acid), contrary to enzymes of other sources only 2 subunits | 72—74 |
| Cytochromes | | | |
| b | 450 | Spectrum reduced: 557.5/527/425 nm, $E_m = -0.204$ V, n = 1, $pH_{iso} = 4.2$ | 75,76 |
| c$_2$ | 12.8 | Spectrum reduced: 550/521/415 nm, $E_m = +0.32$ V, n = 1, $pH_{iso} = 6.0$ | |
| cc' | | Spectrum reduced: 550/—/423 nm, $E_m = 0.01$ V, n = 1, $pH_{iso} = 5.6$, 2 heme/molecule | |
| Nitrogenase | | Consists of 3 components: | |

## Table 10 (continued)
## PROTEINS RELEVANT TO LIGHT ENERGY CONVERSION ISOLATED FROM *R. RUBRUM*

| Name | Molecular weight (× 1000) | Characterization | Ref. |
|---|---|---|---|
| MoFe protein | 230 | 4 subunits, 2 Mo, 20 Fe, similar to other known MoFe proteins | 77—80 |
| Fe protein | 30 and 31.5 | 2 subunits, 4 Fe, similar to other Fe proteins in molecular weight metal composition and e.p.r. signal, but contains bound phosphate, ribose and adenine | 80 |
| Activating factor | — | Needed for activation after isolation of the nitrogenase [possibly because of the content of phosphate, ribose and adenine(?)] | 78 |

### Other Light-Dependent Reactions

A similar "uphill" electron transport operates in light-dependent $N_2$ fixation and $H_2$ evolution. While hydrogen evolution in the dark (e.g., from pyruvate) seems to be coupled to ferredoxin and hydrogenase, hydrogen evolution in the light, as well as nitrogen fixation, is catalyzed by nitrogenase, a system where ATP from photophosphorylation is necessary for activity. Both processes may be important in a concept of utilization of organisms for solar energy conversion. However, very little is known about the actual mechanisms involved in the energy coupling in these pathways.[40,41]

### Action Spectra and Quantum Efficiency

As seen from the absorption spectra of the chromatophores, *R. rubrum,* like other photosynthetic bacteria, absorbs light in a very broad band ranging from the UV (350 nm) to the IR (900 nm). Light energy absorbed by bacteriochlorophyll, as well as the accessory carotenoids, is converted into chemical energy with high efficiency.

Data available on the quantum yield for $CO_2$ assimilation of whole cells vary greatly, ranging from 1 to 12 quanta/$CO_2$ assimilated.[42,43] However, most authors agree on a value of 4 to 5 quanta/$CO_2$ assimilated.

## MUTANTS

Several mutants of *R. rubrum* have been produced. While some cells show a change in shape which may not be relevant for photosynthesis (see Table 11A), others have mutations in the pigment composition or the membrane structure and clearly are defective in the process of light energy transduction (see Table 11B).[44-57]

## Table 11
### DESCRIBED MUTANTS OF *R. RUBRUM*[17]

#### A. Changes in shape

| Strains | Observations | Ref. |
|---|---|---|
| S4, S10, S7F1, SF21 | Bacilliform structure (metabolism of D-alanine impaired) | 45,46 |
| V 1, V 3 | Vibrioid-like structure | 47 |

#### B. Changes in metabolism

| Strains | Formation of | | | Excretion of pigments | Growth phototr. | Electr. transp. | Other | Ref. |
|---|---|---|---|---|---|---|---|---|
| | Bacterio-chlorophyll | Carotine | Internal membranes | | | | | |
| C | + | + | Altered | | + | Changed | | 48 |
| G₁ | − | | Altered | | − | | Light sensitive | 49,54 |
| F 1, F 3, F 6, F 10, F 14, M 46 | + | + | + | Precursors of bacterio-chlorophyll | + | | | 49 |
| F 4, F 5, F 8 | − | − | | Precursors of bacterio-chlorophyll | − | | | 49 |
| F 9 | − | | | Precursors of bacterio-chlorophyll | − | | | 50 |
| F 12, F 13 | − | | | Precursors of bacterio-chlorophyll | − | | | 49 |
| M₂B | + | − | + | | + | | | 51 |
| BG-1 | + | − | + | | + | | | 52 |
| G-9 | + | − | + | | + | | | |
| B 14 | + | +/− | | | + | Yes | *Temperature sensitive | 53 |

Table 11 (continued)
DESCRIBED MUTANTS OF *R. RUBRUM*[17]

| | | | Ref. |
|---|---|---|---|
| 13, 14, D 20, K 1, G 9-6 | − | − | 53 |
| F 11, F 24 | − | − | 55,56 |
| F 24.1 | + | + RC changed | 57 |

# REFERENCES

1. Gest, H., San Pietro, A., and Vernon, L. P., Eds., *Bacterial Photosynthesis,* Antioch Press, Yellow Springs, Ohio, 1963.
2. Kondrat'eva, E. N., *Photosynthetic Bacteria,* Israel Program for Scientific Translations, Jerusalem, 1965.
3. Lascelles, J., *Microbial Photosynthesis,* Benchmark Papers in Microbiology, Dowden, Hutchinson, and Ross, Stroudsburg, Pa., 1973.
4. Clayton, R. and Sistrom, W. R., Eds., *The Photosynthetic Bacteria,* Plenum Press, New York, 1978.
5. Elsden, S. R., Photosynthesis and lithotrophic carbon dioxide fixation, in *The Bacteria,* Vol. 3, Gunsalus, I. C. and Stanier, R. Y., Eds., Academic Press, New York, 1962, chap. 1.
6. Clayton, R. K., Primary processes in bacterial photosynthesis, *Annu. Rev. Biophys. Bioeng.,* 2, 131, 1973.
7. Parson, W. W., Bacterial photosynthesis, *Annu. Rev. Microbiol.,* 28, 41, 1974.
8. Jones, O. T. G., Electron transport and ATP synthesis in the photosynthetic bacteria, in *Microbial Energetics,* 27th Symp. Soc. General Microbiology, Cambridge University Press, Cambridge, 1977, 151.
9. Molisch, H., *Die Purpurbakterien,* Fischer, Jena, 1907.
10. VanNiel, C. B., The culture, general physiology, morphology and classification of the nonsulfur purple and brown bacteria, *Bacteriol. Rev.,* 8, 1, 1944.
11. Pfennig, N. and Trüper, H. G., The phototrophic bacteria, in *Bergey's Manual of Determinative Bacteriology,* 8th Ed., Buchanan, R. E. and Gibbons, N. E., Eds., Williams & Wilkins, Baltimore, 1974, 24.
12. Pfennig, N., Phototrophic green and purple bacteria: a comparative systematic survey, *Annu. Rev. Microbiol.,* 31, 275, 1977.
13. Ormerod, J. G., Ormerod, K. S., and Gest, H., Light-dependant utilization of organic compounds and photoproduction of molecular hydrogen by photosynthetic bacteria; relationships with nitrogen metabolism, *Arch. Biochem. Biophys.,* 94, 449, 1961.
14. Cohen-Bazire, G., Sistron, W. R., and Stanier, R. Y., Kinetic studies of pigment synthesis by non-sulfur purple bacteria, *J. Cell. Comp. Physiol.,* 49, 25, 1957.
15. Schön, G., Fructoseverwertung und Bacteriochlorophyllsynthese in anaeroben Dunkel- und Licht-kulturen von *Rhodospirillum rubrum, Arch. Mikrobiol.,* 63, 362, 1968.
16. Oelze, J. and Drews, G., Membranes in photosynthetic bacteria, *Biochim. Biophys. Acta,* 265, 209, 1972.
17. Frenkel, A. W., Light-induced reactions of chromatophores of *Rhodospirillum rubrum, Brookhaven Symp. Biol.,* 11, 276, 1959.
18. Takacs, B. J. and Holt, S. C., *Thiocapsa florodana:* A cytological, physical and chemical character-isation. I. Cytology of whole cells and isolated chromatophore membranes, *Biochim. Biophys. Acta,* 233, 258, 1971.
19. Clayton, R. K., Absorption spectra of photosynthetic bacteria and their chlorophylls, in *Bacterial Photosynthesis,* Gest, H., San Pietro, A., and Vernon, L. P., Eds., Antioch Press, Yellow Springs, Ohio, 1963.
20. Walter, E., Die chemische Natur der Pigmente aus photosynthetischen Reaktionszentren von *Rho-dospirillum rubrum* G-9⁺, Dissertation ETH Zürich, 1978.
21. Holt, S. C. and Marr, A. G., Effect of light intensity on the formation of intracytoplasmic membrane in *Rhodospirillum rubrum, J. Bacteriol.,* 89, 1421, 1965.
22. Cohen-Bazire, G. and Kunisawa, R., Some observations on the synthesis and function of the photo-synthetic apparatus in *Rhodospirillum rubrum, Proc. Natl. Acad. Sci. U.S.A.,* 46, 1543, 1960.
23. Stanier, R. Y., Formation and function of the photosynthetic pigment system in purple bacteria, *Brookhaven Symp. Biol.,* 11, 43, 1959.
24. Steiner, S., Sojka, G. A., Conti, S. F., Gest, H., and Lester, R. L., Modification of membrane composition in growing photosynthetic bacteria, *Biochim. Biophys. Acta,* 203, 571, 1970.
25. Oelze, J. and Drews, G., Der Einfluss der Lichtintensität und der Sauerstoffspannung auf die Differ-enzierung der Membranen von *Rhodospirillum rubrum, Biochim. Biophys. Acta,* 203, 189, 1970.
26. Schön, G. and Ladwig, R., Bacteriochlorophyllsynthese und Thylakoidmorphogenese in anaerober Dunkelkultur von *Rhodospirillum rubrum, Arch. Mikrobiol.,* 74, 356, 1970.
27. Biedermann, M., Drews, G., Mark, R., and Schröder, J., Der Einfluss des Sauerstoffpartialdruckes und der Antibiotica Actinomycin und Puromycin auf das Wachstum, die Synthese von Bacteriochl-lorophyll und die Thylakoidmorphogenese in Dunkelkulturen von *Rhodospirillum rubrum, Arch. Mikrobiol.,* 56, 133, 1967.
28. Goodwin, T. W., The carotenoids of photosynthetic bacteria. II. The carotenoids of a number of non-sulfur photosynthetic bacteria (Athiorhodaceae), *Arch. Mikrobiol.,* 24, 313, 1956.

29. Jensen, S. L., Cohen-Bazire, G., and Stanier, R. Y., Biosynthesis of carotenoids in purple bacteria, a reevaluation based on consideration of chemical structure, *Nature (London)*, 192, 1168, 1961.
30. Shipman, R. H., Fan, L. T., and Kao, I. C., Single cell production by photosynthetic bacteria, *Adv. Appl. Microbiol.*, 21, 161, 1977.
31. Shipman, R. H., Kao, I. C., and Fan, L. T., Single cell production by photosynthetic bacteria cultivation in agricultural by-products, *Biotechnol. Bioeng.*, 17, 1561, 1975.
32. Kobayashi, M., *Chem. Biol.*, 8, 604, 1970; as cited in Stanier, R. Y., *The Photochemical Apparatus, Its Structure and Function*, Fuller, R. C., Ed., Brookhaven Symposia in Biology 11, Upton, N. Y., 1959, 43.
33. Oelze, J., Characterisation of two cell-envelope fractions from chemotrophically grown *Rhodospirillum rubrum*, *Antonie van Leeuwenhoek J. Microbiol. Scrol.*, 41, 273, 1975.
34. Fuller, R. C., The comparative biochemistry of carbon dioxide fixation in photosynthetic bacteria, in *Progress in Photosynthetic Research*, Vol. 3, Metzner, H., Ed., 1969, 1579.
35. Buchanan, B. B., Evans, M. C. W., and Arnon, D. I., Ferredoxin dependent carbon assimilation in *Rhodospirillum rubrum*, *Arch. Mikrobiol.*, 59, 32, 1967.
36. Gromet-Elhanan, Z., Electron transport and phosphorylation in photosynthetic bacteria, in *Encyclopedia of Plant physiology*, Vol. 5, new series, Pirson, A. and Zimmermann, M. H., Eds., Springer-Verlag, Heidelberg, 1977, 637.
37. Baltscheffsky, H., Van Stedingk, L., Heldt, H. W., and Klingenberg, M., Inorganic pyrophosphate: formation in bacterial photophosphorylation, *Science*, 153, 1120, 1966.
38. Baltscheffsky, H. and Van Stedingk, L., Bacterial phosphorylation in the absence of added nucleotide. A second intermediate stage of energy transfer in light induced formation of ATP, *Biochem. Biophys. Res. Commun.*, 22, 722, 1966.
39. Gest, H., Energy conversion and generation of reducing power in bacterial photosynthesis, *Adv. Microb. Physiol.*, 7, 243, 1972.
40. Meyer, J., Kelley, B. C., and Vignais, P. M., Nitrogen fixation and hydrogen metabolism in photosynthetic bacteria, *Biochimie*, 60, 245, 1978.
41. Weare, N. M. and Shanmugam, K. T., Photoproduction of ammonium ion from $N_2$ in *Rhodospirillum rubrum*, *Arch. Microbiol.*, 110, 207, 1976.
42. Katz, E., Wassink, E. C., and Dorrestein, R., On some methodological problems in the study of photosynthesis of unicellular organisms, *Enzymologia*, 10, 269, 1942.
43. French, C. S., The quantum yield of hydrogen and carbon dioxide assimilation in purple bacteria, *J. Gen. Physiol.*, 20, 711, 1937.
44. Saunders, V. A., Genetics of Rhodospirillaceae, *Microbiol. Rev.*, 42, 357, 1978.
45. Newton, J. W., Bacilliform mutants of *Rhodospirillum rubrum*, *Biochim. Biophys. Acta*, 141, 633, 1967.
46. Newton, J. W., Metabolism of D-alanine in *Rhodospirillum rubrum* and its bacilliform mutants, *Nature (London)*, 228, 1100, 1970.
47. Newton, J. W., Vibrio mutants of *Rhodospirillum rubrum*, *Biochim. Biophys. Acta*, 244, 478, 1971.
48. Uffen, R. L., Sybesma, C., and Wolfe, R. S., Mutants of *Rhodospirillum rubrum* obtained after long-term anaerobic dark grown, *J. Bacteriol.*, 108, 1348, 1971.
49. Oelze, J., Schroeder, J., and Drews, G., Bacteriochlorophyll, fatty-acid and protein synthesis in relation to thylakoid formation in mutant strains of *Rhodospirillum rubrum*, *J. Bacteriol.*, 101, 669, 1970.
50. Oelze, J. and Drews, G., The production of particle bound bacteriochlorophyll precursors by the mutant F9 of *Rhodospirillum rubrum*, *Arch. Mikrobiol.*, 73, 19, 1970.
51. Kuhn, P. J. and Holt, S. C., Characterisation of a blue mutant of *Rhodospirillum rubrum*, *Biochim. Biophys. Acta*, 261, 267, 1972.
52. Hsi, E. S. P. and Bolton, J. R., Flash photolysis-electron spin resonance study of the effect of o-phenantroline and temperature on the decay time of the ESR signal B1 in reaction center preparations and chromatophores of mutant and wild type strains of *Rhodopseudomonas spheroides* and *Rhodospirillum rubrum*, *Biochim. Biophys. Acta*, 347, 126, 1974.
53. Weaver, P., Temperature sensitive mutants of the photosynthetic apparatus of *Rhodospirillum rubrum*, *Proc. Natl. Acad. Sci. U.S.A.*, 68, 136, 1971.
54. Schick, J. and Drews, G., The morphogenesis of the bacterial photosynthetic apparatus. III. The features of a pheophytin-protein-carbohydrate complex excreted by the mutant M46 of *Rhodospirillum rubrum*, *Biochim. Biophys. Acta*, 183, 215, 1969.
55. Del Valle-Tascon, S., Gimenez-Gallego, and Ramirez, J. M., Light-dependent ATP-formation in a non-phototrophic mutant of *Rhodospirillum rubrum* deficient in oxygen photoreduction, *Biochim. Biophys. Res. Commun.*, 66, 514, 1975.
56. Del Valle-Tascon, S., Gimenez-Gallego, G., and Ramirez, J. M., Photooxidase system of *Rhodospirillum rubrum*. I. Photooxidations catalyzed by chromatophores isolated from a mutant deficient in photooxidase activity, *Biochim. Biophys. Acta*, 459, 277, 1977.

57. Picorel, R., Del Valle-Tascon, S., and Ramirez, J. M., Isolation of a photosynthetic strain of *Rhodospirillum rubrum* with an altered reaction center, *Arch. Biochem. Biophys.*, 181, 665, 1977.

58. Gorchein, A., Neuberger, A., and Tait, G. H., The isolation and characterisation of subcellular fractions from pigmented and unpigmented cells of *Rhodopseudomonas spheroides*, *Proc. R. Soc. London Ser. B*, 170, 229, 1968.

59. Adams, M. W. W. and Hall, D. O., Isolation of the membrane-bound hydrogenase from *Rhodospirillum rubrum*, *Biochem. Biophys. Res. Commun.*, 77, 730, 1977.

60. Yoch, D. C., Arnon, D. I., and Sweeney, W. V., Characterisation of two soluble ferredoxins as distinct from bound iron-sulfur proteins in the photosynthetic bacterium *Rhodospirillum rubrum*, *J. Biol. Chem.*, 250, 8330, 1975.

61. Shanmugan, K. T., Buchanan, B. B., and Arnon, D. I., Ferredoxins in light- and dark-grown photosynthetic cells with special reference to *Rhodospirillum rubrum*, *Biochim. Biophys. Acta*, 256, 477, 1972.

62. Yoch, D. C., Carithers, R. P., and Arnon, D. I., Isolation and characterisation of bound iron-sulfur proteins from bacterial photosynthetic membranes. I. Ferredoxins III and IV from *Rhodospirillum rubrum*, *J. Biol. Chem.*, 252, 7453, 1977.

63. Noel, H., Van der Rest, M., and Gingras, G., Isolation and partial characterisation of P 870 reaction center complex from wild type *Rhodospirillum rubrum*, *Biochim. Biophys. Acta*, 275, 219, 1972.

64. Van der Rest, M. and Gingras, G., The pigment complement of the photosynthetic reaction center isolated from *Rhodospirillum rubrum*, *J. Biol. Chem.*, 249, 6446, 1974.

65. Snozzi, M., Isolierung und Charakterisierung von Reaktionszentren aus *Rhodospirillum rubrum*, *Ber. Dtsch. Bot. Ges.*, 90, 485, 1977; also *Biochim. Biophys. Acta*, 546, 236, 1979.

66. Cuendet, P. A. and Zuber, H., Isolation and characterization of a bacteriochlorophyll associated cbromatophore protein from *Rhodospirillum rubrum* G-9, *FEBS Lett.*, 79, 96, 1977.

67. Tonn, S. J., Gogel, G. E. and Loach, P. A., Isolation and characterization of an organic solvent soluble polypeptide component from photoreceptor complexes of *Rhodospirillum rubrum*, *Biochemistry*, 16, 877, 1977.

68. Johannson, B. C., Baltscheffsky, M., Baltscheffsky, H., Baccharini-Melandri, A., and Melandri, B. A., Purification and properties of a coupling factor ($Ca^{2+}$ dependant adenosine triphosphatase) from *Rhodospirillum rubrum*, *Eur. J. Biochem.*, 40, 109, 1973.

69. Philosoph, S., Binder, A., and Gromet-Elhanan, Z., Coupling factor ATPase complex of *Rhodospirillum rubrum*. Purification and properties of a reconstitutively active single subunit, *J. Biol. Chem.*, 252, 8747, 1977.

70. Konings, A. W. T. and Guillory, R. J., Resolution of enzymes catalyzing energy linked transhydrogenation. IV. Reconstitution of ATP driven transhydrogenase in depleted chromatophores of *Rhodospirillum rubrum* by the transhydrogenase factor and a soluble oligomycin sensitive $Mg^{2+}$ adenosine triphosphatase, *J. Biol. Chem.*, 248, 1045, 1973.

71. Oren, R. and Gromet-Elhanan, Z., Coupling factor adenosine triphosphatase complex of *Rhodospirillum rubrum*. Isolation of an oligomycin-sensitive $Ca^{2+}$, $Mg^{2+}$-ATPase, *FEBS Lett.*, 79, 147, 1977.

72. Anderson, L. E. and Fuller, R. C., Photosynthesis in *Rhodospirillum rubrum*. IV. Isolation and characterisation of ribulose-1,5-diphosphate carboxylase, *J. Biol. Chem.*, 244, 3105, 1969.

73. Tabita, F. R. and McFadden, B. A., D-Ribulose-1,5-diphosphate carboxylase from Rhodospirillum rubrum. I. Levels, purification and effect of metallic ions, *J. Biol. Chem.*, 249, 3453, 1974.

74. Tabita, R. F. and McFadden, B. A., D-Ribulose-1,5-diphosphate carboxylase from *Rhodospirillum rubrum*. II. Quarternary structure, composition, catalytic and immunological properties, *J. Biol. Chem.*, 249, 3459, 1974.

75. Kamen, M. D. and Horio, T., Bacterial cytochromes. I. Structural aspects, *Annu. Rev. Biochem.*, 39, 673, 1970.

76. Horio, T. and Kamen, M. D., Bacterial cytochromes. II. Functional aspects, *Annu. Rev. Microbiol.*, 24, 399, 1970.

77. Biggins, D. R., Kelly, M., and Postgate, J. R., Resolution of nitrogenase of *Mycobacterium flavum* 301 into two components and crossreaction with nitrogenase components from other bacteria, *Eur. J. Biochem.*, 20, 140, 1971.

78. Nordlund, S., Eriksson, U., and Baltscheffsky, H., Necessity of a membrane component for nitrogenase activity in *Rhodospirillum rubrum*, *Biochim. Biophys. Acta*, 462, 187, 1977.

79. Nordlund, S., Eriksson, U., and Baltscheffsky, H., Properties of the nitrogenase system from a photosynthetic bacterium *Rhodospirillum rubrum*, *Biochim. Biophys. Acta*, 504, 248, 1978.

80. Ludden, P. W. and Burris, R. H., Purification and properties of nitrogenase from *Rhodospirillum rubrum* amd evidence for phosphate, ribose and an adenine-like unit covalently bound to the iron protein, *Biochem. J.*, 175, 251, 1978.

81. Collins, M. L. P. and Niederman, R. A., Membranes of *Rhodospirillum rubrum*: physiochemical properties of chromatophore fractions isolated from osmotically and mechanically disrupted cells, *J. Bacteriol.*, 126, 1326, 1976.

82. **Drews, G.,** Nachweis von Zucker in den Thylakoiden von *Rhodospirillum rubrum* und *Rhodopseudomonas viridis, Z. Naturforsch.,* 23b, 671, 1968.

# GENERAL CHARACTERISTICS OF BLUE-GREEN ALGAE (CYANOBACTERIA): *SPIRULINA*

## Hubert Durand-Chastel

Blue-green algae (Cyanobacteria, Cyanophytes) are distinct among prokaryotic organisms in their ability of carry out photosynthesis through the photolysis of water in a two-photosystem reaction analogous to that found in eukaryotic organisms.[1] However, unlike eukaryotes, their cells do not contain an organized nucleus, and DNA is dispersed in the form of a filamentous system within the cytoplasm, which contains chlorophyll *a*, carotenoids, and biliproteins. From an evolutionary standpoint, blue-green algae are among the oldest oxygen-evolving photosynthetic organisms. Recent fossil evidence indicates that they may date back as far as 3.5 billion years.[2] Today, they are distributed throughout the oceans, fresh water, and soil environments, with about 1200 species of strains being known.

One important property of many blue-green algae is the ability to fix elemental nitrogen. Thus, these species make a contribution to the fertility of some crops, especially in the paddyfields of the Philippines, Indochina, and India. The beneficial effects can represent increased rice yields of up to 600% in laboratory plots and 100% in the field. Extensive studies were made in Japan by Watanabe[3-5] and in India by Venkataraman.[6] Another interesting property of blue-green algae is the ability of some species to produce hydrogen gas. Among the many new concepts now being explored in solar energy bioconversions, is their use for the production of hydrogen for fuel purposes.[7,8] A third important aspect of blue-green algae is their use as food and feed. Since ancient time, some blue-green algae have been eaten (*Aphanothece sacrum, Nostoc verrucosum, Nostoc commune*, and *Blachytrichia quoyi*) in Japan, but the quantities have been small. In the past 15 years, a great interest in *Spirulina maxima* and *Spirulina platensis* has arisen. These species grow naturally in various countries and are now cultured and processed industrially in Mexico.[9]

*Spirulina maxima* grows as a helicoidal filament, with a length of 200 or 300 $\mu$m and a diameter of 5 to 10 $\mu$m. A complete *S. maxima* forms a filament of 7 spires which measures about 1000 $\mu$m when developed and includes from 100 to 250 cells.[10] Reproduction is accomplished by binary fission in the longitudinal plane. Necrosis of older cells fragments the filaments. *Spirulinae* has no heterocyst and does not fix nitrogen. Because of gas vacuoles, characteristic of this species, filaments are buoyant and float on the surface of the water where they may form clumps. If photosynthesis is very intense, turgor pressure becomes too high, and the gas vesicles collapse. The cells then lose their buoyancy and sink.

*Spirulina* grows in alkaline to very alkaline waters. The nutritional requirements of *Spirulina* include nitrogen, phosphorus, sulfur, potassium, sodium, magnesium, calcium, and iron; trace amounts of other elements are also necessary. The high pH of the culture medium changes the solubility of many salts causing deficiencies, which, in turn, frequently result in failure of *Spirulina* crops. Table 1 gives the ash analysis of *Spirulina*.

Chemical composition studies on *Spirulina* reveal a high protein content. Table 2 indicates the composition of essential and nonessential amino acids. On the average, the available lysine is about 85% of total lysine. The lipid content is moderate, with fatty acids predominating at 83%. δ-Linolenic acid represents 20% of the total fatty acids (see Table 3). Details on the sterols and the carotenoids are given in Table 4. A moderate amount of carbohydrates with low caloric value is present. Among the carbohydrates is a phosphorylated cyclitol which is of nutritional interest because it is an

Table 1
CHEMICAL ANALYSIS OF
*SPIRULINA* ASH

| Weight % | Minimum 6.4 (mg/kg) | Maximum 9.0 (mg/kg) |
|---|---|---|
| Calcium | 1,045 | 1,315 |
| Phosphorus | 7,617 | 8,942 |
| Iron | 475 | 580 |
| Sodium | 275 | 412 |
| Chloride | 4,000 | 4,400 |
| Magnesium | 1,410 | 1,915 |
| Manganese | 18 | 25 |
| Zinc | 27 | 39 |
| Potassium | 13,305 | 15,400 |
| Others | 36,000 | 57,000 |

Table 2
AMINO ACID COMPOSITION OF *SPIRULINA*

| | Minimum (%) | Maximum (%) |
|---|---|---|
| Total organic nitrogen | 10.85 | 13.35 |
| Nitrogen from proteins | 9.60 | 11.36 |
| Crude protein (% N × 6.25) | 60.0 | 71.0 |
| | | |
| Essential amino acids | | |
| Isoleucine | 4.69 | 4.13 |
| Leucine | 5.56 | 5.80 |
| Lysine | 2.96 | 4.00 |
| Methionine | 1.59 | 2.17 |
| Phenylalanine | 2.77 | 3.95 |
| Threonine | 3.18 | 4.17 |
| Tryptophan | 0.82 | 1.13 |
| Valine | 4.20 | 6.00 |
| | | |
| Nonessential Amino acids | | |
| Alanine | 4.97 | 5.82 |
| Arginine | 4.46 | 5.98 |
| Aspartic acid | 5.97 | 6.43 |
| Cystine | 0.56 | 0.67 |
| Glutamic acid | 8.29 | 8.94 |
| Glycine | 3.17 | 3.46 |
| Histidine | 0.89 | 1.08 |
| Proline | 2.68 | 2.97 |
| Serine | 3.18 | 4.00 |

iron chelating agent (see Table 5). The nucleic acid content is low: the RNA content is from 2.20 to 3.50%, and the DNA content is from 0.63 to 1%.[11] The vitamins of *Spirulina* are given in Table 6, and the heavy metal content is given in Table 7. The protein efficiency ratio (PER) is between 2.2 and 2.6 (74 to 87% of casein). Net protein utilization (NPU) is between 53 and 61% (85 to 92% of casein). Digestibility is between 83 and 84%. All these characteristics explain why *Spirulina* has been eaten for centuries in some countries, e.g., Chad and Mexico.[12-14] Presently, the main use of *Spirulina* is for feeding animals — from crustacean larvae to monogastrics to ruminants.[15,16]

Table 3
LIPID COMPOSITION OF *SPIRULINA*

|  | Minimum | Maximum |
|---|---|---|
| Total lipids | 6.0% | 7.0% |
| Fatty acids | 4.9% | 5.7% |
|  | (mg/kg) | (mg/kg) |
| Lauric ($C_{12}$) | 180 | 229 |
| Myristic ($C_{14}$) | 520 | 644 |
| Palmitic ($C_{16}$) | 16,500 | 21,141 |
| Palmitoleic ($C_{16}$) | 1,490 | 2,035 |
| Palmitolinoleic ($C_{16}$) | 1,750 | 2,565 |
| Heptadecanoic ($C_{17}$) | 90 | 142 |
| Stearic ($C_{18}$) | trace | 353 |
| Oleic ($C_{18}$) | 1,970 | 3,009 |
| Linoleic ($C_{18}$) | 10,920 | 13,784 |
| Linolenic ($C_{18}$) | 8,750 | 11,970 |
| Linolenic ($C_{18}$) | 160 | 427 |
| Others | 7,000 | 699 |
|  |  |  |
| Nonsaponifiable | 1.1% | 1.3% |
|  | (mg/kg) | (mg/kg) |
| Sterols | 100 | 325 |
| Triterpene alcohols | 500 | 800 |
| Carotenoids | 2,900 | 4,000 |
| Chlorophyll *a* | 6,100 | 7,600 |
| Others | 1,400 | 150 |
| 3,4-Benzypyrene | 2,600 | 3,600 |

Table 4
STEROLS AND CAROTENOIDS OF *SPIRULINA*

|  | Minimum (mg/kg) | Average (mg/kg) | Maximum (mg/kg) |
|---|---|---|---|
| Sterols | 100 | — | 325 |
| Cholesterol | 60 | — | 196 |
| Sistosterol | 30 | — | 97 |
| Dihidro-7-cholesterol |  |  |  |
| Cholesten-7-ol-3 |  |  |  |
| Stigmasterol | 10 | — | 32 |
| Carotenoids | 2900 | — | 4000 |
| Carotene | — | (trace) | — |
| Carotene | — | 1700 | — |
| Xanthophylls | — | 1600 | — |
| Cryptoxanthin | — | 556 | — |
| Echinenone | — | 439 | — |
| Zeaxanthin | — | 316 | — |
| Lutein and euclenanone | — | 289 | — |

## Table 5
## CARBOHYDRATES OF *SPIRULINA*

| | Average | |
| --- | --- | --- |
| | Minimum (%) | Maximum (%) |
| Total carbohydrates | 13.0 | 16.5 |
| Rhamnose | — | 9.0 |
| Glucan | — | 1.5 |
| Phosphoryled cyclitols | — | 2.5 |
| Glucosamine and muramic acid | — | 2.0 |
| Glycogen | — | 0.5 |
| Sialic Acid and others | — | 0.5 |

## Table 6
## VITAMINS OF *SPIRULINA*

Average values (mg/kg)

| | | |
| --- | --- | --- |
| Biotin (H) | — | 0.4 |
| Cyanocobalamin (B$_{12}$) | — | 2 |
| d-Ca-pantothenate | — | 11 |
| Folic acid | — | 0.5 |
| Inositol | — | 350 |
| Nicotinic acid (PP) | — | 118 |
| Pyridoxine (B$_6$) | — | 3 |
| Riboflavine (B$_2$) | — | 40 |
| Thiamine (B$_1$) | — | 55 |
| Tocopherol (E) | — | 190 |

## Table 7
## METALS AND CYANIDE CONCENTRATIONS IN *SPIRULINA*

| Heavy metals and cyanide | Amount (ppm) |
| --- | --- |
| Arsenic (as As$_2$O$_3$) | 1.10 |
| Cadmiun | >0.10 |
| Lead | 0.40 |
| Mercury | 0.24 |
| Selenium | 0.40 |
| Cyanide | 0.20 |

## REFERENCES

1. Fogg, G. E., Steward, W. D. P., Fay, P., and Walsby, A. E., Eds., *The Blue-green Algae,* Academic Press, London, 1973.
2. Knoll, A. and Barghoorn, E., Archean microfossils showing cell division from the Swaziland system of South Africa, *Science,* 198, 396, 1977.
3. Watanabe, A., Distribution of nitrogen fixing blue-green algae in various areas of South and East Asia, *J. Gen. Appl. Microbiol.,* 5, 21, 1959.
4. Watanabe, A., Effect of nitrogen fixing blue-green alga, *Tolypothrix tenuis* on the nitrogenous fertility of paddy soils and on the crop yield of rice plant, *J. Gen. Appl. Microbiol.,* 8, 85, 1962.
5. Watanabe, A., Studies on the blue-green algae as green manure in Japan, *Proc. Natl. Acad. Sci. India,* 35A, 361, 1965.
6. Venkataraman, G. S., The role of blue-green algae in tropical rice cultivation, in *Nitrogen Fixation by Free-Living Microorganisms,* Stewart, W. D. P., Ed., Cambridge University Press, Cambridge, 1975, 207.
7. Mitsui, A., Applications of photosynthetic hydrogen production and nitrogen fixation research, in *Proc. Conf. Capturing the Sun Through Bioconversion,* Washington Metropolitan Studies, Ed., Washington, D.C., 1976, 653.
8. Hall, D. O., Photobiological energy conversion, *FEBS Lett.,* 64, 6, 1976.
9. Durand-Chastel, H. and David, M., The *Spirulina* algae, in European Seminar on Biological Solar Energy Conversion Systems, Grenoble, France, 1977.

10. **Chanthavong, P.**, Ultraestructure de *Spirulina platensis, Diplome d'etudes aprofondies, Universite de Dijon, France, 1974.*
11. **Jassey, Y., Berlot, J. P., and Barón, C.**, Etude Comparee des Acides Nucleiques de Deux Especes de *Spirulina platensis geitleri* et *Spirulina maxima, C. R. Acad. Sci. Ser. D,* 1971.
12. **Furst, P. T.**, *Spirulina,* nutritious algae, once a staple of Aztec diets, could feed many of the world's hungry people, *Human Nature,* March, 1978.
13. **Sautier, C. and Tremoliere, Y.**, Valeur alimentaire des algues *Spiruline chez l'homme, Ann. Nutr. Aliment.,* 29, 517, 1975.
14. **Busson, F.**, *Spirulina platensis* et *Spirulina geitleri,* Cyanophycees alimentaires, Service de Santé, Parc du Pharo, Marseille, 1977.
15. **Durand-Chastel, H., Santillan Claudio,** The Tecuitlatl and the aquaculture, in FAO Technical Conference on Aquaculture, Kyoto, Japan, 1975.
16. **Durand-Chastel, H.**, Production and use of *Spirulina* in Mexico, in Seminar on Production and Use of Micro-Algae Biomass, Acre, Israel, September 1978.

# GENERAL CHARACTERISTICS OF GREEN MICROALGAE:
## *CHLORELLA*

**Akio Kamiya and Shigetoh Miyachi**

## CULTURE CONDITIONS

Most green algae grow photoautotrophically, and growth is maintained by the products of photosynthesis. Under suitable conditions, green algae exhibit growth rates dependent on light intensities and temperatures. Table 1 shows the growth rates of *Chlorella ellipsoidea* under different culture conditions.[1] The following inorganic elements (in addition to C, H, and O) are required for algal growth: N, P, K, Mg, Ca, S, Fe, Cu, Mn, Mo, Na, Co, V, and Si. Of these, N, P, Mg, S, Fe, Cu, Mn, and Mo are considered to be required by all algae and not replaceable by other elements.[2-4]

The most common sources of nitrogen used by green algae are nitrate and ammonium ions. It is generally accepted that nitrate is reduced to the ammonium level before being incorporated into organic compounds. Although most green algae are able to synthesize all the organic N compounds required for growth from the inorganic N source, such as nitrate and ammonium, the colorless alga *Prototheca zopfii* utilizes adenine and ammonium, but not nitrate.[5] Some are able to use certain organic N compounds as sole sources of nitrogen. Urea can be used as an N source for growth by many kinds of algae. Birdsey and Lynch[6] observed that eight species of the Chlorophyceae were able to use urea and five used uric acid and xanthine as N sources (see Table 2). Roon and Levenberg[7] described a new enzyme (ATP, urea amido-lyase) from urea-grown *C. ellipsoidea* and *C. pyrenoidosa* which catalyzes the ATP-dependent cleavage of urea to $CO_2$ and ammonia:

$$\text{Urea} + \text{ATP} \rightarrow CO_2 + 2\,NH_3 + \text{ADP} + \text{Pi}$$

Phosphate plays a significant role in most algae, especially in energy transformation reactions. Concentrations below 50 $\mu$g/L of phosphorus limit the growth of green algae, those above 20 mg/L are inhibitory, and 100 to 2000 $\mu$g/L are optimum.[8] The biggest difference between phosphate metabolism in higher plants and algae is the formation of large amounts of inorganic polyphosphate in algae. The polyphosphate of *C. pyrenoidosa* has been separated into an acid-soluble fraction (polyphosphate A) and three acid-insoluble fractions (polyphosphate B, C, and D).[9] Insoluble polyphosphate accumulates at a constant rate in *C. pyrenoidosa* during cell growth, but can be used as a source of phosphorus for nucleic acid and phosphoprotein.[10]

Many algae require vitamins: vitamin $B_{12}$, thiamine, and biotin. However, the members of the Chlorophyceae do not require biotin, and the number of species which require other vitamins is also smaller than in other families of algae. The contents of vitamin $B_{12}$, thiamine, and biotin of the Chlorophyceae are 12 to 150 ng/g, 0.9 to 23 $\mu$g/g, and 115 to 230 ng/g, respectively.[11] Table 3 shows the vitamin requirement of the chlorophyceae.[12]

## COMPOSITION OF ALGAL CELL MATERIALS

Table 4 shows the elemental composition of the green algae *Chlorella* and *Scenedesmus*[13] and indicates that the following chemical elements are dominant: C, O, H, N, P, K, Mg, S, Fe, and Ca (see also Table 6). The chlorophyll *a* content of algal cells ranges from 0.3 to 2.0% of the dry weight[14] (see also Table 6). Chlorophyll *b* is the

Table 1
## LINEAR GROWTH RATE OF *CHLORELLA ELLIPSOIDEA* UNDER DIFFERENT LIGHT INTENSITIES AND AT DIFFERENT TEMPERATURES (BUBBLING CULTURE)

| Light Intensity (klx) | Temperature (C°) | Rate of linear growth (g/m²/12 hr) |
|:---:|:---:|:---:|
| 50 | 25 | 32 |
| | 15 | 14 |
| | 7 | 0 |
| 25 | 25 | 23 |
| | 15 | 11 |
| | 7 | 1.0 |
| 10 | 25 | 14 |
| | 15 | 7.5 |
| | 7 | 2.5 |
| 5 | 25 | 8.2 |
| | 15 | 6.5 |
| | 7 | 1.8 |
| 2 | 25 | 3.9 |
| | 15 | 3.6 |
| | 7 | 1.3 |

Reproduced with permission from *Annu. Rev. Plant Physiol.*, 8, 309, 1957. Copyright 1957 by Annual Reviews, Inc.

Table 2
## UTILIZATION OF NITROGEN COMPOUNDS BY GREEN ALGAE

| Alga | Urea | Uric acid | Xanthine | Allantoin | Creatinine | KNO₃ | NH₄Cl |
|---|:---:|:---:|:---:|:---:|:---:|:---:|:---:|
| *Chlorella pyrenoidosa* | + | + | + | − | − | + | + |
| *Chlorella vulgaris* | + | + | + | − | − | + | + |
| *Scenedesmus obliquus* | + | + | + | − | − | + | + |
| *Chlamydomonas reinhardi* | + | + | + | − | − | + | + |
| *Asterococcus superbus* | + | − | | − | − | ± | + |
| *Euglena gracilis* | − | − | | − | − | − | + |
| *Scenedesmus* sp. | + | + | + | − | − | + | + |
| *Chlorella* sp. | + | − | | − | − | + | + |
| *Chlorella* sp. | + | − | | − | − | + | + |

From Birdsey, E. C. and Lynch, V. H., *Science,* 137, 763, 1962. Copyright 1962 by the American Association for the Advancement of Science.

minor component distributed in the Chlorophyceae, Euglenophyceae, and Prasinophyceae. The chlorophyll *a/b* ratio in the Chlorophyceae generally ranges from 2 to 3.[15] Mineral nutrition affects the chlorophyll content. In numerous algae, deficiencies of Fe, Mg, and N, essential constituents of heme or chlorophyll, have pronounced effects on chlorophyll synthesis and its content. The chlorophyll content of green algae is inversely proportional to the light intensity[16] (see Figure 1). The main carotenoids of green algae are β-carotene and lutein. Table 5 shows the distribution of major carotenoids in *C. pyrenoidosa*.[17] Zeaxanthin was found to be a constant component of green

Table 3
## VITAMIN REQUIREMENTS OF THE CHLOROPHYCEAE[12]

| Species | Vitamin requirements[a] | | |
| --- | --- | --- | --- |
| | $B_{12}$ | Thiamin | Biotin |
| *Astrephomene gubernaculifera, Bracteacoccus cinnabarinus, B. engadiensis, B. minor, B. terrestris, Chlamydomonas agloeformis, C. eugametos, C. moewusii, C. pichinchae, C. reinhardtii, Chlorella autotrophica, C. candida, C. ellipsoidea, C. emersonii, C. emersonii globosa, C. fusca, C. fusca vacuolata, C. infusionum, C. infusionum auxenophila, C. miniata, C. mutabilis, C. nocturna, C. photophila, C. pringsheimii, C. pyrenoidosa (Emerson), C. regularis, C. regularis aprica, C. regularis imbricata, C. saccharophila, C. simplex, C. sorokiniana, C. vannielii, C. variabilis, C. vulgaris, C. vulgaris luteoviridis, Chlorococcum aplanosporum, C. diplobionticum, C. echinozygotum, C. ellipsoideum, C. hypnosporum, C. macrostigmatum, C. minitum, C. multinucleatum, C. oleofaciens, C. perforatum, C. pinguideum, C. punctatum, C. scabellum, C. tetrasporum, C. vacuolatum, C. wimmeri, Chlorosarcinopsis auxotrophica, C. eremi, Cosmarium botrytis, C. impressulum, C. laeve, C. lundelli, C. meneghini, Cylindrocystis brebissonnii, Dictyochloris fragrans, Dunaliella primolecta, D. salina, D. viridis, Eudorina elegans, Haematococcus pluvialis, Lobomonas pyriformis, Micrasterias cruxmelitensis, Nautococcus pyriformis, Neochloris alveolaris, N. aquatica, N. gelatinosa, N. minuta, N. pseudoalveolaris, Pandorina morum, P. unicocca, Polytoma obtusum, P. uvella, Radiosphaera dissecta, Scenedesmus obliquus, Selenastrum minutum, Spongiochloris excentrica, S. lamellata, S. spongiosis, Spongiococcum alabamense, S. excentricum, S. multinucleatum, S. tetrasporum, Staurastrum gladiosum, S. paradoxum, S. sebaldii ornatum, Staurodesmus pachyrhynchus, Stichococcus cylindricus, Stigeoclonium aestivale, S. helveticum, S. subsecundum, S. tenue, S. pascheri, S. variabile, S. farctum* | O | O | O |
| *Balticola buetschlii, B. droebakensis, Chlamydomonas chlamydogama, C. mundana, C. mundana astigmata, C. pallens, C. pulsatilla, Chlorosarcinopsis sempervirens, Chlorosphaera consociata, Closterium braunii Z 951/2a, C. macilentum, C. peracerosum, C. pusillum, C. siliqua, C. strigosum, C. turgidum, Cylindrocystis brebissonnii, Cosmarium turpini, Desmidium swartzii, Docidium manubrium, Gonium pectorale, Haematococcus droebakensis, H. buetschlii, H. pluvialis, Hyalotheca dissiliens, Lobomonas rostrata, L. sphaerica, Nannochloris sp., Platydorina caudata, Pleurotaenium trabecula, Stichococcus sp., (cylindricus?), Volvox aureus, V. globator, V. tertius, Volvulina steinii, V. pringsheimii, Xanthidium cristatum* | R | O | O |
| *Chlorella protothecoides, C. protothecoides communis, C. protothecoides galactophila, C. protothecoides mannophila, Coelastrum morus, Gloeocystis gigas, Gonium multicoccum, Polytoma caudatum, P. ocellatum, Polytomella caeca, Prototheca zopfii, Raphidonema nivale, Selenastrum minutum* | O | R | O |
| *Acetabularia mediterranea, Astrephomene gubernaculifera, Brachiomonas submarina, Eremosphaera viridis, Haematococcus capensis typ. H. capensis borealis, H. zimbabwiensis, Stephanosphaera pluvialis, Volvox globator* | R | R | O |

[a]  R = required; O = not required.

Table 4
## THE ELEMENTAL COMPOSITION OF GREEN ALGAE[13]

| Alga | Element | Percent of dry weight |
|---|---|---|
| *Chlorella* | C | 51.4—72.6 |
| | O | 28.5—11.6 |
| | H | 7.0—10.0 |
| *Scenedesmus* | N | 7.7—6.2 |
| | P | 1.0—2.0 |
| | K | 0.85—1.62 |
| | Mg | 0.36—0.80 |
| | S | 0.28—0.39 |
| | Fe | 0.04—0.55 |
| | Ca | 0.005—0.08 |
| | Zn | 0.0006—0.005 |
| | Cu | 0.001—0.004 |
| | Co | 0.00003—0.0003 |
| | Mn | 0.002—0.01 |

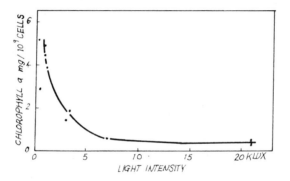

FIGURE 1.    The content of chlorophyll *a* per cell of *Chlorella pyrenoidosa* as a function of the light intensity at which the alga was grown; continuous incandescent light, 20°C.[16]

Table 5
## DISTRIBUTION OF MAJOR CAROTENOIDS IN *CHLORELLA PYRENOIDOSA*[17]

| Carotenoid | Concentration (mg/g dry weight) | Percent of total carotenoids |
|---|---|---|
| α-Carotene | 0.09 | 3.6 |
| β-Carotene | 0.33 | 13.5 |
| Luteine | 1.11 | 54.8 |
| Violaxanthin | 0.22 | 9.0 |
| Neoxanthin | 0.21 | 8.8 |
| Others | 0.25 | 10.3 |
| Total | 2.21 | — |

algae.[18] Recently, a new carotenoid, loroxanthin, was found in *Chlorella vulgaris, Cladophora* spp., and *Scenedesmus obliquus.*[19]

Three polysaccharides have been isolated and purified from the alkali-soluble poly-

## Table 6
### CHARACTERISTICS OF DARK AND LIGHT CELLS[22]

| | Average cell diameter (µm) | Light-saturated photosynthetic rate at 25°C[a] | Respiratory activity ($Qo_2$) at 25°C | Chlorophyll content (%) | Nitrogen content (%) | Phosphorus content (%) |
|---|---|---|---|---|---|---|
| **Dark cells** | | | | | | |
| Active | 3.1—3.4 | 5.2—5.8 | 4.6—6.1 | 2.4—5.2 | 7.0—9.5 | 1.7—1.8 |
| Nascent | 2.3—2.5 | 3.1—4.9 | 2.9—4.6 | 0.9—2.0 | 5.5—7.0 | 1.8 |
| **Light cells** | 5.5—5.9 | 0.8—1.0 | 7.7—9.3 | 0.8—1.3 | 5.2—5.7 | 1.2—1.8 |

[a]  $O_2$ µL/cm³ · 10 min; $CO_2$ concentration, 5%.

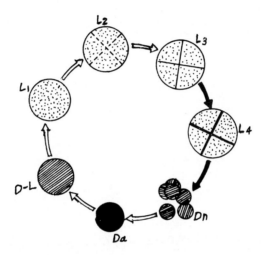

FIGURE 2.    Schematic representation of the life cycle of *Chlorella ellipsoidea.*[24]

saccharides of *C. pyrenoidosa.*[20] These polysaccharides consist of hemicellulose A, hemicellulose B, and starch which are in the ratio of 9:2:1. Pure starch from *C. pyrenoidosa* consisted of amylose (7%) and amylopectin. Acetate-grown *Polytoma uvella* produced a starch-type polysaccharide consisting of 16% amylose and amylopectin similar to that of higher plants.[21]

Changes in carbohydrate, protein, and lipid content in green algae have been studied using the technique of synchronous culture. With *C. ellipsoidea*, Tamiya et al.[22-23] observed that the cells show, in the course of their growth, two distinct forms possessing widely different characters. The young form, which they named "dark cells", is much smaller in size, richer in chlorophyll, and more photosynthetically active than the other developed form called "light cells" (see Table 6). The overall developmental process of *C. ellipsoidea* is shown in Figure 2. Figure 3 shows the change in content of carbohydrate, protein, lipids, and ash during the course of synchronous growth of *C. ellipsoidea.*[24] The composition of carbohydrate, protein, and lipid in *C. pyrenoidosa* grown on glucose has been estimated to be 36, 49, and 15%, respectively.[25] With heterotrophically grown *C. vulgaris*, Bergmann[26] reported a different composition: 70% carbohydrate, 15% protein, and 15% lipid. It has been reported that specific wavelengths of light influence the basic composition of algal cells.[27] Cells grown in blue light contain more total nitrogen and less carbohydrate than those grown in red light.

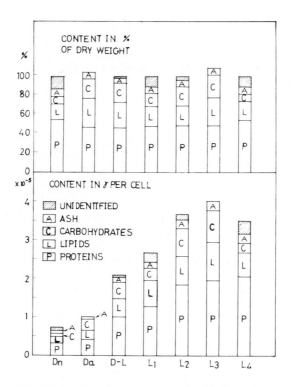

FIGURE 3.   Content of crude proteins, lipids, carbohydrates, and ash in *Chlorella ellipsoidea* cells of different developmental stages.[24]

Although the total protein of *Chlorella* cells is dependent on the age of the culture and nutritive value, the effect of age on the amino acid composition of the bulk protein in the cells is only slight. Table 7 shows the amino acid composition of the proteins isolated from *C. vulgaris*.[28]

Table 7
AMINO ACID COMPOSITION OF
PROTEINS ISOLATED FROM
*CHLORELLA VULGARIS*[28]

| Amino acid | Percent of total protein nitrogen |
|---|---|
| Aspartic acid | 6.4 |
| Glycine | 6.2 |
| Threonine | 2.9 |
| Alanine | 7.7 |
| Tyrosine | 2.8 |
| Valine | 5.5 |
| Phenylalanine | 2.8 |
| Serine | 3.3 |
| Glutamic acid | 7.8 |
| Leucine | 6.1 |
| Isoleucine | 3.5 |
| Proline | 5.8 |
| Arginine | 15.8 |
| Histidine | 3.3 |
| Lysine | 10.2 |
| Tryptophan | 2.1 |
| Methionine | 1.4 |
| Cysteine | 0.2 |
| Amide-N | 6.1 |
| Total-N | 99.9 |

# REFERENCES

1. **Tamiya, H.**, Mass culture of algae, *Annu. Rev. Plant Physiol.*, 8, 309, 1957.
2. **Hutner, S. H. and Provasoli, L.**, Nutrition of algae, *Annu. Rev. Plant Physiol.*, 15, 37, 1964.
3. **O'Kelley, J. C.**, Mineral nutrition of algae, *Annu. Rev. Plant Physiol.*, 19, 89, 1968.
4. **O'Kelley, J. C.**, Inorganic nutrients, in *Algal Physiology and Biochemistry*, Stewart, W. D. P., Ed., Blackwell Scientific, Oxford, 1974, chap. 22.
5. **Stacey, J. L. and Casselton, P. J.**, Utilization of adenine but not nitrate as nitrogen source by *Prototheca zopfii*, *Nature (London)*, 211, 862, 1966.
6. **Birdsey, E. C. and Lynch, V. H.**, Utilization of nitrogen compounds by unicellular algae, *Science*, 137, 763, 1962.
7. **Roon, R. J. and Levenberg, B.**, An adenosine triphosphate-dependent, avidin-sensitive enzymatic cleavage of urea in yeast and green algae, *J. Biol. Chem.*, 243, 5213, 1968.
8. **Chu, S. P.**, The influence of the mineral composition of the medium on the growth of planktonic algae. II. The influenc of the concentration of inorganic nitrogen and phosphate phosphorus, *J. Ecol.*, 31, 109, 1943.
9. **Kanai, R., Aoki, S., and Miyachi, S.**, Quantitative separation of inorganic polyphosphates in *Chlorella* cells, *Plant Cell Physiol.*, 6, 467, 1965.
10. **Baker, A. L. and Schmidt, R. R.**, Polyphosphate metabolism during nuclear division in synchronously growing *Chlorella, Biochim. Biophs. Acta*, 82, 624, 1964.
11. **Kanazawa, A.**, Vitamins in algae, *Bull. Jpn. Soc. Sci. Fish.*, 29, 713, 1963.
12. **Provasoli, L. and Carlucci, A. F.**, Vitamins and growth regulators, in *Algal Physiology and Biochemistry*, Stewart, W. D. P., Ed., Blackwell Scientific, Oxford, 1974, chap. 27.
13. **Krauss, R. W.**, paper presented at the Conference Solar Energy, Tucson, Ariz., October 31 to November 1, 1955.
14. **Rabinowitch, E. I.**, *Photosynthesis and Related Processes*, Vol. 1, Interscience, New York, 1945.
15. **Strain, H. H., Cope, B. T., and Svec, W. A.**, Procedures for the isolation, identification, estimation and investigation of the chlorophylls, in *Methods Enzymol.*, San Pietro, A., Ed., 23A, 452, 1971.

16. Steemann Nielsen, E. and Jorgensen, E. G., The adaptation of plankton algae. I. General part, *Physiol. Plant.*, 21, 401, 1968.

17. Allen, M. B., Goodwin, T. W., and Phagpolngarm, S., Carotenoid distribution in certain naturally occurring algae and in some artificially induced mutants of *Chlorella pyrenoidosa, J. Gen. Microbiol.*, 23, 93, 1960.

18. Hager, A. and Stransky, H., Das carotenoidmuster und die Verbreitung des lichtinduzierten Xanthophyllcyclus in Verschiedenen Algenklassen. III. Grünalgen, *Arch. Mikrobiol.*, 72, 68, 1970.

19. Aitzetmüller, K., Strain, H. H., Svec, W. A., Grandolfo, M., and Katz, J. J., Loroxanthin, a unique xanthophyll from *Scenedesmus obliquus* and *Chlorella vulgaris*, *Phytochem.*, 8, 1761, 1969.

20. Olaitan, S. A. and Northcote, D. H., Polysaccharides of *Chlorella pyrenoidosa*, *Biochem. J.*, 82, 509, 1962.

21. Manners, D. J., Mercer, G. A., Stark, J. R., and Ryley, J. F., Studies on the metabolism of the Protozoa, the molecular structure of a starch-type polysaccharide from *Polytoma uvella*, *Biochem. J.*, 96, 530, 1965.

22. Tamiya, H., Iwamura, T., Shibata, K., and Nihei, T., Correlation between photosynthesis and light-independent metabolism in the growth of *Chlorella*, *Biochim. Biophys. Acta*, 12, 23, 1953.

23. Tamiya, H., Synchronous cultures of algae, *Annu. Rev. Plant Physiol.*, 17, 1, 1966.

24. Hase, E., Morimura, Y., and Tamiya, H., Some data on the growth physiology of *Chlorella* studied by the technique of synchronous culture, *Arch. Biochem. Biophys.*, 69, 149, 1957.

25. Samejima, H. and Myers, J., On the heterotrophic growth of *Chlorella pyrenoidosa*, *J. Gen. Microbiol.*, 18, 107, 1958.

26. Bergmann, L., Stoffwechsel und Mineralsalzernährung einzelliger Grünalgen. II. Vergleichende Untersuchen uber den Einfluss mineralischen bei heterotropher und mixotropher Ernahrung, *Flora*, 142, 493, 1955.

27. Pirson, A. and Kowallik, W., Spectral responses to light by unicellular plants, *Photochem. Photobiol.*, 3, 489, 1964.

28. Fowden, L., A comparison of the compositions of some algal proteins, *Ann. Bot. (London)*, 18, 257, 1954.

# GENERAL CHARACTERISTICS OF PHYTOPLANKTON: DIATOMS

## W. Marshall Darley

Diatoms are a well-defined group of wide-spread, unicellular or colonial, yellow-brown algae. Various aspects of the physiology and ecology of the group have been treated in more detail in a recent monograph.[1] Approximately 12,000 valid species[1] are grouped in some 200 genera.[2] Some authorities include all diatoms in the class Bacillariophyceae within the division Chrysophyta,[2-4] while others recognize divisional status (Bacillariophyta) for the diatoms.[5] On the basis of the chemical form of the stored food reserve, pigment composition, flagellar structure, and ultrastructural features, the diatoms show a fairly close phylogenetic relationship to the Xanthophyceae (yellow-green algae), Chrysophyceae (golden-brown algae), and Phaeophyceae (brown algae) and do not appear closely related to the other groups of algae.

Perhaps the most characteristic feature of diatoms is the highly ornamented siliceous cell wall, or frustule, whose structure and pattern of ornamentation serve as the most important character for classification within the group. The ornamentation usually takes the form of rows of thin places in the cell wall, although slits, thickenings, internal partitions, and various knobs and spines are also common. The resistance of the cell wall to dissolution has resulted in the accumulation of large deposits which are mined as diatomaceous earth (or diatomite) and used commercially as a filtration aid and mild abrasive. The frustule consists of two overlapping halves (or valves) which fit together not unlike the top (or epivalve) and bottom (or hypovalve) halves of a Petri dish.[6] Additional siliceous bands (girdle bands) attached to the free margin of the valves encircle the cell in the region of the overlap. It is the girdle bands, in fact, which overlap to hold the valves together. Vegetative cell division occurs parallel to the plane of the valve, after which each daughter protoplast synthesizes a new valve inside the existing valve. In other words, the protoplast in the epivalve (Petri dish top) synthesizes a new hypovalve (Petri dish bottom), while the daughter protoplast in the hypovalve of the parent frustule produces a new valve inside the existing hypovalve. In the process, the original hypovalve becomes the epivalve (outer valve) of this latter daughter cell. When the new valves are complete, the daughter cells separate, except in the case of colonial species in which they may remain attached to form a filament. As a result of the formation of new valves, one of the two daughter cells is slightly smaller than the parent cell. The species eventually regains its full size through the sexual cycle, although vegetative enlargement is also known to occur.[7] Not all species undergo a diminution in cell size during vegetative cell division, presumably because of flexible frustule components.

The presence of the siliceous frustule gives diatoms an unusual nutritional requirement for silicon.[8,9] Diatoms have an absolute requirement for silicon in the form of silicic acid, $Si(OH)_4$. In addition to the Si requirement for cell wall formation, diatoms require Si for net DNA synthesis a requirement which resides in the translation step in the synthesis of DNA polymerase.[10] Little is known of the silicification process in frustule formation except that silicic acid uptake is by active transport and that it is polymerized to hydrated amorphous silica ($SiO_2 \cdot nH_2O$) within a membranous structure which lies close to, but within, the plasma membrane. There is no ultrastructural evidence as yet of the morphogenetic factors which control the intricate, species specific ornamentation of the frustule. Once the valve is complete, a new plasma membrane appears underneath the new cell wall, thereby leaving it outside the protoplast. A complex organic cell wall layer,[11] derived in part from the membranous deposition vesicle, continues to surround and tightly adhere to the siliceous component of the cell wall.

Silicic acid uptake appears to be limited to the period of cell wall formation and is therefore a discontinuous process during the cell cycle. Germanic acid, $Ge(OH)_4$, with chemical properties similar to those of silicic acid, is a potent and apparently specific growth inhibitor in diatoms at a Ge to Si molar ratio of 0.1 or higher.[9] At lower concentrations, the radioisotope [68]Ge has proved useful as a tracer for silicification[12] and ecological[13] studies, since it has a longer halflife than [31]Si.

The principal storage polysaccharide in diatoms is chrysolaminaran, a water-soluble $\beta$-(1 $\rightarrow$ 3) glucan with a small degree of branching at $C_6$ and containing an average of 21 glucose units.[14] It is similar to laminaran found in the brown algae, but lacks the terminal mannitol residues. Under conditions of Si starvation, diatoms will also accumulate lipid. The pigments found in diatoms include chlorophylls *a*, $c_1$, and $c_2$ and the carotenoids$\beta$-carotene, fucoxanthin (the major xanthophyll), diadinoxanthin, diatoxanthin, and neofucoxanthin.[14] Jørgensen[15] has recently reviewed the function of some of these pigments and other aspects of photosynthesis in the group. A few species of the pennate diatom genus *Nitzschia* are colorless obligate heterotrophs. The only flagellated cell known to occur in the group is the male gamete of the centric diatoms. The single anteriorly directed flagellum has a bilateral array of mastigonemes as in related groups of algae, but is unusual in having a 9 + 0 arrangement of microtubules.[16] In recent years, it is becoming increasingly evident that ultrastructural features, especially those of the chloroplast, are proving to be very useful taxonomic characters at the class and divisional level.[17] As in related groups of algae, the thylakoids are arranged in groups of threes, with a band of girdle lamellae encircling the periphery of the chloroplast just inside the chloroplast envelope. The chloroplast envelope is surrounded by two additional membranes of the endoplasmic reticulum (chloroplast endoplasmic reticulum) which are continuous with the nuclear envelope. The dominant vegetative phase in the diatom life cycle is diploid with meiosis occurring during gametogenesis.

The remainder of this section focuses on the characteristics and common genera of the two major groups of diatoms, the orders Centrales and Pennales. In centric diatoms, there is a radial symmetry to the ornamentation on the valve; the cells usually contain numerous small chloroplasts and a large vacuole. Sexual reproduction is oogamous, involving an egg and flagellated sperm. Centric diatoms are usually planktonic and are more common in marine and brackish waters than in freshwater. They appear to be the more primitive of the two groups, appearing in the fossil record for the first time in the Jurassic Period.[5] Unicellular genera, such as *Coscinodiscus* and *Cyclotella*, look much like miniature Petri dishes; in valve view, they appear circular, while in girdle view, they appear rectangular. In other unicellular genera, the cell is elongated in the valve to valve axis so as to appear square in girdle view. In *Rhizosolenia* the cell is highly elongated and appears long and needle-like in girdle view. Most of the length of the frustule is composed of scale-like intercalary bands, while the valves are small, pointed structures at either end. Many centric diatoms form long filaments in which the valve of one cell remains attached, by various structures, to the valve of the adjacent cell following cell division. In *Skeletonema*, a small, very common, and widely used experimental organism, the cells are joined by a ring of siliceous spines oriented perpendicular to the valve face. In the large and common filamentous genus *Chaetoceros*, each valve has two long spines which extend laterally on opposite sides of the cell. Each spine overlaps with the spine on the valve of the adjacent cell to hold the filament together. In *Thalassiosira* the *Coscinodiscus*-shaped cells are held together in a filament by a single, central strand of organic material. Filament formation and the associated spines in many genera of centric diatoms are thought to aid in flotation of these planktonic forms.

Frustule symmetry in pennate diatoms is bilateral; each valve is shaped something like a canoe rather than circular dish. The cells often appear oval in valve view and rectangular in girdle view. Since it is not easy to identify living pennate diatoms (even to genus), most taxonomic work is carried out on acid-cleaned material in which the features of the frustule ornamentation may be more easily discerned.[2-4,6] Genera commonly used as experimental organisms include *Amphora, Asterionella, Cocconeis, Cylindrotheca, Gomphonema, Navicula,* and *Nitzschia.* Mention should be made of *Phaeodactylum tricornutum* (sometimes referred to as *Nitzschia closterium* f. *minutissima).* This marine pennate diatom is easily cultured and has been used extensively as an example of a "typical" diatom in many biochemical, physiological, and ecological studies. In fact, this species is a very atypical pennate diatom. It occurs in three distinct growth forms: oval, fusiform, and triradiate.[18] The cell wall of all three forms is largely organic, with several narrow siliceous bands in the girdle region. The oval form may have one siliceous valve. It is not known if the morphological peculiarities of this organism extend to physiological parameters.

Pennate diatoms usually contain two chloroplasts or one large H-shaped chloroplast and small vacuoles at the ends of the cell. Sexual reproduction is accomplished through the formation and fusion of morphological identical amoeboid gametes. Pennate diatoms are abundant in both freshwater and marine habitats and are usually associated with a substrate of some sort. They may be attached to plants, animals, rocks, or sand grains by a mucilage pad or stalk, or they may be freely mobile on these surfaces or in muddy sediments. Motility in the vegetative state is restricted to pennate diatoms which possess a special longitudinal slit, called the raphe, in the cell wall. The mechanism of the characteristic gliding movement is not fully understood, although several theories have been proposed.[19] The genus *Asterionella* lacks a raphe and is a common component of the phytoplankton in both lakes and coastal waters. The cells are associated in stellate or spiral colonies.

# REFERENCES

1. Werner, D., Ed., *The Biology of Diatoms,* Blackwell Scientific, Oxford, 1977.
2. Bold, H. C. and Wynne, M. J., *Introduction to the Algae: Structure and Reproduction,* Prentice-Hall, Englewood Cliffs, N.J., 1978.
3. Patrick, R. and Reimer, C. W., *The Diatoms of the United States exclusive of Alaska and Hawaii, Vol. 1, Fragilariaceae, Eunotiaceae, Achnanthaceae, Naviculaceae,* Monograph of the Academy of Natural Science of Philadelphia, No. 13, 1966.
4. Patrick, R. and Reimer, C. W., *The Diatoms of the United States exclusive of Alaska and Hawaii, Vol. 2 (Part 1), Entomoneidaceae, Cymbellaceae, Gomphonemaceae, Epithemiaceae,* Monograph of the Academy of Natural Science of Philadelphia, No. 13, 1975.
5. Lewin, J. C. and Guillard, R. R. L., Diatoms, *Ann. Rev. Microbiol.,* 17, 373, 1963.
6. Trainor, F. R., *Introductory Phycology,* John Wiley & Sons, New York, 1978.
7. von Stosch, H. A. Manipulierung der Zellgrösse von Diatomeen im Experiment, *Phycologia,* 5, 21, 1965.
8. Darley, W. M., Silicification and calcification, in *Algal Physiology and Biochemistry,* Stewart, W. D. P., Ed., Blackwell Scientific, Oxford, 1974, chap. 24.
9. Werner, D., Silicate metabolism, in *The Biology of Diatoms,* Werner, D., Ed., Blackwell Scientific, Oxford, 1977, chap. 4.
10. Sullivan, C. W. and Volcani, B. E., The effects of silicic acid on DNA polymerase, thymidylate kinase and DNA synthesis in *Cylindrotheca fusiformis, Biochim. Biophys. Acta,* 308, 212, 1973.
11. Hecky, R. E., Mopper, K., Kilham, P., and Degens, E. T., The amino acid and sugar composition of diatom cell walls, *Mar. Biol.,* 19, 323, 1973.
12. Sullivan, C. W., Diatom mineralization of silicic acid. II. Regulation of $Si(OH)_4$ transport rates during the cell cycle of *Navicula pelliculosa, J. Phycol.,* 13, 86, 1977.

13. **Azam, F. and Chisholm, S. W.,** Silicic acid uptake and incorporation by natural marine phytoplankton populations, *Limnol. Oceanogr.,* 21, 427, 1976.
14. **Darley, W. M.,** Biochemical composition, in *The Biology of Diatoms,* Werner D., Ed., Blackwell Scientific, Oxford, 1977, chap. 7.
15. **Jørgensen, E. G.,** Photosynthesis, in *The Biology of Diatoms,* Werner, D., Ed., Blackwell Scientific, Oxford, 1977, chap. 5.
16. **Heath, I. B. and Darley, W. M.,** Observations on the ultrastructure of the male gametes of *Biddulphia levis* Ehr., *J. Phycol.,* 8, 51, 1972.
17. **Dodge, J. D.,** *The Fine Structure of Algal Cells,* Academic Press, New York, 1973.
18. **Borowitzka, M. A. and Volcani, B. E.,** The polymorphic diatom *Phaeodactylum tricornutum:* ultrastructure of its morphotypes, *J. Phycol.,* 14, 10, 1978.
19. **Harper, M. A.,** Movements, in *The Biology of Diatoms,* Werner, D., Ed., Blackwell Scientific, Oxford, 1977, chap. 8.

# GENERAL CHARACTERISTICS OF PHYTOPLANKTON: DINOFLAGELLATES

## A. R. Loeblich, III

## INTRODUCTION

Dinoflagellates (Division Pyrrhophyta), along with the diatoms and haptophytes, compose the bulk of the marine phytoplankton. They are rarely a major component of the freshwater phytoplankton community. There are thousands of described species, both extant and extinct, with an excellent fossil record of preserved cysts (presumably zygotes) back to the Triassic.[1] The dinoflagellates occupy more diverse types of niches in nature than any other algal division. They occur as free-living autotrophs or auxotrophs (for vitamins) and as saprophytes, phagotrophs, parasites, symbionts, and epiphytes. Undoubtably this is a reflection of a long evolutionary history.[2] The greatest diversity of these forms occurs in the sea.

They are predominantly unicellular, asymmetric in shape, and biflagellate, the exception being a few filamentous and parasitic multicellular forms. The pyrrhophytes are distinguished from other algae by the architecture of their cell covering (amphiesma),[3] the permanent condensation[4] and fibrous nature[5] of their chromosomes, high degree of DNA base substitution,[6] unique chromatin,[7] flagellar insertion and orientation, structure of the transverse flagella,[8] the xanthophyll peridinin,[9] unusual meiosis,[10] and the presence of bioluminescence in some marine forms.[11,12] The abundance of carotenoids gives the photosynthetic species a yellow-brown or reddish-brown color which suggested the term pyrrhophyte or flame plant.[13]

## SYSTEMATICS AND ARCHITECTURE OF THE AMPHIESMA

The cellular covering (amphiesma) of dinoflagellates[3,14] is unique and perhaps the most complicated among the algae. Because of the complexity of the amphiesma and variations in degree of development in different dinoflagellates, this character, in conjunction with cell shape, is relied on heavily by systematists in developing a classification scheme for dinoflagellates.[2,3] However, no adequate monograph exists. Several indexes of living taxa have been published recently. Reference 15 and its earlier editions index the genera, subgenera, and sections of the Pyrrhophyta, both fossil and recent. Reference 16 and two earlier parts catalog the living marine species.

The amphiesma of the motile stage can be described as a series of appressed membranes of which the outermost is interpreted to be the plasmalemma,[17] as it is reported to be continuous with the flagellar membrane[18] (see Figure 1). Body scales of organic composition have been detected covering the plasmalemma in a few marine species (see Figure 2). The *Oxyrrhis marina* scale (diameter 0.25 $\mu$m) is a flat coil of four or five turns with radial spokes connecting the coils.[19] *Heterocapsa triquetra* and a *Glenodinium* sp. (Cambridge 1116/2) both have scales (diameter of base 0.3 $\mu$m) of open framework construction.[20] Beneath the plasma membrane lies a series of flattened vesicles. In those species that are called armored or thecate, there are structures termed plates composed of cellulose[3] or cellulose-like glucan[21] located in the lumen of these vesicles. These cellulosic structures are called thecal plates, and the thecal plates collectively comprise the theca. The thecal plates overlap one another along their margins in a precise fashion (Reference 22 discusses these studies). The regions of plate overlap are termed sutures. The thecal plates have perforations through which trichocysts (also present in nonthecate species) are discharged (see Figure 3). Beneath the thecal vesicles,

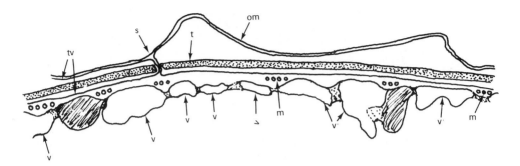

FIGURE 1. *Zooxanthella microadriatica* (isolate 395). The amphiesma (cross section) in the region of a suture(s) consists of an outer membrane (om), thecal vesicles (tv), and thecal plates (t). Sets of microtubules (m) are located beneath the thecal vesicles and above the peripheral vacuolar system (v).

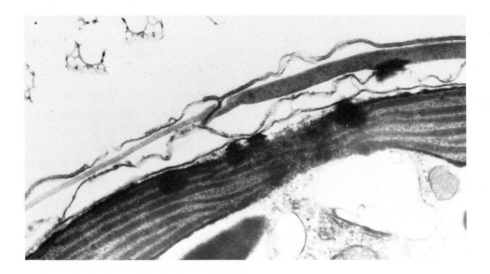

FIGURE 2. *Heterocapsa niei* (isolate 87-1). Electron micrograph of a portion of sectioned cell. The amphiesma, with newly formed thecal plate (thin) on left in contact with an older thecal plate (thick) on right. Body scales seen sectioned in various orientations to upper left of cell. Scale line = 1 μm.

various other structures may occur: membranes, a pellicle[3] (composed of a sporopollenin-like material), a system of vesicles, and arrays of microtubules (perhaps serving a skeletal function).

In armored forms, depending on the species, the daughter cells at cell division may either split the parental theca and each reform the missing complement,[23] or the parental wall may be discarded and each daughter cell synthesize an entirely new cellular covering.[3] The dinoflagellate theca forms within the plasmalemma.[17] Two patterns of thecal growth have been suggested.[23,24] Originally the theca was interpreted to grow in thickness centripetally by laying down new layers on the outside in contrast to the vascular plants which deposit new wall layers internal to the existing layer.[24] More recently wall growth has been found to occur at the base of the plate.[23] The surface area of the theca increases as a consequence of the deposition of new thecal material on the plate margin. The region of this secondary deposition is termed an intercalary band and frequently differs in surface ornamentation from the rest of the plate. Only certain margins of the thecal plates grow; depending on the species, either the overlapping or the underlapping or both plates may grow.[3,25]

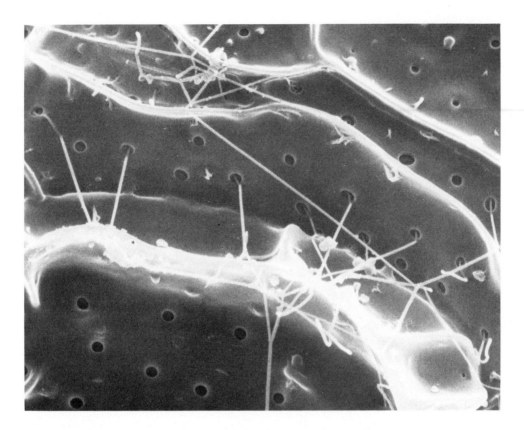

FIGURE 3. *Ceratocorys horrida* (isolate 496). Scanning electron micrograph of critical point dried cell. Close up of epitheca with showing trichocysts discharged through pores. Scale line = 5 μm.

The rigidity of the theca when present in the amphiesma, the facility of preservation by planktonologists of this layer, and the rich variety of thecal surface ornamentations has resulted in far more emphasis being devoted to studies of armored or thecate forms than the naked or unarmored species. For the paleontologists, the cyst wall is practically the only fossilizable portion of a dinoflagellate cell. The exception being silicified intracellular skeletal elements in a few species that are readily preserved.[26] Paleontologists have developed a well-conceived terminology[27] for the anatomy of the cyst wall in contrast to the variety of nonuniform terms applied by biologists to the amphiesma[3] of the living cell.

Dinoflagellates with a vegetative phase that is nonmotile are found in both marine (*Zooxanthella, Pyrocystis,* and others) and freshwater (*Cystodinium, and Stylodinium,* and others) habitats. The cell wall covering the nonmotile stage of the above-mentioned marine dinoflagellates is produced by the motile cell (zoospore) undergoing ecdysis, resulting in the thecal layer being shed and a cell wall that lacks sutures being formed.[28] In *Pyrocystis* this new wall appears to be an amplification of the pellicular layer, as this recently developed wall contains sporopollenin-like material.[29]

## PIGMENTS

All photosynthetic dinoflagellates contain chlorophyll *a* and chlorophyllide $c_2$.[30] The one exception, identified as *Prorocentrum cassubicum*, contains chlorophyll *a*, and chlorophyllides $c_1$, and $c_2$. It should be reanalyzed and the identity of the strain reconfirmed.

Carotenoids have been reported from two nonphotosynthetic dinoflagellates. *Cryp-thecodinium cohnii* contains only the hydrocarbons γ-carotene (major carotenoid present) and β-carotene.[31] *Protoperidinium ovatum* is reported to contain xanthophylls with conjugated carbonyl groups.[32] An array of carotenes have been reported from the symbiotic *Peridinium foliaceum* which are presumably located in the nonplastidic oil droplets of the dinoflagellate partner[32] (see Figure 4).

Numerous studies of photosynthetic dinoflagellates have revealed β-carotene to be the major carotene and peridinin, diadinoxanthin, and dinoxanthin to be major xanthophylls.[2,30] Peridinin is photosynthetically active and is the most abundant carotenoid of photosynthetic species, comprising up to 76% of the carotenoids of the cells.[4] Other reported[2] xanthophylls are astaxanthin, diatoxanthin, fucoxanthin, peridininol, pyrrhoxanthin, and pyrrhoxanthinol. As *Gyrodinium galatheanum* has recently been detected to have 19′- hexanoyloxyfucoxanthin as its major carotenoid,[74] those species previously reported to contain fucoxanthin should be reanalyzed.

In some dinoflagellates, the major carotenoid, peridinin, which transfers energy to the photosynthetic reaction center II with a high degree of efficiency, is located in a peridinin-chlorophyll protein (PCP) complex with a molecular weight of 34,500 to 39,000. The molar ratio of peridinin, chlorophyll *a*, and protein in the complex of *Amphidinium carterae* (Plymouth 450) is 9:2:1, and in *Heterocapsa pygmaea*, sp., it is 4:1:1.[34] Isoelectric focusing of the PCP proteins has revealed that some dinoflagellates have one protein dominant and at least five other forms with different isoelectric points present.[34,35] The protein may consist of one or two subunits. *Heterocapsa niei* is unique in having PCP proteins present in 12 isoelectric forms. Cytological and morphological evidence suggests that *H. niei* arose by a polyploidization which could account for it having twice as many of this protein as other species analyzed.[36]

## SYMBIOSIS

The only symbiotic associations known in this algal division involve marine dinoflagellates.[37] Dinoflagellates occur in symbiotic associations with a variety of animals who utilize carbon compounds excreted by the plant, e.g., photosynthate transported into the sea anemone *Anthopleura elegantissima* by *Zooxanthella* sp.[38] Members of the genus *Zooxanthella* occur as symbionts of all reef building corals and play a major role as primary producers in this tropical community.[39] Algal photosynthate translocated to the tip of coral branches greatly stimulates calcification by the coelenterate.[40] Zooxanthellae also occur in turbellarians, anemones, zoantharians, gorgonians, and scyphomedusans and in the mantle of giant clams. Zooxanthellae also occur in protozoa, e.g., in foraminifera, radiolaria, and acantharia. *Pyrocystis* species are reported to exist in the mesh work of the reticulopodia of planktonic foraminifera.[41] Some *Blastodinium* species infesting copepods may be photosynthetically active as they possess plastids.[42]

Nonphotosynthetic dinoflagellates occur in several symbiotic associations with photosynthetic unicellular algae. *Peridinium balticum* and *P. foliaceum* both harbor a photosynthetic unicellular organism which has been related to the Chrysophyta, Bacillariophyta, or Chloromonadophyta[43,44] (see Figure 5). A report of what appears to be a second nucleus in *Pseliodinium vaubanii* requires more investigation.[45] *Noctiluca scintillans* in the tropics harbors[46] the prasinophyte *Predinomonas noctilucae.*[47]

Two unicellular blue green algae (phaeosomes), *Synechococcus carcerarius* and *Synechocystis consortia*, occur symbiotically with oceanic dinoflagellates. *S. carcerarius* has been found in the cingular region of *Ornithocercus* spp., *Parahistioneis* sp., and *Histioneis* spp. and endocytically in *Amphisolenia globifera*. *S. consortia* was found

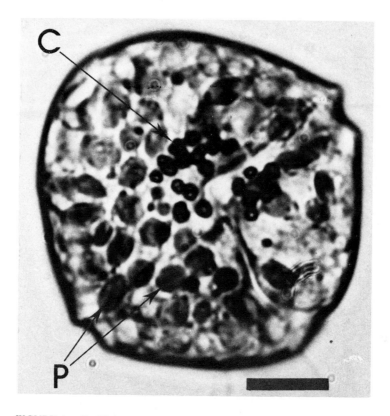

FIGURE 4.    *Peridinium foliaceum* (isolate 129). A photomicrograph of living cell with numerous small elliptical-shaped plastids (P) occurring in the periphery of the cell. Carotenoid granules (C) contained within the cytoplasm, appear as a dark bodies grouped in the sulcal region. Scale line = 10 μm.

in the cingular region of a *Parahistioneis* and *Histioneis* spp.[48] Phaeosomes have also been found in a specialized chamber in *Citharistes.*

## TOXINS

Dinoflagellates are known to produce a wide range of compounds toxic to animals. This property coupled with their ability to form water "blooms" has resulted in incidences of human poisoning and fish kills in various coastal regions of the world. Members of the genus *Gessnerium* (formerly in the genus *Gonyaulax*) produce saxitoxin and up to six derivatives of this compound.[49] These poisons are concentrated by filter feeding organisms and are responsible for paralytic poisoning. *Gymnodinium breve* produces an ichthyotoxic compound of unknown structure. Recently the ciguatera poison has been determined to originate in a dinoflagellate that is an epiphyte on the surface marine algae in the tropics. Herbivorous fish eating infested plants become ciguatoxic when the dinoflagellate is abundant.[50]

## NUTRITION

Most photosynthetic dinoflagellates are auxotrophs for vitamin $B_{12}$, thiamin, and biotin. Some species require all three, some require two, and others require only one of these vitamins. A few true autotrophs do exist. A freshwater *Gymnodinium* sp. (strain 160),[5] *Zooxanthella microadriatica* (strain 395),[28] and *Pyrocystis lunula* (strain

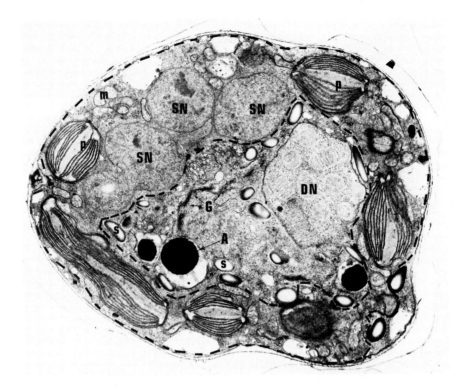

FIGURE 5.    Longitudinal section of a *P. foliaceum* cell illustrating the relationship of the various organelles. As a result of its amorphous nature, the symbiont nucleus (SN) appears, in thin section, as one or several spheres or multi-lobed bodies. The dinoflagellate nucleus (DN) is more regular in shape and contains chromosomes exhibiting the arc-like chromatin striation patterns characteristic of dinoflagellates. Plastids (P) are restricted to the outer portion of the cell. The ring-shaped genophore appears as white patches at either long end of the plastid. Tubules lace through the single, lens-shaped pyrenoid. Starch granules (S), accumulation bodies (A), and Golgi bodies (G) can be seen within the dinoflagellate cytoplasm. The path of the membrane enclosing the cytoplasm of the photosynthetic partner has been traced (dotted line). Scale line = 5 μm.

167)[51] have no vitamin requirements. Most of the nutritional data available for this group of organisms has been obtained with neritic forms which so far have proved the easiest to culture.[52] Other than *Pyrocystis* species the nutrition of truly oceanic forms has not been studied, consequently little can be said concerning their nutrient requirements. A particularly successful culture medium for marine species is GPM,[53] with salinities varied from 2.5 to 3.5%, depending on the preference of the dinoflagellate.[54] The *Pyrocystis* species grow well in medium f/2.[55] Freshwater photosynthetic species grow well in the *Navicula pelliculosa* culture medium.[55]

The nutrition of two nonphotosynthetic species, the saprophyte *Crypthecodinium cohnii*[57] and the phagotroph *Oxyrrhis marina*, have been intensively studied. *O. marina* has an obligate requirement for the light-labile compound ubiquinone.[58]

## CYTOLOGY

Reference 59 reviews the fine structure of dinoflagellates. The motile cell of almost all dinoflagellates has two unequal flagella. A single longitudinal flagellum occurs on the zoospores of *Pyrocystis lunula*[51] and the gametes of *Noctiluca scintillans*.[60] The transverse flagellum contains a paraflagellar rod within the flagellar membrane that is shorter than the axonemal microtubules which are in a helical configuration exterior to the paraflagellar rod.[8] Both flagella may have a coating of fine flexible hairs[18] and even scales as in *O. marina*.[19] A few species have a peduncle arising from the ventral area in close proximity to the point of flagellar insertion. The peduncle contains microtubules and is reminiscent of the haptonema of haptophytes. The peduncle microtubules continue into the cell for some distance and form a rosette-like structure in cross section.[61] Skeletal microtubules frequently occur above the interconnected vacuolar system of the amphiesma.[14]

The plastids have three limiting membranes[62] in common with the plastids of the englenoids. A variety of pyrenoids have been found.[63] Trichocysts have been found in all dinoflagellates[64] except for *Zooxanthella*.[28] Subcellular structures termed nematocysts are found in *Polykrikos* and *Nematodinium*. An organelle termed a pusule has been interpreted[65] as an osmoregulatory structure on the basis of its ultrastructure. An alternate explanation that it is an excretory organelle has recently been proposed.[65]

## NUCLEAR CYTOLOGY, GENETICS, AND LIFE CYCLES

Almost all dinoflagellates have permanently condensed chromosomes, even in the interphase. Exceptions are *Noctiluca*[67] and *Oodinium*,[68] in which the chromosomes decondense in the vegetative or feeding stage and recondense during the reproductive stage. Basic proteins in the chromatin of free-living dinoflagellates are undetectable histochemically. The chromatin of three dinoflagellates with permanently condensed chromosomes has been analyzed and lacks the typical array of eukaryotic histones present in both vertebrates and vascular plants. There is one histone-like protein present in about one tenth the amount of DNA present. This protein migrates in gel electrophoresis similarly to corn histone IV, but is 45% larger and contains fewer basic amino acids.[7] Histochemical studies of the chromosomes of the parasite *Syndinium* sp.[69] and other parasitic dinoflagellates[70] indicate that basic proteins do occur. The present inability to grow these parasites obviates a biochemical analysis. Correlated, perhaps, with the lack of eukaryotic histones and the presence of a low amount of chromosomal proteins is absence of nu-bodies[71] and the configuration[72] assumed by the chromatin which closely resembles $\psi$-DNA (see Figure 6).

A renaturation kinetic study of the DNA of *Crypthecodinium cohnii* reveals that about 55% of the DNA is repeated and 45% of the DNA is unique. An analysis of the repeated DNA sequences indicates that a variety of sequences are present and repeated to varying degrees and interspersed with unique DNA.[5]

Probably all dinoflagellates have a sexual cycle. Most species are homothallic.[2] Meiosis occurs at zygote germination, thus the vegetative cell is haploid.[73] Genetic analysis of flagellar mutants has shown only ditypic segregation at meiosis.[10] It is believed that dinoflagellate meiosis lacks DNA synthesis and that, as a consequence, the chromosomes are single stranded at pairing, resulting in a one-step meiotic division. All subsequent divisions are mitotic.

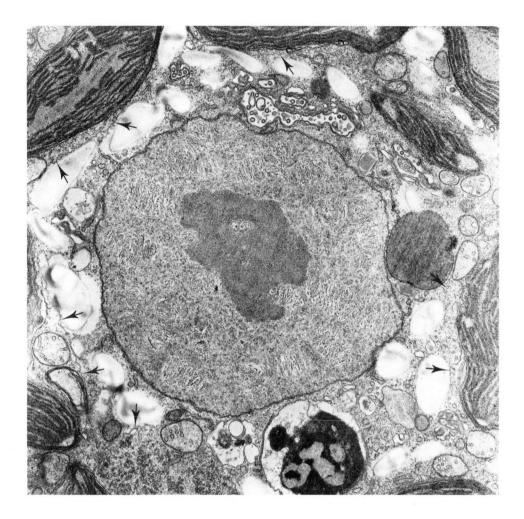

FIGURE 6.    *Peridinium foliaceum.* An electron micrograph of a section through the dinoflagellate nucleus. Arrows indicate bounding membrane of the symbiont encloses all the plastids, while excluding the trichocysts, starch granules, pusule, and dinoflagellate nucleus. Scale line = 2 μm.

# REFERENCES

1. Stover, L. E. and Evitt, W. R., Analyses of pre-Pleistocene organic-walled dinoflagellates, *Stanford Univ. Publ. Geol. Sci.,* 15, 1, 1978.
2. Loeblich, A. R., III, Dinoflagellate evolution: speculation and evidence, *J. Protozool.,* 23, 13, 1976.
3. Loeblich, A. R., III, The amphiesma or dinoflagellate cell covering, *Proc. N. Am. Paleont. Conv.,* G, 867, 1970.
4. Loeblich, A. R., III, Studies on synchronously dividing populations of *Cachonina niei,* a marine dinoflagellate, *Bull. Jpn. Soc. Phycol.,* 25 (Suppl.) (Mem. Iss. Yamada), 119, 1977.
5. Allen, J. R., Roberts, T. M., Loeblich, A. R., III, and Klotz, L. C., Characterization of the DNA from the dinoflagellate *Crypthecodinium cohnii* and implications for nuclear organization, *Cell,* 6, 161, 1975.
6. Rae, P. M. M., and Steele, R. E., Modified bases in the DNAs of unicellular eukaryotes: an examination of distributions and possible roles, with emphasis on hydroxymethyluracil in dinoflagellates, *Biosystems,* 10, 37, 1978.

7. Rizzo, P. J. and Noodén, L. D., Partial characterization of dinoflagellate chromosomal proteins, *Biochim. Biophys. Acta,* 349, 415, 1974.

8. Berdach, J. T., In situ preservation of the transverse flagellum of *Peridinium cinctum* (Dinophyceae) for scanning electron microscopy, *J. Phycol.,* 13, 243, 1977.

9. Strain, H. H., Svec, W. A., Aitzetmüller, K., Grandolfo, M. C., Katz, J. J., Kjøsen, H., Norgård, S., Liaaen-Jensen, S., Haxo, F. T., Wegfahrt, P., and Rapoport, H., The structure of peridinin, the characteristic dinoflagellate carotenoid, *J. Am. Chem. Soc.,* 93, 1823, 1971.

10. Beam, C. A., Himes, M., Himelfarb, J., Link, C., and Shaw, K., Genetic evidence of unusual meiosis in the dinoflagellate *Crypthecodinium cohnii, Genetics,* 87, 19, 1977.

11. Hastings, J. W., Dinoflagellate bioluminescence: molecular mechanisms and circadian control, in *Proc. 1st Int. Conf. Toxic Dinoflagellate Blooms,* LoCicero, V. R., Ed., Massachusetts Science Technology Foundation, Wakefield, Mass., 1975, 235.

12. Schmidt, R. J., Gooch, V. D., Loeblich, A. R., III, and Hastings, J. W., Comparative study of luminescent and nonluminescent strains of *Gonyaulax excavata* (Pyrrhophyta), *J. Phycol.,* 14, 5, 1978.

13. Pascher, A., Über Flagellaten und Algen, *Ber. Dtsch. Bot. Ges.,* 32, 136, 1914.

14. Dodge, J. D. and Crawford, R. M., A survey of thecal fine structure in the Dinophyceae, *Bot. J. Linn. Soc.,* 63, 53, 1970.

15. Loeblich, A. R., Jr. and Loeblich, A. R., III, Index to the genera, subgenera and sections of the Pyrrhophyta. VII, *Phycologia,* 13, 57, 1974.

16. Sournia, A., Catalogue des espéces et taxons infraspecifiques de dinoflagellés marins actuels publiés depuis la revision de J. Schiller, *Rev. Algol.,* n.s., 13, 3, 1978.

17. Wetherbee, R., The fine structure of *Ceratium tripos,* a marine armored dinoflagellate I. The cell covering (theca), *J. Ultrastruct. Res.,* 50, 58, 1975.

18. Dodge, J. D. and Crawford, R. M., Fine structure of the dinoflagellate *Oxyrrhis marina.* II. The flagellar system, *Protistologica,* 7, 399, 1971.

19. Clarke, K. J. and Pennick, N. C., The occurrence of body scales in *Oxyrrhis marina* Dujardin, *Br. Phycol. J.,* 11, 345, 1976.

20. Pennick, N. C. and Clarke, K. J., The occurrence of scales in the peridinian dinoflagellate *Heterocapsa triquetra* (Ehrenb. Stein, *Br. Phycol. J.,* 12, 63, 1977.

21. Nevo, Z. and Sharon, N., The cell wall of *Peridinium westii,* a non cellulosic glucan, *Biochim. Biophys. Acta,* 173, 161, 1969.

22. Loeblich, A. R., III and Loeblich, L. A., The systematics of *Gonyaulax* with special reference to the toxic species. Toxic dinoflaggellate blooms, in *Proc. 2nd Int. Conf. Toxic Dinoflagellate Blooms, Developments in Marine Biology,* Vol. 1, Taylor, D. L. and Seliger, H. H., Eds., Elsevier North-Holland, New York, 1979, 41.

23. Wetherbee, R., The fine structure of *Ceratium tripos,* a marine armored dinoflagellate. III. Thecal plate formation, *J. Ultrastruct. Res.,* 50, 77, 1975.

24. Schütt, F., Die Erklärung des centrifugalen Dickenwachsthums der Membran, *Bot. Zeit.,* 58, 245, 1900.

25. Gocht, H. and Netzel, H., Rasterelektronenmikroskopische Untersuchungen an Panzer von *Peridinium* (Dinoflagellata), *Arch. Protistenkd.,* 116, 381, 1974.

26. Loeblich, A. R., III, Loeblich, L. A., Tappan, H. N., and Loeblich, A. R., Jr., Annotated index of fossil and recent silicoflagellates and ebridians with descriptions and illustrations of validly proposed taxa, *Mem. Geol. Soc. Am.,* 106, 1, 1968.

27. Evitt, W. R., Lentin, J. K., Millioud, M. E., Stover, L. E., and Williams, G. L., Dinoflagellate Cyst Terminology, Canada Geological Survey Paper, 76-24, 1977, 1.

28. Loeblich, A. R., III and Sherley, J. L., Observations on the theca of the motile phase of freeliving and symbiotic isolates of *Zooxanthella microadriatica* (Freudenthal) comb. nov., *J. Mar. Biol. Assoc. U. K.,* 59, 195, 1979.

29. Swift, E. and Remsen, C. C., The cell wall of *Pyrocystis* spp. Dinococcales, *J. Phycol.,* 6, 79, 1970.

30. Jeffrey, S. W., Sielicki, M., and Haxo, F. T., Chloroplast pigment patterns in dinoflagellates, *J. Phycol.,* 11, 374, 1975.

31. Tuttle, R. C. and Loeblich, A. R., III, N-methyl-N'-nitrosoquanidine and UV induced mutants of the dinoflagellate *Crypthecodinium cohnii, J. Protozool.,* 24, 313, 1977.

32. Neveux, J. and Soyer, M.-O., Caracterisation des pigments et structure fine de *Protoperidinium ovatum* Pouchet (Dinoflagellata), *Vie Milieu Ser. A,* 26, 175, 1976.

33. Withers, N. and Haxo, F. T., Chlorophyll $c_1$ and $c_2$ and extraplastidic carotenoids in the dinoflagellate, *Peridinium foliaceum* Stein, *Plant Sci. Lett.,* 5, 7, 1975.

34. Siegelman, H. W., Kycia, J. H., and Haxo, F. T., Peridinin-Chlorophyll *a* — proteins of dinoflagellate algae, *Brookhaven Symp. Biol.* , 28, 162, 1977.

35. Prezelin, B. B. and Haxo, F. T., Purification and characterization of peridinin-chlorophyll *a*-proteins from the marine dinoflagellates *Glenodinium* sp. and *Gonyaulax polyedra, Planta,* 128, 133, 1976.

36. Loeblich, A. R., III, Schmidt, R. J., and Sherley, J. L., Scanning electron microscopy of *Heterocapsa pygmaea* sp. nov., and evidence for polyploidy as a speciation mechanism in dinoflagellates, *J. Plankton Res.*, in press.

37. Taylor, D. L., Symbiotic dinoflagellates, *Symp. Soc., Exp. Biol.*, 29, 267, 1975.

38. Muscatine, L. and Hand, C., Direct evidence for the transfer of materials from symbiotic algae to the tissues of a coelenterate, *Proc. Natl. Acad. Sci. U.S.A.*, 44, 1259, 1958.

39. Muscatine, L. and Cernichiari, E., Assimilation of photosynthetic products of zooxanthellae by a reef coral, *Biol. Bull.*, 137, 506, 1969.

40. Pearse, V. B. and Muscatine, L., Role of symbiotic algae zooxanthellae) in coral calcification, *Biol. Bull.*, 141, 350, 1971.

41. Bé, A. W. H., Cover, *Science*, 192, 829, 1976.

42. Soyer, M.-O., Etude ultrastructurale de l'endoplasme et des vacuoles chez deux types de Dinoflagellés appartenant aux genres *Noctiluca* (Suriray) et *Blastodinium* (Chatton), *Z. Zellforsch. Mikrosk. Anat.*, 105, 350, 1970.

43. Tomas, R. N., Cox, E. R., and Steidinger, K. A., *Peridinium balticum* (Levander) Lemmermann, and unusual dinoflagellate with a mesocaryotic and an eucaryotic nucleus, *J. Phycol.*, 9, 91, 1973.

44. Jeffrey, S. W. and Vesk, M., Further evidence for a membrane bound endosymbiont within the dinoflagellate *Peridinium foliaceum*, *J. Phycol.*, 12, 450, 1976.

45. Jacques, G. and Soyer, M.-O., Nouvelles observations sur *Pseliodinium vaubanii* (Sournia) dinoflagellé libre planctonique, *Vie Milieu Ser. A*, 27, 83, 1977.

46. Sweeney, B. M., *Pedinomonas noctilucae* (Prasinophyceae), the flagellate symbiotic in *Noctiluca* (Dinophyceae) in Southeast Asia, *J. Phycol.*, 12, 460, 1976.

47. Subrahmanyan, R., A new member of the Euglenineae, *Protoeuglena noctilucae* gen. et sp. nov., occurring in *Noctiluca miliaris* Suriray, causing green discoloration of the Sea off Calicut, *Proc. Indian Acad. Sci.*, 39, 118, 1954.

48. Norris, R. E., Algal consortisms in marine plankton, in *Proc. Seminar Sea, Salt and Plants*, Krishnamurthy, V., Ed., Central Salt and Marine Chemicals Research Institute, Bhavnagar, India, 1966, 178.

49. Shimizu, Y., Alam, M., Oshima, Y., Buckley, L. J., Fallon, W. E., Kasai, H., Miura, I. Gullo, V. P., and Nakanishi, K., Chemistry and distribution of deleterious dinoflagellate toxins, in *Marine Natural Products Chemistry*, Faulkner, D. J., and Fenical, W. H., Eds., Plenum Press, New York, 1977, 261.

50. Yasumoto, T., Inoue, A., and Bagnis, R., Ecological survey on a toxic dinoflagellate associated with Ciguatera, Toxic dinoflagellate blooms, *Proc. 2nd Int. Conf. on Toxic Dinoflagellate Blooms, Developments in Marine Biology*, Vol. 1, Taylor, D. L. and Seliger, H. H., Eds., Elsevier North-Holland, New York, 1979, 221.

51. Loeblich, A. R., III, *Pyrocystis lunula* and its relationship to *Sporodinium*, *J. Protozool.*, 21, 435, 1974.

52. Loeblich, A. R., III, Aspects of the physiology and biochemistry of the Pyrrhophyta, *Phykos*, 5, 216, 1967.

53. Loeblich, A. R., III, A seawater medium for dinoflagellates and the nutrition of *Cachonina niei*, *J. Phycol.*, 11, 80, 1975.

54. Schmidt, R. J. and Loeblich, A. R., III, Distribution of paralytic shellfish poison among Pyrrhophyta, *J. Mar. Biol. Assoc. U.K.*, 59, 479, 1979.

55. Guillard, R. R. L. and Ryther, J. H., Studies of marine planktonic diatoms. I. *Cyclotella nana* Hustedt, and *Detonula confervacea* (Cleve) Gran, *Can. J. Microbiol.*, 8, 229, 1962.

56. Lee, R. F. and Loeblich, A. R., III, Distribution of 21:6 hydrocarbon and its relationship to 22:6 fatty acid in algae, *Phytochemistry*, 10, 593, 1971.

57. Tuttle, R. C. and Loeblich, A. R., III, An optimal growth medium for the dinoflagellate *Crypthecodinium cohnii*, *Phycologia*, 14, 1, 1975.

58. Droop, M. R. and Pennock, J. F., Terpenoid quinones and steroids in the nutrition of *Oxyrrhis marina*, *J. Mar. Biol. Assoc. U.K.*, 51, 455, 1971.

59. Dodge, J. D., Fine structure of the Pyrrhophyta, *Bot. Rev.*, 37, 481, 1971.

60. Soyer, M.-O., Les ultrastructures liees aux fonctions de relation chez *Noctiluca miliaris* S. (Dinoflagellata), *Z. Zellforsch Mikrosk. Anat.*, 104, 29, 1970.

61. Lee, R. E., Saprophytic and phagocytic isolates of the colourless heterotrophic dinoflagellate *Gyrodinium lebouriae* Herdman, *J. Mar. Biol. Assoc. U.K.*, 57, 303, 1977.

62. Dodge, J. D., A survey of chloroplast ultrastructure in the Dinophyceae, *Phycologia*, 14, 253, 1975.

63. Dodge, J. D. and Crawford, R. M., A fine-structural survey of dinoflagellate pyrenoids and food-reserves, *Bot. J. Linn. Soc.*, 64, 105, 1971.

64. Bouck, G. B. and Sweeney, B. M., The fine structure and ontogeny of trichocysts in marine dinoflagellates, *Protoplasma*, 61, 205, 1966.

65. Dodge, J. D., The ultrastructure of the dinoflagellate pusule: a unique osmo-regulatory organelle, *Protoplasma*, 75, 285, 1972.
66. Loeblich, A. R., III, Sherley, J. L., and Schmidt, R. J., The correct position of flagellar insertion in *Prorocentrum* and description of *Prorocentrum rhathymum* sp. nov. (Pyrrhophyta), *J. Plankton Res.*, 1, 113, 1979.
67. Soyer, M.-O., Les ultrastructures nucleaires de la noctiluque (Dinoflagellé libre) au cours de la sporogenèse, *Chromosoma*, 39, 419, 1972.
68. Cachon, J. and Cachon, M., Observations on the mitosis and on the life cycle of *Oodinium*, a parastic dinoflagellate, *Chromosoma*, 60, 237, 1977.
69. Ris, H. and Kubai, D. F., An unusual mitotic mechanism in the parasitic protozoan *Syndinium* sp., *J. Cell Biol.*, 60, 702, 1974.
70. Hollande, A., Etude comparée de la mitose syndinienne et de celle des péridiniens libres et des hypermastigines infrastucture et cycle évolutif des syndinides parasites de radiolaires, *Protistologica*, 10, 413, 1975.
71. Hamkalo, B. A. and Rattner, J. B., The structure of a mesokaryote chromosome, *Chromosoma*, 60, 39, 1977.
72. Livolant, F., Girand, M.-M., and Bouligand, Y., A goniometric effect observed in sections of twisted fibrous materials, *Biol. Cell*, 31, 159, 1978.
73. von Stosch, H. A., Observations on vegetative reproduction and sexual life cycles of two freshwater dinoflagellates, *Gymnodinium pseudopalustre* Schiller and *Woloszynskia apiculata* sp. nov., *Br. Phycol. J.*, 8, 105, 1973.
74. Bjornland, T. and Tangen, K., Pigmentation and morphology of a marine Gyrodinium (Dinophyceae) with a major carotenoid different from peridinin and fucoxanthin, *J. Phycol.*, 15, 457, 1979.

# GENERAL CHARACTERISTICS OF RED MACROALGAE: *PORPHYRA*

## Akio Miura

## MORPHOLOGY AND TAXONOMY

*Porphyra* is a genus (belonging to the Bangiaceae, Bangiales, Bangiophycideae, Rhodophyceae, and Rhodophycophyta) with nearly 100 species in the world.[1] *Porphyra* has two morphologically differing stages during its life cycle. The stage usually seen on shore is macroscopic and leafy (see Figure 1A); the other stage, not usually seen, is microscopic and filamentous (see Figure 1B). Filamentous-stage plants are known as "Conchocelis" filaments. Leafy-stage plants have a simple blade upon a minute holdfast and vary in shape from oblanceolate to orbiculate to ovate, attaining 2 to 100 cm in length and 0.5 to 30 cm in width. The color of leafy plants varies according to the species and environmental conditions. Most species are purplish black when fresh; the shade of leafy plants depends mainly on the ratio of phycobilins to chlorophyll contained in chloroplasts. Vegetative cells which compose a leafy plant are polygonal or spherical and are arranged pavement-like in surface view (see Figure 2A), but appear rectangular in sectional view (see Figure 2D).

In sectional view, the leafy plants are 20 to 250 $\mu$m in thickness and are composed of monostromatic, distromatic, or partially distromatic cell layers according to species. Monostromatic plants contain one or two chloroplasts in each cell; distromatic plants contain a single chloroplast in each cell. Partially distromatic plants contain a single chloroplast in each cell at the distromatic portion and two chloroplasts in each cell at monostromatic the portion. The genus is subdivided into three subgenera based on the structure in sectional view and the number of chloroplasts in each cell. Monostromatic plants containing only one chloroplast in each cell belong to subgenus *Porphyra*. Monostromatic plants containing two chloroplasts in each cell and partially distromatic plants containing two chloroplasts in each cell belong to subgenus *Diplastida*. Distromatic plants containing a single chloroplast in each cell belong to subgenus *Diploderma*.[2] The basal part of a leafy plant forms a minute holdfast which is composed of rhizoidal cells (see Figure 2B).

The plant at the filamentous stage is composed of filamentous cells 2 to 3 $\mu$m thick (see Figure 1B). Filamentous plants penetrate calcareous matrix, e.g., mollusk shells, egg shells, and calcite grains, but can grow without calcareous matrices. When filamentous plants inhabit the calcareous matrix, to the naked eye they are discernible as blackish or reddish spots on the surface of the matrix and appear pinnately branched under a microscope. When the filamentous plants grow in a free-living condition, they form an intertwining mass.

## REPRODUCTION AND LIFE CYCLE

Leafy plants of some species propagate asexually by neutral spores formed on the sexually immature small leafy plants (see Figure 3). The mature leafy plants release spermatia (see Figure 2G) as male gametes and carpogonia (see Figure 2E) as female gametes. They are formed at the margin of the blade. Spermatangial portions are a yellowish color, and carpogonial portions are a deeper color than the vegetative portions. Some species are monoecious; others are dioecious, androdioecious, or trioecious according to the species.

Spermatangia and carpogonia are converted from ordinal vegetative cells. Spermatangia repeat periclinal and anticlinal divisions until mature spermatangia have been

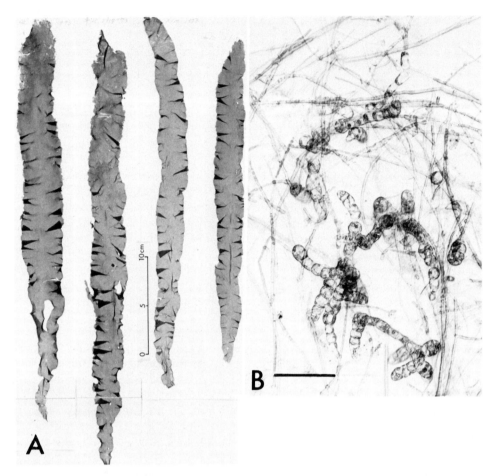

FIGURE 1.   *P. yezoensis* Ueda. (A) Habit of leafy plants, (B) photomicrograph of filamentous plant, "Conchocelis" stage, bearing conchosporangial branch in free-living condition. Scale bar = 50 μm.

formed. Consequently, 32, 64, 128, or 256 spermatia are produced from a single sper-matangium; the maximum number of spermatia produced in a single spermatangium is definite according to the species. Spermatia are released into the water by deteriora-tion of a spermatangial wall. The carpogonium is ellipsoidal, with a trichogyne at one or both sides of the blade (see Figure 2E). The spermatia do not have flagella, but are transported by water movement and become attached to the trichogynes; fertilization between the carpogonia and spermatia then is completed.[3]

Fertilization is followed by repeated periclinal and anticlinal divisions of the fertil-ized carpogonia forming carpospores. Depending on the species, 2, 4, 8, 16, or 32 carpospores are produced in a single carposporangium (see Figure 2F). Carpospores are released by deterioration of carposporangial wall; they have a single stellate chlo-roplast. Released carpospores immediately germinate into unipolar sporlings which give rise to the filamentous plant "Conchocelis". The filamentous plants form con-chosporangial branches (see Figure 1B) which consist of uniserial or multiserial quad-rate cells thicker than the filamentous vegetative cells. The filamentous vegetative cells contain a single parietal chloroplast, while the cells of conchosporangial branches con-tain a single stellate chloroplast. Each conchosporangium asexually produces one, two, or four conchospores. They are released through an opening at the distal end of the conchosporangial cell row after deterioration of the spermatangial wall, particularly

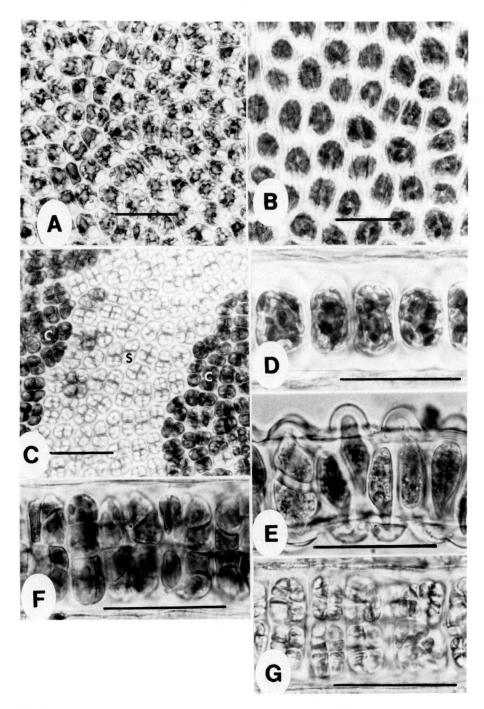

FIGURE 2.    *P. yezoensis* Ueda. Photomicrographs of leafy plant. Scale bar = 50 μm. (A) Surface view of vegetative cells, (B) surface view of rhizoidal cells, (C) surface view of spermatangia (s) and carposporangia (c), (D) sectional view of vegetative cells, showing monostromatic structure and containing a single chloroplast in each cell, (E) sectional view of carposporangia, (G) sectional view of spermatangia.

the transverse septum. Released conchospores immediately germinate into bipolar sporlings which give rise to the leafy plants.

Most of the species which have been studied showed the development of the filamen-

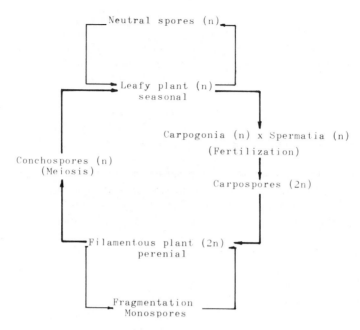

FIGURE 3.   Diagrammatic representation of life cycle in *Porphyra*. The thick line represents the basic course of the life cycle, and the thin line represents the accessory course which some species follow.

tous plants. However, two species from the west coast of North America did not show this development.[4,5] In some species, a bud-like structure formed on the filamentous plant directly gives rise to a leafy plant without producing conchospores.[6,7] The filamentous plants propagate by fragmenting vegetative filaments, in some species by monospores formed on the filaments (see Figure 3). There are variations of the life cycle in the *Porphyra* species. In Figure 3, a representative life cycle is diagrammatically shown.

The leafy plant has a haploid chromosome number; the filamentous plant has a diploid chromosome number. Meiosis occurs during the formation of conchospores.[8-10] The haploid chromosome number of the species of *Porphyra* varies from two to seven according to the species.[10,11] The species with three in the haploid chromosome number of 3 predominates in the number of species examined.

## ECOLOGY

*Porphyra* ranges widely from polar to tropical seas. In cold-current seas, leafy plants are seasonally luxuriant from spring to summer, and in warm-current seas, from autumn to spring through winter. Leafy plants of most species grow in the intertidal zone, but some species occur in the subtidal zone.[2,5] The leafy plants attach themselves by a minute holdfast to a substratum, such as rocks, pebbles, mollusk shells, and other seaweeds and also to woods, bamboo, natural and synthetic fibers, and rubber and concrete structures. Filamentous plants inhabit and penetrate in a calcareous matrix of dead and living mollusk shells. Under the most favorable cultivating conditions in Japan, the maximum yield attains 15 to 30 kg (in dry weight)/18 × 1.2 m Nori net/6 months (October to March).

## CULTURE IN LABORATORY

Provasoli's ES formula medium[1] gives satisfactory results for the culture of leafy and filamentous plants. Sensitivity of the leafy plant and the filamentous plant to light and temperature is different according to the species. The formation of conchospores in the filamentous plant of *Porphyra tenera*, autumn to spring species in Japan, is regulated by a definite photoperiod. A treatment of 7 days under short-day conditions is effective for the formation of conchospores. In *P. yezoensis* and *P. tenera* (autumn to spring species in Japan), for the germination and vegetative growth of filamentous plants, culture conditions at 20 to 24°C for 14 light:10 dark photoperiod under 500 to 1000 lx, using fluorescent cool white illumination, are favorable. When the culture is kept under this condition, vegetative growth is maintained without the formation of conchospores. For the formation of conchospores, culture conditions are 20 to 24°C for 10 light:14 dark photoperiod under 4000 lx. For conchospore release and germination, culture conditions at 13 to 15°C for 10 light:14 dark photoperiod under 6000 lx are favorable. When the conchosporangia are mature, their release takes place on the fourth day after the time the culture is transferred from 20 to 24 to 18°C. The leafy plants grow well under the same conditions and produce carpospores on the 30th to 40th day after the conchospore germination. In *P. miniata*, a summer annual species in the west coast of North America, the formation of conchospores takes place at 13 to 15°C for 10 light:14 dark photoperiod under 200 to 400 lx, and the conchospore release takes place at a low temperature of 3 to 7°C.[14] In *P. purpurea* (as *P. umbilicalis*), a summer annual species in the eastern district of Hokkaido, Japan, the formation and release of conchospores occurs under a long-day condition of 12 to 14 light:12 to 10 dark photoperiod and a low temperature below 20°C.[15] In *P. linearis*, a winter annual species of the north Atlantic coast, the photoperiodic control of the formation and release of conchospores is not observed; conchospore release occurs only at 13°C.[16]

## REFERENCES

1. Bold, H. C. and Wynne, M. J., *Introduction to the Algae,* Prentice-Hall, Englewood Cliffs, N.J., 1978, 477 and 577.
2. Kurogi, M., Systemetics of *Porphyra* in Japan, in *Contribution to the Systematics of Benthic Marine Algae of the North Pacific,* Abbott, I. A. and Kurogi, M., Eds., Japanese Society of Phycology, Kobe, Japan, 1972, 167.
3. Hawkes, M. W., Sexual reproduction in *Porphyra gardneri* (Smith et Hollenberg) Hawkes (Bangiales, Rhodophyta), *Phycologia,* 17, 329, 1978.
4. Conway, E. and Cole, K., Studies in the Bangiaceae: structure and reproduction of the conchocelis of *Porphyra* and *Bangia* in culture (Bangiales, Rhodophyceae), *Phycologia,* 16, 14, 1977.
5. Conway, E., Mumford, T. F., Jr., and Scagel, R. F., The genus *Porphyra* in British Columbia and Washington, *Seysis,* 8, 185, 1975.
6. Miura, A., A new species *Porphyra* and its *Conchocelis*-phase in nature, *J. Tokyo Univ. Fish.,* 47, 305, 1961.
7. Krishnamurthy, V., The *Conchocelis* phase of three species of *Porphyra* in culture, *J. Phycol.,* 5, 42, 1969.
8. Migita, S., Cytological studies on *Porphyra yezoensis* Ueda *Bull. Fac. Fish. Nagasaki Univ.,* 24, 55, 1967.
9. Giraud, A. and Magne, F., La place de la méiose dans le cycle de développment de *Porphyra umbilicalis, C. R. Acad. Sci. Ser. D,* 267, 586, 1968.
10. Kito, H., Cytological studies on genus *Porphyra, Bull. Tohoku Reg. Fish. Res. Lab.,* 39, 29, 1978.

11. **Mumford, T. F., Jr., and Cole, K.,** Chromosome numbers for fifteen species in the genus *Porphyra* (Bangiales, Rhodophyta) from the west of North America, *Phycologia,* 16, 373, 1977.

12. **Dring, M. J.,** Effects of daylength on growth and reproduction of the conchocelis of *Porphyra tenera, J. Mar. Biol. Assoc. U.K.,* 47, 501, 1967.

13. **Rentschler, H. G.,** Photoperiodische Induktion der Monosporenbildung bei *Porphyra tenera* Kjellm. (Rhodophyta-Bangiophyceae), *Planta,* 6, 216, 1967.

14. **Chen, L. C-M., Edelstein, T. E., Ogata, E., and MacLachlan, J.,** The life history of *Porphyra miniata, Can. J. Bot.,* 48, 385, 1970.

15. **Kurogi, M., Sato, S., and Yoshida, T.,** Effects of water temperature on the liberation of monospores from the conchocelis of *Porphyra umbilicalis* (L.) Kütz., *Bull. Tohoku Reg. Fish. Res. Lab.,* 27, 131, 1967.

16. **Bird, C. J., Chen, L. C-M., and McLachlan, J.,** The culture of *Porphyra linearis* (Bangiales, Rhodophyceae), *Can. J. Bot.,* 50, 1859, 1972.

# GENERAL CHARACTERISTICS OF RED MACROALGAE: *EUCHEUMA*

## Clinton J. Dawes

## INTRODUCTION

*Eucheuma* is a pantropical red algal genus belonging to the family Solieriaceae and the order Gigartinales. The primary taxonomic feature of *Eucheuma* is the presence of a nonprocarpic central fusion cell producing the carposporophyte, zonale tetrasporangia, and a filamentous or rhizoidal medulla.[1] The genus is of special interest because of the phycocolloid, carrageenan, that is present in the cell wall.[2] The species are typically large (0.5 to 2 m tall), usually red to orange-yellow in color, and fleshy to cartilaginous in texture. The branching pattern is subdichotomous to irregular, with relatively thick (0.5 to 2 + cm in diameter) branches that may be spiny or covered with knobs and irregular protrusions (see Figures 1 and 2).

## TAXONOMY

Forty eight species have been ascribed to the genus.[3] Weber-van Bossee[4] placed over 30 species in either the axifera (plants that have a distinct central medulla of filaments) or the anaxifera (plants that lack a distinct central filamentous medulla). Six species are known from the Caribbean (see Table 1), and all of these species are axiferous and yield iota carrageenan. Cheney,[3] in a biosystematic study of Floridian *Eucheuma*, demonstrated that only *Eucheuma isiforme* had a fusion cell, the primary taxonomic feature of the genus. He suggested that the other four species be removed to a new genus, *Meristiella*, in which the sterile tissue cystocarp construction is used as the primary taxonomic feature. Furthermore, Cheney compared *E. isiforme* with *Eucheuma nudum* using electrophoretic and morphological characters and concluded that these are varieties of a single species *E. isiforme*.

One species, *Eucheuma uncinatum* is endemic to the central region of the Gulf of California.[5] It strongly resembles *E. isiforme* in morphology and the presence of a dense medulla and iota carrageenan.[6] Doty and Santos[7] suggest that there are probably 15 species of *Eucheuma* present in the central and western tropical Pacific, with most species being anaxiferous and yielding *x* carrageenan. However, *Eucheuma denticulatum* and *Eucheuma gelatinae* are both axiferous and yield *ι* carrageenan.[7,8] The one exception to the axiferous and *ι* carrageenan appears to be *Eucheuma arnoldii* (see Table 1) in which the *ι* carrageenan-producing plant is anaxiferous.

## CARRAGEENANS FROM *EUCHEUMA*

Carrageenans are sulfated galactans that can be extracted with hot water. They are used as thickeners, stabilizers, and gelling agents for food processing, as well as cosmetic and industrial purposes.[9,10] Various species of *Eucheuma* have been shown to yield between 60 and 70% carrageenan from sun-dried plants.[6,7,11-13] Although a variety of carrageenans are known from various members of the Gigartinales, apparently only *ι* and *x* carrageenan occur in species of *Eucheuma*. Only *ι* carrageenan has been found in the Caribbean and Gulf of California species,[6,12] while the majority of Pacific species produce *x* carrageenan.[7] Of the *ι* carrageenan-producing species studied, there appear to be two forms based on level of sulfation, a "typical" high sulfated form extracted from *E. denticulatum* and *E. uncinatum* of the Pacific and Gulf of California

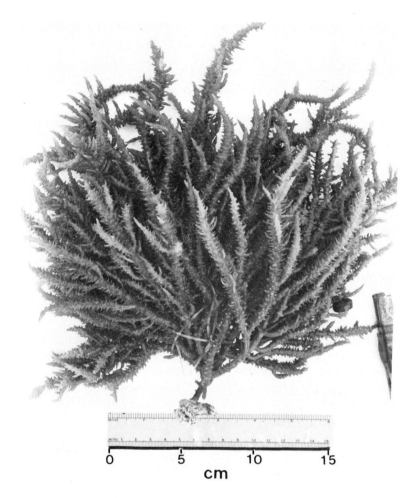

FIGURE 1.    A mature plant of *E. isiforme* from the Florida Keys growing in 1 m of water. Note the heavy spinose appearance and dense branching.

and a lower sulfated form extracted from the Caribbean species of *Eucheuma*.[6] Thus far, all species that were found to be reproductive show no distinction in carrageenan chemistry between the tetrasporic and cystocarpic plants in contrast to other carrageenan producing members of the Gigartinales.[6,7,12]

Farming of *Eucheuma straitum* (in the trade called "cottonii")[13] is extensive in the Philippines,[14,15] with over 4000 metric ton exported in 1977 at a value of $40 to $60 U.S. per metric ton.[16] The predicted exportation of around 6000 metric ton in 1978 would indicate that the farming is still expanding. Table 2 summarizes the export levels of *Eucheuma* from the Philippines and indicates a rapid rise to 1974 and then a major fall back in production because of overproduction.[16] The farming procedure is simple, using a series of monofilament lines (approximately 200-lb test) tied to wooden stakes. The plant cuttings of *Eucheuma* are tied to the 10-m long nylon line spaced every 0.2 m with a series of lines about 0.5 m apart. A farm usually has about 100,000 plants per hectare. In 1977 it was estimated that there were over 350 family farms averaging about 1.5 million plants each of *E. denticulatum* and about 600 family farms growing about 3.1 million plants of *E. striatum*. The Tamblang strain of *E. striatum* is the most-desired form in farming, and it has been found to have the broadest tolerances and most-rapid growth rates of all species grown in the Philippines Philippines.[15] Be-

FIGURE 2. A mature, cystocarpic plant of *E. nudum* from the west coast of Florida collected in about 30 m of water. The cystocarps are small bulbous branchlets.

cause of the high value placed on ι carrageenan, there is general agreement that cultivation of plants producing ι carrageenan is necessary.[10]

## ECOLOGY AND LIFE HISTORIES

The plants are usually attached to firm substances in areas of high water movement.[13,17] The depths are usually shallow (1 to 5 m) and in clear water. Although most species tend to be bushy, secondary attachments will result in a prostrate form, and the plant is entirely prostrate in a few Pacific species.[13] All the Caribbean species appear to have a typical triphasic red algal life history,[3,17] with the plants becoming reproductive in the fall. The most commonly found stage in Caribbean populations are tetrasporic plants (diploid), with cystocarpic (female) plants common and spermatangial (male) plants being rare. *E. uncinatum* from the Gulf of California is reproductive in late March,[5] and female and tetrasporic plants have been collected.[6] Both the Caribbean and the Gulf of California species tend to break down after spore production.[6,12] The majority of species of Pacific *Eucheuma* appear to have incomplete or modified life histories, with only tetrasporic or sterile plants present.[7]

Table 1
PARTIAL LIST OF SPECIES OF *EUCHEUMA* AND ASSOCIATED
CHARACTERISTICS

| Species | Medullary type | | Carrageenan type | | Ref. |
|---|---|---|---|---|---|
| | Axiferous | Anaxiferous | ι | x | |
| Caribbean | | | | | 12 |
| E. acanthocladum[a] (Harvey) J. Agardh | x | | x | | |
| E. echinocarpum[a] Areschoug | x | | x | | |
| E. gelidium[a] (J. Agardh) J. Agardh | x | | x | | |
| E. isiforme (C. Agardh) J. Agardh | x | | x | | |
| E. nudum J. Agardh | x | | x | | |
| E. schrammii[a] (Crouan) J. Agardh | x | | x | | |
| | | | | | |
| Gulf of California | | | | | 5 |
| E. uncinatum Setchell and Gardner | x | | x | | |
| | | | | | |
| Central and Western Pacific | | | | | 7 |
| E. arnoldii Weber-van Bossee | | x | | x | |
| E. cottonii Weber-van Bossee | | x | | x | |
| E. denticulatum (Burmann) Coll. and Herv. | x | | x | | |
| E. gelatinae (Esper) J. Agardh | x | | | | |
| E. odontophorum Boergesen | | x | | x | |
| E. platycladum Schmitz | | x | | x | |
| E. procrustaeanum Kraft | | x | | x | |
| E. striatum Schmitz | | x | | x | |

    [a]   Cheney[3] suggests that these species be combined and placed in a new genus, *Meristiella*.

## PHYSIOLOGICAL ECOLOGY

Photosynthetic and respiratory rates of ι carrageenan-producing species of *Eucheuma* from the Caribbean (*E. isiforme, E. nudum, and E. gelidium*), the Gulf of California (*E. uncinatum*), and the Pacific (*E. denticulatum*) have been compared.[18] All species except *E. gelidium* have higher photosynthetic rates with increasing light levels (see Figure 3). With regard to temperature (see Figure 4), photosynthetic responses were highest and respiration rates were lowest at temperatures from 20 to 24°C. Higher temperature did produce high photosynthetic rates, but when the plants were allowed to equilibrate back to 25°C, the final rates were substantially lower, possibly because of enzymatic inactivation (see *E. denticulatum*, Figure 4). After being held for 3 days in various salinities, all species examined had high photosynthetic rates of 30 to 35 °/$_{oo}$. Freshly sampled branches of *E. denticulatum* had higher rates than material held for 3 days (see arrows, Figure 5, *E. denticulatum*), indicating a sensitivity of *Eucheuma* to tank culture. The study demonstrated that at least the ι-producing species of *Eucheuma* show strong tropical adaptations, being stenohaline (30 to 35 °/$_{oo}$) with tolerances to cooler water (20 to 25°C) and no inhibition to high light intensity (14,000 to 18,000 µW/cm²/white light). The sensitivity to higher temperatures may indicate why Floridian plants of shallow water (1 to 5 m), but not deep water (20 to 80 m), become reproductive in the fall and then rapidly deteriorate. In the Gulf of California, *E. uncinatum* also becomes reproductive in the late spring, but deteriorates in the summer with the rapid rise in temperature.[5]

## CONCLUSIONS

Future studies on *Eucheuma* should emphasize the cultivation of ι carrageenan-yielding forms. The elucidation of the life histories, especially for species showing modified or sterile patterns, should also be carried out. Practically nothing is known about the biosynthesis of carrageenan and the value of the product to the plant, although it represents up to 70% of the cell wall dry weight in Caribbean *Eucheuma*.

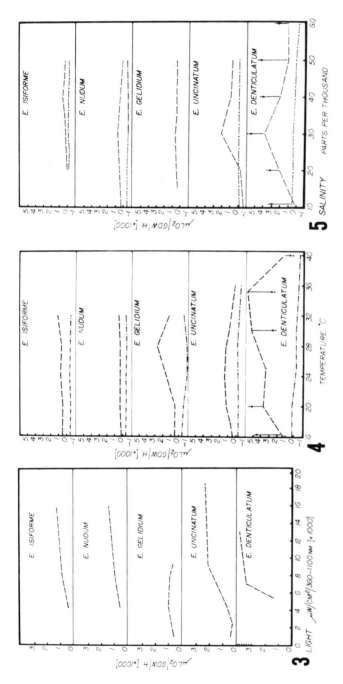

FIGURE 3-5. Photosynthetic and respiratory responses of *Eucheuma* from the Caribbean (*E. isiforme*, *E. nudum*, and *E. gelidium*), Mexico (*E. uncinatum*) and the Pacific (*E. denticulatum*) to light, temperature, and salinity. All measurements are expressed in microliters oxygen per gram dry weight per hour. Photosynthetic rates are expressed by a dashed line, and respiratory rates are expressed by a dash-dot line. The scales expressing rates of oxygen evolution or consumption are identical for all species in a given figure.

Table 2
## EXPORTATION OF *EUCHEUMA* FROM THE PHILIPPINES[16] (IN METRIC TONS)

| Year | *E. striatum*[a] (κ carrageenan) | *E. denticulatum* (ι carrageenan) | Total |
|------|------|------|------|
| 1966 | 565 | 240 | 805 |
| 1967 | 430 | 245 | 675 |
| 1968 | 185 | 80 | 265 |
| 1969 | 306 | 122 | 428 |
| 1970 | 230 | 88 | 318 |
| 1971 | 195 | 145 | 340 |
| 1972 | 330 | 155 | 485 |
| 1973 | 751 | 214 | 965 |
| 1974 | 6286 | 304 | 6590 |
| 1975 | 2670 | 58 | 2728 |
| 1976 | 3277 | 253 | 3530 |
| 1977[b] | 4400 | 250 | 4650 |

[a]  A number of species collected under the general term "cottonii".

[b]  Exports estimated from projected harvests.


## REFERENCES

1. **Kylin, H.,** *Die Gattungen der Rhodophyceen,* Gleerups Forlag., Lund, 1956.
2. **LaClaire, J. W., II. and Dawes, C. J.,** An autoradiographic and histochemical localization of sulfated polysaccharides in *Eucheuma nudum,* J. Agardh, *J. Phycol.,* 12, 368, 1976.
3. **Cheney, D. P.,** A Biosystematic Investigation of the Red Algal Genus *Eucheuma* (Solieriaceae) in Florida, Doctoral dissertation, University of South Florida, Tampa, 1975.
4. **Weber-van Bossee, A.,** Liste des algues du Siboga, *Siboga Exped. Mongr.,* 59, 404, 1928.
5. **Norris, J. N.,** The Marine Algae of the Northern Gulf of California, Doctoral dissertation, University of California, Santa Barbara, 1975.
6. **Dawes, C. J., Stanley, N. F., and Moon, R. E.,** Physiological and biochemical studies on the iota carrageenan producing red alga *Eucheuma uncinatum,* Setchell and Gardner from the Gulf of California, *Bot. Mar.,* 20, 437, 1977.
7. **Doty, M. W. and Santos, G. A.,** Carrageenans from tetrasporic and cystocarpic *Eucheuma* species, *Aquat. Bot.,* 4, 143, 1978.
8. **Kraft, G. T.,** The Red Algal Genus *Eucheuma* in the Philippines, Masters thesis, University of Hawaii, Honolulu, 1969.
9. **Silverthorne, W. and Sorensen, P. E.,** Marine Algae as an Economic Resource, 7th Marine Technology Society Ann. Meeting, Washington, D.C., 7, 523, 1971.
10. **Dawes, C. J.,** On the Mariculture of the Florida Seaweed, *Eucheuma isiforme,* Florida Sea Grant Office, Gainesville, Fla., Florida Sea Grant Progress Rep. #5, 1974.
11. **Dawes, C. J., Lawrence, J. M., Cheney, D. P., and Mathieson, A. C.,** Ecological studies of Floridian *Eucheuma* (Rhodophyta, Gigartinales). III. Seasonal variations in carrageenan, total carbohydrate, protein and lipid, *Bull. Mar. Sci.,* 24, 286, 1974.
12. **Dawes, C. J., Stanley, N. F., and Stancioff, D. J.,** Seasonal and reproductive aspects of plant chemistry, and iota carrageenan from Floridian *Eucheuma* (Rhodophyta, Gigartinales), *Bot. Mar.,* 20, 137, 1977.
13. **Doty, M. S.,** Farming the red seaweed, *Eucheuma* for carrageenans, *Micronesia (J. Coll. Guam),* 9, 59, 1973.
14. **Parker, H. S.,** The culture of the red algal genus *Eucheuma* in the Philippines, *Aquaculture,* 3, 425, 1975.
15. **Doty, M. S. and Alvarez, V. B.,** Status, problems, advances and economics of *Eucheuma* farms, *Mar. Technol. Soc. J.,* 9(4), 30, 1975.

16. **Ricohermoso, M. A. and Deveau, L. E.,** Commercial Propagation of *Eucheuma* spp. Clones in the South China Sea , A Discussion of Trends in Cultivation Technology and Commercial Production Pattern, handout at 9th Int. Seaweed Symp., Santa Barbara, Calif., 1977.
17. **Dawes, C. J., Mathieson, A. C., and Cheney, D. P.,** Ecological studies of Floridian *Eucheuma* (Rhodophyta, Gigartinales). I. Seasonal growth and reproduction, *Bull. Mar. Sci.,* 24, 235, 1974.
18. **Dawes, C. J.,** Physiological and biochemical comparisons of species of *Eucheuma* yielding iota carrageenan from Florida and the Gulf of California with *E. denticulatum* from the Pacific (Rhodophyceae), 9th Int. Seaweed Symp., Santa Barbara, Calif., 1977, Science Press, Princeton, 199.

# GENERAL CHARACTERISTICS OF BROWN MACROALGAE: *LAMINARIA*

Yoshio Hasegawa

## INTRODUCTION

Botanically, *Laminaria* belongs to the Phaeophyta. Of the 14 genera Engler[1] proposed, 13 are cryptogams and the 14th is a phanerogam. *Laminaria*'s genus is one of the cryptogams. Worldwide there are 45 to 50 species in the genus *Laminaria*, with the majority found in the northern hemisphere.

## MORPHOLOGY

A *Laminaria* plant can generally be divided into three parts: a root, stem, and blade, but they are different from those of higher plants, so that it would be correct to claim only root and blade. The root is normally fibrous. The tissue in the lower parts of the stem extends fibrous branches, and the extensions form the root: the end points adhere to the substrates. Roots are finer if *Laminaria* grows in deep quiet places. Certain *Laminaria* species have deformed roots, others have flat roots; in rare cases, they have an underground stem creeping on the substrates, with the blades growing out. There are also *Laminaria* with plate-like roots, the ends becoming flat discs with the full underside adhering to the substrates.

The external morphology of roots is one of the distinguishing features in the classification of *Laminaria*. Many have long, belt-like stems, and the midportion is especially thick. Towards both ends, the stem gradually shows wavelike contractions. The juvenile fronds have a dragon pattern on the surface not present in the fully developed frond. In some species, the midportion shows no particular thickening, but the whole stem becomes uniformly thick. Some species have branching with palmate blades. Figure 1 shows the photograph of external morphology of *Laminara japonica* as an example.

The internal morphology of *Laminaria* has been reported. Killian[2] reported on *L. hyperborea*, Yendo[3] described *L. angustata*, and more recently Kain[4] described *L. hyperborea*.

The blade surface is made up of a number of thin layers forming the epidermis. Assimilation is supported by the pigment in these cells. The cells in the inner epidermal layers are generally larger than those in the outer layers. From the inner cells, long thin, threadlike cells grow and intertwine with each other to make up the pith. At the contact points, some of these threadlike cells show swellings with openings in the surface from which a phloem-like structure develops as in higher plants. According to recent work by Luning,[5] the products of photosynthesis are accumulated in such vascular strands. There are slime glands in the tissue and cavities which keep the surface on the plant viscous and give leaves elasticity when they adhere to foreign objects.

## ASSIMILATION

There have been many reports on the classification of *Laminaria* and the early stages of development of zoospores. However, only a few reports have appeared on the physiological mechanism. Kain[6] reported that in the sexual reproduction of zoospores of *L. hyperborea*, 2 days after discharge the rate of photosynthesis at saturation irradiation was equal to respiration. Lüning[5] reported that the growth of new blades early in

**c m**

FIGURE 1.    *L. japonica* Ares, adult size, Uchiura
Bay of Hokkaido, Japan.[9]

the year was supported not only by their own assimilated products, but also by those
accumulated in old blades and stems.

## RESPIRATION

Lüning[5] used *L. hyperborea* to study the respiration of blades throughout different
periods of the year. The rate of production of oxygen per unit blade area was minute
from March to August. For new blades, the March value was 0.2 mL·$O_2$/g/hr and
increased to 0.7 in May. From August to December, the value was shown to be lower
than in March. Lüning also reported that the stored reserves in the *Laminaria* increased
substantially during the summer.

## NUTRITION

It is generally known that *Laminaria* utilizes inorganic substances in seawater. Experiments with nutrients effective for the cultivation of *Laminaria* zoospores were first
carried out by Schreiber.[7] This nutrient seawater is ordinary filtered seawater with
$NaNO_3$ and $Na_2NO_4$ added. Until some 10 years ago, this "Schreiber's Solution" was
used throughout the world. However, since Provasoli[8] developed an artificial seawater,

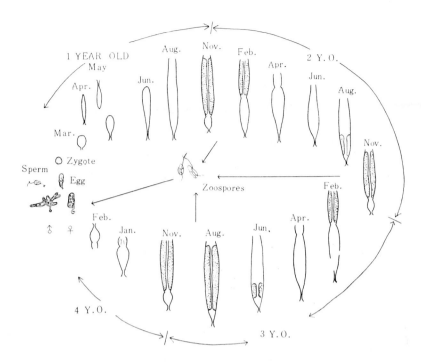

FIGURE 2.    Scheme showing life span of *L. angustata* Kjellman. Dotted parts of the fronds show the sporangia.[9]

nutrition experiments with seaweeds in laboratories have developed rapidly. This seawater contains chelated metal salts in addition to inorganic substances.

## GROWTH

Because *Laminaria* has both a sexual and asexual generation, there is an alternation of generations[9] (see Figures 2 and 3). The asexual generation grows and germinates a zoospore; the sexual gametophyte then develops with the germination of the zoospore. With this sexual gametophyte comes the zygote from spermatozooa and egg, growing into the asexual generation with blades. Since Sauvageau,[10] much research has been done on the former, and all show the results. In autumn zoospores 8 to 9 × 4 to 5 μM in size are released from the zoosporangia on the surface of the mature sporophytes. The zoospores attach themselves to the substrates and immediately start sporeling. After 1 to 2 weeks, the sprout becomes the tiny, microscopical gametophyte of the sexual generation which is a dioecious gametophyte.[9]

When the zoospore has grown for 1 to 2 weeks, the female gametophyte develops an oogonium, and the male gametophyte develops a spermatangium. The female gametophyte is generally larger than its male counterpart and consists of one or more cells. The male gametophyte, however, becomes threadlike or "grapeshaped", and at maturity, one whole cell becomes the spermatangium, each producing a single spermatozoon. On the other hand, when the egg is mature, it emerges from the oogonium and is fertilized while attached at the entrance of oogonium by a spermatozoon swimming in the vicinity. The zygote formed soon starts to divide and after 20 days divides into 1000 cells.[6] The asexual generation or sporophyte continues to grow and develop. The male-female fertilization takes place from late summer into the autumn. In the year of fertilization, the sporophyte does not grow very much. However, from spring to summer of the following year, it grows into fully developed blades. There are species

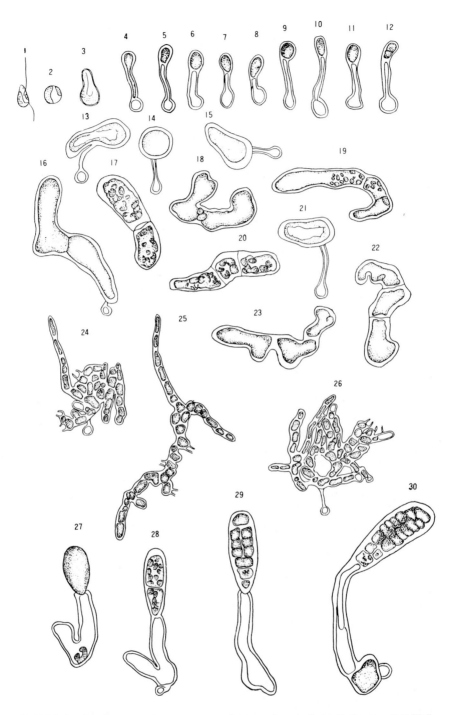

FIGURE 3.    Development of spore, gametophytes, and sporophytes in *L. angustata* Kjell-man.[9] (1) Zoospore (× ca. 670); (2) embryospore (× ca. 670); (3) germination tube after 5 min (× ca. 670); (4-12) germination of the embryospores from 6 days of culture. (The contents of the original cells migrated into germination tubes.) (× ca. 670); (13-23) various types of the matured female gametophytes from 47 days of culture (× ca. 670); (24-26) various types of the matured male gametophytes from 47 days of culture (× ca. 370); (27) egg discharged from one celled female gametophyte from 47 days of culture (× ca. 370); (28) young sporophyte from 47 days of culture, showing the first cell division of the fertilized egg (× ca. 370); (29-30) young sporophytes from 47 days of culture (× ca. 370).

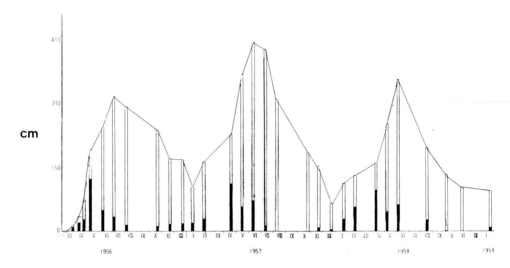

FIGURE 4.   Average actual blade length of the population at the time of measurement. The black block at the base of each column shows the increment in length of blades made by the population since the previous measurement.[9]

which mature and complete their lifespan in this time, as well as species that stop their growth sometime from fall to winter and in the following spring use the stored metabolic products to grow very actively. In some species, plants live up to 10 years. *Laminaria* blade elongation takes place by so-called intercalary growth, in which the main growth points exists between the blade and the stem.

In general, the period when the blades show the greatest elongation is May to July (see Figure 4), and the period of the greatest increase in weight is between July and August.

## USE

From early times, *Laminaria* has been used as raw material for human food, animal fodder, manure, alginic acid, and industrial material for potassium iodide. Except for *Laminaria* used for human food, organized gathering has practically never taken place, and only what have been washed ashore has been used. Only a few speices grown in the seas around Japan have been used for human food. In no other area has *Laminaria* ever been used for food. In Japan, these species have been actively increased and successfully used for human food for many years.

## REFERENCES

1. **Engler, A.**, Die natürlichen Pflanzenfamilien, Leipzig, 1897.
2. **Killian, K.**, Beiträge zur Kenntnis der Laminarien, *Z. Bot.*, 3, 433, 1911.
3. **Yendo, K.**, A monograph of the genus *Alaria*, *J. Coll. Sci. Imp. Univ. Tokyo.*, 43, 1, 1919.
4. **Kain, J. M.**, Synopsis of Biological Data on *Laminaria hyperborea*, *FAO (F.A.O.U.N.) Fish. Synop.*, 87, 1971.
5. **Lüning, K.**, Growth of amputed and dark-exposed individuals of the brown alga *Laminaria hyperborea*, *Mar. Biol.*, 2, 218, 1969.
6. **Kain, J. M.**, The biology of *Laminaria hyperborea*. V. Comparison with early stage of competitors, *J. Mar. Biol. Assoc. U.K.*, 49(2), 455, 1969.

7. **Schreiber, E.,** Untersuchungen über Partenogenesis, Geschlechtsbestimmung und Bastardierungs vermögen bei Laminarien, *Planta,* Bd. 12, 1930.
8. **Provasoli, L., McLaughlin, J. J. A., and Droop, M. R.,** The development of artificial media for marine algae, *Arch. Mikrobiol.,* 25, 392, 1975.
9. **Hasegawa, Y.,** An ecological study of *Laminaria angustata* Kjellman, on the coast of Hidaka Prov., *Hokkaido, Bull. Hokkaido Reg. Fish. Res. Lab.,* 24, 116, 1962.
10. **Sauvageau, C.,** Sur la sexualité hétérgamiques d'une Laminaire, *C. R. Acad. Sci.,* 161, 1915.

# GENERAL CHARACTERISTICS OF BROWN MACROALGAE: *MACROCYSTIS*

W. J. North

## INTRODUCTION

The giant kelp, *Macrocystis*, is a member of the division Phaeophyta (brown algae) and the order Laminariales which includes most of the large kelps used for commercial purposes. *Macrocystis* beds along the Pacific shores of North America range from Punta San Hipolito in Baja California to Sitka, Alaska. Plants most commonly attach to rocky substrates lying within a depth range of about 8 to 20 m for coastal populations. Island populations may experience clearer waters, permitting development to depths of 30 m or more. Locations where wave exposure is reduced sometimes support *Macrocystis* attached only to sedimentary bottom.

The total area of all *Macrocystis* beds in southern California has been officially estimated at about 285 km² by the State Department of Fish and Game.[1] Declines have occurred in many beds since the Department's earliest survey, so that only about 70% of the total area still remains. Ownership resides in the state, which derives revenue from leases, taxes, and licensing fees that allow harvesting and fishing in the kelp beds. A number of other kelp species occur in the California beds, but *Macrocystis* is the primary support of the harvesting industry. South of Point Conception, *Macrocystis* is usually the dominant species in the depth range it occupies. North of Point Conception, dominance is shared by *Macrocystis* and *Nereocystis leutkeana*, the bull kelp. Decline of major kelp beds near Los Angeles and San Diego in the 1940s and early 1950s aroused great concern among conservationists, fishermen, and the kelp harvesting industry.

## BIOLOGY OF *MACROCYSTIS*

*Macrocystis* probably originated in the southern hemisphere, where it occurs in association with every major land mass where temperate waters occur (i.e., South Africa, Australia, New Zealand, and both sides of South America). Migration to temperate-boreal shores in the northeast Pacific Ocean may have occurred during a recent ice age. The genus has been successfully cultured on several occasions at locations in the north Atlantic Ocean. Thus, giant kelp could probably be grown anywhere provided environmental circumstances, such as water temperature, protection from waves, etc., are favorable.

Womersley[2] has recognized three species in the genus: *Macrocystis pyrifera*, *Macrocystis angustifolia*, and *Macrocystis integrifolia*. Intensive work in California and Baja California has revealed presence of morphologically distinct populations in various geographic areas.[3,4] Intergradations also occur, and a clearcut taxonomy is difficult.

Adult specimens of giant kelp consist of a mixture of fronds of varying lengths and ages emanating from the holdfast attachment at the plant base (see Figure 1). Frond lifespan is of the order of half a year, so this upper foliage must be continually renewed. Fronds arise from basal meristems situated just above the holdfast apex. Each frond is comprised of a vinelike stipe, a number of gas floats or pneumatocysts distributed along the stipe, and leaflike blades subtended from each pneumatocyst. Usually more than half of all blades on a mature frond occur in the top meter of the water column (i.e., the canopy region). An apical meristem occurs at the frond tip during its growth phase. This apical blade continually produces new stipe, pneumatocysts,

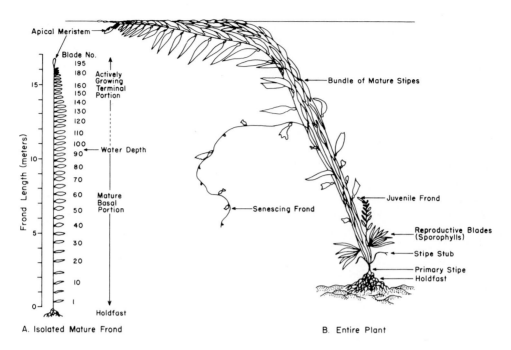

FIGURE 1. Diagrams of *Macrocystis*. (A) An isolated *Macrocystis* frond showing distributions of blades along the stipe and relations to tissue maturity and to water depth from which the specimen was taken, (B) a young adult plant, about 2 years old, standing in water about 10 m deep.

and blades. Vigorous development is usually confined to the terminal 2 to 3 m of fronds. The remainder represents mature tissues that grow slowly or not at all. The lowermost two or three blades at the frond base are specialized for spore production. Fronds elongate rapidly (i.e., the length increases from 8 to 12%/day) during the first month or two of their existence. Overall growth falls to 3 to 5%/day and decreases as the frond apex nears the surface.[5] Some evidence indicates fronds in deeper water may elongate more rapidly.[6] Growth continues to decline in the fully mature canopy frond until senescence (which probably progresses upward from the oldest basal tissues) overwhelms the organ. Senescent fronds may sink and be destroyed by wave surge grinding them across the bottom, or they can be pruned from the stipe bundle by storms. When tops of mature fronds are cut by harvesters, the lower attached portion continues its normal existence, but does not contribute significantly to canopy regeneration because the apical meristem is now gone. Canopies are replaced by new growth of fully intact fronds lying below the depth of cutting.

Continuous canopies can develop when frond densities in kelp beds reach two per square meter or greater. The blades spread more or less horizontally on the surface. As greater numbers of fronds grow into the canopy region, blades overlap to form loose layering that can be a meter or more in thickness. Incoming sunlight is effectively absorbed, so that only one percent or less is transmitted to the underlying waters.[7] Nearly all bottom vegetation disappears beneath well-developed *Macrocystis* canopies, and the underlying rocky surfaces become encrusted with sponges, tunicates, bryozoans, and other sessile animals. Under such conditions, even the short, rapidly growing younger fronds on the *Macrocystis* plants cannot supply their growth requirements from their own photosynthesis.[8,9] These growth needs are primarily derived from translocation processes that move products of photosynthesis downward from the well-illuminated canopies.[10,11] The self-shading problem is thus solved, so that *Macrocystis* plants can aggregate to produce dense forest-like associations. The conducting elements in which translocation occurs are sieve tubes situated between the cortex and medulla within the stipes.[12]

Developing *Macrocystis* fronds also require mineral nutrients, such as trace metals, nitrogen, and phosphorus. These materials initially enter plants by absorption from the surrounding seawater. All exposed surfaces are believed to participate in nutrient accumulation. There are no known organs, such as roots among higher plants, specialized for mineral uptake. When plentiful in the environment, nitrogen (and possibly other nutrients) appears to be accumulated in excess, stored, and then utilized at times when the element is scarce. Uptake of nitrate follows rate-saturating Michaelis-Menten kinetics, yielding values of $K_s$ = 13.1 $\mu M/L$ and $V_{max}$ = 3.1 $\mu M/g$ wet weight/hr.[13] Similar values for ammonium at concentrations below 22 $\mu M$ were $K_s$ = 5.3 $\mu M/L$ and $V_{max}$ = 2.4 $\mu M/g$ wet weight/hr. Nitrogen appears to be translocated in *Macrocystis*[33] probably in the form of amino acids.[14] Healthy mature blades, believed to be lacking nutrient reserves, yield N contents in the range of 0.8 to 1.2% dry weight. N contents of such blades may rise almost to 5% dry weight during seasons when nutrients are plentiful. The N content of meristems, sporophylls, and juvenile blades usually exceeds those of mature blades. Senescing blades may contain only 0.2 to 0.5% dry weight. Possibly N is withdrawn from senescent tissues and translocated to young fronds, or it may be returned to the water. Thus, meaningful interpretations of N determinations in *Macrocystis* tissues should take account of the age and history of the sample, as well as history of the surrounding water.

Reported biomass from established *Macrocystis* stands are variable and influenced, to some extent, by measuring techniques.[15,16] North[16] listed a range of 3 to 22 kg/m² for determinations among beds in California and Baja California. The mean value for the sampled areas was about 6 kg/m² for sample sizes 100 m² or larger. "Thickets" occur in *Macrocystis* stands, so that small areas may display much higher values than cited above. Productivity estimates for *Macrocystis* beds also vary somewhat according to the investigator and the method. North[17] summarized results from seven determinations yielding estimates from 16 to about 130 metric ton dry weight per hectare per year. Plant size (fronds per plant) may vary seasonally,[15,18] so presumably biomass and productivity may fluctuate throughout the year.

The *Macrocystis* life cycle involves a heteromorphic alternation of generations.[19,20] The large, familiar, diploidal sporophytes liberate biflagellated spores that settle and germinate into haploid gametophytes. Gametophytes in nature are probably always microscopic and may be comprised of only a few cells. Much larger individuals can be raised in laboratory culture, consisting of many hundreds of cells in tiny mats one or two mm in diameter. Maturation leads to production of swimming biflagellated antherozooids (sperm) by male gametophytes that fertilize ova arising from females. The resulting zygote divides rapidly, becoming an embryonic sporophyte. The developing sporophyte exists first as a single flat blade.[21] The blade is evenly divided in two by a cleft arising basally when the length reaches around 4 to 20 cm. Further clefts then appear that divide the ensuing blades in a complex pattern which finally produces the basal branching system of the adult form. Time from spore release to appearance of juvenile sporophytes just visible to the unaided eye is 3 to 6 months in the ocean or about 1 to 2 months under continuous illumination in the laboratory. Tiny sporophytes in the sea require 6 to 12 months, depending on depth, to develop to sizes that reach the surface. A year later, they may be contributing significantly to the commercial harvest. Mortality is usually high among small sporophytes in nature. Even populations of young adults display half-lives of only 6 to 9 months along exposed coasts. Year-old plants bear perhaps 10 to 20 fronds, followed by roughly a twofold increase during the second year. Where protection from wave damage exists, very large plants

Table 1
GROSS COMPOSITION OF
*MACROCYSTIS*[24,25]

| Substance or compound | Composition (%) |
|---|---|
| Water | 88.2 |
| Solids | 11.8 ± 1.0[a] |
| Ash (550°C) | 38.96 ± 6.24[b] |
| KCL | 28.7 |
| NaCl | 7.5 |
| $Na_2SO_4$ | 4.3 |
| Volatiles (550°C) | 66—53 |
| Simple carbohydrate | 28.0 |
| Mannitol | 10—22 |
| Laminarin | 1—2 |
| Fucoidan | 0.5—2 |
| Alginic acid | 8—23 |
| Cellulose | 3—8 |
| Protein | 5—13 |
| Fat | 0.6 |

[a]   Based on 42 samples, percent of wet weight.
[b]   Based on 44 samples, percent of dry weight.

may occur that display hundreds of fronds. Presumably such specimens are many years old.

Grazing by urchins, competition from other seaweeds, water temperatures much above 20°C, nutrient availability, and wave exposure are factors that significantly limit *Macrocystis* at times. Photosynthesis declines when current is less than 6 cm/sec across blade surfaces.[22]

## THE *MACROCYSTIS* RESOURCE

The gross and elemental components of *Macrocystis* resemble constituents occurring among other Laminarian kelps (see Tables 1 and 2). Most of the *Macrocystis* harvest in the U.S. is processed for production of alginic acid. Moss[23] reported a world alginate production of 16,100 metric tons for 1976, valued at $88 million. The U.S. production was 5500 metric tons in 1976. Small proportions of the *Macrocystis* harvest are used for food additives and fertilizer. Alginic acid derivatives find varied uses as thickening, emulsifying, and suspending agents among hundreds of commercial products, including foods and beverages. A mean of 15.7 ± 2.6% of the dry weight of *Macrosystis* (35 samples) occurred as alginic acid, but large seasonal variations were reported.[24,25] Other products and extractives have been marketed from *Macrocystis*, including potash during World War I.[27]

*Macrocystis* was harvested sporadically in California during the early 1900s. Records of the harvests have been maintained by the State Department of Fish and Game for about 70 years,[28] except for the period 1921 to 1930, when no industry existed. Annual yields ranged from 236 metric tons in 1931 to 358,000 metric tons in 1918. Demand for kelp and its extractives strongly influenced amounts harvested during the early years. More recently, the California resource has been fully utilized and foreign kelp has been imported for processing. Consequently, fluctuations in the annual take have stabilized. Harvests since 1970 ranged from 115,000 to 155,000 metric tons. The *Macrocystis* beds are classified by the state as open (freely harvestable) or closed (harvestable only by lessees). Harvesting companies pay license and leasing fees, as well as

## Table 2
### ELEMENTAL COMPOSITION OF *MACROCYSTIS*[26]

| Element | Concentration (μM/g dry weight) | Element | Concentration (μM/g dry weight) |
|---------|----------------|---------|----------------|
| Hydrogen | 56,000 | Vanadium | 0.3 |
| Carbon | 26,000 | Beryllium | 0.2 |
| Chlorine | 1,100 | Antimony | 0.1 |
| Nitrogen | 840 | Barium | 0.1 |
| Calcium | 240 | Manganese | 0.1 |
| Magnesium | 240 | Copper | 0.08 |
| Sulfur | 220 | Titanium | <0.07 |
| Potassium | 120 | Fluorine | 0.05 |
| Phosphorus | 93 | Nickel | 0.05 |
| Boron | 25 | Cobalt | 0.03 |
| Iodine | 4.7 | Molybdenum | 0.01 |
| Strontium | 3.8 | Rubidium | 0.01 |
| Bromine | 3.2 | Chromium | <0.01 |
| Silicon | 2.8 | Rhodium | <0.01 |
| Lithium | 1.5 | Selenium | <0.01 |
| Arsenic | 1.4 | Lead | 0.004 |
| Iron | 1.4 | Gold | <0.002 |
| Aluminum | 0.96 | Mercury | <0.002 |
| Zinc | 0.55 | Yttrium | <0.001 |

royalties per wet ton of kelp collected. Leases cover 20-yr periods. Depth of cutting is limited to 1.2 m. Various management techniques are utilized by state and industry biologists to ensure integrity of the resource.[29] These include aerial monitoring of kelp canopies, control of predators such as grazing urchins, and transplanting both juveniles and adults to assist restoring depleted stocks.

Many marine animals, including edible varieties, seek food, shelter, and settling substrate among *Macrocystis* forests. Kelp tissues drifting away from kelp beds as debris on the sea floor provide nourishment for organisms not closely associated with kelp beds.[15] Roles of kelp beds in commercial and sport fisheries have been discussed by Davies,[1] Quast,[30] Feder et al.,[31] and Miller and Geibel.[32]

## REFERENCES

1. **Davies, D. H.,** Statistical analysis of the relations between kelp harvesting and sportfishing in the California kelp beds, in Utilization of Kelp Bed Resources in Southern California, North, W. J. and Hubbs, C. L., Eds., Fish Bulletin 139, Department of Fish and Game, Sacramento, Calif., 1968, chap. 10.

2. **Womersley, H. B. S.,** The species of *Macrocystis* with reference to those on south Australian coasts, *Univ. Calif. Berkeley Publ. Bot.,* 27, 109, 1954.

3. **Neushul, M.,** Studies on the Growth and Reproduction of the Giant Kelp *Macrocystis*, Ph.D. thesis, University of California, Los Angeles, 1959.

4. **Brostoff, W. N.,** A Taxonomic Revision of the Giant Kelp *Macrocystis* in Southern California Based on Morphometric and Transplant Studies, M.S. thesis, San Diego State University, San Diego, 1977.

5. **North, W. J.,** Growth of individual fronds of the mature Giant Kelp, *Macrocystis*, in *Biology of Giant Kelp Beds (Macrocystis) in California*, North, W. J., Ed., Cramer, Lehre, Germany, 1971, chap. 3.

6. **North, W. J.,** Observations on populations of *Macrocystis*, in *Contributions to the Systematics of Benthic Marine Algae of the North Pacific*, Kurogi, M. and Abbott, I. A., Eds., Japanese Society of Phycology, Kobe, 1972, 75.

7. **Neushul, M.,** Submarine illumination in *Macrosystis* beds, in *Biology of Giant Kelp Beds (Macrosystis) in California,* North, W. J., Ed., Cramer, Lehre Germany, 1971, chap. 10.
8. **Sargent, M. C. and Lantrip, L. W.,** Photosynthesis, growth, and translocation in the Giant Kelp, *Am. J. Bot.,* 39, 99, 1952.
9. **Clendenning, K. A. and Sargent, M. C.,** Photosynthesis and general development in *Macrocystis,* in *Biology of Giant Kelp Beds (Macrocystis) in California,* North, W. J., Ed., Cramer, Lehre Germany, 1971, chap. 4.
10. **Parker, B. C.,** Studies of translocation in *Macrocystis,* in *Biology of Giant Kelp Beds (Macrocystis) in California,* North, W. J., Ed., Cramer, Lehre Germany, 1971, chap. 5.
11. **Lobban, C. S.,** Translocation of $^{14}$C in *Macrocystis pyrifera* (Giant Kelp), *Plant Physiol.,* 61, 585, 1978.
12. **Parker, B. C.,** The internal structure of *Macrocystis,* in *Biology of Giant Kelp Beds (Macrocystis) in California,* North, W. J., Ed., Cramer, Lehre Germany, 1971, chap. 2.
13. **Haines, K. C. and Wheeler, P. A.,** Ammonium and nitrate uptake by the marine macrophytes *Hypnea musciformis* (Rhodophyta) and *Macrocystis pyrifera, J. Phycol.,* 14, 319, 1978.
14. **Jackson, G. A.,** Nutrients and production of Giant Kelp, *Macrocystis pyrifera* off southern California, *Limnol. Oceanogr.,* 22, 979, 1977.
15. **Gerard, V. A.,** Some Aspects of Material Dynamics and Energy Flow in a Kelp Forest in Monterey Bay, California, Ph.D. thesis, University of California, Santa Cruz, 1976.
16. **North, W. J.,** Introduction and background, in *Biology of Giant Kelp Beds (Macrocystis) in California,* North, W. J., Ed., Cramer, Lehre Germany, 1971, chap. 1.
17. **North, W. J.,** The oceanic setting as a site for biomass production, in *Proc. Fuels from Biomass Symp.,* Pfeffer, J. T. and Stukel, J. J., Eds., University of Illinois, Urbana-Champaign, 1977, 99.
18. **Kirkwood, P. D.,** Seasonal Patterns in the Growth of the Giant Kelp, *Macrocystis pyrifera,* Ph.D. thesis, California Institute of Technology, Pasadena, 1977.
19. **Papenfuss, G. F.,** Studies of South African Phaeophyceae. I. *Ecklonia maxima, Laminaria pallida, Macrocystis pyrifera, Am. J. Bot.,* 29, 15, 1942.
20. **Cole, K.,** Gametophytic development and fertilization in *Macrocystis integrifolia, Can. J. Bot.,* 46, 777, 1968.
21. **Skottsberg, C.,** Zur Kenntnis der subantarktischen und anarktischen Meeresalgen. I. Phaeophyceen, *Wiss. Ergeb. Schwedischen Sudpolar-Expedition,* 4, 1, 1907.
22. **Wheeler, W. N.,** Transport Limitation of Photosynthesis in *Macrocystis,* M.A. thesis, University of California, Santa Barbara, 1976.
23. **Moss, J. R.,** Essential considerations for establishing seaweed extraction factories, in *The Marine Plant Biomass of the Pacific Northwest Coast,* Krauss, R. W., Ed., Oregon State University Press, Corvallis, 1977, chap. 17.
24. **Hart, M. R., de Fremery, D., Lyon, C. K., Kuzmicky, D. D., and Kohler, G. O.,** Ocean Food and Energy Farm Kelp Pretreatment and Separation Processes, Rep. under ARS agreement No. 12-14-5001-6402, Western Regional Research Center, U.S. Department of Agriculture, Berkeley, Calif., 1976.
25. **Lindner, E., Dooley, C. A., and Wade, R. H.,** Chemical Variation of Chemical Constituents in *Macrocystis pyrifera,* Final Report, U.S. Naval Undersea Center, San Diego, Calif., 1977.
26. **North, W. J.,** Trace metals in Giant Kelp, *Macrocystis, Am. J. Bot.,* 67, 1097, 1980.
27. **Scofield, W. L.,** History of kelp harvesting in California, *Calif. Fish Game,* 45, 135, 1959.
28. **Pinkas, L.,** California Marine Fish Landings for 1975, Department of Fish and Game Fish Bulletin 168, Sacramento, Calif., 1977.
29. **Wilson, K. C., Haaker, P. L., and Hanan, D. A.,** Kelp restoration in southern California, in The Marine Plant Biomass of the Pacific Northwest Coast, Krauss, R. W., Ed., Oregon State University Press, Corvallis, 1977, chap. 10.
30. **Quast, J. C.,** Fish fauna of the rocky inshore zone; estimates of the populations and the standing crop of fishes; observations on the food and biology of the Kelp Bass, *Paralabrax clathratus,* with notes on the sportfishery at San Diego California; observations on the food of the kelp-bed fishes; the effects of kep harvesting on the fishes of the kelp beds, in Utilization of Kelp-Bed Resources in Southern California, North, W. J. and Hubbs, C. L., Eds., Department of Fish and Game Fish Bulletin 139, Sacramento, Calif. 1968, chaps. 5 to 9.
31. **Feder, H. M., Turner, C. H., and Limbaugh, C.,** Observations on Fishes Associated with Kelp Beds in Southern California, Department of Fish and Game Fish Bulletin, 160, 1974.
32. **Miller, D. J. and Geibel, J. J.,** Summary of Blue Rockfish and Lingcod Life Histories; a Reef Ecology Study; and Giant Kelp *Macrocystis pyrifera* Experiments in Monterey Bay, Calif., Department of Fish and Game Fish Bulletin 158, 1973.
33. **Wheeler, P. A.,** personal communication.

# GENERAL CHARACTERISTICS OF FRESHWATER VASCULAR PLANTS

## L. A. Garrard and T. K. Van

## INTRODUCTION

A number of terms have been used to describe vascular plant species that have become adapted for growth in aquatic habitats. Sculthorpe[1] suggests that the difficulty in selecting an acceptable definition of this group arises because terrestrial and aquatic habitats cannot be sharply divided. This author uses the term "vascular hydrophyte" throughout his major work on the biology of these species. Hutchinson[2] applies the term "aquatic tracheophyte" to this same group of plants.

## SYSTEMATIC CONSIDERATIONS

The taxonomic grouping of many vascular hydrophytes has caused considerable controversy, and no attempt will be made here to deal with problems arising in this area. For treatment of this subject, the reader is referred to discussions by Sculthorpe[1] and Hutchinson[2] and to taxonomic organizations by Fassett[3] and Cook et al.[4].

Sculthorpe[1] lists 33 families that are characteristically aquatic; of these families, 30 have less than 10 genera, 17 have but a single genus, while 3 are monotypic. Also, the number of species per family is generally much smaller than the number found in terrestrial families. In addition to the members of aquatic families, many other vascular hydrophytes are scattered throughout taxa of terrestrial plants. Table 1 presents a list of families which are either totally aquatic or have some aquatic members. Also given are the common life forms or growth habits of some better-known vascular hydrophytes. It is interesting to note that with the exceptions of a comparatively few aquatic ferns (e.g., Azollaceae and Salviniaceae) and a few gymnosperms that grow rooted in swamps, all vascular hydrophytes are angiosperms. Further, contrary to what is observed with terrestrial plant communities, aquatic plant communities have an equal or greater number of monocotyledonous species than dicotyledenous species.[1,2]

Vascular hydrophytes must spend significant portions of their lives in an aquatic habitat. The characters of each habitat will vary to some degree, as will the adaptations of plants to each aquatic site. In general, emergent species rooted in aerial (shore) or submersed soils resemble in many respects their terrestrial relatives both morphologically and physiologically. The root and rhizome systems of these plants, however, may be permanently existing in an anaerobic substratum, and the presence of a well-developed intercellular lacunae system provides a pathway for the exchange and transport of photosynthetic and respiratory gases throughout the plant.[1,2,5] Some of these emersed species utilize the best conditions of both the aquatic and terrestrial environments to become the most productive plants in the world.

The submersed vascular hydrophytes display the greatest degree of adaptation to the aquatic environment, and it is generally very difficult to point out their terrestrial progenitors. However, the fact that these aquatic species did evolve from terrestrials is clearly demonstrated by the presence of such features as poorly lignified xylem components, thin leaf cuticles, and functionless stomata.[1,2,5]

Submersed vascular hydrophytes move toward structural reduction with little dry matter utilized for the formation of mechanical tissue. Suspension of the plants within the water column is by flotation phenomena which in turn depends on a vast complex of internal air passages (lacunae, aerenchyma tissue). The lacunal system also provides

## Table 1
## EXAMPLES OF COMMON AQUATIC ANGIOSPERMS AND THEIR LIFE FORMS

| Family name[a] | Family common name | Number of genera[b] | Common examples | Life forms[c] |
|---|---|---|---|---|
| **Dicotyledonae** | | | | |
| Acanthaceae | Water-willow | 2 | *Justicia americana* (Water-willow) | B |
| | | | *J. ovata* | B |
| Amaranthaceae | Amaranth | 2 | *Alternanthera philoxeroides* (Alligator weed) | B, A |
| Apiaceae | Parsley | 15 | *Hydrocotyle umbellata* (Water-pennywort) | B, A |
| | | | *Sium suave* (Water-parsnip) | A |
| Brassicaceae | Mustard | 3 | *Nasturtium officinales* (Watercress) | B, D |
| | | | *Cardamine clematitis* (Bitter cress) | B, D |
| Cabombaceae | Fanwort | 2 | *Cabomba caroliniana* (Fanwort) | D |
| | | | *Bresenia schreberi* (Water shield) | C |
| Ceratophyllaceae* | Hornwort | 1 | *Ceratophyllum demersum* (Coontail) | E |
| | | | *C. echinatum* (Hornwort) | E |
| Haloragaceae* | Water milfoil | 4 | *Myriophyllum spicatum* (Eurasian milfoil) | D |
| | | | *Proserpinaca palustris* (Mermaid-weed) | B, D, A |
| Lentibulariaceae | Bladderwort | 3 | *Utricularia purpurea* (Purple bladderwort) | E |
| | | | *U. inflata* (Big floating-bladderwort) | E |
| Nelumbonaceae* | American lotus | 1 | *Nelumbo lutea* (American lotus) | C |
| | | | *N. nucifera* (Sacred lotus) | C |
| Nymphaceae* | Waterlily | 6 | *Nuphar luteum* (Spatter-dock) | C, B |
| | | | *Nymphaea odorata* (Fragrant waterlily) | C |
| Primulaceae | Primrose | 4 | *Hottonia inflata* (Water-violet) | F |
| | | | *Lysimachia terrestris* (Loosestrife) | B, A |
| **Monocotyledonae** | | | | |
| Alismataceae* | Water plantain | 11 | *Sagittaria latifolia* (Common arrowhead) | B, A |
| | | | *Echinodorus cordifolius* (Bur-head) | B, A |
| Araceae | Arum | 16 | *Pistia stratiotes* (Water-lettuce) | F, B |
| | | | *Peltandra virginica* (Arrow-arum) | A, B |
| Cyperaceae | Sedge | 31 | *Cladium jamaicense* (Sawgrass) | B, A |
| | | | *Scirpus californicus* (Giant bulrush) | B, A |
| Hydrocharitaceae* | Frogbit | 15 | *Hydrilla verticillata* (Hydrilla) | D |
| | | | *Vallisneria americana* (Tapegrass) | D |
| Juncaceae | Rush | 2 | *Juncus effusus* (Soft rush) | A, B |
| | | | *J. coriaceus* | A, B |
| Lemnaceae* | Duckweed | 6 | *Lemna minor* (Common duckweed) | F |
| | | | *Spirodela polyrhiza* (Giant duckweed) | F |
| Najadaceae* | Water nymph | 1 | *Najas guadalupensis* (Southern naiad) | D |
| | | | *N. minor* (Slender naiad) | D |
| Poaceae | Grass | 17 | *Panicum hemitomon* (Maidencane) | A, B |
| | | | *Phragmites australis* (Common reed) | A, B |
| Pontederiaceae* | Pickerel weed | 9 | *Eichhornia crassipes* (Floating water-hyacinth) | F, B, A |
| | | | *Pontederia lanceolata* (Pickeralweed) | B |
| Potamogetonaceae* | Pondweed | 2 | *Potamogeton pectinatus* (Sago pondweed) | D |
| | | | *P. diversifolius* (Variable-leaf pondweed) | D, C |
| Typhaceae* | Cattail | 1 | *Typha latifolia* (Common cattail) | B, A |
| | | | *T. angustifolia* (Narrow-leaved cattail) | B, A |

[a]   Families containing aquatic genera; asterisk (*) indicates family composed solely of aquatic members.

[b]   Number of genera containing aquatic representatives.

[c]   Life forms adapted from similar schemes given by other authors:[1,2,5] (A) rooted in wet soil, with aerial stems and leaves (marsh, ditchbank, or shoreline plants); (B) rooted in shallow hydrosoil and/or dense mats, with aerial (emersed) leaves; (C) rooted in hydrosoil, with floating leaves; (D) rooted in hydrosoil, with submersed leaves; (E) nonrooted, submersed plants; (F) free-floating plants with submersed and/or aerial leaves.

for the storage and recycling of oxygen and carbon dioxide which, to some extent, alleviates the effect of the low diffusiveness of gases through water.[5] Thus, the ratio of photosynthesis and respiration may be influenced by the capacity of submersed plants to provide internal supplies of gaseous reactants. Other adaptations that appear to favor photosynthesis and uptake of materials from the aqueous medium include the presence of a very thin cuticle, the concentration of chloroplasts in the epidermal cells of extremely thin leaves (often without any well-defined mesophyll), leaf morphology designed for the maximum surface area for absorbing efficiently the available light and nutrients (may be entire, fenestrated, or dissected), the presence of hydropoten, and the existence of a mechanism for the utilization of the bicarbonate ion.[1,2,5]

## PHOTOSYNTHESIS AND PRODUCTIVITY

### Emergent Plants

Comparisons of productivity among freshwater ecosystems are difficult because of the variations in analytical and sampling techniques. Westlake[6,7] discussed the problems of standardizing published data for the purpose of making valid comparisons of productivity. Tables 2 and 3 give some representative estimates of photosynthetic capacity, maximum seasonal biomass, and annual productivity of freshwater vascular plants. In aquatics, emergent plant communities are considered among the most productive. Westlake[6] estimated the organic productivity of the common emergents *Typha* and *Phragmites* to be 30 to 45 t/ha·yr in temperate climates, while even higher values of 65 to 85 t/ha·yr or more may be reached in subtropical and tropical habitats. It has been speculated that the high growth rate in the emergent angiosperms results from ready nutrient and water availability from the aqueous sediment, combined with much greater availability of $CO_2$ from the atmosphere.[1] A number of authors have reported net photosynthetic rates of 18 to 42 mg $CO_2$/g·hr for emergent macrophytes (see Table 3). The exceptionally high photosynthetic rates on leaf area basis (69 mg $CO_2$/dm²·hr, Table 3) reported for *Typha latifolia* may have been an artifact of measurement techniques. Westlake[8] indicated that it may be better to use areas of both leaf surfaces for photosynthesis determinations in the case of thick erect leaves.

A few emergent angiosperms (e.g., *Panicum repens, P. purpurascens, P. maximum, P. dichotomiflorum, Cyperus esculentus,* and *Echinochloa crusgalli*) have been shown to have "Kranz"-type leaf anatomy and low $CO_2$ compensation points,[9] characteristics usually associated with the $C_4$ photosynthetic pathway. These plants grow well in marshlands and very moist sites,[10] indicating that high-efficiency carbon metabolism may have been a sufficient advantage for evolution of $C_4$ photosynthetic metabolism in the invasion of aquatic environments.

### Floating Plants

Very few quantitative studies of floating plant communities are available. Penfound[11] stated that *Eichhornia crassipes* accumulated 14.6 g dry matter/m²·day and *Pistia stratiotes* 15.3 g dry matter/m²·day during the growing season. Westlake[6] estimated that a community of *E. crassipes* may produce 15 to 45 t dry matter/ha·yr in warmer climates and may even achieve greater productivity with a longer growing season under optimal tropical conditions. Chen et al.[12] measured a photosynthetic rate of 18 mg $CO_2$/dm²·hr in *E. crassipes*. Haller and Knipling[13] determined that maximum leaf photosynthetic rates of *Alternanthera philoxeroides, Hydrocotyle umbellata, E. crassipes, Pontederia cordata,* and *Myriophyllum brasiliense* were from 14 to 40 mg $CO_2$/dm²·hr.

## Table 2
## REPRESENTATIVE ESTIMATES OF SEASONAL MAXIMUM BIOMASS, MAXIMUM GROWTH RATE, AND NET ANNUAL PRODUCTIVITY OF AQUATIC MACROPHYTES

| Species and location | Seasonal maximum biomass (kg dry wt/m$^2$) | Maximum growth rate (g dry wt/m$^2$·day) | Net annual productivity (t dry wt/ha·yr) | Ref. |
|---|---|---|---|---|
| Emergents | | | | |
| *Typha* (hybrid), Minnesota, U.S. | 1.68 (aerial) 2.96 (underground) | 14 | 25 | 42 |
| *T. latifolia*, Oklahoma, U.S. | 1.53 (aerial) | 31 | — | 11 |
| *Glyceria, Phragmites*, Surlingham Broad, England | 0.8—1.2 (aerial) | — | — | 43 |
| *Phragmites*, Czechoslovakia | 1.1—2.2 (aerial) 6.0—8.6 (underground) | — | — | 5 |
| Floatings | | | | |
| *Eichhornia crassipes*, New Orleans, U.S. | 1.5 | 7.4—22 | 15—44 | 11, 6 |
| *E. crassipes*, Florida, U.S. | — | 14.6 | — | 11 |
| *Pistia stratiotes*, Florida, U.S. | — | 15.3 | — | 44 |
| *Alternanthera philoxeroides*, Alabama, U.S. | 0.84 | 17 | — | 45 |
| Submersed | | | | |
| *Ceratophyllum demersum*, Osbysjon, Sweden | 0.78 | 5.7 | 9 | 14, 6 |
| *Myriophyllum verticillatum*, Osbysjon, Sweden | 0.24 | 2.8 | — | 14 |
| *Myriophyllum spicatum*, Tennessee, U.S. | 0.18—0.36 | 2.8—4.5 | — | 46 |
| *Scirpus subterminalis*, Michigan, U.S. | 0.40 | — | 5.6 | 16 |
| *Callitriche, Ranunculus, Potamogeton*, England | 0.52 | 4.2—6.0 | 5.2 | 47 |
| *Hydrilla verticillata*, Florida, U.S. | 0.18 | — | — | 17 |
| *Vallisneria neotropicalis*, Florida, U.S. | 0.40 | — | — | 17 |
| *V. densesurrulata*, Tokyo, Japan | 0.2—0.3 | 5.5 | — | 15 |

Table 3

REPRESENTATIVE RATES OF NET PHOTOSYNTHESIS OF AQUATIC MACROPHYTES

| Species | Net photosynthesis | | Conditions | Ref. |
|---|---|---|---|---|
| | (mg $CO_2$/g · hr) | (mg $CO_2$/dm² · hr) | | |
| **Emergents** | | | | |
| Phragmites communis | 32 | 22 | 300 W/m², 20°C | 48 |
| P. communis | 42 | 32 | July 1970, Death Valley, Calif. | 49 |
| Typha latifolia | 18 | 69 | 2 × 10⁶ erg/cm² · sec, 25°C | 50 |
| Pontederia cordata | 21 | 14 | 1500 μE/m² · s, 25°C | 13 |
| **Floating** | | | | |
| Eichhornia crassipes | 41 | 62 | 8000 fc, 25°C | 12 |
| E. crassipes | 91 | 40 | 1500 μE/m² · sec, 25°C | 13 |
| Alternanthera philoxeroides | 47 | 18 | 1500 μE/m² · sec, 25°C | 13 |
| **Submersed** | | | | |
| Myriophyllum spicatum | 8.7 | — | 300 W/m², 25°C | 51 |
| M. spicatum | 9.0 | — | 105 erg/cm², 20°C, 300 μℓ $CO_2$/ℓ | 32 |
| Elodea canadensis | 11.7 | — | Optimum conditions | 52 |
| E. occidentalis | 11 | — | 18 Klx, 27°C | 27 |
| Ceratophyllum demersum | 5.4 | — | 1000 μE/m² · sec, 30°C, 340 μℓ $CO_2$/ℓ | 19 |
| Vallismeria asiatica | 11 | — | 43 Klx, 27°C | 27 |
| V. densesurralata | 15—22 | 1.9—3.2 | 62 Klx, 22—30°C | 15 |
| Hydrilla verticillata | 7—14 | — | 18 Klx, 27°C | 27 |
| H. verticillata | 5.2 | — | 1000 μE/m² · sec, 30°C, 340 μℓ $CO_2$/ℓ | 19 |

## Submersed Plants

Submersed freshwater plant communities are much less productive than floating and emergent communities. Westlake[6] stated that the organic productivity of submersed freshwater macrophytes in the temperate region exceeded only that of desert plants and phytoplankton. Temperate, submersed communities could produce about 1 to 7 t/ha·yr, while tropical communities might well produce 10 to 20 t/ha·yr under favorable conditions for carbon fixation. Forsberg[14] estimated the maximum net production of *Ceratophyllum demersum* under normal lake conditions to be 5.7 g dry matter/m²·day or about 9 t/ha·yr. Ikusima[15], working with *Vallisneria denseserrulata*, observed a net daily production of 1 to 5.5 g dry matter/m² at the height of the growing season, depending on environmental conditions. He also observed a maximum standing crop during the summer of 2 to 3 t dry matter per hectare, a value considerably lower than those of most terrestrial species. Rich et al.[16] found the seasonal maximum biomass of *Scirpus subterminalis* to be 0.4 kg dry matter per square meter. They also estimated an annual productivity of 5.6 t dry matter per hectare for this species. Haller and Sutton,[17] working with *Hydrilla verticillata* and *Vallisneria neotropicalis* grown in artificial ponds in Central Florida, reported standing crops of only 1.6 t dry matter per hectare and 4.0 t dry matter per hectare, respectively.

The lower productivities of submersed plants may be explained in terms of their low photosynthetic capacities and the characteristics of their aquatic environment (e.g., low total incident radiation, slow diffusion rates of gases in solution). Numerous investigations have reported photosynthesis rates of 5 to 15 mg $CO_2$/g·hr (see Table 3). Ikusima[18] stated that the maximum rate of photosynthesis of submersed macrophytes could potentially amount to 5 to 6 mg $CO_2$/dm²·hr at light saturation during their period of maximum growth, but the mean value might be approximately 1 mg $CO_2$/dm²·hr. Van et al.[19] found the light- and $CO_2$-saturated photosynthetic rates of *H. verticillata*, *C. demersum*, and *Myriophyllum spicatum* to be 50 to 60 μmol $O_2$/mg Chl·hr at 30°C. However, these rates were considerably higher than those observed at atmospheric levels of $CO_2$ (0.42 mg $CO_2$ per liter in the aqueous phase, pH 5.5). Rates at light saturation and natural lake levels of $CO_2$ were approximately 3 to 7 μmol $CO_2$/mg Chl·hr, or less than 5% of those achieved by terrestrial plants at comparable $CO_2$ levels. Zieman[20] has shown that the measurement of photosynthetic rates of submersed macrophytes by gas exchange methods may be underestimated because of the storage of $CO_2$ and oxygen in the internal lacunae of the plant tissues. However, since in most species of submersed angiosperms, chloroplasts are found predominantly in the epidermis,[1,10] gas exchange with the surrounding solution during photosynthesis appears likely to be more important than with the lacunae.[21] The low rates of photosynthesis in submersed macrophytes may be attributed, in part, to the low diffusion rate of $CO_2$ in solution.[22] Lloyd et al.[21] reported that photosynthesis rates in *M. spicatum* and *Potamogeton amphifolius* increased with increasing levels of $CO_2$ in the gas phase up to 3500 μℓ $CO_2$ per liter. Similarly, Van et al.[19] found the apparent $K_m$ ($CO_2$) for photosynthesis in three submersed macrophytes to be as high as 150 to 170 μM $CO_2$ at pH 4 and 70 to 95 μM at pH 8. In contrast, the $K_m$ ($CO_2$) of photosynthesis in terrestrial plant species is near 9 μM.[23] It is unlikely that the excessively high $K_m$ ($CO_2$) for photosynthesis in the submersed macrophytes was because of RuBP carboxylase kinetics, as *Hydrilla* and *Spinacia oleracea* RuBP carboxylases exhibited similar $K_m$ ($CO_2$) values.[19] The low rates of photosynthesis in submersed freshwater macrophytes are more likely related to the low activities of the carboxylase enzyme systems found in these plants.

*Light and Temperature*

Photosynthetically active radiation is an important factor in the distribution of submersed aquatic macrophytes.[24] Brown et al.[25] also reported a correlation between the photosynthetic responses to light by a number of exotic submersed macrophytes and their ability to displace native vegetation in New Zealand lakes.

For terrestrial plants, shade-type $C_3$ species usually reach light saturation at about 25 to 50% full sunlight (500 to 1000 $\mu E/m^2 \cdot sec$), whereas sun-adapted $C_3$ species and $C_4$ plants tend to saturate at values near full sunlight.[26] In regard to submersed macrophytes, Van et al.[19] reported light saturation levels of 600 to 700 $\mu E/m^2 \cdot sec$ for *H. verticillata, C. demersum, M. spicatum,* and *Cabomba caroliniana.* Further, Ikusima[27] reported a light saturation value of approximately 25 klx (about 25% full sunlight) for *Elodea occidentalis, H. verticillata,* and *Vallisneria asiatica.* The light compensation points of terrestrial plants and submersed macrophytes are highly variable and depend mostly on the light regimes under which the plants are grown.[28] Since water often significantly reduces the radiation reaching submersed plants, it is not uncommon to find some plants with comparatively low light compensation points. Van et al.[19] measured the light compensation points of 15, 35, 35, and 55 $\mu E/m^2 \cdot sec$, respectively, for *H. verticillata, C. demersum, M. spicatum,* and *C. caroliniana.* Further, the dominance of *Hydrilla* among submersed species of southern freshwater ecosystems was attributed, in part, to its low light compensation point.

*Carbon Assimilation*

Because of their aquatic environment, submersed plants are exposed to free $CO_2$, $HCO^-_3$, and $CO^=_3$ ions in the surrounding medium.[1] Steemann-Nielsen[22] noted that *Elodea* uses free $CO_2$ and $HCO^-_3$ for photosynthesis, whereas *Fontinalis antipyretica* assimilates only free $CO_2$. Carr[29] reported that $CO_2$ was the preferred carbon source for photosynthesis in *C. demersum,* and Brown et al.[25] could find no evidence for the utilization of $HCO^-_3$ by *Egeria densa, Lagarosiphon major,* or *Elodea canadensis.* The ability to utilize $HCO^-_3$ ions has been suggested as a factor influencing the distribution of certain aquatic species.[30] However, the importance of $HCO^-_3$ use for the growth of aquatic plants is difficult to estimate and has yet to be unequivocally demonstrated. The enzyme carbonic anhydrase which functions to convert $CO_2$ to $HCO^-_3$ and vice versa has been shown to be widespread in the plant kingdom.[31] Chen et al.[12] detected a very active carbonic anhydrase in *E. crassipes.* Van et al.[19] also reported the activity of this enzyme in leaf extracts of three submersed plants in amounts far exceeding those required to support $HCO^-_3$ utilization. Recent studies have indicated that free $CO_2$ is used preferentially by most aquatic plants. Stanley and Naylor[32] found the affinity of *M. spicatum* for $CO_2$ to be more than three times that for $HCO^-_3$. Similar observations for *H. verticillata, C. demersum,* and *M. spicatum* were made by Van et al.[19] These authors also noted that the efficiency of $HCO^-_3$ utilization may depend on the relative level of free $CO_2$ present in the system, and that no $HCO^-_3$ utilization was observed at saturating levels of free $CO_2$.

Investigations of the carbon metabolism pathways in submersed angiosperms have produced a variety of results. Stanley and Naylor[32] found 3-PGA, glycolate, and glucose-6-P as early photosynthetic products in *M. spicatum,* indicating a $C_3$-type photosynthesis in this species. Similar findings were reported by Hough[33] for *Najas flexilis* and *Scirpus subterminalis.* Browse et al.[34] found that the initial product of photosynthesis in *E. densa* was 3-PGA, with malate being only 5% of the total product and not further metabolized actively. Brown et al.[25] also reported 3-PGA to be the major initial product in *E. densa* and *Lagarosiphon major;* however, up to 30% of the early product was $C_4$ acids. DeGroote and Kennedy[35] studied $^{14}CO_2$ fixation in *E. canaden-*

*sis.* Exposure of the plant to $^{14}/CO_2$ for 2 sec resulted in 45% of label in $C_4$ acids and 15% in 3-PGA. This relatively high level of $C_4$ acid synthesis is not typical in $C_3$ species. However, pulse-chase studies of *Elodea*[35] and also of *Egeria*[34] showed very slow turnover rates of label in the $C_4$ acids, suggesting the lack of a true $C_4$ photosynthesis mechanism in these plant species.

Photorespiration, a consequence of altered $C_3$-type carbon assimilation, is potentially a significant factor in aquatic plant primary productivity. Van et al.[19] presented field data suggesting that aquatic species are frequently exposed to conditions that are most conducive to accelerated photorespiration at the expense of $CO_2$ fixation. Such conditions also might be expected to select for plants with reduced photorespiratory capacity. A recent survey by Hough and Wetzel,[10] however, indicated that the presence of a true $C_4$ pathway in the totally aquatic system is unlikely, despite the many $C_4$ characteristics reported for submersed macrophytes.[36,37]

Numerous studies suggested the presence of photorespiration in submersed plant species. Stanley and Naylor,[38] by supplying radioactive glycolate, glyoxylate, and $CO_2$, showed glycolate synthesis and glycolate oxidase activity in *M. spicatum.* Van et al.[19] found the glycolate oxidase activities of *H. verticillata, M. spicatum,* and *C. demersum* to be lower than in *S. oleracea,* but considerably higher than the level in *Zea mays.* Also, photosynthesis rates of the three submersed macrophytes were inhibited 9 to 20% by atmospheric levels of oxygen. Direct measurements of photorespiration as release of labeled $CO_2$ in the light were made by Hough and Wetzel[39] for *N. flexilis.* Amount of $CO_2$ released in the light increased with increasing levels of dissolved oxygen. However, the rate of photorespiration at 21% oxygen was never appreciably greater than dark respiration. Similar results were obtained by Bowes et al.[28] for photorespiration in *H. verticillata.* They measured the rate of $CO_2$ release into $CO_2$-free atmosphere. The limited rate of photorespiration in aquatic angiosperms compared to terrestrial $C_3$ plants was thought to be the result of extensive refixation of photorespired $CO_2$ and low solubility of oxygen in water.[10] However, for aquatic and terrestrial plants, oxygen must dissolve to become available to the cells. Also, data by Van et al.[19] suggested that aquatic plants under certain conditions may be exposed to dissolved oxygen levels far in excess of those to which terrestrial plants are exposed. It appears more likely that reduced photorespiration in aquatic plants is the result of generally lower photoactive enzyme levels. When the low photosynthetic rate and reduced activities of RuBP carboxylase and glycolate oxidase were taken into account, the ratio of photosynthesis to photorespiration may be similar to that occurring in more typical $C_3$ plants.

Conflicting reports of submersed angiosperm $CO_2$ compensation points have been published, although most of the recent investigations give values within the range of $C_3$ plants.[19,21,25] The near zero value reported for *M. spicatum* by Stanley and Naylor[32] may have been an artifact of their measurement techniques.[25] Van et al.[19] found $CO_2$ compensation points for *H. verticillata, M. spicatum,* and *C. demersum* to be 44, 19, and 41 in 21% oxygen and 17, 9, and 11, respectively, in 1% oxygen. The high $CO_2$ compensation points of the aquatic plants in 1% oxygen indicate that dark respiration may continue in the light. The authors also noted some intraspecific variations in $CO_2$ compensation point for a number of submersed angiosperms, suggesting some variations of the photosynthesis/photorespiration ratio in these species. Similarly, Hough[33] measured a seasonal fluctuation in photorespiration rates of *N. flexilis* and *S. subterminalis* based on *in situ* lake measurements. Bowes et al.[36] reported seasonal changes for *H. verticillata* $CO_2$ compensation points which were high in winter and low in summer. The high winter values were within the range for $C_3$ plants, but the low summer values were typical of the $C_3$-$C_4$ intermediate group. The summer decline in $CO_2$ compensation point was associated with an increase in net photosynthesis and/or a decrease in photorespiration.[36] The changes in $CO_2$ compensation point for submersed aquatic angiosperms were apparently not related to the ontogeny of the plants, but were environmentally induced. Long photoperiods and higher temperatures were shown to induce a decrease in $CO_2$ compensation points under growth chamber conditions for a number of submersed macrophytes.[36]

Holaday and Bowes[40] compared the photosynthetic characteristics of *Hydrilla* plants with high and low $CO_2$ compensation points. Summer plants which had a low $CO_2$ compensation point synthesized large amounts of malate as initial photosynthetic product. The PEP and RuBP carboxylase activities of these plants were 150 and 11 $\mu$mol $CO_2$ per mg Chl·hr, respectively. Oxygen-induced inhibition of photosynthesis was reduced to 10%. These plants also were capable of fixing $CO_2$ at night at a rate equivalent to 33% of light fixation, resulting in marked diurnal fluctuations in titrable acidity. Cold-water *Hydrilla*, however, had a higher $CO_2$ compensation point, low PEP carboxylase activity, and low $^{14}CO_2$ dark fixation. Photosynthesis in this plant was inhibited over 20% by atmospheric oxygen.

The variation of photosynthetic characteristics with changing growth parameters makes it difficult to classify submersed macrophytes in terms of $C_3$, $C_4$, or CAM species[40,41] and may explain the conflicting data reported. Submersed aquatic plants may belong to a new, distinct group based on their carbon metabolism.

# REFERENCES

1. **Sculthorpe, C. D.,** *The Biology of Aquatic Vascular Plants,* St. Martin's Press, New York, 1967.
2. **Hutchinson, G. E.,** A Treatise on Limnology, Vol. 3, Limnological Botany, John Wiley & Sons, New York, 1975.
3. **Fassett, N. C.,** *A Manual of Aquatic Plants,* University of Wisconsin Press, Madison, 1957.
4. **Cook, C. D. K., Gut, B. J., Rix, E. M., Schneller, J., and Seitz, M.,** *Water Plants of the World,* W. Junk b.v., The Hague, 1974.
5. **Wetzel, R. G.,** *Limnology,* W. B. Saunders, Philadelphia, 1975.
6. **Westlake, D. F.,** Comparisons of plant productivity, *Biol. Rev.,* 38, 385, 1963.
7. **Westlake, D. F.,** Some basic data for investigations of the productivity of aquatic macrophytes, in *Primary Productivity in Aquatic Environments,* Goldman, C. R., Ed., University of California Press, Berkley, 1966, 231.
8. **Westlake, D. F.,** Primary production of freshwater macrophytes, in *Photosynthesis and Productivity in Different Environments,* Cooper, J. P., Ed., Cambridge University Press, Cambridge, 1975, 189.
9. **Downton, W. J. S.,** The occurrence of $C_4$ photosynthesis among plants, *Photosynthetica,* 9, 96, 1975.
10. **Hough, R. A. and Wetzel, R. G.,** Photosynthetic pathways of some aquatic plants, *Aquat. Bot.,* 3, 291, 1977.
11. **Penfound, W. T.,** Primary production of vascular aquatic plants, *Limnol. Oceanogr.,* 2, 92, 1956.
12. **Chen, T. M., Brown, R. H., and Black, C. C.,** $CO_2$ compensation concentration, rate of photosynthesis, and carbonic anhydrase activity of plants, *Weed Sci.,* 18, 399, 1970.
13. **Haller, W. T. and Knipling, E. B.,** unpublished data, 1977.
14. **Forsberg, C.,** Subaquatic macrovegetation in Osbysjon, Djursholm, *Oikos,* 11, 183, 1960.
15. **Ikusima, I.,** Ecological studies on the productivity of aquatic plant communities. II. Seasonal changes in standing crop and productivity of a natural submersed community of *Vallisneria denseserrulata, Bot. Mag. Tokyo,* 79, 7, 1966.
16. **Rich, P. H., Wetzel, R. G., and Thuy, N. V.,** Distribution, production, and role of aquatic macrophytes in a southern Michigan marl lake, *Freshwater Biol.,* 1, 3, 1971.
17. **Haller, W. T. and Sutton, D. L.,** Community structure and competition between hydrilla and vallisneria, *Hyacinth Control J.,* 13, 48, 1975.
18. **Ikusima, I.,** Ecological studies on the productivity of aquatic plant communities. I. Measurement of photosynthetic activity, *Bot. Mag.,* 78, 202, 1965.
19. **Van, T. K., Haller, W. T., and Bowes, G.,** Comparison of the photosynthetic characteristics of three submersed aquatic plants, *Plant Physiol.,* 58, 761, 1976.

20. **Zieman, J. C.,** Methods for the study of the growth and production of turtle grass *Thalassia testudinum* Konig, *Aquaculture,* 4, 139, 1974.

21. **Lloyd, N. D. H., Canvin, D. T., and Bristow, J. M.,** Photosynthesis and photorespiration in submersed aquatic vascular plants, *Can. J. Bot.,* 55, 3001, 1977.

22. **Steemann-Nielsen, E.,** Photosynthesis of aquatic plants with special reference to the carbon sources, *Dansk. Bot. Ark.,* 12, 3, 1947.

23. **Zelitch, I.,** *Photosynthesis, Photorespiration, and Plant Productivity,* Academic Press, New York, 1971.

24. **Spence, D. H. N. and Chrystal, J.,** Photosynthesis and zonation of freshwater macrophytes. I. Depth distribution and shade tolerance, *New Phytol.,* 69, 205, 1970.

25. **Brown, J. M. A., Dromgoole, F. I., Towsey, M. W., and Browse, J.,** Photosynthesis and photorespiration in aquatic macrophytes, in *Mechanisms of Regulation of Plant Growth,* Bulletin 12, Bieleske, R. L., Gerguson, A. R., and Cresswell, M. M., Eds., The Royal Society of New Zealand, Wellington, 1974, 243.

26. **Black, C. C.,** Photosynthetic carbon fixation in relation to net $CO_2$ uptake, *Annu. Rev. Plant Physiol.,* 24, 253, 1973.

27. **Ikusima, I.,** Ecological studies on the productivity of aquatic plant communities. III. Effect of depth on daily photosynthesis in submersed macrophytes, *Bot. Mag. Tokyo,* 80, 57, 1967.

28. **Bowes, G., Van, T. K., Garrard, L. A., and Haller, W. T.,** Adaptation to low light levels by hydrilla, *J. Aquat. Plant Manage.,* 15, 32, 1977.

29. **Carr, J. L.,** The primary productivity and physiology of *Ceratophyllum demersum.* II. Micro primary productivity, pH, and the P/R ratio, *Aust. J. Mar. Freshwater Res.,* 20, 115, 1969.

30. **Hutchinson, G. E.,** The chemical ecology of three species of *Myriophyllum* (Angiospermae, Haloragaceae), *Limnol. Oceanogr.,* 15, 1, 1970.

31. **Atkins, C. A., Patterson, B. D., and Graham, D.,** Plant carbonic anhydrases. I. Distribution of types among species, *Plant Physiol.,* 50, 214, 1972.

32. **Stanley, R. A. and Naylor, A. W.,** Photosynthesis in Eurasian water-milfoil (*Myriophyllum spicatum* L.), *Plant Physiol.,* 50, 149, 1972.

33. **Hough, R. A.,** Photorespiration and productivity in submersed aquatic vascular plants, *Limnol. Oceanogr.,* 19, 912, 1974.

34. **Browse, J. A., Dromgoole, F. I., and Brown, J. M. A.,** Photosynthesis in the aquatic macrophyte *Egeria densa.* I. $^{14}CO_2$ fixation at natural $CO_2$ concentrations, *Aust. J. Plant Physiol.,* 4, 169, 1977.

35. **DeGroote, D. and Kennedy, R. A.,** Photosynthesis in *Elodea canadensis* Michx. Four-carbon acid synthesis, *Plant Physiol.,* 59, 1133, 1977.

36. **Bowes, G., Holaday, A. S., Van, T. K., and Haller, W. T.,** Photosynthetic and photorespiratory carbon metabolism in aquatic plants, in *Proc. 4th Int. Congr. Photosynthesis,* Hall, D. O., Coombe, J., and Goodwin, T. W., Eds., The Biochemical Society, London, 1977, 289.

37. **Newton, R. J., Scott, R. J., Benedict, C. R.,** Leaf structure and $^{13}C$ values of the aquatic *Hydrilla verticillata, Plant Physiol.,* Suppl. 59, 65, 1977.

38. **Stanley, R. A. and Naylor, A. W.,** Glycolate metabolism in Eurasian watermilfoil (*Myriophyllum spicatum*), *Physiol. Plant.,* 29, 60, 1973.

39. **Hough, R. A. and Wetzel, R. G.,** A $^{14}C$-assay for photorespiration in aquatic plants, *Plant Physiol.,* 49, 987, 1972.

40. **Holaday, A. S. and Bowes, G.,** Photosynthetic/photorespiratory variation and dark fixation in submersed aquatic plants, *Plant Physiol.,* Suppl. 61, 8, 1978.

41. **Benedict, C. R.,** Nature of obligate photoautotrophy, *Annu. Rev. Plant Physiol.,* 29, 67, 1978.

42. **Bray, J. R., Lawrence, D. B., and Pearson, L. C.,** Primary production in some Minnesota terrestrial communities for 1957, *Oikos,* 10, 38, 1959.

43. **Buttery, B. R. and Lambert, J. M.,** Competition between *Glyceria maxima* and *Phragmites communis* in the region of Surlingham Broad. I. The competition mechanism, *J. Ecol.,* 53, 163, 1965.

44. **Odum, H. T.,** Trophic structure and productivity in Silver Springs, Florida, *Ecol. Monogr.,* 27, 55, 1957.

45. **Boyd, C. E.,** Production, mineral nutrient absorption, and biochemical assimilation by *Justicia americana* and *Alternanthera philoxeroides, Arch. Hydrobiol.,* 66, 139, 1969.

46. **Stanley, R. A., Shackelford, E., Wade, D., and Warren, C.,** Effects of season and water depth on Eurasian watermilfoil, *J. Aquat. Plant Manage.,* 14, 32, 1976.

47. **Edwards, R. W. and Owens, M.,** The effects of plants on river conditions. I. Summer crops and estimates of net productivity of macrophytes in a chalk stream, *J. Ecol.,* 48, 159, 1960.

48. **Gloser, J.,** Characteristics of $CO_2$ exchange in *Phragmites communis* Trin. derived from measurements *in situ, Photosynthetica,* 11, 139, 1977.

49. **Pearcy, R. W., Berry, J. A., and Bartholomew, B.,** Field photosynthetic performance and leaf temperatures of *Phragmites communis* under summer conditions in Death Valley, California, *Photosynthetica,* 8, 104, 1974.

50. **McNaughton, S. J. and Fullem, L. W.**, Photosynthesis and photorespiration in *Typha latifolia*, *Plant Physiol.*, 45, 703, 1970.
51. **McGahee, C. F. and Davis, G. J.**, Photosynthesis and respiration in *Myriophyllum spicatum* L. as related to salinity, *Limnol. Oceanogr.*, 16, 826, 1971.
52. **Hartman, R. T. and Brown, D. L.**, Changes in internal atmosphere of submersed vascular hydrophytes in relation to photosynthesis, *Ecology*, 48, 252, 1967.

# GENERAL CHARACTERISTICS OF MARINE VASCULAR PLANTS: MANGROVES

Howard J. Teas

Mangroves are the trees and shrubs that grow between the level of the high water of spring tides and mean sea level.[1] Mangroves occupy the major part of the coast lines of the world between 25° north and south latitude.[2] One or another species of mangrove is found as far north as 35° latitude on Kyushu Island, Japan[3] and as far south as 37° latitude at New Zealand.[4] Mangroves are characteristic of gradually sloping shorelines and estuaries, although they are not limited to such areas. They are generally found in habitats in the subtropics and tropics that along cooler shores would be occupied by salt marshes.

Mangroves are botanically diverse, including 14 plant families and 19 genera with at least 59 species. The principal mangrove families and genera are compared with nonmangroves in Table 1. Their close relatives indicate that mangroves are land plants that have adapted to growth at the edge of the sea, rather than marine plants that have evolved in the sea and adapted to growth along the shore.

Mangroves are often found in well-defined zones of particular species or groups of species. Zonation has been attributed to frequency of tidal inundation and to interstitial soil salinity.[1]

Mangroves change the soil where they grow. Their roots add organic matter to the soil and serve to collect sand, litter, etc. which raises the soil elevation and can, in time, cause an area to become more suitable for other species of mangroves or for upland plants. Changes in mangrove soils are illustrated by analyses of organic matter: freshly deposited alluvium that has been colonized by mangroves may contain only 5 to 15% organic matter, whereas mature mangrove soils may be 65% or more organic matter.[1] Because of the gradual changes which result in species succession, the mangrove forests are not climax communities. There have been claims that mangroves recover land from the sea. However, the evidence indicates that mangroves advance seaward along accreting shorelines, but very little along stable shorelines.[1]

Many species of mangroves show vivapary, i.e., the seeds germinate inside the fruit before it separates from the parent plant. Vivapary has been cited as an adaptation to life in the mangrove forest, but there are equally successful mangroves species that have seeds with a resting stage.

Mangroves are important biologically as producers of food material and as a habitat for a variety of animals. Mangroves have been shown to be involved in a food chain in which leaves, twigs, fruit, flowers, trunks, roots, etc. form detritus that is fed upon by an array of microorganisms and invertebrates that in turn are consumed by higher forms, and those are eaten by still others.[7,8] Crabs, prawns, mollusks, insects, and other invertebrates, as well as birds, fishes, amphibians, reptiles, and mammals, are found in mangrove swamps. Commercially important species use mangrove waterways as nursery grounds: in Florida, the mangrove habitat has been identified as a nursery and feeding ground for such commercially valuable species as the pink shrimp (*Penaeus duorarum*), mullet (*Mugil cephalus*), grey snapper (*Lutjianus griseus), red drum (Sciaenops ocellata*), sea trout (*Cynoscion nebulosus*), and blue crab (*Callinectes sapidus*).[7] The commercial value of mangroves to fisheries is indicated by Macnae,[9] who tabulated production of prawns in the Indo-Pacific area with mangrove forests. He found that for Mozambique, Madagascar, West Thailand, Malaya, and combination of West Iran, Papua, and north Australia, there was a general correlation of man-

Table 1

**PRINCIPAL FAMILIES AND GENERA OF
MANGROVES COMPARED WITH RELATED
NONMANGROVES**

| Mangrove family | Nonmangroves[5] | | Mangroves[6] | |
|---|---|---|---|---|
| | Number of genera | Number of species | Genera | Number of species |
| Acanthaceae | 259 | 2498 | *Acanthus* | 2 |
| Avicenniaceae | 0 | 0 | *Avicennia* | 11 |
| Bombacaceae | 19 | 178 | *Camptostemon* | 2 |
| Combretaceae | 16 | 596 | *Conocarpus* | 1 |
| | | | *Laguncularia* | 1 |
| | | | *Lumnitzera* | 2 |
| Euphorbiaceae | 299 | 4999 | *Excoecaria* | 1 |
| Meliaceae | 49 | 1390 | *Xylocarpus* | 10 |
| Myrtaceae | 99 | 2999 | *Osbornia* | 1 |
| Palmae | 216 | 2499 | *Nypa* | 1 |
| Pellicieriaceae | 0 | 0 | *Pelliciera* | 1 |
| Plumbaginaceae | 9 | 498 | *Aegialitis* | 2 |
| Rhizophoraceae | 12 | 104 | *Rhizophora* | 7 |
| | | | *Bruguiera* | 6 |
| | | | *Ceriops* | 2 |
| | | | *Kandelia* | 1 |
| Rubiaceae | 499 | 5999 | *Scyphiphora* | 1 |
| Sonneratiaceae | 1 | 2 | *Sonneratia* | 5 |
| Sterculiaceae | 59 | 698 | *Heritiera* | 2 |

groves and prawns. His data showed an average of about 4 tons of prawns produced per year per square kilometer of mangroves.

Mangroves are widely recognized as playing an important role in the protection of shorelines from storms and erosion. *Rhizophora mangle* was planted among the ballast stones along the overseas railway in the Florida Keys to control erosion,[10] and was introduced into Hawaii on the lee shore of Molokai for erosion control early in this century.[11] Elsewhere, e.g., in Sri Lanka, mangroves have been planted in order to reclaim land from the sea and to stabilize the dikes of fish ponds and canals.[1] Fosburg[12] has suggested that earlier cutting of thousands of hectares of mangroves in Bangladesh may have been partly responsible for the heavy loss of life in the 1970 storm and tidal wave in that country. Recently mangroves have been planted in Florida because of appreciation of their ecological role.[13]

Mangroves are frequently found growing in water-logged anaerobic soils that are saline or hypersaline. Mangroves species have evolved several systems for aerating their roots. These involve snorkels or ventilator tubes which connect the roots with small above-ground openings called lenticels that are located on the prop roots, pneumatophores, or on the trunks of the trees. Scholander and co-workers[14] demonstrated that the lenticels of *Avicennia* pneumatophores and the prop roots or stilt roots of *Rhizophora* were functionally connected with the underground roots. When the lenticels of breathing organs (pneumatophores or prop roots) were sealed off with petroleum jelly, oxygen concentrations in the roots decreased and carbon dioxide concentrations increased. Because of these adaptations for root aeration, mangroves are very vulnerable to conditions that clog or block lenticels, such as flooding, siltation, or heavy petroleum residues from oil spills. An example of mangrove susceptibility to lenticel blocking is the many acres of mangroves killed by hurricanes in Everglades National Park in Florida when layers of fine sediment were deposited by the storms.[15]

Mangroves live where the waters are brackish or saline, but do not appear to be obligate halophytes. A number of species of mangroves have been grown in freshwater for many years.[1] Laboratory experiments indicate also that *Rhizophora* and *Avicennia* are facultative halophytes, i.e., they grow better when supplied with salt than without salt.[16,17] This is consistent with the report that mangroves grow most luxuriously in the middle reaches of estuaries rather than where they are exposed to full strength seawater or to freshwater.[18] The salinity of the soil, i.e., the interstitial salinity, is the level of salt to which the plant is exposed. The salinities of surface and interstitial water can differ markedly.[18] Some soils, such as fine marl or heavy clays, are slow to equilibrate their interstitial salinity with that of the surface water.[19,20]

Mangrove species differ considerably in their salinity preference or tolerance. For example, *Sonneratia caseolaris* grows only if the salinity is less than approximately 10 $^o/_{oo}$ (10 parts per thousand), *Bruguiera parviflora* grows best at about 20 $^o/_{oo}$, and *Rhizophora mucronata* prefers water approximately the salinity of seawater, 35 $^o/_{oo}$. Some species are quite salt tolerant. *Ceriops tagal* is often found growing at 60 $^o/_{oo}$ or greater and *Avicennia marina* and *Lumnitzera racemosa* can grow where soil salinities are greater than 90 $^o/_{oo}$.[1]

Mangroves grow in saline waters, but ordinarily have internal salt concentrations that are a fraction of those in the surrounding soil medium.[21,22] Mangroves have adapted to the saline environment by several means: excluding salt, excretion of salt by glands on their leaves, accumulation of salt in thickened leaves (succulence), and by discard of salt-laden leaves or other parts. An effective salt-excluding species is *R. mangle*,[22] salt-secreting genera include *Avicennia* and *Aegialitis*; succulent genera include *Sonneratia* and *Lumnitzera*,[23] genera that shed salt-laden leaves include *Ceriops, Avicennia, Rhizophora, Excoecaria,* and *Lumnitzera*.[24] Probably all mangroves carry on some degree of salt exclusion which is a physical process that operates essentially as reverse osmosis in which pressure for the ultrafiltration is supplied by transpiration.[22] The process by which mangroves separate salt for excretion through salt glands is very different physiologically from salt exclusion. Secretion by salt-secreting gland is inhibited by uncouplers of oxidative phosphorylation and other respiratory poisons and is markedly temperature sensitive.[22,25,26] Thus, salt excretion appears to require expenditure of metabolic energy, probably in the form of high-energy phosphate.

Mangroves are poor competitors in the upland, but salinity tolerance enables mangroves to compete effectively with upland plants in saline environments. There appears to be a physiological "price" for such salt tolerance. The transpiration of mangrove leaves is less than that of ordinary plants. According to experiments of Bowman,[10] zero transpiration for *R. mangle* would be expected to occur in the range of 60 to 75 o/oo in the root medium salinity. Hicks and Burns[27] found that respiration and photosynthesis of mangroves in Florida were affected by salinity. They reported that, although total photosynthesis rose with increases in salinity, at a community level the net photosynthetic activity dropped because of a marked increase in respiration.

The principal use of mangroves is for wood, either for stakes and timber or to burn as wood or charcoal.[28] Mangroves have a variety of other uses, such as for production of tannin, smoking sheet rubber and fish, chipboard, newspaper, rayon production, and occasionally for food (propagules) or medicines.[29] The leaves of mangroves are also used as food for water buffalo, cattle, and camels.[30]

# REFERENCES

1. Macnae, W., A general account of the fauna and flora of mangrove swamps in the Indo-West Pacific region, *Adv. Mar. Biol.,* 6, 73, 1968.
2. McGill, J. G., Coastal landforms of the world, map supplement in Russell, R. J., Second Coastal Geography Conference, Coastal Studies Institute, Louisiana State University, Baton Rouge, 1959, 472.
3. Steenis, C. G. and Van, G. J., The distribution of mangrove plant genera and its significance for paleogeography., *K. Ned. Akad. Wet. Versl. Gewone Vergad. Afd., Natuurkd. Ser. C,* 65, 164, 1962.
4. Chapman, V. J. and Ronaldson, J. W., The mangrove and salt-grass flats of the Auckland Isthmus, *N. Z. Dep. Sci. Ind. Res. Bull.,* 125, 79, 1958.
5. Willis, J. C., *A Dictionary of the Flowering Plants and Ferns,* 7th ed., Cambridge University Press, Cambridge, 1966, 53 and 1214.
6. Chapman, V. J., Mangrove phytosociology, *Trop. Ecol.,* 11, 1, 1970.
7. Odum, W. E. and Heald, E. J., Trophic analyses of an estuarine mangrove community, *Bull. Mar. Sci.,* 22, 671, 1972.
8. Untawale, A. G., Balasubramian, T., and Wafar, M. V. M., Structure and production in a detritus rich estuarine mangrove swamp, *Mahasagar-Bull. Natl. Inst. Oceanogr.,* 10, 173, 1977.
9. Macnae, W., Mangrove Forests and Fisheries, Reports IOFC, International Indian Ocean Fisheries Survey and Development Programme, No. 74/34, 1974.
10. Bowman, H. H. M., Ecology and physiology of red mangroves, *Proc. Am. Philos. Soc.,* 56, 589, 1917.
11. MacCaughey, V., The mangrove in the Hawaiian Islands, *Hawaii. For. Agric.,* 14, 361, 1917.
12. Fosberg, F. R., Mangroves versus tidal waves, *Biol. Conserv.,* 4, 38, 1971.
13. Teas, H. J., Ecology and restoration of mangrove shorelines in Florida, *Environ. Conserv.,* 4, 51, 1977.
14. Scholander, P. F., Dam, L. van, and Scholander, S. I., Gas exchange in the roots of mangroves, *Am. J. Bot.,* 42, 92, 1955.
15. Craighead, F. C., Sr., *The Trees of South Florida,* University of Miami Press, Miami, 1971.
16. Pannier, P. F., El efecto de distintas concentraciones salinas sobre el desarrolo de *Rhizophora mangle* L., *Acta Cient. Venoz.* 10, 68, 1959.
17. Connor, D. J., Growth of grey mangrove (*Avicennia marina*) in nutrient culture, *Biotropica,* 1, 36, 1969.
18. Davis, J. H., Jr., The ecology and geological role of mangroves in Florida, *Carnegie Inst. Washington Publ.,* 32, 305, 1940.
19. Scholl, D. W., High interstitial water chlorinity in estuarine mangrove swamps, Florida, *Nature (London),* 207, 284, 1965.
20. Giglioli, M. E. C. and King, D. F., The mangrove swamps of Keneba, Lower Gambia River basin. III, *J. Appl. Ecol.,* 3, 1, 1966.
21. Scholander, P. F., Bradstreet, E. D., Hammel, H. T., and Hemmingsen, E. A., Sap concentrations in halophytes and some other plants, *Plant Physiol.,* 41, 529, 1966.
22. Scholander, P. F., How mangroves desalinate seawater, *Physiol. Plant.,* 21, 258, 1968.
23. Joshi, G. V., Jamale, B. B., and Bhosale, L. J., Ion regulation in mangroves, in Int. Symp. Biology and Management of Mangroves, Institute for Food and Agricultural Sciences, University of Florida, Gainesville, 1975, 595.
24. Jamale, B. B. and Joshi, G. V., Physiological studies in senescent leaves of mangroves, *Indian J. Exp. Biol.,* 14, 697, 1976.
25. Atkinson, M. R., Findlay, G. P., Hope, A. B., Pitman, M. G., Saddler, H. D. W., and West, H. R., Salt regulation in the mangroves *Rhizophora mangle* Lam. and *Aegialitis annulata* R., *Aust. J. Biol. Sci.,* 20, 589, 1967.
26. Arisz, W. H., Chapman, I. J., Heikens, H., and van Tooren, A. J., The secretion of the salt glands of *Limonium latifolium* Ktze, *Acta Bot. Neerl.,* 4, 322, 1975.
27. Hicks, D. B. and Burns, L. A., Mangrove metabolic response to alteration of natural freshwater drainage in southwestern Florida estuaries, in Proc. Int. Symp. on Biology and Management Mangroves, Institute for Food and Agricultural Sciences, University of Florida, Gainesville, 1975, 238.
28. Watson, J. D., Mangrove forests of the Malaya peninsula, *Malay. For.,* 6, 1, 1928.
29. Morton, F. J., Can the red mangrove provide food, feed, and fertilizer?, *Econ. Bot.,* 19, 113, 1965.
30. Bhosale, L. J., *Ecophysiological Studies of the Mangroves from the Western Coast of India,* Department of Botany, Shivaji University, Kolhapur, India, 1978.

# GENERAL CHARACTERISTICS OF MARINE VASCULAR PLANTS: SEAGRASSES

## Anitra Thorhaug

Many phyla of terrestrial and freshwater plants are completely missing from the marine environment. Notably absent are bryophytes, pteridophytes, and gymnosperms. According to den Hartog,[1] there is "not a single indication that they have ever inhabited the sea". There are no marine dicotyledons and only one order of monocotyledons (Helobiae) having marine representatives. Nine genera belong to the family Potamogetonaceae, and three belong to the Hydrocaritaceae. These 12 genera contain 50 species of seagrasses. Thus, an important question about the marine ecosystem, which has abundant water, light, carbon dioxide and most other qualities necessary for plant growth, is "Why have phyla gaining dominance in land plant communities not evolved strategies to exploit marine resources?."

The seagrasses are the only group of higher plants that have evolved strategies for complete marine submersion for extended times. They have done so with remarkable success, usually dominating, in terms of primary production and biomass abundance in the shallow water communities they inhabit.

In estuarine and very nearshore coastal waters, one of the most important and dominant plant groups throughout the world is the seagrasses. Their importance lies not only in their biomass dominance of the nearshore marine plant community, but as a source of nutrition and shelter for the food web animals. These are not true grasses, but rather members of two families of angiosperms which have completely adapted over the course of evolution to life in marine waters. The small number of seagrass species (50) is not proportional to their high production of plant material from solar energy as a contribution to the estuaries of the world.

Several species are universally distributed throughout the oceans, and others are either temperate or tropical. In general, the Pacific estuaries have more species, which often share dominance within an area, than the Atlantic or Mediterranean. The Atlantic area is often dominated by the genera *Zostera* in temperate areas or *Thalassia testudinum* in the tropics. (See den Hartog[1] for distribution diagrams, and the section on seagrass primary productivity in this volume for spatial biomass values.)

Geologically these plants have secondarily readapted to the marine environment in the fairly recent geological past. They have probably descended from marsh and other brackish water species and become completely adapted to total submersion and saline rather than fresh water media so that their generative cycle occurs when submerged in saline water.

Distribution nearshore or zonation depends in part on the sediment type and on the climatic zone, as well as on the light penetration of the particular area. den Hartog[1,2] has an extensive discussion of zonation of seagrass forms. Some groups of seagrasses are more likely to be intertidal, and others are more strictly sublittoral. In the Indo-Pacific, *Thalassia hemprichii*, *Cymodosea serulata*, and *Cymodosea rotunda* are the most important species. The East African coast is dominated in sandy bottoms and sand-covered bogs by *Thalassodendron ciliata*. In areas that are frequently disturbed at the mouth of estuaries where inundations of fresh water occur, the usual dominance may be replaced by *Halodule* or other successional stages.

The distribution of these plants in the oceans of the world is limited by the bottom type. Most of the species of seagrasses do not appear on rock, but need a sufficient amount of sediment, either sand, mud, or peat, for their rhizomal and root system to take hold. When light penetration is sufficient for growth, seagrasses frequently become the dominant plant. This is usually the case in estuaries or shallow areas on continental shelves such as the Bahama Banks. In clear tropical waters, the seagrasses are found at considerable depths, such as 30 m in Cuba[3] and 37 m in Jamaica.[4] Seagrasses often form dominant or climax communities and compete well with benthic marine algae for space on the benthic sediment. Several species do well in the highly fluctuating estuarine environment, while others require more stable conditions. Distribution of the seagrasses and speciation have been covered by den Hartog.[5]

Seagrasses have not elicited much attention because general botanists confined themselves to terrestrial species and marine botanists confined themselves to algae. Recently the importance of seagrasses has been recognized. The seagrasses often form a major part of the nutrition of estuarine and nearshore species. The seagrass "meadows" can produce up to 2000 g dry w/m² year.[1,6,7] Although in certain locations large amounts of direct grazing have been found,[8] much of the nutritional pathway for seagrasses is through a detrital cycle.[9] The blades grow very rapidly. (Each 3 to 6 weeks a turnover occurs in tropical species.) In addition to providing direct carbon fixation, the blades provide a substrate onto which micro- and macroepiphytes can grow which also contribute substantial amounts of primary productivity. Thus, the entire plant population in the seagrass community fixes high amounts of solar energy and ranks with that of agricultural crops, such as *Zea mays* or a tropical rain forest. Seagrass values can be higher than production values associated with phytoplankton in the marine upwellings.

The vegetative morphology of representative seagrasses is shown in Figures 1, 2, and 3 which have been adapted from Tomlinson.[10]

The question of whether seagrasses are $C_4$ or $C_3$ plants has been controversial. Evidence from the first products by Benedict and Scott[11] and from morphological evidence (see Wetzel and Hough[12]) would tend to support the argument that seagrasses are $C_4$ plants. Parker[13] analyzed $C-12/C-13$ ratios in 40 seagrass species from field conditions where these ratios were in the range of $C_4$ plants. However, $^{14}CO_2$ studies show they are $C_3$ plants.[14]

Another important function of seagrasses beyond fixation of solar energy is that the seagrasses have blades which can reach from several inches to many feet and thus form an important habitat and shelter for the many juvenile forms of fisheries and food web organisms which utilize seagrass beds as a nursery. The flat large blades also are excellent places for attachment of egg cases and sessile organisms. In the subtropics and tropics with flat shallow bottom topography, the seagrasses are particularly important as a sheltering nursery for many commercial and sports fishes, such as the pink shrimp, the rock lobster, the stone crab, and other organisms. Another important aspect of seagrasses is their sediment-binding and producing capacity. The seagrasses have an extensive rhizomal system (see Figures 1, 2, and 3), in some cases going down 1 to 2 m into the sediment and forming an extremely thick mat. Some of the residual rhizomal material persists through geological time. This consolidates and binds sediments.[16] The baffling effect of their blades also adds to sediment production in areas where sediment suspension is frequent. Seagrasses are often associated with sediment-producing organisms, such as foraminifera and micromollusks, which grow directly on the blade and calcium carbonate-producing algae, which can grow within the seagrass population or on the blades, such as *Halimeda and Penicillus*. A removal of seagrasses often results in fine sediments washing away and a complete change in sediment properties.[16]

The chemical composition of seagrass for ash, protein, and carbohydrate is given in Table 1.[17,18] The ash content is approximately 20% for the seagrass *Thalassia testudinum* (for which most data are available). Protein content in the tropical Puerto Rican seagrasses was 13% and in the north Gulf of Mexico was about 26%. Amino acid

THALASSIA

POSIDONIA

ZOSTERA

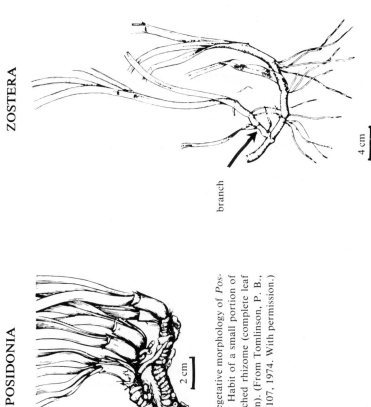

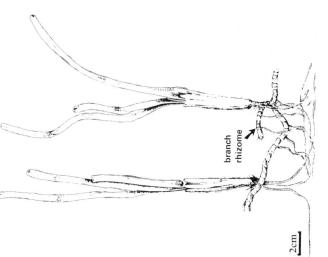

FIGURE 1. Vegetative morphology of *Thalassia testudinum*. (From Tomlinson, P. B., Aquaculture, 4, 107, 1974. With permission.)

FIGURE 2. Vegetative morphology of *Posidonia oceanica*. Habit of a small portion of plant with branched rhizome (complete leaf blades not shown). (From Tomlinson, P. B., Aquaculture, 4, 107, 1974. With permission.)

FIGURE 3. Vegetative morphology of *Zostera marina*. (From Tomlinson, P. B., Aquaculture, 4, 107, 1974. With permission.)

Table 1
CHEMICAL COMPOSITION OF
*THALASSIA* BLADES FROM PUERTO
RICO AND THE GULF OF MEXICO[21]

| Composition | Puerto Rico | Gulf of Mexico |
|---|---|---|
| Protein (N × 6.25) | | |
| Blades | 13.1 | 25.7 |
| Rhizomes | — | 11.0 |
| Fat | 0.5 | — |
| Ash | | |
| Blades | 24.8 | 24.5 |
| Rhizomes | — | 23.8 |
| Crude fiber | 16.4 | — |
| Other carbohydrates | 35.6 | — |
| Total carbohydrates | | |
| Blades | 51.0 | 23.6 |
| Rhizomes | — | 72.1 |
| Calories/100 g | | |
| Blades | 199.0 | 46.6 |
| Rhizomes | — | 48.8 |

*Note:* Data in percent dry weight, except for caloric values.

content in Puerto Rico is shown in Table 2. Fat content was about 0.5%. Walsh and Grow[18] looked at seasonal distribution of these components and found that there was no annual variation in ash content; however, there was considerable variation in the protein content, with March, April, and May, the peak growing season, being the time of peak protein.[18,19] There was some annual variation of high carbohydrate content, with August being a peak carbohydrate month and September, October, and November, when the plants were in a more senescent period, being the low carbohydrate period. Carbohydrates ranged between 18 and 25% for blades and between 55 and 80% for rhizomes. The rhizome peak of carbohydrate was between August and December, and the low carbohydrate in the rhizomes was during March through July.

Figure 4 shows the annual variation in the major elements (Na, K, Mg, Fe, Mn, and Zn), which appears to be associated with functions of the material analyzed and growth processes for the leaves and rhizomes.[18] Mg, Fe, and Zn appear to have bimodal yearly distributions, while the rest go on an annual cycle, notably winter being the peak for Mg, whereas summer and spring are peaks for Na, K, Mn, and Zn in leaves.

Trace metal content has been examined in some detail from a grid of extensive sampling, and it was found in a subtropical estuary on the edge of the tropics (see Table 3) that most of the trace metal inventory in the estuary resided in two compartments: the sediment and the seagrass *Thalassia* as the chief component of the biotic system.[20] Extensive cycling of trace metals from the sediment through the seagrasses to the other members of the food chain has been hypothesized by Segar and Gilio[20] and established experimentally by Schroeder and Thorhaug,[21] carried out by radiotracer work in microcosms. A cycling model in terms of trace metals and of energy flow has been given by Schroeder and Thorhaug.[21] *Thalassia* appears to concentrate certain elements preferentially and store them at various concentrations in different parts of the plant, for instance, Sr. However, some of the elements, including Ca, were at approximately the same concentrations in the substrate and in the plant. Certain substances such as Fe appeared to be taken up chiefly from sediment. On the other hand, monovalent cations appear to be taken up from the water. Translocation was noted between blades and roots in most of the elements studied. Since much of the rapid productivity and turn-

## Table 2
### AMINO ACID COMPOSITION OF
### *T. TESTUDINUM* LEAVES EXPRESSED IN
### PERCENT (GRAMS AMINO ACID PER 100
### G OF DRIED MATERIAL)[21]

| Amino acid | Percent of dry matter |
|---|---|
| Arginine | 0.702 |
| Aspartic acid | 1.120 |
| Glutamic acid | 1.090 |
| Histidine | 0.310 |
| Isoleucine | 0.249 |
| Leucine | 0.693 |
| Lysine | 0.720 |
| Methionine | 0.187 |
| Phenylalanine | 0.465 |
| Threonine | 0.204 |
| Tryptophan | 0.049 |
| Valine | 0.317 |

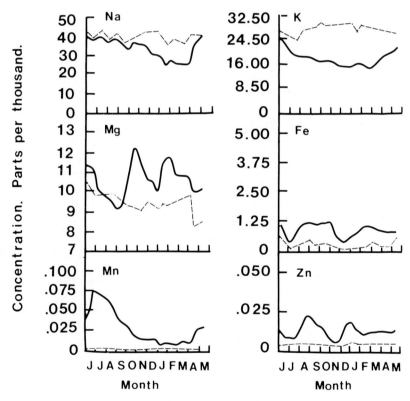

FIGURE 4. Annual variations of some elements in *T. testudinum*. The solid line ( —— ) indicates leaves, and the dotted line ( - - - ) indicates rhizomes.

Table 3

FOUR COMPARTMENT MODEL OF TRACE ELEMENTS IN CARD
SOUND[20]

| Compartment | Elements (mg/m²) | | | | | |
|---|---|---|---|---|---|---|
| | V | Fe | Cu | Zn | Cd | Pb |
| Sediment | $8.0 \times 10^3$ | $6.3 \times 10^5$ | $6.7 \times 10^2$ | $1.4 \times 10^3$ | 23 | $3.4 \times 10^2$ |
| Water | 2.6 | $5.2 \times 10^2$ | $1.2 \times 10^2$ | $2.6 \times 10^2$ | 2.1 | 15 |
| Thalassia | 1.4 | 53 | 0.27 | 3.0 | 0.033 | 0.12 |
| Biota | 2.7 | $1.5 \times 10^2$ | 0.76 | 7.0 | 0.068 | 0.16 |

Note: All values are total element concentrations.

over of blades is consumed directly by invertebrates or fish or indirectly by the detrital cycle, there is an active cycling of material by seagrasses in nearshore water.

Seagrass is consumed directly by humans and perhaps has some potential as a harvestable fodder for animals. Reports of seagrass consumption by native tribes in Mexico[22] show that the fruit is eaten and used as a bread. Use of the seagrass fruit by peoples in the Malaysian and Indonesian archipelagoes is common. Experiments for using seagrass as a fodder or a compost have been undertaken by the Cuban government, as well as others, but to date have not proved commercially successful.

# REFERENCES

1. Hartog, C. den, *The Sea-grasses of the World,* North-Holland, Amsterdam, 1970, chap. 1.
2. Hartog, C. den, Structure, function, and classification of seagrass communities, in *Seagrass Ecosystems,* McRoy, C. P. and Helfferich, C., Eds., Marcel Dekker, New York, 1977, chap. 3.
3. Buesa, R. J., Population and biological data on turtle grass (*Thalassia testudinum* Konig, 1801) on the northwestern Cuban shelf, *Aquaculture,* 4, 207, 1974.
4. Thorhaug, A., unpublished data, 1978.
5. Hartog, C. den, Klassifikate van Zeegrasgezelschappen, *Jaarb. Versl. Meded. Kon. Ned. Bot. Ver.,* 32, 1971.
6. Thorhaug, A. and Roessler, M. A., Seagrass community dynamics in a subtropical estuarine lagoon, *Aquaculture,* 12, 253, 1977.
7. Thorhaug, A., Teas, H., and Penhale, P., Total primary productivity in a subtropical estuarine lagoon, *J. Phycol.,* 13, 67, 1977.
8. Greenway, M., The effects of cropping on the growth of *Thalassia testudinum* (Konig) in Jamaica, *Aquaculture,* 4, 199, 1974.
9. Fenchel, T., Aspects of the decomposition of seagrasses, in *Seagrass Ecosystems,* McRoy, C. P. and Helfferich, C., Eds., Marcel Dekker, New York, 1977, chap. 4.
10. Tomlinson, P. B., Vegetative morphology and meristem dependence — the foundation of productivity in seagrasses, *Aquaculture,* 4, 107, 1974.
11. Benedict, C. R. and Scott, J. R., Photosynthetic carbon metabolism of a marine grass, *Plant Physiol.,* 57, 876, 1976.
12. Wetzel, R. G. and Hough, R. A., personal communication to C. P. McRoy and C. McMillan; as quoted in McRoy, C. P. and McMillan, C., Ecology and physiology of seagrasses, in *Seagrass Ecosystems,* McRoy, C. P. and Helfferich, C., Eds., Marcel Dekker, New York, 1977, chap. 2.
13. Parker, P. L., The biogeochemistry of the stable isotopes of carbon in a marine bay, *Geochem. Cosmochim. Acta,* 28, 1155, 1964.
14. Benedict, C. R., Wong, W. W. L., and Wong, J. H. H., The fractionation of stable carbon isotopes in seagrasses, *Plant Physiol.,* 63, S-1, 1979.

15. Thorhaug, A., Roessler, M. A., and Segar, D., Impact of a power plant on a subtropical estuarine environment, *Bull. Mar. Poll.,* 7, 166, 1973.
16. Wanless, H. R., Sedimentary dynamics and significance of sea-grass beds, *Fl. Sci.,* 38, 29, 1975.
17. Burkholder, P. R., Burkholder, L. M., and Rivero, J. A., Some chemical constituents of the turtle grass *Thalassia testudinum, Bull. Torrey Bot. Club,* 86, 88, 1959.
18. Walsh, G. E. and Grow, T. E., Composition of *Thalassia testudinum* and *Ruppia maritima, Q. J. Fl. Acad. Sci.,* 35, 96, 1972.
19. Ford, E., Moore, S., and Humm, H. J., *A Study of the Seagrass Beds in the Anclote Estuary During the First Year of Operation of the Anclote Power Plant,* Florida Power Corporation, St. Petersburg, Fla., 1975.
20. Segar, J. L. and Segar, D. A., Biogeochemistry of trace elements in Card Sound, Florida, Proc. Biscayne Bay Symp. I, April 2 to 3, 1976.
21. Schroeder, P. and Thorhaug, A., A model of heavy metal cycling through subtropical and tropical estuarine systems with energy related industry, Waste Heat Symposium II, 1978.
22. Felger, R. and Moser, M. B., Eelgrass (*Zostera marina* L.) in the Gulf of California: discovery of its nutritional value by the Seri Indians, *Science,* 181, 355, 1973.

# GENERAL CHARACTERISTICS OF TERRESTRIAL PLANTS (AGRONOMIC AND FORESTS) — C₃, C₄, AND CRASSULACEAN ACID METABOLISM PLANTS

## Bruce N. Smith

Solar energy is trapped and transformed into chemical energy by green plants in the form of adenosine triphosphate (ATP) and reducing power (NADPH). This chemical energy (ATP and NADPH) is then used to reduce $CO_2$ to the level of carbohydrate. All photosynthetic organisms on earth use similar mechanisms to trap light energy and transform it to chemical energy. However, some striking differences exist in higher plants as to the mechanism for using light-generated chemical energy to fix $CO_2$. This chapter will briefly consider the basis for these differences and will then outline the distribution among terrestrial cultivated and native plants of the principal modes of $CO_2$ fixation.

The overall reaction for plant photosynthesis can be written:

$$CO_2 + 2H_2O \xrightarrow[\text{chloroplasts}]{\text{light}} (CH_2O) + O_2 + H_2O$$

Calvin and co-workers[1] have elucidated very elegantly the mechanism by which inorganic carbon is synthesized into complex organic molecules. The initial fixation step involves the enzyme ribulose bisphosphate carboxylase (RuBP carboxylase) which can make up as much as half of the soluble protein in some leaves. The reaction requires $CO_2$ and a five-carbon sugar (ribulose-1,5-bisphosphate) and the product is two molecules of a three-carbon acid (3-phosphoglyceric acid). The three-carbon acid is then reduced with the aid of ATP and NADPH to the level of a sugar. Some of this sugar then goes to make starch, while some goes to regenerate the reactant ribulose-1,5-bisphosphate. This mechanism for carbon fixation is found in virtually all green tissues. It is thought to be the only mode of carbon fixation used by algae, mosses, and lycopods. It is also by far the most common type of photosynthesis in ferns, gymnosperms, and flowering plants.

RuBP carboxylase not only combines $CO_2$ and RuBP to produce PGA, but it also reacts with oxygen.[2] The products with oxygen are phosphoglyceric acid and phosphoglycolate. Oxygen and $CO_2$ thus compete for the same active site on the enzyme — that is, oxygen competes with $CO_2$ in photosynthesis. The phosphoglycolate is degraded eventually to $CO_2$ which is then released into the atmosphere. Thus, the oxygenase reaction results in a loss of carbon through photorespiration that in essence greatly decreases the efficiency of photosynthesis.

An impressive body of evidence now exists for other modes of carbon fixation in a relatively small number of land plant species. One mechanism, $C_4$ photosynthesis, includes a syndrome of distinguishing anatomical and physiological characters all associated with photosynthetic fixation of carbon.[3-5] Anatomical features include a chlorenchymatous sheath of large, thick-walled cells (Kranz cells) surrounding vascular bundles of leaves. Bundle sheath (Kranz) cells have an ultrastructural specialization which includes extensive cytoplasm and in some groups, reduced grana.[6] In addition, mesophyll cells surrounding the bundle sheath cells are often radially arranged. Functional aspects of the syndrome include initial carbon fixation into four-carbon acids — oxaloacetic acid then malic or aspartic acid in the mesophyll cells. The four-carbon acids pass into the bundle sheath cells where they are decarboxylated and the $CO_2$ released is then fixed into PGA.[3] The syndrome is also characterized by having a much

Table 1
## DISTINGUISHING CHARACTERISTICS OF THREE TYPES OF PHOTOSYNTHETIC PLANTS

| | Plants | | |
|---|---|---|---|
| Characteristics | $C_4$ | $C_3$ | CAM |
| Initial $CO_2$-fixing enzyme | PEP carboxylase | RuBP carboxylase | Both |
| Kranz anatomy | Present | Absent | Absent |
| Succulence | No | Yes and no | Yes |
| $CO_2$ conc. required for fixation | <10 ppm | 30—150 ppm | ? |
| Light saturation | Full sunlight | <40 klx | ∿10 klx |
| Optimum temperature for $CO_2$ fixation | 30—35°C | 15—20°C | ∿35°C |
| Photosynthetic rate (mg $CO_2$ dm$^{-2}$ hr$^{-1}$) | 40—80 | 15—35 | 0.5—0.7 |
| Photorespiration | Very low | Active | Very low |
| $^{13}C/^{12}C$ ratio (δ$^{12}C$ o/oo) | −9 to −18 | −23 to −36 | −9 to −36 |

higher temperature optimum and light saturation than other plants (see Table 1). This phenomenon is now called $C_4$ photosynthesis, and plants possessing the syndrome are termed $C_4$ plants. Other plants are called $C_3$ plants, as the first product of photosynthesis is a three-carbon acid — phosphoglyceric acid.

In $C_4$ plants, atmospheric $CO_2$ diffuses into the leaf through open stomates and enters the mesophyll cell. Once in the mesophyll, the $CO_2$ reacts with phosphoenolpyruvate (PEP) to yield oxaloacetic acid (OAA). The enzyme which catalyzes the reaction is PEP carboxylase. PEP carboxylase has a high affinity for $CO_2$ and a low affinity for $O_2$ — i.e., there is no competition between oxygen and $CO_2$ on this enzyme. All of the carbon fixed by $C_4$ plants is fixed via PEP carboxylase in the mesophyll cells. Oxaloacetic acid is rapidly transformed into malic acid (in some species into aspartic acid) for transport into the bundle sheath cells where the malic acid is decarboxylated. The $CO_2$ liberated in the bundle sheath cells is in turn fixed into PGA by the ubiquitous RuBP carboxylase.[4] The PGA can then go into starch or other organic substances for storage or transport.

Table 1 lists a number of distinguishing characteristics between $C_4$ and $C_3$ plants. As a consequence of more efficient $CO_2$ fixation, $C_4$ plants make less of a distinction between the naturally occurring stable isotopes of carbon than does RuBP carboxylase in $C_3$ plants.[7] $C_4$ photosynthesis does not occur in algae, bryophytes, pteridophytes, gymnosperms, and the great majority of the families of the angiosperms. Only 16 families of the approximately 300 families of flowering plants contain $C_4$ plants (see Table 2). Within none of the families do $C_4$ plants constitute a majority. $C_4$ photosynthesis appears to be a relatively recent adaptation to warm, arid, semitropical to tropical conditions. The oldest $C_4$ fossil known is from the Pliocene.[8]

Succulent desert plants often exhibit a mode of carbon fixation called crassulacean acid metabolism (CAM). CAM plants under conditions of drought, hot days, and cool nights will fix $CO_2$ at night via PEP carboxylase and store it as malic acid in the vacuole. During the day, the malic acid is decarboxylated and the $CO_2$ is fixed into PGA via RuBP carboxylase. While many desert plants must fix their carbon in this way ("obligate" CAM plants), other desert succulents apparently use the CAM mode only when under stress due to drought or salt. These plants ("facultative" CAM) in unstressed, mesic conditions may fix all of their carbon in the daytime through normal $C_3$ mechanisms.[9] Although certain superficial similarities do exist between $C_4$ photo-

## Table 2
## FAMILIES IN WHICH C₄
## PHOTOSYNTHESIS IS
## KNOWN TO OCCUR

Monocotyledonae
Cyperaceae
Gramineae
Liliaceae

Dicotyledonae
Aizoaceae
Amaranthaceae
Asteraceae
Boraginaceae
Capparadaceae
Caryophyllaceae
Chenopodiaceae
Euphorbiaceae
Molluginaceae
Nyctaginaceae
Polygonaceae
Portulacaceae
Zygophyllaceae

## Table 3
## FAMILIES IN WHICH CAM IS KNOWN
## TO OCCUR

Filicinae
Polypodiaceae
Gymnospermae
Welwitschiaceae
Angiospermae
Monocotyledonae

| | |
|---|---|
| Agavaceae | Liliaceae |
| Bromeliaceae | Orchidaceae |

Dicotyledonae

| | |
|---|---|
| Aizoaceae | Convolvulaceae |
| Asclepiadaceae | Crassulaceae |
| Asteraceae | Euphorbiaceae |
| Bataceae | Plantaginaceae |
| Cactaceae | Portulacaceae |
| Caryophyllaceae | Vitaceae |
| Chenopodiaceae | |

synthesis and CAM, the differences (see Table 1) are fundamental enough that C₄ plants and CAM plants probably had a separate origin.

CAM metabolism has now been reported from a fern, a gymnosperm, and from 19 families of flowering plants (see Table 3). *Drymoglossum* is a succulent, epiphytic fern which probably occupies a xeric microhabitat.[10] The desert gymnosperm, *Welwitschia mirabilis* has also been reported to be a CAM plant.[11] CAM metabolism thus exists in a diverse group of plants united more by adaptation to xeric environments than by phylogeny based on morphological consideration. It may have evolved more than once. Since CAM plants keep their stomates closed in the daytime, water loss as well as gas exchange is greatly restricted. Under these conditions, rates of photosynthesis are very low, but the strategy of CAM plants seems to be survival in extreme aridity rather than rapid growth.

Table 4

**REPRESENTATIVE AGRONOMIC PLANTS BY
PHOTOSYNTHETIC TYPE**

**$C_3{}^a$**

Grains
  Wheat
  Rice
  Barley
  Rye

Composites
  Sunflower
  Safflower

Legumes
  Soybeans
  Beans
  Peas
  Alfalfa
  Clover

Chenopods
  Cabbage
  Cauliflower
  Spinach
  Chard
  Brussels sprouts

Root crops
  Carrot
  Turnip
  Taro
  Cassava

Solanaceae
  Potato
  Tomato
  Pepper

Tree fruits
  Apple
  Pear
  Cherry
  Apricot
  Plum
  Peach
  Almond
  Mango
  Papaya

Forb fruits
  Squash
  Watermelon
  Cantelope
  Strawberry

**$C_4$**

Crops
  Corn
  Sorghum
  Sugarcane
  Millet
  Amaranthus
  Portulaca

Forage, hay, and pasture
  Several $C_4$ grass species
  are important:
    *Bouteloua* sp.
    *Panicum* sp.
    *Eragrostis* sp.
    etc.

**CAM**
  Pineapple
  *Agave* — sisal and tequila
  Houseplants — kalanchoe,
  aloe, cactus

    [a]  Includes most crop plants.

The great majority of crop plants world-wide are $C_3$ plants. A few representative $C_3$ crop plants are listed in Table 4. Thus, of plants we raise for food, all legumes, composites, chenopods, and solonaceae are $C_3$. Tree fruits, most grains, root crops, etc. are $C_3$. $C_4$ plants under cultivation are quite important, but represent a small number of species (see Table 4). Thus, corn, sorghum, and sugarcane probably represent most of the $C_4$ agronomic food plants. In addition, a number of $C_4$ grasses are grown for hay, pasture, or used for forage. CAM plants are similarly limited as crop plants: pineapple and *Agave* (sisal and tequila). A variety of CAM plants are cultivated and sold as house and garden plants.

At this time, any survey of $C_4$ plants is bound to be incomplete. For example, 5 years ago we knew of only 10 families containing $C_4$ species. Today we know of 16 families containing $C_4$ plants. However, Table 5 may prove to be a useful interim checklist. This is certainly more complete than earlier published checklists.[12,13]

CAM plants have been of interest for a long time, but mostly to a relatively small group of specialists. The enumeration of CAM plants in Table 6 is also certainly incomplete, but may serve a useful purpose as a checklist of species known to be either "facultative" or "obligate" CAM. This is somewhat more complete than the excellent checklist published by Black and Williams.[14]

Table 5
## C₄ PLANTS LISTED BY FAMILY

| Plant | Ref. | Plant | Ref. |
|---|---|---|---|
| Monocotyledonae | | | |
| | | | |
| Cyperaceae[a] | 15 | *Anthaenantia rufa* | 21 |
| *Abildgaardia* sp. | 15 | *Anthephora cristata* | 21 |
| *Ascolipis* sp. | 15 | *A. elegans* | 21 |
| *Bulbostylis* sp. | 15 | *A. elongata* | 21 |
| *Crosslandia* sp. | 15 | *A. hermaphrodita* | 21 |
| *Cyperus albomarginatus* | 16 | *A. pubescens* | 18, 21 |
| *C. eragrostis* | 17 | *Apluda mutica* | 24 |
| *C. esculentus* | 18 | *Aristida acutiflora* | 23 |
| *C. filiculmis* | 19 | *A. adscensionis* | 21, 23 |
| *C. odoratus* | 19 | *A. armata* | 17 |
| *C. papyrus* | 18 | *A. biglandulosa* | 20 |
| *C. rotundus* | 17 | *A. caerulescens* | 23 |
| *C. ustalatus* | 17 | *A. ciliata* | 23 |
| *Fimbristylis* sp. | 15 | *A. glauca* | 21 |
| *Hemicarpha* sp. | 15 | *A. hamulosa* | 21 |
| *Kyllinga* sp. | 15 | *A. obtusa* | 23 |
| *Lipocarpa* sp. | 15 | *A. plumosa* | 23 |
| *Mariscus* sp. | 15 | *A. pungens* | 23 |
| *Nelmesia* sp. | 15 | *A. purpurea* | 21 |
| *Nemum* sp. | 15 | *A. ramosa* | 24 |
| *Pycreus* sp. | 15 | *A. uniplumis* | 21 |
| *Queenslandiella* sp. | 15 | *A. vagans* | 24 |
| *Remirea* sp. | 15 | *Arthraxon hispidus* | 18 |
| *Scirpus* sp. | 15 | *A. quartinianus* | 20 |
| *Torulinum* sp. | 15 | *Anthropogon scaber* | 21 |
| *Volkiella* sp. | 15 | *A. villosus* | 21 |
| | | *A. xerachne* | 21 |
| Gramineae | | *Arundinella anomala* | 21 |
| *Acrachne verticillata* | 20 | *A. berteroniana* | 21 |
| *Achlaena piptostachya* | 21 | *A. deppeana* | 21 |
| *Alloteropsis angusta* | 22 | *A. hirta* | 21 |
| *A. cimicina* | 22 | *A. metzii* | 20 |
| *A. gwebonsis* | 22 | *A. nepalensis* | 24 |
| *A. paniculata* | 22 | *A. villosa* | 20 |
| *A. semialata* | 22 | *Astrebla pectinata* | 21 |
| *Amphilophis affinis* | 20 | *A. squarrosa* | 20 |
| *A. intermedia* | 20 | *Austrochloris dichanthioides* | 24 |
| *Andropogon amplectens* | 20 | *Axonopus affinis* | 21 |
| *A. cirratus* | 18 | *A. argentinus* | 21 |
| *A. condensatus* | 18 | *A. compressus* | 21 |
| *A. distachys* | 23 | *Beckeropsis uniseta* | 20 |
| *A. gerardi* | 18 | *Bothriochloa alta* | 21 |
| *A. hassleri* | 18 | *B. barbinodis* | 18 |
| *A. hirtiflorus* | 18 | *B. caucasia* | 20 |
| *A. lateralis* | 18 | *B. decipiens* | 18 |
| *A. papillosus* | 18 | *B. ewartiana* | 18 |
| *A. perforatus* | 21 | *B. glabra* | 18 |
| *A. saccharoides* | 21 | *B. insculpta* | 18 |
| *A. schirensis* | 20 | *B. intermedia* | 18 |
| *A. scoparius* | 21 | *B. ischaemum* | 18 |
| *A. selloanus* | 18 | *B. laguroides* | 18 |
| *A. ternatus* | 18 | *B. macra* | 24 |
| *A. venustus* | 20 | *B. pertusa* | 18 |
| *A. virgatus* | 18 | *B. springfieldii* | 18 |
| *A. virginicus* | 21 | *Bouteloua aracilis* | 20 |

## Table 5 (continued)
## C₄ PLANTS LISTED BY FAMILY

| Plant | Ref. | Plant | Ref. |
|---|---|---|---|
| *B. aristidoides* | 25 | *C. virgata* | 17 |
| *B. barbata* | 25 | *Chrysopogon fallax* | 24 |
| *B. curtipendula* | 21 | *C. gryllus* | 21 |
| *B. filiformis* | 21 | *C. montanus* | 21 |
| *Bouteloua foliosa* | 20 | *C. serrulatus* | 18 |
| *B. gracilis* | 21 | *C. zeylanicus* | 20 |
| *B. hirsuta* | 18 | *Cleistachne sorghoides* | 20 |
| *Brachiaria deflexa* | 20 | *Coelorachis cimicina* | 18 |
| *B. distichophylla* | 20 | *C. rottboellioides* | 24 |
| *B. crucaetormis* | 18 | *C. selloana* | 18 |
| *B. foliosa* | 24 | *Coix lacryma-jobi* | 21 |
| *B. jubata* | 20 | *Cymbopogon citratus* | 21, 20 |
| *B. laeta* | 18 | *C. giganteus* | 20 |
| *B. mutica* | 21 | *C. martini* | 18, 20 |
| *B. paspaloides* | 20 | *C. nardus* | 20 |
| *B. platyphylla* | 21 | *C. refractus* | 24 |
| *B. ramosa* | 18 | *C. schoenanthus* | 23 |
| *B. reptans* | 20 | *C. tortilus* | 21 |
| *B. subquadripara* | 21 | *C. validus* | 20 |
| *Brachyachne convergens* | 21 | *Cymnogopon refractus* | 17 |
| *Buchloë dactyloides* | 21 | *Cynodon arcuatus* | 18 |
| *Calamovilfa longifolia* | 20 | *C. dactylon* | 21 |
| *Capillipedium parviflorum* | 20 | *Dactyloctenium aegyptiacum* | 21 |
| *C. spicigerum* | 21 | *D. giganteum* | 20 |
| *Cenchrus biflorus* | 18 | *Danthoniopsis dinteri* | 21 |
| *C. calycalatus* | 17 | *D. humbertii* | 20 |
| *C. ciliaris* | 21 | *D. minor* | 20 |
| *Cenchrus echinatus* | 21 | *D. stocksii* | 20 |
| *C. incertus* | 21 | *D. viridis* | 20 |
| *C. myosuroides* | 21 | *Desmostachya bipinnata* | 20 |
| *C. pauciflorus* | 18 | *Dichanthium annulatum* | 18 |
| *C. pilosus* | 18 | *D. aristatum* | 20, 21 |
| *C. setigerus* | 18 | *D. polyptychum* | 20 |
| *Chamaeraphis hordeacea* | 24 | *D. sericeum* | 18, 20 |
| *Chionachne cyathopoda* | 24 | *D. superciliatum* | 18 |
| *Chloris acicularis* | 18 | *Digitaria adscendens* | 21 |
| *C. argentina* | 18 | *D. argyrograpta* | 21 |
| *C. barbata* | 20 | *D. barbonica* | 20 |
| *C. canterai* | 18 | *D. bicornis* | 18 |
| *C. caribaea* | 18 | *D. brazzae* | 20 |
| *C. cucullata* | 21 | *D. brownii* | 18 |
| *C. distichophylla* | 21 | *D. decumbens* | 21 |
| *C. filiformis* | 20 | *D. diagonalis* | 18 |
| *C. gayana* | 21 | *D. eriantha* | 18 |
| *C. inflata* | 18 | *D. eriostachya* | 18 |
| *C. pectinata* | 18 | *D. gazensis* | 18 |
| *C. petraea* | 18 | *D. glauca* | 18 |
| *C. pilosa* | 18 | *D. horizontalis* | 18, 20 |
| *C. polydactyla* | 18 | *D. iburua* | 18 |
| *C. pycnothrix* | 18 | *D. ischaemum* | 18 |
| *C. radiata* | 18 | *D. kilimandscharica* | 18 |
| *C. robusta* | 20 | *D. milanjiana* | 18, 20 |
| *C. submutica* | 18 | *D. pentzii* | 20 |
| *C. truncata* | 18 | *D. phaeothrix* | 18 |
| *C. uliginosa* | 18 | *D. rhopaloricha* | 21 |
| *C. ventricosa* | 17 | *D. sanguinalis* | 21 |

## Table 5 (continued)
## C₄ PLANTS LISTED BY FAMILY

| Plant | Ref. | Plant | Ref. |
|---|---|---|---|
| D. seriata | 18 | E. flaccida | 18 |
| D. smutsii | 21 | E. gangetica | 20 |
| D. swazilandensis | 18 | E. gummiflua | 18 |
| D. valida | 18 | E. heteromera | 18 |
| Dimeria thwaitesii | 20 | E. horizontalis | 18 |
| Dinebra retroflexa | 24 | E. intermedia | 21 |
| Diplachne parviflora | 24 | E. lappula | 18 |
| Distichlis distichophylla | 24 | E. lehmanniana | 18 |
| D. spicata | 21 | E. margaritacea | 18 |
| D. stricta | 26 | E. mexicana | 21 |
| Echinochloa colonum | 21 | E. nigra | 18 |
| E. crusgalli | 21 | E. obtusa | 18 |
| E. frumentacea | 18 | E. oxylepis | 18 |
| E. haploclada | 18 | E. papposa | 18 |
| E. holubii | 18 | E. parviflora | 17 |
| E. pyramidalis | 18 | E. patentissima | 18 |
| E. spiralis | 18 | E. pilosa | 21 |
| E. stagnina | 21 | E. plana | 18 |
| Eleusine compressa | 18 | E. poaeoides | 18 |
| E. coracana | 21 | E. polytricha | 18 |
| E. flagellifera | 18 | E. rigidior | 21 |
| E. floccifolia | 18 | E. robusta | 18 |
| E. indica | 21 | E. rufescens | 18 |
| E. jaegeri | 18 | E. secundiflora | 18 |
| E. multiflora | 18 | E. spectabilis | 26 |
| E. tristachya | 18 | E. starosselsky | 18 |
| Elionurus chevalieri | 20 | E. superba | 18 |
| E. citreus | 20 | E. tremula | 18 |
| E. hirtifolius | 20 | E. trichodes | 18 |
| Enneapogon cenchroides | 20 | E. truncata | 18 |
| E. nigricans | 24 | E. unioloides | 18 |
| E. scaber | 23 | E. virescens | 18 |
| Enteropogon acicularis | 24 | Eremochloa bimaculata | 24 |
| Eragrostiella bifaria | 20 | E. muricata | 20 |
| Eragrostis acutiflora | 18 | E. ophiuroides | 21 |
| E. acutiglumis | 18 | Erianthus fastigiatus | 24 |
| E. airoides | 18 | E. hostii | 20 |
| E. aspera | 20 | E. maximus | 21 |
| E. atherstonei | 18 | Eriochloa australiensis | 20 |
| E. bahiensis | 18 | E. gracilis | 21 |
| E. bethamii | 24 | Euchlaena mexicana | 18, 21 |
| E. bicolor | 18 | E. perennis | 21 |
| E. brasiliensis | 18 | Euchlaezea mertonensis | 20 |
| E. brownii | 21 | Euclasta condylotricha | 20 |
| E. chalcantha | 18 | Eulalia fulva | 24 |
| E. chariis | 18 | E. geniculata | 20 |
| E. cholormelas | 21, 20 | E. phaeothrix | 20 |
| E. cilianensis | 21 | Eustachys distichophylla | 24 |
| E. collocarpa | 18 | E. paspaloides | 20 |
| E. curvula | 21 | Fingerhuthia africana | 20 |
| E. denudata | 18 | Garnotia adscendens | 21 |
| E. dielsii | 18 | G. courtallensis | 20 |
| E. diffusa | 18 | G. stricta | 21 |
| E. diplachnoides | 20 | Gymnopogon ambiguus | 21 |
| E. echinochloidea | 20 | G. delicatulus | 20 |
| E. ferruginea | 18 | G. foliosus | 20 |

## Table 5 (continued)
## C₄ PLANTS LISTED BY FAMILY

| Plant | Ref. | Plant | Ref. |
|---|---|---|---|
| *G. spicatus* | 20 | *Manisuris altissima* | 21 |
| *Hackelochloa granularis* | 24 | *Melinis minutiflora* | 20, 21 |
| *Hemarthria uncinata* | 24 | *Mirochloa caffra* | 20 |
| *Heterachne brownii* | 24 | *Microchloa indica* | 24 |
| *Hetropogon contortus* | 18, 21, 27 | *Microsteguim ciliatum* | 20 |
| *Hilaria belangeri* | 21 | *Miscanthes condensatus* | 20 |
| *H. mutica* | 21 | *M. sacchariflorus* | 20, 21 |
| *Hyparrhenia filipendula* | 24 | *M. sinensis* | 24 |
| *H. dissoluta* | 20 | *Monanthochloë littoralis* | 21 |
| *H. hirta* | 18 | *Monelytrum luederitzianum* | 20 |
| *H. rufa* | 18 | *Muhlenbergia arisanensis* | 24 |
| *Imperata arundinacea* | 21 | *M. emersleyi* | 21 |
| *I. chesemanii* | 17 | *M. lindheimeri* | 21 |
| *I. cylindrica* | 17 | *M. racemosa* | 21 |
| *Indopoa paupercula* | 20 | *M. schreberi* | 21 |
| *Ischaemum australe* | 20 | *Munroa mendoana* | 20 |
| *I. commutatum* | 20 | *M. squarrosa* | 20 |
| *I. laxum* | 20 | *Neostapfia colusana* | 21 |
| *I. santapaui* | 20 | *Ophiurus exaltatus* | 24 |
| *Iseilema membranacea* | 18, 20 | *Orcuttia californica* | 21 |
| *I. vaginiflora* | 18, 24 | *Oropetium africanum* | 20 |
| *I. wightii* | 18 | *O. thomaeum* | 20 |
| *Jouvea pilosa* | 21 | *Panicum amarulum* | 18 |
| *Lasiurus hirsutus* | 23 | *P. ambiguum* | 18 |
| *L. sindicus* | 31 | *P. anceps* | 21 |
| *Latipes senegalensis* | 20 | *P. antidotale* | 21 |
| *Leptocarydion vulpiastrum* | 20 | *P. australiense* | 24 |
| *Leptochloa coerulescens* | 20 | *P. bergii* | 21 |
| *L. digitata* | 24 | *P. capillare* | 21 |
| *L. dubia* | 21 | *P. bulbosum* | 21 |
| *L. fascicularis* | 18 | *P. coloratum* | 21 |
| *L. fusca* | 21 | *P. cymbiforme* | 21 |
| *L. monstachya* | 21 | *P. decompositum* | 21 |
| *L. uniflora* | 20 | *P. deustum* | 21 |
| *Leptochloopsis condensata* | 28 | *P. dichotomiflorum* | 21 |
| *L. virgata* | 28 | *P. dilatatum* | 17 |
| *Leptoloma cognatum* | 21, 26 | *P. distichum* | 17 |
| *Lepturella aristata* | 20 | *P. effusum* | 18 |
| *L. capensis* | 20 | *P. filipes* | 21 |
| *Lepturus hildebrandtii* | 20 | *P. firmulum* | 18 |
| *L. mildbraedianus* | 20 | *P. geminatum* | 21 |
| *L. radicans* | 21 | *P. hallii* | 21 |
| *L. repens* | 21 | *P. havardii* | 18 |
| *L. xerophilus* | 20 | *P. indicum* | 21 |
| *Loudetia superba* | 20 | *P. laevifolium* | 21 |
| *Loudetiopsis ambiens* | 20 | *P. lanipes* | 18 |
| *L. capillipes* | 20 | *P. larcomianum* | 18 |
| *L. chevalieri* | 20 | *P. longijubatum* | 18 |
| *L. chrysothrix* | 20 | *P. makarikariensis* | 21 |
| *L. glabrata* | 20 | *P. maximum* | 21 |
| *L. kerstingii* | 20 | *P. miliaceum* | 21 |
| *L. purpurea* | 20 | *P. minus* | 21 |
| *L. ternata* | 20 | *P. molle* | 18 |
| *L. tristachyoides* | 20 | *P. notatum* | 17 |
| *L. villosipes* | 20 | *P. obtusum* | 21 |
| *Lycurus phleoides* | 21 | *P. paspalioides* | 17 |

## Table 5 (continued)
## C₄ PLANTS LISTED BY FAMILY

| Plant | Ref. | Plant | Ref. |
|---|---|---|---|
| *P. plenum* | 21 | *Pennisetuna alopecuroides* | 24 |
| *P. polygonatum* | 21 | *Pennisetum ciliare* | 21 |
| *P. prolutum* | 21 | *P. clandestinum* | 27 |
| *P. pumilum* | 17 | *P. dichotomum* | 23 |
| *P. purpurascens* | 21 | *P. flaccidum* | 18 |
| *P. queenslandicum* | 18 | *P. glaucum* | 21 |
| *P. ramisetum* | 21 | *P. macrourum* | 18, 20 |
| *P. repens* | 20, 21 | *P. massaicum* | 18 |
| *P. reptans* | 20, 21 | *P. orientale* | 18 |
| *P. reverchoni* | 21 | *P. pedicellatum* | 21 |
| *P. seminudum* | 20 | *P. polystachyum* | 18 |
| *P. simile* | 24 | *P. purpureum* | 21 |
| *P. stapfianum* | 21 | *P. setaceum* | 23 |
| *P. tenerum* | 21 | *P. spicatum* | 18 |
| *P. texanum* | 21 | *P. typhoides* | 18, 20 |
| *P. trachyrhachis* | 18 | *P. villosum* | 20, 24 |
| *P. turgidum* | 21 | *Perotis indica* | 20 |
| *P. urvilleanum* | 21 | *P. patens* | 20 |
| *P. virgatum* | 21 | *P. rara* | 17, 20 |
| *P. whitei* | 18 | *Pheidochloa gracilis* | 24 |
| *Pappophorum alopecuroideum* | 20 | *Plectrachne schinzii* | 24 |
| *P. bicolor* | 21 | *Pogonarthria squarrosa* | 20 |
| *P. pappiferum* | 20 | *P. tuberculata* | 20 |
| *Paraneurachne muelleri* | 24 | *Pogonatherum paniceum* | 24 |
| *Paspalidium geminatum* | 20 | *Pollinia fulva* | 20 |
| *P. gracile* | 24 | *Pommerculla cornucopiae* | 20 |
| *P. jubiflorum* | 24 | *Polytoca macrophylla* | 24 |
| *P. rarum* | 24 | *Polytrias amaura* | 24 |
| *P. almum* | 18 | *Pseudochaetochloa australiensis* | 24 |
| *P. boscianum* | 18 | *Pseudanthistiria umbellata* | 20 |
| *P. brunneum* | 18 | *Pseudopogonatherum iritans* | 24 |
| *P. ciliatifolium* | 18 | *Rattraya petiolata* | 21 |
| *P. commersonii* | 21 | *Reimarochloa acuta* | 21 |
| *Paspalum conjagatum* | 18, 21 | *Reynaudia filiformis* | 21 |
| *P. dilatatum* | 21 | *Rhynchelytrum grandiflorum* | 21 |
| *P. distichum* | 21 | *R. repens* | 24 |
| *P. geminiflorum* | 18 | *R. roseum* | 21 |
| *P. hartwegianum* | 21 | *Rottboellia exaltata* | 24 |
| *P. intermedium* | 18 | *Saccharum benghalense* | 20 |
| *P. juegensii* | 18 | *S. officinarum* | 21 |
| *P. mandiocanum* | 18 | *S. robustum* | 21 |
| *P. nicorae* | 18 | *S. sinensis* | 21 |
| *P. notatum* | 21 | *S. spontaneum* | 21 |
| *P. paniculatum* | 18, 20 | *Schedonnardus paniculatus* | 20, 21 |
| *P. paucispicatum* | 18 | *Schizachyrium jeffreysii* | 20 |
| *P. platyphyllum* | 18 | *S. obliquiberbe* | 20 |
| *P. plicatulum* | 18 | *Sclerandrium truncatiglume* | 24 |
| *P. polystachyum* | 18 | *Scleropogon brevifolius* | 20 |
| *P. publiflorum* | 21 | *Setaria adhaerens* | 18 |
| *P. pumilum* | 18 | *S. almaspicata* | 18 |
| *P. quadrifarium* | 18 | *S. anceps* | 27 |
| *P. rojasii* | 18 | *S. argentina* | 18 |
| *P. scrobiculatum* | 18 | *S. barbata* | 21 |
| *P. umbrosum* | 18 | *S. faberii* | 18 |
| *P. urvillei* | 21 | *S. geniculata* | 24 |
| *P. virgatum* | 18 | *S. glauca* | 18 |
| *P. yaguaronense* | 18 | *S. holstii* | 18 |

## Table 5 (continued)
## C₄ PLANTS LISTED BY FAMILY

| Plant | Ref. | Plant | Ref. |
| --- | --- | --- | --- |
| *S. italica* | 21 | *S. heterolepsis* | 26 |
| *S. lutescens* | 21 | *S. indicus* | 18 |
| *S. neglecta* | 18 | *S. ioclados* | 18 |
| *S. pallidifusca* | 18 | *S. jacquemontii* | 18 |
| *S. palmifolia* | 18 | *S. molleri* | 20 |
| *S. phanerococca* | 18 | *S. phyllotrichus* | 18 |
| *S. plicata* | 20 | *S. poiretii* | 21 |
| *S. scheelei* | 21 | *S. pyramidatus* | 18 |
| *S. sphacelata* | 21 | *S. usitatus* | 18 |
| *S. verticillata* | 18 | *S. wrightii* | 21 |
| *S. viridis* | 21 | *Stenotaphrum dimidiatum* | 20 |
| *Snowdenia polystachya* | 21 | *S. secundatum* | 21, 24 |
| *Sorghastrum nutans* | 21 | *Tetrapogon villosus* | 20 |
| *S. pellitum* | 18 | *Thaumastochloa* sp. | 24 |
| *Sorghum almum* | 21 | *Themeda arenacea* | 20 |
| *S. arundinaceum* | 18 | *T. australis* | 21 |
| *S. bicolor* | 21 | *T. quadrivalvis* | 20 |
| *S. caffrorum* | 18 | *T. tremula* | 20 |
| *S. caudatum* | 18 | *T. triandra* | 21 |
| *S. controversum* | 18 | *Thuarea involuta* | 20, 21 |
| *S. dochna* | 18 | *Trachys muricata* | 21 |
| *S. drummondii* | 18 | *Tragus australianus* | 21, 24 |
| *S. gambicum* | 18 | *T. berteronianus* | 20 |
| *S. halepense* | 21 | *Trichachne california* | 21 |
| *S. hewisonii* | 18 | *T. insularis* | 21 |
| *S. japonicum* | 18 | *T. sacchariflora* | 21 |
| *S. leiocladum* | 24 | *Trichloris crinita* | 18 |
| *S. nigricans* | 18 | *Tricholaena monachne* | 18 |
| *S. pellitum* | 21 | *T. repens* | 18 |
| *S. propinquum* | 21 | *Tridens albescens* | 21 |
| *S. saccharatum* | 18 | *T. flava* | 20 |
| *S. sudanense* | 21 | *T. grandiflorus* | 20 |
| *S. technicum* | 18 | *T. muticus* | 20 |
| *S. verticilliflorum* | 18 | *T. pilosus* | 21 |
| *S. virgatum* | 18 | *Triodia pungens* | 24 |
| *S. vulgare* | 21 | *Triplopogon spathiflorus* | 20 |
| *Spartina alterniflora* | 21 | *Tripogon loliiformis* | 24 |
| *S. cynosuroides* | 21 | *Triraphis mollis* | 24 |
| *S. foliosa* | 21 | *T. pumilio* | 20 |
| *S. maritima* | 24 | *Tripsacum dactyloides* | 20, 21 |
| *S. pectinata* | 21 | *Tristachya hispida* | 20 |
| *S. spartinae* | 21 | *T. inamoena* | 20 |
| *S. townsendii* | 20 | *Uniola paniculata* | 28 |
| *Spinifex hirsutus* | 17 | *U. pittieri* | 28 |
| *S. littoralis* | 4 | *Urochloa mosambicensis* | 21 |
| *Sporobolus africanus* | 20 | *U. panicoides* | 20 |
| *S. airoides* | 20 | *U. pullulans* | 20, 21 |
| *S. asper* | 21 | *Vaseyochloa multinervosa* | 21 |
| *S. capensis* | 18 | *Vetiveria elongata* | 20 |
| *S. caroli* | 21 | *V. zizanioides* | 18, 20 |
| *S. contractis* | 18 | *Xerochloa barbata* | 24 |
| *S. cryptandrus* | 21 | *X. chirboa* | 21 |
| *S. diander* | 20 | *X. imberbis* | 24 |
| *S. elongatus* | 18 | *X. lanifera* | 24 |
| *S. fimbriatus* | 18 | *Zea mays* | 21 |
| *S. helvola* | 18 | *Zoysia japonica* | 21 |

## Table 5 (continued)
## C₄ PLANTS LISTED BY FAMILY

| Plant | Ref. | Plant | Ref. |
|---|---|---|---|
| Z. macrantha | 24 | G. haegeana | 29 |
| Z. matrella | 21 | G. maritina | 32 |
| Z. minima | 17 | G. nitida | 29 |
| | | G. sonorae | 29 |
| Liliaceae | | Gossypianthus lanuginosus | 29 |
| Poellnitzia rubiflora | 17 | G. rigidiflorus | 29 |
| | | Guilleminea densa | 32 |
| Dicotyledonae | | Lithophila vermiculares | 29 |
| Aizoaceae | | Philoxerus vermiculares | 32 |
| Cypselea humifusa | 29 | Tidestromia lanuginosa | 29 |
| Gisekia pharnacoides | 30 | T. oblongifolia | 29 |
| Trianthema portulacastrum | 18, 29, 30 | T. suffruticosa | 29 |
| T. triquetra | 18 | Asteraceae | |
| | | Flaveria australasia | 36 |
| Amaranthaceae | | F. bidentis | 36 |
| Acanthochiton wrightii | 29 | F. intermedia | 36 |
| Aerva javanica | 19 | F. trinervia | 36 |
| A. pseudomentosa | 19 | Pectis ambigua | 36 |
| Alternanthera caracosana | 32 | P. arenaria | 36 |
| A. pugens | 30 | P. berlandieri | 36 |
| A. repens | 29 | P. capillaris | 36 |
| Amaranthus acanchochitan | 29 | P. depressa | 36 |
| A. albus | 29, 33 | P. dichotoma | 36 |
| A. arenicola | 29 | P. filipes | 36 |
| A. blitoides | 33 | P. haenkeana | 36 |
| A. californicus | 29 | P. humifusa | 36 |
| A. caudatus | | P. incisifolia | 36 |
| A. chlorostachys | 29 | P. latisquama | 36 |
| A. cruentus | 29 | P. leibmannii | 36 |
| A. deflexus | 29 | P. linifolia | 36 |
| A. edulis | 33, 34 | P. multiseta | 36 |
| A. fimbriatus | 29 | P. papposa | 36 |
| A. graecizans | 29 | P. propetes | 36 |
| A. hybridus | 29 | P. prostrata | 36 |
| A. hypochondriacus | 33 | P. repens | 36 |
| A. lividus | 29 | P. saturejoides | 36 |
| A. melancholicus | 33 | P. tenella | 32 |
| A. palmeri | 29 | P. uniaristata | 36 |
| A. patulus | 33 | Chrysanthellum americanum | 36 |
| A. paniculatus | 29 | C. integrifolium | 36 |
| A. powellii | 29 | C. involutum | 36 |
| A. retroflexus | 17, 29, 22 | C. mexicanum | 36 |
| A. spinosus | 29 | Eryngiophyllum pennatisectum | 36 |
| A. tamariscinus | 29 | Glossocardia bosvallea | 36 |
| A. tricolor | 33 | Glossogyne tenuifolia | 36 |
| A. tuberculatus | 29 | Isostigma acaule | 36 |
| A. viridus | 33 | I. brasiliense | 36 |
| Brayulinea densa | 29 | I. impala | 36 |
| Froelichia arizonica | 29 | I. scorzonerifolium | 36 |
| F. campestris | 29 | I. speciosa | 36 |
| F. floridona | 29 | | |
| F. gracilis | 29 | Boraginaceae | |
| Gomphrena caespitosa | 29 | Heliotropium scabrum | 30 |
| G. celosoides | 35 | | |
| G. dispersa | 29 | Capparadaceae | |
| G. globosa | 17 | Gynandropsis speciosa | 18 |

## Table 5 (continued)
## C₄ PLANTS LISTED BY FAMILY

| Plant | Ref. | Plant | Ref. |
|---|---|---|---|
| Caryophyllaceae | | *Halogeton glomeratus* | 29 |
| *Polycarpaea corymbosa* | 30 | *Haloxylon salicornicum* | 19 |
| | | *Kochia childsii* | 16, 33, 34 |
| Chenopodiaceae | | | |
| *Atriplex argenta* | 17, 29 | *K. scoparia* | 16, 29, 33, 34 |
| *A. bonnevillensis* | 29 | | |
| *A. breweri* | 29 | *Sasola kali* | 18, 29, 33 |
| *A. buchananii* | 17 | | |
| *A. canescens* | 29, 34 | *Sasola foetida* | 19 |
| *A. confertifolia* | 29 | *Suaeda californica* | 29 |
| *A. corrugata* | 29 | *S. inermis* | 17 |
| *A. coulteri* | 29 | *S. nigra* | 29 |
| *A. cuneata* | 29 | *S. suffrutescens* | 29 |
| *A. decumbens* | 20 | *S. taxifolia* | 17 |
| *A. elegens* | 29 | *S. torreyana* | 17, 29 |
| *A. falcata* | 29 | *Theleophyton billardieri* | 17 |
| *A. gardneri* | 29 | | |
| *A. garrettii* | 29 | Euphorbiaceae | |
| *A. halimus* | 16, 29, 34 | *Chamaesyce hirta* | 30 |
| | | *C. maculata* | 37 |
| *A. holocarpa* | 16 | *C. supina* | 37 |
| *A. hymenelytra* | 29 | *Euphorbia albomarginata* | 29 |
| *A. inflata* | 16 | *E. arizonica* | 29 |
| *A. lentiformis* | 29, 34 | *E. arnottiana* | 38 |
| *A. leucocladium* | 29 | *E. articulata* | 16 |
| *A. leucophylla* | 29 | *E. atoto* | 38 |
| *A. linearis* | 29 | *E. atrococca* | 38 |
| *A. muricata* | 29 | *E. buxifolia* | 29 |
| *A. navajoensis* | 29 | *E. capitellata* | 29 |
| *A. nummularia* | 16, 17, 34 | *E. carunculata* | 39 |
| | | *E. celastroides* | 30 |
| *A. obovata* | 29 | *E. chamaesyce* | 29 |
| *A. parryi* | 29 | *E. cinerascens* | 29 |
| *A. pentandra* | 29 | *E. clusiaefolia* | 38 |
| *A. polycarpa* | 29, 34 | *E. degeneri* | 38 |
| *A. powellii* | 29 | *E. drummondii* | 39 |
| *A. pusilla* | 29 | *E. fenleri* | 29 |
| *A. rhagadioides* | 16 | *E. florida* | 29 |
| *A. rosea* | 16, 17, 29, 33 | *E. forbesii* | 38, 39 |
| | | *E. geyeri* | 29 |
| *A. sabulosa* | 17 | *E. glyptosperma* | 29, 39 |
| *A. saccaria* | 29 | *E. granulata* | 39 |
| *A. semibaccata* | 29, 34 | *E. hillebrandii* | 29, 38 |
| *A. serenana* | 29 | *E. hirsuta* | 29 |
| *A. sibirica* | 16 | *E. hirta* | 29, 30 |
| *A. spongiosa* | 16, 17 | *E. hirtula* | 29 |
| *A. suberecta* | 16 | *E. hooveri* | 29 |
| *A. tatarica* | 16 | *E. humistrata* | 29 |
| *A. tenuissima* | 29 | *E. hypericifolia* | 39 |
| *A. torreyi* | 29 | *E. hyssopifolia* | 29 |
| *A. tridentata* | 29 | *E. lata* | 29, 39 |
| *A. truncata* | 29 | *E. maculata* | 29, 39 |
| *A. vallicola* | 29 | *E. melanadenia* | 29 |
| *A. vesicaria* | 34 | *E. mesembrianthemifolia* | 30 |
| *A. wolfii* | 29 | *E. micromera* | 29 |
| *Bassia hyssopifolia* | 16, 29, 32 | *E. missurica* | 29, 39 |
| | | *E. multiformis* | 38 |

## Table 5 (continued)
## C₄ PLANTS LISTED BY FAMILY

| Plant | Ref. | Plant | Ref. |
|---|---|---|---|
| *E. neomexicana* | 29 | Portulacaceae | |
| *E. nutans* | 29, 39 | *Portulaca grandiflora* | 17, 33 |
| *E. olowaluana* | 38 | *P. mundula* | 29 |
| *E. parryi* | 29 | *P. oleracea* | 17, 29, |
| *E. pediculifera* | 29 | | 33 |
| *E. peplis* | 39 | *P. parvula* | 29 |
| *E. petaloidea* | 29 | *P. pilosa* | 29 |
| *E. polygonifolia* | 29 | *P. suffrutescens* | 29 |
| *E. remyi* | 29, 38 | | |
| *E. revoluta* | | Zygophyllaceae | |
| *E. rockii* | 38, 39 | *Kallstroemia adscendens* | 32 |
| *E. serpyllifolia* | 29 | *K. boliviana* | 32 |
| *E. serpens* | 29 | *K. Californica* | 32 |
| *E. serrula* | 29 | *K. curta* | 32 |
| *E. setiloba* | 29 | *K. grandiflora* | 29 |
| *E. skottsbergii* | 38 | *K. hintonii* | 32 |
| *E. stictospora* | 29 | *K. hirsutissima* | 32 |
| *E. supina* | 29 | *K. maxima* | 18 |
| *E. tatitensis* | 38 | *K. parviflora* | 32 |
| *E. tirucalli* | 26 | *K. peninsularis* | 32 |
| *E. tomestulosa* | 29 | *K. pennellii* | 32 |
| *E. vermiculata* | 29 | *K. perennans* | 32 |
| *E. villifera* | 29 | *K. pubescens* | 18 |
| *E. viscoides* | 39 | *K. rosei* | 18 |
| *E. wheeleri* | 39 | *K. standleyi* | 32 |
| *E. zygophylloides* | 29 | *K. tribuloides* | 32 |
| *Pedilanthus tithymaloides* | 19 | *K. turcumanensis* | 32 |
| | | *Tribulus alatus* | 32 |
| Molluginaceae | | *T. cistoides* | 18 |
| *Mollugo cerviana* | 22 | *T. hystrix* | 18 |
| *M. nudicaulis* | 30 | *T. terrestris* | 29, 31 |
| | | *Zygophyllum simplex* | 18 |
| Nyctaginaceae | | | |
| *Allionia choisyi* | 29 | ᵃ   At least 950 species are C₄. | |
| *A. cristata* | 29 | | |
| *A. incarnata* | 22, 29 | | |
| *Boerhaavia carbaea* | 29 | | |
| *B. coccinea* | 22 | | |
| *B. coulteri* | 29 | | |
| *B. diffusa* | 22 | | |
| *B. elegans* | 32 | | |
| *B. erecta* | 22, 29 | | |
| *B. gracillima* | 29 | | |
| *B. intermedia* | 22, 29 | | |
| *B. linearifolia* | 29 | | |
| *B. megaptera* | 29 | | |
| *B. paniculata* | 31 | | |
| *B. purpurascens* | 29 | | |
| *B. specata* | 29 | | |
| *B. tenuifolia* | 29 | | |
| *B. torreyana* | 29 | | |
| *B. wrightii* | 22, 29 | | |
| | | | |
| Polygonaceae | | | |
| *Calligonum persicum* | 40 | | |

## Table 6
## CAM PLANTS LISTED BY FAMILY

| Plant | Ref. | Plant | Ref. |
|---|---|---|---|
| Filicinae | | *A. millotii* | 17 |
| Polypodiaceae | | *A. spinulosa* | 48 |
| | | *A. variegata* | 17 |
| *Drymoglossum piloselloides* | 10 | *A. vera* | 14 |
| | | *Apiera spiralis* | 42 |
| Gymnospermae | | *Gasteria carinata* | 17 |
| | | *G. verrucosa* | 45 |
| Welwitschiaceae | | *Haworthia attenuata* | 46 |
| | | *H. cuspidata* | 17 |
| *Welwitschia mirabilis* | 11 | *Sanseveria fasiculata* | 19 |
| | | *S. hahnii* | 42 |
| Angiospermae | | *S. liberica* | 43 |
| Monocotyledonae | | *S. trifasciata* | 45 |
| Agavaceae | | *Yucca brevifolia* | 25 |
| *Agave americana* | 17 | *Y. filamentosa* | 46 |
| *A. desertii* | 25 | *Y. schidigera* | 25 |
| *A. lurida* | 25 | | |
| | | Orchidaceae | |
| Bromeliaceae | | *Arachnis hookeriana* | 45 |
| *Aechmea tillandsioides* | 41 | *Ascocentrum ampullaceum* | 44 |
| *A. bromelifolia* | 41 | *Brassolacliocattelya* sp. "Mauna- | 44 |
| *A. gigantea* | 42 | *lani"* | |
| *Ananas comosus* | 42 | *Cattelya labiata* | 45 |
| *A. sativus* | 43 | *Dendrobium taurinum* | 45 |
| *Billbergia nutans* | 44 | *Encyclia atropurpurea* | 45 |
| *B. saundersii* | 42 | *Epidendrum alatum* | 44 |
| *Bromelia humilis* | 41 | *E. ellipticum* | 45 |
| *Cryptanthus bivattatas* | 42 | *E. radicans* | 44 |
| *C. diversifolius* | 42 | *E. schomburgkii* | 45 |
| *Dickya brevifolia* | 44 | *Phalaenopsis schilleniana* | 44 |
| *D. fosteriana* | 44 | *Pleurothrus ophiocephalus* | 17 |
| *D. tuberosa* | 41 | *Schomburgkia crispa* | 45 |
| *Guzmania lingulata* | 42 | *Vanilla aromatica* | 45 |
| *G. monostachia* | 41 | *V. fragrans* | 44 |
| *Neoregelia ampullaceae* | 40 | | |
| *N. carolinia* | 17 | Dicotyledonae | |
| *N. cruenta* | 44 | Aizoaceae | |
| *Nidularium mayendorffii* | 45 | *Aptenia cordifolia* | 45 |
| *Puya alpestris* | 17 | *Bergeranthus multiceps* | 45 |
| *Tillandsia andreana* | 41 | *Carpobrotus edulis* | 50 |
| *T. balbisiana* | 41 | *Disphyma australa* | 17 |
| *T. circinnata* | 41 | *D. blackii* | 17 |
| *T. gardneri* | 41 | *D. papillatum* | 17 |
| *T. incarnata* | 41 | *Drossanthemum floribundum* | 17 |
| *T. juncea* | 42 | *Faucaria hybrida* | 45 |
| *T. polystachia* | 41 | *Glottiphyllum linguiforme* | 53 |
| *T. recurvata* | 41 | *Lithops leslei* | 53 |
| *T. tenuifolia* | 41 | *L. salicola* | 17 |
| *T. usneoides* | 41 | *L. venteri* | 17 |
| *T. utriculata* | 41 | *Mesembryanthemum chilense* | 46 |
| | | *M. crystallinum* | 51 |
| Liliaceae | | *M. edule* | 52 |
| *Aloe arborescens* | 17 | *M. forsskalii* | 51 |
| *A. aristata* | 45 | *M. nodiflorum* | 51 |
| *A. broomii* | 46 | | |
| *A. cymbaefolia* | 47 | Asclepiadaceae | |
| *A. juvenna* | 17 | *Caralluma negevensis* | 51 |

## Table 6 (continued)
## CAM PLANTS LISTED BY FAMILY

| Plant | Ref. | Plant | Ref. |
|---|---|---|---|
| *Hoya carnosa* | 19, 42 | *O. ramosissima* | 25 |
| *Huernia aspera* | 53 | *O. strobiliformis* | 19 |
| *H. macracarpa* | 42 | *O. versicolor* | 61 |
| *Stapelia gigantea* | 42 | *Pachycereus pringlei* | 17 |
| *S. nobilis* | 43 | *Phyllocactus pfersdorffii* | 45 |
| *S. semota* | 18 | *Rhipsalis cassutha* | 42 |
| *S. variegata* | 45, 54 | *Zygocactus truncatus* | 42, 45 |
| | | | |
| Asteraceae | | Caryophyllaceae | |
| *Borrichia frutescens* | 55 | *Arenaria peploides* | 52 |
| *Hoplophytum grande* | 43 | | |
| *Kleinia articulata* | 14 | Chenopodiaceae | |
| *K. radicans* | 14 | *Salicornia annua* | 52 |
| *K. repens* | 42 | *S. europaea* | 55 |
| *K. tomentosa* | 53 | *S. herbacea* | 53 |
| *Notonia petraea* | 43 | *S. virginica* | 55 |
| *Senecio gregori* | 19, 46 | *Suaeda maritima* | 52 |
| *S. herreianus* | 45 | | |
| | | Convolvulaceae | |
| Bataceae | | *Alona rostrata* | 67 |
| *Batis maritima* | 55 | | |
| | | Crassulaceae | |
| Cactaceae | | *Adromischus cristatus* | 14 |
| *Anhalonium lewinii* | 53 | *Aeonium haworthi* | 68 |
| *Astrophytum myriostigma* | 53 | *Bryophyllum calycinum* | 69 |
| *Carnegiea gigantea* | 56 | *B. crenatum* | 69 |
| *Cereus peruvianus* | 19, 46 | *B. daigremontianum* | 46 |
| *Chamaecereus sylvestris* | 17 | *B. fedtschenkoi* | 46 |
| *Crypotocereus anthocyanus* | 14 | *B. proliferum* | 46 |
| *Echinocactus acanthoides* | 25 | *Cotyledon agavoides* | 53 |
| *Echinocereus engelmanii* | 25 | *C. orpiculata* | 14 |
| *E. ledigii* | 57 | *C. peacockii* | 70 |
| *Echinopsis eriesii* | 45 | *Crassula arborea* | 62 |
| *Ferocactus acanthodes* | 57 | *C. arborescens* | 64 |
| *Hatiora salicornoides* | 14 | *C. argentea* | 46 |
| *Lobivia* sp. | 17 | *C. lactea* | 62 |
| *Mamillaria graham* | 58 | *C. pallida* | 62 |
| *M. rhodantha* | 45 | *C. tomentosa* | 19, 46 |
| *M. tetrancistra* | 58 | *Dudleya farionosa* | 71 |
| *Nopalea cochinellifera* | 59 | *D. lanceolata* | 58 |
| *N. dejecta* | 60 | *D. saxosa* | 58 |
| *Nothocactus mammulosus* | 17 | *Echeveria cilva* | 18, 46 |
| *Opuntia acanthocarpa* | 57 | *E. glauca* | 72 |
| *O. basilaris* | 25, 42 | *E. kircheriana* | 45 |
| *O. bigelovii* | 25 | *E. metallica* | 48 |
| *O. blakeana* | 61 | *E. pumila* | 73 |
| *O. camanchica* | 53 | *E. secunda* | 53 |
| *O. cylindrica* | 62 | *Kalanchoe beharensis* | 14 |
| *O. disctata* | 61 | *K. blossfeldiana* | 42, 45, 70 |
| *O. echinocarpa* | 25 | | |
| *O. fiscus-indica* | 63 | *K. longifolis* | 53 |
| *O. humifusa* | 19 | *K. maculata* | 17 |
| *O. leptocaulis* | 61 | *K. marmorata* | 70 |
| *O. megacantha* | 64 | *K. tubiflora* | 18, 46 |
| *O. polyacantha* | 65 | *K. verticillata* | 46 |
| *O. puberula* | 66 | *Nanathus malherbi* | 46 |

## Table 6 (continued)
## CAM PLANTS LISTED BY FAMILY

| Plant | Ref. | Plant | Ref. |
|---|---|---|---|
| *Rochea falcata* | 48 | *E. drupifera* | 39 |
| *Sedum acre* | 62 | *E. grandidens* | 39, 45 |
| *S. adolfe* | 74 | *E. lactiflua* | 67 |
| *S. guatamalense* | 53 | *E. millisplendens* | 14 |
| *S. kamschatkicum* | 62 | *E. nivulia* | 39 |
| *S. praeltum* | 46 | *E. splendens* | 48 |
| *S. pulchellum* | 75 | *E. submammilaris* | 39, 44 |
| *S. purpurescens* | 62 | *E. tirucalli* | 39, 46 |
| *S. purpureum* | 62 | *E. trigona* | 39 |
| *S. rubrotinctum* | 18, 46 | *E. xylophyloides* | 39, 44 |
| *S. sexangulare* | 64 | *Monodenium lugardoe* | 44 |
| *S. spectabile* | 46 | *Pedilanthus tithymaloides* | 46 |
| *S. spurium* | 46 | *Synadenium capulare* | 44 |
| *S. tectorum* | 72 | *S. grantii* | 44 |
| *S. telephium* | 46 | | |
| *S. telephoides* | 46 | Plantaginaceae | |
| *S. weinbegii* | 42 | *Plantago maritima* | 52 |
| *Sempervivum calcarum* | 18, 46 | | |
| *S. chlorochrysum* | 48 | Portulacaceae | |
| *S. glaucum* | 69 | *Calandrinia maritima* | 67 |
| *S. guiseppii* | 53 | *Portulacaria afra* | 14 |
| *S. tectorum* | 45 | | |
| *Titanopsis calcarum* | 46 | Vitaceae | |
| Euphorbiaceae | | *Cissus digitata* | 42 |
| *Euphorbia bubalina* | 39 | *C. quadrangularis* | 43 |
| *E. caducifolia* | 39 | | |

# REFERENCES

1. **Calvin, M.**, The path of carbon in photosynthesis, *Science,* 135, 879, 1962.
2. **Bowes, C. and Ogren, W. L.**, Oxygen inhibition and other properties of soybean ribulose-1,5-diphosphate carboxylase, *J. Biol. Chem.*, 247, 2171, 1972.
3. **Black, C. C., Jr.**, Photosynthetic carbon fixation in relation to net $CO_2$ uptake, *Annu. Rev. Plant Physiol.*, 24, 253, 1973.
4. **Hatch, M. D. and Osmond, C. B.**, Compartmentation and transport in $C_4$ photosynthesis, in *Encyclopedia of Plant Physiology*, New Series, Vol. 3, Stocking, C. P. and Heber, U., Eds., Springer-Verlag, Berlin, 1976, 144.
5. **Laetsch, W. M.**, The $C_4$ syndrome: a structural analysis, *Annu. Rev. Plant Physiol.*, 25, 27, 1974.
6. **Johnson, C., Sr. and Brown, W. V.**, Grass leaf ultrastructural variations, *Am. J. Bot.*, 60, 727, 1973.
7. **Whelan, T., Sackett, W. M., and Benedict, C. R.**, Enzymatic fractionation of carbon isotopes by phosphoenol pyruvate carboxylase from $C_4$ plants, *Plant Physiol.*, 51, 1051, 1973.
8. **Nambudiri, E. M. V., Tidwell, W. D., Smith, B. N., and Hebbert, N. P.**, A $C_4$ plant from the Pliocene, *Nature (London)*, 276, 816, 1978.
9. **Osmond, C. B., Allaway, W. G., Sutton, B. G., Troughton, J. H., Queiroz, O., Luttge, U., and Winter, K.**, Carbon isotope discrimination in photosynthesis of CAM plants, *Nature (London)*, 246, 41, 1973.
10. **Hew, C-S. and Wong, Y. S.**, Photosynthesis and respiration of ferns in relation to their habit, *Am. Fern J.*, 64, 40, 1974.
11. **Smith, B. N.**, Evolution of $C_4$ photosynthesis in response to changes in carbon and oxygen concentrations in the atmosphere through time, *Biosystems*, 8, 24, 1976.

12. **Downton, J. S.**, Check list of C₄ species, in *Photosynthesis and Photorespiration*, Hatch, M. D., Osmond, C. B., and Slatyer, R. O., Eds., Wiley-Interscience, New York, 1971, 554.
13. **Brown, W. V.**, The Kranz syndrome and its subtypes in grass systematics, *Mem. Torrey Bot. Club*, 23(3), 1, 1977.
14. **Black, C. C. and Williams, S.**, Plants exhibiting characteristics common to Crassulacean acid metabolism, in *CO₂ Metabolism and Plant Productivity*, Burris, R. H. and Black, C. C., Eds., University Park Press, Baltimore, 1976, 407.
15. **Lerman, J. C. and Raynal, J.**, La teneur en isotopes stables du carbone chez les Cyperacees: sa valeur taxonomique, *C. R. Acad. Sci.*, 275, 1391, 1972.
16. **Tregunna, E. B., Smith, B. N., Berry, J. A., and Downton, W. J. S.**, Some methods for studying the photosynthetic taxonomy of the angiosperms, *Can. J. Bot.*, 48, 1209, 1970.
17. **Troughton, J. H., Card, K. A., and Hendy, C. H.**, Photosynthetic pathways and carbon isotope discrimination by plants, *Carnegie Inst. Washington Yearb.*, 73, 768, 1974.
18. **Krenzer, E. G., Moss, D. N., and Crookston, R. K.**, Carbon dioxide compensation points of flowering plants, *Plant Physiol.*, 56, 194, 1975.
19. **Bender, M. M.**, Variations in the ¹³C/¹²C ratios of plants in relation to the pathway of photosynthetic carbon dioxide fixation, *Phytochemistry*, 10, 1239, 1971.
20. **Hattersley, P. W. and Watson, L.**, C₄ grasses: an anatomical criterion for distinguishing between NADP-malic enzyme species and PCK or NAD-malic enzyme species, *Aust. J. Bot.*, 24, 297, 1976.
21. **Smith, B. N. and Brown, W. V.**, The Kranz syndrome in the Gramineae as indicated by carbon isotopic ratios, *Am. J. Bot.*, 60, 505, 1973.
22. **Smith, B. N. and Robbins, M. J.**, Evolution of C₄ photosynthesis: an assessment based on ¹³C/¹²C ratios and Kranz anatomy, in *Proc. 3rd Int. Congr. Photosynthesis*, Avron, M., Ed., Elsevier, Amsterdam, 1974, 1579.
23. **Winter, K., Troughton, J. H., and Card, K. A.**, δ¹³C values of grass species collected in the northern Sahara desert, *Oecologia*, 25, 115, 1976.
24. **Hattersley, P. W. and Watson, L.**, Anatomical parameters for predicting photosynthetic pathways of grass leaves: the 'maximum lateral cell count' and the 'maximum cells distant count', *Phytomorphology*, 25, 325, 1975.
25. **Johnson, H. B.**, Gas-exchange strategies in desert plants, in *Perspectives of Biophysical Ecology*, Gates, D. M. and Schmerl, R. B., Eds., Springer-Verlag, New York, 1975, 105.
26. **Bender, M. M. and Smith, D.**, Classification of starch and fructosan-accumulating grasses as C-3 or C-4 species by carbon isotope analysis, *J. Br. Grassl. Soc.*, 28, 97, 1973.
27. **Ludlow, M. M., Troughton, J. H., and Jones, R. J.**, A technique for determining the proportion of C₃ and C₄ species in plant samples using stable natural isotopes of carbon, *J. Agric. Sci.*, 87, 625, 1976.
28. **Brown, W. V. and Smith, B. N.**, The Kranz syndrome in *Uniola* (Gramineae), *Bull. Torrey Bot. Club*, 101, 117, 1974.
29. **Welkie, G. W. and Caldwell, M.**, Leaf anatomy of species in some dicotyledon families as related to the C₃ and C₄ pathways of carbon fixation, *Can. J. Bot.*, 48, 2135, 1970.
30. **Rathnam, C. K. M., Raghavendra, A. S., and Das, V. S. R.**, Diversity in the arrangements of mesophyll cells among leaves of certain C₄ dicotyledons in relation to C₄ physiology, *Z. Pflanzenphysiol.*, 77, 283, 1976.
31. **Crookston, R. K. and Moss, D. N.**, The relation of carbon dioxide compensation and chlorenchymatous vascular bundle sheaths in leaves of dicots, *Plant Physiol.*, 46, 564, 1970.
32. **Robbins, M. J.**, Origin and Number of Times the Kranz Syndrome and C₄ Photosynthesis Evolved in Each of Ten Families of the Angiosperms, M. S. thesis, University of Texas, Austin, 1974, 1.
33. **Tregunna, E. B. and Downton, J.**, Carbon dioxide compensation in members of the Amarantheceae and some related families, *Can. J. Biol.*, 45, 2385, 1967.
34. **Smith, B. N. and Epstein, S.**, Two categories of ¹³C/¹²C ratios for higher plants, *Plant Physiol.*, 47, 380, 1971.
35. **Hatch, M. D., Kagawa, T., and Craig, S.**, Subdivisions of C₄-pathway species based on differing C₄ acid decarboxylating systems and ultrastructural features, *Aust. J. Plant Physiol.*, 2, 111, 1975.
36. **Smith, B. N. and Turner, B. L.**, Distribution of Kranz syndrome among Asteraceae, *Am. J. Bot.*, 62, 541, 1975.
37. **Gutierrez, M., Gracen, V. E., and Edwards, G. E.**, Biochemical and cytological relationships in C₄ plants, *Planta*, 119, 279, 1974.
38. **Pearcy, R. W. and Troughton, J.**, C₄ photosynthesis in tree form *Euphorbia* species from Hawaiian rainforest sites, *Plant Physiol.*, 55, 1054, 1975.
39. **Webster, G. L., Brown, W. V., and Smith, B. N.**, Systematics of photosynthetic carbon fixation pathways in *Euphorbia*, *Taxon*, 24, 27, 1975.

40. **Winter, K., Kramer, D., Troughton, J. H., Card, K. H., and Fischer, K.**, C₄ pathway of photosynthesis in a member of the Polygonaceae: *Calligonum persicum* (Boiss. & Buhse) Boiss., *Z. Pflanzenphysiol.*, 81, 341, 1977.

41. **Mendina, E. and Troughton, J. H.**, Photosynthetic patterns in the Bromeliaceae, *Carnegie Inst. Washington Yearb.*, 73, 805, 1974.

42. **Dittrich, P., Campbell, W. H., and Black, C. C.**, Phosphoenol pyruvate carboxykinase in plants exhibiting Crassulacean acid metabolism, *Plant Physiol.*, 52, 357, 1973.

43. **Milburn, T. R., Pearson, D. J., and Ndegwe, N. A.**, Crassulacean acid metabolism under natural tropical conditions, *New Phytol.*, 67, 883, 1968.

44. **McWilliams, E. L.**, Comparative rates of dark CO₂ uptake and acidification in the Bromeliaceae, Orchidaceae, and Euphorbiaceae, *Bot. Gaz.*, 131, 285, 1970.

45. **Nuernbergk, E. L.**, Endogener Rhythmus and CO₂-stoffwechsel bei Pflanzen mit diurnalem Saurerhythmus, *Planta*, 56, 28, 1961.

46. **Bender, M. M., Rouhani, I., Vines, H. M., and Black, C. C.**, ¹³C:¹²C ratio changes in Crassulacean acid metabolism plants, *Plant Physiol.*, 52, 427, 1973.

47. **Hempel, J.**, Buffer processes in the metabolism of succulent plants, *C. R. Trav. Lab. Carlsberg*, 13, 1, 1917.

48. **de Vries, H.**, Ueber Periodicität und Säuregehalt der Fettplanzen, Verslag, *Med. K. Ak. Wet. Amsterdam*, 1, 58, 1884.

49. **Treichel, S.**, Crassulacean acid metabolism in a salt-tolerant member of the Aizoaceae: *Apteria cordifolia, Plant Sci. Lett.*, 4, 141, 1975.

50. **Winter, K.**, Sodium chloride-induced Crassulacean acid metabolism (CAM) in a secondary member of the Aizoaceae family, *Planta*, 115, 187, 1973.

51. **Winter, K., Troughton, J. H., Evenari, M., Läuchli, A., and Lüttge, U.**, Mineral ion composition and occurrence of CAM-like diurnal malate fluctuations in plants of coastal and desert habitats of Israel and the Sinai, *Oecologia*, 25, 125, 1976.

52. **Delf, E. M.**, Transpiration in succulent plants, *Ann. Bot.*, 27, 409, 1912.

53. **James, W.O.**, Succulent plants, *Endeavor*, 17, 90, 1958.

54. **de Saussure, T.**, *Recherches Chimiques Sur la Vegetation*, Paris, 1804.

55. **Webb, K. L. and Burley, J. W. A.**, Dark fixation of ¹⁴CO₂ by obligate and facultative salt marsh halophytes, *Can. J. Bot.*, 43, 281, 1965.

56. **Despain, D. G., Bliss, L. C., and Boyer, J. C.**, Carbon dioxide exchange in *Saguaro* seedlings, *Ecology*, 51, 912, 1970.

57. **Patten, D. T. and Dinger, B. E.**, Carbon dioxide exchange patterns of cacti from different environments, *Ecology*, 50, 686, 1969.

58. **Ting, I. P. and Dugger, W. M.**, Non-autotrophic carbon dioxide metabolism in cacti, *Bot. Gaz.*, 129, 9, 1968.

59. **Master, R. W. P.**, Organic acid and carbohydrate metabolism in *Nopalea cochirellifera, Experientia*, 15, 30, 1959.

60. **Mukerji, S. K.**, Four-hourly variations in the activities of malate dehydrogenase (decarboxylating) and phosphopyruvate carboxylase in the cactus (*Nopalea dejecta*) plant, *Ind. J. Biochem.*, 5, 62, 1968.

61. **Richards, H. M.**, *Acidity and Gas Interchange in Cacti*, Carnegie Institute Publication #209, 1915.

62. **Thomas, M. and Ranson, S. L.**, Physiological studies on acid metabolism in green plants. III. Further evidence of CO₂-fixation during dark acidification of plants showing Crassulacean acid metabolism, *New Phytol.*, 53, 1, 1954.

63. **Mukerji, S. K. and Ting, I. P.**, Intracellular localization of CO₂ metabolism in cactus phylloclades, *Phytochemistry*, 7, 903, 1968.

64. **Mayer, A.**, Uber die Säuerstoffausscheidung eininger Crassulacean, *Landev. Versuchst.*, 21, 277, 1878.

65. **Troughton, J. H., Wells, P. V., and Mooney, H. A.**, Photosynthetic mechanisms in ancient C₄ and CAM species, *Carnegie Inst. Washington Yearb.*, 73, 812, 1974.

66. **Kausch, W.**, Relationships between root growth, transpiration, and CO₂ exchange in several cacti, *Planta*, 66, 229, 1965.

67. **Mooney, H. A., Troughton, J. H., and Berry, J. A.**, Arid climates and photosynthetic systems, *Carnegie Inst. Washington Yearb.*, 73, 793, 1974.

68. **Neales, T. F., Patterson, A. A., and Hartney, V. J.**, Physiological adaptation to drought in the carbon assimilation and water loss of xerophytes, *Nature (London)*, 219, 469, 1968.

69. **Wolf, J.**, Beitrag zer Kenntris des Säurestoffwechsels succulenter Crassulacean, *Planta*, 15, 572, 1932.

70. **Nishida, K.**, Studies on stomatal movement of Crassulacean plants in relation to the acid metabolism, *Physiol. Plant.*, 16, 281, 1963.

71. **Bartholomew, B.**, Drought response in the gas exchange of *Dudleya farinosa* (Crassulaceae) grown under natural conditions, *Photosynthetica*, 7, 117, 1973.

72. **Astruc, A.,** Recherches sur l'acidité végétale, *Ann. Sci. Nat. (Bot.),* Ser. 8, 17, 1, 1903.
73. **Meinzer, F. C. and Rundel, P. W.,** Crassulacean acid metabolism and water use efficiency in *Echeveria pumila, Photosynthetica,* 7, 358, 1973.
74. **Sutton, B. G. and Osmond, C. B.,** Dark fixation of $CO_2$ by Crassulacean plants, *Plant Physiol.,* 50, 360, 1972.
75. **Franzen, H. and Osterkag, R.,** Über Nichtexistenz der Crassulaceenäpfelsäure, *Z. Physiol. Chem.,* 122, 263, 1922.

*Section 6*
*Response of Photosynthetic Organisms to*
*Major Environmental Factors*

# RESPONSE OF MICROALGAE TO LIGHT INTENSITY AND LIGHT QUALITY

## Horst Senger

The photosynthetic response of microalgae to the intensity of incident light is directly proportional at low values. Under such conditions, only the photochemical reactions of the photosynthetic apparatus are limiting. Differences in the light-dependent response of the photochemical reactions, when based on the same chlorophyll amount, can be due to changes at various levels in excitation energy distribution among antenna pigments or between antenna pigments and reaction centers; excitation energy transformation between reaction centers and primary acceptors; or electron transport in the electron transport chain.[1] Little or no effect of external factors such as temperature and $CO_2$ concentration on pH is found in the light-limiting region of the photosynthetic response curve.

Compensation points of respiratory $O_2$ uptake by photosynthetic $O_2$ evolution are reached at low light intensities, rarely surpassing 10 $W/m^2$. For most algae, the dark respiration remains below 10% of the maximal $O_2$ evolution under saturating light conditions.

The question of whether the respiration value measured in the dark remains the same in light can be evaluated by adding the value of $O_2$ uptake by dark respiration to the values of photosynthetically produced $O_2$. When such a corrected curve extrapolates through zero, one can assume that respiration values do not change in the dark and light. All wild-type microalgae checked under normal conditions demonstrate this behavior. Extreme nutritional conditions and mutations [2,3] may show differences in respiration in dark and light. So far, definitive evidence has not been presented on the occurence of photorespiration in microalgae.[4]

Light intensity curves of photosynthetic $O_2$ evolution of various microalgae are computed on the basis of the same chlorophyll content and shown in Figure 1. Although all cells were measured polarographically under red light above 620 nm in an oxygraph, they can be compared only relatively, since developmental stages, preculturing, and suspension media were different. Nevertheless, the comparison of the different species demonstrates a remarkably small region (100 to 500 $W/m^2$) in which a saturating light intensity is reached.

The level of saturation of photosynthesis in microalgae depends on many internal and external factors when compared on the basis of equal chlorophyll content. Internal factors, such as the concentration of other light-harvesting pigments, the effectiveness of the dark reactions, age and developmental stage of the algae, and preculturing conditions influence the photosynthetic capacity. External factors influencing the photosynthetic capacity most effectively are temperature, $CO_2$ concentration, and enzyme inhibitors. It also has been shown that growing microalgae under low or high intensities of light cause the same phenomena as demonstrated by shade and sun plants, i.e., considerable differences in the compensation point and in the light saturating region of the light intensity curves of photosynthesis.[5]

Among photosynthetic organisms, the algae demonstrate the largest variety in the compostion of photosynthetic pigments. In addition to the different chlorophylls, carotenoids and biliproteins are found. Accordingly, the variation in light absorption is manifold. Many action spectra for photosynthetic reactions of microalgae have been reported, including *Chlorella*,[6] *Chroococcus*,[7] *Cyanidium*,[8] *Gonyaulax*,[8] *Hemiselmis*,[8] *Navicula*,[9] and *Porphyridium*.[8,10,11] These action spectra reflect the different compositions of photosynthetically active pigments, i.e., their absorption spectra. This absorp-

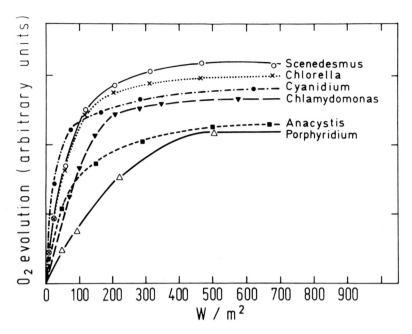

FIGURE 1.    Light intensity curves for photosynthetic $O_2$ evolution of various microalgae. Oxygen evolution was measured polarographically in a Gilson ® oxygraph under red light above 620 nm. Preculturing and suspension media during measurements were different for various algae.

tion is taken into account in measurements of quantum yield under different wavelengths; thus, it is not surprising that the quantum yield of photosynthesis does not change considerably over the entire range of absorption.[6,9,10] One has also to keep in mind that the wavelength dependence of photosynthesis is specific to low intensities. At saturating intensities, photosynthesis becomes more or less independent of wavelengths.[12]

Since saturating intensities are not reached in some ecosystems, the pigment composition of the algae is important for photosynthetic productivity. Some algae have developed complementary chromatic adaptation, i.e., the ability to synthesize pigments according to light availability.[13-16]

# REFERENCES

1. Strasser, R., The grouping model of plant photosynthesis, in *Chloroplast Development,* Akoyunoglou, G., Ed, Elsevier/North-Holland, 1978, 513.
2. Kowallik, U. and Kowallik, W., Eine wellenlängenäbhangige Atmungssteigerung während der Photosynthese von *Chlorella, Planta,* 84, 141, 1969.
3. Senger, H. and Bishop, N. I., Changes in fluorescence and absorbance during synchronous growth of *Scenedesmus,* in *Proc. 2nd Int. Congr. Photosynthesis Research* Vol. 1, Forti, G., Avron, M., and Melandri, A. Eds., DR. W. Junk, The Hague, The Netherlands,1972, 723.
4. Kowallik, W., Light effects on carbohydrate and protein metabolism in algae, in *Photobiology of Microorganisms,* Halldal, P., Ed, Wiley - Interscience, London, 1970, 165.
5. Senger, H. and Fleischhacker, Ph., Adaptation of the photosynthetic apparatus of *Scenedesmus obliquus* to strong and weak light conditions. I. Differences in pigments, photosynthetic capacity, quantum yield and dark reactions, *Physiol. Plant.,* 43, 35, 1978.

6. **Emerson, R. and Lewis, C. M.,** The dependence of the quantum yield of *Chlorella* photosynthesis on wave length of light, *Am. J. Bot.,* 30, 165, 1943.

7. **Emerson, R. and Lewis, C. M.,** The photosynthetic efficiency of phycocyanin in *Chroococcus* and the problem of carotenoid participation in photosynthesis, *J. Gen. Physiol.,* 25, 579, 1942.

8. **Haxo, F. T.,** The wavelength dependence of photosynthesis and the role of accessory pigments, in *Comparative Biochemistry of Photoreactive Systems,* Vol. 1, Allen, M. B., Ed, Academic Press, New York, 1960, 339.

9. **Tanada, T.,** The photosynthetic efficiency of carotenoid pigments in *Navicula minima, Am. J. Bot.,* 38, 276, 1951.

10. **Duysens, L. N. M.,** Transfer of Excitation Energy in Photosynthesis, Doctoral thesis, University of Utrecht, Holland, 1952.

11. **Brody, M. and Emerson, R.,** The quantum yield of photosynthesis in *Porphyridium cruentum,* and the role of chlorophyll *a* in the photosynthesis of red algae, *J. Gen. Physiol.,* 43, 251, 1959a.

12. **Gabrielsen, E. K.,** Lichtwellenlange und Photosynthese, in *Handbuch der Pflanzenphysiologie,* Vol. 5 (Part 2), Ruhland, W., Ed, Springer-Verlag, New York, 1960, 49.

13. **Bogorad, L,** Phycobiliproteins and complementary chromatic adaptation, *Annu. Rev. Plant Physiol.,* 26, 369, 1975.

14. **Tandeau de Marsac, N.,** Occurence and nature of chromatic adaptation in cyanobacteria, *J. Bacteriol.,* 130, 82, 1977.

15. **Hanri, J. F. and Bogorad, L.,** Action spectra for phycobiliprotein synthesis in a chromatically adapting cyanophyte, *Fremyella diplosiphon, Plant Physiol.,* 60, 835, 1977.

16. **Ohki, K. and Fujita, Y.,** Photocontrol of phycoerythrin formation in the blue-green alga *Tolyphthrix tenuis* growing in the dark, *Plant Cell Physiol.,* 19, 7, 1978.

# RESPONSE OF MICROALGAE TO $CO_2$, $HCO_3^-$, $O_2$, AND pH

## Kurt Schneider

## RESPONSE TO $CO_2$, $HCO_3^-$

The liquid culture of a photosynthetic microorganism is an extremely complex system. $CO_2$ in the air must be physically dissolved in the liquid phase; a process governed by the general laws of mass transfer. After solubilization in the medium, a series of reactions occurs between $CO_2$ (aq), $HCO_3^-$ and $CO_3^{2-}$. The absorbed light energy per cell or unit volume of the culture depends on the geometry and on the density of the culture. Hence, the culture must be defined as completely as possible. At least the following parameters should be defined:

1. Type of culture (batch, continuous)
2. Dynamic state of the culture (stationary,* nonstationary)
3. Homogeneity of the culture (homogeneous, ** nonhomogeneous)
4. Input and state variables of the culture (definition, measurement, and control)

The system shown in Figure 1 is described extensively in the literature.[1-5] The principle of mass transfer between gas and liquid phase is given in several textbooks,[6,7] whereas only a few works are available on the quantitative exchange of $CO_2$ between the gaseous and liquid phases.[8,9]

Figure 2 shows the driving potentials for high and low $CO_2$ exchange areas. For reasons of mass balance, the mass flow through the boundary areas I and II must be the same and is proportional to the exchange area multiplied by the corresponding driving potential.[8,9] It is evident that a low turbulence in an aerated culture can be compensated by a high inlet $CO_2$ concentration and vice versa. The exchange area varies by about a factor of 100 regarding a simple bubble column and a high turbulent fermentor. Therefore, in a low turbulent culture, a $CO_2$ concentration of about 3% is needed to attain the same productivity as in the high turbulent fementor aerated with air (0.03% $CO_2$), where the stirrer speed directly influences the apparent photosynthesis.[10]

Figure 1 shows the way to use $HCO_3^-$ or $CO_3^{2-}$ as a carbon source. The direct uptake of $HCO_3^-$ through the cell wall is still a controversial issue.[5,11-14] The reaction rate from the bicarbonate pool to the $CO_2$ seems to be sufficient for all the measured photosynthetic rates.[15] The $HCO_3^-$ and $CO_3^{2-}$ concentrations are very much dependent on the pH.

Seven concentrations are defined in Figure 2, describing the complete $CO_2$ system: inlet and outlet $CO_2$ concentration, gas phase concentration, $CO_2$ (aq), $HCO_3^-$, $CO_3^{2-}$, and the concentration at the reaction site C*. From the aspect of reaction kinetics, the concentration C* is the most important. The dependency of photosynthesis on C* can be described in the form of Monod kinetics;[8] the apparent photosynthesis rate is APS $= APS_{max} \cdot C^*/(K_s + C^*)$. A very low value for the saturation constant $K_s$ is expected (as it is known for most microorganism-substrate relations). This is confirmed for *Chlorella vulgaris*, where $K_s$ has been calculated to be $4.77 \cdot 10^{-8}$ mol/l $CO_2$ (aq) corresponding to a partial pressure of only 1.55 ppm $CO_2$. This means that a medium in

---

* If the n variables describing the culture state are denoted by $V_i$, i = 1, n then the stationary state (steady state) is defined by $dV_i/dt = 0$, i = 1, n.

** The intensive properties of the culture are independent of the measuring position except for the light intensity. The light absorption profile is near homogeneity only for very dilute cultures.

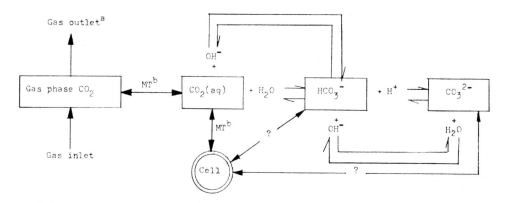

FIGURE 1.    Carbon components in a culture. (a) In a homogeneous culture, the outlet concentration corresponds to the gas phase concentration. (b) MT: mass transfer depending on the driving potential and the exchange area.

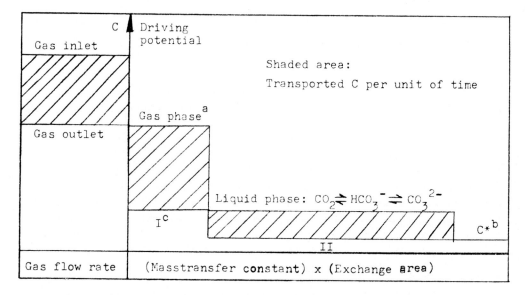

FIGURE 2.    Driving potentials and chemical reactions in an aerated, homogeneous stationary culture. (a) In a homogeneous culture, the outlet concentration corresponds to the gas phase concentration. (b) C* represents the concentration at the reaction site. (c) For reasons of mass balance, the mass flow through the boundary areas I and II must be the same.

equilibrium with 1.55 ppm $CO_2$ is enough for half maximum photosynthetic rate. On the other hand, for the production of plants, a knowledge of C* is irrelevant. In this case, the essential questions are the following: (1) How much additional (and costly) $CO_2$ is necessary for an optimal production for a given mass transfer, and (2) how can one increase productivity by influencing mass transfer for a given $CO_2$ feed?

The responses of photosynthesis to $CO_2$ and $HCO_3^-$ given in the literature are valid only for the cultivation device used. In most cases, the mass transfer coefficient, $K_L a$, is not known and often the description does not allow comparisons between different types of cultures. Consequently, the data found in the literature can be compared only with great caution. Therefore Table 1 serves primarily as a reference list.

## Table 1
## RESPONSES OF MICROALGAE TO $CO_2$ AND $HCO_3^-$

| Species[a] | Data[b] | Ref. |
|---|---|---|
| **Cyanophyta** | | |
| *Anabaena variabilis* | 25°C, light saturation | 16 |
| | $APS_{max} = 140 \, \mu L \cdot g \, dw^{-1} \cdot min^{-1}$, $K_s = 2.5 \, mM HCO_3^-$ | |
| *Anabaena flos-aquae* | APS linearly dependent on $CO_2$ in the concentration range of 6—14 | 17 |
| | $\mu L/L \, CO_2$; $CO_2$ compensation point $= 3 \, \mu L/L$ | |
| *Anacystis nidulans* | Stationary, homogeneous turbidostat culture; APS as function of | 10 |
| | outlet concentration, pH and temperature as parameter | |
| **Chromophyta** | | |
| *Nitzschia palea* | 26°C, Warburg manometry, 75-W lamps at 9 cm | 18 |
| | $APS_{max} = 100 \, mm^3 \, O_2 \cdot h^{-1} \cdot (10 \, mm^3 \, of \, cells)^{-1}$, $K_s = 0.5 \, \mu mol/L$ $CO_2$ | |
| | *N. palea* is compared to *Chlorella*. | |
| *Navicula pelliculosa* | 25°C, 270 $\mu E \cdot m^{-2} \cdot sec^{-1}$ quantum flux | 17 |
| | $APS_{max} = 280 \, \mu mol \cdot mg^{-1} Chl \cdot h^{-1}$ $K_s = 100 \, \mu L \, CO_2/L$ | |
| *Thalassiosira fluviatilis* | 25°C, 270 $\mu E \cdot m^{-2} \cdot sec^{-1}$ quantum flux | 17 |
| | $APS = 17.3 \, \mu mol \cdot mg^{-1} Chl \cdot h^{-1}$ at 350 $\mu L/L \, CO_2$ | |
| **Rhodophyta** | | |
| *Porphyridium* sp. | 25°C, 270 $\mu E \cdot m^{-2} \cdot sec^{-1}$ quantum flux | 17 |
| | $APS = 27 \, \mu mol \cdot mg^{-1} Chl \cdot h^{-1}$ | |
| **Chlorophyta** | | |
| *Chlamydomonas reinhardtii* | 25°C, air bubbled, pH 8, grown at low $CO_2$ concentration, $K_s(CO_2)$ $= 3.7 \, \mu M$ | 12 |
| | 25°C, 270 $\mu E \cdot m^{-2} \cdot sec^{-1}$, 350 $\mu L/L \, CO_2$ | 17 |
| | $APS = 55 \, \mu mol \cdot mg^{-1} Chl \cdot h^{-1}$ | |
| *Chlorella pyrenoidosa* | Laboratory cultures, 6% $CO_2$: 86 g dw $\cdot m^{-2} \cdot d^{-1}$; outdoor cultures: | 19 |
| | 1—43 g dw $\cdot m^{-2} \cdot d^{-1}$ | 20-24 |
| | Diagrams APS as function of $CO_2$ concentration | 25 |
| | Production of *Chlorella* in Asian countries | |
| *Chlorella vulgaris* | Continuous fermentor culture, 28°C, pH 6.5 | 8 |
| | $K_s = 5.10^{-8} \, mol \, CO_2 \cdot L^{-1}$ | |
| *Dunaliella salina* | Air-bubbled cultures, saturated light | 12 |
| | $K_s(CO_2) = 1.5 \, \mu M CO_2$ | |
| *Gonium pectorale* | 20°C, 390 ppm $CO_2$, air-bubbled cultures | 26 |
| | $APS_{max} = 320 \, \mu g \, CO_2 \cdot min^{-1} \cdot g^{-1}$, $K_s = 100 \, ppm$ | |
| | Comparison to *Chlorella* | |
| *Mougeotia* sp. | 25°C, 270 $\mu E \cdot m^{-2} \cdot sec^{-1}$ quantum flux, 350 $\mu L \cdot L^{-1} \, CO_2$ | 17 |
| | $APS = 32 \, \mu mol \cdot mg^{-1} Chl \cdot h^{-1}$ | |
| *Scenedesmus quadricauda* | Inorganic C-limited chemostat; calculations using the Monod model, 27°C | 4 |
| | $K_s(C_{total}) = 0.14(pH \, 7.1) - 0.54 \, (pH \, 7.61) \, mg/L$ | 25,28 |
| | APS as a function of $CO_2$ concentration | |
| *Scenedesmus obliquus* | Adaptation of the algae to high and low $CO_2$ concentrations | 14 |
| *Selenastrum capricornutum* | Inorganic limited chemostat; calculations using the Monod model, 27°C | 4 |
| | $K_s(C_{total}) = 0.41 \, (pH \, 7.05) - 1.49 \, (pH \, 7.59) \, mg/L$ | |
| | $\mu_{max} = 2.45 \, day^{-1}$ (independent of pH) | |

[a] Taxonomy according to Fott.[28]

[b] Abbreviations: APS, apparent photosynthesis rate; $K_s$, saturation constant; $\mu_{max}$, maximum specific growth rate.

## RESPONSE TO $O_2$

The concentration of dissolved $O_2$ in the medium influences apparent photosynthesis in two ways:

1.  An increasing $O_2$ concentration enhances the respiration which decreases apparent photosynthesis.
2.  High $O_2$ concentration acts as a product inhibitor on the $O_2$ evolution process.

The mathematical description of the $O_2$ mass balance is simpler than for $CO_2$. There are no further chemical reactions of the dissolved gas with the medium. For the mass transfer between gaseous and liquid phases, the same considerations are valid as shown previously for $CO_2$, except for the direction of the driving potential. In dense cultures (e.g., mass production ponds), the transfer of $O_2$ from liquid to gas phase can be critical if the turbulence is low. In this case, the $O_2$ concentration can be extremely high, supersaturation, and severely depresses the photosynthetic productivity.

The question of the response of photosynthesis to $O_2$ is closely related to the problems of photorespiration. Generally, microalgal photosynthesis is mostly influenced by $O_2$ concentrations higher than 21%, whereas the influence between 0 and 21% is small or not even measurable (Warburg effect). Table 2 summarizes responses of microalgae to $O_2$.

## RESPONSE TO pH

Microalgae, as well as other microorganisms, have upper and lower pH limits for normal physiological activity. The situation is complicated by the fact that the $HCO_3^-$ and $CO_3^{2-}$ concentrations are extremely dependent on pH, whereas the concentration of dissolved $CO_2$ remains constant as shown in Figure 3. Therefore pH can influence the photosynthetic activity indirectly via the $HCO_3^-$ concentration. This influence is excluded for stationary aerated cultures.

The growth of the algae itself influences the pH of a culture. In a stationary continuous culture using gaseous $CO_2$ as a C source, the pH increases when $NO_3^-$ is used as a N source (nitrate assimilation). If $HCO_3^-$ is the C source, via the reaction $HCO_3^- + H^+ \rightleftharpoons CO_2 \text{ (aq)} + H_2O$, hydrogen ions are removed from the medium and the pH increases. Figure 4 presents the pH ranges for several algae.

## Table 2
## RESPONSES OF MICROALGAE TO O$_2$

| Species[a] | Data[b] | Ref. |
|---|---|---|
| **Cyanophyta** | | |
| *Anabaena flos-aquae* | 25°C, 270 $\mu$E · m$^{-2}$ · sec$^{-1}$ quantum flux | |
| | 150 $\mu$L/L CO$_2$: rel. APS in 2/21/50% O$_2$ $\sim$ 103/100/108% | |
| | 350 $\mu$L/L CO$_2$: rel. APS in 2/21/50% O$_2$ $\sim$ 100/100/102% | 17 |
| | 1000 $\mu$L/L CO$_2$: rel. APS in 2/21/50% O$_2$ $\sim$ 100/100/100% | |
| *Anacystis nidulans* | 25°C, 270 $\mu$E · m$^{-2}$ · sec$^{-1}$ quantum flux | 17 |
| | 350 $\mu$L/L CO$_2$: rel. APS in 2/21/50% O$_2$ $\sim$ 100/100/102% | |
| | Continuous turbidostat culture, full range of physiological conditions: APS not influenced between 0 and 21% O$_2$ | 10 |
| **Chromophyta** | | |
| *Navicula pelliculosa* | 25°C, 270 $\mu$E · m$^{-2}$ · sec$^{-1}$ quantum flux | |
| | 150 $\mu$L/L CO$_2$: rel. APS in 2/21/50% O$_2$ $\sim$ 103/100/113% | |
| | 350 $\mu$L/L CO$_2$: rel. APS in 2/21/50% O$_2$ $\sim$ 97/100/105% | 17 |
| | 1000 $\mu$L/L CO$_2$: rel. APS in 2/21/50% O$_x$ $\sim$ 91/100/91% | |
| *Phaeodactylum tricornutum* | 100% O$_2$ inhibits APS up to 70%, depending on the HCO$_3^-$ concentration | 29 |
| *Thalassiosira fluviatilis* | 25°C, 270 $\mu$E · m$^{-2}$ · sec quantum flux | 17 |
| | 350 $\mu$L/L CO$_2$: rel. APS in 2/21/50% O$_2$ $\sim$ 104/100/108% | |
| *Thalassiosira pseudonana* | 24°C, 1.3 × 10$^{17}$ quanta/sec/cm$^2$ | 30 |
| | APS in N$_2$/air/O$_2$ $\sim$ 93.9/74.6/56.2 $\mu$mol $\sim$ C/mg Chl a/h | |
| Zooxanthellae (sep. from *Pocillopora capitata*) | 24°C, 1.3 × 10$^{17}$ quanta/sec/cm$^2$ | 30 |
| | APS in N$_2$/air/O$_2$ $\sim$ 15.7/16.5/13.8 $\mu$mol C/mg Chl a/h | |
| **Rhodophyta** | | |
| *Porphyridium* sp. | 25°C, 270 $\mu$E · m$^{-2}$ · sec$^{-1}$ quantum flux | 17 |
| **Chlorophyta** | | |
| *Chlamydomonas reinhardtii* | 0% O$_2$: K$_s$[d] = 0.7 m$M$, APS $_{max}$ = 350 $\mu$mol · mg Chl$^{-1}$ · h$^{-1}$ | 31 |
| | 100% O$_2$: K$_s$ = 2.9 m$M$, APS$_{max}$ = 350 $\mu$mol · mg Chl$^{-1}$ · h$^{-1}$ | |
| | 3% CO$_2$: 21/97% O$_2$ $\sim$ 9.9/25.3 h t$_d$ | 11 |
| | 350 ppm CO$_2$: 21/97% CO$_2$ $\sim$ 9.9/15.6 h t$_d$ | |
| | 25°C, 270 $\mu$E · m$^{-2}$ · sec$^{-1}$ quantum flux | 17 |
| | 350 $\mu$L/L CO$_2$: rel. APS in 2/21/50% O$_2$ $\sim$ 103/100/100% | |
| *Chlorella pyrenoidosa* | 24°C, 1.3 × 10$^{17}$ quanta/sec/cm$^2$ | 30 |
| | APS in N$_2$/air/O$_2$ $\sim$ 13.0/9.9/3.6 $\mu$mol C/mg Chl a/h | |
| | 25°C, 270 $\mu$E · m$^{-2}$ · sec$^{-1}$ quantum flux | |
| | 150 $\mu$L/L CO$_2$: rel. APS in 2/21/50% O$_2$ $\sim$ 109/100/104% | |
| | 350 $\mu$L/L CO$_2$: rel. APS in 2/21/50% O$_2$ $\sim$ 102/100/96% | 17 |
| | 1000$\mu$L/LCO$_2$: rel. APS in 2/21,/50% O$_2$ $\sim$95/100/95% | |
| *Dunaliella salina* | 25°C, 270 $\mu$E · m$^{-2}$ · sec$^{-1}$ | 17 |
| | 350 $\mu$L/L CO$_2$: rel. APS in 2/21/50% O$_2$ $\sim$ 100/100/100% | |
| *Mougeotia* sp. | 25°C, 270 $\mu$E · m$^{-2}$ · sec$^{-1}$ | 17 |
| | 350 $\mu$L/L CO$_2$: rel. APS in 2/21/50% O$_2$ $\sim$ 100/100/102% | |
| *Scenedesmus quadricauda* | 25°C, 270 $\mu$E · m$^{-2}$ · sec$^{-1}$ | 17 |
| | 350 $\mu$L/L CO$_2$: rel. APS in 2/21/50% O$_2$ $\sim$ 104/100/99% | |

[a]  Taxonomy after Fott.[28]

[b]  Abbreviations: APS, apparent photosynthesis rate; $\sim$, corresponding to; K$_s$, saturation constant.

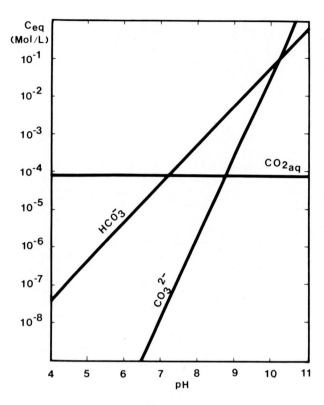

FIGURE 3.   Equilibrium concentrations of $CO_2$ (aq), $HCO_3^-$ and $CO_3^{2-}$ with air (350 ppm $CO_2$, 25°C) as function of pH.

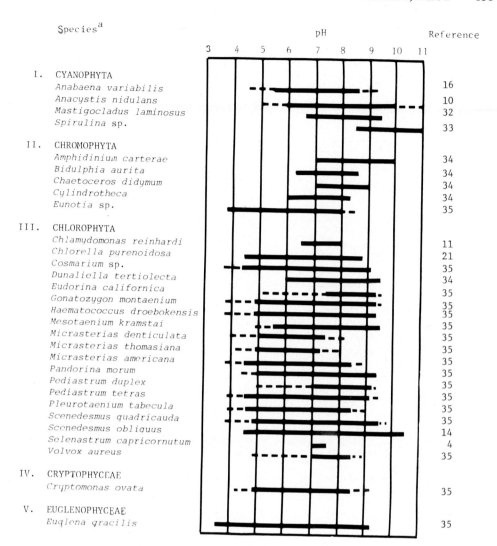

FIGURE 4.   Response of microalgae to pH, where —, normal physiological activity, – – –, decreasing activity, damage at the extreme values. Taxonomy after Fott.[28]

# REFERENCES

1. **Baczko, v., K.,** Wasserhaltige Lösungen von Kohlensäure, Carbonat-Ionen, Peroxokohlensäuren, in *Gmelin Handbuch der anorganischen Chemie,* Kohlenstoff, Teil C 3, Gmelin, L., Verlag Chemie, Weinheim, 1973, 117.
2. **Skirrow, G.,** The dissolved gases — carbon dioxide, in *Chemical Oceanography,* Vol. 1, Riley, J. P. and Skirrow, G., Eds., Academic Press, London, 1965, chap. 7.
3. **Stumm, W., and Morgan, J. J.,** *Aquatic Chemistry,* Wiley Interscience, New York, 1970.
4. **Goldman, J. C., Oswald, W. J., and Jenkins, D.,** The kinetics of inorganic carbon limited algal growth, *J. Water Polut. Control Fed.,* 46, 554, 1974.
5. **Wetzel, R. G.,** Inorganic carbon, in *Limnology,* W. B. Saunders, Philadelphia, 1975, chap. 10.
6. **Aiba, S., Humphrey, A. E., and Millis, N. F.,** *Biochemical Engineering,* 2nd ed., Academic Press, New York 1973.
7. **Brauer, H.,** *Stoffaustausch einschliesslich chemischer Reaktionen,* Sauerländer AG, Aarau, 1971.
8. **Märkl, H.,** $CO_2$ transport and photosynthetic productivity of a continuous culture of algae, *Biotechnol. Bioeng.,* 19, 1851, 1977.
9. **Schneider, K. and Frischknecht, K.,** Determination of the $O_2$ and $CO_2$ $K_L$ a values in fermenters with the dynamic method measuring the step responses in the gas phase, *J. Appl. Chem. Biotechnol.,* 27, 631, 1977.
10. **Frischknecht, K., and Schneider, K.,** Physiological performances of the blue-green alga *Anacystis nidulans* in continuous turbiodostat fermenter culture, *Arch. Microbiol.,* in press.
11. **Berry, J., Boynton, J., Kaplan, A., and Badger, M.,** Growth and photosynthesis of *Chlamydomonas reinhardtii* as a function of $CO_2$ concentration, *Carnegie Inst. Washington Yearb.* 75, 423, 1975.
12. **Badger, M. R., Kaplan, A., and Berry, J. A.,** The internal $CO_2$ pool of *Chlamydomonas reinhardtii:* response to external $CO_2$, *Carnegie Inst. Washington Yearb.,* 76, 362, 1976.
13. **Raven, J. A.,** Exogenous inorganic carbon sources in plant photosynthesis, *Biol. Rev.,* 45, 167, 1970.
14. **Findenegg, G. R.,** Correlations between accessibility of carbonic anhydrase for external substrate and regulation of photosynthetic use of $CO_2$ and $HCO_3^-$ by *Scenedesmus obliquus, Z. Pflanzenphysiol.,* 79, 428, 1976.
15. **King, D. L.,** The role of carbon in eutrophication, *J. Water Polut. Control Fed.,* 42, 2035, 1970.
16. **Kratz, W. A. and Myers, J.,** Photosynthesis and respiration of the blue-green algae, *Plant Physiol.,* 30, 275, 1955.
17. **Lloyd, N. D. H., Canvin, D. T., and Culver, D. A.,** Photosynthesis and photorespiration in algae, *Plant Physiol.,* 59, 936, 1977.
18. **Barker, H. A.,** Photosynthesis in diatoms, *Arch. Mikrobiol.,* 6, 141, 1935.
19. **Thomas, M. D.,** Algal Culture, in *Plant Physiolog* Vol. A, Steward, F. C., Ed., Academic Press, 1965, 168.
20. **Warburg, O.,** Ueber die Geschwindigkeit der photochemischen Kohlensäurezersetzung in lebenden Zellen, *Biochem. Z.,* 100, 230, 1919.
21. **Emerson, R., Green, L.,** Effect of hydrogen-ion on *Chlorella* photosynthesis, *Plant Physiol.,* 13, 157, 1938.
22. **Briggs, G. E. and Whittingham, C. P.,** Factors affecting the rate of photosynthesis of *Chlorella* at low concentrations of carbon dioxide and in high illumination, *New Phytol.,* 51, 236, 1952.
23. **Fock, H., Canvin, D. T., and Grant, B. R.,** Effects of oxygen and carbon dioxide on photosynthetic $O_2$ evolution and $CO_2$ uptake in sunflower and *Chlorella, Photosynthetica,* 5, 389, 1971.
24. **Metzner, H. und Lorenzen, H.,** Untersuchungen über den Photosynthesegaswechsel an vollsynchronen *Chlorella-*Kulturen, *Ber. Dtsch. Bot. Ges.,* 73, 410, 1960.
25. **Tsukada, O. and Kawahara, T.,** Mass culture of *Chlorella* in Asian countries, in *Biological Solar Energy Conversion,* Mitsui, A., Miyachi, S., San Pietro, A., and Tamura, S., Eds., Academic Press, New York, 1977, 363.
26. **Brown, D. L. and Tregunna, E. B.,** Inhibition of respiration during photosynthesis by some algae, *Can. J. Bot.,* 45, 1135, 1967.
27. **Rabinowitch, E. L.,** *Photosynthesis and Related Processes,* Vol. 2 (Part 2), Interscience Publishers, New York, 1956, 1886.
28. **Fott, B.,** *Algenkunde,* 2nd ed., VEB Gustav Fischer, Jena, 1971.
29. **Beardall, J. and Morris, L.,** Effects of environmental factors on photosynthesis patterns in *Phaeodactylum tricornutum* (Bacillariophyceae). II. Effect of oxygen, *J. Phycol.,* 11, 430, 1975.
30. **Burris, J. E.,** Photosynthesis, photorespiration, and dark respiration in eight species of algae, *Mar. Biol.,* 39, 371, 1977.
31. **Bowes, G. and Berry, J. A.,** The Effect of oxygen on photosynthesis and glycolate excretion in *Chlamydomonas reinhardtii, Carnegie Inst. Washington Yearb.,* 71, 148, 1971.

32. **Binder, A., Locher, P., and Zuber, H.,** Concerning the large scale cultivation of thermophilic cosmopolitan *Mastigocladus laminosus* COHN in Icelandic hot springs, *Arch. Hydrobiol.*, 70, 541, 1972.
33. **Clement, G. und Landeghem, H., van,** *Spirulina*: ein günstiges Objekt für die Massenkultur von Mikroalgen, *Ber. Dtsch. Bot. Ges.*, 83, 559, 1970.
34. **Humphrey, G. F.,** The photosynthesis: respiration ratio of some unicellular marine algae, *J. Exp. Mar. Biol. Ecol.*, 18, 111, 1975.
35. **Moss, B.,** The influence of environmental factors of the distribution of freshwater algae: an experimental study. II. The role of pH and the carbon dioxide-bicarbonate system, *J. Ecol.*, 61, 157, 1973.

# RESPONSE OF PHOTOSYNTHETIC BACTERIA TO MINERAL NUTRIENTS

Johannes F. Imhoff

## SULFIDE

Sulfide is the most important photosynthetic electron donor for Chlorobiaceae and Chromatiaceae. It is also used by *Chloroflexus aurantiacus* strains[1] and some Rhodospirillaceae species.[2,3] The sulfide tolerance is quite different among the species, and sulfide is regarded as the main regulatory factor. In addition to light intensity, duration of light phase, temperature, and pH, it controls the occurrence, distribution, and mass development of the different species in natural environments.[4,5] Depending on the sulfide concentration and light penetration, the motile forms show diurnal vertical movements. The organic matter produced by the photosynthetic sulfur bacteria is about 3 to 5% of the total annual production in lakes poor in sulfide,[6] 9 to 25% in lakes rich in sulfide,[6] and reaches values of 83% in Solar Lake and Fayetteville Green Lake.[7,8] The percentage of the total daily production under stratified conditions is even higher and ranges from 20 to 91% for different lakes, with an absolute carbon fixation rate of 45 to 8015 mg carbon per m²/day.[8-10]

Chromatiaceae and Chlorobiaceae species are generally found in sulfide-rich habitats and form mass developments predominantly in marine coastal areas and in lakes with a well-established chemocline, located in the photic zone. Rhodospirillaceae species in general are less sulfide tolerant and typically inhabit sulfide-poor freshwater habitats. In general, the Chlorobiaceae are more sulfide tolerant than the Chromatiaceae.[4,11] *Chlorobium* and *Prosthecochloris* strains grow well with up to 4.1 to 8.3 m$M$ sulfide in mineral salts medium; in stratified environments, they are usually found in deeper layers than the purple sulfur bacteria. The species of the genera *Pelodictyon* and *Ancalochloris,* containing gas vacuoles, need low sulfide concentrations of 0.4 to 2.0 m$M$ and additionally require low light intensity and temperature.[10,12] In mixed pure cultures with the sulfur-reducing *Desulfuromonas acetoxidans,* optimal growth of *Chlorobium* and *Prosthecochloris* is possible with only 0.2 to 0.23 m$M$ sulfide, while no growth was observed below 0.12 m$M$.[13] Among the Chromatiaceae, medium to high concentrations of sulfide (2 to 6 m$M$ ) are tolerated by the small cell *Chromatium* species, *Thiocapsa roseopersicina, Thiocystis violacea,* and *Ectothiorhodospira* species. Growth inhibition is more serious in the large-cell *Chromatium* species, *Thiospirillum jenense, Thiocystis gelatinosa,* and the species containing gas vacuoles which are found in the uppermost layers of sulfide-containing stratified environments. The sulfide concentration must be reduced to 0.4 to 2.0 m$M$ to allow growth of these species.[4,14] Most sensitive to sulfide is *Thiopedia rosea* which often shows mass development in environments with very low sulfide production.

In the small-cell *Chromatium vinosum,* the saturation constant for sulfide was determined in continuous culture to be $7 \times 10^{-6}$ $M$.[15] At sulfide concentrations above 0.02 m$M$, the growth rate was reduced, and no growth was observed above 2.2 m$M$ sulfide. At continuous illumination with sulfide as the growth-limiting factor, the growth rate of *Chromatium vinosum* (maximum specific growth rate of 0.115/hr) exceeds that of the large cell species *Chromatium weissei* (maximum specific growth rate of 0.040/hr) regardless of the sulfide concentration. With intermittent light-dark periods, however, both organisms show balanced coexistence, because sulfide oxidation occurs more rapidly in *Chromatium weissei* (41.5 $\mu$mol S$^{2-}$ per milligram cell nitrogen/hr) than in *Chromatium vinosum* (18.5 $\mu$mol S$^{2-}$ per milligram cell nitrogen/hr). Consequently, in

*Chromatium vinosum,* 54 to 57% and in *Chromatium weissei,* 85-89% of the added sulfide was recovered as elemental sulfur in the cells.[5]

The intermediary storage of elemental sulfur in the cells during sulfide oxidation to sulfate is a general feature of the Chromatiaceae. The oxidation of sulfur is partially repressed by sulfide, and upon sulfide depletion, an increased rate of sulfur oxidation was observed in *Chromatium vinosum, Chr. okenii,* and *Thiocapsa roseopersicina.*[16-20] In species of the genus *Ectothiorhodospira, Rhodopseudomonas* sp. strain "51", and in the Chlorobiaceae, all of which accumulate elemental sulfur outside the cells, sulfide oxidation and sulfur oxidation are clearly sequential processes. Sulfur oxidation does not start before sulfide has completely dissappeared.[2,20]

In *Thiocapsa roseopersicina, Chromatium vinosum,* and *Thiocystis violacea* growing photoheterotrophically on fructose as a sole carbon source, growth is inhibited by sulfide. Mixotrophic growth on $CO_2$ and fructose as carbon sources and sulfide or thiosulfate as electron donors is possible.[21]

Sulfide tolerance of most Rhodospirillaceae species is very low. *Rhodopseudomonas sphaeroides, Rps. palustris* and *Rhodospirillum rubrum* do not tolerate more than 0.4 m$M$ sulfide in mineral salts medium.[3] The sulfide tolerance of *Rps. capsulata* and *Rhodomicrobium vannielii* is 2 to 3 m$M$, with a maximum specific growth rate of 0.14/hr for *Rps. capsulata.* The highest sulfide tolerance is observed in *Rps. sulfidophila;* it is 5.2 to 6.3 m$M$ in a mineral medium and increases to 7 to 8 m$M$ with the addition of 0.01% yeast extract.[2] Sulfide tolerance of *Rps. sp.* strain "51" is about 5 m$M$. Both *Rps. palustris* and *Rps. sulfidophila* are able to grow photolithotrophically with sulfide or thiosulfate as electron donors. The oxidation of both compounds is inducible, and they are oxidized to sulfate without intermediary accumulation of other sulfur compounds.[2,22,23] *Rm. vannielii* can grow at the expense of sulfide as oxidizable substrate which is oxidized to tetrathionate.[2] All other species tested oxidize sulfide to elemental sulfur only, but *Rps. sp.* strain "51" is the only member of the Rhodospirillaceae which is able to oxidize elemental sulfur to sulfate.[2]

Some strains of *Chloroflexus aurantiacus,* which were routinely grown with 2 to 4 m$M$ sulfide, tolerate more than 6 m$M$ sulfide. Autotrophic growth with sulfide in the light was obtained only with strain OK-70fl.[24]

## SULFATE

In the Chlorobiaceae sp. and in most species of the Chromatiaceae which are well adapted to anaerobic life in sulfide-rich habitats, assimilation of sulfate is not possible. Only some species of the genera *Thiocapsa, Thiocystis, Ectothiorhodospira,* and the small-cell *Chromatium* sp. are able to grow with sulfate as the sole source of sulfur.[24] Growth of halophilic *Ectothiorhodospira* sp. was observed in natural brines with up to 500 m$M$ sulfate, routinely grown in laboratory cultures between 100 to 200 m$M$.[26,27] Most Rhodospirillaceae sp are able to assimilate sulfate. *Rps. sulfoviridis,*[28] *Rps. sp.* strain "51",[2] and some isolates of marine habitats[29] require reduced sulfur compounds for growth and are not able to assimilate sulfate or show reduced capacity of sulfate reduction. *Rps. globiformis* needs 0.02% thiosulfate for normal growth.[30] Growth is apparently inhibited by normal sulfate concentrations of about 1 m$M$ and is enhanced if the sulfate content is reduced to about 100 $\mu M$.[29] The photoorganotrophic growth of *Rps. sulfidophila* is limited by sulfate below 0.25 m$M$. The sulfate uptake of growing cells is inhibited by reduced sulfur compounds like sulfide, sulfite, cysteine, methionine, or glutathione.[29]

## OTHER SULFUR COMPOUNDS

Elemental sulfur is oxidized by all Chlorobiaceae and Chromatiaceae, but *Chloroflexus aurantiacus* and the Rhodospirillaceae, with the exception of *Rps. sp.* strain "51", do not use sulfur. Thiosulfate is utilized by many Chromatiaceae sp., two subspecies of *Chlorobium limicola* and *Chlorobium vibrioforme*, as well as *Rps. palustris*[22,31] and *Rps. sulfidophila*.[32,33] The thiosulfate oxidizing enzyme system is inducible in *Rps. palustris*[34] and *Rps. sulfidophila*,[2] but is present in sulfide grown cells of *Chr. vinosum* and *Thiocapsa roseopersicina*.[16] Many strains of Chlorobiaceae and Chromatiaceae also use sulfite, but at low concentrations.[19] *Chlorobium limicola* f. *thiosulfatophilum* also uses tetrathionate as a photosynthetic electron donor.[35]

## AMMONIA

Ammonia is assimilated by all phototrophic bacteria and enhances growth much better than other inorganic nitrogen sources. Good growth of most species in batch culture is observed at 7.5 to 18.7 m$M$ ammonium chloride. *Chloroflexus aurantiacus* was routinely grown with 3.7 m$M$ ammonium chloride,[24] and *Chromatium vinosum* was grown with only 1.12 m$M$ ammonia in sulfide-limited continuous culture.[15]

Although there are generally two enzyme systems present in the phototrophic bacteria which are possibly involved in ammonia assimilation, both respond differently to low and high concentrations of ammonia. The glutamine synthetase/glutamate synthase sequence normally represents the primary pathway of nitrogen assimilation, if cells are grown on $N_2$, $NO_3^-$, or low ammonia concentrations (0.71 m$M$). Growth on nitrogen or nitrate is, in this respect, similar to growth on low ammonia concentrations, as both are reduced to ammonia and the ammonia concentration remains low. In cultures of *Rps. capsulata* with high nitrogenase activity, the ammonia concentration is 10 to 20 $\mu M$.[36] The $K_m$ of the glutamine synthetase for ammonia is low, 0.3 to 0.6 m$M$.[37] In the Rhodospirillaceae sp. *R. fulvum*, *R. molischianum*, *Rm. vannielii*, *Rps. sphaeroides*, *Rps. palustris*, and *Rps. acidophila*, glutamine synthetase and NADH- or NADPH-dependent glutamate synthase activities were high in cultures grown on nitrate (0.7 and 14.3 m$M$), nitrogen, or low ammonia concentrations (0.7 and 3.6 m$M$).[38,39] These activities were also found in *R. rubrum*[40-42] and *Rps. capsulata*.[43] Also in the Chromatiaceae and Chlorobiaceae species *Amoebobacter roseus*, *Chr. gracile*, *Chr. minus*, *Chr. vinosum*, *Chr. warmingii*, *Thiocapsa roseopersicina*, *Thiocystis violacea*, *Chlorobium limicola*, and *Chlorobium vibrioforme*, glutamine synthetase and NADH-dependent glutamate synthase are present.[37,41] They are generally more active in cells grown on low ammonia concentrations and *Thiocapsa roseopersicina* completely lacks glutamate synthase activity in cells grown on high ammonia concentrations (28.6 m$M$).

Assimilation of ammonia via glutamate dehydrogenase is only favored at high ammonia concentrations, as the $K_m$ for ammonia is between 6 to 16 m$M$.[37] Even at high ammonia concentration (28.6 m$M$), the role of the glutamate dehydrogenase is not clear. If the glutamine synthetase is active under these conditions, both enzymes will compete for ammonia. Glutamate dehydrogenase is found in all Rhodospirillaceae so far tested,[38,43,44] except *Rps capsulata* (even if grown on glutamate or high ammonia concentrations) and *Rps. acidophila* which does not grow on glutamate as the sole nitrogen source.[46] The activity is nearly unaffected by the ammonia concentration, but glutamate-grown cells of *R. rubrum* show a several-fold increase in the glutamate dehydrogenase activity in the direction of glutamate degradation.[38,45] In the Chromatiaceae and Chlorobiaceae species, glutamate dehydrogenase is present only if cells are grown on high ammonia concentrations. Only in *Chromatium vinosum* is the enzyme also present at low ammonia concentrations.[37,41]

## NITROGEN

Considerable activities of nitrogenase are present in most species of Rhodospirillaceae and Chromatiaceae. Only in *Rhodocyclus purpureus, Thiocystis gelatinosa*, and *Amoebobacter pendens* are no activities found. Minor activities were measured in Chlorobiaceae species. From 62 strains of phototrophic bacteria, only 9 showed no activities and, 7 showed very low activities.[46] Most data on nitrogen responses are available from Rhodospirillaceae species.

In *R. rubrum* and *Rps. acidophila*, the synthesis of nitrogenase is initiated at ammonium limitation. As synthesis is also induced in nitrogen-free medium under an atmosphere lacking $N_2$ and the $N_2$-fixing system is present under nitrogen limitation regardless of the form of nitrogen used, it is concluded that nitrogenase is not positively induced by $N_2$.[46-48] The $K_m$ ($N_2$) for a cell-free system of *R. rubrum* was determined to be 0.07 atm.[48] In whole cells of *R. rubrum, Rps. capsulata*, and *Rps. palustris*, the nitrogenase activity is immediatly inactivated by one or more of the following nitrogen compounds, ammonia, L-glutamine, L-asparagine, and urea.[47,49,50] The inactivation depends on the ammonia concentration, and nitrogenase activity is switched off by an ammonia concentration of less than 0.36 m$M$ in *R. rubrum*.[51] In *Rps. capsulata* 0.1 to 0.5 m$M$ ammonia inhibits nitrogenase (photoproduction of hydrogen);[50] acetylene reduction is completely inhibited by 10 to 20 $\mu M$ ammonia in *Rps. capsulata*,[52] and nitrogenase activity resumes its initial activity after the ammonia concentration has fallen below 0.1 m$M$ in the medium in *Rps. palustris*.[49] Nitrogenase activity at low partial pressures of $O_2$ was shown in *R. rubrum* (0.01 atm),[53] *Rps. capsulata*[54] and *Rps. acidophila*.[46] Optimal $pO_2$ for nitrogen fixation by *Rps. acidophila* grown semiaerobically in the dark is 0.08 atm.[46] In the light, nitrogenase is inhibited at lower $pO_2$, probably as light inactivates respiration, which might be responsible for keeping the actual $pO_2$ low. From these data, it is concluded that nitrogenase activity is not dependent on electron transport driven by light. Reports like those by Meyer et al.[52] which report a rapid stop of nitrogen fixation, if photosynthetically grown cells are suddenly exposed to darkness, possibly reflect a rapid depletion of ATP which does not allow further action of nitrogenase. After a period of darkness, nitrogenase activity resumes in cells that are incubated under microaerophilic conditions and are thus able to maintain a high ATP pool by respiratory activity.[52]

## NITRATE

Nitrate is not generally used as a nitrogen source by photosynthetic bacteria and growth is less pronounced than with other nitrogen sources. The ability to use nitrate is reported for a number of Rhodospirillaceae species, namely *R. rubrum, R. fulvum, R. molischianum, Rm. vannielii, Rps. palustris, Rps. sphaeroides, Rps. capsulata*, and *Rps. acidophila*[38,39,46,55-58] and the Chromatiaceae *Chromatium vinosum, Thiocapsa roseopersicina*, and *Ectothiorhodospira shaposhnikovii*.[59-61] The concentrations of nitrate usually applied are 6 to 10 m$M$ for nitrate assimilation and 20 m$M$ for nitrate dissimilation.[62] Considerable activity of nitrate reductase in *E. shaposhnikovii* and *Rps. capsulata* is found only after growth on nitrate, not on ammonia nor glutamate.[59,63] The nitrate reducing enzyme system is inducible by nitrate in *R. rubrum*,[55,57] *Rps. capsulata*,[63] *E. shaposhnikovii*,[59] *Rps. palustris, Rps. sphaeroides*, and *Thiocapsa roseopersicina*.[56] Nitrate reductase is repressed in *Rps. capsulata* to 30% of the activity of nitrate grown cells by ammonia and to 10% by glutamate.[63] Nitrite accumulates in the medium up to 2 m$M$ during nitrate reduction.[63] Nitrate inhibits the ability of *Rps. capsulata* to grow anaerobically in the dark with dimethylsulfoxide as the electron acceptor.[63] Nitrate does not repress nitrogenase synthesis and nitrogen is also assimilated in the presence of nitrate. In *Rps. acidophila*, nitrate is not used under aerobic conditions in the dark, if active nitrogenase is absent.[46]

There is only one report on the denitrification in phototrophic bacteria: *Rps. sphaeroides* f. *denitrificans* grows well anaerobically in the dark with nitrate as the electron acceptor.[62] $N_2$ is evolved during nitrate reduction, and growth is not possible with nitrate as sole nitrogen source. Denitrification is induced by nitrate, either in the dark or in the light, and is suppressed by $O_2$.

The energy and quantum requirement of phototrophic bacteria is far greater for growth on nitrogen and nitrate than for growth on ammonia as the sole nitrogen source. In *Rps. capsulata* and *Rps. acidophila*, 47 and 75 quanta/$N_2$, corresponding to 31 and 50 ATP/$N_2$, are required, respectively. If growth on ammonia and nitrogen is compared, the quantum requirement (per gram dry weight) increases by 88% in *Rps. acidophila* and by 57% in *Rps. capsulata*.[64] As a consequence of the increased energy demand, bacteriochlorophyll synthesis increases if cells are grown on $N_2$ or $NO_3^-$ as sole nitrogen sources. The bacteriochlorophyll content of nitrogen fixing *Rps. acidophila* is 1.3-fold compared to ammonia-grown cells.[46] On the other hand, if ammonia is added to a nitrogen-fixing culture, nitrogen fixation stops and bacteriochlorophyll synthesis is inhibited until the level is adapted to the now lower energy requirement. Concomitant with the increase of the energy requirement, the doubling time of *Rps. acidophila* increases. Under anaerobic light conditions with lactate as substrate, the minimal doubling times are 3.0 hr with ammonia, 3.7 hr with nitrogen, and 6.7 hr with nitrate as the sole nitrogen source.[39,46] Additionally, the growth yield decreases by 17% if ammonia-grown *Rps. acidophila* cells are adapted to nitrogen fixation. The turnover rate of the substrate is also reduced under nitrogen-fixing conditions. Under nitrogen limitation with increasing C/N ratio, the production of storage material and photoproduction of $H_2$ become dominant.[65]

## PHOSPHATE

Little is known about the response to phosphate. The usually applied concentrations for enrichment and growth are 7 to 12 m$M$ for Rhodospirillaceae sp.[66-68] and 2 to 4 m$M$ for the lithotrophically growing Chromatiaceae and Chlorobiaceae.[74,26,69] In sulfide-limited continuous culture, *Chromatium vinosum* grows well with 0.07 m$M$ phosphate.[15] Studies on phosphate limitations are not known. In *Thiocapsa roseopersicina*, growth on fructose in the presence of 20 m$M$ NaHCO$_3$ and 1 m$M$ sulfide is strongly inhibited at 30 m$M$ phosphate, and no growth occurred at 40 m$M$. Optimal phosphate concentration under these conditions was between 10 to 20 m$M$.[21] Similar inhibition was not observed in *Rps. capsulata*[70] and *Rps. sphaeroides*[71] with as much as 100 m$M$ phosphate in the medium. Pyruvate kinase in *R. rubrum* is inhibited by phosphate as an allosteric inhibitor with $K_i = 0.05$ m$M$ phosphate.[72]

## POTASSIUM

The requirement for potassium for growth was demonstrated for *Rps. capsulata*,[72] *Rps. sphaeroides*,[74,75] and *R. rubrum*.[76] The optimal concentration of $K^+$ for *Rps. sphaeroides* is 0.5 to 1.3 m$M$, whereas cells of *Rps. capsulata* grow as rapidly in 0.1 m$M$ as in 15 m$M$ $K^+$ and reach the same yield. The cell yield decreased below 0.1 m$M$ $K^+$. $K^+$ is accumulated actively in the cells by an energy-dependent transport system, while $Na^+$ is extruded from the cells.[73] In cells of *R. rubrum* and *Rps. capsulata* growing anaerobically in the light with 15 to 16 m$M$ K$^+$, K$^+$ accumulates to 110 to 125 m$M$, but the Na$^+$ concentration is nearly the same in the cells as in the medium (37 and 34 m$M$). The K$^+$ content was highest at late logarithmic growth in *Rps. capsulata* and decreased during stationary phase. Similarly, in *R. rubrum*, the K$^+$ level falls from 129 to 46 m$M$ within 72 hr if light-grown cells are incubated in the dark; simultaneously, Na$^+$ rises from 31 to 68 m$M$.

Rb$^+$ and Cs$^+$ can replace K$^+$ during growth in *Rps. capsulata*, but are less efficient, and the doubling time decreases from 2.7 to 4.8 hr with 1 m$M$ Rb$^+$ and to 15.5 hr with 1 m$M$ Cs$^+$ instead of K$^+$. The K$_m$ values of the transport system are 0.2 m$M$ for K$^+$, 0.5 m$M$ for Rb$^+$, and 2.6 m$M$ for Cs$^+$.[73] In *Rps. sphaeroides*, only Rb$^+$, but not Cs$^+$ or Li$^+$, can replace K$^+$ for at least three or four generations.[74]

## SODIUM

Many species of Chromatiaceae and Chlorobiaceae (but only a few Rhodospirilla-ceae) have been isolated from marine or brackish habitats. Some isolates of *Ectothior-hodospira* sp. have also been obtained from extreme saline habitats. The Chlorobi-aceae sp. of the genus *Chlorobium* and *Prosthecochloris aestuarii* are found in brackish and marine environments.[4,23,37,77,78] The Chromatiaceae sp. *Chr. vinosum*, *Chr. gracile*, *Chr. buderi*, *Chr. warmingii*, *Chr. violascens*, *Chr. minutissimum*,[23,33,37,78-81] as well as *Thiocystis violacea*,[78] *Thiocapsa roseopersicina*,[37,78,81,82] *Thiocapsa pfennigii*,[78] and *Lamprocystis roseopersicina*[83,84] were isolated from marine habitats. Mass developments of *Thiopedia rosea* were also often observed in marine habitats.[78,83-85] *Ectothiorhodospira mobilis* was isolated from marine and extreme saline habitats,[27,33,78,86,87] and the extreme halophilic species *E. halophila*,[23,27,69,88] and *E. halochloris*[26] were found in habitats with salt concentrations up to a saturated level. The only Rhodospirillaceae species which are reportedly iso-lated from marine habitats are *Rm. vannielii*,[89,90] some unidentified Rhodospirilla-ceae,[91] *Rps. sulfidophila*,[2,23,32,33] and *Rps. sp.*.[29,33]

In most of the marine isolates, the salt requirement was not established. The strains tested for their salt requirement most frequently prefer salinities of 1 to 3% NaCl, but do not necessarily require it.[78] A requirement of 1 to 3% was shown for a strain of *Thiocapsa roseopersicina*[82] and for the species *Chromatium buderi*[79] and *E. mobilis*.[78,86] *Chromatium buderi* was isolated from a habitat with 26% NaCl. It showed 100% growth at 2%, 92 at 1% NaCl, and only 30% without added NaCl in the me-dium. In NaCl-free medium, the cells show deformations.[79] Similar deformations were regularly found in the marine forms of *Chromatium* species in media with low NaCl concentrations. The genus *Rhabdochromatium*[92] was later recognized as *Chromatium* which showed abnormal cell shapes under unfavorable growth conditions.[79,93-95]

Strains of *E. mobilis* were also isolated from habitats with high salinity.[84,86] The type strain of *E. halophila* grows well between 11 to 22% NaCl, with a doubling time of 7.5 hr. Cells can be adapted to growth at 8 to 9% NaCl; with less than 8%, no growth occurred. At 9 and 30% NaCl, the doubling time is 18 hr. Below 2%, the cells lyse.[69] *E. halochloris* is optimally adapted to growth between 14 to 27% and requires at least 10% total salts, present as NaCl, Na$_2$SO$_4$, and Na$_2$CO$_3$.[26]

The marine isolates of *Chlorobium* sp. require at least 1% NaCl.[4,78] *Prosthecochloris aestuarii* grows well between 2 to 5% NaCl and tolerates 1 to 8% NaCl.

Among numerous isolates of Rhodospirillaceae, the strains isolated from fresh water sources showed growth optima at 0 to 1% NaCl, whereas the strains from marine sources (*Rps. sulfidophila* and *Rps. sp.*) showed optima between 2 to 4% NaCl or are nearly unaffected by NaCl concentrations between 1 to 4% NaCl.[2,23]

A low sodium requirement was also found in some fresh water isolates. Three strains of *Rps. sphaeroides* and one strain of *Rps. palustris* require sodium for growth; the optimum concentration is about 4.4 m$M$.[74] One strain of *R. rubrum* has been shown to grow well without added sodium to the medium. It was also found that Na$^+$ deficiency inhibited the formation of organic phosphorus compounds like nucleic acids and lipids and reduced the bacteriochlorophyll content of *Rps. sphaeroides* cells.[96]

## DI- AND POLYVALENT CATIONS

Magnesium is the major divalent cation in all living cells and plays an important role in the regulation of metabolic key reactions in photosynthetic bacteria. The requirement is usually met by 0.4 to 2.0 m$M$,[26,66,97-99] but some marine isolates need about 12 m$M$.[100] Some *Ectothiorhodospira* sp. were isolated from strongly alkaline natural brines, with extremely low Mg$^{2+}$ concentrations.[26,27,101] Mg$^{2+}$ is accumulated in the cells by an energy-dependent transport system. The K$_m$ for Mg$^{2+}$ uptake is 0.5 $\mu M$ in *Rps. capsulata* with v$_{max}$ = 0.6 to 1.8 $\mu$mol Mg$^{2+}$ /min · g dry weight.[102] Mg$^{2+}$ plays an important role in some general reactions, such as ribosome subunit association or ATP dependent reactions. It accelerates ATPase and PPase reactions, but seems not to be necessary for the formation of ATP in *R. rubrum*.[103] The key enzyme of the autotrophic CO$_2$ fixation pathway, ribulose 1,5-bisphosphate carboxylase, is strongly dependent on the presence of Mg$^{2+}$ in *R. rubrum*.[104] The key enzyme of nitrogen assimilation, glutamine synthetase, is strongly regulated by a mechanism involving Mg$^{2+}$ and Mn$^{2+}$. At low ammonia concentrations, it is deadenylated and Mg$^{2+}$-dependent; at high ammonia concentrations, it is adenylated and Mg$^{2+}$-inhibited (but Mn$^{2+}$-dependent in *Rps. capsulata*).[105] Mg$^{2+}$ is essential for the synthesis of bacteriochlorophyll and the photosynthetic apparatus.

Calcium is another major cation in the environment. The requirement of Ca$^{2+}$ is generally met by 0.5 to 4.4 m$M$ and is higher in some marine *Chlorobium* strains.[100] In alkaline media, its concentration must be reduced and in some natural alkaline brines, the Ca$^{2+}$ content is below the analytical detection limit.[101] It appears to be mainly extracellular in the growth media. In *Rps. capsulata*, it was not taken up, but normally excreted by the cells.[102] Thus, its function seems to be to protect the integrity of the cell envelope.

The requirement for other metal ions is usually met by trace element solutions.[26,106] Accordingly, the media contain 7 to 9 $\mu M$ Fe$^{2+}$, 150 to 350 n$M$ Mn$^{2+}$, 120 n$M$ MoO$_4^{2-}$, 840 to 1050 n$M$ Co$^{2+}$, 40 to 80 n$M$ Ni$^{2+}$, 60 n$M$ Cu$^{2+}$, 700 n$M$ Zn$^{2+}$, and 40 n$M$ SeO$_3^{2-}$. The content of trace metals has been determined in some species and is expressed as micromoles per gram dry weight: 0.055 Mn$^{2+}$, 8.16 Fe$^{2+}$, 0.105 Cu$^{2+}$, and 0.597 Zn$^{2+}$ in *Chr. vinosum*, 0.105 Mn$^{2+}$, 3.62 Fe$^{2+}$, 0.368 Cu$^{2+}$, and 0.346 Zn$^{2+}$ in *R. rubrum*, and 0.695 Mn$^{2+}$, 2.93 Fe$^{2+}$, 0.26 Cu$^{2+}$, and 0.47 Zn$^{2+}$ in *Rps. sphaeroides* grown phototrophically.[107] A similar Fe$^{2+}$ content was found in a *Rps. sp.* and *Chr. minutissimum*. The Mn$^{2+}$ content of these strains was 0.182 for *Rps. sp.* and 0.728 to 2.84 $\mu$mol/g dry weight for *Chr. minutissimum*.[108]

In Mn$^{2+}$-deficient media, the Mn$^{2+}$ content of the cells was 1/25 to 1/100 of normal cells, but nearly the same yield was obtained.[107] In *Chr. vinosum*, the cellular content of Mn$^{2+}$ is eightfold, and chromium increased twofold during autotrophic culture conditions compared with a culture in the presence of acetate.[109]

The optimal concentration of Fe$^{2+}$ in *Rps. sphaeroides* was 6 to 9 $\mu M$; only slight inhibition effects were observed between 2 to 6 $\mu M$, but inhibition increased markedly below 2 $\mu M$ Fe$^{2+}$. Iron deficiency causes excretion of porphyrins by whole cells, inhibition of bacteriochlorophyll synthesis, a decrease in the total heme content by a factor of 20 and the accumulation of poly-$\beta$-hydroxybutyrate.[110-112]

The effect of molybdate concentration on growth and nitrogenase activity was investigated in *R. rubrum,*[113] *Rps. acidophila,*[46] and *Chlorobium limicola* f. *thiosulfatophilum.*[46] Growth inhibition was observed at increased molybdate concentration above 600 $\mu$g Mo/$\ell$ in *Rps. acidophila* and above 200 $\mu$g Mo/$\ell$ in *Chlorobium limicola* grown with ammonia as the nitrogen source. Under nitrogen-fixing conditions, the growth rate of *Rps. acidophila* decreased from 0.19 to 0.17, as the molybdate concentration was increased from 30 to 150 $\mu$g Mo/$\ell$.[46] Calculated from the maximum specific nitrogenase activity, the maximum available cell density in carbon-limited culture and the Mo content of nitrogenase, the minimal requirement was 0.056 nmol Mo/m$\ell$ compared with 0.12 nmol Mo/m$\ell$ present in the medium.[46] In *R. rubrum,* 10 $\mu M$ was sufficient for the normal requirement of the cells. The Mo content of these cells was 39 to 66 $\mu$g Mo/g dry weight, but less than 0.2 $\mu$g Mo/g dry weight in Mo-deficient cells. Mo deficiency reduces nitrogenase activity to about 20%. Tungsten can substitute for Mo in the nitrogenase activity, but not in nitrate reductase activity. Substitution is far better in the $H_2$-evolving than in the $N_2$-fixing reaction.[113]

One striking problem for photosynthetic bacteria living in anaerobic and sulfide-containing environment is the availability of heavy metal ions which are precipitated by sulfide. The concentrations of the ions $Co^{2+}$, $Ni^{2+}$, $Zn^{2+}$, $Mn^{2+}$, and $Fe^{2+}$ are between $10^{-10}$ to $10^{-20}$ $M$ in 1 m$M$ sulfide. $Cu^{2+}$ concentration is far below $10^{-20}$ $M$. This problem may be resolved by the excretion of chelating agents which can help to raise the available metal ions for the cells or by extremely efficient transport systems for the metal ions.

# REFERENCES

1. **Madigan, M. T. and Brock, T. D.,** Photosynthetic sulfide oxidation by *Chloroflexus aurantiacus,* a filamentous photosynthetic gliding bacterium, *J. Bacteriol.,* 122, 782, 1975.
2. **Hansen, T. A.,** Sulfide als Electronendonor voor Rhodospirillaceae, Doctoral thesis, University of Groningen, The Netherlands, 1974.
3. **Hansen, T. A. and van Gemerden, H.,** Sulfide utilization by purple nonsulfur bacteria, *Arch. Mikrobiol.,* 86, 49, 1972.
4. **Pfennig, N.,** Photosynthetic bacteria, *Annu. Rev. Microbiol.,* 21, 285, 1967.
5. **van Gemerden, H.,** Coexistence of organisms competing for the same substrate: an example among the purple sulfur bacteria, *Microb. Ecol.,* 1, 104, 1974.
6. **Takahashi, M. and Ichimura, S.,** Vertical distribution and organic matter production of photosynthetic sulfur bacteria in Japanese lakes, *Limnol. Oceanogr.,* 13, 644, 1968.
7. **Culver, D. A. and Brunskill, G. J.,** Fayetteville Green Lake, New York. V. Studies of primary production and zooplankton in a meromictic lake, *Limnol. Oceanogr.,* 14, 862, 1969.
8. **Cohen, Y., Krumbein, W. E., and Shilo, M.,** Solar Lake (Sinai). II. Distribution of photosynthetic microorganisms and primary production, *Limnol. Oceanogr.,* 22, 609, 1977.
9. **Czeczuga, B.,** Primary production of the green hydrosulphuric bacteria, *Chlorobium limicola* Nads., *Photosynthetica,* 2, 11, 1968.
10. **Pfennig, N.,** General physiology and ecology of photosynthetic bacteria, in *The Photosynthetic Bacteria,* Clayton, R. E. and Sistrom, W. R., Eds., Plenum Press, New York, 1978, chap. 1.
11. **Pfennig, N.,** Phototrophic green and purple bacteria: a comparative systematic survey, *Annu. Rev. Microbiol.,* 31, 275, 1977.
12. **van Niel, C. B.,** Techniques for the enrichment, isolation, and maintenance of the photosynthetic bacteria, in *Methods in Enzymology,* Vol. 23A, San Pietro, A., Ed., Academic Press, New York, 1971, 3.
13. **Biebl. H. and Pfennig, N.,** Growth yields of green sulfur bacteria in mixed cultures with sulfur and sulfate reducing bacteria, *Arch. Microbiol.,* 117, 9, 1978.
14. **Pfennig, N.,** Anreicherungskulturen für rote und grüne Schwefelbakterien, *Zentralbl. Bakteriol. Parasitenkd. Hyg. Abt. 1, Suppl.,* 1, 179, 1965.

15. van Gemerden, H. and Jannasch, H. W., Continuous culture of Thiorhodaceae: sulfide and sulfur limited growth of *Chromatium vinosum*, *Arch. Mikrobiol.*, 79, 345, 1971.

16. Trüper, H. G. and Schlegel, H. G., Sulphur metabolism in Thiorhodaceae. I. Quantitative measurements on growing cells of *Chromatium okenii*, *Antonie van Leeuwenhoek J. Microbiol. Serol.*, 30, 225, 1964.

17. van Gemerden, H., On the ATP generation by *Chromatium* in darkness, *Arch. Mikrobiol.*, 64, 118, 1968.

18. Schmidt, G. L. and Kamen, M. D., Variable cellular composition of *Chromatium* in growing cultures, *Arch. Mikrobiol.*, 73, 1, 1970.

19. van Gemerden, H. and Beeftink, H. H., Specific rates of substrate oxidation and product formation in autotrophically growing *Chromatium vinosum* cultures, *Arch. Microbiol.*, 119, 135, 1978.

20. Trüper, H. G., Sulfur metabolism, in *The Photosynthetic Bacteria*, Clayton, R. E. and Sistrom, W. R., Eds., Plenum Press, New York, 1978, chap. 35.

21. Conrad, R. and Schlegel, H. G., Metabolism of fructose in *Thiocapsa roseopersicina*, *Z. Allg. Mikrobiol.*, 18, 309, 1978.

22. Rolls, J. P. and Lindstrom, E. S., Effect of thiosulfate on the photosynthetic growth of *Rhodopseudomonas palustris*, *J. Bacteriol.*, 94, 860, 1967.

23. Imhoff, J. F., Phototrophe Bakterien salzhaltiger Standorte: Ökologische und Taxonomische Aspekte, Diploma thesis, University of Bonn, 1976.

24. Madigan, M. T., Petersen, S. R., and Brock, T. D., Nutritional studies on *Chloroflexus*, a filamentous photosynthetic, gliding bacterium, *Arch. Microbiol.*, 100, 97, 1974.

25. Pfennig, N. and Trüper, H. G., The phototrophic bacteria, in *Bergey's Manual of Determinative Bacteriology*, 8th ed., Buchanan, R. E. and Gibbons, N. E., Eds., Williams & Wilkins, Baltimore, 1974, 24.

26. Imhoff, J. F. and Trüper, H. G., *Ectothiorhodospira halochloris* sp. nov., a new extremely halophilic phototrophic bacterium containing bacteriochlorophyll b, *Arch. Microbiol.*, 114, 115, 1977.

27. Imhoff, J. F., Hashwa, F., and Truper, H. G., Isolation of extremely halophilic phototrophic bacteria from the alkaline Wadi Natrun, Egypt, *Arch. Hydrobiol.*, 84, 24, 1978.

28. Keppen, O. I. and Gorlenko, V. M., A new species of purple budding bacteria containing bacteriochlorophyll b, *Mikrobiologiya*, 44, 258, 1975.

29. Imhoff, J. F., Aspekte des assimilatorischen Schwefelstoffwechsels in Rhodospirillaceae, Doctoral thesis, University of Bonn, 1980.

30. Pfennig, N., *Rhodopseudomonas globiformis*, sp. n., a new species of the Rhodospirillaceae, *Arch. Microbiol.*, 100, 197, 1974.

31. van Niel, C. B., The culture, general physiology, morphology and classification of the non-sulfur purple and brown bacteria, *Bacteriol. Rev.*, 8, 1, 1944.

32. Hansen, T. A. and Veldkamp, H., *Rhodopseudomonas sulfidophila*, nov. spec., a new species of the purple nonsulfur bacteria, *Arch. Mikrobiol.*, 92, 45, 1973.

33. Imhoff, J. F. and Trüper, H. G., Marine sponges as habitats of anaerobic phototrophic bacteria, *Microb. Ecol.*, 3, 1, 1976.

34. Rolls, J. P. and Lindstom, E. S., Induction of a thiosulfate-oxidizing enzyme in *Rhodopseudomonas palustris*, *J. Bacteriol.*, 94, 784, 1967.

35. Larsen, H., On the culture and general physiology of the green sulfur bacteria, *J. Bacteriol.*, 64, 187, 1952.

36. Meyer, J., Kelley, B. C., and Vignais, P. M., Effect of light on nitrogenase function and synthesis in *Rhodopseudomonas capsulata*, *J. Bacteriol.*, 136, 201, 1978.

37. Brown, C. M. and Herbert, R. A., Ammonia assimilation in purple and green sulphur bacteria, *FEBS Lett.*, 1, 39, 1977.

38. Brown, C. M. and Herbert, R. A., Ammonia assimilation in members of the Rhodospirillaceae, *FEBS Lett.*, 1, 43, 1977.

39. Herbert, R. A., Siefert, E., and Pfennig, N., Nitrogen assimilation in *Rhodopseudomonas acidophila*, *Arch. Microbiol.*, 119, 1, 1978.

40. Weare, N. M. and Shanmugam, K. T., Photoproduction of ammonium ion from $N_2$ in *Rhodospirillum rubrum*, *Arch. Microbiol.*, 110, 207, 1976.

41. Nagatani, H., Shimizu, M., and Valentine, R. C., The mechanism of ammonia assimilation in nitrogen fixing bacteria, *Arch. Mikrobiol.*, 79, 164, 1971.

42. Slater, J. H. and Morris, J., Light dependent synthesis of glutamate in *Rhodospirillum rubrum*, *Arch. Microbiol.*, 95, 337, 1974.

43. Johannsen, B. C. and Gest, H., Inorganic nitrogen assimilation by the photosynthetic bacterium *Rhodopseudomonas capsulata*, *J. Bacteriol.*, 128, 683, 1976.

44. Engelhardt, H., Reinigung und Charakterisierung der Glutamatdehydrogenase aus *Rhodopseudomonas sphaeroides*, Diploma thesis, University of Bonn, 1977.

45. Bachofen, R. and Neeracher, H., Glutamatdehydrogenase im photosynthetischen Bakterium *Rhodospirillum rubrum*, *Arch. Mikrobiol.*, 60, 235, 1968.
46. Siefert, E., Die Fixierung von molekularem Stickstoff bei phototrophen Bakterien am Beispiel von *Rhodopseudomonas acidophila*, Doctoral thesis, University of Göttingen, 1976.
47. Neilson, A. H. and Nordlund, S., Regulation of nitrogenase synthesis in intact cells of *Rhodospirillum rubrum* — inactivation of nitrogen fixation by ammonia, L-glutamine and L-asparagine, *J. Gen. Microbiol.*, 91, 53, 1975.
48. Munson, T. O., Burris, R. H., Nitrogen fixation by *Rhodospirillum rubrum* grown in nitrogen-limited continuous culture, *J. Bacteriol.*, 97, 1093, 1969.
49. Zumpft, W. G. and Castillo, F., Regulatory properties of the nitrogenase from *Rhodopseudomonas palustris*, *Arch. Mikrobiol.*, 117, 53, 1978.
50. Hilmer, P. and Gest, H., $H_2$ metabolism in the photosynthetic bacterium *Rhodopseudomonas capsulata*: production and utilization of $H_2$ by resting cells, *J. Bacteriol.*, 129, 732, 1977.
51. Schick, H. -J., Substrate and light dependent fixation of molecular nitrogen in *Rhodospirillum rubrum*, *Arch. Mikrobiol.*, 75, 89, 1971.
52. Meyer, J., Kelley, C. B., and Vignais, P. M., Nitrogen fixation and hydrogen metabolism in photosynthetic bacteria, *Biochimie*, 60, 245, 1978.
53. Gest, H., Kamen, M. D., and Bregoff, H. M., Studies on the metabolism of photosynthetic bacteria. V. Photoproduction of hydrogen and nitrogen fixation by *Rhodospirillum rubrum*, *J. Biol. Chem.*, 182, 153, 1950.
54. Meyer, J., Kelley, B. C., and Vignais, P. M., Aerobic nitrogen fixation by *Rhodopseudomonas capsulata*, *FEBS Lett.*, 85, 224, 1978.
55. Katoh, T., Nitrate reductase in the photosynthetic bacterium *Rhodospirillum rubrum*. Purification and properties of nitrate reductase in nitrate adapted cells, *Plant Cell Physiol.*, 4, 13, 1963.
56. Malofeeva, I. V., Bogorov, L. V., and Gogotov, I. N., Utilization of nitrates by purple bacteria, *Mikrobiologiya*, 43, 821, 1974.
57. Taniguchi, S. and Kamen, M. D., On the nitrate metabolism of facultative photoheterotrophs, in *Microalgae and Photosynthetic Bacteria*, Japanese Society of Plant Physiologists, Ed., University Press, Tokyo, 1963, 465.
58. Alef, K. and Klemme, J. -H., Characterization of a soluble NADH-independent nitrate reductase from the photosynthetic bacterium *Rhodopseudomonas capsulata*, *Z. Naturforsch.*, 32c, 954, 1977.
59. Malofeeva, I. V., Kondratieva, E. N., and Rubin, A. B., Ferrodexin-linked nitrate reductase from the phototrophic bacterium *Ectothiorhodospira shaposhnikovii*, *FEBS Lett.*, 53, 188, 1975.
60. Malofeeva, I. V. and Laush, D., Utilization of various nitrogen compounds by phototrophic bacteria, *Mikrobiologiya*, 45, 512, 1976.
61. Bast, E., Utilization of nitrogen compounds and ammonia assimilation by Chromatiaceae, *Arch. Microbiol.*, 113, 91, 1977.
62. Satoh, T., Hoshina, Y., and Kitamura, H., *Rhodopseudomonas sphaeroides f. sp. denitrificans*, a denitrifying strain as a subspecies of *Rhodopseudomonas sphaeroides*, *Arch. Microbiol.*, 108, 265, 1976.
63. Alef, K., Assimilatorische Nitratreduktion in dem phototrophen Bakterium *Rhodopseudomonas capsulata*: Struktur und Funktion der Nitratreduktase in neu isolierten Stämmen, Doctoral thesis, University of Bonn, 1978.
64. Göbel, F., Quantum efficiencies of growth, in *The Photosynthetic Bacteria*, Clayton, R. K., Sistrom, W. R., Eds., Plenum Press, New York, 1978, chap. 50.
65. Dierstein, R. and Drews, G., Nitrogen-limited continuous culture of *Rhodopseudomonas capsulata* growing photosynthetically or heterotrophically under low oxygen tensions, *Arch. Microbiol.*, 99, 117, 1974.
66. Drews, G., Die Isolierung schwefelfreier Purpurbakterien, *Zentralbl. Bakteriol. Parasitenkd. Hyg. Abt. 1, Suppl.*, 1, 170, 1965.
67. Ormerod, J. G., Ormerod, K. S., and Gest, H., Light dependent utilization of organic compounds and photoproduction of molecular hydrogen by photosynthetic bacteria; relationships with nitrogen metabolism, *Arch. Biochem. Biophys.*, 94, 449, 1961.
68. Pfennig, N., *Rhodopseudomonas acidophila*, sp. n., a new species of the budding purple nonsulfur bacteria, *J. Bacteriol.*, 99, 597, 1969.
69. Raymond, J. C. and Sistrom, W. R., *Ectothiorhodospira halophila*: a new species of the genus *Ectothiorhodospira*, *Arch. Mikrobiol.*, 69, 121, 1969.
70. Eidels, L. and Preiss, J., Carbohydrate metabolism in *Rhodopseudomonas capsulata*: enzyme titers, glucose metabolism, and glucose polymer synthesis, *Arch. Biochem. Biophys.*, 140, 75, 1970.
71. Engelhardt, H., personal communication, 1978.
72. Klemme, J. -H., Allosterische Kontrolle der Pyruvatkinase aus *Rhodospirillum rubrum* durch anorganisches Phosphat und Zuckerphosphatester, *Arch. Microbiol.*, 90, 305, 1973.

73. Jasper, P., Potassium transport system of *Rhodopseudomonas capsulata, J. Bacteriol.,* 133, 1314, 1978.

74. Sistrom, W. R., A requirement for sodium in the growth of *Rhodopseudomonas sphaeroides, J. Gen. Microbiol.,* 22, 778, 1960.

75. Grosse, W., Wachstum und Phosphathaushalt von *Rhodopseudomonas sphaeroides* unter dem Einfluss der Natrium-, Kalium- und Magnesiumversorgung, *Flora,* 153, 157, 1963.

76. Stenn, K. S., Cation transport in a photosynthetic bacterium, *J. Bacteriol.,* 96, 862, 1968.

77. Larsen, H., On the microbiology and biochemistry of the photosynthetic green sulfur bacteria, *K. Nor. Vidensk. Selsk. Skr.,* 1, 1, 1953.

78. Trüper, H. G., Culture and isolation of phototrophic sulfur bacteria from the marine environment, *Helgol. Wiss. Meeresunters.,* 20, 6, 1970.

79. Trüper, H. G. and Jannasch, H. W., *Chromatium buderi* nov. spec., eine neue Art der "großen" Thiorhodaceae, *Arch. Mikrobiol.,* 61, 363, 1968.

80. Trüper, H. G. and Genovese, S., Characterization of photosynthetic sulfur bacteria causing red water in Lake Faro (Messina, Sicily), *Limnol. Oceanogr.,* 13, 225, 1968.

81. Hashwa, F. and Trüper, H. G., Viable phototrophic sulfur bacteria from the Black-Sea bottom, *Helgol. Wiss. Meeresunters.,* 31, 249, 1978.

82. Bogorov, L. V., About the properties of *Thiocapsa roseopersicina* strain BBS, isolated from estuaria of white Sea, *Mikrobiologiya,* 43, 326, 1974.

83. Gietzen, J., Untersuchungen über marine Thiorhodaceen, *Zentralbl. Bakteriol. Parasitenkd. Infektionskr. Hyg. Abt. 2,* 83, 183, 1931.

84. Imhoff, J. F., unpublished data, 1978.

85. Hirsch, P., Ecology and morphogenesis of *Thiopedia* spp. in ponds, lakes and laboratory cultures, *Proc. 2nd Int. Symp. Photosynthetic Procaryotes,* Codd, G. A. and Stewart, W. D. P., Eds., Dundee, Scotland, 1976, 13.

86. Trüper, H. G., *Ectothiorhodospira mobilis* Pelsh, a photosynthetic bacterium depositing sulfur outside the cells, *J. Bacteriol.,* 95, 1910, 1968.

87. Tew, R. W., Photosynthetic Halophiles from Owens Lake, NASA report CR 361, 1966.

88. Raymond, J. C. and Sistrom, W. R., The isolation and preliminary characterization of a halophilic photosynthetic bacterium, *Arch. Mikrobiol.,* 59, 255, 1967.

89. Douchow, E., Douglas, H. C., *Rhodomicrobium vannielii,* a new photoheterotrophic bacterium, *J. Bacteriol.,* 58, 409, 1949.

90. Hirsch, P. and Conti, S. F., Enrichment and isolation of stalked and budding bacteria, *Zentralbl. Bakteriol. Parasitenk. Hyg. Abt. 1, Suppl.,* 1, 100, 1965.

91. Wynn-Williams, D. D. and Rhodes, M. E., Nitrogen fixation of marine photosynthetic bacteria, *J. Appl. Bacteriol.,* 37, 217, 1974.

92. Cohn, F., Untersuchungen über Bakterien II, *Beitr. Biol. Pflanz.,* 1, 141, 1875.

93. van Niel, C. B., On the morphology and physiology of the purple and green sulphur bacteria, *Arch. Mikrobiol.,* 3, 1, 1931.

94. Petrova, E. A., The morphology of purple sulfur bacteria of the genus *Chromatium* in relation to the medium, *Mikrobiologiya,* 28, 414. 1959.

95. Schlegel, H. G. and Pfennig, N., Die Anreicherungskultur einiger Schwefelpurpurbakterien, *Arch. Mikrobiol.,* 55, 245, 1966.

96. Pirson, A. and Grosse, W., Zur Rolle des Natriums im anaeroben Lichtstoffwechsel von *Rhodopseudomonas, Naturwissenschaften,* 50, 359, 1963.

97. Pfennig, N., Eine vollsynthetische Nährlosung zur selektiven Anreicherung einiger Schwefelpurpurbakterien, *Naturwissenschaften,* 48, 136, 1961.

98. Maximov, V. N., A study on optimal composition of the medium for cultivation of green sulphur bacteria, *Mikrobiologiya,* 40, 258, 1971.

99. Trentini, W. C., Defined medium allowing maximal growth of *Rhodomicrobium vannielii, J. Bacteriol.,* 94, 1260, 1967.

100. Pfennig, N. and Trüper, H. G., The Chlorobiaceae and Chromatiaceae. Habitats and isolation, in *The Procaryotes. A Handbook on Habitats, Isolation and Identification of Bacteria,* Starr, M. P., Stolp, H., Trüper, H. G., Balows, A., and Schlegel, H. G., Eds., Springer, New York, 1981.

101. Imhoff, J. F., Sahl, H. G., Soliman, G. S. H., and Trüper, H. G., The Wadi Natrun: chemical composition and microbial mass developments in alkaline brines of eutrophic desert lakes, *Geomicrobiology J.,* 1, 219, 1978.

102. Jasper, P. and Silver, S., Divalent cation transport systems of *Rhodopseudomonas capsulata, J. Bacteriol.,* 133, 1323, 1978.

103. Suter, W., Lutz, H. U., and Bachofen, R., Phosphate binding to chromatophores of *Rhodospirillum rubrum, Eur. J. Biochem.,* 67, 57, 1976.

104. Tabita, F. R. and McFadden, B. A., D-Ribulose 1,5-diphosphate carboxylase from *Rhodospirillum rubrum, J. Biol. Chem.,* 249, 3453, 1974.

105. Johannson, B. C. and Gest, H., Adenylation/deadenylation control of the glutamine synthetase of *Rhodopseudomonas capsulata*, *Eur. J. Biochem.*, 81, 365, 1977.
106. Pfennig, N. and Lippert, K. D., Über das Vitamin $B_{12}$-Bedurfnis phototropher Schwefelbakterien, *Arch. Mikrobiol.*, 55, 245, 1966.
107. Kassner, R. J. and Kamen, M. D., Trace metal composition of photosynthetic bacteria, *Biochim. Biophys. Acta*, 153, 270, 1968.
108. Udelnova, T. M., Kondratieva, E. N., and Boichenko, E. A., Iron and manganese content in various photosynthetizing microorganisms, *Mikrobiologiya*, 37, 197, 1968.
109. Udelnova, T. M., Chudina, V. I., Osnitskaya, L. K., Boichenko, E. A., Chernogorova, S. M., and Karyakin, A. V., Content of polyvalent metals in the presence of a change in metabolism of *Chromatium vinosum*, *Mikrobiologiya*, 46, 333, 1977.
110. Wiesner, W., Wachstum und Stoffwechsel von *Rhodopseudomonas sphaeroides* in Abhängigkeit von der Versorgung mit Mangan und Eisen, *Flora*, 149, 1, 1960.
111. Reiss-Husson, F., De Klerk, H., Jolchine, G., Jauneau, E., and Kamen, M. D., Some effects of iron deficiency on *Rhodopseudomonas sphaeroides* strain Y, *Biochim. Biophys. Acta*, 234, 73, 1971.
112. Jones, O. T. G., The production of magnesium protoporphyrin monomethyl ester by *Rhodopseudomonas sphaeroides*, *Biochem. J.*, 86, 429, 1963.
113. Paschinger, H., A changed nitrogenase activity in *Rhodospirillum rubrum* after substitution of tungsten for molybdenum, *Arch. Microbiol.*, 101, 379, 1974.

# RESPONSE OF PHOTOSYNTHETIC BACTERIA TO LIGHT QUALITY, LIGHT INTENSITY, TEMPERATURE, $CO_2$, $HCO_3^-$, $O_2$, AND pH

S. Morita

## LIGHT QUALITY

The near IR wavelengths of light absorbed by bacteriochlorophyll, 780 to 920 nm in Rhodospirillaceae and Chromatiaceae and 700 to 770 nm in Chlorobiaceae, are effectively utilized by photosynthetic bacteria.[1,2] Red light, 620 to 770 nm in Rhodospirillaceae and Chromatiaceae and 550 to 700 nm in Chlorobiaceae, is less effective for photosynthesis. Green light, absorbed by carotenoids from 420 to 540 nm, is utilized, but the efficiency varies from 27 to 90% among bacterial species.[3-7]

## LIGHT INTENSITY

Total photosynthetic production, such as with $CO_2$ fixation, shows a simple relationship to light intensity. The relationship between light intensity and the rate of $CO_2$-fixation fits a rectangular hyperbola.[1] The rate of $CO_2$ fixation or growth is light saturated at values near 0.5 to 4 $mW/cm^2$.[8-13] The responses of light-induced cytochrome oxidation to increasing light intensity are complicated and difficult to interpret.[8-10]

## TEMPERATURE

The photosynthetic productivity of photosynthetic bacteria is representative of the growth of bacteria in general. The responses of photosynthetic bacteria to temperature are listed in Table 1 (expressed by a range of temperature and the optimum temperature for growth).

## $CO_2$, $HCO_3^-$

All species of photosynthetic bacteria can assimilate $CO_2$ and/or $HCO_3^-$ in the light upon the addition of an electron donor, mostly under anaerobic conditions.[15-17]

## OXYGEN

Photosynthetic bacteria do not produce $O_2$ during photosynthesis, and some species of photosynthetic bacteria cannot grow in the presence of $O_2$. These grow as obligate phototrophic organisms under strictly anaerobic conditions and do not grow in dark aerobic conditions. These are all species of the family Chlorobiaceae: all species of the genera *Chromatium, Thiocystis, Thiosarcina, Thiospirillum, Lamprocystis, Thiodictyon, Thiopedia, Amoebobacter,* and *Ectothiorhodospira,* and the species *Thiocapsa pfennigii.*[14,18]

Other species of photosynthetic bacteria grow anaerobically in the light. They also can grow under microaerophilic to aerobic conditions in the dark. These species are *Rhodopseudomonas palustris, Rps. acidophila, Rps. gelatinosa, Rps. capsulata, Rps. sphaeroides, Rhodospirillum rubrum,* and *R. tenue.*[14,18]

Other species grow anaerobically in the light and under microaerophilic conditions in the dark. These species are *Rhodospirillum fulvum, R. molischianum, R. photometricum, Rhodomicrobium vannielii,* and *Thiocapsa roseopersicina.*[14,18]

## Table 1
## RESPONSE OF GROWTH TO TEMPERATURE[14]

| Organism | Range or optimum temperature (°C) |
|---|---|
| Rhodospirillaceae | |
| *Rhodospirillum* | |
| R. rubrum | 30—35 |
| R. tenue | 30 |
| R. fulvum | 30 |
| R. molischianum | 30 |
| | |
| *Rhodopseudomonas* | |
| R. palustris | 30—37 |
| R. viridis | 25—30 |
| R. acidophila | 25—30 |
| R. gelatinosa | 25—30 |
| R. capsulata | 25—30 |
| R. sphaeroides | 25—30 |
| | |
| *Rhodomicrobium* | |
| R. vannielii | 30 |
| | |
| Chromatiaceae | |
| *Chromatium* | |
| C. okenii | 25—30 |
| C. weissei | 25—30 |
| C. warmingii | 25—30 |
| C. buderi | 25—30 |
| C. minus | 25—30 |
| C. violascens | 25—30 |
| C. vinosum | 25—30 |
| C. gracile | 25—30 |
| C. minutissimum | 25—30 |
| | |
| *Thiocystis* | |
| T. violacea | 25—30 |
| T. gelatinosa | 25—30 |
| | |
| *Thiospirillum jenense* | 20—25 |
| | |
| *Thiocapsa* | |
| T. roseopersicina | 25—30 |
| T. pfennigii | 25 |
| | |
| *Lamprocystis roseopersicina* | 20—25 |
| | |
| *Thiodictyon* | |
| T. elegans | *20* |
| T. bacillosum | 20—25 |
| | |
| *Thiopedia rosea* | 20—25 |
| | |
| *Amoebobacter* | |
| | |
| A. roseus | 25—30 |
| A. pendens | 25—30 |

Table 1 (continued)
## RESPONSE OF GROWTH TO TEMPERATURE[14]

| Organism | Range or optimum temperature (°C) |
|---|---|
| *Ectothiorhodospira* | |
| *E. mobilis* | 25—30 |
| *E. shaposhnikovii* | 30—35 |
| *E. halophila* | 25—47, 44 |
| | |
| Chlorobiaceae | |
| *Chlorobium* | |
| *C. limicola* | 25—30 |
| *C. vibrioforme* | 25—30 |
| *C. phaeobacteroides* | 25—30 |
| *C. phaeovibrioides* | 25—30 |
| | |
| *Prosthecochloris aestuarii* | 25—30 |
| | |
| *Pelodictyon* | |
| *P. clathratiforme* | 20—25 |
| *P. luteolum* | 20—25 |

A blue-green mutant of Rhodospirillaceae grows under anaerobic-light or aerobic-dark conditions, but cannot grow in aerobic-light conditions. Indeed these strains are killed by aerobic-light conditions.[19]

The photosynthetic productivity of bacteria can be expressed by the rate of growth in light under anaerobic conditions.

## pH

The effects of culture media pH are listed in Table 2 as a range of pH values and as the optimum pH for growth.[14] General reviews of these environmental influences on photosynthetic bacteria are given in References 1, 12, 16, and 20.

## Table 2
## RESPONSE OF GROWTH TO pH

| Organism | pH range | pH optimum |
|---|---|---|
| Rhodospirillaceae | | |
| *Rhodospirillum* | | |
| R. rubrum | 6.0—8.5 | 6.8—7.0 |
| R. tenue | — | 6.6—7.4 |
| R. fulvum | 6.0—8.5 | 7.3 |
| R. molischianum | 6.0—8.5 | 7.3 |
| | | |
| *Rhodopseudomonas* | | |
| R. palustris | 5.5—8.5 | — |
| R. viridis | 6.3—8.0 | 6.5—7.0 |
| R. acidophila | 4.8—7.0 | 5.8 |
| R. gelantinosa | 5.5—8.5 | 7.0 |
| R. capsulata | 5.5—8.5 | 7.0 |
| R. sphaeroides | 6.0—8.5 | 7.0 |
| | | |
| *Rhodomicrobium* | | |
| R. vannielii | 5.2—7.5 | 6.0 |
| | | |
| Chromaticeae | | |
| *Chromatium* | | |
| C. okenii | 6.5—7.6 | — |
| C. weissei | 6.5—7.6 | — |
| C. warmingii | 6.5—7.6 | — |
| C. buderi | 6.5—7.6 | — |
| C. minus | 6.5—7.6 | — |
| C. violascens | 6.5—7.6 | — |
| C. vinosum | 6.5—7.6 | — |
| C. gracile | 6.5—7.6 | — |
| C. minutissimum | — | 7—8 |
| | | |
| *Thiocystis* | | |
| T. violacea | 4.3—8 | 7—7.5 |
| T. gelatinosa | 6.5—7.6 | — |
| | | |
| *Thiospirillum jenense* | 7.0—7.5 | — |
| | | |
| *Thiocapsa* | | |
| T. roseopersicina | — | 7.0—7.5 |
| T. pfennigii | 6.5—7.5 | — |
| | | |
| *Lamprocystis roseopersi* cina | 6.5—7.6 | — |
| | | |
| *Thiodictyon* | | |
| T. elegans | 6.5—7.6 | — |
| T. bacillosum | 6.5—7.5 | — |
| | | |
| *Amoebobacter* | | |
| A. roseus | 6.5—7.5 | — |
| A. pendens | 7.0—7.5 | — |
| | | |
| *Ectothiorhodospira* | | |
| E. mobilis | — | 7.5—8.0 |
| E. shaposhnikovii | 8.0—8.5 | — |
| E. halophila | 7.6—8.0 | |

## Table 2 (continued)
## RESPONSE OF GROWTH TO pH

| Organism | pH range | pH optimum |
|---|---|---|
| Chlorobiaceae | | |
| *Chlorobium* | | |
| *C. limicola* | 6.0—7.0 | 6.8 |
| *C. vibrioforme* | 6.0—7.5 | — |
| *C. phaeobacteroides* | 6.0—7.5 | — |
| *C. phaeovibrioides* | 6.0—7.5 | — |
| *Prosthecochloris aestuarii* | — | 6.7—7.0 |
| *Pelodictyon* | | |
| *P. clathratiforme* | 6.5—7.0 | — |
| *P. luteolum* | 6.5—7.0 | — |

## REFERENCES

1. **Gobel, F.,** Quantum efficiency of growth, in *The Photosynthetic Bacteria,* Clayton, R. K. and Sistrom, W. R., Eds., Planum Press, New York, 1979, 907.
2. **French, C. S.,** The rate of $CO_2$ assimilation by purple bacteria at various wavelengths of light, *J. Gen. Physiol.,* 21, 711, 1937.
3. **Duysens, L. N. M.,** Transfer of Excitation Energy in Photosynthesis, Thesis, Kemink, Utrecht, 1952.
4. **Goedheer, J. C.,** Energy transfer between carotenoids and bacteriochlorophyll in chromatophores of purple bacteria, *Biochim. Biophys. Acta.,* 35, 1, 1959.
5. **Olson, J. M. and Nadler, K. D.,** Energy transfer and cytochrome function in a new type of photosynthetic bacterium, *Photochem. Photobiol.,* 4, 783, 1965.
6. **Olson, J. M. and Sybesma, C.,** Energy transfer and cytochrome oxidation in green bacteria, in *Bacterial Photosynthesis,* Gest, H., San Pietro, A., and Vernon, L. P., Eds., Antioch Press, Yellow Springs, Ohio, 1963, 413.
7. **Nishimura, M. and Takamiya, A.,** Analyses of light-induced bacteriochlorophyll absorbance change and fluorescence emission in purple bacteria, *Biochim. Biophys. Acta.,* 120, 34, 1966.
8. **Olson, J. M. and Chance, B.,** Oxidation-reduction reactions in the photosynthetic bacterium *Chromatium* I. Absorption spectrum changes in whole cells, *Arch. Biochem. Biophys.,* 88, 26, 1960.
9. **Morita, S., Edwards, M. L., and Gibson, J.,** Influence of metabolic conditions of light-induced absorbancy changes in *Chromatium* D, *Biochim. Biophys. Acta,* 109, 45, 1965.
10. **Morita, S., Olson, J. M., and Conti, S. F.,** Light-induced reactions of cytochromes and carotenoids in *Rhodomicrobium vannielii Arch. Biochem. Biophys.,* 104, 346, 1964.
11. **Morita, S., Suzuki, K., and Takashima, S.,** On the cause of the S-shaped rate - light intensity - relationship in the photosynthesis of purple bacteria, *J. Biochem.,* 38, 255, 1951.
12. **Pfennig, N.,** General physiology and ecology of photosynthetic bacteria, in *The Photosynthetic Bacteria,* Clayton, R. K. and Sistrom, W. R., Eds., Plenum Press, New York, 1979, 3.
13. **Uemura, T., Suzuki, K., Nagano, K., and Morita, S.,** Comparative studies on growth, respiration, photosynthesis, and pigment content in *Rhodopseudomonas palustris, Plant Cell Physiol.,* 2, 451, 1961.
14. **Pfennig, N. and Truper, H. G.,** The phototrophic bacteria, in *Bergey's Manual of Determinative Bacteriology,* 8th ed., Buchanan, R. E. and Gibbons, N. E., Eds., Williams & Wilkins, Baltimore, 1974, 24.
15. **Fuller, R. C.,** Photosynthetic carbon metabolism in the green and purple bacteria, in *The Photosynthetic Bacteria,* Clayton, R. K. and Sistrom, W. R., Eds., Plenum Press, New York, 1979, 691.
16. **Pfennig, N.,** Photosynthetic bacteria, *Annu. Rev. Microbiol.,* 21, 285, 1967.
17. **Bose, S. K.,** Media for anaerobic growth of photosynthetic bacteria, in *Bacterial Photosynthesis,* Gest, H., San Pietro, A., and Vernon, L. P., Eds., Antioch Press, Yellow Springs, Ohio, 1963, 501.
18. **Keister, D. L.,** Respiration vs. photosynthesis, in *The Photosynthetic Bacteria,* Clayton, R. K. and Sistrom, W. R., Eds., Plenum Press, New York, 1979, 849.

19. **Griffiths, M., Sistrom, W. R., Cohn-Bazire, G., and Stanier, R. Y.,** Function of carotenoids in photosynthesis, *Nature (London),* 176, 1211, 1955.
20. **van Niel, C. B.,** The culture, general physiology, morphology, and classification of the non-sulfur purple and brown bacteria, *Bacteriol. Rev.,* 8, 1, 1944.

# CULTURE CONDITIONS OF PHOTOSYNTHETIC MICROORGANISMS

## H. Senger

The growth of photosynthetic microorganisms under autotrophic conditions depends on several parameters: light, $CO_2$, temperature, inorganic nutrients, and pH. It is relatively easy to keep a constant temperature and $CO_2$ content, but the intensity of the light and the concentration of nutrients available to the cells, together with the pH, change considerably during cell growth. The average light intensity available to a cell depends on the shading by other cells, i.e., the density of the culture. During a 14-hr growth period of a *Scenedesmus* culture, the cell mass increased about ten times, and the light intensity measured behind the culture vessel decreased 60% at the same time.[1] During the same period, other growth parameters did not become limiting factors. Accordingly, the selection of the light regime and the density of the cultures determine the three general types of cultures: the batch culture, the homocontinuous culture, and the synchronous culture.[2,3]

A batch culture is most commonly used for photosynthetic studies, but its cell population is most difficult to define. This type of culture can be obtained easily by inoculation of a culture vessel containing fresh nutrient medium with an undefined amount of cells of any developmental stage. If the intervals of subculturing and the inoculum are kept constant, the cultures are fairly well reproducible. Frequent subculturing also reduces the accumulation of metabolites in the medium, the formation of foam, and contamination of the culture. The culture will proceed through the different growth stages, reach the stationary phase after either light intensity or nutrients become the limiting factor, and finally the activity will decay. The advantage of such cultures is the easy handling. When the cultures are used at the beginning of the stationary growth phase, the cells of the cultures yield relatively good reproducible results. Batch cultures are suitable to study growth kinetics and parameters influencing cell growth. In addition, the only way to study aging of microorganisms is to use batch cultures beyond the stationary growth phase.[4,5]

To obtain a homocontinuous culture, all growth factors have to be kept constant. This requires that the cell density of the culture be controlled and always diluted to the same concentration. This can be achieved with either a chemostat[6,7] or a turbidostat.[8,9] The chemostat measures and controls one component of the nutrient solution, mostly pH. Changes are compensated for by a regulated continuous dilution. The turbidostat operates via a photoelectrical control of the density of the culture. When a present density is reached, a solenoid valve will be operated, and fresh nutrient medium flows into the culture vessel. In addition to the regulation of the density by dilution, the chemostat or turbidostat prevent major changes in the culture medium and the pH.

In such a homocontinuous culture, the beginnings of the individual life cycles of the cells depend only upon statistical distribution. The culture is completely randomized and the whole population kept in a steady state and continuously in the logarithmic growth phase. The advantage of a homocontinuous culture is that samples taken at any time are identical and most reproducible over a long period.[2] However, one has to be aware that not all microorganisms survive continuous photosynthetically saturating illumination. Much consideration has been given to the continuous cultivation of microorganisms, and several symposia with a wealth of literature have been dedicated to this topic.[10-15]

A synchronous culture requires that all life cycles of the individual cells be initiated simultaneously. This requires at least one discontinuous growth factor which causes synchronization. In unicellular algae, the synchronizing factor is represented by the dark-light change combined with the dilution to a preset cell number. This dilution can either be carried out by hand[16] or by a turbidostat.[1] Some algae, like *Bumilleriopsis*,[17] *Dunaliella*, and *Cyanidium*,[18,19] need a combination of dark-light regime and temperature change to become synchronized. *Anacystis* can either be synchronized by dark-light regime[20] or by a temperature change.[21] Additional methods for synchronization of algae are cell starvation[22] and mechanical fractionation of cell populations.[23] Bacteria generally can be synchronized by temperature changes or temporary application of inhibitors.

The synchronizing procedure must be adapted to each species, particularly the lengths of the "free running" life cycle has to be considered. The quality of the synchronization achieved should be evaluated by the following criteria:[2,24]

1.    Complete and permanent synchronization
2.    Homogeneity of the developmental stages
3.    Shortest life cycle and high productivity
4.    Nonsusceptibility of the life cycle to the synchronizing procedure

The advantages of using synchronized cultures are manifold. They have excellent reproducibility. The whole culture represents a single cell during its life cycle. Activity changes, biosynthetic pathways, sensitive stages, and other physiological parameters linked to the cell cycle can be studied. Last, but not least, synchronized cultures are an excellent tool to study some phenomena of the endogenous rhythms.[25,26] Detailed methods for cultivation and synchronization of photosynthetic microorganisms have appeared.[27]

## REFERENCES

1. Pfau, J., Werthmüller, K., and Senger, H., Permanent automatic synchronization of microalgae achieved by photoelectrically controlled dilution, *Arch. Mikrobiol.*, 75, 338, 1971.
2. Senger, H., Pfau, J., and Werthmüller, K., Continuous automatic cultivation of homocontinuous and synchronized microalgae, in *Methods in Enzymology,* Vol. 5, San Pietro, A., Ed., Academic Press, New York, 1972, 301.
3. Bishop, N. I. and Senger, H., Preparation and photosynthetic properties of synchronous cultures of *Scenedesmus*, in *Methods in Enzymology*, Vol 23A, San Pietro, A., Ed., Academic Press, New York, 1971, 53.
4. Kulandaivelu, G. and Senger, H., Changes in the reactivity of the photosynthetic apparatus in heterotrophic ageing cultures of *Scenedesmus obliquus*. I. Changes in the photochemical activities, *Physiol. Plant.*, 36, 157, 1976.
5. Kulandaivelu, G. and Senger, H., Changes in the reactivity of the photosynthetic apparatus in heterotrophic ageing cultures of *Scenedesmus obliquus*. II. Changes in ultrastructure and pigment composition, *Physiol. Plant.*, 36, 165, 1976.
6. Monod, J., La technique de culture continue, *Ann. Inst. Pasteur,* 79, 390, 1950.
7. Myers, J. and Graham, I. R., On the mass culture of algae. II. Yield as a function of cell concentration under continuous sunlight irradiance, *Plant Physiol.*, 34, 345, 1959.
8. Myers, J. and Clark, L. B., Apparatus for the continuous culture of *Chlorella*, *J. Gen. Physiol.*, 28, 103, 1944.
9. Herbert, D., Some principles of continuous culture, in *Recent Progress in Microbiology,* Blackwell Scientific, Oxford, 1958, 381.
10. Burlew, J. S., Algal culture, from laboratory to pilot plant, *Carnegie Inst. Washington Publ.*, Nr. 600, Washington, D.C., 1953.
11. Malek, I., Beran, K., and Hospodka, J., *Continuous Cultivation of Microorganisms,* Academic Press, New York, 1964.

12. Malek, I. and Fenel, Z., *Theoretical and Methodical Basis of Continuous Cultivation of Microorganisms,* Academic Press, New York, 1966.

13. Wattanabe, A. and Hattori, A., *Cultures collections Algae,* Proc. U.S.-Jap. Conf., Japanese Soc. of Plant Physiologists, Kyoto, Japan, 1966.

14. Powell, E. D., Evans, C. G. T., Strange, R. E., and Tempest, D. W., in Proc. 3rd Int. Symp. Microbial Physiology Continuous Culture, Her Majesty's Stationary Office, London, 1967.

15. Malek, I., Beran, K., and Hospodka, J., *Continuous Cultivation of Microorganisms,* Academic Press, New York, 1969.

16. Lorenzen, H., Synchrone Zellteilungen von *Chlorella* bei verschiedenen Licht-Dunkel-Wechseln, *Flora,* 144, 32, 1957.

17. Hesse, M., Wachstum und Synchronisierung der Alge *Bumilleriopsis filiformis* Vischer (Xanthophyceae), *Planta,* 120, 135, 1974.

18. Wegmann, K. and Metzner, H., Synchronization of *Dunaliella* cultures, *Arch. Mikrobiol.,* 78, 360, 1971.

19. Möller, M. and Senger, H., Photosyntheseleistung synchroner Kulturen von *Cyanidium caldarium,* *Ber. Dtsch. Bot. Ges.,* 85, 391, 1972.

20. Lorenzen, H. and Konshik, B. D., Experiments with synchronous *Anacystis nidulans,* *Ber. Deutsch. Bot. Ges.,* 89, 491, 1976.

21. Lorenzen, H. and Venkataraman, G. S., Synchronous cell division in *Anacystis nidulans* Richter, *Arch. Mikrobiol.,* 67, 251, 1969.

22. Coombs, J., Halicki, P. J., Holm-Hansen, O., and Volcani, B. E., Studies on the biochemistry and fine structure of silica shell formation in diatoms. II. Changes in concentration of nucleoside triphosphates in silicon-starvation synchrony of *Navicula pelliculosa* (Brĕb.) Hilse, *Exp. Cell Res.,* 47, 315, 1967.

23. Tamiya, H., Iwamura, T., Shibata, K., Hase, H., and Nihei, T., Correlation between photosynthesis and light-independent metabolism in the growth of *Chlorella,* Biochim. Biophys. Acta, 12, 23-1953.

24. Lorenzen, H., in *Photobiology of Microorganisms,* Per Halldal, Ed., Wiley Interscience, London, 1970, 187.

25. Prĕzelin, B. B., Meeson, B. W., and Sweeney, B. M., Characterization of photosynthetic rhythms in marine dinoflagellates. I. Pigmentation, photosynthetic capacity and respiration, *Plant Physiol.,* 60, 384, 1977.

26. Prĕzelin, B. B. and Sweeney, B. M., Characterization of photosynthetic rhythms in marine dinoflagellates. II. Photosynthesis-irradiance curves and *in vivo* chlorophyll *a* flourescence, *Plant Physiol.,* 60, 388, 1977.

27. Photosynthesis, in *Methods in Enzymology,* Vol. 23A, San Pietro, A., Ed., Academic Press, New York, London, 1971.

# RESPONSE OF MACROALGAE TO LIGHT QUALITY, LIGHT INTENSITY, TEMPERATURE, $CO_2$, $HCO_3^-$, $O_2$, MINERAL NUTRIENTS, AND pH

William N. Wheeler

## RESPONSE TO LIGHT QUALITY AND LIGHT INTENSITY

The optical properties of seawater in which macroalgae grow are determined by the optical properties of water and by various dissolved and particulate substances.[1] Scattering and absorption of light by these substances are wavelength dependent.[1] The result is that with increasing depth in the sea, light becomes increasingly monochromatic (either blue 450- to 475-nm peak transmittance) in places like the eastern Mediterranean or blue-green, green in temperate coastal waters (500- to 575-nm peak transmittance).[1] Algae which live several meters beneath the surface must utilize light of wavelengths which are not efficiently absorbed by chlorophyll$a$ (chl$a$; see Table 1). These algae must depend on accessory pigments to capture the incident irradiation. The red algae (Rhodophyta) contain phycocyanin and phycoerythrin (see Table 1), both of which absorb light in the green region of the spectrum. The brown algae (Phaeophyta) contain fucoxanthin, a brown pigment which absorbs in the blue-green region (see Table 1), and chlorophyll$c$ (chl$c$) which absorbs in the yellow-red (see Table 1). These accessory pigments have been shown to transfer captured energy efficiently to chl$a$ (see Figure 1).[2-4] For example, the transfer of energy between fucoxanthin and chl$a$ is greater than 90% and between chl$c$ and chl$a$, 55%.[2]

The adjustment of these accessory pigments and chl$a$ in the photosynthetic unit to produce an optimum rate of apparent photosynthesis (AP) (net photosynthesis = AP) under conditions of changing light quality and quantity can be termed qualitative and quantitative adaptation. Uniform changes in the size or number of these units can be called quantitative adaptation. If, however, the ratios of the pigments within these units change, qualitative adaptation has occurred. *Ulva, Codium, Porphyra*, and *Chondrus* have been shown to increase pigment content with increasing depth.[5] Further, the ratio of phycobiliproteins and chlorophyll$b$ (chl$b$) to chl$a$ increases with depth.[5] Plants transferred from depths to the surface decrease pigment concentration and resynthesize them upon transfer back to a low irradiance environment.[5] These pigment changes enhance the AP capacity as well as the photosynthetic efficiency[6-9] in limiting irradiance environments. Seaweeds transplanted from the intertidal to low irradiance environments were found to change only the quantity of pigment.[5] However, in another study,[10] algae from grottoes and shaded environments were also found to increase pigmentation with increasing shade, although shaded species were also found to increase ratios of chl$b$ and phycobiliproteins to chl$a$, carotenoids to chl$a$ and chl$c$ to chl$a$ decreased with increasing shade. *Fucus* and *Ascophyllum* have also been shown to decrease carotenoids to chl$a$ ratios with increasing depth in the sea, but no clear photosynthetic advantage was found.[11] These pigment changes can be rapid, occurring in less than 4 hr,[12] and need not occur simultaneously with cell division.[5] Although these pigments are alterable, the limits of the alterations are probably genetically fixed.[13]

Photosynthesis in algae is to a certain extent proportional to light intensity. When irradiance levels are low, the AP rate is directly proportional to the light intensity. However, as the intensity increases, this proportionality becomes less as other factors become more important. The efficiency with which a plant can utilize low-level irradiance can be approximated from the initial slope of the AP-light intensity curve, and

Table 1
IMPORTANT MACROALGA PIGMENTS AND THEIR IN
VITRO ABSORPTION PEAKS[100]

| Pigment | In vitro peaks (nm) | Chlorophyta | Phaeophyta | Rhodophyta |
|---|---|---|---|---|
| Chlorophylls | | | | |
| Chl *a* | 665,432[a] | X | X | X |
| Chl *b* | 650, 475[a] | X | | |
| Chl *c* | 624, 585, 450[a] | | X | |
| | | | | |
| Phycobiliproteins | | | | |
| Phycocyanin | 553, 615[b] | | | X |
| Phycoerythrin | 498, 540, 566[b] | | | X |
| | | | | |
| Xanthophylls | | | | |
| Fucoxanthin | 453[c] | | X | |
| Lutein | 446, 476[c] | X | | X |
| Neoxanthin | 437, 467[c] | X | | |
| Siphonein | 460[c] | X | | |
| Violaxanthin | 443, 472[c] | X | X | |
| Zeaxanthin | 452, 483[c] | X | | X |

[a]   Extracted with methanol.
[b]   Extracted with ethanol.
[c]   Extracted with water.

indirectly from the $K_s$ which is the light intensity which gives half of the maximum AP response. The smaller the $K_s$, the more efficient the plant is at harvesting the low-level incident irradiation. Plants which normally live in high irradiance environments, where light is not normally limiting, usually have a rather high $K_s$ (see Table 2). Plants which live in shade environments, or deep water, must be efficient and have correspondingly smaller $K_s$ values (see Table 2). The sublittoral species generally have lower $K_s$ values than corresponding eulittoral species.[7-9] This is, however, not always the case. *Macrocystis* fronds have blades which span the entire water column (usually < 20 m). Blades just above the holdfast in 12 m of water should behave as "shade" plants. Blades near the surface should behave as "sun" plants. The data in Table 3 indicate that this is not the case. Here, the photosynthetic characteristics appear to be controlled almost entirely by ontogeny. Ontogenetic responses are not limited to this giant kelp, but are a general phenomenon.[14-16]

The compensation point is that irradiance for which the AP rate is just large enough to compensate for the respiration rate. This point varies with the morphology of the alga, and, because it is linked to respiration, with the water temperature (see Table 2). Seasonal studies on a number of different algae indicate that temperature adaptation of respiration does occur.

The maximum photosynthetic rate, $P_{max}$, is not light dependent. This rate is generally controlled by the rate of fixation of carbon which is a light-independent enzyme-mediated process. Seasonal variations noted in the photosynthetic capacity of some *Laminaria* species[17,18] can be partially explained by the seasonal variation in the activity of ribulose 1,5-bisphosphate carboxylase. The activity of RuBP-Case in *Laminaria* appears to be a function of the external temperature and nitrogen nutrition.[19] Availability of inorganic carbon can also play a role in the determination of $P_{max}$ in some species of algae, such as *Macrocystis* and *Carpophyllum*.[20,101]

The structure of the thallus may play a very important role in the photon capture process. Sheet-like and finely divided forms show much higher productivity than crus-

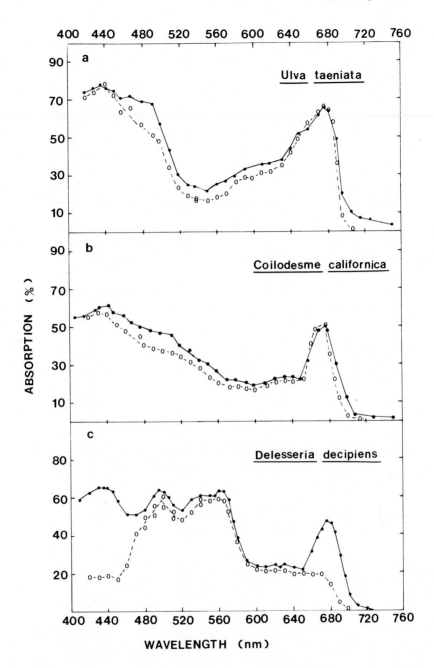

FIGURE 1.   Absorption spectra (●) and action spectra (○) for (A) a green alga (*Ulva taeniata*), (B) a brown alga (*Coilodesme californica*), (C) a red alga (*Delesseria decipiens*). The absorption and action spectra for A and B are drawn to coincide at 675 nm.[3]

tose forms.[21,22] Short-lived annual species rather than perennials, eulittoral rather than sublittoral, and sheet-like or filamentous forms rather than coarsely branched forms also have higher productivities.[21] Individuals which are optically opaque are better suited to low-light environments than translucent ones.[23] *Codium*, an opaque green, may also possess light guides in the form of air-filled utricles which may enhance the capture of incident light.[23] Deep water individuals of *Halimeda tuna*, as compared to

Table 2
## PHOTOSYNTHESIS — IRRADIANCE RELATIONSHIPS FOR SELECTED MACROALGAE

| Alga | Depth (m) | $I_{sat}$ | $K_s$ | Comp. $(\mu Ein\ m^{-2}sec^{-1})^a$ | $p_{max}$ | Ref. |
|---|---|---|---|---|---|---|
| **Chlorophyta** | | | | | | |
| *Monostroma nitidum* | 0 | 560 | 120 | <10 | 1.1 $\mu mol\ O_2\ cm^{-2}\ hr^{-1}$ | 8 |
| *Cladophora* sp. | 0 | 300 | 60 | <10 | 625 $\mu mol\ O_2\ gdw^{-1}\ hr^{-1}$ | 8 |
| *Ulva pertusa* | −1 | 300 | 100 | <10 | 1.3 $\mu mol\ O_2\ cm^{-2}\ hr^{-1}$ | 8 |
| *Cladophora wrightiana* | −5 | 100 | 20 | <10 | 134 $\mu mol\ O_2\ gdw^{-1}\ hr^{-1}$ | 8 |
| *Ulva japonica* | −10 | 160 | 40 | <10 | 1.3 $\mu mol\ O_2\ cm^{-2}\ hr^{-1}$ | 8 |
| **Phaeophyta** | | | | | | |
| *Fucus* sp. | 0 | 720 | 310 | — | 410 $\mu mol\ O_2\ gdw^{-1}\ hr^{-1}$ | 26 |
| *Fucus* sp. | 0 | 700 | 75 | 2 | 194 $\mu mol\ O_2\ gdw^{-1}\ hr^{-1}$ | 21 |
| *Sargassum ringgoldianum* | −1 | 300 | 55 | <10 | — | 7 |
| *Undaria pinnatifida* | −1 | 175 | 60 | 5 | — | 7 |
| *L. hyperborea* | −2 | 190 | 48 | — | 1.3 $\mu mol\ C\ cm^{-2}\ hr^{-1}$ | 27 |
| *Undaria peterseniana* | −15 | 140 | 50 | 5 | — | 7 |
| **Rhodophyta** | | | | | | |
| *Porphyra suborbiculata* | 0 | 360 | 110 | — | 1.6 $\mu mol\ O_2\ cm^{-2}\ hr^{-1}$ | 9 |
| *Chondrus verrucosa* | −1 | 250 | 70 | 5 | — | 9 |
| *Ptilota serrata* | −6 | 182 | 42 | <7 | 126 $\mu mol\ O_2\ gdw^{-1}\ hr^{-1}$ | 28 |
| *P. serrata* | −14 | 116 | 42 | <7 | 96 $\mu mol\ O_2\ gdw^{-1}\ hr^{-1}$ | 28 |
| *P. serrata* | −27 | 116 | 20 | <7 | 54 $\mu mol\ O_2\ gdw^{-1}\ hr^{-1}$ | 28 |
| *Plocamium telfairiae* | −8 | 100 | 15 | 5 | — | 9 |
| *Meristotheca papulosa* | −10 | 150 | 40 | 5 | — | 9 |
| *Phyllophora truncata* | −14 | 116 | 42 | <7 | 129 $\mu mol\ O_2\ gdw^{-1}\ hr^{-1}$ | 28 |

[a]   Irradiances converted using the relation: 250 lx = 1 W m$^{-2}$ = 5 $\mu Ein\ m^{-2}\ sec^{-1}$. [29]

Table 3
## PHOTOSYNTHETIC CHARACTERISTICS OF BLADES LOCATED ALONG A SINGLE *M. PYRIFERA* FROND[16]

| Distance above holdfast (m) | Initial slope of PS curve (nmol $O_2 h^{-1} \mu Ein^{-1}$ m$^2$ sec) | | $K_s$ ($\mu mol\ P\ m^{-1}\ sec^{-1}$) | | $P_{max}$ (nmol $O_2\ hr^{-1})^a$ | | Chl $a$ (nmol cm$^{-2}$) |
|---|---|---|---|---|---|---|---|
| | cm$^{-2}$ | chl $a^{-1}$ | cm$^{-2}$ | chl $a^{-1}$ | cm$^{-2}$ | chl $a^{-1}$ | |
| 4.5 | 14 | 1.4 | 85 | 89 | 825 | 88.7 | 10.4 |
| 5.5 | 17.8 | 1.7 | 41 | 39 | 744 | — | — |
| 6.5 | 16.2 | 1.6 | 62 | 61 | 684 | 77.7 | 9.9 |
| 8.5 | 21.9 | 1.9 | 23 | 23 | 675 | 74.2 | 10.2 |
| 10.5 | 18.1 | 2.0 | 41 | 40 | 819 | 99.9 | 9.2 |
| 12.2 | 22.2 | 2.4 | 33 | 34 | 1040 | 129 | 9.1 |
| 13.5 | 21.8 | 2.6 | 21 | 23 | 959 | 120 | 8.3 |
| 15.5 | 14.4 | 3.7 | 26 | 26 | 463 | 132 | 3.9 |

[a]   Measured at 500 $\mu Ein\ m^{-2}\ sec^{-1}$.

those individuals living near the surface, have been shown to have thinner segments, larger utricles, and thinner cell walls, allowing a larger exposed surface area.[24] Chloroplast structure of these individuals remained unchanged. Many of the Phaeophyta apparently have the ability to orient their chloroplasts to obtain an optimum surface area in shaded environments and a minimum surface area in bright environments.[25]

## Table 4
## $Q_{10}$ VALUES OF APPARENT
## PHOTOSYNTHESIS FOR SELECTED ALGAL
## SPECIES FROM WINTER AND SUMMER[a]

| Alga | $Q_{10}$[b] summer | $Q_{10}$[b] winter |
|------|------|------|
| Chlorophyta | | |
| *U. pertusa* | 2.7 | 2.0 |
| *Enteromorpha compressa* | 1.8 | 1.6 |
| | | |
| Phaeophyta | | |
| *Ishige okamurai* | 1.9 | 1.4 |
| *Hizikai fuciformis* | 1.9 | 1.4 |
| *Padina arborescens* | 2.5 | 2.0 |
| | | |
| Rhodophyta | | |
| *Gymnogongrus flabelliformis* | 5.0 | 2.1 |
| *Gloiopeltis complanata* | 2.9 | 1.6 |
| *Gelidium amansii* | 3.3 | 2.1 |

[a]   Approximated from the data of Yokohama.[30]
[b]   $Q_{10} = AP_{20}/AP_{10}$.

Although light is probably the most important single factor in the regulation of growth, reproduction and morphogenesis in macroalgae, it is by no means the only factor. For *Laminaria longicruris* only, 61% of the variability in the *in situ* AP could be accounted for by light.[18]

## RESPONSE TO TEMPERATURE

Photosynthesis involves light reactions which are not temperature dependent. The products of the light reactions are utilized in enzymatic reactions which are temperature sensitive. This temperature-dependent phase can limit the overall photosynthetic process. Changes in the photosynthetic capacity and the compensation point are end results of the action of temperature on enzyme-catalyzed steps such as ribulose bisphosphate carboxylase. Changes in the compensation point may be due to temperature-enhanced metabolic processes such as changes in the respiration rate.

Metabolic processes, over a limited temperature range, generally increase logarithmically with temperature, approximately doubling with an increase of 10°C. This increase is called the $Q_{10}$, and its value gives a reasonable indication of the temperature sensitivity of a metabolic process. Changes in the $Q_{10}$ with corresponding changes in the ambient temperature infer a change in the metabolic process and may indicate an adaptive response. Table 4 shows photosynthetic $Q_{10}$ for a number of different algae during both summer and winter. The $Q_{10}$ in these cases was approximated from increases in the AP rate between 10 and 20°C. The $Q_{10}$ values were consistently higher in summer than in winter, demonstrating a higher thermal sensitivity during summer.[30] Temperature optima for the algae shown were between 28 and 32°C in summer and for the red and green algae were between 25 and 30°C in winter.[30] The brown algae in the winter were more sensitive, with optima between 20 and 23°C.[30] The difference in optimum temperature between winter and summer algae was in the range of 5 to 10°C in the brown algae and less than 5°C in the green and most of the red algae. The difference in temperature of the seawater between winter and summer was usually more than 10°C.[30]

Comparison of algal temperature responses from colder and warmer regions[31] indi-

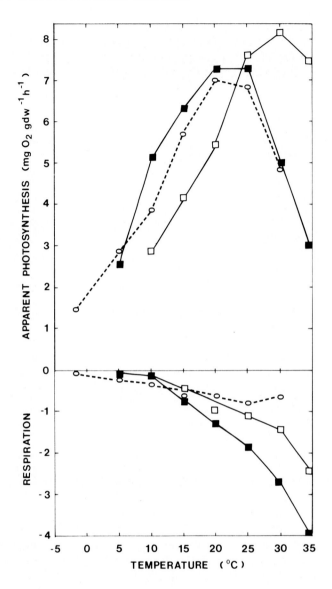

FIGURE 2.    AP  and  respiration  rates  for  an  arctic  *Fucus*  sp.[32]  (−O−) and a temperate *Ischige* sp.[30] in both summer (□) and winter (■).

cates that green algae are generally more heat resistant, possibly because of the tropical origin of the green algae,[31] whereas brown algae are more heat susceptible, again possibly because of their more temperate origin.[31] In general, optimum temperature for AP is shifted in the direction of the growth temperature. In the colder region of Japan,[31] optimum temperatures generally lay 2 to 4°C lower than for the same species in the southern parts of Japan, which is generally 5°C warmer.

Artic algae collected beneath the ice in the Bering Sea[32] showed temperature responses similar to those from more temperate climates.[26] Figure 2 shows the AP response to temperature of *Fucus* sp. from the Arctic and a brown alga from southern Japan. Algae from the arctic were able to carry out a considerable amount of AP, even at temperatures as low as −1.5°C. Similar results were found by Kanwisher with *Fucus* and *Laminaria*.[26]

### Table 5
### $Q_{10}$ VALUES OF RESPIRATION FOR
### SELECTED ALGAL SPECIES FROM WINTER
### AND SUMMER[33]

| Alga | $Q_{10}^a$ summer | $Q_{10}^a$ winter |
|---|---|---|
| Chlorophyta | | |
| *Enteromorpha intestinalis* | 1.09 | 2.83 |
| *U. lactuca* | 1.12 | 1.92 |
| Phaeophyta | | |
| *Fucus (ceranoides?)* | 1.17 | 2.75 |
| Rhodophyta | | |
| *C. crispus* | 1.23 | 2.6 |
| *Porphyra umbilicalis* | 1.06 | — |
| *Griffithsia flosculosa* | 1.22 | 2.63 |

$^a$ $Q_{10} = R_{20}/R_{10}$

The compensation point is that light level at which AP approaches the respiration rate. For algae growing under light-limited conditions, an adaptation in the respiration rate is as important as an adaptation in AP. The $Q_{10}$ for respiration for a number of seaweeds is higher in winter than in summer,[26,30,33] with the exception of *Laminaria saccharina* in which this trend is reversed[34] (see Table 5). This increase in $Q_{10}$ has been interpreted as an adaptation to summer temperatures.[33] Relatively low $Q_{10}$ values (1.2) in summer demonstrate that the respiration in these seaweeds is almost unaffected by fluctuations in temperature.[33] Newell and Pye[33] also concluded that the shallowest portion of the respiration-temperature curves appeared in the ambient temperature region. Algae in winter respire at about the same rate as at 10°C higher in the summer. Since summer temperatures are near 10 to 20°C warmer, respiration is still lower in winter than in summer. Respiration rates for arctic species are unusually low, less than half that shown by more temperate species.[32] Long winter months under the ice without light demand a low respiration rate.[26]

Since chemical reaction rates decrease with lowered temperatures, the photosynthetic enzyme activity is also likely to decrease. Ribulose bisphosphate carboxylase, which has been found to be closely associated with AP capacity in higher plants,[35] increases in activity with increases in assay and growth temperature.[36] The low activities of this enzyme in the winter may be a factor in the limitation of AP rate at low temperatures encountered in the winter. Ribulose bisphosphate carboxylase activity for *Laminaria hyperborea* follows an annual cycle (see Figure 3). Activities are much higher in the winter months than in summer. Küppers[19] has concluded that this annual cycle is predominantly controlled by ambient temperature. To maintain a reasonably high AP rate in winter when water temperature approaches 3°C, the plant must increase synthesis of the enzyme to compensate. In another study with field-grown *L. longicruris* plants, Hatcher et al.[18] found that 56% of the variability in light-saturated AP could be accounted for by temperature.

The effect of temperature on the AP of marine macroalgae has not been fully studied. We can say at present that there is AP adaptation as well as respiration adaptation to temperature. However, the biochemical, physiological, or genetic mechanisms for this adaptation have not been studied. The degree of adaptation shown, however, by summer and winter plants and arctic and temperate plants is no greater than expected for physiological acclimation as opposed to genetic adaptation.

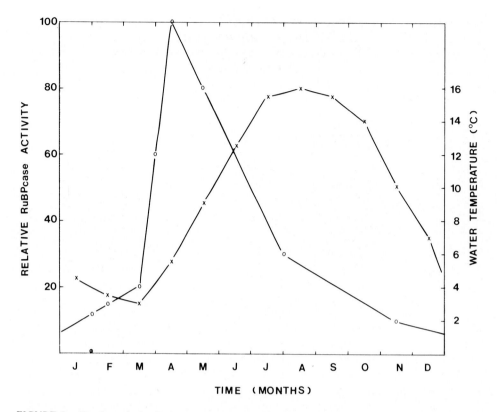

FIGURE 3.   Yearly cycle in ribulose bisphosphate carboxylase activity[19] (O) as measured in vitro at 25°C (X) and Helgoland water temperatures[37] throughout the year. Monthly means for (1965 to 1975) are shown.

## RESPONSE TO $CO_2$, $HCO_3^-$

As the substrate for photosynthesis, $CO_2$ is one of the most important factors governing the rate of AP. In the ocean, not only $CO_2$, but two other forms of inorganic carbon, $HCO_3^-$ and $CO_3^{2-}$, are present. At pH 8, the normal pH for seawater, $HCO_3^-$ is the predominant form, comprising well over 99% of the inorganic carbon. Thus, it is not surprising to find that many marine algae are able to utilize $HCO_3^-$ as a carbon source for photosynthesis. Table 6 lists a number of species which have been shown to utilize $HCO_3^-$ and those which have been shown to utilize only $CO_2$. Those algae which must rely on $CO_2$ alone have only about 10 $\mu$mol $L^{-1}$ available for photosynthesis.[38] Those which can utilize all three forms have about 2 mmol $L^{-1}$ inorganic carbon available.[39] Land plants must be content with 0.03% which at 20°C and 1 bar is about 13 $\mu$mol $L^{-1}$, and under light-saturating conditions, this concentration is usually limiting. A plant which can utilize bicarbonate in the sea should, then, have no such problem. In air, $O_2$ and $CO_2$ have diffusion constants on the order of 0.2 and 0.16 cm$^2$ sec$^{-1}$ (20°C), respectively, while in water these same molecules have constants of 1.7 and 2.0 × 10$^{-5}$ cm$^2$ sec$^{-1}$ (20°C), respectively.[40] This difference of 10$^4$ more than compensates for any advantage that a bicarbonate user has over a land plant.

The fixation of carbon depends upon molecular transport through the external fluid to the sites of carbon fixation. The greatest transport distance in marine algae is through the boundary layer which is a thin layer of fluid through which momentum, heat, and mass are transported from the fluid stream to the plant surface.

Biologists tend to look at the boundary layer as a resistance to molecular transport.

**Table 6**
## CARBON SOURCE FOR
## PHOTOSYNTHESIS IN MACROALGAE

| Alga | $HCO_3^-$ | $CO_2$ | Ref. |
|---|---|---|---|
| Chlorophyta | | | |
| *Enteromorpha* sp. | | X | 39 |
| *Hydrodictyon africanum* | X | | 57 |
| *Ulva* sp. | X | | 39 |
| *U. pertusa* | X | | 46, 58 |
| | | | |
| Phaeophyta | | | |
| *Alaria* sp. | X | | 39 |
| *Carpophyllum flexuosum* | | X | 59 |
| *C. maschalocarpum* | | X | 59 |
| *C. plumosum* | | X | 59 |
| *C. angustifolium* | | X | 59 |
| *Costaria costata* | X | | 39 |
| *Desmarestia munda* | | X | 39 |
| *Fucus* sp. | X | | 39 |
| *L. saccharina* | X | | 39 |
| *M. pyrifera* | X | | 16 |
| *Nereocystis luetkeana* | X | | 39 |
| *Sargassum muticum* | X | | 60 |
| | | | |
| Rhodophyta | | | |
| *G. amansii* | X | | 46 |
| *G. cartilagineum* | | X | 61 |
| *Gigartina (cristata?)* | X | | 39 |
| *G. harveyana* | | X | 59 |
| *Iridaea cordata* | X | | 58 |
| *Palmeria palmata* | X | | 62 |
| *Porphyra schizophylla* | | X | 39 |
| *Porphyra tenera* | X | | 46 |

This resistance can be defined as the gradient developed between the fluid stream and the plant surface divided by the transport rate,

$$R = (C - C_o) J^{-1} \qquad (1)$$

where C is the concentration of carbon in the fluid stream, and $C_o$ is the concentration at the plant surface; J is the boundary layer transport rate. This flux or rate can be estimated from Fick's first law:

$$J = D \, dC \, dY^{-1} \qquad (2)$$

Assuming that an alga has a flat morphology (for simplicity) such as the blade of *Macrocystis* which is two-dimensional with a zero-pressure gradient and further assuming that the water movement in the ocean or created in the laboratory contains low levels of turbulence, the following equation gives a rough approximation of the laminar diffusion boundary layer thickness on such a surface,[41]

$$T_1 = 3 \, Sc^{-1/3} \, Re_x^{-1/2} \, X \qquad (3)$$

Here, Sc is the Schmidt number, a dimensionless number which for seawater is constant at 522. The Reynolds number, Re, is another dimensionless number which for seawater is equal to 100 X v, X being the characteristic dimension of the algal thallus and v being the water velocity.

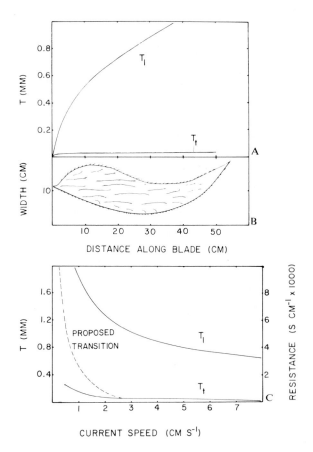

FIGURE 4.   Theoretical curves for boundary layer thicknesses and diffusion resistances. (A) Diffusion boundary layer thickness over a blade of *Macrocystis* with a water flow rate of 5 cm sec$^{-1}$, the laminar boundary layer thickness ($T_l$) and the turbulent boundary layer thickness ($T_t$) are computed from Equations 3 and 4, (B) a representation of the *Macrocystis* blade, and (C) laminar ($T_l$) and turbulent ($T_t$) diffusion boundary layer thicknesses at a point 25 cm along the blade, plotted for different water velocities. The boundary layer resistance (R) is calculated on the right. The dotted line is the estimated transition from a laminar to a turbulent boundary layer.

If, on the other hand, the water motion contains high levels of turbulence or the surface is not strictly a two-dimensional, flat, smooth surface, the boundary layer may be turbulent. In such a case, the diffusion thickness can be approximated by the relation:[41]

$$T_t = 2.5 \ Sc^{-1/4} \ Re_x^{-7/8} \ X \tag{4}$$

This thickness is shown to vary with both the distance along a theoretical (flat) *Macrocystis* blade (see Figure 4) and with water speed.[101]

The resistance equation (1) can be rewritten as

$$R = T \ D^{-1} \tag{5}$$

and $T_t$ and $T_l$ substituted for T. The resistance and boundary layer thicknesses for a number of characteristic algal dimensions are shown in Table 7. With a laminar bound-

## Table 7
## BOUNDARY LAYER THICKNESS AND RESISTANCE FOR A NUMBER OF CHARACTERISTIC MACROALGAL DIMENSIONS (THICKNESS IN CM)

| Dimension | Laminar thickness | R | Turbulent thickness | R |
|---|---|---|---|---|
| 100 cm | 0.12 | 24,000 | 0.021 | 4,200 |
| 50 cm | 0.083 | 16,600 | 0.0020 | 400 |
| 10 cm | 0.037 | 7,400 | 0.0017 | 340 |
| 1 cm | 0.012 | 2,360 | 0.0012 | 240 |
| 5 mm | 0.008 | 1,600 | 0.0011 | 220 |
| 1 mm | 0.004 | 800 | 0.0009 | 180 |

*Note:* Calculated using laminar and turbulent boundary layer equations in text. All values assuming a two-dimensional, flat surface with a zero-pressure gradient. R units (sec cm$^{-1}$); velocity = 10 cm sec$^{-1}$; D = $2.0 \times 10^{-5}$ cm$^2$sec$^{-1}$.

ary layer, the resistance (at 10 cm sec$^{-1}$) for a large kelp surface (1 m) can be as high as 24,000 sec cm$^{-1}$. One might compare this with a desert succulent with closed stomates (100 to 300 sec cm$^{-1}$) and to crop plants (1 to 5 sec cm$^{-1}$).[42] The importance of these high resistances is underlined by the correlation between water motion and algal productivity.[43]

The resistance equation can be further modified to provide

$$J = D(C - C_o) T^{-1} = (C - C_o) R^{-1} \qquad (6)$$

Again, $T_t$ and $T_l$ can be substituted for T. The result is graphed for the theoretical *Macrocystis* blade in Figure 5.[101]

It is not realistic to assume that the uptake remains linear. At a certain speed, the enzymatic uptake systems of the plant will become saturated and show no further response to water velocity. The chemical process of carbon fixation has been described in the form of the Michaelis-Menten equations which is used here only in the operational sense and does not imply any particular knowledge of the actual mechanism. Thus,

$$J = (V_m C_o) (C_o + K_s)^{-1} \qquad (7)$$

The relationship between boundary layer flux and enzymatic flux has been described in detail elsewhere for higher plants, phytoplankton, and larger algae,[20,44,45] with the result that

$$J = ( (C + K_s + RV_m) - ( (C + K_s + RV_m)^2 - 4RCV_m)^{1/2} ) 2R^{-1} \qquad (8)$$

The uptake of inorganic carbon by an aquatic plant such as *Macrocystis* should be determinable for a given C, R, $V_m$, and $K_s$. Figure 6 shows data taken from *Macrocystis pyrifera* by the author for the calculation of the above variables.[101] Figure 7 shows how Equation 8 can be used to assess the various effects of the boundary layer on the productivity of *Macrocystis*.[101]

Although the $V_m$ for inorganic carbon uptake is or can be controlled by a number

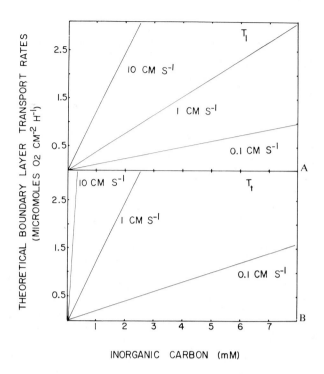

FIGURE 5.    Calculated rates (Equation 6) of boundary layer transport through (A) laminar and (B) turbulent boundary layers, expressed as oxygen evolved ($\mu$mol $O_2$ $cm^{-2}$ $hr^{-1}$).

of other factors including mineral nutrition, the $K_s$ generally lies between 1 and 3 mmol $L^{-1}$.[39,46,101] Saturation concentrations for bicarbonate users generally lie in the range between 6 and 8 mmol $L^{-1}$.[39,46,101]

Ribulose 1,5-bisphosphate carboxylase (RuBP-Case), the primary carbon fixation enzyme, has been shown to require $CO_2$ as a substrate.[47] Algae which utilize bicarbonate must be able to convert $HCO_3^-$ to $CO_2$. The presence of carbonic anhydrase (C-Aase) has been demonstrated in macroalgae.[48-50]

The entrance of $CO_2$ into algae and higher plant carbon compounds is mostly via RuBP-Case. However, in a number of higher plants, $CO_2$ enters through phosphoenolpyruvate carboxylase (PEP-Case) as well. A characteristic of this enzyme in CAM plants is that carbon can be fixed in the absence of light. Dark fixation also has been demonstrated in a number of marine algae (see Table 8).[14,51,52] Although a search of marine algae has failed to turn up appreciable amounts of PEP-Case, high activities of phosphoenolpyruvate carboxykinase (PEP-CKase) have been demonstrated (see Table 9).[14,53-55] This enzyme is usually associated with the growing regions of the alga, where it can contribute as much as 50% of the fixed carbon.[56] Its activity is, however, much lower in older parts of the plant.[56] This enzyme can contribute significantly to the carbon balance of *Laminaria*,[56] but its ecological significance is still largely unknown.

## RESPONSE TO $O_2$

From work with higher plants, it is now clear that the major receptor of $CO_2$ is ribulose 1,5-bisphosphate carboxylase (RuBP-Case). This enzyme also has an affinity for $O_2$.[63] Under conditions of high $O_2$ concentration and low $CO_2$ concentration, a large percentage of the activity of this enzyme is devoted to producing phosphoglyco-

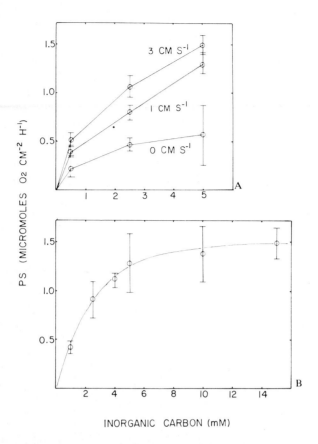

FIGURE 6.   Photosynthetic oxygen production ($\mu$mol $O_2$ cm$^{-2}$ hr$^{-1}$) as a function of inorganic carbon concentration (mmol $\ell^{-1}$). (A) Upper curves show the effect of water speed on plant disc AP under nonsaturating inorganic carbon concentrations. (B) The lower curve shows the effect of concentration under saturating water motion (7 cm sec$^{-1}$). The mean and SD for 15 blade discs are shown.

late.[63] Therefore, high $O_2$ concentrations increasingly inhibit the fixation of $CO_2$ and thus photosynthesis (see Figure 8). Evidence for the oxygenase activity of RuBP-Case in algae can be seen when the AP under high $O_2$ concentration is compared with the AP under low $O_2$ concentration.[18,64-66] A list of selected algae and their degree of $O_2$ inhibited AP is shown in Table 10.

The $CO_2$ equilibrium level is another indication of photorespiration. This compensation point is maintained by an equilibrium between the rate of $CO_2$ fixation and the rate of $CO_2$ lost from photorespiration and dark respiration. This compensation point is nearly 0 for $C_4$ plants without photorespiration because of the efficient $CO_2$ fixation through the $C_4$ enzymes in the mesophyll cells. A large $CO_2$ compensation point, then, indicates normal $C_3$ metabolism and photorespiration.[67] *Boodlea composita, Halimeda cylindracea*, and *Enteromorpha* sp. (Chlorophyta) all were found to have $CO_2$ compensation points increased with increasing $O_2$ concentration.[68]

The primary steps in the glycolate pathway are catalyzed by a number of enzymes which are specific for the pathway, and their presence in algae demonstrates existence of the pathway in algae. Ribulose 1,5-bisphosphate and $O_2$ are transformed through RuBP oxygenase into phosphoglycolate. Phosphoglycolate in the presence of NAD and phosphoglycolate phosphatase results in the formation of glycolate. Glycolate is

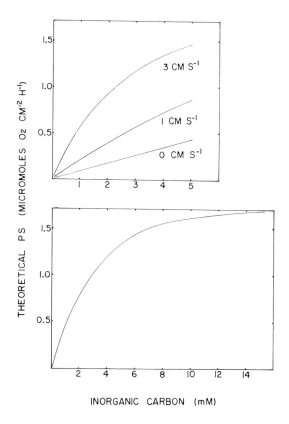

FIGURE 7.   Theoretical AP rate expressed as $\mu$mol $O_2$ cm$^{-2}$ hr$^{-1}$ as a function of inorganic carbon. Curves calculated from Equation 8. Upper curves show the effect of nonsaturating current velocities, while the lower curve is calculated for saturating water velocity over the same range presented in Figure 6.

then either oxidized (with $O_2$ to form $H_2O_2$ and glyoxylate as in higher plants or oxidized without $O_2$ (but with DCPIP) to form glyoxylate as in unicellular algae.[63] The former reaction is catalyzed by glycolate oxidase and the latter by glycolate dehydrogenase.[63] Further steps produce intermediates such as glycine and serine.

The RuBP-Case/Oase has been shown to actively accept $O_2$ in vitro in both *H. cylindracea* and *Chaetomorpha* (Chlorophyta), where the oxygenase activity was found to be about 1% that of the carboxylase activity.[69] No activity could be detected in either the phaeophytes or rhodophytes tested. The amount of $CO_2$ fixed through the carboxylase was, on the other hand, enough to account for the measured AP. Semipurified enzymes from *H. cylindracea* were similar to those enzymes purified from spinach and *Chlorella*.[69]

Phosphoglycolate phosphatase is an especially good indicator of photorespiration because of its specificity for phosphoglycolate. PGPase activity has been found to be greater than or equal to that found in terrestrial plants (on a chl *a* basis) in 13 of 14 chlorophytes, 2 of 3 phaeophytes, and all 4 rhodophytes examined (see Table 11).[70] The semipurified enzyme from *H. cylindracea* was found to be very similar to the same enzyme from spinach, verifying its substrate specificity for phosphoglycolate in algae.[70]

Efforts to measure the activities of glycolate oxidase (GOx) and/or glycolate dehydrogenase (GDH) in marine algae have met with some success (see Table 11).[71] Members of the Chlorophyta appear to contain only glycolate dehydrogenase as in unicel-

Table 8
COMPARISON OF RATES OF
PHOTOSYNTHESIS (PS) AND LIGHT-
INDEPENDENT CO₂-FIXATION (DF) IN
MACROALGAE

| Alga | nmol $CO_2$ gdw$^{-1}$ hr$^{-1}$ | | DF in % of PS |
| --- | --- | --- | --- |
| | DF | PS | |
| Chlorophyta | | | |
| *Acrosiphonia arcta* | 3.5 | 635.8 | 0.55 |
| *Codium fragile* | 9.2 | 683.0 | 1.34 |
| *Enteromorpha compressa* | 5.7 | 288.2 | 1.98 |
| *Enteromorpha linza* | 2.8 | 324.9 | 0.85 |
| *Ulva lactuca* | 2.7 | 93.9 | 2.94 |
| *Urospora* sp. | 2.9 | 115.0 | 2.50 |
| | | | |
| Phaeophyta | | | |
| *Ascophyllum nodosum* | 1.4 | 8.0 | 16.70 |
| *Bifurcaria bifurcata* | 6.8 | 108.4 | 6.25 |
| *Cystoseira ericoides* | 7.6 | 180.0 | 4.35 |
| *Dictyota dichotoma* | 19.9 | 464.5 | 4.35 |
| *Fucus serratus* | 9.4 | 235.8 | 4.00 |
| *F. spiralis* | 22.3 | 223.9 | 10.00 |
| *F. vesiculosus* | 22.2 | 292.3 | 7.60 |
| *Himanthalia elongata* | 3.8 | 91.4 | 4.17 |
| *Laminaria digitata* | 9.4 | 252.6 | 3.70 |
| *L. hyperborea* | 29.3 | 271.4 | 10.80 |
| *L. saccharina* | 10.1 | 185.4 | 5.40 |
| *Pelvetia canaliculata* | 3.4 | 87.1 | 4.00 |
| | | | |
| Rhodophyta | | | |
| *Bostrychia scorpioides* | 0.61 | 132.8 | 1.06 |
| *Calliblepharis jubata* | 1.8 | 352.6 | 0.51 |
| *C. crispus* | 0.97 | 212.4 | 0.46 |
| *Dumontia incrassata* | 3.3 | 389.7 | 0.85 |
| *Gastroclonium ovatum* | 3.0 | 391.2 | 0.76 |
| *Griffithsia flosculosa* | 0.66 | 132.8 | 0.50 |
| *Laurencia pinnatifida* | 2.3 | 312.2 | 0.72 |
| *Lomentaria articulata* | 4.3 | 498.2 | 0.87 |
| *P. palmata* | 1.1 | 449.4 | 0.24 |

From Kremer, B. and Küppers, U., *Planta*, 133, 191, 1977.
With permission.

lular greens. Problems with extraction techniques have made assessment of the activity in phaeophytes and rhodophytes difficult.[71]

The relative amounts of the intermediates glycine and serine have also been shown to increase with $O_2$ concentration in a number of marine algae.[72] For example, glycine and serine increased from 48 to 73% of the total amino acid pool when $O_2$ concentrations increased from 0 to 100% saturation.[72] These numbers were typical for the two rhodophytes, one phaeophyte, and two chlorophytes tested.[72]

The loss of organic compounds with increasing $O_2$ has been observed in *Caulerpa verticillata, Chlorodesmis fastigiata, Halimeda cylindracea,* and *Sargassum.*[73] This loss of organic carbon was accompanied by increasing respiration rates in the light which may be indicative of the oxidation steps in the glycolate pathway.[73]

Many species of microalgae are known to excrete glycolate under high $O_2$ concentra-

Table 9
## RELATIVE PHOTOSYNTHETIC ENZYME RATES IN
## MARINE ALGAE

| Alga | Relative enzyme activities | | |
|------|-----------|----------|---------|
|      | RuBP-Case | PEP-CKase | PEP-Case |
| **Chlorophyta** | | | |
| *Cladophora rupestris* | 98.1 | 1.5 | 0.4 |
| *U. lactuca* | 97.5 | 1.5 | 1.0 |
| | | | |
| **Phaeophyta** | | | |
| *Fucus serratus* | 45.2 | 54.8 | 0.0 |
| *Fucus spiralis* | 50.1 | 49.9 | 0.0 |
| *Fucus vesiculosus* | 52.8 | 47.1 | 0.1 |
| *Laminaria digitata* | 85.6 | 14.4 | 0.0 |
| *Laminaria hyperborea* | 27.5 | 72.5 | 0.0 |
| *Laminaria saccharina* | 60.5 | 39.5 | 0.0 |
| *Petalonia zosterifolia* | 91.0 | 8.5 | 0.5 |
| | | | |
| **Rhodophyta** | | | |
| *Chondrus crispus* | 97.3 | 1.8 | 0.9 |
| *Corallina officinalis* | 91.7 | 8.3 | 0.0 |
| *Delesseria sanguinea* | 100.0 | 0.0 | 0.0 |
| *Phyllophora membranifolia* | 98.7 | 1.3 | 0.0 |
| *Rhodomela subfusca* | 96.9 | 2.4 | 0.7 |

From Kremer, B. and Küppers, U., *Planta*, 133, 191, 1977. With permission.

tions and high illumination.[63] Although the larger macroalgae excrete a number of dissolved organic compounds (above), glycolate has been measured as exudate from only a small number of marine algae (see Table 12).[74] This may be due to the fact that many of the algae tested seemed to exude compounds which interfere with the normal glycolate test.[74]

$C_4$ plants with phosphoenolpyruvate carboxylase (PEP-Case) and the Kranz anatomy are able to utilize $CO_2$ more efficiently. As such, light respiration rates are almost nonexistent, and $CO_2$ compensation points are almost 0 in these plants. An extensive search of the marine algae has not turned up solid proof for the existence of PEP-Case in the marine algae.[53] However, in the Phaeophyta, PEP carboxykinase has been demonstrated.[54] This enzyme performs basically the same function as PEP-Case in higher plants. Indeed, in those plants were high activities of PEP carboxykinase have been found, photorespiration (as indicated by the Warburg effect) is minor, accounting for less than a 10% change in the AP rates in the case of *Laminaria longicruris*.[18]

$\delta^{13}$ C values have been used as an indicator of the presence of PEP-Case in higher plants. This is because PEP-Case discriminates only slightly between the two forms of carbon ($^{12}C$ and $^{13}C$). On the other hand, RuBP-Case discriminates heavily in favor of $^{12}C$ carbon. While this method is reasonably good for higher plants, seawater has an entirely different ratio from air making extrapolations more difficult. Of the 13 species of Chlorophyta tested, only 3 (*H. cylindracea, Neomeris vambossae,* and *Acetabularia*) had ratios indicative of $C_4$ plants.[75] Among the rhodophytes tested, only *Laurencia, Galoraing dermanema,* and *Acanthophora spicifera* had low ratios.[75] The only phaeophyte found to have a low ratio was *Turbinaria ornata*. Although these results are of physiological interest, such low ratios do not mean the existence of $C_4$ enzymes. *Halimeda* has been in many other ways proven to use primarily the $C_3$ pathway.

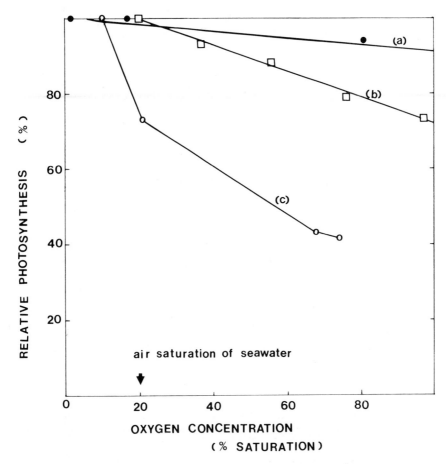

FIGURE 8.   Relative photosynthetic rate as a function of oxygen concentration (expressed as % saturation). (A) *L. longicruris* data,[18] (B) *C. maschalocarpum* data,[20] and (C) *H. cylindracea* data.[65]

Although photorespiration has been convincingly demonstrated in marine algae, its ecological significance to the overall productivity of these algae is still in question.[76] In a study with *Carpophyllum*, Dromgoole[20] estimated that in high tidepools, the AP could be inhibited as much as 54%. Almost all of this inhibition could, however, be accounted for by changes in pH and not by the changes in $O_2$ saturation which also occurred. Seasonal carbon budgets for both *Laminaria longicruris*[18] and *L. saccharina*[77] do not include terms for photorespiration because this term was considered insignificant.

## RESPONSE TO MINERAL NUTRIENTS

The effects of mineral nutrition on photosynthesis of macroalgae are difficult to measure directly. Mineral nutrition is generally considered in relation to protein or enzyme formation and degradation and the effects of these processes on photosynthetic productivity. Mineral nutrients are usually divided into two groups, macro- and micronutrients. In the aquatic environment, the availability of the macronutrients is probably one of the main factors limiting algal growth. However, under cultivation conditions, where optimum growth is the goal, micronutrients also can play an important role. Table 13 indicates the mineral nutrients which have been shown to be required by at least one algal species and the concentration ranges of these elements in

Table 10
WARBURG EFFECT ON MACROALGAE

| Alga | $O_2$ conc. (% Saturation) | Inhibition (%) | Ref. |
|---|---|---|---|
| Chlorophyta | | | |
| *Enteromorpha* sp. | 20—100 | 18 | 66 |
| *Enteromorpha* sp. | 21—62 | 14 | 65 |
| *Enteromorpha* sp. | 0—100 | 19[a] | 64 |
| | 0—21 | 24[a] | 64 |
| *Ulva lobata* | 1—21 | 0 | 67 |
| *Cladophora* sp. | 20—100 | 43 | 66 |
| *Codium fragile* | 20—100 | 35 | 20 |
| *Bryopsis* sp. | 20—100 | 5 | 66 |
| *Chaetomorpha crassa* | 5—21 | 6 | 65 |
| | 5—80 | 51 | 65 |
| *Boergesenia forbesii* | 21—65 | 64 | 65 |
| *Boodlea composita* | 21—80 | 75 | 65 |
| *Halimeda cylindracea* | 10—74 | 58 | 65 |
| | 0—100 | 0[b] | 64 |
| | | | |
| Phaeophyta | | | |
| *Carpophyllum plumosum* | 20—100 | 17—43 | 20 |
| *Dictyota* sp. | 21—90 | 61 | 65 |
| *Ecklonia radiata* | 20—100 | 33 | 20 |
| *Laminaria longicruris* | 2—108 | 10 | 18 |
| *Nereocytis luetkeana* | 5—14 | 47 | 76 |
| *Padina tennuis* | 21—75 | 69 | 65 |
| *Sargassum sinclarii* | 20—100 | 29 | 20 |
| *Sargassum* sp. | 21—55 | 31 | 65 |
| | 21—85 | 53 | 65 |
| *Scytothamnus australis* | 21—100 | 30 | 20 |
| | | | |
| Rhodophyta | | | |
| *Chyclocladia muelleri* | 5—100 | 34 | 66 |
| *Ceramium* sp. | 21—85 | 55 | 65 |
| *Halymenia durvillaei* | 0—100 | 83[a] | 64 |
| *Laurencia* sp. | 0—100 | 0[b] | 64 |
| *Nitzemia* sp. | 5—100 | 20 | 66 |
| *Phacelocarpus complanatus* | 5—100 | 36 | 66 |
| *Polysiphonia abscissa* | 5—100 | 34 | 66 |

[a]    After an incubation of 30 min.
[b]    Authors suggest methods failure.

algae. Table 14 indicates how much water must pass by a *Macrocystis* plant to satisfy its requirements for 1 day's growth.[78] One can see that the macronutrients, N and P, require the largest water volumes and are therefore most likely to be limiting under natural conditions. In culture, a number of algae have been shown to require larger than trace amounts of certain elements. For example, many red and brown algae accumulate large quantities of I and Br and require substantial amounts for minimum growth.[79-81]

In the sea, N is the main nutrient limiting growth, while in freshwater, P is the main limiting nutrient.[82] Lack of sufficient N in the marine environment in summer produces several important physiological as well as ecological effects.

N-starved algal cells have been shown to degrade pigments.[83,84] Chlorophylls and carotenoids in green algae,[85] fucoxanthin and chlorophylls in brown algae,[84] and chlorophylls and phycobilins in the red algae[83] all are degraded during N starvation. In

### Table 11
### DISTRIBUTION OF GLYCOLATE ENZYMES IN
### MARINE ALGAE[70,71]

| Alga | PGPase[a] | GOxase | GDHase |
|---|---|---|---|
| **Chlorophyta** | | | |
| *Avrainvillea erecta* | − | − | −[b] |
| *Boergesenia forbesii* | + | − | + |
| *Caulerpa sertularioides* | + | − | + |
| *C. verticillata* | + | − | + |
| *Caulerpa* sp. | n | − | + |
| *Chaetomorpha crassa* | + | − | + |
| *Chlorodesmis fastigiata* | + | − | −[b] |
| *Cladophora fascicularis* | n | − | + |
| *Dictyosphaeria versluysii* | n | − | + |
| *Entermorpha flexuosa* | + | − | + |
| *Halimeda cylindracea* | + | − | + |
| *H. macroloba* | + | − | + |
| *H. opuntia* | + | − | + |
| *Neomeris vanbossae* | + | − | + |
| *Udotea argentea* | n | − | + |
| *Valonia* sp. | n | − | + |
| | | | |
| **Phaeophyta** | | | |
| *Dictyota* sp. | + | − | −[b] |
| *Padina tennuis* | − | − | −[b] |
| *Sargassum* sp. | + | − | −[b] |
| | | | |
| **Rhodophyta** | | | |
| *Acanthophora spicifera* | n | − | −[b] |
| *Ceratodictyon spongiosum* | + | − | −[b] |
| *Halymenia durvillaei* | n | − | −[b] |
| *Hypnea pannosa* | n | − | −[b] |
| *Laurencia* sp. | + | − | −[b] |

[a]   n, not tested.
[b]   Authors suggest unsatisfactory extraction technique.

### Table 12
### GLYCOLLATE EXCRETION
### BY MARINE ALGAE

**Alga**

Chlorophyta
   *Chlorodesmis fastigiata*
   *Halimeda opuntia*
   *H. cylindracea*
   *Boodlea composita*
   *Neomeris vanbossae*

Phaeophyta
   *Sargassum* sp.
   *Padina tennuis*

Rhodophyta
   *Laurencia obtusa*
   *Galaxaura fastigiata*

*Note:* Algae tested and found to excrete
glycollate by Fogg.[74]

Table 13
ELEMENTARY COMPOSITION OF
ALGAE, INCLUDING ALL THOSE
ELEMENTS KNOWN TO BE
REQUIRED BY AT LEAST ONE
ALGA[81]

| Element | mg gdw⁻¹ | |
| --- | --- | --- |
| | Average | Range |
| H | 65 | 29—100 |
| C | 430 | 175—650 |
| O | 275 | 205—330 |
| N | 55 | 10—140 |
| Si | 55 | 0—230 |
| K | 17.3 | 1—75 |
| P | 11 | 0.5—33 |
| Na | 6.1 | 0.4—47 |
| Mg | 5.6 | 0.5—75 |
| Ca | 8.7 | 0—80 |
| S | 5.9 | 1.5—16 |
| Fe | 5.9 | 0.2—34 |
| Zn | 0.28 | 0.005—1.0 |
| B | 0.03 | 0.001—0.25 |
| Cu | 0.1 | 0.006—0.3 |
| Mn | 0.06 | 0.002—0.24 |
| Co | 0.06 | 0.0001—0.2 |
| Mo | 0.0008 | 0.0002—0.001 |
| I | 0.60[a] | |
| Br | 0.1[a] | |

[a]   Data from North.[78]

Table 14
COMPUTATIONS INDICATING THE
VOLUMES OF SEAWATER NEEDED TO
SUPPLY THE AMOUNTS OF SIX
NUTRIENT ELEMENTS SUFFICIENT TO
SUSTAIN A DAILY *MACROCYSTIS*
PRODUCTION OF 5.5 gdw m⁻²

| Element | Water conc. (ngat L⁻¹) | Plant conc. (mg gdw⁻¹) | Liters d⁻¹ |
| --- | --- | --- | --- |
| Cu | 3—10 | 0.08 | 50—150 |
| Zn | 10—50 | 0.55 | 60—300 |
| Fe | 10—30 | 1.40 | 30—800 |
| Mn | 5—40 | 0.10 | 30—220 |
| N | 0—500 | 850.0 | 10,000 |
| P | 200—600 | 93.0 | 1,000—3,000 |

Data from North.[78]

some members of the Chlorophyta, N starvation enhances the production of secondary carotenoids.[85] This "bleaching effect" is, however, not restricted to just pigments. Most of the enzymes associated with nitrogen metabolism and photosynthesis also are degraded.[36] This general degradation is correlated with decreased protein content and growth rates in a number of marine algae.[18,77,86] The decreased rate of energy capture

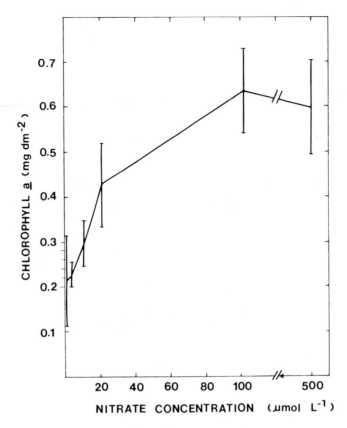

FIGURE 9.   Chl *a* content of *L. longicruris* sporophytes at various
NO₃⁻ nitrogen concentrations (±95% confidence limits). (From Chapman, A. R. O., Markham, J. W., and Lüning, K., *J. Phycol.*, 14,
195, 1978. With permission.)

by pigments is thus a natural result of decreased need for photosynthate. Decreased
pigment concentrations at a time of high irradiance in summer may also have a protective effect, in that potentially damaging peroxides can form when the rate of production of photoreductant exceeds that of its use.[87]

An increase in carbohydrate production in summer accompanies N starvation. Pho-
tosynthates produced in the absence of N are shunted into mannitol and laminarin
synthesis.[84] Mannitol and laminarin produced at this time can be used later in the
winter months when photosynthesis cannot meet the demands of growth.[86,88] Mannitol
metabolic enzymes such as mannitol-1-P dehydrogenase are, in contrast to most of the
photosynthetic and nitrogen enzymes, quite active in summer.[19]

Addition of nitrate or ammonia and phosphate to algal cultures reestablishes pigment synthesis and increases photosynthetic rates.[84] Chapman et al.[84] grew *L. saccharina* (after a period of starvation) on different concentrations of NO₃⁻ nitrogen. Synthesis of chl*a* continued up to 100 μmol NO₃⁻ L⁻¹ (see Figure 9). In contrast,
photosynthesis and growth were N saturated at or before 10 μmol NO₃⁻ L⁻¹ (see Figure
10). The synthesis of chl*a* thus continues well past the N level required for maximum
photosynthesis or growth. This might be considered a form of "luxury synthesis" and
might be a source of storage for nitrogenous compounds.

Addition of N and P to *Ulva lactuca* cultures also has been shown to have a marked
effect on the photosynthetic rate. *Ulva* grown under chemostat conditions responded
to increasing concentrations of NH₃ nitrogen and phosphate by increasing carbon fix-

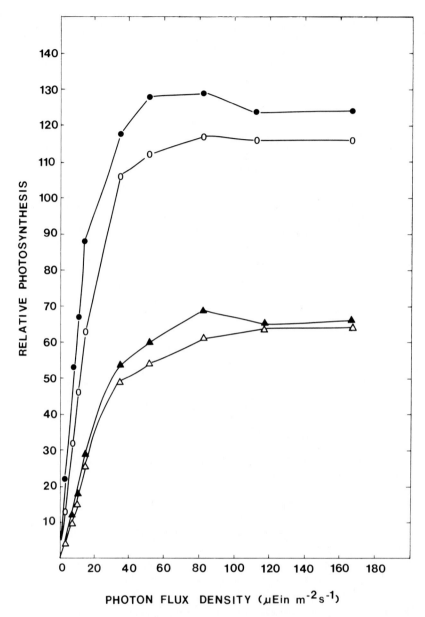

FIGURE 10.    Relative gross photosynthesis of *L. longicruris* sporophytes vs. irradiance.
(●) 500 μmol NO₃⁻ L⁻¹; (○) 10 μmol NO₃ L⁻¹; (▲) 3 μmol NO₃ L⁻¹; (△) 0 μmol NO₃ L⁻¹.
(From Chapman, A. R. O., Markham, J. W., and Lüning, K., *J. Phycol.*, 14, 195, 1978.
With permission.)

ation after only a 12-hr incubation period.[89] Carbon fixation was further stimulated
by addition of both elements.[89] The productivity of *Ulva, Enteromorpha, Gracilaria,
Fucus,* and *Chondrus* as well as a number of other algae is similarly enhanced with
the addition of domestic sewage to the cultivation tank.[90-92]

## RESPONSE TO pH

The effect of pH on the apparent photosynthesis of macroalgae is difficult to determine. This is because pH influences the AP in two ways:

Table 15

## pH VALUE AT WHICH THE AP APPROACHES ZERO FOR A NUMBER OF MARINE ALGAE[94]

| Alga | pH |
|---|---|
| Chlorophyta | |
| *Cladophora trichotoma* | >10 |
| *Cladophora graminea* | 9.8 |
| *Codium* sp. | >10 |
| *Enteromorpha* sp. | >10 |
| *Ulva* sp. | >10 |
| | |
| Phaeophyta | |
| *Carpophyllum maschalocarpum* | 8.6[a] |
| *Desmarestia* sp. | 9.5 |
| *Dictyoneurum* sp. | 9.4 |
| *Egregia* sp. | 9.2 |
| *Fucus* sp. | >10 |
| *Macrocystis* sp. | 9.5 |
| *Pelvetia* sp. | >10 |
| | |
| Rhodophyta | |
| *Bossiella* sp. | >10 |
| *Botryocladia* sp. | >10 |
| *Botryoglossum* sp. | 9.8 |
| *Aghardiella* sp. | 9.8 |
| *Centroceros* sp. | >10 |
| *Corallina* sp. | >10 |
| *Endocladia* sp. | 9.8 |
| *Gastroclonium sp.* | >10 |
| *Gigartina harveyana* | 9.5 |
| *Gigartina papillata* | 9.8 |
| *Gracilaria* sp. | 9.1 |
| *Iridaea* sp. | 9.8 |
| *Schizymenia* sp. | 9.2 |

[a]   Data from Dromgoole.[97]

1.    By affecting the protoplasm and ion balance which is particularly sensitive above pH 9.8[93]
2.    By changing the $CO_2 - HCO_3^- - CO_3^{2-}$ balance of the medium

Increasing the pH of seawater with NaOH decreases AP (see Table 15).[94] The response of an alga to increasing pH is probably one of tolerance. Those algae living in high tide pools where AP can drive the pH as high as 10 must be able to tolerate high alkalinity.[95] Those algae living lower in the tidal region or subtidally are never exposed to such drastic changes and are likely to be more sensitive.

Free $CO_2$ concentrations decrease from about 10 $\mu$mol $L^{-1}$ at pH 8.2 to negligible quantities at pH 9.[96] An alga which is dependent on free $CO_2$ as a carbon source should, then, be unable to carry out AP above pH 9. Thus, AP should drop with increasing pH. Most of the algae in Table 15 appear to carry out AP well above pH 9 which is an indication that they are able to use, at least partially, another source of inorganic carbon. Dromgoole[97] found the AP of *Carpophyllum* dropped to 0 at pH 8.6 (see Figure 11). He correlated the drop in AP with increasing pH and falling $CO_2$ concentrations. On the other hand, Jolliffe and Tregunna,[39] analyzing the AP response of *Sargassum* to different concentrations of inorganic carbon, found little or no re-

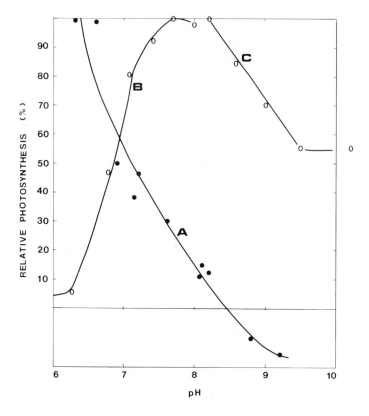

FIGURE 11.   The relative photosynthetic rate of (A) *Carpophyllum*[97] and (B)[93] and (C),[94] *Ulva* vs. pH. *Carpophyllum* is an obligate $CO_2$ user, while *Ulva* is capable of using bicarbonate.

sponse to pH or to free $CO_2$ between pH 8.0 and 9.2. *Ulva* has been shown repeatedly to utilize bicarbonate.[93] Its pH response is shown in Figure 11 and demonstrates probably more realistically the pH response in algae.

At more acidic pH, the concentration of free $CO_2$ increases. At pH 7.4, the concentration of $CO_2$ is 10 times higher than at pH 8.2 (100 $\mu$mol $L^{-1}$).[96] $CO_2$-dependent species should then increase AP with decreasing pH. The AP rate of *Carpophyllum*, a typical $CO_2$ species, increases with decreasing pH (see Figure 11).[97] The AP rate of *Palmeria*, a probable bicarbonate user, increases down to pH 6.5, at which point the rate begins to decline.[62] Respiration rates of *Carpophyllum* and *Palmeria* also increase with decreasing pH.[62] Using a phosphate buffer with a constant supply of bubbled $CO_2$, the optimum rates of AP for *Porphyra*, *Ulva*, and *Gelidium* were found to be between pH 7.5 and 8 (see Figure 11 for *Ulva*). Growth studies of *Chondrus crispus* indicate that the optimum pH for growth is between 7 and 8.[98]

The cellular pH of a few selected algae lies between 6.5 and 7.5.[99] The pH optima for photosynthesis enzymes such as RuBP carboxylase also lie in this region, 7.8 for *Halimeda cylindracea*[69] and 8.5 for *Laminaria hyperborea*.[19] The pH optima for PEP carboxykinase are 7.6 for *L. hyperborea*[19] and 7.0 to 7.3 for other brown algae.[54]

# REFERENCES

1. Jerlov, N. G., *Optical Oceanography,* Elsevier, New York, 1968.
2. Inoue, H., Sugahara, K., and Nishimura, M., Energy-and-electron transfer reactions in brown algae, *Proc. Int. Seaweed Symp.,* 7, 515, 1972.
3. Haxo, F. T. and Blinks, L. R., Photosynthetic action spectra of marine algae, *J. Gen. Physiol.,* 33, 389, 1955.
4. Yocum, C. S. and Blinks, L. R., Photosynthetic efficiency of marine plants, *J. Gen. Physiol.,* 38, 1, 1954.
5. Ramus, J., Beale, S. I., Mauzerall, D., and Howard, K. L., Changes in the photosynthetic pigment concentration in seaweeds as a function of water depth, *Mar. Biol.,* 37, 223, 1976.
6. Ramus, J., Beale, S. I., and Mauzerall, D., Correlation of changes in pigment content with photosynthetic capacity of seaweeds as a function of water depth, *Mar. Biol.,* 37, 231, 1976.
7. Kageyama, A. and Yokohama, Y., Photosynthetic properties of marine benthic brown algae from different depths in coastal area, *Bull. Jpn. Soc. Phycol.,* 22, 119, 1974.
8. Yokohama, Y., Photosynthetic properties of marine benthic green algae from different depths in the coastal area, *Bull. Jpn. Soc. Phycol.,* 21, 70, 1973.
9. Yokohama, Y., Photosynthetic properties of marine benthic red algae from different depths in coastal area, *Bull. Jpn. Soc. Phycol.,* 21, 119, 1973.
10. Li, B. D., and Titlyanov, E. A., Adaptation of benthic plants to light. III. Content of photosynthetic pigments in marine macrophytes from differently illuminated habitats, *Biol. Morya,* 4, 597, 1978.
11. Ramus, J., Lemons, F., and Zimmerman, C., Adaptation of light harvesting pigments to downwelling light and the consequent photosynthetic performance of the eulittoral rockweeds *Ascophyllum nodosum* and *Fucus vesiculosus, Mar. Biol.,* 42, 293, 1977.
12. Duncan, M. J., *In situ* studies of growth and pigmentation of the phaeophycean *Nereocystis luetkeana, Hegol. Wiss. Meeresunters.,* 24, 510, 1973.
13. Keast, J. F. and Grant, B. R., Chlorophyll *a:b* ratios in some siphonous green algae in relation to species and environment, *J. Phycol.,* 12, 328, 1976.
14. Küppers, U. and Kremer, B. P., Longitudinal profiles of carbon dioxide fixation capacities in marine macroalgae, *Plant Physiol.,* 62, 49, 1978.
15. King, R. J. and Schramm, W., Determination of photosynthetic rates for the marine algae *Fucus vesiculosus* and *Laminaria digitata, Mar. Biol.,* 37, 209, 1976.
16. Wheeler, W. N., Pigment content and photosynthetic rate of the fronds of *Macrocystis pyrifera, Mar. Biol.,* 56, 97, 1980.
17. Lüning, K., Seasonal growth of *Laminaria hyperborea* under recorded underwater light conditions near Helgoland, in *4th European Marine Biology Symposium,* Crisp, D. J., Ed., Cambridge University Press, London, 1971, 347.
18. Hatcher, B. G., Chapman, A. R. O., and Mann, K. H., An annual carbon budget for the kelp, *Laminaria longicruris, Mar. Biol.,* 44, 85, 1977.
19. Küppers, U., Enzymologie der $CO_2$- Fixerung bei *Laminaria hyperborea*: jahresperiodische, umweltbezogene Veranderungen von Enzymaktivitaten, Ph.D. thesis, University of Koln, 1978.
20. Dromgoole, F. I., The effects of oxygen on dark respiration and apparent photosynthesis of marine macro-algae, *Aquat. Bot.,* 4, 281, 1978.
21. King, R. J. and Schramm, W., Photosynthetic rates of benthic marine algae in relation to light intensity and seasonal variations, *Mar. Biol.,* 37, 215, 1976.
22. Littler, M. M. and Murray, S. N., The primary productivity of marine macrophytes from a rocky intertidal community, *Mar. Biol.,* 27, 131, 1974.
23. Ramus, J., Seaweed anatomy and photosynthetic performance: the ecological significance of light guides, heterogeneous absorption and multiple scatter, *J. Phycol.,* 14, 352, 1978.
24. Colombo, P. M. and Orsenigo, M., Sea depth effects on the algal photosynthetic apparatus. II. An electron microscopic study of the photosynthetic apparatus of *Halimeda tuna* (Chlorophyta, Siphonales) at −0.5 m and −6.0 m sea depths, *Phycologia,* 16, 9, 1977.
25. Nultsch, W. and Pfau, J., Occurrence and biological role of light-induced chromatophore displacements in seaweeds, *Mar. Biol.,* 51, 77, 1979.
26. Kanwisher, J. W., Photosynthesis and respiration in some seaweeds, in *Some Contemporary Studies in Marine Science,* Barnes, H., Ed., George Allen and Unwin, London, 1966, 407.
27. Kain, J. M., Drew, E. A., and Jupp, B. P., Light and the ecology of *Laminaria hyperborea.* II, in *Light as an Ecological Factor. II,* Evans, G. C., Bainbridge, R., and Rackham, O., Eds., Blackwell Scientific, Oxford, 1976, 63.
28. Mathieson, A. C. and Norall, T. L., Physiological studies of subtidal red algae, *J. Exp. Mar. Biol. Ecol.,* 20, 237, 1975.
29. Lüning, K., Critical levels of light and temperature regulating the gametogenesis of three *Laminaria* species (Phaeophyceae), *J. Phycol.,* 16, 1, 1980.

30. **Yokohama, Y.**, A comparative study on photosynthesis-temperature relationships and their seasonal changes in marine benthic algae, *Int. Rev. Ges. Hydrobiol.*, 58, 463, 1973.
31. **Hata, M. and Yokohama, Y.**, Photosynthesis-temperature relationships in seaweeds and their seasonal changes in the colder region of Japan, *Bull. Jpn. Soc. Phycol.*, 24, 1, 1976.
32. **Healey, F. P.**, Photosynthesis and respiration of some arctic seaweeds, *Phycologia*, 11, 267, 1972.
33. **Newell, R. C. and Pye, V. I.**, Seasonal variations in the effect of temperature on the respiration of certain intertidal algae, *J. Mar. Biol. Assoc. U.K.*, 48, 341, 1968.
34. **Grintal, A. R.**, The effect of temperature on the rate or respiration in *Laminaria saccharina*, *Bot. Zh. SSSR.*, 61, 1068, 1976.
35. **Björkman, O.**, Carboxydismutase activity in shade-adapted and sunadapted species of higher plants, *Physiol. Plant.*, 21, 1, 1968.
36. **Wheeler, W. N.**, unpublished data, 1979.
37. **Weigel, H. -P.**, Temperature and salinity observations from Helgoland Reede in 1976, *Ann. Biol.*, 33, 35, 1978.
38. **Bidwell, R. G. S.**, The carbon dioxide machine or plants on the make, *Proc. N.S. Inst. Sci.*, 28, 27, 1977.
39. **Jolliffe, E. A. and Tregunna, E. B.**, Studies on $HCO_3^-$ ion uptake during photosynthesis in benthic marine algae, *Phycologia*, 9, 293, 1970.
40. **Leyton, L.**, *Fluid Behaviour in Biological Systems*, Clarendon Press, Oxford, 1975.
41. **Levich, V. G.**, *Physiochemical Hydrodynamics*, Prentice-Hall, Englewood Cliffs, N.J., 1962.
42. **Nobel, P. S.**, *Introduction to Biophysical Plant Physiology*, W. H. Freeman, San Francisco, 1974.
43. **Schwenke, H.**, Water movement — plants, in *Marine Ecology*, Vol. 1(Part 2), Kinne, O., Ed., Wiley Interscience, New York, 1971, 1091.
44. **Lommen, P. W., Schwintzer, C. R., Yocum, C. S., and Gates, D. M.**, A model describing photosynthesis in terms of gas diffusion and enzyme kinetics, *Planta*, 98, 195, 1971.
45. **Pasciak, W. J. and Gavis, J.**, Transport limitation of nutrient uptake in phytoplankton, *Limnol. Oceanogr.*, 19, 881, 1974.
46. **Ogata, E. and Matsui, T.**, Photosynthesis in several marine plants of Japan in relation to carbon dioxide supply, light and inhibitors, *Jpn. J. Bot.*, 19, 83, 1965.
47. **Raven, J. A.**, Exogenous inorganic carbon sources in plant photosynthesis, *Biol. Rev.*, 45, 167, 1970.
48. **Bowes, G. W.**, Carbonic anhydrase in marine algae, *Plant Physiol.*, 44, 726, 1969.
49. **Okazaki, M. and Furuya, K.**, Carbonic anhydrase in algae, *Proc. Int. Seaweed Symp.*, 7, 522, 1972.
50. **Graham, D. and Smillie, R. M.**, Carbonate dehydratase in marine organisms of the Great Barrier Reef, *Aust. J. Plant Physiol.*, 3, 113, 1976.
51. **Weidner, M. and Küppers, U.**, Phosphoenolpyruvat-Carboxykinase und Ribulose-1.5-diphosphat-Carboxylase von *Laminaria hyperborea* (Gunn.) Fosl.: Das Verteilungsmuster der Enzymaktivitäten in Thallus, *Planta*, 114, 365, 1973.
52. **Willenbrink, J., Rangoni-Kübbeler, M., and Tersky, B.**, Frond development and $CO_2$-fixation in *Laminaria hyperborea*, *Planta*, 125, 161, 1975.
53. **Kremer, B. and Küppers, U.**, Carboxylating enzymes and pathway of photosynthetic carbon assimilation in different marine algae — evidence for the $C_4$ pathway?, *Planta*, 133, 191, 1977.
54. **Akagawa, H., Ikawa, T., and Nisizawa, K.**, The enzyme system for the entrance of $^{14}CO_2$ in the dark fixation of brown algae, *Plant Cell Physiol.*, 13, 999, 1972.
55. **Craigie, J. S.**, Dark fixation of $C^{14}$-bicarbonate by marine algae, *Can. J. Bot.*, 41, 317, 1963.
56. **Akagawa, H., Ikawa, T., and Nisizawa, K.**, $^{14}CO_2$-fixation in marine algae with special reference to the dark fixation in brown algae, *Bot. Mar.*, 15, 126, 1972.
57. **Raven, J. A.**, The mechanism of photosynthetic use of bicarbonate by *Hydrodictyon africanum*, *J. Exp. Bot.*, 19, 193, 1978.
58. **Ikemori, M. and Nishida, K.**, Inorganic carbon source and the inhibition of diamox on the photosynthesis of marine algae — *Ulva pertusa*, *Ann. Rep. Noto Mar. Lab. (Kanazawa Daigaku Rigakubu Rinkai Jikkenjo Nenpo)*, 7, 1, 1966.
59. **Emerson, R., and Green, L.**, Manometric measurements of photosynthesis in the marine alga *Gigartina*, *J. Gen. Physiol.*, 18, 817, 1934.
60. **Thomas, E. A. and Tregunna, E. B.**, Bicarbonate ion assimilation in photosynthesis by *Sargassum muticum*, *Can. J. Bot.*, 46, 411, 1968.
61. **Tseng, C. K. and Sweeney, B. M.**, Physiological studies of *Gelidium cartilagineum*. I. Photosynthesis, with special reference to the carbon dioxide factor, *Am. J. Bot.*, 33, 706, 1946.
62. **Robbins, J. V.**, Effects of physical and chemical factors on photosynthetic and respiratory rates of *Palmeria palmata* (Florideophyceae), *Proc. Int. Seaweed Symp.*, 9, 273, 1979.
63. **Tolbert, N. E.**, Photorespiration, in *Algal Physiology and Biochemistry*, Stewart, W. D. P., Ed., Blackwell Scientific, Oxford, 1974, 474.
64. **Black, C. C., Jr., Burris, J. E., and Everson, R. G.**, Influence of oxygen concentration on photosynthesis in marine plants, *Aust. J. Plant Physiol.*, 3, 81, 1976.

65. Downton, W. J. S., Bishop, D. G., Larkum, A. W. D., and Osmond, C. B., Oxygen inhibition of photosynthetic oxygen evolution in marine plants, *Aust. J. Plant Physiol.,* 3, 73, 1976.

66. Turner, J. S., Todd, M., and Brittain, E. G., The inhibition of photosynthesis by oxygen. I. Comparative physiology of the effect, *Aust. J. Biol. Sci.,* 9, 494, 1956.

67. Björkman, O., The effect of oxygen concentration on photosynthesis in higher plants, *Physiol. Plant.,* 19, 618, 1966.

68. Tolbert, N. E. and Garey, W., Apparent total $CO_2$ equilibrium point in marine algae during photosynthesis in seawater, *Aust. J. Plant Physiol.,* 3, 69, 1976.

69. Akazawa, T. and Osmond, C. B., Structural properties and Ribulose bisphosphate carboxylase and oxygenase activity of fraction-1 protein from the marine alga *Halimeda cylindracea* (Chlorophyta), *Aust. J. Plant Physiol.,* 3, 93, 1976.

70. Randall, D. D., Phosphoglycollate phosphatase in marine algae: isolation and characterization from *Halimeda cylindracea, Aust. J. Plant Physiol.,* 3, 105, 1976.

71. Tolbert, N. E., Glycollate oxidase and glycollate dehydrogenase in marine algae and plants, *Aust. J. Plant Physiol.,* 3, 93, 1976.

72. Burris, J. E., Holm-Hansen, O., and Black, C. C., Jr., Glycine and serine production in marine plants as a measure of photorespiration, *Aust. J. Plant Physiol.,* 3, 87, 1976.

73. Haugh, R. A., Light and dark respiration and release of organic carbon in marine macrophytes of the Great Barrier Reef region, *Aust. J. Plant Physiol.,* 3, 63, 1976.

74. Fogg, G. E., Release of glycollate from tropical marine plants, *Aust. J. Plant Physiol.,* 3, 57, 1976.

75. Black, C. C., Jr. and Bender, M. M., $\delta^{13}C$ values in marine organisms from the Great Barrier Reef, *Aust. J. Plant Physiol.,* 3, 25, 1976.

76. Surbeck, I. E., Holt, V., and Lund, E. J., Effect of oxygen and carbon dioxide concentration on inhibition of respiration and photosynthesis by KCN, *Proc. Soc. Exp. Biol. Med.,* 23, 681, 1962.

77. Johnston, C. S., Jones, R. G., and Hunt, R. D., A seasonal carbon budget for a lamarian population in a Scottish Sealoch, *Helgol. Wiss. Meeresunters.,* 30, 527, 1977.

78. North, W. J., The Role of Trace Metals in Nutrition of Giant Kelp, *Macrocystis,* presented at Symposium on Trace Metal Cycling in the Coastal Zone, Blacksburg, Va., June 27, 1978, 1.

79. Fries, L., Influence of iodine and bromine on growth of some red algae in axenic culture, *Physiol. Plant.,* 19, 800, 1966.

80. Hsio, S. I. C., Life history and iodine nutrition of the marine brown alga, *Petalonia fascia* (O. F. Müll.) Kuntze, *Can. J. Bot.,* 47, 1611, 1969.

81. Pedersen, M., The demand for iodine and bromine of three brown algae grown in bacteria-free cultures, *Physiol. Plant.,* 22, 680, 1969.

82. Ryther, J. H. and Dunstan, W. M., Nitrogen, phosphorus and eutrophication in the coastal marine environment, *Science,* 171, 1008, 1971.

83. Haxo, F. and Strout, P., Nitrogen deficiency and coloration in red algae, *Biol. Bull.,* 99, 360, 1950.

84. Chapman, A. R. O., Markham, J. W., and Lüning, K., Effects of nitrate concentration on the growth and physiology of *Laminaria saccharina* (Phaeophyta) in culture, *J. Phycol.,* 14, 195, 1978.

85. Czygan, F. -C., Seundär-Carotinoide in Grünalgen.I. Chemie, Vorkommen und Faktoren, welche die Bildung dieser Polyene beeinflussen, *Arch. Mikrobiol.,* 61, 81, 1968.

86. Chapman, A. R. O. and Craigie, J. S., Seasonal growth in *Laminaria longicruris:* relations with dissolved inorganic nutrients and internal reserves of nitrogen, *Mar. Biol.,* 40, 197, 1977.

87. Healey, F. P., Inorganic nutrient uptake and deficiency in algae, *CRC Crit. Rev. Microbiol.,* 3, 69, 1973.

88. Chapman, A. R. O. and Craigie, J. S., Seasonal growth in *Laminaria longicruris:* relations with reserve carbohydrate storage and laminarin production, *Mar. Biol.,* 40, 197, 1977.

89. Waite, T. and Mitchell, R., The effect of nutrient fertilization on the benthic alga *Ulva lactuca, Bot. Mar.,* 15, 152, 1972.

90. Prince, J. S., Nutrient assimilation and growth of some seaweeds in mixtures of seawater and secondary sewage treatment effluents, *Aquaculture,* 4, 69, 1974.

91. Ryther, J. H., DeBoer, J. A., and Lapointe, B. E., Cultivation of seaweeds for hydrocolloids, waste treatment and biomass for energy conversion, *Proc. Int. Seaweed Symp.,* 9, 1, 1979.

92. Harlin, M. M., Nitrate uptake by *Enteromorpha* spp. (Chlorophyceae): applications to aquaculture system, *Aquaculture,* 15, 373, 1978.

93. Ogata, E. and Matsui, T., Photosynthesis in several marine plants as affected by salinity, drying and pH, with attention given to their growth habits, *Bot. Mar.,* 8, 199, 1965.

94. Blinks, L. R., The effect of pH upon the photosynthesis of littoral marine algae, *Protoplasma,* 57, 126, 1963.

95. Biebl, R., Physiological aspects of ecology: seaweeds, in *Physiology and Biochemistry of Algae,* Lewin, R. A., Ed., Academic Press, New York, 1962, 799.

96. Skirrow, G., The dissolved gases — carbon dioxide, in *Chemical Oceanography,* Vol. 1, Riley, J. P. and Skirrow, G., Eds., Academic Press, New York, 1965, 227.

97. **Dromgoole, F. I.,** The effects of pH and inorganic carbon on photosynthesis and dark respiration of *Carpophyllum* (Fucales, Phaeophyceae), *Aquat. Bot.,* 4, 11, 1978.

98. **Simpson, F. J., Neish, A. C., Shacklock, P. F., and Robson, D. R.,** The cultivation of *Chondrus crispus.* Effect of pH on growth and production of carrageenan, *Bot. Mar.,* 21, 229, 1978.

99. **Atkins, W. R. G.,** The hydrogen ion concentration of the cells of some marine algae, *J. Mar. Biol. Assoc. U.K.,* 12, 785, 1922.

100. **Strain, H. H.,** The pigments of algae, in *Manual of Phycology,* Smith, G. M., Eds., Ronald Press, New York, 1951, chap. 13.

101. **Wheeler, W. N.,** Effect of boundary layer transport on the fixation of carbon by the giant kelp *Macrocystis pyrifera, Mar. Biol.,* 56, 103, 1980.

# RESPONSE OF TERRESTRIAL PLANTS TO LIGHT QUALITY, LIGHT INTENSITY, TEMPERATURE, CO₂, AND O₂

### R. H. Brown

## RESPONSE TO LIGHT QUALITY AND LIGHT INTENSITY

The quality of light* available for photosynthesis in higher plants varies to a small degree compared to most other controlling factors. The composition of light from the sun varies very little in the wavelengths to which photosynthesis is sensitive, about 350 to 700 nm. There may be some variance, however, in the quality of light received by leaves at different levels in plant canopies. Those at lower levels are increasingly exposed to wavelengths in the mid-range and less effective portion of the photosynthetically active region.

Extracted chlorophyll shows sharp absorption peaks at wavelengths between 650 and 700 nm and below 500 nm, with almost no absorption between 500 and 600 nm. Leaves exhibit no such sharp absorption peaks, but have maximum absorption from the UV up to about 500 nm and a rather broad peak at about 680 nm (see Figure 1). There is a drop in absorption centered on about 550 nm which in Figure 1 reaches about 75% of incident light. For plant communities, the spectral absorption curve flattens even more.[2] In the upper part of Figure 1, the line represents absorption by a corn canopy.

Absorption spectra of individual leaves are relatively unaffected by environment or genetics. Twenty-two species of plants are represented by the data in Figure 1, and there was little variation in absorption among them. Absorption at most wavelengths and particularly those near 550 nm was decreased by growing plants at low nitrogen levels so that leaves were yellow.[1] A similar effect on absorption was demonstrated by Björkman[3] in leaves with low chlorophyll content.

Photosynthesis is not as constant at wavelengths from 400 to 700 nm as is light absorption. In Figure 1, the relative quantum yield (photosynthesis per absorbed quanta) is shown to drop greatly below 400 nm and above 675 nm. Maximum quantum yield occurs in a broad band from about 550 to 675 nm. A second peak occurs at about 425 nm, in which quantum yield is about 75% of maximum. At the lower wavelengths, quantum yield is somewhat higher for plants grown in growth chambers than those grown in the field. This difference is attributed to greater absorption of short wavelengths by nonphotosynthetic tissue in leaves of field grown plants than in growth chamber plants.[1] Field-grown plants had thicker leaves. A similar increase in relative quantum yield at short wavelengths was observed when dicot leaves were illuminated on the lower rather than the upper side.[1] As with the differential response of field- and chamber-grown plants, the higher quantum yield was attributed to lower absorption of UV wavelengths by nonphotosynthetic tissue in leaves illuminated from beneath. Although temperature and CO₂ affect the quantum yield, the relative effects are similar at all wavelengths, so that the spectrum is not changed.[1]

Since the energy used in photosynthetic reduction of CO₂ in plants comes from sunlight, it follows that the rate of CO₂ fixation is proportional to the intensity of sunlight. This proportionality holds, however, only when other environmental factors or physiological processes are nonlimiting. Therefore, leaf photosynthesis is proportional to light intensity only when the radiant energy levels are low (see Figure 2). At higher

---

* In this section, the terms light and photosynthetically active radiation (PAR) are used interchangeably, although PAR is the preferable scientific term denoting quanta of energy available for photosynthesis. PAR units used will be milli-Einsteins per square meter per second⁻¹ (mEin m⁻² sec⁻¹).

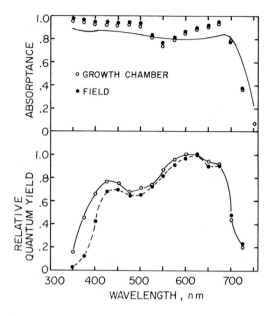

FIGURE 1.    (A) Absorptance of light by leaves (circles) of crop plants and by a whole corn (*Zea mays*) crop (solid line) and (B) relative quantum yield of leaves as a function of wavelength. Quantum yield was measured in terms of $CO_2$ uptake. Open circles represent averages of 20 species grown in growth chambers, and solid circles represent 8 species grown in the field. Data for leaves from McCree[1] and for the corn crop from Lemon.[2] (From McCree, K. J., *Agric. Meteorol.*, 9, 211, 1972. With permission.)

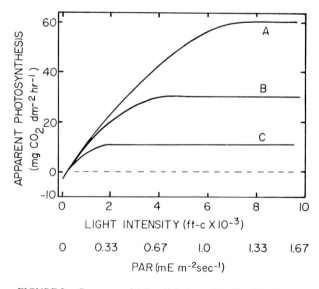

FIGURE 2.    Response of AP to light intensity. The light intensity scale is in foot candles (fc) and also in PAR units mEin m⁻² sec⁻¹. The generalized curves (A, B, and C) represent conditions given in Table 1.

levels of light, other factors such as diffusion of $CO_2$ to the chloroplast become limiting.

At low light intensity the response is steepest, representing maximum efficiency of radiant energy conversion (quantum efficiency). This efficiency of energy conversion under 30°C and atmospheric $CO_2$ and $O_2$ levels is 0.05 to 0.06 mol of $CO_2$ reduced per absorbed quantum of light.[4,5] The efficiency of conversion is relatively constant in $C_4$ species over a range of temperatures and oxygen concentrations, but it is reduced by high temperatures and high $O_2$ levels in $C_3$ species.[4,5] The constancy of maximum quantum efficiency in $C_4$ plants apparently results from near elimination of photorespiration which in $C_3$ plants is increased by high $O_2$ and temperature.

A small amount of radiant energy (.01 to 0.03 mE m$^{-2}$ sec$^{-1}$) is needed to offset the $CO_2$ lost in respiration, and this intensity is referred to as the "light compensation point". It is an important characteristic in the adaptation of leaves to low-light environments, since plants must be above the compensation point for a considerable part of the day to insure survival. Survival in low light environments is favored by low respiration rates and by high quantum efficiency.[6]

As the radiant energy absorbed by the leaf is increased, the response of apparent photosynthesis (AP)* becomes less steep, and when other factors limit photosynthesis to the degree that light intensity no longer affects the rate, leaf photosynthesis is "light saturated" (see Figure 2). It is characteristic of photosynthetic light response that leaves with high AP rates require high light intensities for saturation (see Figure 2). The number of environmental and physiological factors affecting maximum AP and light saturation intensities are too numerous to discuss here in any detail, but some of the more important are discussed below and listed in Table 1.

Plant species differ greatly in their response to light. Species which are adapted to areas of high insolation generally require high light levels for maximum AP. Even among sun plants, however, there is great variation in light response. Plants with the $C_4$ pathway of $CO_2$ assimilation require higher intensities for saturation than do $C_3$ species (see Figure 3). An extreme example of the high light response of $C_4$ species is the desert species, *Tidestroma oblongifolia*, which has a linear response of AP up to full sunlight.[8] Most woody forest species are at the other extreme of light response and saturate at less than one third of full sunlight.[9]

The most prevalent underlying causes of light saturation are probably the low supply of $CO_2$ at the fixation site in the chloroplast and reduced enzyme activity. The light response curves for different $CO_2$ concentrations (see Figure 2) have a similar slope at low light intensities, but at high $CO_2$ concentrations, AP rates and saturating light intensities are increased. The more nearly linear light response curves of $C_4$ plants may be due to the greater affinity for $CO_2$ by phosphoenolpyruvate carboxylase and the concentration of $CO_2$ in bundle sheath cells, the site of ribulose biphosphate (RuBP) carboxylase. Thus, $CO_2$ is not as limiting to AP in $C_4$ as in $C_3$ species, and higher light intensity is required for saturation. The effects of several environmental factors on the light response curve may be indirect effects of $CO_2$ and similar to those shown in Figure 2. Since $CO_2$ supply to the chloroplast is chiefly by diffusion, partial stomatal closure will reduce the $CO_2$ supply and result in light response represented by the lower curve in Figure 2.

Temperatures below or above optimum reduce both maximum AP and the light intensity required for saturation. In addition, dark respiration is increased as temperature rises, and in $C_3$ species the initial slope of the light response is decreased.[4,5] Therefore, the steepest portions of the light response curves representing 15, 30, and 45°C (see Figure 2) would be slightly different. The reduction in AP and lowering of saturation light levels caused by supra- and suboptimum temperatures may be partially caused by reduced $CO_2$ supply, through stomatal closure, but may also be direct effects

* AP is the net $CO_2$ exchange and is not corrected for respiration.

Table 1
FACTORS AFFECTING PHOTOSYNTHETIC LIGHT
RESPONSE OF LEAVES

| Parameter | Light response curves in Figure 2 | | |
|---|---|---|---|
| | A | B | C |
| 1. Maximum AP (mg $CO_2$ $dm^{-2}$ $hr^{-1}$) | 60 | 30 | 10 |
| 2. Saturating PAR (mEin $m^{-2}$ $sec^{-1}$) | 1.4 | 1.0 | 0.5 |
| 3. $CO_2$ concentration ($\mu L$ $L^{-1}$) | 600 | 300 | 100 |
| 4. Stomatal conductance (cm $sec^{-1}$) | — | 0.5 | 0.2 |
| 5. Temperature (°C) | — | 30 | 15,45 |
| 6. Leaf age beyond full expansion days) | | 0 | + 20 |
| 7. Sunlight during development | | Full sunlight | 1/3 sunlight |
| 8. Leaf water potential (bars) | | −4 | −15 |
| 9. Leaf nitrogen content (% of dry weight) | | 5 | 2 |

*Note:* Data are approximate for parameters associated with the generalized curves
for a $C_3$ plant in Figure 2. In consideration of Parameters 3 through 9, other
conditions are assumed optimum for apparent photosynthesis.

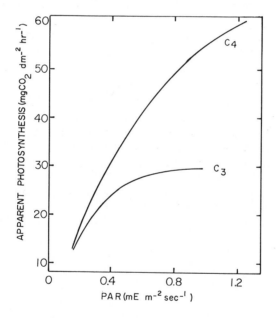

FIGURE 3.   Response of AP to light intensity in a $C_4$
species (johnsongrass — *Sorghum halepense*) and a $C_3$
species (tall fescue — *Festuca arundinacea*).[7]

on enzymatic reaction rates associated with $CO_2$ fixation and at high temperatures to
the increase in respiration, particularly photorespiration in $C_3$ species.

Sunlight intensity under which leaves develop also determines their light response.
Leaves which develop in full sunlight require much higher intensities for saturation
than do those developing in shade.[6,10] Björkman[11] and Singh et al.[12] found that the
higher AP in "sun" leaves was associated with higher levels of RuBP carboxylase.
Sun leaves are also much thicker and have more compact mesophyll tissue which may
help account for their higher light response. Plants adapted to deep shade appear to
have a slightly higher initial slope of the light response curve than do leaves of sun
plants[6] which may be an adaptation for survival in deep shade.

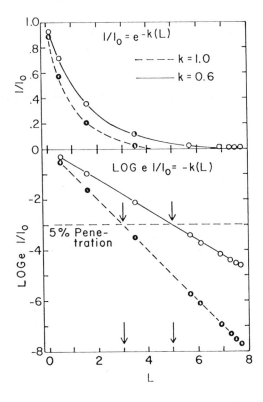

FIGURE 4.   Light penetration (I) through a plant
stand expressed as a fraction of light at the top of
the stand ($I_o$) and plotted against L. The Naperian
logarithm of the light penetrating the stand is plotted
in the lower part of the graph. Arrows indicate L at
95% light interception.

Photosynthetic response to light by plant communities is more complex than for
individual leaves because communities are made up of many leaves which differ in
age, orientation, and degree of shading. In addition, plant communities have nonphotosynthetic organs which contribute variable amounts to the respiratory load. The absorption of light by plant communities is also more complex than absorption by individual leaves further complicating light responses.

Two main factors which control light penetration and AP of plant communities are
the amount and orientation of leaf surfaces. Leaf surface in plant communities is usually expressed as the ratio of leaf surface (one side only) to ground surface and is
referred to as leaf area index (L). The relationship of light penetration to L is shown
in Figure 4 and conforms generally to the Beer-Lambert law as follows:

$$\text{Log}_e I/I_o = -kL \qquad (1)$$

where $I_o$ = light intensity above the canopy, I = light, intensity below the canopy, L
= the leaf area index, and k = the extinction coefficient of the foliage. Figure 4
represents two foliage types having extinction coefficients of 0.6 and 1.0. The foliage
with the lower k value requires a higher L to intercept a given percentage of the incident
radiation. For example, the L required to intercept 95% of the incident radiation
(sometimes referred to as "critical L") is 5.0 and 3.0, respectively, for foliages with k
of 0.6 and 1.0. Foliages with low extinction coefficients are usually characterized by
(1) erect leaves (leaf surfaces tending to parallel the light rays) or (2) considerable
clumping of leaves in the plane perpendicular to the direction of the radiation.

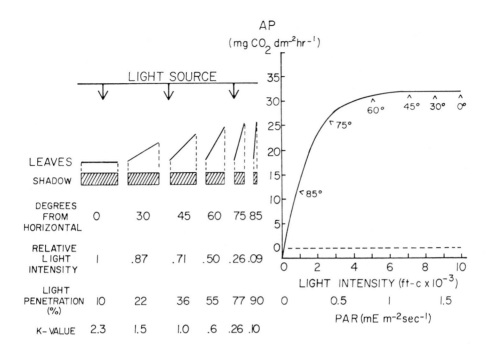

FIGURE 5.   Diagrammatic illustration of the influence of leaf angle on relative light intensity at the leaf surface, light penetration through a single leaf layer, and the extinction coefficient (K value). The curve at right shows AP of a leaf plotted against light intensity (fc) and PAR up to nearly full sunlight. Positions marked on the curve show AP in full sunlight for leaves at different angles from horizontal (assuming sun is directly overhead).

The difference in light response of individual leaves and plant communities may be visualized with the aid of Figure 5 in which leaves are represented by lines drawn at several angles from horizontal. Light response curves for individual leaves reported in the literature are developed with light rays predominantly perpendicular to the leaf surface (represented in Figure 5 as horizontal). As the leaf surface becomes more nearly parallel to the light rays, the effective intensity at the leaf surface decreases. A leaf inclined 85% from horizontal would be irradiated with only about 10% of the intensity at the horizontal plane, and its AP would be about one third of maximum. Thus, leaves whose surfaces are at oblique angles to the sun may never reach light saturation. By reference to the light response curve in Figure 5, it may be seen that leaves may be inclined 60° from horizontal with very little effect on their AP, if light intensity on the horizontal plane is equivalent to full sunlight. Leaves inclined more than 75° show a steep decline in AP. Although steeply inclined leaves have low AP rates, they also cast small shadows, so that leaves lower in the canopy are better illuminated. Lower leaves in such a canopy may contribute nearly as much to total AP as upper, erect leaves. In a canopy with much of the leaf surface at angles oblique to the sun's rays, several leaf layers may receive low enough light to be on a nearly linear portion of the response curve.

In plant canopies with low L, a horizontal leaf arrangement is most advantageous for maximum light interception. As L increases, mutual shading of leaves occurs and is greater with greater with horizontal leaf orientation. In dense plant canopies, increasingly vertical leaf angles become beneficial because more of the leaf surfaces can be illuminated. The advantage of erect leaf angle in AP has been shown experimentally with barley seedlings.[13] The results in Figure 6 show that at low L values (about 2) there was no effect of leaf angle on AP. At high L, however, there was much higher

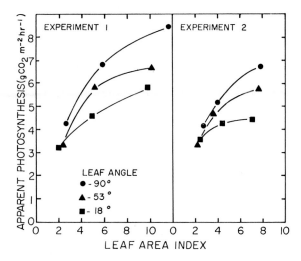

FIGURE 6. The influence of leaf angle from horizontal on AP of barley (*Hordeum vulgare*) seedling communities at different L values. (From Pearce, R. B., Brown, R. H., and Blaser, R. E., *Crop Sci.*, 7, 321, 1967. By permission of Crop Science Society of America.)

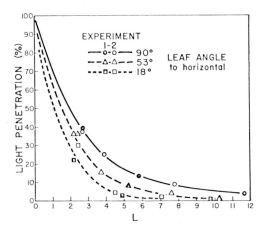

FIGURE 7. Light penetration through barley seedling communities as a function of L and leaf angle. Data are from Pearce et al.[13] and the experiments shown in Figure 6. (From Pearce, R. B., Brown, R. H., and Blaser, R. E., *Crop Sci.*, 7, 324, 1967. By permission of Crop Science Society of America.)

AP in communities with erect leaves. It may be seen in Figure 7 that those seedlings with erect leaves allowed greater light penetration to the soil at a given L than leaves at 53 or 18° from horizontal. The higher AP in dense canopies of erect leaves must have been less mutual shading of leaves, since canopies with more erect leaves intercepted less light. Leaf angle effects in the field may be less pronounced than in the barley seedling experiments because sunlight direction changes throughout the day, and leaf angles may also change because of wind movement or phototropic responses.

Generally, plant communities require much higher irradiance levels for saturation than individual leaves. In fact, many dense plant canopies show linear increases in AP

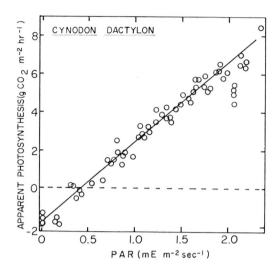

FIGURE 8.    Response of AP of a coastal Bermuda-grass (*Cynodon dactylon*) sod to light intensity. L = 5.[92]

up to the highest sunlight intensities (see Figure 8). In communities with low L and particularly those with horizontal leaves, the response curve may saturate because all leaves are well enough illuminated to be saturated at less than full sunlight. In addition, AP in plant canopies may be light saturated if environmental or physiological conditions are substantially suboptimal for AP, as, for example, in the case of advanced plant maturity.

Dark respiration is much more variable in plant communities than in individual leaves, and therefore its influence on the light response is quite variable. Plant communities with high respiratory loads because of mutual shading of leaves or a high percentage of nonphotosynthetic tissue have less steep light response curves and require rather high light levels for photosynthesis to compensate for respiration, i.e., high light compensation values (see Figure 9). In canopies with very high relative respiration rates, photosynthesis may not compensate for the loss of $CO_2$ at any light intensity. For example, orchardgrass sods soon after mowing (see Figure 9) were light saturated at about one third of full sunlight, but even at saturation AP was $-0.34$ g $CO_2$ m$^{-2}$ hr$^{-1}$.[14] Leaf area index of the orchardgrass sods after mowing from 33 to 10 cm was about 1, but the plants were composed mostly of roots and culm bases.

Although the light response of photosynthesis in higher plants is one of the most thoroughly studied responses, it is still not well enough characterized to predict responses in the widely variable environments in which photosynthesis occurs. Leaves of some species are especially suited for photosynthesis in desert environments and others for survival in the dim light of the forest floor. In many environments such as those under which crops grow, the leaf may be fully exposed early in its ontogeny, but become shaded as it ages. As the sun moves during the day, leaves may be alternately sunlit and shaded. Thus, the light response of plant communities is complicated by variability in leaf age, position, and aging patterns, as well as by many other factors discussed earlier. Photosynthesis by organs other than leaves may also contribute to the photosynthetic complexity of the plant community. In cereal crops, for example, the ears, especially if awned, may contribute significantly to photosynthesis of the crop.

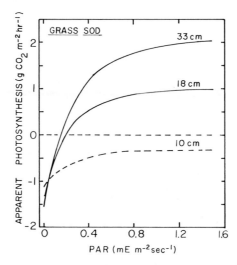

FIGURE 9.   Response of an orchardgrass (*Dactylis glomerata*) sod to light intensity before cutting (33 cm) and after mowing back to 18 and 10 cm. L was approximately 7, 3.5, and 1 for sods of 33, 18, and 10 cm, respectively. (From Pearce, R. B., Brown, R. H., and Blaser, R. E., *Crop Sci.*, 5, 553, 1965. By permission of Crop Science Society of America.)

## Response to Temperature

Photosynthetic $CO_2$ assimilation of terrestrial plants is influenced by environmental temperatures through several mechanisms, and therefore the effects are complex. The diffusion of $CO_2$ into the leaf is controlled by stomatal opening which responds to temperature. By raising the water-holding capacity of air surrounding the leaf and the kinetic activity of water molecules, high temperatures increase transpiration rates and make water stress more likely. Thus, temperature may affect AP indirectly through effects on water status. Direct effects of temperature on enzymatic reaction rates cause increases in AP rates as temperature is raised up to an optimum, beyond which rates decrease because of increases in respiration and/or inactivation of photosynthetic enzymes. In addition to the effects of temperature on stomatal opening and enzymatic reaction rates, membrane permeability and solubility of the gases $O_2$ and $CO_2$ in cytoplasm are influenced by temperature.

The large number of characteristics involved in response of AP to temperature allows plants to adapt to a wide range of climatic temperature regimes. An example of this adaptation is given in Figure 10 which shows the response of AP to temperature for a $C_3$ species adapted to a cool coastal environment and a $C_4$ species which grows during summer in a hot desert.[8] In most species adapted to warm environments, AP rises steeply as temperature is increased from near 5 to about 30°C. For the $C_4$ species, *Tidestromia oblongifolia* (see Figure 10), AP drops to near zero at 10°C, and the optimum temperature is about 47°C. On the other hand, many plants adapted to cool climates have rather constant AP rates over the range of 10 to 30°C as illustrated by *Atriplex glabriuscula* ($C_3$) in Figure 10.

The $C_4$ pathway of $CO_2$ fixation appears to be one mechanism by which plants adapt to warm environments. In early studies, it became clear that $C_4$ species have higher temperature optima for AP than do $C_3$ plants.[15] This is particularly true when $C_4$ plants are compared to $C_3$ species adapted to cool climates, but even tropical legumes ($C_3$)

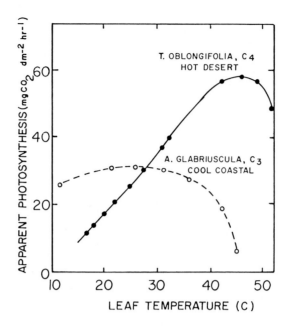

FIGURE 10. Effects of temperature on AP of a $C_4$ species from a hot desert (*Tidestromia oblongifolia*) and a $C_3$ species from a cool coastal climate (*Atriplex glabriuscula*). (From Björkman, O., in *Environmental and Biological Control of Photosynthesis*, Marcelle, R., Ed., Dr. W. Junk, The Hague, 1975, 10. With permission.)

have lower temperature optima than tropical grasses (see Table 2).[16] The tropical legumes in Table 2 also had lower maximum temperatures (temperature at which AP dropped to zero) than the grasses, although minimum temperatures for AP were similar. The higher temperature optimum for $C_4$ plants apparently derives in part from their lack of photorespiration. Photorespiration in $C_3$ species has been shown to be higher at temperatures between 15 and 30°C,[17-19] and therefore if photorespiration in $C_3$ plants is repressed, as when leaves are exposed to low $O_2$ concentrations, the optimum for AP shifts to higher temperatures.[20] Although $C_4$ species generally have higher temperature optima than $C_3$ species, Björkman[8] has shown that $C_4$ *Atriplex* species adapted to cool climates may have similar temperature responses to those of $C_3$ species.

Photosynthetic response varies greatly among species, but variation among ecotypes or cultivars within species also may be considerable. Ecotypes of temperate grasses from Mediterranean and continental climates have been shown to differ in the response of AP to temperature. When *Dactylis glomerata* ecotypes from Norway and Portugal were grown at 25°C, AP rate and its response to measurement temperature were similar in the two ecotypes.[21] When grown at 5°C, the ecotype from Norway had twice the AP rate and a much greater respones to temperature during measurement than the Portuguese ecotype. In coastal and desert ecotypes of the $C_4$ plant *Atriplex lentiformis*, response of AP to temperature was related to the environment to which they were adapted (see Figure 11).[22] The two ecotypes had similar AP rates, and short-term response to temperature was also similar when plants were grown under a 23-18°C, day-night regime. The coastal ecotypes, however, had much lower rates than those from the desert under a 43-30°C regime.

The temperature under which plants are grown thus influences the short-term response of AP to temperature. In general, the optimum temperature for AP is shifted in the direction of the growth temperature. For example, ecotypes of *A. lentiformis*

### Table 2
### OPTIMUM, MAXIMUM, AND MINIMUM LEAF
### TEMPERATURES FOR APPARENT
### PHOTOSYNTHESIS IN C₃ AND C₄ SPECIES[16]

| | Leaf temperature (°C) | | |
| --- | --- | --- | --- |
| | Minimum | Optimum | Maximum |
| C₃ Legumes | 6.3 | 31.5 | 50.1 |
| C₄ Grasses | 7.0 | 38.3 | 57.3 |

*Note:* Data are averages for four C₃ legumes and six C₄ grass species.

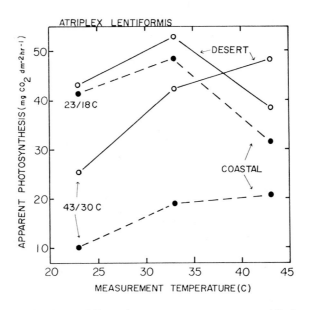

FIGURE 11.   Effects of measurement temperature on AP of leaves of coastal and desert races of *Atriplex lentiformis* grown at day/night temperatures of 23/18°C or 43/30°C.[22]

(Figure 11) had optimum temperatures of near 43°C when grown under a daytime temperature of 43°C. However, at a temperature of 23°C during the day, the optimum for AP was near 33°C. Similar shifts in temperature optima have been observed in temperate grasses,[20] apricot,[23] the desert shrubs *Hammada scoparia*[23] and *Larrea divaricata*,[24] and pine.[25] It was found by Strain et al.[25] that the optimum temperature for AP in *Pinus taeda* shifted from approximately 25°C during the summer months to 10°C in winter.

The response of leaf AP to temperature may be mediated by changes in biochemical and/or diffusive characteristics. The differential responses of plants with C₃ and C₄ pathways have already been referred to, but other biochemical mechanisms must also be involved in temperature response. *Atriplex sabulosa*, a C₄ plant adapted to a cool climate, responds to temperature like the C₃ species, *A. glabriuscula*, while the C₄ desert species *Tidestromia oblongifolia* has a 4°C temperature optimum and is very sensitive to low temperature.[8] The differences in temperature response among these species are therefore not closely tied to the C₄ mechanism. The different responses were not caused by stomatal conductance changes either, since stomatal conductance was nearly constant over the range of temperature tested. Greater sensitivity of AP to high temperature in *A. sabulosa* than in *T. oblongifolia* was related to greater heat stability of "photosystem II-driven electron transport and noncyclic photophosphorylation" in *T. oblongifolia* and of some photosynthetic enzymes, notably NADP glyceraldehyde 3-P dehydrogenase and ribulose 5-P kinase.[26] Decreases in photosynthetic capacity of the desert shrub, *Larrea divaricata*, at high temperature has been attributed to a blockage of excitation energy transfer from chlorophyll *b* to chlorophyll *a* and changes in distribution of excitation energy between photosystems I and II.[27] Acclimation of this species to higher temperatures involves increased stability of the interaction between the light harvesting pigments and the photosystem reaction centers.

The mechanism of control of AP by temperature may differ above and below the optimum. Since chemical reactions decrease in rate at low temperatures, AP is more likely to be limited by enzymatic activity than diffusion processes at temperatures below optimum. In Sitka spruce, mesophyll conductance ($K_m$) was much lower than stomatal conductance ($K_s$) at low temperatures, and therefore changes in $K_m$ with temperature exerted the main control of AP.[28] A similar response was found in bean[29] and can be deduced from results of Pearcy[30] on *A. lentiformis*. The lower $K_m$ relative to $K_s$ at low temperatures indicates control of AP by enzymatic activity, since $K_m$ contains metabolic as well as physical diffusion components. Ribulose bisphosphate carboxylase which has been found to be closely associated with AP is increased in activity by raising the assay temperature in the range of 5 to 30°C,[20,21] and the low activities of this enzyme may be a factor in the limitation of AP at low temperatures. This low activity of RuBP carboxylase in vitro is sometimes offset by higher activity and presumably increased synthesis of the enzyme inplants grown at low temperature.[31] Thus, acclimation of AP to low temperature may be via increased synthesis of RuBP carboxylase.

A low temperature inhibition of photosynthesis which may be fairly common in tropical grasses is the exposure to cool nights. Overnight temperatures of 10°C have been shown to decrease AP on the following day compared to 30°C night temperatures.[32-34] The "cool night" effect on AP was associated with the inhibition of starch breakdown in chloroplasts during the dark period.[34,35] Karbassi et al.[35] found that amylase activity was reduced at temperatures which prevented starch breakdown and that gibberellic acid overcame the inhibition of amylase. The accumulation of starch and soluble carbohydrate in leaves and culms of cool season grasses and other plants[36] at low temperatures may indicate that AP is reduced at low temperatures in many plants by photosynthate accumulation.

Decreases in AP above the optimum temperature have been associated with decreased $K_s$ in several studies, but not in all. Corresponding decreases in $K_s$ and AP at high temperatures have been reported in tall fescue,[20] bean,[29] potato,[37] spruce,[28] and other species.[38,39] In a number of experiments in which $K_s$ was decreased at high temperature, $K_m$ was also decreased,[28-30] so it is questionable whether the effect was exerted through $K_s$ or $K_m$. A decrease in $K_s$ because of high temperature would lower intercellular $CO_2$ concentrations in the leaf unless some biochemical or physical component of $K_m$ were also changed so that assimilation capacity is decreased. In experiments with *A. lentiformis*,[31] decreases in AP at high temperatures were associated with increased rather than decreased intercellular $CO_2$ (see Figure 12). Therefore, some component of $K_m$ was decreased to a greater degree than $K_s$ by high temperature. Since $K_m$ includes both biochemical and physical diffusion components, it is uncertain whether changes in $K_m$ at high temperature represent biochemical or physical alterations of the leaf. Studies which show high temperature sensitivity of electron transport in leaves[26,27] indicate that in many environments AP is reduced at high temperatures by direct effects on biochemistry of photosynthesis.

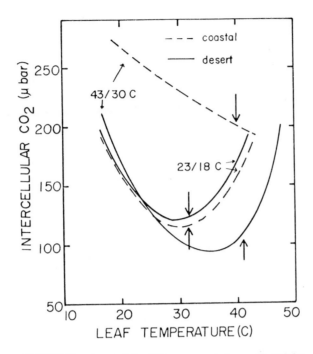

FIGURE 12. Intercellular $CO_2$ pressures in leaves of *Atriplex lentiformis* as a function of leaf temperature. Vertical arrows indicate optimum temperature for AP. Plants were from coastal and desert habitats and were grown at 43/30°C and 23/18°C day/night temperatures. (From Pearcy, R. W., *Plant Physiol.*, 59, 797, 1977. With permission.)

The interaction of $O_2$ and $CO_2$ concentrations on AP of $C_3$ plants is widely known, and Ku and Edwards[40] have proposed that the oxygen response of AP in $C_3$ species at moderate to high temperatures is related to the differential effect of temperature on the solubility of $O_2$ and $CO_2$. Since AP is favored by increased $CO_2$ and decreased $O_2$ concentrations, it may decrease with temperature if the solubility of $O_2$ relative to $CO_2$ is greater at higher temperatures. The solubility of $CO_2$ decreases at a faster rate than $O_2$ with increased temperature; therefore, an increase in leaf temperature would lower the dissolved $CO_2/O_2$ ratio and increase photorespiration relative to photosynthesis. The significance of changes in this ratio to temperature response above the optimum is uncertain, however, since AP at low $O_2$ concentrations (<2%) also decreases at temperatures not much higher than the optimum at 21% $O_2$.

As mentioned at the outset, control of AP by temperature is complex and mediated by biochemical and diffusive processes. The biochemical processes appear to dominate especially at temperatures near the lower and upper limits of AP. Even in those experiments where stomatal conductance is correlated with AP, it may not be controlling, since, in several cases, the same temperature conditions which reduce $K_s$ also reduce $K_m$, implying reduced biochemical activity and perhaps a feedback control of $K_s$ through changes in photosynthetic processes.

## RESPONSE TO CARBON DIOXIDE

It is axiomatic that the concentration of $CO_2$, a substrate in photosynthesis, controls the rate of assimilation. The concentration in the atmosphere, which is approximately 320 $\mu$L L$^{-1}$ limits AP under saturating light intensities. Most studies of effects of $CO_2$ concentrations have shown a linear dependence of AP on $CO_2$, at least up to the at-

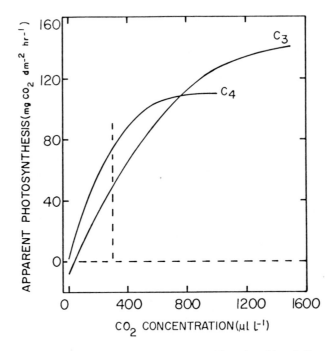

FIGURE 13.    Response of AP of a $C_3$ (*Vigna luteola*) and $C_4$ (*Sorghum almum*) species to $CO_2$ concentration. The vertical broken line indicates 300 $\mu L$ $L^{-1}$ of $CO_2$.[16]

mospheric level. In $C_4$ species, the response tends to level off just above this concentration (Figure 13), but leaves of $C_3$ species may respond in a linear fashion at up to twice atmospheric $CO_2$ concentrations.

The diffusion model for leaf photosynthesis put forward by Gaastra[41] depicts $CO_2$ uptake as proportional to the $CO_2$ concentration gradient from atmosphere to chloroplast and inversely proportional to the sum of resistances to $CO_2$ diffusion down the concentration gradient. Therefore,

$$AP = \frac{C_a - C_{chl}}{r_a + r_s + r_m} \qquad (2)$$

where $C_a$ and $C_{chl}$ are $CO_2$ concentrations outside the leaf and in the chloroplast, respectively, and $r_a$, $r_s$, and $r_m$ are resistances imposed by the leaf surface-air boundary layer, stomata, and mesophyll tissue, respectively. Mesophyll resistance includes metabolic and physical components and is poorly defined in relation to either class of components. In early use of the model, $C_{chl}$ was assumed to be zero, but the $CO_2$ compensation concentration (see below) is more usually taken as $C_{chl}$. It is also more common in recently published papers to express diffusive characteristics as conductances (reciprocals of resistance) primarily because AP is more linearly related to conductance than to resistance.

The generalized curves in Figure 14 show several characteristics of the response of AP to $CO_2$. Over the range of $CO_2$ concentrations shown, the response for $C_3$ species is linear. Line A—B represents the $CO_2$ response of a $C_3$ leaf at 21% $O_2$. The intersect of line A—B at AP = 0 is the $CO_2$ compensation concentration ($\Gamma$), that concentration at which photosynthesis equals respiration. In $C_3$ species, $\Gamma$ is near 50 $\mu L$ $L^{-1}$ at 30°C. Extrapolation of line A—B to a $CO_2$ concentration of zero gives an estimate of pho-

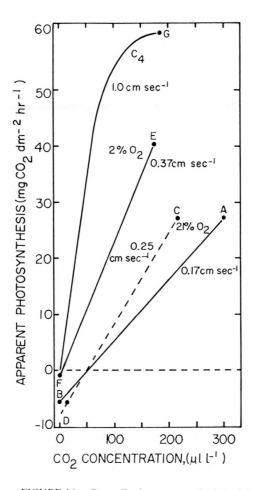

FIGURE 14. Generalized responses of AP to $CO_2$ concentration. Line A—B is for a $C_3$ species at 21% $O_2$ and external $CO_2$ concentrations; C—D is the same AP values as for line A—B plotted against intercellular $CO_2$ concentrations; E—F represents AP of a $C_3$ species at 2% $O_2$ plotted against intercellular $CO_2$ concentrations; and, G—F is the response of AP of a $C_4$ species to intercellular $CO_2$ concentration. Numbers beside lines are the slopes or conductance values.

torespiration. The slope of line A—B represents the conductance of the leaf to $CO_2$ and includes the conductance of the leaf-air boundary layer ($K_a$), stomata ($K_s$), and mesophyll tissue ($K_m$).

Line C—D shows the response of AP to intercellular $CO_2$ concentration. This internal $CO_2$ concentration is calculated from the diffusion model for photosynthesis as,

$$C_i = C_a - \frac{AP}{K_1} \tag{3}$$

where $C_i$ and $C_a$ are the intercellular and external $CO_2$ concentrations, respectively, and $1/K_1 = 1/K_a + 1/K_s$. In construction of Figure 14, $K_1$ was assumed to be constant, although it usually decreases slightly with increased $CO_2$ concentration in the range of 0 to 300 $\mu L\ L^{-1}$. When the calculated $C_i$ is plotted against AP (Line C—D), the slope gives an estimate of $K_m$. It will be noted that lines A—B and C—D cross at the $CO_2$ compensation point, and since C—D is steeper, it intercepts the AP-axis at a lower point, giving a greater and more nearly correct estimate of photorespiration than line A—B. The difference in the AP-axis intercepts of lines A—B and C—D is an estimate of the amount of $CO_2$ evolved from the mesophyll and recycled before escaping through the stomata.

If photorespiration is eliminated or greatly reduced in $C_3$ leaves by reducing $O_2$ concentration from 21 to 1 or 2%, the response to $CO_2$ changes as shown by line E—F in Figure 14. The $CO_2$ compensation point is reduced to less than 10 $\mu L$ $L^{-1}$ and the AP-axis intercept (photorespiration) is about 1 mg $CO_2$ $dm^{-2}$ hr or less. In addition, the slope or $K_m$ is increased compared to line C—D. This increase in slope represents mainly the removal of $O_2$ inhibition of carboxylation of RuBP to form 3-phosphoglyceric acid.

The response of AP to $CO_2$ at low levels is much greater in $C_4$ than in $C_3$ species.[16,43,44] In Figure 14, line G—F represents the response of AP in a $C_4$ leaf to intercellular $CO_2$, and the slope of the linear portion is much steeper than for the $C_3$ species at 21% $O_2$ (Line C—D). It is also steeper than the line for the $C_3$ species at 2% $O_2$ (Line E—F), indicating the greater affinity for $CO_2$ in the carboxylation of phosphoenolpyruvate in $C_4$ species than in the carboxylation of RuBP by $C_3$ species, even when $O_2$ inhibition is suppressed. Line G—F is similar at 21 or 2% $O_2$ because of the lack of $O_2$ inhibition of photosynthesis in $C_4$ species.

Apparent photosynthesis begins to level off at lower intercellular $CO_2$ concentrations in $C_4$ than in $C_3$ plants (Figure 14), and therefore increases in $CO_2$ concentration above atmospheric have much less influence on AP in $C_4$ than in $C_3$ species. If lines C—D and G—F in Figure 14 are extrapolated to intercellular $CO_2$ levels of approximately 400 or 450 $\mu L$ $L^{-1}$, AP rates are equal for the $C_3$ and $C_4$ species. Hesketh[45] showed that AP of maize ($C_4$) exceeded that of sunflower ($C_3$) at $CO_2$ concentrations below 600 $\mu L$ $L^{-1}$, but was less than that of sunflower at between 800 and 1000 $\mu L$ $L^{-1}$. Likewise, Ludlow and Wilson[16] found AP of several tropical grasses ($C_4$) to be higher than those of tropical legumes ($C_3$) at atmospheric levels of $CO_2$, but AP of the $C_3$ species was similar to or higher than $C_4$ at 1200 $\mu L$ $L^{-1}$ of $CO_2$ (see Figure 13). Therefore, $C_4$ species make efficient use of $CO_2$ at concentrations near atmospheric and below, but their response to higher $CO_2$ levels is less than that of $C_3$ species.

Interactions with temperature and light intensity complicate the response of AP to $CO_2$ concentrations. Linear portions of the $CO_2$ response curves of $C_3$ plants are less steep when light intensity is reduced below saturating levels.[46,47] Light intensity has little influence on $\Gamma$, so the $CO_2$ response curves at low light intensities extrapolate to lower photorespiration rates. In $C_4$ species, the $CO_2$ response curves extrapolate to near zero AP at zero $CO_2$, and the initial slopes are similar for a wide range of light intensities.[48] The generalized curves in Figure 15 show the difference in interactions of $CO_2$ with light intensity in $C_3$ and $C_4$ species. The concentration of $CO_2$ required to saturate AP is increased as light intensity is raised in $C_3$ and $C_4$ species, and this has led to the conclusion that $CO_2$ enrichment enhances plant growth to a greater extent as light intensity is increased. The response of AP to $CO_2$ concentration is steepest at temperatures optimum for AP and tends to become less steep at temperatures below and above.[48] At low temperatures, AP is $CO_2$ saturated at low concentrations, but at high temperatures, it may increase in as near linear fashion as at optimum temperature (see Figure 16).[17]

Increase in $CO_2$ concentrations cause partial stomatal closure, and therefore $K_1$ decreases with $CO_2$ enrichment. The decrease in $K_1$ results in decreased transpiration at higher $CO_2$ levels, and since AP is increased, the efficiency of water use is increased, i.e., more $CO_2$ is fixed per unit of water transpired.

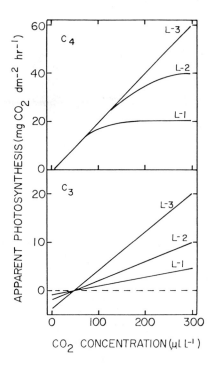

FIGURE 15.   Generalized $CO_2$ response curves for $C_3$ and $C_4$ species as affected by light intensity. L-1, L-2, and L-3 represent increasing light intensities. Patterned after data for $C_4$ [48] and $C_3$ plants.[46,47]

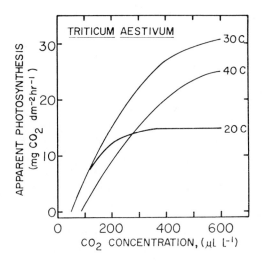

FIGURE 16.   The influence of temperature on the $CO_2$ response of AP in wheat leaves.[17]

Because of the response of AP to $CO_2$ concentration, many attempts have been made to increase yields by $CO_2$ enrichment. Raising $CO_2$ levels above atmospheric has generally increased yields of greenhouse crops,[49] but attempts at $CO_2$ fertilization of field crops have been few and generally unsuccessful. Harper et al.[50] estimated that a

Table 3
EFFECTS OF $CO_2$ CONCENTRATION ON
THE LEAF AREA/PLANT WEIGHT
RATIO (cm$^2$ g$^{-1}$) OF FOUR SPECIES[52]

| Species | $CO_2$ concentration ($\mu$L L$^{-1}$) | | |
|---|---|---|---|
| | 300 | 1000 | 3300 |
| Sugarbeet | 99 | 86 | 68 |
| Kale | 125 | 93 | 81 |
| Barley | 222 | 181 | 138 |
| Corn | 148 | 144 | 165 |

35% increase in daily net photosynthate production resulted from releasing 222.6 kg of $CO_2$/ha/hr in a cotton crop. At the top of the crop, $CO_2$ concentration was raised by about 100 $\mu$L L$^{-1}$ in the enriched plots. Recovery of the released $CO_2$ ranged from 7 to 38%, depending mainly on light intensity.

Response of plant growth or yield to $CO_2$ concentration has been variable and generally less than would be expected from response of AP. As noted by Neales and Nicholls,[51] increases in AP due to $CO_2$ enrichment, if sustained, should cause exponential increases in plant growth. That is, the increased AP should cause increased production of leaf area which in turn increases the total photosynthetic capacity of the plant, especially isolated plants or seedlings. In a summary of AP responses to $CO_2$ reported in the literature, Neales and Nicholls[51] list increases which average 2.5-fold for 8 species when $CO_2$ was raised from 200 to 800 $\mu$L L$^{-1}$. Relative dry matter growth rates, however, were increased by only 1.2-fold over a similar $CO_2$ range in experiments with tomato and wheat seedlings. The fact that growth response to $CO_2$ is similar to or less than AP response is apparently because of a shift in partitioning of dry weight. In several experiments and as shown in Table 3, $CO_2$ enrichment reduced the leaf area/plant weight ratio.[51-53] This reduction in leaf area increase per unit dry weight increase at high $CO_2$ levels partially offsets the advantage of higher AP. Leaf area/plant weight ratio was not reduced by high $CO_2$ in corn, and as will be seen later, growth was not stimulated in corn nor other $C_4$ species by $CO_2$ enrichment.

A second indirect effect of $CO_2$ on AP may be the accumulation of surplus carbohydrate in the leaves which partially inhibits AP. Since AP is stimulated by high $CO_2$ levels, products of photosynthesis may accumulate unless carbohydrate translocation and utilization keeps pace with the increased AP. Nafziger and Koller[54] obtained evidence that starch accumulation caused by exposing plants to elevated $CO_2$ levels (Table 4) reduced AP when measured at 300 $\mu$L L$^{-1}$. Others have also shown that exposing plants to levels of $CO_2$ above atmospheric increases leaf and stem carbohydrate levels,[55,56] and Ford and Thorne[52] observed increased sucrose concentration in sugar beet roots. Therefore the degree of stimulation of growth by high $CO_2$ levels may depend partially upon the degree to which photosynthesis is the limiting factor. If photosynthetic sink capacity is limiting, then $CO_2$ enrichment may increase initial AP rates substantially, but accumulation of photosynthetic products may reduce the effect of $CO_2$ on AP and growth in the longer term. In some experiments, however, no long-term effects of $CO_2$ concentration were observed on AP of individual leaves measured at atmospheric concentrations.[52,57]

Yields of grain have been increased by $CO_2$ enrichment, but the yield response depends on the stage of plant development during enrichment. Enrichment prior to reproductive development usually does not influence yield. Rice yield was increased by about 30% when enrichment with 700 to 1100 $\mu$L L$^{-1}$ occurred during 30 days prior to

Table 4

EFFECTS OF $CO_2$ CONCENTRATION ON LEAF
STARCH LEVELS IN SOYBEAN[54]

| Experiment designation | $CO_2$ concentration ($\mu$L L$^{-1}$) | Starch[a] | |
|---|---|---|---|
| | | mg dm$^{-2}$ | % dry weight |
| Gas exchange | 50 | 64 | 14 |
| | 300 | 129 | 24 |
| | 2000 | 181 | 30 |
| $CO_2$ compensation | 50 | 67 | 15 |
| | 300 | 129 | 25 |
| | 2000 | 205 | 35 |

[a] Leaf starch levels are expressed on leaf area and dry weight bases and rounded to whole numbers.

heading in two experiments[55] and by enrichment to 900 $\mu$L L$^{-1}$ in a third experiment.[58] Enrichment for 30 days following heading increased grain yield by 15 to 20%. The enrichment before heading increased yield mainly through an increase in grain numbers, whereas postheading treatment increased the percentage of grains filled and individual grain weight. Krenzer and Moss[59] reported increases in grain yield of wheat by $CO_2$ enrichment to 600 $\mu$L L$^{-1}$. The increases were in the range of 10 to 20% when enrichment occurred 30 days prior to or after anthesis, except that in one cultivar, yield was increased 38% when fertilized with $CO_2$ after anthesis. Enrichment during the vegetative stage had no effect on grain yield. In an experiment in which $CO_2$ was raised 200 $\mu$L L$^{-1}$ above ambient throughout the growth cycle, Gifford[57] found grain yield to be increased 44% mainly because of a greater number of grain-bearing tillers. Kernel weight was barely affected.

Interactions of $CO_2$, temperature, and light intensity on AP indicate that environmental conditions may influence the growth response to $CO_2$ enrichment, and many of the cases in which little or no response was reported may have involved suboptimum levels of other factors. Soybeans cultured at 22/17°C day/night temperatures did not grow any faster at 1000 $\mu$L L$^{-1}$ $CO_2$ than at 300 $\mu$L L$^{-1}$. At 27/22°C, enrichment nearly doubled the growth rate, and at 36/31°C, the $CO_2$ response was intermediate between that in the other two temperature regimes.[60] In another experiment in which soybeans were grown at temperatures from 15 to 35°C, $CO_2$ enrichment had no effect below 25°C.

The differential growth response of $C_3$ and $C_4$ species to $CO_2$ shown in Figure 17 indicates much less response of growth to $CO_2$ enrichment in $C_4$ species. In fact, the few experiments with $C_4$ plants show no increase in growth at $CO_2$ levels above atmospheric. In Figure 17, results of Akita and Tanaka[61] show that yield of $C_3$ species increased with $CO_2$ enrichment up to 1000 $\mu$L L$^{-1}$ and in cool season species up to 2500 $\mu$L L$^{-1}$, whereas $C_4$ species did not respond. In a comparison of barley, sugar beet, and corn, Ford and Thorne[52] found no response in corn, but significant increases in growth in the other two species (Figure 17). As discussed earlier in this section, the lack of responses of $C_4$ plants to high $CO_2$ concentrations is probably the $CO_2$ concentrating mechanism of the $C_4$ cycle of $CO_2$ fixation.

Even though $CO_2$ is a substrate for photosynthesis, its effects on plant growth are not as straightforward as might be expected. Therefore, $CO_2$ fertilization in the field and greenhouse may produce variable results, depending on environmental conditions and indirect effects of $CO_2$ on partitioning of dry weight during plant growth. In addition, differences in $CO_2$ response exist among species, and a $CO_2$ concentrating

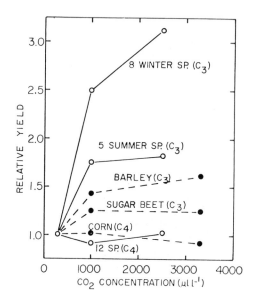

FIGURE 17.   Relative yield of seedlings of several species grown at different $CO_2$ concentrations. Yield at 300 $\mu L$ $L^{-1}$ equals 1. Data represented by solid lines from Akita and Tanaka[61] and by broken lines from Ford and Thorne.[52]

mechanism in $C_4$ plants may make them unresponsive. Certainly yield increases from $CO_2$ enrichment are possible only when photosynthesis is the process limiting yield. The variability in results of $CO_2$ enrichment makes uncertain predictions of increased productivity on a global scale due to rising atmospheric $CO_2$ levels.

## RESPONSE TO OXYGEN

It has been known since the earliest studies of photosynthesis that $O_2$ was a product. Warburg[62] discovered early in this century that $O_2$ also inhibits photosynthesis; however, systematic studies of $O_2$ effects were not conducted with higher plants until the mid-1960s.[63,64] This inhibition of photosynthesis by $O_2$ became known as the "Warburg effect" from his discovery of the inhibition. It is now recognized as a major factor limiting productivity of $C_3$ species.[65]

A close tie between $O_2$ effects and photorespiration in $C_3$ plants was suspected for several years because atmospheric $O_2$ concentrations reduced AP by about 30% and stimulated photorespiration compared to 1 or 2% $O_2$. Stimulated photorespiration at 21% $O_2$ was exhibited as

1.   Increased $CO_2$ evolution from illuminated leaves[66,67]
2.   An increased surge of respiration immediately after darkening of a leaf[63,67]
3.   An increase in the $CO_2$ compensation concentration[64,67]

This association between the $O_2$ inhibition of AP and stimulation of photorespiration in $C_3$ species was strengthened by the fact that $O_2$ had neither effect in $C_4$ species.[19,44,68] The decrease in AP by 21% $O_2$ in $C_3$ species could then be explained by increased photorespiration, but other factors, such as light intensity[64,67] and leaf age,[69] caused parallel changes in AP and photorespiration. In addition, there were positive correlations between AP and photorespiration in different plant types or cultivars.[70-71] There were therefore apparent anomalies in the relationship between AP and photorespiration, with $O_2$ concentration having opposite effects on the two processes and other factors having parallel effects. The relationship between these two processes was, however, greatly clarified by the discovery that $O_2$ was a competitive substrate with $CO_2$ for RuBP carboxylase in the Calvin cycle.[42,72] Oxygen was shown to react with RuBP to produce phosphoglycolate, a substrate for photorespiration. Since $O_2$ is also required in the oxidation of glycolate,[65] it has two inhibitory effects on AP:

1.   It competes with $CO_2$ for the substrate, RuBP, thus reducing the amount of $CO_2$ assimilated.
2.   It reacts with glycolate in the pathway which produces photorespired $CO_2$.

The effects of $O_2$ on leaf AP are shown by the $CO_2$ response curves in Figure 14. Line E—F represents the $CO_2$ response of a leaf at 2% $O_2$. Raising $O_2$ to 21% has three effects shown by line C—D:

1.   The intercept at zero $CO_2$ is changed from $-1.2$ to $-8.1$ mg $CO_2$ dm$^{-2}$ hr$^{-1}$.
2.   The slope of the line (and likewise $K_m$) is decreased from 0.37 cm sec$^{-1}$ to 0.25 cm sec$^{-1}$.
3.   $\Gamma$ is increased from 5 to 50 $\mu$L L$^{-1}$.

If the difference in $CO_2$ evolution at points F and D is taken as $O_2$ stimulation of photorespiration (6.9 mg dm$^{-2}$ hr$^{-1}$) and the difference between AP at points E and C (13.5 mg $CO_2$ dm$^{-2}$ hr$^{-1}$) is the total inhibition at 21% $O_2$, then the remaining $O_2$ effect, which is assumed to be mainly $O_2$ inhibition of RuBP carboxylation, is $13.5 - 6.9$ or 6.6 mg $CO_2$ dm$^{-2}$ hr$^{-1}$. So at $CO_2$ and $O_2$ concentrations equivalent to atmospheric, AP is estimated to be inhibited 17% by photorespiration and 16% by $O_2$ effects on carboxylation, for a total inhibition by $O_2$ of 33%. This degree of total inhibition is similar to that reported by several authors.[63-64]

The above analysis shows that the effect of 21% $O_2$ on AP is about equally divided between stimulation of photorespiration and inhibition of "true photosynthesis". A similar effect at atmospheric $CO_2$ levels has been reported by D'Aoust and Canvin[73] and Ludwig and Canvin,[74] although other reports show relatively less effect on photorespiration.[18,75] Ku and Edwards[18] found only 20 to 25% of the total inhibition due to photorespiration at atmospheric levels of $O_2$ and $CO_2$, but an increased proportion as the solubility ratio of $O_2/CO_2$ increased, i.e., as $O_2$ concentration was raised or $CO_2$ lowered.

Results with soybean[64] shown in Figure 18 extend the relationships of Figure 14 to higher $O_2$ concentrations. Although the $CO_2$ concentrations shown are those external to the leaf, they show the same features discussed above. An increase in $O_2$ from 1 to 100% causes a linear increase in $\Gamma$ from near zero to 170 $\mu$L L$^{-1}$. Extrapolation of the lines to the AP axis shows increases in photorespiration from near 0 to 12 $\mu$g leaf$^{-1}$ min$^{-1}$. In addition, the slope of the $CO_2$ response lines decrease with each increase of $O_2$ concentration, indicating progressively greater $O_2$ inhibition of carboxylation. The lines in Figure 18 diverge as $CO_2$ concentration is increased, but at higher levels than those shown, AP becomes $CO_2$ saturated and $O_2$ inhibition is decreased, apparently because of the competitive advantage of $CO_2$ in the reaction with RuBP.

Inhibition of AP by 21% $O_2$ is not observed in $C_4$ species, apparently because of the $CO_2$ concentrating characteristic of the $C_4$ pathway. The decarboxylation of the 4-carbon acids in bundle sheath cells proceeds at a rapid enough rate to raise the $CO_2$ concentration, so that RuBP carboxylase is apparently not inhibited by $O_2$. Estimates by Hatch[76] of 60 $\mu M$ $CO_2$ (equivalent to about 10 times atmospheric) in bundle sheath

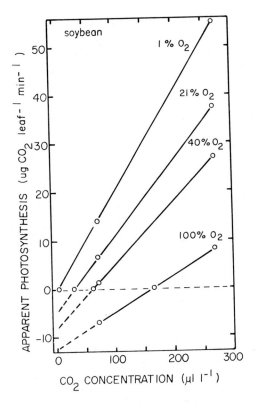

FIGURE 18.    Response of AP in soybean leaves
to $CO_2$ and $O_2$ concentrations. Broken lines are
extrapolations of experimental data to zero $CO_2$
concentration. (From Forrester, M. L. et al.,
*Plant Physiol.*, 41, 426, 1966. With permission.

cells during photosynthesis in normal air are equivalent to concentrations which elim-
inate the inhibition of AP by 21% $O_2$ in $C_3$ species.[17,37] So while the RuBP carboxylase
in $C_3$ species is usually exposed to less than 300 $\mu$L $L^{-1}$ of $CO_2$ during photosynthesis,
that in $C_4$ plants may be exposed to saturating levels and is therefore insensitive to
atmospheric $O_2$ concentration. Photosynthesis in $C_4$ plants has been shown to be reduced by very high $O_2$ levels,[77] but the effect is not readily reversible as in $C_3$ plants
and therefore is probably upon some other process than carboxylation or photorespiration. At any rate, 21% $O_2$ appears to have no appreciable influence on AP in $C_4$
species.

It may be assumed that a reduction in the $O_2$ inhibition of AP through genetic variation would result in increased plant productivity. Species with the $C_4$ pathway are
more productive than those with the $C_3$ pathway, and experimental reduction of $O_2$
concentration has resulted in increased growth.[78,79] The consistent effects of 21% $O_2$
observed in many experiments with $C_3$ species and the chemical characteristics of the
RuBP oxygenase/carboxylase reaction has led to the speculation that a change in $O_2$
response was not possible through changes in the RuBP carboxylase protein.[80] Recent
discovery of plant species with reduced $O_2$ response[81,82] shows some promise of manip-
ulation of this characteristic, though perhaps not through a change in the RuBP oxy-
genase/carboxylase reaction.

*Panicum milioides*, a grass species adapted to tropical climates, exhibits several characteristics intermediate between $C_3$ and $C_4$ photosynthesis. Reduction of AP in this
species at 21% $O_2$ is about 20 to 25% compared to about 33% in $C_3$ species.[81,83] In

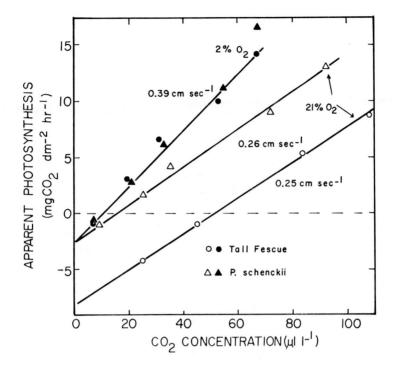

FIGURE 19. Response of AP in *P. schenckii* (intermediate between C₃ and
C₄) and tall fescue (C₃) to intercellular $CO_2$ concentrations at 2 and 21% $O_2$.
Figures beside the lines are $K_m$ values.[93]

addition its $\Gamma$ value is 15 to 25 $\mu$L L⁻¹ compared to 40 to 60 $\mu$L L⁻¹ for C₃ species.[81]
Two other species, *P. schenckii* and *P. decipiens*, have been found by the author to
exhibit $O_2$ responses similar to *P. milioides*. Figure 19 shows the reduced $O_2$ effect on
AP of *P. schenckii* compared to tall fescue, a C₃ species. *P. schenckii* exhibits a $\Gamma$
value at 21% $O_2$ of 16 $\mu$L L⁻¹ as shown in Figure 19 compared to 50 $\mu$L L⁻¹ for tall
fescue. Likewise at 21% $O_2$, photorespiration of *P. schenckii* is much less than for tall
fescue. At 2% $O_2$, photorespiration of *P. schenckii* is nearly identical to that at 21%,
but in tall fescue, photorespiration is reduced from about 8 to 2.5 mg $CO_2$ dm⁻² hr⁻¹.
The increase in $K_m$ (slope of the $CO_2$ response lines) at 2% $O_2$ is similar, however, in
*P. schenckii* and tall fescue. These and similar unpublished data indicate that the reduced $O_2$ effect in *P. schenckii* (and in *P. milioides* and *P. decipiens*) is due to suppres-
sion of photorespiration rather than a change in the $O_2$ inhibition of carboxylation.
Biochemical and leaf anatomical characteristics of the above *Panicum* species also
show some degree of intermediacy between C₃ and C₄ plants.[84,85]

The relative effect of $O_2$ on AP is nearly independent of light intensity, but is
strongly dependent on temperature. Björkman[63] reported that the initial slope of the
light response curve was reduced by 21% $O_2$ to the same extent (30%) as AP at saturating light intensity. Ku et al.,[37] however, showed a decrease in $O_2$ inhibition of AP
in potato with increased light intensity from 50% at 0.15mEin m⁻² sec⁻¹ to 37% at 1.5
mEin m⁻² sec⁻¹. Interactions of temperature and $O_2$ on AP are strong, with $O_2$ effects
being absent at low temperatures and maximal at optimum or higher temperatures.
Data from studies with wheat[17] show no effect of 21% $O_2$ on AP at temperatures
below about 22°C (see Figure 20). The maximum effect of 21% $O_2$ was at 32°C which
was the optimum temperature for AP. The optimum temperature for AP is usually
shifted to lower values by raising $O_2$ concentration. The effect is not noticeable for

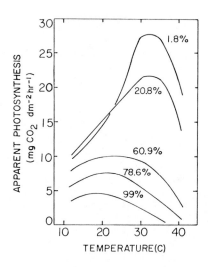

FIGURE 20.    Response of AP in wheat ($C_3$) to temperature at different $O_2$ concentrations. (From Jolliffe, P. A. and Tregunna, E. B., *Can. J. Bot.*, 51, 841, 1973. With permission.)

21% $O_2$ in Figure 20, but is evident at higher $O_2$ levels. In several other reports, optimum temperature for AP was shown to be lower at 21% than at 1 or 2% $O_2$.[19,66,86]

Photorespiration as determined by $CO_2$ efflux in the light has been shown to increase with temperature,[19] and elimination of this process by low $O_2$ may be the reason for the upward shift of the temperature optimum. Peisker and Apel[87] concluded that the decrease in AP with temperature increase from 20 to 30°C was mainly photorespiration and much less to a decrease in $K_m$. Ku and Edwards,[18] in an extensive study of $O_2$ and $CO_2$ interactions with temperature, showed very little effect of temperature on the competitive inhibition of AP by $O_2$, so that the increased $O_2$ effects at high temperatures were due to increased photorespiration.

As might be expected from the results of $O_2$ level on AP, dry weight accumulation by $C_3$ plants is increased with decreased $O_2$ levels,[78,88] while that of $C_4$ species is unaffected. Low $O_2$ concentration (2.5%) increased growth of wheat, even at temperatures of 5 and 10°C at which $O_2$ effects on AP have been observed to disappear.[17,86] This effect of low $O_2$ level on growth apparently is similar to effects of increased $CO_2$ concentrations. As in the case of increased $CO_2$, the advantage of increased AP at low $O_2$ is offset to some degree by a decrease in growth of leaves relative to stems and roots.[89-91] In experiments with rice and wheat (Table 5), Fukuyama et al.[90] found that, although net assimilation rate was stimulated by low $O_2$, the relative expansion of leaf area was reduced. Others have reported data which show the leaf area/plant weight ratio to be reduced by low $O_2$.[89,91] Fukuyama et al. concluded, however, that the reduced leaf growth at low $O_2$ was not related to $O_2$ inhibition of AP, since leaf expansion was inhibited by low $O_2$ in $C_4$ species[89] and when administered to $C_3$ species during the dark rather than the light period. In addition, leaf growth was not influenced at $O_2$ levels above 10%, whereas AP is inhibited by $O_2$ at all levels above 10%. Besides the inhibition of leaf growth, low $O_2$ has been found to reduce or prevent fruiting in soybean, a $C_3$ species, and sorghum, a $C_4$ species.[79] Therefore, low $O_2$ atmospheres are not entirely suitable for plant growth, even though photosynthesis is stimulated and photorespiration is suppressed.

### Table 5
## THE INFLUENCE OF OXYGEN CONCENTRATION ON NET ASSIMILATION RATE (NAR) AND RELATIVE LEAF GROWTH RATE (RLGR) OF WHEAT AND RICE AT DIFFERENT TEMPERATURES[90]

| Species | Temperature (°C) | Oxygen concentration (%) | NAR $gm^{-2} wk^{-1}$ | RLGR $cm^2 dm^{-2} wk^{-1}$ |
|---|---|---|---|---|
| Wheat | 10 | 21 | 36 | 49 |
| | | 2.5 | 50[a] | 32[a] |
| | 20 | 21 | 62 | 82 |
| | | 2.5 | 86[a] | 48[a] |
| Rice | 20 | 21 | 71 | 47 |
| | | 2.5 | 88[a] | 48 |
| | 30 | 21 | 42 | 55 |
| | | 2.5 | 81[a] | 42[a] |

[a] Significantly different from the value at 21% $O_2$ — 1% probability level.

## REFERENCES

1. **McCree, K. J.**, The action spectrum, absorptance, and quantum yield of photosynthesis in crop plants, *Agric. Meteorol.*, 9, 191, 1972.
2. **Lemon, E.**, Aerodynamic studies of $CO_2$ exchange between the atmosphere and the plant, in *Harvesting the Sun-Photosynthesis in Plant Life*, San Pietro, A., Greer, F. A., and Army, F. J. Eds., Academic Press, New York, 1967, 263.
3. **Björkman, O.**, Further studies on differentiation of photosynthetic properties of sun and shade ecotypes of *Solidago virgaurea, Physiol. Plant.*, 21, 84, 1968.
4. **Ehleringer, J. and Björkman, O.**, Quantum yields for $CO_2$ uptake in $C_3$ and $C_4$ plants. Dependence of temperature, $CO_2$ and $O_2$ concentration, *Plant Physiol.*, 59, 86, 1977.
5. **Ku, S. B. and Edwards, G. E.**, Oxygen inhibition of photosynthesis. III. Temperature dependence of quantum yield and its relation to $O_2/CO_2$ solubility ratio, *Planta*, 140, 1, 1978.
6. **Björkman, O. and Holmgren, P.**, Adaptability of the photosynthetic apparatus to light intensity in ecotypes from exposed and shaded habitats. *Physiol. Plant.*, 16, 889, 1963.
7. **Chen, T. M., Brown, R. H., and Black, C. C.**, $CO_2$ compensation concentration, rate of photosynthesis, and carbonic anhydrase activity of plants, *Weed Sci.*, 18, 399, 1970.
8. **Björkman, O.**, Environmental and biological control of photosynthesis: inaugural address, in *Environmental and Biological Control of Photosynthesis*, Marcelle, R., Ed., Dr. W. Junk, The Hague, 1975, 1.
9. **Hesketh, J. D. and Moss, D. N.**, Variation in response of photosynthesis to light, *Crop Sci.*, 3, 107, 1963.
10. **Björkman, O. and Holmgren, P.**, Photosynthetic adaptation to light intensity in plants native to shaded and exposed habitats, *Physiol. Plant.*, 19, 854, 1966.
11. **Björkman, O.**, Carboxydismutase activity in shade-adapted and sun-adapted species of higher plants, *Physiol. Plant.*, 21, 1, 1968.
12. **Singh, M., Ogren, W. L., and Widholm, J. M.**, Photosynthetic characteristics of several $C_3$ and $C_4$ plant species grown under different light intensities, *Crop Sci.*, 14, 563, 1974.
13. **Pearce, R. B., Brown, R. H., and Blaser, R. E.**, Photosynthesis in plant communities as influenced by leaf angle, *Crop Sci.*, 7, 321, 1967.
14. **Pearce, R. B., Brown, R. H., and Blaser, R. E.**, Relationship between leaf area index, light interception and net photosynthesis in orchardgrass, *Crop Sci.*, 5, 553, 1965.
15. **Murata, Y., Iyama, J., and Honma, T.**, Studies on the photosynthesis of forage crops.IV. Influence of air-temperature upon the photosynthesis and respiration of alfalfa and several southern type forage crops, *Proc. Crop Sci. Soc. Jpn.*, 34, 154, 1965.

16. **Ludlow, M. M. and Wilson, G. L.,** Photosynthesis of tropical pasture plants. I. Illuminance, carbon dioxide concentration, leaf temperature, and leaf-air vapor pressure difference, *Aust. J. Biol. Sci.,* 24, 449, 1971.

17. **Jolliffe, P. A. and Tregunna, E. B.,** Environmental regulation of the oxygen effect on apparent photosynthesis in wheat, *Can. J. Bot.,* 51, 841, 1973.

18. **Ku, S. B. and Edwards, G. E.,** Oxygen inhibition of photosynthesis.II. Kinetic characteristics as affected by temperature, *Plant Physiol.,* 59, 991, 1977.

19. **Hofstra, G. and Hesketh, J. D.,** Effects of temperature on the gas exchange of leaves in the light and dark, *Planta,* 85, 228, 1969.

20. **Treharne, K. J. and Nelson, C. J.,** Effect of growth temperature on photosynthetic and photorespiratory activity in tall fescue, in *Environmental and Biological Control of Photosynthesis,* Marcelle, R. Ed., Dr. W. Junk, The Hague, 1975, 71.

21. **Treharne, K. J. and Eagles, C. F.,** Effect of temperature on photosynthetic activity of climatic races of *Dactylis glomerata* L., *Photosynthetica,* 4, 107, 1970.

22. **Pearcy, R. W.,** Temperature responses of growth and photosynthetic $CO_2$ exchange rates in coastal and desert races of *Atriplex lentiformis, Oecologia (Berlin),* 26, 245, 1976.

23. **Lange, O. L., Schulze, E. -D., Evenari, M., Kappen, L., and Buschbom, U.,** The temperature-related photosynthetic capacity of plants under desert conditions.I. Seasonal changes of the photosynthetic response to temperature, *Oecologia (Berlin),* 17, 97, 1974.

24. **Mooney, H. A., Björkman, O., and Collatz, G. J.,** Photosynthetic acclimation to temperature in the desert shrub, *Larrea divaricata.* I. Carbon dioxide exchange characteristics of intact leaves, *Plant Physiol.,* 61, 406, 1978.

25. **Strain, B. R., Higginbotham, K. O., and Mulroy, J. C.,** Temperature preconditioning and photosynthetic capacity of *Pinus taeda* L., *Photosynthetica,* 10, 47, 1976.

26. **Björkman, O. and Badger, M.,** Thermal stability of photosynthetic enzymes in heat- and cool-adapted $C_4$ species, *Carnegie Inst. Washington Yearb.,* 76, 400, 1977.

27. **Armond, P. A., Schreiber, U., and Björkman, O.,** Photosynthetic acclimation to temperature in the desert shrub, *Larrea divaricata.* II. Light harvesting efficiency and electron transport, *Plant Physiol.,* 61, 411, 1978.

28. **Neilson, R. E., Ludlow, M. M., and Jarvis, P. G.,** Photosynthesis in Sitka spruce (*Picea sitchensis* (Bong.) Carr.). II. Response to temperature, *J. Appl. Ecol.,* 9, 721, 1972.

29. **Kuiper, P. J. C.,** Temperature dependence of photosynthesis of bean plants as affected by decenyl-succinic acid, *Plant Physiol.,* 40, 915, 1965.

30. **Pearcy, R. W.,** Acclimation of photosynthetic and respiratory carbon dioxide exchange to growth temperature in *Atriplex lentiformis* (Torr.) Wats., *Plant Physiol.,* 59, 795, 1977.

31. **Pearcy, R. W., Berry, J. A., and Fork, D. C.,** Effects of growth temperature on the thermal stability of the photosynthetic apparatus of *Atriplex lentiformis* (Torr.) Wats., *Plant Physiol.,* 59, 873, 1977.

32. **West, S. H.,** Biochemical mechanism of photosynthesis and growth depression in *Digitaria decumbens* when exposed to low temperatures, *Proc. 11th Int. Grassld. Congr.,* 1970, 514.

33. **Chatterton, N. J., Carlson, G. E., Hungerford, W. E., and Lee, D. R.,** Effect of tillering and cool nights on photosynthesis and chloroplast starch in pangola, *Crop Sci.,* 12, 206, 1972.

34. **Hilliard, J. H. and West, S. H.,** Starch accumulation associated with growth reduction at low temperatures in a tropical plant, *Science,* 168, 494, 1970.

35. **Karbassi, P., Garrard, L. A., and West, S. H.,** Reversal of low temperature effects on a tropical plant by gibberellic acid, *Crop Sci.,* 11, 755, 1971.

36. **Blaser, R. E., Brown, R. H., and Dunton, H. L.,** The relationship between carbohydrate accumulation and growth of grasses under different microclimates, *Proc. 10th Int. Grassld. Congr.,* 1966, 147.

37. **Ku, S. B., Edwards, G. E., and Tanner, C. B.,** Effects of light, carbon dioxide and temperature on photosynthesis, oxygen inhibition of photosynthesis, and transpiration in *Solanum tuberosum, Plant Physiol.,* 59, 868, 1977.

38. **Doley, D. and Yates, D. J.,** Gas exchange of Mitchell grass (*Astrebla lappacea* (Lindl.) Domin) in relation to irradiance, carbon dioxide supply, leaf temperature and temperature history, *Aust. J. Plant Physiol.,* 3, 471, 1976.

39. **Williams, G. J. and Kemp, P. R.,** Temperature relations of photosynthetic response in populations of *Verbascum thapsus* L., *Oecologia (Berlin),* 25, 47, 1976.

40. **Ku, S. B. and Edwards, G. E.,** Oxygen inhibition of photosynthesis. I. Temperature dependence and relation to $O_2/CO_2$ solubility ratio, *Plant Physiol.,* 59, 986, 1977.

41. **Gaastra, P.,** Photosynthesis of crop plants as influenced by light, carbon dioxide, temperature and stomatal resistance, *Meded. Landbouwhogesch. Wageningen,* 59, 1, 1959.

42. **Bowes, G. and Ogren, W. L.,** Oxygen inhibition and other properties of soybean ribulose 1,5-diphosphate carboxylase, *J. Biol. Chem.,* 247, 2171, 1972.

43. **Björkman, O.**, Adaptive and genetic aspects of photosynthesis, in *CO₂ Metabolism and Plant Productivity*, Burris, R. H. and Black, C. C., Eds., University Park Press, Baltimore, Md., 1976, 287.
44. **Slatyer, R. O.**, Comparative photosynthesis, growth and transpiration of two species of *Atriplex, Planta*, 93, 175, 1970.
45. **Hesketh, J. D.**, Limitations to photosynthesis responsible for differences among species, *Crop Sci.*, 3, 493, 1963.
46. **Whiteman, P. C. and Koller, D.**, Interactions of carbon dioxide concentration, light intensity and temperature on plant resistances to water vapor and carbon dioxide diffusion, *New Phytol.*, 66, 463, 1967.
47. **Heath, O. V. S.**, *The Physiological Aspects of Photosynthesis*, Stanford University Press, Stanford, Calif., 1969.
48. **El-Sharkawy, M. A., Loomis, R. S., and Williams, W. A.**, Photosynthetic and respiratory exchanges of carbon dioxide by leaves of the grain amaranth, *J. Appl. Ecol.*, 5, 243, 1968.
49. **Wittwer, S. H. and Robb, W.**, Carbon dioxide enrichment of greenhouse atmospheres for food crop production, *Econ. Bot.*, 18, 34, 1964.
50. **Harper, L. A., Baker, D. N., Box, J. E., Jr., and Hesketh, J. D.**, Carbon dioxide and the photosynthesis of field crops: a metered carbon dioxide release in cotton under field conditions, *Agron. J.*, 65, 7, 1973.
51. **Neales, T. F. and Nicholls, A. O.**, Growth responses of young wheat plants to a range of ambient CO₂ levels, *Aust. J. Plant Physiol.*, 5, 45, 1978.
52. **Ford, M. A. and Thorne, G. N.**, Effect of CO₂ concentration on growth of sugar beet, barley, kale and maize, *Ann. (London) Bot.*, 31, 629, 1957.
53. **Cooper, R. L. and Brun, W. A.**, Response of soybeans to a CO₂-enriched atmosphere, *Crop Sci.*, 7, 455, 1967.
54. **Nafziger, E. D. and Koller, H. R.**, Influence of leaf starch concentration on CO₂ assimilation in soybean, *Plant Physiol.*, 57, 560, 1976.
55. **Cock, J. H. and Yoshida, S.**, Changing sink and source relations in rice (*Oryza sativa* L.) using carbon dioxide enrichment in the field, *Soil Sci. Plant Nutr. (Tokyo)*, 19, 229, 1973.
56. **Madsen, E.**, Effect of CO₂ concentration on the accumulation of starch and sugar in tomato leaves, *Physiol. Plant.*, 21, 168, 1968.
57. **Gifford, R. M.**, Growth pattern, carbon dioxide exchange and dry weight distribution in wheat growing under differing photosynthetic environments, *Aust. J. Plant Physiol.*, 4, 99, 1977.
58. **Yoshida, S.**, Effects of CO₂ enrichment at different stages of panicle development on yield components and yield of rice (*Oryza sativa* L.), *Soil Sci. Plant Nutr. (Tokyo)*, 19, 311, 1973.
59. **Krenzer, E. G., Jr. and Moss, D. N.**, Carbon dioxide enrichment effects upon yield and yield components in wheat, *Crop Sci.*, 15, 71, 1975.
60. **Hofstra, G. and Hesketh, J. D.**, The effects of temperature and CO₂ enrichment on photosynthesis in soybeans, in *Environmental and Biological Control of Photosynthesis*, Marcelle, R. Ed., Dr. W. Junk, The Hague, 1975, 71.
61. **Akita, S. and Tanaka, I.**, Studies on the mechanism of differences in photosynthesis among species. IV. The differential response in dry matter production between C₃ and C₄ species to atmospheric carbon dioxide enrichment, *Proc. Crop Sci. Soc. Jpn.*, 42, 288, 1973.
62. **Warburg, O.**, Uber die geschwindigkeit der photokemischen kohlensaurezersetzung in lebenden zellen II, *Biochem. Z.*, 103, 188, 1920.
63. **Björkman, O.**, The effect of oxygen concentration on photosynthesis in higher plants, *Physiol. Plant.*, 19, 618, 1966.
64. **Forrester, M. L., Krotkov, G., and Nelson, C. D.**, Effect of oxygen on photosynthesis, photorespiration and respiration in detached leaves. I. Soybean, *Plant Physiol.*, 41, 422, 1966.
65. **Zelitch, I.**, *Photosynthesis, Photorespiration and Plant Productivity*, Academic Press, New York, 1971.
66. **Holmgren, P. and Jarvis, P. G.**, Carbon dioxide efflux from leaves in light and darkness, *Physiol. Plant.*, 20, 1045, 1967.
67. **Tregunna, E. B., Krotkov, G., and Nelson, C. D.**, Effect of oxygen on the rate of photorespiration in detached tobacco leaves, *Physiol. Plant.*, 19, 723, 1966.
68. **El-Sharkawy, M. and Hesketh, J.**, Photosynthesis among species in relation to characteristics of leaf anatomy and CO₂ diffusion resistances, *Crop Sci.*, 5, 517, 1965.
69. **Hodgkinson, K. C.**, Influence of partial defoliation on photosynthesis, photorespiration and transpiration by lucerne leaves of different ages, *Aust. J. Plant Physiol.*, 1, 561, 1974.
70. **Carlson, G. E., Pearce, R. B., Lee, D. R., and Hart, R. H.**, Photosynthesis and photorespiration in two clones of orchardgrass, *Crop Sci.*, 11, 35, 1971.
71. **Criswell, J. G. and Shibles, R. M.**, Physiological basis for genotypic variation in net photosynthesis of oat leaves, *Crop Sci.*, 11, 550, 1971.

72. Bowes, G., Ogren, W. L., and Hageman, R. H., Phosphoglycolate production catalyzed by ribulose diphosphate carboxylase, *Biochem. Biophys. Res. Commun.*, 45, 716, 1971.

73. D'Aoust, A. L. and Canvin, D. T., Effect of oxygen concentration on the rates of photosynthesis and photorespiration of some higher plants, *Can. J. Bot.*, 51, 457, 1973.

74. Ludwig, L. J. and Canvin, D. T., An open gas-exchange system for the simultaneous measurement of the $CO_2$ and $^{14}CO_2$ fluxes from leaves, *Can. J. Bot.*, 49, 1299, 1970.

75. Ludwig, L. J. and Canvin, D. T., The rate of photorespiration during photosynthesis and the relationship of the substrate of light respiration to the products of photosynthesis in sunflower leaves, *Plant Physiol.*, 48, 712, 1971.

76. Hatch, M. D., The $C_4$ pathway of photosynthesis: mechanism and function, in *$CO_2$ Metabolism and Plant Productivity*, Burris, R. H. and Black, C. C., Eds. University Park Press, Baltimore, Md., 1976, 59.

77. Forrester, M. L., Krotkov, G., and Nelson, C. D., Effect of oxygen on photosynthesis, photorespiration and respiration in detached leaves. II. Corn and other monocotyledons, *Plant Physiol.*, 41, 428, 1966.

78. Björkman, O., Hiesey, W. M., Nobs, M. A., Nicholson, F., and Hart, R. W., Effect of oxygen concentration in higher plants, *Carnegie Inst. Washington Yearb.*, 66, 228, 1968.

79. Quebedeaux, B. and Hardy, R. W. F., Oxygen concentration: regulation of crop growth and productivity, in *$CO_2$ Metabolism and Plant Productivity*, Burris, R. H. and Black, C. C., Eds., University Park Press, Baltimore, Md., 1976, 185.

80. Lorimer, G. H. and Andrews, T. J., Plant photorespiration — an inevitable consequence of the existence of atmospheric oxygen, *Nature (London)*, 243, 359, 1973.

81. Brown, R. H. and Brown, W. V., Photosynthetic characteristics of *Panicum milioides*, a species with reduced photorespiration, *Crop Sci.*, 15, 681, 1975.

82. Kanai, R. and Kashiwagi, M., *Panicum milioides*, a Gramineae plant having Kranz leaf anatomy without $C_4$-photosynthesis, *Plant Cell Physiol.*, 16, 669, 1975.

83. Keck, R. W. and Ogren, W. L., Differential oxygen response of photosynthesis in soybean and *Panicum milioides*, *Plant Physiol.*, 58, 552, 1976.

84. Black, C. C., Goldstein, L. D., Ray, T. B., Kestler, D. P., and Mayne, B. C., The relationship of plant metabolism to internal leaf and cell morphology and to the efficiency of $CO_2$ assimilation, in *$CO_2$ Metabolism and Plant Productivity*, Burris, R. H. and Black, C. C., Eds., University Park Press, Baltimore, Md., 1976, 113.

85. Brown, R. H., Characteristics related to photosynthesis and photorespiration of *Panicum milioides*, in *$CO_2$ Metabolism and Plant Productivity*, Burris, R. H. and Black, C. C., Eds., University Park Press, Baltimore, Md., 1976, 311.

86. Akita, S. and Miyasaka, A., Studies on the differences of photosynthesis among species. II. Effect of oxygen-free air on photosynthesis, *Proc. Crop Sci. Soc. Jpn.*, 38, 525, 1969.

87. Peisker, M. and Apel, P., Influence of oxygen on photosynthesis and photorespiration in leaves of *Triticum aestivum* L. III. Response of $CO_2$ gas exchange to oxygen at various temperatures, *Photosynthetica*, 11, 29, 1977.

88. Quebedeaux, B. and Chollet, R., Comparative growth analyses of *Panicum* species with differing rates of photorespiration, *Plant Physiol.*, 59, 42, 1977.

89. Fukuyama, M., Takeda, T., and Maeda, H., Studies on the photosynthesis and the growth of crop plants. II. Relationship between photorespiration and expansion of leaf area, *Proc. Crop Sci. Soc. Jpn.*, 43, 453, 1974.

90. Fukuyama, M., Takeda, T., and Taniyama, T. Studies on the photosynthesis and the growth of crop plants, I. The effects of oxygen concentration under various growth temperatures on the growth of wheat and rice plants, *Proc. Crop Sci. Soc. Jpn.*, 43, 267, 1974.

91. Parkinson, K. J., Penman, H. L., and Tregunna, E. B., Growth of plants in different oxygen concentrations, *J. Exp. Bot.*, 25, 132, 1974.

92. Brown, R. H. and Morgan, J. A., unpublished data.

93. Brown, R. H., unpublished data.

# RESPONSE OF TERRESTRIAL PLANTS TO MINERAL NUTRIENTS

## B. Clifford Gerwick

## INTRODUCTION

Most higher plants require 16 elements for normal growth and completion of their life cycles. A severe deficiency in any one of these 16 will impair photosynthesis. The relationship between mineral nutrition and photosynthesis is complex, however, since mineral nutrition affects many physiological processes and frequently affects these processes differently, depending upon the stage of plant growth, the environmental conditions during plant growth, and the form in which they are made available to the plant. Further, the deficiency of one nutrient typically affects the status of other nutrients in the plant making direct and indirect effects difficult to resolve. The response of photosynthesis to mineral nutrients is presented within these limitations, and, where possible, the effects at the whole plant level are related to specific physiological and biochemical processes. A list of the physiological and biochemical roles of mineral nutrients related to photosynthesis is provided in Table 1, along with the concentration and amount of these nutrients considered adequate for normal plant growth.

## NITROGEN

While plant growth has long been known to respond markedly to N applications, only recently has a relationship between photosynthesis and leaf N levels been established. This relationahip is shown for *Panicum maximum* and *Lolium perenne* in Figure 1. Such a relationship is not unexpected, since as much as 75% of leaf N is located in the chloroplasts.[1] A similar correlation between photosynthetic rate and leaf N has been established for a wide number of species, including wheat,[2] barley,[2] and rice.[3,4] In rice, increasing leaf N results in a near linear increase in photosynthesis up to leaf N concentrations of 6%.[3] This high level of leaf N could only be obtained after intensive cultivation which suggests that under many field conditions photosynthesis may be limited by N.

N-use efficiency, or the photosynthetic rate per unit of leaf N, varies considerably among higher plants. Plants with the $C_4$ pathway of carbon assimilation, however, have a higher nitrogen use efficiency than plants with the $C_3$ pathway.[5] This relationship emerges most clearly in the Gramineae, as comparatively little information is available for other groups of plants. In Figure 1, photosynthetic rates as a function of leaf N levels are shown for *P. maximum*, a $C_4$ species, and *L. perenne*, a $C_3$ species. It is apparent that at all leaf N levels the $C_4$ species achieves a greater rate of photosynthesis than the $C_3$ species. Similar conclusions can be drawn from the work of Dirven.[6] Here five $C_4$ species and the $C_3$ species *Panicum laxum* and *Lolium multiflorum* were grown for varying lengths of time and analyzed for N content (see Figure 2). The $C_4$ species all had lower N contents than the $C_3$ species and during the growth period achieved a greater fresh weight production.

The response of photosynthesis to N appears to involve changes in both mesophyll and stomatal resistances. The relative importance of these resistances in restricting photosynthesis varies with the magnitude of stress, species, and growth conditions. Most workers have attributed to a major role to mesophyll resistance which is consistent with the reduced chlorophyll levels and levels of Fraction I protein (RuBP carboxylase/oxygenase) found in N-deficient plants.[3] Considerable evidence is accumulating that photosynthetic rates can be directly related to the levels of this protein.[5,7-9] In-

Table 1
# ROLE OF MINERAL NUTRIENTS IN PHOTOSYNTHETICALLY RELATED PROCESSES

| Element[a] | Conc. in plant[50] | | Role in photosynthesis and associated metabolism | Ref. |
|---|---|---|---|---|
| | μmol/gdw | ppm or % dw | | |
| N | 1000 | 1.5% | Constituent of proteins and nucleic acids. Nitrogen Effects on Photosynthesis Result Largely From Changes in The Levels of Fraction I protein. | 5 |
| K | 250 | 1.0% | A key role in the generation of osmotic potentials for stomatal opening and closing. Necessary for chlorophyll biosynthesis, photosynthetic translocation, metabolite transport in $C_4$ plants, and photosynthetic electron transport. Cofactor for pyruvate kinase and starch synthetase. | 17,24,35,36,37,38, 39,40,41,42,43 |
| Ca | 125 | 0.5% | General role in membrane stability, including the chloroplast envelope. Constituent of amylase. | 41,44 |
| Mg | 80 | 0.2% | Cofactor for a large number of phosphate transfer reactions and photophosphorylation. Involved in control of energy transfer between photosystem II and I. Constituent of chlorophyll molecule and cofactor and activator of RuBP carboxylase-oxygenase. Cofactor for PEP carboxylase and numerous other enzumes. | 45,46,47,48,49 |
| P | 60 | 0.2% | A key role in energy conservation and transfer. | 41,50,51 |
| S | 30 | 0.1% | Component of primary electron acceptor of photosystem I and chloroplast membrane lipids. Constituent of ferredoxin, biotin, and Coenzyme A. | 50,52 |
| Cl | 3 | 100 ppm | Component of photosystem II. Associated with restoration of electrical potential equilibrium during proton pumping. | 48,53,54,55,56 |
| Fe | 2 | 100 ppm | Component of primary electron accepter of photosystem I, ferredoxin, cytochrome, cytochrome oxidase, catalase, and necessary for chlorophyll biosynthesis. | 41,50,52,57 |
| B | 2 | 20 ppm | A possible role in sugar translocation. | 51 |
| Mn | 1 | 50 ppm | Associated with photosystem II. Cofactor for some forms of NAD malic enzyme and dehydrogenases of the TCA cycle. | 50,51,53,58,59,60, 61 |
| Zn | 0.3 | 20 ppm | Constituent of superoxide dismutase and carbonic anhydrase. | 41,62,63 |
| Cu | 0.1 | 6 ppm | Constituent of plastocyanin and superoxide dismutase. | 51,64 |
| Mo | 0.001 | 0.1 ppm | Component of nitrate reductase. | 65 |

[a]    The elements C, H, and O are not considered in this table. These elements are obtained by plants primarily as $CO_2$ and $H_2O$. Plants typically are 45% C, 45% O, and 6% H on a dry weight basis.[50]

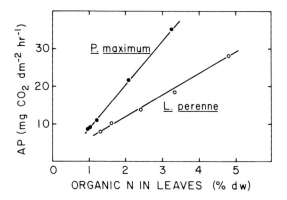

FIGURE 1. Photosynthetic response of *P. maximum* (C₄) and *L. perenne* (C₃) to leaf organic N levels.[5,66]

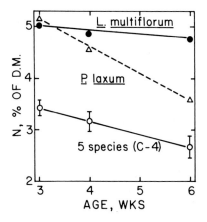

FIGURE 2. The N content of $C_3$ and $C_4$ grass species in relation to plant age.[6] *L. multiflorum* and *P. laxum* are $C_3$ species. Mean date ± 1 SD has been plotted for five $C_4$ species. D.M. = dry matter.

creased protein levels in the leaf result in an increased percentage of protein in Fraction I (see Figure 3) which suggests that the effects of N on whole-plant photosynthesis may result largely from changes in Fraction I.

Brown[5] has put forward the hypothesis that high N-use efficiency in $C_4$ plants may be related to the levels of Fraction I protein. In $C_4$ plants, this enzyme is compartmentalized to the bundle sheath cells of the leaf, where elevated $CO_2$ concentrations increase its efficiency in $CO_2$ fixation. Hence, while in $C_3$ species this protein accounts for 25 to 60% of the leaf protein, in $C_4$ species, this percentage ranges from 8 to 23%.[5,10] A reduced investment in Fraction I, unless largely offset by N investment in proteins of the $C_4$ cycle, may account for the high N-use efficiency of $C_4$ plants.

## POTASSIUM

K is known to have an important role in guard cell opening and closing, as well as in photosynthetic electron transport, phloem loading, and chlorophyll biosynthesis (see

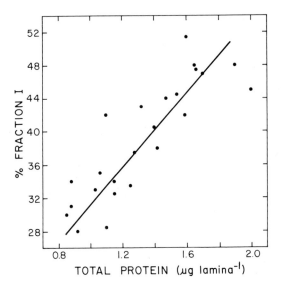

FIGURE 3.   The percentage of leaf protein in Fraction I as influenced by the total amount of leaf protein in barley.[67]

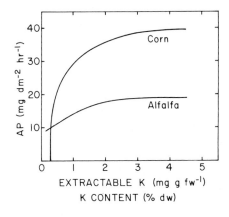

FIGURE 4.   Effect of leaf K content on AP. Data for alfalfa from Cooper et al.,[12] where K content is expressed as percentage of D.M. Data for corn from Peaslee and Moss,[11] where K content is extractable K in mg gfw[-1] (gfw = gram of fresh weight).

Table 1). K concentrations in the leaf below 1 to 2% are correlated with reduced photosynthesis. The response of photosynthesis to K concentrations in maize and alfalfa is shown in Figure 4. The reduction of photosynthesis at low K levels in the leaf results in part from impaired stomatal functioning. K deficiency in maize is correlated with reduced leaf porosity,[11] an indication of increased stomatal resistance. In alfalfa, both stomatal aperature and the number of stomata may decrease with decreasing K levels.[12]

In other work on alfalfa, both stomatal and mesophyll resistance were found to increase with K deficiency.[13] The increase in mesophyll resistance was not attributable

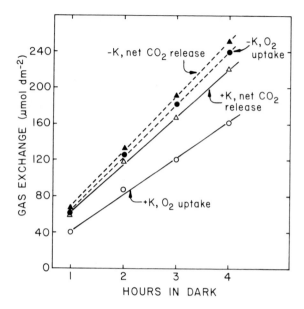

FIGURE 5.   Effects of K deficiency on net $CO_2$ and $O_2$ exchange in the dark in bean leaves.[14]

to changes in PSI or II, but appeared to result from changes in the levels of Fraction I.

Plants deficient in K also demonstrate increased dark respiration[13-16] (see Figure 5). A respiratory quotient (RQ) near unity in K-deficient plants indicates reductants produced during respiration were being oxidized almost exclusively by way of oxidative phosphorylation. In contrast, the reductants produced in nonstressed plants serve to reduce other metabolites resulting in higher RQs.[16] Based on the change in RQ, K deficiency may impair a number of metabolic processes requiring reduced pyridine nucleotide or reduced electron transport intermediates.

In plants with the $C_4$ pathway of carbon assimilation, K deficiency alters the labeling pattern observed with $^{14}CO_2$.[17] In K-deficient maize plants, label accumulated in the organic acid and amino acid fractions at the expense of soluble sugars, sugar phosphates, and starch. It seems probable that K deficiency impairs the intercellular transport of photosynthetic intermediates between the mesophyll and bundle sheath cells.[17]

## PHOSPHORUS

The important role of P in energy conservation and transfer suggests that deficiencies of this element would greatly affect photosynthesis. However, the effects of P deficiency on photosynthesis are variable, and several workers have concluded that leaf P concentrations have little direct relationship to leaf photosynthetic rates.[18-20] P deficiency has been reported to reduce photosynthesis in spinach,[21] tobacco,[22] and sugar beet,[23] but no effects have been observed by other workers for a variety of other species.[18-20] Andreeva and Persanov[20] found P deficiency in bean to inhibit growth primarily through a reduction in leaf area. In sugar beets, however, P deficiency reduces photosynthesis, photorespiration, and dark respiration.[23] After 26 days of growth on P-free medium, the rate of photosynthesis was approximately 35% of the control (see Figure 6). At low light intensities, leaf diffusive resistance (stomatal) was markedly higher in the P-deficient plants. This effect was overcome at higher light

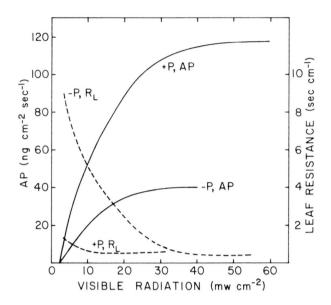

FIGURE 6. AP and leaf resistance (RL) changes with light intensity in control and phosphorus deprived sugar beet.[23]

intensities which suggests that P deficiency, by way of an effect on ATP concentrations, was altering stomatal response.[23] The decrease in photosynthesis in P-deficient plants at high light intensities was accounted for largely by increases in mesophyll resistance. Such increases are consistent with the pronounced inhibitory effect of P deficiency on ATP synthesis and the Hill reaction in isolated chloroplasts.[24]

## OTHER ELEMENTS

S deficiencies typically result in reductions of photosynthesis which are paralleled by reductions in chlorophyll content.[21] S deficiencies also impair protein synthesis and N assimilation. Much of the S requirement of higher plants, however, can be met through the assimilation of $SO_2$. The stimulation of plant growth by low concentrations of $SO_2$ is frequently over and above that expected by meeting the S requirement alone. $SO_2$ moves into the tissues primarily as $SO_3^=$ and $HSO_3^-$. Low to moderate concentrations of $SO_3^= - HSO_3^-$ stimulate photosynthesis.[25-27] These forms of S are potent inhibitors of glycolate oxidase. A stimulation of photosynthesis by $SO_3^= - HSO_3^-$ in isolated poppy cells results, at least in part, from the inhibition of glycolate oxidase.[25] Sulfite also can be oxidized to the $HSO_3$ radical in chloroplasts and accept electrons from photosystem I.[26] This has been suggested to stimulate photosynthesis by allowing photophosphorylation to proceed under NADPH saturing conditions.[26] The relative contribution of these two processes in stimulating photosynthesis has yet to be evaluated.

The key role of Mg as a constituent of the chlorophyll molecule and as a cofactor for numerous enzymes associated with photosynthesis and respiration (see Table 1) suggests that deficiencies would have pronounced effects on these processes. Mg deficiency in maize inhibits photosynthesis at concentrations less than 200 ug/gfw,[11] and in sugar beet, the critical concentration is very nearly the same.[28] An increase in mesophyll resistance largely accounts for the inhibition.[28] Peaslee and Moss[11] found that the inhibition of photosynthesis in maize was paralleled by decreasing chlorophyll concentrations. However, in sugar beet,[28] photosynthesis was inhibited prior to changes

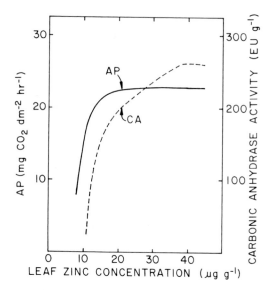

FIGURE 7.    Effects of leaf Zn concentration on AP
and carbonic anhydrase (CA) activity in soybean.[31]

in chlorophyll concentration. The inhibition of photosynthesis under Mg deficiency undoubtedly results from the impairment of numerous processes including chlorophyll synthesis.

Fe deficiencies have a very pronounced effect of chlorophyll levels of the leaf.[29,30] This finding is consistent with a role of this element in chlorophyll biosynthesis (see Table 1). Photosynthesis is limited under Fe deficiency, however, even when rates of photosynthesis are expressed on a chlorophyll basis substantial reductions of photosynthesis occur.[29] Fe deficiencies also limit the amount of ferredoxin and Fraction I protein in the leaf.[30]

Zn deficiencies retard growth, reduce net photosynthesis, and stimulate dark respiration.[31] Carbonic anhydrase activity is also reduced[32] and in a manner which can parallel the reductions in photosynthesis[31] (see Figure 7). This relationship has led to the suggestion that a restriction in the hydration of $CO_2$ might impair photosynthesis at low leaf Zn levels. However, Randall and Bouma[32] found substantial reductions in carbonic anhydrase activity in spinach, with little or no effect on net photosynthesis. Plants deficient in Zn also show reduced levels of Fraction I protein[33] which suggests that reductions in photosynthesis result from several biochemical limitations.[32]

A deficiency of Ca appears to limit photosynthesis only when prolonged and severe.[34] Gallaher et al.[34] found Ca deficiency to reduce photosynthesis in soybean, peanut, and cotton, only when the deficiency was evident in the leaves. In spinach, Ca deficiency had no effect on net photosynthesis.[21]

Deficiencies in B and Mn have been reported to limit photosynthesis in spinach.[21]

The limitations on plant growth imposed by nutrient deficiencies generally result from the impairment of several to numerous metabolic processes, including photosynthesis. In general, photosynthesis responds to these essential elements in a curvilinear fashion such that saturation is achieved at some level of that element. The exception to this response is that observed with N. Photosynthesis appears to increase in concert with all N increases in the leaf. Nitrogen, primarily through its effect on the levels of Fraction I protein, provides a general constraint to both photosynthesis and plant productivity.

# REFERENCES

1. Stocking, C. R. and Ongun, A., The intercellular distribution of some metallic elements in leaves, *Am. J. Bot.*, 49, 284, 1962.
2. Natŕ, L., Influence of mineral nutrition of photosynthesis and the use of assimilates, in *Photosynthesis and Productivity in Different Environments*, Cooper, J. P., Ed., Cambridge University Press, Cambridge, 1975.
3. Wananabe, H. and Yoshida, S., Effects of nitrogen, phosphorus, and potassium on photophosphorylation in rice in relation to the photosynthetic rate of single leaves, *Soil Sci. Plant Nutr. (Tokyo)*, 16, 163, 1970.
4. Yoshida, S. and Coronel, V., Nitrogen nutrition, leaf resistance, and leaf photosynthetic rate of the rice plant, *Soil Sci. Plant Nutr. (Tokyo)*, 22, 207, 1976.
5. Brown, R. H., A difference in N use efficiency in $C_3$ and $C_4$ plants and its implications in adaptation and evolution, *Crop Sci.*, 18, 93, 1978.
6. Dirven, J. G. P., The chemical composition of some temperate and tropical grasses, cultivated in Surinam, De Surinaamse landbarrw, 19, 5, 1971.
7. Björkman, O., Carboxydismutase activity in shade-adapted and sun-adapted species of higher plants, *Physiol. Plant.*, 21, 1, 1968.
8. Björkman, O., Further studies on differentiation of photosynthetic properties in sun and shade ecotypes of *Solidago virgaurea*, *Physiol. Plant.*, 21, 84, 1968.
9. Medina, E., Effect of nitrogen supply and light intensity during growth on the photosynthetic capacity and carboxydismutase activity of leaves of *Atriplex patula* spp. *hastata*, *Carnegie Inst. Washington Yearb.*, 69, 551, 1970.
10. Ku, M. S. B., Schmitt, M. R., and Edwards, G. E., Quantitative determination of RuBP carboxylase-oxygenase protein in leaves of several $C_3$ and $C_4$ plants, *J. Exp. Bot.*, 30, 89, 1979.
11. Peaslee, D. E. and Moss, D. N., Photosynthesis in K- and Mg-deficient maize (*Zea mays* L.) leaves, *Soil Sci. Soc. Am. Proc.*, 30, 220, 1966.
12. Cooper, R. B., Blazer, R. E., and Brown, R. H., Potassium nutrition effects on net photosynthesis and morphology of alfalfa, *Soil Sci. Soc. Am. Proc.*, 31, 231, 1967.
13. Peoples, T. R. and Koch, D. W., Role of potassium in carbon dioxide assimilation in *Medicago sativa* L., *Plant Physiol.*, 63, 878, 1979.
14. Osbun, J. L., Volk, R. J., and Jackson, W. A., Effects of potassium deficiency on photosynthesis, respiration, and the utilization of photosynthetic reductant by immature bean leaves, *Crop Sci.*, 5, 69, 1965.
15. Okamoto, S., The respiration in leaf discs from younger taro plants under a moderate potassium deficiency, *Soil Sci. Plant Nutr. (Tokyo)*, 15, 274, 1969.
16. Jackson, W. A. and Volk, R. J., Role of potassium in photosynthesis and respiration, in *The Role of Potassium in Agriculture*, Kilmer, V. J., Younts, S. E., and Brady, N. C., Eds., American Society of Agronomy, Madison, Wis., 1968.
17. Barankiewicz, T. J., $CO_2$ exchange rates and $^{14}C$ photosynthetic products of maize leaves as affected by potassium deficiency, *Z. Pflanzenphysiol.*, 89, 11, 1978.
18. Natŕ, L., Influence of mineral nutrients on photosynthesis of higher plants, *Photosynthetica*, 6, 80, 1972.
19. Natŕ, L. and Purš, J., The relation between rate of photosynthesis and N, P, K concentrations in barley leaves. II. Phosphorus absent from the nutrient solution, *Photosynthetica*, 4, 31, 1970.
20. Andreeva, T. F. and Personov, U. M., Effect of duration of phosphorus deficiency on rate of photosynthesis and growth in connection with plant productivity, *Fiziol. Rast.*, 17, 478, 1970.
21. Bottrill, D. E., Possingham, J. V., and Kriedemann, P. E., The effect of nutrient deficiencies on photosynthesis and respiration in spinach, *Plant Soil*, 32, 424, 1970.
22. Kakie, T., Effect of a phosphorus deficiency on the photosynthetic carbon fixation-products in tobacco, *Soil Sci. Plant Nutr. (Tokyo)*, 15, 245, 1970.
23. Terry, N. and Ulrich, A., Effects of phosphorus deficiency on the photosynthesis and respiration of leaves of sugar beet, *Plant Physiol.*, 51, 43, 1973.
24. Tombesi, L., Calé, M. T., and Tiborné, B., Effects of nitrogen, phosphorus and potassium fertilizer on the assimilation capacity of *Beta vulgaris* chloroplasts, *Plant Soil*, 31, 65, 1969.
25. Paul, J. S. and Bassham, J. A., Effect of sulfite on metabolism in isolated mesophyll cells from *Papaver somniferum*, *Plant Physiol.*, 62, 210, 1978.
26. Ziegler, I. and Libera, W., The enhancement of $CO_2$ fixation in isolated chloroplasts by low sulfite concentrations and by ascorbate, *Z. Naturforsch.*, 30, 634, 1975.
27. Libera, W., Ziegler, I., and Ziegler, H., The action of sulfite on the $HCO_3^-$ fixation and the fixation pattern of isolated chloroplasts and leaf tissue slices, *Z. Pflanzenphysiol.*, 74, 420, 1975.
28. Terry, N. and Ulrich, A., Effects of magnesium deficiency on the photosynthesis and respiration of leaves of sugar beet, *Plant Physiol.*, 54, 379, 1974.

29. **Stocking, R. C.,** Iron deficiency and the structure and physiology of maize chloroplasts, *Plant Physiol.,* 55, 626, 1975.
30. **Marsh, H. V., Jr., Evans, H. J., and Matrone, G.,** Investigations on the role of iron deficiency on chlorophyll and heme content and on the activities of certain enzymes in leaves, *Plant Physiol.,* 38, 632, 1963.
31. **Ohki, K.,** Zinc concentration in soybeans as related to growth, photosynthesis, and carbonic anhydrase activity, *Crop Sci.,* 18, 79, 1978.
32. **Randall, P. J. and Bouma, D.,** Zinc deficiency, carbonic anhydrase, and photosynthesis in leaves of spinach, *Plant Physiol.,* 52, 229, 1973.
33. **Jyung, W. H., Camp, M. E., Polson, D. E., Adams, M. W., and Wittwer, S. N.,** Differential response of two bean varities to zinc as revealed by electrophoretic protein pattern, *Crop Sci.,* 12, 26, 1971.
34. **Gallaher, R. N., Brown, R. H., Ashley, D. A., and Jones, J. B., Jr.,** Photosynthesis of and $^{14}CO_2$-photosynthate translocation from calcium deficient leaves of crops, *Crop Sci.,* 16, 116, 1976.
35. **Willmer, C. M. and Pallas, J. E., Jr.,** A survey of stomatal movements and associated potassium fluxes in the plant kingdom, *Can. J. Bot.,* 51, 37, 1973.
36. **Humble, G. D. and Hsiao, T. C.,** Light dependent influx and efflux of potassium of guard cells during stomatal opening and closing, *Plant Physiol.,* 46, 483, 1970.
37. **Fisher, R. A.,** Stomatal opening: role of potassium uptake by guard cells, *Science,* 160, 284, 1968.
38. **Peaslee, D. E. and Moss, D. N.,** Stomatal conductivities in K-deficient leaves of maize (*Zea mays,* L.), *Crop Sci.,* 8, 427, 1968.
39. **Marschner, H. and Possingham, J. V.,** Effect of $K^+$ and $Na^+$ on growth of leaf discs of sugar beet and spinach, *Z. Pflanzenphysiol.,* 75, 6, 1975.
40. **Pflüger, R. and Mengel, K.,** Die photochemische aktivitat von chloroplasten aus unterschieldlich mit kalium ernahrten pflanzen, *Plant Soil,* 36, 417, 1972.
41. **Hewitt, E. J. and Smith, T. A.,** *Plant Mineral Nutrition,* John Wiley & Sons, New York, 1974.
42. **Hartt, C. E.,** Effect of potassium deficiency upon translocation of $^{14}C$ in attached blades and entire plants of sugarcane, *Plant Physiol.,* 44, 1461, 1969.
43. **Spanner, D. C.,** The translocation of sugar in sieve tubes, *J. Exp. Bot.,* 9, 332, 1958.
44. **Hall, J. D., Barr, R., Al-Abbas, A. H., and Crane, F. L.,** The ultrastructure of chloroplasts in mineral-deficient maize leaves, *Plant Physiol.,* 50, 404, 1972.
45. **Jagendorf, A. T. and Smith, M.,** Uncoupling phosphorylation in spinach chloroplasts by absence of cations, *Plant Physiol.,* 37, 135, 1962.
46. **Butler, W. L. and Kitajima,** Energy transfer between photosystem II and photosystem I in chloroplasts, *Biochim. Biophys. Acta,* 396, 72, 1975.
47. **Murata, N.,** Control of excitation transfer in photosynthesis. II. Magnesium ion-dependent distribution of excitation energy between two pigment systems in spinach chloroplasts, *Biochim. Biophys. Acta,* 189, 171, 1969.
48. **Hind, G., Nakatani, H. Y., and Izawa, S.,** Light-dependent redistribution of ions in suspensions of chloroplast thylakoid membranes, *Proc. Natl. Acad. Sci. U.S.A.,* 71, 1484, 1974.
49. **Weissbach, A., Horecker, B. L., and Hurwitz, J.,** The enzymatic formation of phosphoglyceric acid from ribulose diphosphate and carbon dioxide, *J. Biol. Chem.,* 218, 795, 1956.
50. **Epstein, E.,** *Mineral Nutrition of Plants: Principles and Perspectives,* John Wiley & Sons, New York, 1972.
51. **Mengel, K. and Kirkby, E. A.,** *Principles of Plant Nutrition,* International Potash Institute, Bern, Switzerland, 1978.
52. **Malkin, R., Bearden, A. J., and Hunter, F. A.,** Properties of the low-temperature photosystem I primary reaction the P-700-chlorophyll a-protein, *Biochim. Biophys. Acta,* 430, 389, 1976.
53. **Cheniae, G. M.,** Photosystem II and $O_2$ evolution, *Annu. Rev. Plant Physiol.,* 21, 467, 1970.
54. **Izawa, S., Heath, R. L., and Hind, G.,** The role of chloride ion in photosynthesis. III. The effect of artificial electron donors upon electron transport, *Biochim. Biophys. Acta,* 180, 388, 1969.
55. **Schröder, H., Muhle, H., and Rumberg, B.,** Relationship between ion transport phenomena and phosphorylation in chloroplasts, 2nd Int. Congr. Photosynthesis, Stresa, 1971, 919.
56. **Deamer, D. W. and Packer, L.,** Light-dependent anion transport in isolated spinach chloroplasts, *Biochim. Biophys. Acta,* 172, 539, 1969.
57. **Marsh, H. V., Jr., Evans, H. J., and Matrone, G.,** Investigations of the role of iron in chlorophyll metabolism. II. Effect of iron deficiency on chlorophyll synthesis, *Plant Physiol.,* 38, 639, 1963.
58. **Possingham, J. V. and Spencer, D.,** Manganese as a functional component of chloroplasts, *Aust. J. Biol. Sci.,* 15, 58, 1962.
59. **Cheniae, G. M. and Martin, I. F.,** Site of manganese function in photosynthesis, *Biochem. Biophys. Acta,* 153, 819, 1968.
60. **Itoh, Y., Yamashita, K., Nishi, T., Konishi, K., and Shibata, K.,** The site of manganese function in photosynthetic electron transport system, *Biochim. Biophys. Acta,* 180, 509, 1969.

61. **Hatch, M. D. and Kagawa, T.**, NAD malic enzyme in leaves with the $C_4$ pathway photosynthesis and its role in $C_4$ acid decarboxylation, *Arch. Biochem. Biophys.*, 160, 346, 1974.
62. **Atkins, C. A., Patterson, B. D., and Graham, D.**, Plant carbonic anhydrases. I. Distribution of types among species, *Plant Physiol.*, 50, 214, 1972.
63. **Toblin, A. J.**, Carbonic anhydrase from parsley leaves, *J. Biol. Chem.*, 245, 2656, 1970.
64. **Katoh, S.**, A new copper protein from *Chlorella ellipsoidea*, *Nature (London)*, 186, 533, 1960.
65. **Nicholas, D. J. D. and Nason, A.**, Role of molybdenum as a constituent of nitrate reductase from soybean leaves, *Plant Physiol.*, 30, 135, 1955.
66. **Wilson, J. R.**, Comparative response to nitrogen deficiency of a tropical and temperate grass in the interrelation between photosynthesis, growth and the accumulation of non-structural carbohydrate, *Neth. J. Agric. Sci.*, 23, 104, 1975.
67. **Blenkinsop, P. G. and Dale, J. E.**, The effects of nitrate supply and grain reserves on Fraction I protein level in the first leaf of barley, *J. Exp. Bot.*, 25, 913, 1974.

# SALINITY PREFERENCE AND TOLERANCE OF AQUATIC PHOTOSYNTHETIC ORGANISMS

## E. O. Duerr and A. Mitsui

## INTRODUCTION

Aquatic saline environments may be found in virtually any region of the world. Terrestrial saline areas of the world have been described by Waisel[1] and Chapman.[2,3] In addition to the saline lakes and rivers of these terrestrial areas, the oceans and seas of the world comprise upwards of 71% of the surface area of the earth or 98% of the hydrosphere.[4]

Marine waters have been divided into six categories according to the total quantity of salts dissolved in a unit volume of water.[5] This classification has been summarized in Table 1. Oligohaline waters represent the first level above freshwater, containing greater than approximately 0.5 °/$_{oo}$ dissolved salts, but less than approximately 5.0 °/$_{oo}$. In order of increasing salinity, the other categories are termed mesohaline, polyhaline, mixoeuhaline, euhaline, and hyperhaline. Inland saline waters represent a more complex situation because of the diversity of the dominant ionic complexes. They do not lend themselves to such a classification.[5] A number of other physical chemical definitions have been proposed to describe saline waters. Several of these have been summarized and discussed by Remane.[6]

Organisms have also been classified according to their tolerance of, or requirement for inorganic salts (especially NaCl). The biological classification employs only one critical point from the physical chemical classification, and that is the division between freshwater and saltwater. Freshwater species (glycophytes) may be defined as being incapable of completing a life cycle in waters containing greater than 0.5 °/$_{oo}$ salinity.[6,7] Other authors have established this upper salinity limit of freshwater species at different levels. These limits include 0.15,[8] 1,[6] 3,[6] and 5 °/$_{oo}$.[9] Organisms capable of growth in waters of 0.5 °/$_{oo}$ or greater are termed halophytes. Employing terminology put together by Ingram,[10] halophytes can be subdivided into three general types: intolerant, facultative, and obligate. The intolerant halophyte shows optimal growth in freshwater, but is capable of reduced growth in saline environments. The facultative halophyte shows optimal growth in saline habitats, but is capable of maintaining growth in freshwater. The obligate halophyte has an absolute requirement for the presence of inorganic salts and experiences its highest growth rates at salinities greater than 0.5 °/$_{oo}$. A schematic diagram illustrating definitions of salinity tolerance is given in Figure 1.

## GROWTH IN SALINE ENVIRONMENTS

Laboratory studies of plant growth responses to varying salinities have been numerous among the algae, but very sparse among the angiosperms and photosynthetic bacteria. Table 2 provides a partial list of aquatic photosynthetic, halophilic organisms with specific information about each. Because of possible secondary nutrient dilution effects by some methods, the types of media used in the various experiments have been described. Four basic types of saline media have been used:

1. Seawater
2. Enriched seawater media (SWM)
3. Freshwater based artificial seawater media (ASW)
4. NaCl dissolved in distilled water

## Table 1
## THE VENICE SYSTEM FOR THE
## CLASSIFICATION OF MARINE WATERS
## ACCORDING TO SALINITY

|  | Zone | Salinity (°/oo) |
|---|---|---|
| Brackish | Oligohaline | 0.5—∿5 |
|  | Mesohaline | ∿5—∿18 |
|  | Polyhaline | ∿18—∿30 |
|  | Mixoeuhaline | less than adjacent euhaline sea, but greater than ∿30 |
| Marine | Euhaline | ∿30 to ∿40 |
|  | Hyperhaline | Greater than ∿40 |

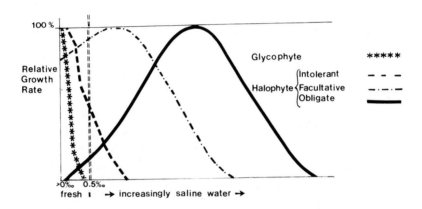

FIGURE 1.    Classifications[10] of growth types according to salinity tolerances.[9]

In those cases where the media salinity was varied solely by the addition or deletion of NaCl (micro-element composition remaining constant), the "+" designation has been added. In each case, the range of NaCl molar strength of the experimental media has been calculated and listed in the third column. Unless otherwise stated in the respective article, seawater was assumed to have 35 °/oo salinity,[11] which converts to a NaCl molar strength of approximately 0.5 $M$. Generally the experiments using seawater achieved different molar strengths by adding distilled water for dilution or distilling the seawater to increase the concentration.

Many specific reactions in terms of growth have been measured in response to varying salinity. The reactions and specific values derived for each of the species listed have been given in Columns 4 and 5 of Table 2. These values have been matched with their corresponding salinities (converted to NaCl molar values). Generally low, high, and intermediate salinities have been chosen to give the range and optimal salinities. An attempt was made to transform the growth data into uniform and therefore comparable units. Salinity preferences of the macroalgae and single angiosperm listed were measured most frequently by $O_2$ evolution rates. This gives an indirect index of the photosynthetic response to the media. Growth measurements of photosynthetic microorganisms have traditionally been given in units of "doubling time". Numbers of microorganisms in log phase nonsynchronous growth can be described according to the formula:

$$N_1 = N_0 e^{kt} \tag{1}$$

where $N_0$ and $N_1$ represent cell numbers at the initiation and conclusion of the experiment, respectively; e is the base of natural logarithms; k is the growth constant (a measure of the efficiency of growth); and t is the time period between initiation and conclusion.[12] From Formula 1, an expression for the growth constant can be derived:

$$k_e = \frac{\log_e N_1 - \log_e N_0}{t} \tag{2}$$

or

$$k_e = \frac{\log_e (N_1/N_0)}{t} \tag{3}$$

Doubling time (G) is defined as

$$G = \frac{\log_e 2}{k_e} \tag{4}$$

Measurements can be taken in $\log_{10}$ ($k_{10}$) or $\log_2$ ($k_2$) units as well. If t is measured in days, G can be brought to an hourly basis by multiplying by 24. Column 7 of Table 2 gives the G values (in hours) for each experiment where such conversions were possible.

Experimental variables other than salinity are summarized in the comment column. A number of factors are listed which were shown to have a synergistic effect with salinity. These factors include temperature,[13,14] light levels,[13-16] $Na^+/K^+$ ratio,[17] $HCO_3^-$ concentration,[14,18] osmotic ionic relation,[19,20] and $Ca^{2+}$ concentration.[16] Several reviews exist which discuss these interactions.[1,21-27]

A number of papers have been published which list the salinity ranges in which particular photosynthetic organisms have been found.[6,24,28-34] The growth characteristics determined in the laboratory have usually confirmed field observation for each of the authors, listed in Table 2, who discussed the subject. It should be noted that the growth rates listed in Table 2 should be considered in a relative sense. It was frequently the case that experiments involving other parameters yielded faster growth rates. Hoogenhout and Amesz[35] have compiled a table of maximum growth rates of photosynthetic microorganisms.

## SALT TOLERANCE AND CONTROL MECHANISMS

Most studies on salt tolerance have concentrated on NaCl. Specific ionic requirements for $Na^+$ have been demonstrated in a few aquatic photosynthetic organisms, i.e., *Dunaliella viridis*[36] and *Aphanothece halophytica*.[37] Other species may display abnormal growth in the absence of $Na^+$. Tolerance of organisms to potassium which substitutes most readily for sodium has been investigated extensively, along or in conjunction with sodium.[6,38-64] Calcium is generally the dominant cation in freshwater.[7] Calcium tolerances and control mechanisms have also been studied for halophytes.[6,15,30,41-44,49,57,60,63,64] The effect of magnesium on aquatic halophytes has received the least attention of the dominant cations.[7,30,36,42-44,49,57,60,63,64] Among the inorganic anions, chloride[41-43,45,47,50,57,63,65-68] and sulphates[16,36,43,60] are the only groups which have been extensively studied. A review of the specific ionic effects of each of the above listed ions on aquatic algae has been given by Raven.[24]

## Table 2
## GROWTH OF VARIOUS PHOTOSYNTHETIC ORGANISMS IN SALINE MEDIA

| Genus, species | Salinity range tested | Converted NaCl molar | Growth potential Units of measure | Growth potential Value | Molar NaCl conc. | Doubling time G (hr) | Comments | Ref. |
|---|---|---|---|---|---|---|---|---|
| Angiosperma *Zostera nana* | 0—2X conc. seawater | 0—1.0 | O₂ evolved (μL/mg dry weight/ hr) | 2 7 4 | 0.0 0.4 1.0 | | Maximum rate increases to 17.5 if $HCO_3^-$ held constant at all salinities | 18 |
| Chlorophyta *Chlamydomonas moewasii* | 0—0.5 *M* ASW + | 0—0.5 | Optical density (at 6 days) (6600 A.U. filter) | 0.45 0.62 0.65 0.08 | 0.0 0.03 0.1 0.4 | | | 90 |
| *Chlamydomonas uva-maris* | 3—100°/₀₀ (SWM) NaCl | 0.07—1.6 | ¹⁴C fixation (relative rate) | 85% 100% 50% | 0.2 0.4 1.6 | | Short-term growth in media of NaCl + H₂O was identical at comparable SWM concentrations | 91 |
| *Chlorella pyrenoidosa* | 0—68 meq NaCl ASW | 0—0.068 | Growth rate (mg dry weight/L/ day) | 70 75 30 40 | 0.0 0.0017 0.0085 0.017 | 32.5 30.4 76.0 57.0 | All NaCl concentrations in iso-osmotic Tamiya's medium, balanced by KNO₃; no osmotic effects were noted at the various polyethylene glycol concentrations tested | 92 |
| *Chlorella sorokiniana* | 0—0.5 *M* ASW + | 0—0.5 | Cell division/day | 9.6 9.3 6.3 0.0 | 0.0 0.2 0.3 0.35 | 2.5 2.6 3.8 — | Values given were measured at 9.4 mW/ cm²; reference gives other information on NaCl tolerance affected by light intensity and Ca²⁺ levels | 16 |

| Species | Medium | Range | Measure (units) | | | | Ref. | Notes |
|---|---|---|---|---|---|---|---|---|
| *Codium decorticatum* | 15—50⁰/₀₀ SWM+ | 0.2—0.7 | Kₑ (wet weight basis) | 0.0 | 0.2 | — | 57 | |
| | | | | 0.06—0.15 | 0.3 | 111—277 | | |
| | | | | 0.06—0.15 | 0.6 | 111—277 | | |
| | | | | 0.04 | 0.7 | 415 | | |
| *Dunaliella parva* | 0.3—8.0 osM ASW+ | 0.15—4.1 | O₂ evolved (μmol/mg chl/hr) | 0 | 0.15 | | 71 | Glucose and glycine can replace NaCl for similar O₂ evolution |
| | | | | 55 | 2.0 | | | |
| | | | | 27 | 4.1 | | | |
| *D. tertiolecta* | 0.17—5.0 M ASW+ | 0.17—5.0 | 24/Gᵃ (E₇₀₀ basis) | 2.4 | 0.17 | 10 | 74 | |
| | | | | 2.0 | 1.0 | 12 | | |
| *D. viridis* | 0.17—5.0 M ASW+ | 0.17—5.0 | 24/Gᵃ (E₇₀₀ basis) | 0.0 | 1.7 | — | 74 | |
| | | | | 0.0 | 1.3 | — | | |
| | | | | 1.0 | 1.7 | 24 | | |
| | | | | 0.8 | 4.8 | 30 | | |
| *Mychonastes ruminatus* | 0—30⁰/₀₀ ASW (?) | 0—0.4 | 24/Gᵃ (E₇₀₀ basis) | 1.1 | 0.02 | 22 | 93 | |
| | | | | 1.6 | 0.2 | 15 | | |
| | | | | 0.8 | 0.4 | 30 | | |
| *Platymonas suecica* | 0.1—3X SWM+ | 0.05—1.4 | k₁₀ (hr) (cell-count basis) | 0.010 | 0.05 | 30 | 77 | |
| | | | | 0.017 | 0.4 | 18 | | |
| | | | | 0.013 | 0.9 | 23 | | |
| | | | | 0.006 | 1.4 | 50 | | |
| *Platymonas* sp. | 2.5—35⁰/₀₀ SWM | 0.03—0.5 | log₂ N₁/N₀ (t = 7 days) (cell-count basis) | 1.2 | 0.03 | 140 | 94 | |
| | | | | 5.5 | 0.2 | 31 | | |
| | | | | 6.0 | 0.5 | 28 | | |
| *Scenedesmus obliquus* | 1—22 atm. SWM | 0.02—0.4 | k₁₀ (E₅₆₀ basis) | 0.82 | 0.02 | 8.8 | 19 | Osmotic inhibition does not begin until 0.2 M NaCl in SWM |
| | | | | 0.15 | 0.2 | 48 | | |
| | | | | 0.05 | 0.3 | 145 | | |
| *S. obliquus* | 2—26⁰/₀₀ ASW+ | 0.03—0.45 | Relative ¹⁴C fixation rate | 100% | 0.03 | | 91 | |
| | | | | 50% | 0.14 | | | |
| | | | | 10% | 0.41 | | | |
| *Ulva pertusa* | 0—2X seawater | 0—1.0 | O₂ evolved (μL/mg dry weight/hr) | 3 | 0.0 | | 18 | Rate shifts at 18 and 0.5 M if HCO₃⁻ held constant at all salinities |
| | | | | 12 | 0.5 | | | |
| | | | | 12 | 1.0 | | | |
| *Valonia macrophysa* | 15—60⁰/₀₀ SWM | 0.2—0.8 | kₑ (wet weight basis) | 0.012 | 0.3 | 1385 | 61 | |
| | | | | 0.052 | 0.4 | 319 | | |
| | | | | 0.053 | 0.7 | 319 | | |
| | | | | 0.002 | 0.8 | 8304 | | |

Table 2 (continued)
## GROWTH OF VARIOUS PHOTOSYNTHETIC ORGANISMS IN SALINE MEDIA

| Genus, species | Salinity range tested | Converted NaCl molar | Units of measure | Growth potential | | | Comments | Ref. |
|---|---|---|---|---|---|---|---|---|
| | | | | Value | Molar NaCl conc. | Doubling time G (hr) | | |
| **Pyrrhophyta** | | | | | | | | |
| *Amphidinium carteri* | 2.5—35 °/oo SWM | 0.03—0.5 | $\log_2 N_t/N_0$ (t = 9 days) (cell-count basis) | 0<br>6.5<br>6.0 | 0.14<br>0.34<br>0.5 | —<br>33<br>36 | | 94 |
| *Amphidinium* sp. | 5—45 °/oo seawater | 0.07—0.62 | Cell division/day | 0.41<br>0.57<br>0.41<br>0.13 | 0.14<br>0.24<br>0.55<br>0.62 | 59<br>42<br>59<br>185 | | 13 |
| *Exuviaella baltica* | 5—45 °/oo seawater | 0.07—0.62 | Cell division/day | 0.22<br>0.30<br>0.22<br>0 | 0.07<br>0.14<br>0.55<br>0.62 | 109<br>80<br>109<br>— | | 13 |
| *Peridinium trochoideum* | 5—60 °/oo seawater | 0.07—0.82 | Cell division/day | 0<br>0.95<br>0.23 | 0.07<br>0.27<br>0.62 | —<br>25<br>104 | Growth rate and maximum salinity tolerance is a function of light and temperature conditions | 13 |
| *Prorocentrum micans* | 10—45 °/oo seawater | 0.14—0.62 | Cell division/day | 0.13<br>0.34<br>0.20 | 0.14<br>0.21<br>0.62 | 185<br>71<br>120 | Moderate changes in salinity tolerance exist in different clones of same species | 13 |
| **Cryptophyta** | | | | | | | | |
| *Cryptomonas* sp. | 5—45 °/oo seawater | 0.07—0.62 | Cell division/day | 0<br>0.39<br>0.10 | 0.07<br>0.27<br>0.48 | —<br>62<br>240 | | 13 |
| **Dinophyta** | | | | | | | | |
| *Gonyaulax excavata* | 6—42.2 °/oo SWM | 0.09—0.62 | Cell division/day | 0<br>0.36<br>0.28 | 0.09<br>0.48<br>0.62 | —<br>67<br>86 | | 95 |

| | | | | | | | Ref. | |
|---|---|---|---|---|---|---|---|---|
| *Gymnodinium breve* | 20—43 %o ASW+ | 0.3—0.6 | $k_r$ (cell-count basis) | 0<br>0.18<br>0 | 0.3<br>0.45<br>0.6 | —<br>92<br>— | 96 | |
| **Chrysophyta** | | | | | | | | |
| *Asterionella japonica* | 10—40 %o SWM+ | 0.14—0.55 | $k_r$ ($E_{680}$/cell-count basis)[b] | 0<br>0.7<br>0.4 | 0.2<br>0.4<br>0.55 | —<br>24<br>42 | 97 | |
| *Cyclotella cryptica* | 0.1—1.5X ASW+ | 0.05—0.7 | Cell count at 4 days ($N_0$ = 1200 cells) | $1.8\times10^5$<br>$2.2\times10^5$<br>$1.3\times10^5$<br>$7.0\times10^3$ | 0.05<br>0.15<br>0.37<br>0.7 | 13.6<br>12.9<br>14.9<br>35.0 | 98 | |
| *Cyclotella meneghiniana* | 0.25X SWM to −38.6 bars SWM+ | 0.05—0.8 | $\Delta E_{510}$ (at 48 hrs) | 0.16<br>0.34<br>0.06 | 0.05<br>0.2<br>0.65 | 145<br>77<br>532 | 85 | |
| *Isochrysis galbana* | 10—40 %o SWM+ | 0.14—0.55 | $k_r$ ($E_{680}$/cell-count basis) | 0.27<br>0.36<br>0.36 | 0.14<br>0.2<br>0.55 | 62<br>46<br>46 | 99 | |
| *Monochrysis lutheri* | 0.013—1.02 *M* SWM+ | 0.013—1.02 | G (hr) (cell-count basis) | 26.3<br>18.1<br>41.0 | 0.013 [Na+]<br>0.46 [Na+]<br>1.02 [Na+] | 18<br>14.9<br>33.8 | 15 | Values given were measured at light intensity of 0.1 g cal/cm²/min; reference gives effects of other light intensities |
| *Olisthodiscus* sp. | 2.5—35 %o SWM | 0.03—0.5 | $\log_2 N_1/N_0$ (t = 9 days) (cell-count basis) | 0<br>5.5<br>5.0 | 0.07<br>0.2<br>0.5 | —<br>39<br>43 | 94 | |
| *Syracosphaera carterae* | 5—45 %o seawater | 0.07—0.62 | Cell division/day | 0.30<br>0.67<br>0.58 | 0.07<br>0.41<br>0.62 | 80<br>36<br>41 | 13 | |
| **Phaeophyta** | | | | | | | | |
| *Dictyopteris membranacea* | 0—75 %o NaCl(?) and seawater | 0—1.0 | $O_2$ evolved (mg/g dry weight/40 min) | 0<br>2.5<br>1.2 | 0.08<br>0.75<br>1.03 | | 100 | $O_2$ evolution measured in seawater after pretreatment for 30 min in dark at given salinities |

## Table 2 (continued)
## GROWTH OF VARIOUS PHOTOSYNTHETIC ORGANISMS IN SALINE MEDIA

| Genus, species | Salinity range tested | Converted NaCl molar | Units of measure | Growth potential | | Doubling time G (hr) | Comments | Ref. |
|---|---|---|---|---|---|---|---|---|
| | | | | Value | Molar NaCl conc. | | | |
| *Fucus serratus* | 5.95—32.06 °/oo seawater | 0.08—0.45 | $O_2$ evolved (μmol/10 min/100 mg dry weight) | 0.97 1.41 1.45 | 0.08 0.3 0.45 | | | 78 |
| *Fucus vesiculosus* | 5.95—32.06 °/oo seawater | 0.08—0.45 | $O_2$ evolved (μmol/10 min/100 mg dry weight) | 1.33 1.41 1.50 | 0.08 0.2 0.45 | | | 78 |
| *Fucus virsoides* | 0—75 °/oo NaCl (?) and seawater | 0—1.0 | $O_2$ evolved (mg/g dry weight/40 min) | 1.8 1.8 1.8 | 0.0 0.4 1.0 | | $O_2$ evolution measured in seawater after pretreatment for 30 min in dark at given salinities | 100 |
| *Porphyridium* sp. | 2.5—35 °/oo SWM | 0.03—0.5 | $log_2 N$ ; $diN_0$ (t = 8 days) (cell-count basis) | 4.5 5.5 5.5 | 0.03 0.14 0.5 | 43 35 35 | | 94 |
| Rhodophyta | | | | | | | | |
| *Gelidium amansii* | 0—2X seawater | 0—1.0 | $O_2$ evolved (μL/mg dry weight/hr) | 1.5 5.8 4.6 | 0.0 0.5 1.0 | | Rate and optima shift to 8.1 and 0.35 $M$ if $HCO_3^-$ held constant at all salinities | 18 |
| *Fucus virsoides* | 0—75 °/oo NaCl (?) and seawater | 0—1.0 | $O_2$ evolved (mg/g dry weight/40 min) | 1.8 1.8 1.8 | 0.0 0.4 1.0 | | $O_2$ evolution measured in seawater after pretreatment for 30 min in dark at given salinities | 100 |
| *Porphyridium* sp. | 2.5—35 °/oo SWM | 0.03—0.5 | $log_2 N$ ; $diN_0$ (t = 8 days) (cell-count basis) | 4.5 5.5 5.5 | 0.03 0.14 0.5 | 43 35 35 | | 94 |

| Organism | Medium | Range | Measurement | | | | Comments | Ref |
|---|---|---|---|---|---|---|---|---|
| **Rhodophyta** | | | | | | | | |
| *Gelidium amansii* | 0—2X seawater | 0—1.0 | O₂ evolved (µL/mg dry weight/hr) | 1.5 / 5.8 / 4.6 | 0.0 / 0.5 / 1.0 | | Rate and optima shift to 8.1 and 0.35 M if HCO₃⁻ held constant at all salinities | 18 |
| *Porphyra tenera* | 0—2X seawater | 0—1.0 | O₂ evolved (µL/mg dry weight/hr) | 3 / 19 / 13 | 0.0 / 0.7 / 1.0 | | Rate and optima shift to 23 and 0.35 M if HCO₃⁻ held constant at all salinities | 18 |
| *Porphyridium cruentum* | 4.5—35 ‰ ASW+ | 0.06—0.5 | $k_c$ (E₇₁₀/cell-count basis) | 0.8 / 0.8 | 0.06 / 0.5 | 20.9 / 20.9 | | 101 |
| **Cyanophyta** | | | | | | | | |
| *Agmenellum quadruplicatum* | 0—1.7 M ASW+ | 0—1.7 | $k_{10}$ (E/cell-count basis) | 2.0 / 2.1 / 0.4 | 0.0 / 0.17 / 1.7 | 3.6 / 3.4 / 18.0 | | 20 |
| *Anabaena* sp. | 0—0.6 M ASW+ | 0—0.6 | $k_{10}$ (E₆₆₀/cell-count basis) | 0.0 / 1.6 / 0.9 | 0.0 / 0.08 / 0.6 | — / 4.5 / 8.0 | Grown in combined N free-media: Na⁺ essential, Cl⁻ not essential | 64 |
| *Aphanothece halophytica* | 0.5—4.0 M ASW+ | 0.5—4.0 | $k_{10}$ (E₇₅₀ basis) | 0.0 / 0.125 / 0.0 | 0.5 / 1—2 / 4.0 | — / 60 / — | Li⁺, Rb⁺, K⁺, and Cs⁺ cannot substitute for Na⁺ in the growth media | 37 |
| *A. halophytica* | 2.0—4.0 M ASW+ | 2.0—4.0 | $k_{10}$ (E₇₅₀ basis) | 0.43 / 0.41 / 0.23 | 2 / 3 / 4 | 14.5 / 17.6 / 31.7 | Values given were measured under 40°C, 4250 lx, and 1% CO₂ add.; reference gives other information, including NaCl tolerance affected by temperature, CO₂ conc., and light intensity | 14 |
| *Anacystis nidulans* | 0—1.7 M ASW+ | 0—1.7 | $k_{10}$ (E/cell-count basis) | 2.4 / 1.0 / 0.3 | 0.0 / 0.25 / 0.34 | 3.0 / 7.2 / 24 | | 20 |
| *Coccochloris elebans* | 0—1.7 M ASW+ | 0—1.7 | $k_{10}$ (E/cell-count basis) | 2.0 / 2.1 / 0.4 | 0.0 / 0.17 / 1.7 | 3.6 / 3.4 / 18.0 | | 20 |

## Table 2 (continued)
## GROWTH OF VARIOUS PHOTOSYNTHETIC ORGANISMS IN SALINE MEDIA

| Genus, species | Salinity range tested | Converted NaCl molar | Growth potential | | | Doubling time G (hr) | Comments | Ref. |
|---|---|---|---|---|---|---|---|---|
| | | | Units of measure | Value | Molar NaCl conc. | | | |
| *Gomphosphaeria aponina* | 20—36 $^o/_{oo}$ ASW+ | 0.3—0.5 | $k_r$ (cell-count basis) | 0.67 | 0.3 | 16 | | 96 |
| | | | | 1.18 | 0.4 | 14 | | |
| | | | | 1.04 | 0.5 | 25 | | |
| Photosynthetic bacteria | | | | | | | | |
| *Rhodopseudomonas spheroides* | 0—250 µg Na+/1 ASW+ | 0—0.11 | $k_r$ (E$_{680}$ dry weight basis) | 0.02 at >0.2 µg Na+/mL | | 34.6 | Na+ requirement and K+ requirement interaction; values given from experiment with K+/Na+ ratio = 2.5 | 17 |
| | | | | 0.11 at >10 µg Na+/mL | | 6.3 | | |
| | | | | 0.15 at >50 µg Na+/mL | | 4.6 | | |
| | | | | 0.175 at >100 µg Na+/mL | | 3.9 | | |
| | | | | 0.155 at >250 µg Na+/mL | | 4.5 | | |

*Note:* Media types: SWM = enriched seawater; SWM+ = salinity of SWM adjusted before addition of enrichment; ASW = artificial seawater media (dist. H$_2$O base); ASW+ = nutrients in distilled H$_2$O with NaCl added to desired level; NaCl = strictly NaCl dissolved in H$_2$O. See text for additional comments.

[a]    Spectrophotometric absorption at 700 nm.

[b]    Spectrophotometric absorption at 680 nm correlated with cell count.

Table 3 gives a partial listing of the halophytic species that have been studied with respect to their behavior in saline media. The optimal NaCl concentration of the growth media was listed when available; otherwise the NaCl molar concentration of the test media was listed in parentheses. When given, the maximum NaCl concentration tolerated by the species was listed. In some cases, these tolerances refer to intermittent exposures. At these maximum concentrations, plants would no longer grow. If mention was made of a higher than optimum salt concentration at which growth occurred, then the maximum concentration was listed as being greater than that number. Maximum salinity tolerances were useful to distinguish species of *Chlorella*.[69] An extensive series of tests for plasmolysis in marine algae in different concentrations of seawater has been reported by Biebl.[28,29] Upper and lower limits of salinity tolerance with respect to plasmolysis are given with extensive discussions on habitat distribution.

Control mechanisms for regulating ionic or osmotic levels in many of the species listed in Table 3 have been studied. These include the regulation of carbohydrates[19,70] and other organic solutes such as glycerol,[71-75] cyclohexanetetrol,[15] mannitol,[76-80] and floridosides.[58,81-83] Amino acid or protein level regulation has been found to vary both directly[62,63,84-86] and inversely[32,78,87-89] with salinity levels. Much attention has been given to the ion pumps common in aquatic organisms. Those studies which indicated the presence of active ion pumps were designated by listing the ion(s) involved. The presence of sodium:potassium exchange pumps was indicated by a "/" between the two ions.

## ENZYME ACTIVITY IN SALINE MEDIA

The activity of enzymes in the presence of NaCl or nonionic osmotic agents (especially glycerol) has received much recent attention. Several of the studies involving photosynthetic aquatic organisms have been presented in Table 4. Only the glucose-6-phosphate dehydrogenase enzymes of *Dunaliella tertiolecta* and *D. viridis* showed highest activity at a salinity other than 0 $^{o}/_{oo}$.[74] All other enzyme systems, including the glucose-6-phosphate dehydrogenase enzyme of *Dunaliella parva*,[71] show maximal activity in the absence of ionic or nonionic osmotic agents.

## Table 3
## SALT TOLERANCE AND CONTROL MECHANISMS OF PHOTOSYNTHETIC AQUATIC HALOPHYTES

| Phyla, Genus Species | Optimal NaCl $M$ conc. | Upper NaCl tolerance | Ionic and/or osmotic controls found | Ref. |
|---|---|---|---|---|
| **Angiosperma** | | | | |
| Diplanthera wrightii | 0.1—0.7 | <1.0 | | 102 |
| D. wrightii | (0.4) | >1.0 | | 103 |
| Halophila engelmanni | (0.4) | <1.0 | | 103 |
| Ruppia maritima | (0.4) | 0.6—>1.0 | | 103 |
| Syringodium filiforme | (0.4) | 0.6—0.8 | | 103 |
| S. filiforme | 0.5 | <0.7 | | 102 |
| Thalassia testudinum | 0.5—0.6 | | | 104 |
| T. testudinum | | | Plasmalemma salt secretion | 105 |
| T. testudinum | (0.4) | 0.8—>1.0 | | 103 |
| Vallisneria spiralis | | | $Cl^-$ | 65 |
| Zostera marina | 0.5 | >1.5 | | 106 |
| Z. nana | 0.4 | >1.0 | | 18 |
| **Chlorophyta** | | | | |
| Acetabularia mediterranea | (0.5) | | $Na^+$, $K^+$, $Cl^-$ | 51 |
| Chaetomorpha linum | (0.4) | | $K^+$, $Cl^-$(?), $Na^+$(?) | 42, 53, 107 |
| Chaetomorpha darwinii | (0.5) | | $K^+/Na^+$ | 46,66 |
| Chlamydomonas moewusii | 0.03—0.1 | 0.4 | Contractile vacuoles eliminate $H_2O$ | 90 |
| C. uva-maris | 0.5 | >1.3 | | 91 |
| Chlamydomonas sp. | 1.71 | | $Na^+$, $H_2O$, $K^+$ induced contraction (?) | 45, 49, 108 |
| Chlorella pyrenoidosa | (0.0) | | Sucrose | 48, 70 |
| C. pyrenoidosa | | >0.4 m$M$[$Na^+$] | $Na^+/K^+$, $H^+/K^+$, $Cl^-$ | 52, 109 |
| C. pyrenoidosa | 1.7 m$M$ | >0.068 | | 92 |
| Chlorella salina | 0.1—0.6 | | Proline, $Na^+$, $K^+$ | 63 |
| C. sorokiniana | 0 | 0.3 | $Na^+$, $Cl^-$ | 16 |
| Codium decorticatum | 0.3—0.6 | >0.7 | $Na^+$, $Cl^-$, $SO_4^{2-}$ | 57 |

| Species | | | | |
|---|---|---|---|---|
| *C. fragile* | (0.4) | | $Na^+$, $Cl^-$, $K^+$ | 43 |
| *C. tomentosum* | (0.7) | | | 21 |
| *Dunaliella parva* | 1.5 | | Large pores in membrane | 110 |
| *D. parva* | 2.0 | >4.0 | Glycerol, $H_2O$ | 71, 72 |
| *D. tertiolecta* | 1.0 | 5.0 | Glycerol | 73 |
| *D. tertiolecta* | ≤0.17 | <1.7 | Glycerol, $K^+$, $Na^+$ | 74 |
| *D. tertiolecta* | 0.5—2.5 | | Glycerol | 75 |
| *D. viridis* | ≤1.7 | >4.8 | Glycerol, $K^+$, $Na^+$ | 74 |
| *D. viridis* | 0.8—2.0 | >3.75 | $Na^+$(?) | 36 |
| *Halicystis osterhoutii* | (0.5) | | $Na^+$, $K^+$, $Cl^-$ | 21 |
| *Halicystis ovalis* | (0.5) | | $Na^+$, $Cl^-$ | 111 |
| *Mychonastes ruminatus* | 0.2 | | | 93 |
| *Nannochloris oculata* | 0.09 [$Na^+$] | >0.4 | | 30 |
| *Nitellopsis obtusa* | (0.03) | 0.52 [$Na^+$] | | 112 |
| *Platymonas subcordiformis* | 0.2 | >0.8 | $Na^+$, $Cl^-$ | 59, 63, 76, 113-115 |
| *P. suecica* | 0.4 | >1.4 | Mannitol, $K^+$, $Na^+$, $Cl^-$, Mannitol, $H_2O$, total cell carbon | 77 |
| *Platymonas* sp. | 0.5 | >>0.5 | $K^+$(?)/$Na^+$, carbohydrates | 94 |
| *Scenedesmus obliquus* | 0.02 | >0.3 | | 19 |
| *S. obliquus* | <0.03 | >0.45 | | 91 |
| *Scenedesmus* sp. | 0.003 | | | 116 |
| *Ulva lactuca* | (0.6) | | $K^+$/$Na^+$ | 38, 39, 117 |
| *U. pertusa* | 0.5—1.0 | >>1.0 | $K^+$, $Na^+$ | 18 |
| *Valonia macrophysa* | 0.4—0.7 | | $K^+$, $Na^+$(?) | 61 |
| *Valonia ventricosa* | (0.5) | >0.8 | $K^+$, $Na^+$ | 47, 118, 119 |
| **Pyrrophyta** | | | | |
| *Amphidinium carteri* | 0.34 | >0.5 | | 44, 94 |
| *A.* sp. | 0.21—0.27 | | | 13 |
| *Exuviaella baltica* | 0.14 | 0.6 | | 13 |
| *Noctiluca miliaris* | | 0.6 | $H_2O$ | 120 |
| *N. miliaris* | (0.5) | | $Na^+$, $Cl^-$ | 121 |
| *Peridinium trochoideum* | 0.21—0.27 | >0.5 | | 13 |
| *Prorocentrum micans* | 0.21—0.27 | >0.6 | | 13 |
| **Cryptophyta** | | | | |
| *Cryptomonas* sp. | 0.03—0.5 | | | 94 |
| *Cryptomonas* sp. | 0.21—0.27 | | | 13 |
| *Hemiselmis virescens* | 0.09 [$Na^+$] | 0.5 [$Na^+$] | | 30 |

Table 3 (continued)
SALT TOLERANCE AND CONTROL MECHANISMS OF PHOTOSYNTHETIC AQUATIC
HALOPHYTES

| Phyla, Genus Species | Optimal NaCl M conc. | Upper NaCl tolerance | Ionic and/or osmotic controls found | Ref. |
|---|---|---|---|---|
| Dinophyta | | | | |
| Gonyaulax excavata | 0.5 | >0.7 | | 95 |
| Gymnodinium breve | 0.45 | 0.58 | | 96 |
| Chrysophyta | | | | |
| Asterionella japonica | 0.4 | >0.55 | | 97 |
| Cyclotella cryptica | 0.15—0.24 | >0.7 | Proline, K$^+$ | 62, 84 |
| C. cryptica | 0.2 | >0.5 | | 44, 94 |
| C. meneghiniana | 0.2 | 0.65 | | 85 |
| Isochrysis galbana | 0.3—0.6 | >>0.6 | Proline, Na$^+$, Cl$^-$ | 99 |
| Monochrysis lutheri | 0.3—0.4 | | | 44 |
| M. lutheri | 0.46 | >1.02 | Cyclohexanetetrol | 15 |
| M. lutheri | 0.09 [Na$^+$] | 0.5 [Na$^+$] | | 30 |
| Ochromonas malhamensis | | >0.1 | Isofloridoside, a.a., H$_2$O, K$^+$ | 58, 81, 82 |
| Olisthodiscus sp. | 0.2 | >0.5 | | 44, 94 |
| Phaedactylum tricornutum | (0.3) | | K$^+$(?) | 60 |
| P. tricornutum | (0.4) | >0.6 | a.a. (proline) | 86 |
| P. tricornutum | 0.13 [Na$^+$] | 0.52 [Na$^+$] | | 30 |
| Skeletonema costatum | 0.4 | | | 44 |
| S. costatum | 0.26 [Na$^+$] | 0.52 [Na$^+$] | | 30 |
| Syracosphaera carterae | 0.3—0.4 | | | 44 |
| S. carterae | 0.41 | >0.6 | | 13 |
| Thalassiosira decipiens | 0.03—0.5 | | | 94 |
| Thalassiosira fluviatilis | 0.3—0.4 | | | 44 |
| Phaeophyta | | | | |
| Ascophyllum nodosum | (0.45) | | Glutamic acid ↓/sal.↑ | 32, 87 |
| Dictyopteris membranacea | 0.75 | >1.0 | | 100 |

| Species | | | Solutes | Ref. |
|---|---|---|---|---|
| *Hormosira banksii* | (0.5) | | $K^+$, $Na^+$ | 40 |
| *Fucus ceranoides* | (0.06) | | Glutamic acid ↓/sal.↑ | 32, 87 |
| *Fucus serratus* | 0.45 | | Mannitol, glutamic acid ↓/ sal.↑ | 32, 78 |
| *F. serratus* | 0.1—0.86 | | | 122 |
| *Fucus vesiculosus* | (0.45) | | Mannitol, crude protein ↓/ sal.↑ | 32, 78, 88 |
| *Fucus virsoides* | 0—1.0 | | | 100 |
| *Pelvetia canaliculata* | (0.5) | ≫1.0 | Volemitol, mannitol | 80 |
| Rhodophyta | | | | |
| *Bryopsis hypnoides* | (0.4) | | $Na^+$, $K^+$, $Cl^-$ | 43 |
| *Bryopsis plumosa* | (0.5) | | $Na^+$, $K^+$, $Cl^-$ | 123 |
| *Delesseria sanguinea* | 0.2—0.5 | | $Na^+$, $K^+$, $Cl^-$, $Br^-$ | 122 |
| *Enteromorpha intestinalis* | (0.5) | | $Na^+$, $K^+$, $Cl^-$ | 124 |
| *Gelidium amansii* | 0.5 | >1.0 | | 18 |
| *Gracilaria foliifera* | (0.5) | | $Na^+$ | 125 |
| *Griffithsia flabelliformis* | (0.5) | | $Cl^-$ | 67 |
| *Griffithsia monile* | (0.5) | | $Cl^-$ | 67 |
| *Griffithsia* sp. | (0.5) | | $Na^+$, $K^+$, $Cl^-$ | 126 |
| *Iridophycus flacidum* | (0.5) | >1.0 | Floridoside | 83 |
| *Porphyra perforata* | (0.5) | >1.0 | Floridoside, isofloridoside | 83 |
| *P. perforata* | (0.5) | >1.2 | $Cl^-$, $Na^+$, $K^+$ | 41, 127 |
| *P. tenera* | 0.7 | >1.0 | | 18 |
| *Porphyridium aerugineum* | (0.001) | >0.01 | | 56 |
| *P. cruentum* | (0.46) | >0.5 | | 101 |
| *Porphyridium* sp. | 0.14—0.5 | | | 44, 94 |
| *Rhodymenia palmata* | (0.5) | | $Na^+$, $Cl^-$, $K^+$(?) | 128 |
| Cyanophyta | | | | |
| *Agmenellum quadruplicatum* | 0.17 | >1.7 | $Na^+$ | 20 |
| *Anabaena flos-aquae* | (0.5) | | | 129 |
| *Anabaena variabilis* | 0.0007 [$Na^+$] | | Low mol wt solute product | 130 |
| *Anabaena* sp. | 0.08 | >0.6 | | 64 |
| *Anacytis nidulans* | 0 | ≥0.34 | | 20 |
| *A. nidulans* | 0.0007 [$Na^+$] | >0.05 | | 56, 130 |

**Table 3 (continued)**

## SALT TOLERANCE AND CONTROL MECHANISMS OF PHOTOSYNTHETIC AQUATIC HALOPHYTES

| Phyla, Genus Species | Optimal NaCl $M$ conc. | Upper NaCl tolerance | Ionic and/or osmotic controls found | Ref. |
|---|---|---|---|---|
| A. nidulans | 1.0—2.0 | | Na$^+$, K$^+$(?), Cl$^-$ | 54, 68 |
| Aphanothece halophytica | 2.0 | | Total protein ↓/sal.↑ | 37, 89 |
| A. halophytica | 0.1—0.3 | >4.0 | | 14 |
| Calothrix scopulorum | 0.17 | 1.1 | | 131 |
| Coccochloris elebans | 0.4 | >1.7 | | 20 |
| Gomphosphaeria aponina | (0.5) | >0.5 | | 96 |
| Lichina pygmaea (a marine lichen) | | | Mannosido-mannitol | 79 |
| Nostoc entophytum | 0.1—0.2 | 0.56 | | 131 |
| Photosynthetic bacteria | | | | |
| Ectothiorhodospira halophila | (1.7—5.1) | | | 132 |
| Ectothiorhodospira mobilis | (0.5) | >4.0 | | 133 |
| Rhodopseudomonas spheroides | 0.002—0.004 [Na$^+$] | >0.011 [Na$^+$] | | 17 |

*Note:* Molar concentrations in parentheses indicate growth media concentrations employed. See text for further explanation.

Table 4

SALT EFFECTS ON IN VITRO ENZYME ACTIVITY OF SOME MICROALGAE

| Genus, species | Enzyme | Relative activity | | Maximum activity | Comment | Ref. |
|---|---|---|---|---|---|---|
| *Dunaliella tertiolecta* | Glucose-6-phosphate dehydrogenase | 100% = | 0.2 $M$(NaCl + KCl) | 0.136 μmol NADP⁺ reduced/min/mg protein | Glycerol added to media decreases activity | 74 |
| | | 50% = | 0.7 $M$(NaCl + KCl) | | | |
| *D. viridis* | Glucose-6-phosphate dehydrogenase | 100% = | 0.2 $M$(NaCl + KCl) | 0.06 μmol NADP⁺ reduced/min/mg protein | Glycerol added to media decreases activity | 74 |
| | | 50% = | 0.7 $M$(NaCl + KCl) | | | |
| *D. tertiolecta* | Glycerol dehydrogenase | 100% = | 0.0 $M$(NaCl + KCl) | 0.23 μmol NADP⁺ reduced/min/mg protein | Glycerol added to media increases activity | 74 |
| | | 50% = | 0.3—0.4 $M$(NaCl + KCl) (glycerol added) | | | |
| *D. viridis* | Glycerol dehydrogenase | 100% = | 0.0 $M$(NaCl + KCl) | 0.39 μmol NADP⁺ reduced/min/mg protein | Glycerol added to media increases activity | 74 |
| | | 50% = | 0.3—0.4 $M$(NaCl + KCl) (glycerol added) | | | |
| *D. parva* | Lactate dehydrogenase | 100% = | 0.0—0.4 $M$NaCl | 10 μmol NADH oxidized/hr/mg chlorophyll | | 71 |
| | | 50% = | 1.2 $M$NaCl | | | |
| *D. parva* | Glucose-6-phosphate dehydrogenase | 100% = | 0.0 $M$NaCl | 15 μmol NADP⁺ reduced/hr/mg chlorophyll | | |
| | | 50% = | 0.5 $M$NaCl | | | |
| *D. viridis* | Phosphoribose isomerase | 100% = | 0.0 $M$NaCl | 3.0 μmol ribulose-5-phosphate/min/mg protein | Similar activity if cells grow at 1.28 $M$ or 3.75 $M$NaCl | |
| | | 50% = | 0.43 $M$NaCl | | | |
| *D. viridis* | Glucose-6-phosphate dehydrogenase | 100% = | 0.0 $M$NaCl | 0.06 μmol NADP⁺ reduced/min/0.74 mg protein | Stimulated by MgSO₄; preincubation of enzyme with NaCl reversed inhibition | |
| | | 77% = | 0.17 $M$NaCl | | | |
| | | 0% = | 1.28 $M$NaCl | | | |
| *D. viridis* | Phosphoglucose isomerase | 100% = | 0.0 $M$NaCl | 0.15—0.20 μmol ketose formed/min/mg protein | Other cations had similar effects | 36 |
| | | 57% = | 0.3 $M$NaCl | | | |
| *Anabaena cylindrica* | Nitrate reductase | 100% = | 0.0 $M$NaCl | 4.3 μmol NO₂⁻ formed/hr/mg dry weight | | 134 |
| | | 50% = | 0.004 m $M$NaCl | | | |

# REFERENCES

1. **Waisel, Y.,** *Biology of Halophytes,* Academic Press, London, 1972.
2. **Chapman, V. J.,** *Salt Marshes and Salt Deserts of the World,* 2nd ed., Lehre: Cramer Verlag, Bremerhaven, 1974.
3. **Chapman, V. J.,** The salinity problem in general, its importance, and distribution with special reference to natural halophytes, in *Plants in Saline Environments,* Poljakoff-Mayber, A. and Gale, J., Eds., Springer-Verlag, Berlin, 1975, 7.
4. **Von Arx, W. S.,** *An Introduction into Physical Oceanography,* Addison-Wesley, Reading, Mass., 1962.
5. Symposium on the classification of brackish waters, Venice, 8-14 April 1958, *Arch. Oceanogr. Limnol.,* 2 (Suppl.), 1, 1959.
6. **Remane, A. and Schlieper, C.,** *Biology of Brackish Water,* John Wiley & Sons, New York, 1971.
7. **Provasoli, L., McLaughlin, J. J. A., and Pintner, I. J.,** Relative and limiting concentrations of major mineral constituents for the growth of algal flagellates, *Trans. N.Y. Acad. Sci., Ser. II,* 16, 412, 1954.
8. **Price, J. B. and Gunter, G.,** Studies of the chemistry of fresh and low salinity waters in Mississippi and the boundary between fresh and brackish water, *Int. Rev. Ges. Hydrobiol.,* 49, 629, 1964.
9. **Barbour, M. G.,** Is any angiosperm an obligate halophyte?, *Am. Midl. Nat.,* 84, 105, 1970.
10. **Ingram, M.,** Micro-organisms resisting high concentrations of sugars or salts, in *Microbial Ecology — 7th Symposium of the Society of General Microbiology,* Cambridge University Press, London, 1957, 90.
11. **Hood, D. and Pytkowicz, R. M.,** Chemical oceanography, in *Handbook of Marine Science,* Vol. 1, Walton Smith, F. G., Ed., CRC Press, Cleveland, 1974, 1.
12. **Fogg, G. E.,** *Algal Cultures and Phytoplankton Ecology,* University of Wisconsin Press, Madison, 1975.
13. **Braarud, T.,** Salinity as an ecological factor in marine phytoplankton, *Physiol. Plant.,* 4, 28, 1951.
14. **Tindall, D. R., Yopp, J. H., Miller, D. M., and Schmid, W. E.,** Physico-chemical parameters governing the growth of *Aphanothece halophytica* (Chroococcales) in hypersaline media, *Phycologia,* 17, 179, 1978.
15. **Craigie, J. S.,** Some salinity-induced changes in growth, pigments, and cyclohexanetetrol content of *Monochrysis lutheri,* *J. Fish. Res. Board Can.,* 26, 2959, 1969.
16. **Chimiklis, P. E. and Karlander, E. P.,** Light and calcium interactions in *Chlorella* inhibited by sodium chloride, *Plant Physiol.,* 51, 48, 1973.
17. **Sistrom, W. R.,** A requirement for sodium in the growth of *Rhodopseudomonas spheroides,* *J. Gen. Microbiol.,* 22, 778, 1960.
18. **Ogata, E. and Matsui, T.,** Photosynthesis in several marine plants of Japan as affected by salinity, drying and pH, with attention to their growth habitats, *Bot. Mar.,* 8, 199, 1965.
19. **Wetherell, D. F.,** Osmotic equilibration and growth of *Scenedesmus obliquus* in saline media, *Physiol. Plant.,* 16, 82, 1963.
20. **Batterton, J. C. and Van Baalen, C.,** Growth responses of blue-green algae to sodium chloride concentration, *Arch. Mikrobiol.,* 76, 151, 1971.
21. **Gutknecht, J. and Dainty, J.,** Ionic relations of marine algae, *Oceanogr. Mar. Biol. Annu. Rev.,* 6, 163, 1968.
22. **Rains, D. W.,** Salt transport by plants in relation to salinity, *Annu. Rev. Plant Physiol.,* 23, 367, 1972.
23. **Kylin, A. and Quatrano, R. S.,** Metabolic and biochemical aspects of salt tolerance, in *Plants in Saline Environments, Ecol. Stud.,* Vol. 15, Poljakoff-Mayber, A. and Gale, J., Eds., Springer-Verlag, Heidelberg, 1975, 147.
24. **Raven, J. A.,** Transport in algal cells, in *Transport in Plants II, Part A, Cells,* Lüttge, U. and Pitman, M. G., Eds., Springer-Verlag, Berlin, 1976, 129.
25. **Cram, W. J.,** Negative feedback regulation of transport in cells. The maintenance of turgor, volume and nutrient supply, in *Transport in Plants II, Part A, Cells,* Lüttge, U. and Pitman, M. G., Eds., Springer-Verlag, Berlin, 1976, 284.
26. **Hellebust, J. A.,** Osmoregulation, *Annu. Rev. Plant Physiol.,* 27, 485, 1976.
27. **Flowers, T. J., Troke, P. F., and Yeo, A. R.,** The mechanism of salt tolerance in halophytes, *Annu. Rev. Plant Physiol.,* 28, 89, 1977.
28. **Biebl, R.,** Ecological and non-environmental constitutional resistance of the protoplasm of marine algae, *J. Mar. Biol. Assoc. U.K.,* 31, 307, 1952.
29. **Biebl, R.,** Vergleichende untersuchungen zur temperaturresistenz von meeresalgen entlang der pazifischen küste nordamerikas, *Protoplasma,* 69, 61, 1970.
30. **Droop, M. R.,** Optimum relative and actual ionic concentrations for growth of some euryhaline algae, *Verh. Int. Verein. Theor. Angew. Limnol.,* 13, 722, 1958.

31. Phillips, R. C., Observations on the ecology and distribution of the Florida sea grasses, *Fl. Bd. Conserv., Prof. Papers Ser.,* 2, 1, 1960.

32. Munda, I., Der einfluss der salinität auf die chemische zusammensetzung, das wachstum und die fruktifikation einiger Fucaceen, *Nova Hedwigia Z. Kryptogamenkd.,* 13, 471, 1967.

33. Munda, I., Changes in the algal vegetation of a part of the deltaic area in the southern Netherlands (Veerse Meer) after its closure, *Bot. Mar.,* 10, 141, 1967.

34. Munda, I. M., Salinity dependent distribution of benthic algae in estuarine areas of Icelandic Fjords, *Bot. Mar.,* 21, 451, 1978.

35. Hoogenhout, H. and Amesz, J., Growth rates of photosynthetic microorganisms in laboratory cultures, *Arch. Mikrobiol.,* 50, 10, 1965.

36. Johnson, M. K., Johnson, E. J., MacElroy, R. D., Speer, H. L., and Bruff, B. S., Effects of salts on the halophilic alga *Dunaliella viridis, J. Bacteriol.,* 95, 1461, 1968.

37. Yopp, J. H., Tindall, D. R., Miller, D. M., and Schmid, W. E., Isolation, purification and evidence for a halophilic nature of the blue-green alga *Aphanothece halophytica* Fremy (Chroococcales), *Phycologia,* 17, 172, 1978.

38. Scott, G. T. and Hayward, H. R., Metabolic factors influencing sodium and potassium distribution in *Ulva lactuca, J. Gen. Physiol.,* 36, 659, 1953.

39. Scott, G. T. and Hayward, H. R., Evidence for the presence of separate mechanisms regulating potassium and sodium distribution in *Ulva lactuca, J. Gen. Physiol.,* 37, 601, 1954.

40. Berquist, P. L., Evidence for separate mechanisms of sodium and potassium regulation in *Hormosira banksii, Physiol. Plant.,* 11, 760, 1958.

41. Eppley, R. W., Sodium exclusion and potassium retention by the red marine alga, *Porphyra perforata, J. Gen. Physiol.,* 41, 901, 1958.

42. Kesseler, H., Die bedeutung einiger anorganischer komponenten des seewassers fur die turgorregulation von *Chaetomorpha linum,* (Cladophorales), *Helgol. Wiss. Meeresunters.,* 10, 73, 1964.

43. Kesseler, H., Zellsaftgewinnung, AFS (apparent free space) und vakuolenkonzentration der osmotisch wichtigsten mineralischen bestandteile einiger Helgoländer meeresalgen, *Helgol. Wiss. Meeresunters.,* 11, 258, 1964.

44. McLachlan, J., Some considerations of the growth of marine algae in artificial media, *Can. J. Microbiol.,* 10, 769, 1964.

45. Okamoto, H. and Suzuki, Y., Intracellular concentration of ions in a halophilic strain of *Chlamydomonas.* I. Concentration of Na, K and Cl in the cell, *Z. Allg. Mikrobiol.,* 4, 350, 1964.

46. Dodd, W. A., Pitman, M. G., and West, K. R., Sodium and potassium transport in the marine alga *Chaetomorpha darwinii, Aust. J. Biol. Sci.,* 19, 341, 1966.

47. Gutknecht, J., Sodium, potassium, and chloride transport and membrane potentials in *Valonia ventricosa, Biol. Bull.,* 130, 331, 1966.

48. Greenway, H. and Hiller, R. G., Effects of low water potentials on respiration and on glucose and acetate uptake, by *Chlorella pyrenoidosa, Planta,* 75, 253, 1967.

49. Yamamoto, M. and Okamoto, H., Osmotic regulation in a halophilic *Chlamydomonas* cell. I. General feature of the response to the change in osmotic pressure, *Z. Allg. Mikrobiol.,* 7, 143, 1967.

50. Barber, J., Light-induced uptake of potassium and chloride by *Chlorella pyrenoidosa, Nature (London),* 217, 876, 1968.

51. Saddler, H. W., Fluxes of sodium and potassium in *Acetabularia mediterranea, J. Exp. Bot.,* 21, 605, 1970.

52. Shieh, Y. J. and Barber, J., Intracellular sodium and potassium concentrations and net cation movements in *Chlorella pyrenoidosa, Biochim. Biophys. Acta,* 233, 594, 1971.

53. Zimmermann, U. and Steudle, E., Effects of potassium concentration and osmotic pressure of sea water on the cell-turgor pressure of *Chaetomorpha linum, Mar. Biol.,* 11, 132, 1971.

54. Dewar, M. A. and Barber, J., Cation regulation in *Anacystis nidulans, Planta,* 113, 143, 1973.

55. Brückner, U., Höfner, W., and Weller, H., Der einfluss steigender kaliumkonzentrationen des nahrmediums auf wachstum und eisenaufnahme von *Chlorella pyrenoidosa, Z. Pflanzenphysiol.,* 74, 35, 1974.

56. Ullrich-Eberius, C. J. and Yingchol, Y., Phosphate uptake and its pH dependence in halophytic and glycophytic algae and higher plants, *Oecologia (Berlin),* 17, 17, 1974.

57. Bisson, M. A. and Gutknecht, J., Osmotic regulation in the marine alga, *Codium decorticatum.* I. Regulation of turgor pressure by control of ionic composition, *J. Membr. Biol.,* 24, 183, 1975.

58. Kauss, H., Lüttge, U., and Krichbaum, R. M., Änderung im kalium — und isofloridosidgehalt wahrend der osmoregulation bei *Ochromonas malhamensis, Z. Pflanzenphysiol.,* 76, 109, 1975.

59. Kirst, G. O., Wirkung unterschiedlicher konzentrationen von NaCl und anderen osmotisch wirksamen substanzen auf die CO$_2$-fixierung der einzelligen alge *Platymonas subcordiformis, Oecologia (Berlin),* 20, 237, 1975.

60. Overnell, J., Potassium and photosynthesis in the marine diatom *Phaeodactylum tricornutum* as related to washes with sodium chloride, *Physiol. Plant.,* 35, 217, 1975.

61. Hastings, D. F. and Gutknecht, J., Ionic relations and the regulation of turgor pressure in the marine alga, *Valonia macrophysa*, *J. Membr. Biol.*, 28, 263, 1976.

62. Liu, M. S. and Hellebust, J. A., Effects of salinity and osmolarity of the medium on amino acid metabolism in *Cyclotella cryptica*, *Can. J. Bot.*, 54, 938, 1976.

63. Kirst, G. O., Ion composition of unicellular marine and fresh-water algae, with special reference to *Platymonas subcordiformis* cultivated in media with different osmotic strengths, *Oecologia (Berlin)*, 28, 177, 1977.

64. Stacey, G., Van Baalen, C., and Tabita, F. R., Isolation and characterization of a marine *Anabaena* sp. capable of rapid growth on molecular nitrogen, *Arch. Microbiol.*, 114, 197, 1977.

65. Arisz, W. H., Active uptake, vacuole secretion, and plasmatic transport of chloride ions in leaves of *Vallisneria spiralis*, *Acta Bot. Neerl.*, 7, 1, 1953.

66. Findlay, G. P., Hope, A. B., Pitman, M. G., Smith, F. A., and Walker, N. A., Ionic relations of marine algae. III. *Chaetomorpha*: membrane electrical properties and Cl⁻ fluxes, *Aust. J. Biol. Sci.*, 24, 731, 1971.

67. Lilley, R. McC. and Hope, A. B., Chloride transport and photosynthesis in cells of *Griffithsia*, *Biochim. Biophys. Acta*, 226, 161, 1971.

68. Dewar, M. A. and Barber, J., Chloride transport in *Anacystis nidulans*, *Planta*, 117, 163, 1974.

69. Kessler, E., Physiologische und biochemische beiträge zur taxonomie der gattung *Chlorella*, *Arch. Microbiol.*, 100, 51, 1974.

70. Hiller, R. G. and Greenway, H., Effects of low water potentials on some aspects of carbohydrate metabolism in *Chlorella pyrenoidosa*, *Planta*, 78, 49, 1968.

71. Ben-Amotz, A. and Avron, M., Photosynthetic activities of the halophilic alga *Dunaliella parva*, *Plant Physiol.*, 49, 240, 1972.

72. Ben-Amotz, A., Adaptation of the unicellular alga *Dunaliella parva* to a saline environment, *J. Phycol.*, 11, 50, 1975.

73. Wegmann, K., Osmotic regulation of photosynthetic glycerol production in *Dunaliella*, *Biochim. Biophys. Acta*, 234, 317, 1971.

74. Borowitzka, L. J. and Brown, A. D., The salt relations of marine and halophilic species of the unicellular green alga, *Dunaliella*, *Arch. Microbiol.*, 96, 37, 1974.

75. Craigie, J. S. and McLachlan, J., Glycerol as a photosynthetic product in *Dunaliella tertiolecta* Butcher, *Can. J. Bot.*, 42, 777, 1964.

76. Kirst, G. O., Beziehung zwischen mannitkonzentration und osmotischer belastung bei der brackwasseralge *Platymonas subcordiformis* Hazen, *Z. Pflanzenphysiol.*, 76, 316, 1975.

77. Hellebust, J. A., Effect of salinity on photosynthesis and mannitol synthesis in the green flagellate *Platymonas suecica*, *Can. J. Bot.*, 54, 1735, 1976.

78. Munda, I. M. and Kremer, B. P., Chemical composition and physiological properties of Fucoids under conditions of reduced salinity, *Mar. Biol.*, 42, 9, 1977.

79. Feige, G. B., Untersuchungen zur okologie und physiologie der marinen blaualgenflechte *Lichina pygmeae*. III. Einige aspekte der photosynthetischen C-fixierung unter osmoregulatorischen bedingungen, *Z. Pflanzenphysiol.*, 77, 1, 1975.

80. Kremer, B. P., Distribution and biochemistry of alditols in the genus *Pelvetia* (Phaeophyceae, Fucales), *Br. Phycol. J.*, 11, 239, 1976.

81. Kauss, H., Turnover of galactosylglycerol and osmotic balance in *Ochromonas*, *Plant Physiol.*, 52, 613, 1973.

82. Quader, H. and Kauss, H., Die rolle einiger zwischenstoffe des galaktosylglyzerinstoffwechsels bei der osmoregulation in *Ochromonas malhamensis*, *Planta*, 124, 61, 1975.

83. Kauss, H., α-Galaktosylglyzeride und osmoregulation in rotalgen, *Z. Pflanzenphysiol.*, 58, 428, 1968.

84. Liu, M. S. and Hellebust, J. A., Regulation of proline metabolism in the marine centric diatom *Cyclotella cryptica*, *Can. J. Bot.*, 54, 949, 1976.

85. Schobert, B., The influence of water stress on the metabolism of diatoms. I. Osmotic resistance and proline accumulation in *Cyclotella meneghiniana*, *Z. Pflanzenphysiol.*, 74, 106, 1974.

86. Besnier, V., Bazin, M., Marchelidon, J., and Genevet, M., Étude de la variation du pool intracellulaire des acides aminés libres d'une diatomée marine en fonction de la salinité, *Bull. Soc. Chim. Biol.*, 51, 1255, 1969.

87. Munda, I. M., Differences in amino acid composition of estuarine and marine fucoids, *Aquat. Bot.*, 3, 273, 1977.

88. Munda, I. M. and Garrasi, C., Salinity-induced changes of nitrogenous constituents in *Fucus vesiculosus* (Phaeophyceae), *Aquat. Bot.*, 4, 347, 1978.

89. Tindall, D. R., Yopp, J. H., Schmid, W. E., and Miller, D. M., Protein and amino acid composition of the obligate halophile *Aphanothece halophytica* (Cyanophyta), *J. Phycol.*, 13, 127, 1977.

90. Guillard, R. R. L., A mutant of *Chlamydomonas moewusii* lacking contractile vacuoles, *J. Protozool.*, 7, 262, 1960.

91. **Vosjan, J. H., Siezen, R. J.**, Relation between primary production and salinity of algal cultures, *Neth. J. Sea Res.*, 4, 11, 1968.

92. **Kalinkina, L. G., Spektorov, K. S., and Bogoslovskaya, V. O.**, Effects of different concentrations of NaCl on growth and development of *Chlorella pyrenoidosa*, *Sov. Plant Physiol.*, 25, 16, 1978.

93. **Simpson, P. D., Karlander, E. P., and Van Valkenburg, S. D.**, The growth rate of *Mychonastes ruminatus* Simpson et Van Valkenburg under various light, temperature and salinity regimes, *Br. Phycol. J.*, 13, 291, 1978.

94. **McLachlan, J.**, The effect of salinity on growth and chlorophyll content in representative classes of unicellular marine algae, *Can. J. Microbiol.*, 7, 399, 1961.

95. **White, A. W.**, Salinity effects on growth and toxin content of *Gonyaulax excavata*, a marine dinoflagellate causing paralytic shellfish poisoning, *J. Phycol.*, 14, 475, 1978.

96. **Martin, D. F. and Gonzalez, M. H.**, Effects of salinity on synthesis of DNA, acidic polysaccharide and growth in the blue-green alga, *Gomphosphaeria aponia*, *Water Res.*, 12, 951, 1978.

97. **Kain, J. M. and Fogg, G. E.**, Studies on the growth of marine phytoplankton, I. *Asterionella japonica* Gran, *J. Mar. Biol. Assoc. U.K.*, 37, 397, 1958.

98. **Liu, M. S. and Hellebust, J. A.**, Effects of salinity changes on growth and metabolism of the marine centric diatom *Cyclotella cryptica*, *Can. J. Bot.*, 54, 930, 1976.

99. **Kain, J. M. and Fogg, G. E.**, Studies on the growth of marine phytoplankton. II. *Isochrysis galbana* Parke, *J. Mar. Biol. Assoc. U.K.*, 37, 781, 1958.

100. **Gessner, F.**, Photosynthesis and ion loss in the brown alga *Dictyopteris membranacea* and *Fucus virsoides*, *Mar. Biol.*, 4, 349, 1969.

101. **Jones, R. F., Speer, H. L., and Kury, W.**, Studies on the growth of the red alga *Porphyridium cruentum*, *Physiol. Plant.*, 16, 636, 1963.

102. **McMahan, C. A.**, Biomass and salinity tolerance of shoalgrass and manateegrass in Lower Laguna Madre, Texas, *J. Wildl. Manage.*, 32, 501, 1968.

103. **McMillan, C. and Moseley, F. N.**, Salinity tolerances of five spermatophytes of Redfish Bay, Texas, *Ecology*, 48, 503, 1967.

104. **Hammer, L.**, Salzgehalt und photosynthese bei marinen pflanzen, *Mar. Biol.*, 1, 185, 1968.

105. **Jagels, R.**, Studies of a marine grass, *Thalassia testudinum*. I. Ultrastructure of the osmoregulatory leaf cells, *Am. J. Bot.*, 60, 1003, 1973.

106. **Biebl, R. and McRoy, C. P.**, Plasmatic resistance and rate of respiration and photosynthesis of *Zostera marina* at different salinities and temperatures, *Mar. Biol.*, 8, 48, 1971.

107. **Kesseler, H.**, Beziehungen zwischen atmung und turgor-regulation von *Chaetomorpha linum*, *Helgol. Wiss. Meeresunters.*, 8, 243, 1962.

108. **Yamamoto, M.**, Osmotic regulation in a halophilic strain of *Chlamydomonas*. II. Effects of sulfhydryl reagents, *Z. Allg. Mikrobiol.*, 7, 267, 1967.

109. **Barber, J. and Shieh, Y. L.**, Sodium transport in Na$^+$-rich *Chlorella* cells, *Planta*, 111, 13, 1973.

110. **Ginzburg, M.**, The unusual membrane permeability of two halophilic unicellular organisms, *Biochim. Biophys. Acta*, 173, 370, 1969.

111. **Blount, R. W. and Levedahl, B. H.**, Active sodium and chloride transport in the single celled marine alga *Halicystis ovalis*, *Acta Physiol. Scand.*, 49, 1, 1960.

112. **MacRobbie, E. A. C. and Dainty, J.**, Ion transport in *Nitellopsis obtusa*, *J. Gen. Physiol.*, 42, 335, 1958.

113. **Kirst, G. O. and Keller, H.-J.**, Der einfluss unterschiedlicher NaCl — konzentrationen auf die atmung der einzelligen alge *Platymonas subcordiformis* Hazen, *Bot. Mar.*, 19, 241, 1976.

114. **Kirst, G. O.**, The cell volume of the unicellular alga, *Platymonas subcordiformis*: effect of the salinity of the culture media and of osmotic stresses, *Z. Pflanzenphysiol.*, 81, 386, 1977.

115. **Kirst, G. O.**, Coordination of ionic relations and mannitol concentrations in the euryhaline unicellular alga, *Platymonas subcordiformis* (Hazen) after osmotic shocks, *Planta*, 135, 69, 1977.

116. **Kylin, A.**, Uptake and loss of Na, Rb and Cs in relation to an active mechanism for extrusion of Na in *Scenedesmus*, *Plant Physiol.*, 41, 579, 1966.

117. **West, K. R. and Pitman, M. G.**, Ionic relations and ultrastructure in *Ulva lactuca*, *Aust. J. Biol. Sci.*, 20, 901, 1967.

118. **Gutknecht, J.**, Salt transport in *Valonia*: inhibition of potassium uptake by small hydrostatic pressures, *Science*, 160, 68, 1968.

119. **Aikman, D. P. and Dainty, J.**, Ionic relations of *Valonia ventricosa*, in *Some Contemporary Studies in Marine Science*, Barnes, H., Ed., Allen & Unwin, London, 1966, 37.

120. **Nawata, T. and Sibaoka, T.**, Ionic composition and pH of the vacuolar sap in marine dinoflagellate *Noctiluca*, *Plant Cell Physiol.*, 17, 265, 1976.

121. **Kesseler, H.**, Beitrag zur kenntnis der chemischen und physikalischen eigenschaft des zellsaftes von *Noctiluca miliaris*, *Veröff. Inst. Meeresforsch. Bremerh. Sonderbd.*, 2, 357, 1966.

122. Nellen, U. R., Über den einfluss des salzgehaltes auf die photosynthetische leistung verschiedener standortformen von *Delesseria sanguinea* und *Fucus serratus*, *Helgol. Wiss. Meeresunters.* 13, 288, 1966.
123. Munday, J. C., Jr., Membrane potentials in *Bryopsis plumosa*, *Bot. Mar.*, 15, 61, 1972.
124. Black, D. R. and Weeks, D. C., Ionic relations of *Enteromorpha intestinalis*, *New Phytol.*, 71, 119, 1972.
125. Gutknecht, J., Ion distribution and transport in the red marine alga *Gracilaria foliifera*, *Biol. Bull.*, 129, 495, 1965.
126. Findlay, G. P., Hope, A. B., and Williams, E. J., Ionic relations of marine algae. I. *Griffithsia*: membrane electrical properties, *Aust. J. Biol. Sci.*, 22, 1163, 1969.
127. Eppley, R. W. and Cyrus, C. C., Cation regulation and survival of the red alga *Porphyra perforata*, in diluted and concentrated sea water, *Biol. Bull.*, 118, 55, 1960.
128. MacRobbie, E. A. C. and Dainty, J., Sodium and potassium distribution and transport in the seaweed *Rhodymenia palmata* (L.) Grev., *Physiol. Plant.*, 11, 782, 1958.
129. Grant, N. G. and Walsby, A. E., The contribution of photosynthate to turgor pressure rise in the planktonic blue-green alga *Anabaena flos-aquae*, *J. Exp. Bot.*, 28, 409, 1977.
130. Kratz, W. A. and Myers, J., Nutrition and growth of several blue-green algae, *Am. J. Bot.*, 42, 282, 1955.
131. Stewart, W. D. P., Nitrogen fixation by myxophyceae from marine environments, *J. Gen. Microbiol.*, 36, 415, 1964.
132. Imhoff, J. F., Hashwa, F., and Trüper, H. G., Isolation of extremely halophilic phototrophic bacteria from the alkaline Wadi Natrun, Egypt, *Arch. Hydrobiol.*, 84, 381, 1978.
133. Trüper, H. G., *Ectothiorhodospira mobilis* Pelsh, a photosynthetic sulfur bacterium depositing sulfur outside the cells, *J. Bacteriol.*, 95, 1910, 1968.
134. Brownell, P. F. and Nicholas, D. J. D., Some effects of sodium on nitrate assimilation and $N_2$ fixation in *Anabaena cylindrica*, *Plant Physiol.*, 42, 915, 1967.

# TERRESTRIAL HALOPHYTES: HABITATS, PRODUCTIVITY, AND USES

## T. P. Croughan and D. W. Rains

Semiarid and arid regions constitute roughly one third ($4.8 \times 10^9$ ha) of the land surface of the earth, and highly saline soils are prevalent in about half this area. In comparison, the entire land-sea interface comprises considerably less than 2% of the land surface.

The extent of irrigated agriculture may be judged by its current consumption of 80% of the total water usage of the world during the farming season.[1] Salinization under irrigated agriculture is a problem of major proportions, as can be gauged from the fact that more than $1.6 \times 10^8$ ha of the arable lands of the world are under irrigation,[2] and one third of this area ($5.3 \times 10^7$ ha) is estimated to be plagued with salinity problems.[3] Further, this problem is increasing, as indicated by the reported loss of at least 40,000 ha a year of irrigated land because of salinization problems in India alone.[1]

Sand dunes, which by virtue of their excellent drainage may offer the greatest promise for irrigation with extremely saline water, cover about $1.3 \times 10^9$ ha (5 million mi$^2$) or about 9% of the surface of the earth.[4] However, the area of dunes which could be economically irrigated with saline ground water has not yet been determined. Coastal deserts, on the other hand, border the oceans for a total of almost 33,000 km, and about half of this area lies at elevations to which it is practical to pump seawater.[5] It is of considerable interest that Epstein and colleagues have successfully raised barley on California coastal sand dunes employing undiluted seawater for irrigation.[6] It appears that this system holds promise for the identification and development of salt tolerance in agricultural species and may find future utility in directly contributing to food production.

The first of the tables that follow presents estimates for the extent of the salt-affected land area in various countries (see Table 1). A partial but representative list of halophytic species will be found in Table 2. We are indebted to Mudie[5] for the compilation of much of this information which includes the scientific and common names of some 150 halophytes, the habitats in which they may be found, their economic uses, and their classification as to salt tolerance.

Table 3 lists biomass and productivity values for a number of halophytes at various locations. The brevity of this list reflects the relatively small number of halophytic species for which biomass and productivity values are available. This lack of information is particularly severe for the inland halophytes from arid habitats, explaining the predominance of salt marsh species in this list. While considerable variability exists in the values in Table 3, the biomass figures are generally higher than the above-ground biomass value for an average wheat field in the U.S. (500 g dry weight/m$^2$).

Table 4 presents salt tolerance information on some 70 agricultural species. An understanding of the salt tolerance of these agricultural species is primarily because of the employment of quantitative and standardized approaches in assessing their tolerance. Application of a similar approach would facilitate a better understanding of the salt tolerance of naturally occurring species.

## Table 1
## SALINE LAND AREA

| Country | Extent (million hectares) | |
|---|---|---|
| U.S. | Saline: At least 1.20 total in Southwest, 0.40 in California; 0.60 agricultural land in North Dakota | |
| | Salt affected: 25% of the irrigated acreage area of coastal marshes in the U.S. | |
| | Atlantic Coastal States (including East Florida) | 0.9423 |
| | West Coast of Florida | |
| |   Tidal Marsh | 0.2140 |
| |   Mangrove | 0.1592 |
| | Alabama | 0.0048 |
| | Mississippi | 0.0271 |
| | Louisiana | 1.0629 |
| | Texas | 0.1611 |
| | West Coast of U.S. (California, Oregon, Washington) | 0.0308 |
| | Total | 2.6022 |
| | Gulf Area | 1.6291 |
| U.S.S.R. | Salt affected: 75 (3.4% of the land area) | |
| | Solonetz: 50—60 | |
| | Saline/sodic: 89% of the irrigated area in central Asia | |
| China | Salt affected: 20 | |
| | Saline: 0.02 in the Sinkiang region | |
| Iraq | Saline: 3.68 (80% of the irrigable land) | |
| India | Saline/sodic: 8.1 | |
| | Sodic ("usar"): >0.8 | |
| | Saline: >25% of the irrigated area of Punjab | |
| | Saline/sodic: ∼0.21 in the Little Rann of Cutch | |
| West Pakistan | Saline: 1.2/5.06 ( = 24%) of the irrigated land; ∼0.50 out of cultivation | |
| Spain | Saline/sodic: >0.1 in the lower valley of Guadalquiver | |
| Yugoslavia | Salt affected: >0.2, mostly in the South Pannonian Plain | |
| Hungary | Salt affected: >0.5 | |
| Egypt | Saline (in 1960): 0.12; reclamation proceeding at rate of $8 \times 10^3$ ha/yr | |
| Canada | Alkali: ∼0.36 in Saskatchewan | |
| | Solonetz: ∼2.84 to south and east of Edmonton | |
| Puerto Rico | Salt affected: 36% of the soils in Lajas Valley | |
| Peru | Saline: 0.004 of the irrigated coastal valley soils | |
| | Saline/sodic: ∼0.051 | |
| | Sodic: ∼0.016 | |
| Northeast Brazil | Saline/sodic: 0.031—0.038 of the irrigated land | |
| Australia | Saline: 20% of the area irrigated by the Murrumbidgee | |
| | Salt-affected: 0.405 in the West Australia wheat belt | |
| | Solonetz: 4.925 in West Australia | |
| | Solonized brown soils ("mallee"): 42.5 in the South | |

*Note:* Estimates of the extent of salt-affected agricultural or potentially arable land in various countries, including a listing of coastal marshland area in the U.S.[5,7]

Table 2
## REPRESENTATIVE LIST OF HALOPHYTIC SPECIES[5]

Partial List of Halophytic Species

| Family | Species and common name | Habitat[a] | Use | Salt tolerance[a] |
|---|---|---|---|---|
| Agavaceae | *Phormium tenax*, J. R. and C. Ford (New Zealand flax) | I,C | Ornamental, fiber | MIO |
| Aizoaceae | *Mesembryanthemum australe* Soland. (Iceplant) | I,C | Horticulture | MES? |
| | *Mesembryanthemum crystallinum* L. | I,C | Horticulture, vegetable | MIO |
| | *Mesembryanthemum nodiflorum* L. | I,C | Horticulture | MIO |
| | *Tetragonia expansa* Murr. (New Zealand spinach) | I,C | Vegetable | MIO |
| Amaryllidaceae | *Pancratium maritimun* L. (Sea lily) | I,C | Ornamental, vegetable, fiber | MIO |
| Apocynaceae | *Cerbera odallum* Gaertn. (Rubber vine) | I | Ornamental | MIO |
| | *Rhabdadenia biflora* (Jacq.) Muell. Arg. | C | Ornamental | MIO |
| | *Urechites lutea* (L.) Britt. var. *lutea* (Wild allemander) | C | Ornamental | MIO |
| Avicenniaceae | *Avicennia alba* B1. | C | Buffalo fodder, timber, medicine | EU |
| | *Avicennia germinans* (L.) L. ( = *Avicennia nitida* L.) (Black mangrove) | C | Cattle fodder, fuel, medicine | EU |
| | *Avicennia marina* (Forsk.) Vierh. (Black mangrove) | C | Camel fodder, timber, fuel, sheep and cattle fodder, tannin | EU |
| | *Avicennia officinalis* L. | C | Edible seedlings | MES? |
| Boraginaceae | *Heliotropium curassavicum* L. (Chinese pusley) | I | Ornamental, medicine | MES |
| Chenopodiaceae | *Atriplex argentea* Nutt. (Salt bush) | I | Forage | MIO? |
| | *Atriplex canescens* (Pursh.) Nutt. | I | Forage | MIO? |
| | *Atriplex confertifolia* (Torr. and Frem.) Wats. (salt bush) | I | Forage | MIO? |
| | *Atriplex elegans* (Moq.) D. Dietr. | I | Forage, vegetable (North American Indian) | MIO? |
| | *Atriplex hortensis* L. (Garden orache) | C | Potherb | MES |
| | *Atriplex lentiformis* (Torr.) Wats. (Lens-scale) | I | Forage | MIO? |
| | *Atriplex nummularia* Lindl. (Salt Bush) | I | Forage | EU |
| | *Atriplex patula* L. ssp. hastata (L.) Hall and Clem. | I,C | Potherb | MES |
| | *Atriplex polycarpa* (Torr.) Wats. (All-scale) | I | Forage | MES |

## Table 2 (continued)
## REPRESENTATIVE LIST OF HALOPHYTIC SPECIES[5]

Partial List of Halophytic Species

| Family | Species and common name | Habitat[a] | Use | Salt tolerance[a] |
|--------|------------------------|-----------|-----|-------------------|
| | *Atriplex semibaccata* R. Br. (Australian salt bush) | I | Forage | MIO |
| | *Atriplex wrightii* S. Wats | I | Grain (North American Indian) | MIO? |
| | *Beta vulgaris* L. (Beet) | I,C | Vegetable, sugar, fodder | MIO |
| | *Kochia scoparia* Schrad. (Summer cypress) | I | Ornamental, fodder | MES? |
| | *Nitraria retusa* (F.) Aschers. | I | Forage | EU |
| | *Salicornia brachiata* Roxb. | I,C | Forage | EU |
| | *Salicornia europaea* L. (Glasswort, samphire) | I,C | Potherb | EU |
| | *Salicornia herbacea* L. ( = *Salicornia stricta*) | I,C | Potherb | EU |
| | *Salicornia virginica* L. (Pickleweed) | I,C | North American Indian Potherb | EU |
| | *Sarcobatus vermiculatus* (Hook.) Torr. | I | Forage | MES? |
| | *Suaeda maritimia* (L.) Dum. (Sea blite) | I,C | Potherb | EU |
| | *Suaeda nudiflora* Moq. | I,C | Potherb | EU |
| Combretaceae | *Lumnitzera littorea* Voigt. | C | Timber | MES? |
| | *Lumnitzera racemosa* Willd. | C | Timber | EU |
| Compositae | *Aster tripolium* L. (Sea aster) | I,C | Ornamental | MES |
| | *Artemisia maritima* L. (Santonin) | C | Insecticide | MIO |
| | *Baccharis halimifolia* L. (Groundsel bush) | C | Ornamental | MIO? |
| | *Cotula cornopifolia* L. (Brass buttons) | I,C | Ornamental | MIO |
| | *Lasthenia glabrata* Lindl. | I,C | Ornamental | MES? |
| | *Grindelia humilis* H. and A. | C | Ornamental | MIO |
| | *Grindelia latifolia* Kell. | I,C | Ornamental | MIO |
| | *Grindelia stricta* D. C. | I,C | Ornamental | MIO |
| | *Jaumea carnosa* (Less.) Gray (Jaumea) | I,C | Ornamental | EU? |
| | *Matricaria chamomilla* L. (False camomile) | I | Medicine, flavoring, perfume | MIO |
| | *Pluchea purpurascens* (Sw.) D.C. (Salt marsh fleabane) | I,C | Ornamental | MIO |
| | *Melilotus indicus* (L.) All. (Sweet clover) | I | Forage | MIO |
| | *Prosopis juliflora* (Sw.) D.C. (Mesquite) | I | Alcohol, forage, resin, legume | MIO |
| | *Prosopis pubescens* Benth. (Screw bean) | I | Legume (North American Indian) | MIO |
| | *Trifolium fragiferum* L. (Strawberry clover) | I | Forage | MIO |
| | *Trifolium repens* L. (White clover) | I | Forage | MIO |
| | *Trifolium resupinatum* L. (Persian clover) | I | Forage | MIO |
| | *Trifolium tomentosum* L. | I | Forage | MIO |

## Table 2 (continued)
## REPRESENTATIVE LIST OF HALOPHYTIC SPECIES[5]

### Partial List of Halophytic Species

| Family | Species and common name | Habitat[a] | Use | Salt tolerance[a] |
|--------|------------------------|-----------|-----|-------------------|
| Convolvulaceae | *Convolvulus sepium* L. (Bindweed) | I,C | Medicine, vegetable, ornamental | MIO |
| | *Convolvulus soldanella* L. (Beach morning glory) | I,C | Medicine, vegetable, ornamental | MES/MIO |
| | *Cressa cretica* L. | I,C | Cattle fodder | MES/MIO? |
| | *Ipomea alba* L. (Moon vine) | I,C | Ornamental | MIO |
| | *Ipomea pes-caprae* (L.) R. Br. | I,C | Ornamental, sandbinder | MIO |
| Cruciferae | *Cakile maritima* Scop. (Sea rocket) | I,C | Vegetable | MES |
| | *Cochlearia anglica* L. (Long-leaved scurvy grass) | C | Medicine, spice | MES/MIO |
| | *Cochlearia officinalis* L. ssp. *officinalis* (Scurvy grass) | C | Vitamin C, medicine, spice | MIO? |
| | *Crambe maritima* L. (Sea kale) | C | Vegetable | MIO? |
| | *Raphanus maritimus* Sm. (Sea radish) | C | Vegetable | MIO? |
| Cyperaceae | *Scirpus americanus* (Bulrush) | I,C | Fiber | MIO? |
| Euphorbiaceae | *Excoecaria agallocha* L. | C | Paper | MIO |
| Frankeniaceae | *Frankenia grandifolia* Cham. and Schlecht. | I,C | Ornamental | EU |
| Goodeniaceae | *Selliera radicans* Cav. | C | Ornamental | MIO? |
| Gramineae | *Agropyron elongatum* (Hort.) Beauv. (Tall wheatgrass) | I,C | Forage | MIO |
| | *Agropyron junecum* (L.) Beauv. | I,C | Forage, sand-binding | MES? |
| | *Agrostis alba* L. var. *palustris* (Red top) | I,C | Forage, hay, turf | MIO |
| | *Agrostis tenuis* Sibth. (Colonial bent) | I,C | Turf | MIO |
| | *Chloris gayana* Kunth (Rhodes grass) | I | Forage | MIO |
| | *Cymbopogon nardus* (L.) Rendle (Citronella grass) | I | Oil, perfume | MIO |
| | *Cynodon dactylon* Pers. (Bermuda grass) | I,C | Lawn, forage | MIO |
| | *Distichlis palmeri* (Vasey) Fassett ex L. M. Johnston (Desert salt grass) | I | Grain (North American Indian) | MES? |
| | *Distichlis spicata* (L.) Greene (Salt grass) | I,C | Forage | EU |
| | *Festuca arundinaceae* Schreb. (Tall fescue) | I,C | Forage, turf | MES/MIO |
| | *Festuca rubra* L. (Red fescue) | I,C | Forage, turf | MES |
| | *Hordeum murinum* Huds. (Sea barley) | I,C | Forage | MIO |
| | *Hordeum murinum* L. (Sterile barley) | I,C | Forage | MIO |
| | *Panicum amarulum* Hitch. and Chase | I,C | Dune binder, forage | MES |
| | *Panicum antidotale* Retz. (Blue panicum) | I,C | Forage, sand-binding | MIO |

## Table 2 (continued)
## REPRESENTATIVE LIST OF HALOPHYTIC SPECIES[5]

### Partial List of Halophytic Species

| Family | Species and common name | Habitat[a] | Use | Salt tolerance[a] |
|---|---|---|---|---|
| | *Panicum virgatum* L. (Switch grass) | C | Forage | MIO? |
| | *Parapholis incurva* (L.) C. E. Hubb (Sickle grass) | I,C | Forage | MIO? |
| | *Paspalum distichum* L. (Knot grass) | I,C | Forage | MIO? |
| | *Paspalum gayanum* E. Desv. | C | Forage | MIO? |
| | *Paspalum vaginatum* Sw. | C | Forage | MIO? |
| | *Puccinellia distans* (L.) Parl. | I | Turf | EU |
| | *Puccinellia maritima* (Huds.) Parl. (Sea poa) | I,C | Forage | EU |
| | *Spartina alterniflora* Loisl. (Smooth cordgrass) | C | Forage, silage | EU |
| | *Spartina cynosuroides* (L.) Roth. (Big cordgrass) | C | Hay | MIO |
| | *Spartina patens* (Ait.) Muhl. (Hay cordgrass) | C | Hay, forage | EU |
| | *Spartina pectinata* Link (Prairie cordgrass) | I,C | Hay, brushes | EU |
| | *Spartina spartinae* (Trin.) Merrill | I,C | Hay, brushes | MIO? |
| | *Spartina townsendii* H. and J. Groves (Cordgrass) | C | Hay, silage, forage | EU |
| | *Sporobolus airoides* (Torr.) Torr. (Alkali sacaton) | I,C | Hay, forage | MIO |
| | *Sporobolus pyramidatus* (Lam.) Hitche. (Rat's tail grass) | I | Forage, grain | MIO |
| | *Stenotaphrum subsecundum* (Walt.) O. Kuntze (St. Augustine grass) | I,C | Lawn, forage | MIO |
| Juncaceae | *Juncus maritimus* (Lam.) (Sea rush) | I,C | Fiber, paper | MES |
| Juncaginaeceae | *Triglochin palustris* L. (Arrow grass) | I,C | Fodder | MES/ MIO? |
| Leguminosae | *Anthyllis vulneraria* L. | I,C | Fodder, ornamental | MIO |
| | *Derris ecastophyllum* (L.) Benth (Derris) | I,C | Insecticide | MIO |
| | *Derris heterophylla* Willd. (Derris) | C | Insecticide (India) | MIO |
| | *Dichrostachys glomerata* Hutch. and Dalz. | C | Fiber, timber, medicine, ornamental | MIO |
| | *Lotus corniculatus* L. var. *tenuifolius* (Narrow leaf bird's foot trefoil) | I,C | Forage | MIO |
| | *Medicago hispida* Gaertn. (Bur clover) | I | Forage | MIO |
| Malvaceae | *Hibiscus tiliaceus* L. (Rose mallow) | I | Ornamental, fiber | MIO |
| | *Sida hederacea* (Doug.) Torr. (Alkali mallow) | I,C | Ornamental | MIO |
| | *Thespesia populnea* Soland. ex. Correa (Portia tree) | C | Ornamental, dye | MIO |
| Meliaceae | *Xylocarpus granatum* Koenig. | C | Tannin, oil | MES |
| | *Xylocarpus moluccensis* Roem. | C | Cabinet wood | MIO? |

Table 2 (continued)
**REPRESENTATIVE LIST OF HALOPHYTIC SPECIES[5]**

Partial List of Halophytic Species

| Family | Species and common name | Habitat[a] | Use | Salt tolerance[a] |
|---|---|---|---|---|
| Nypaceae | *Nypa fruticans* Wurmb. (Nypa palm) | I | Alcohol, sugar, fiber | MIO |
| Palmaceae | *Cocos nucifera* L. (Coconut palm) | I,C | Alcoholic liquors, beverages, nuts, oil, sugar, textile, timber | MIO |
| | *Oncosperma filamentosa* Bl. (Nibung palm) | I,C | Fiber, vegetable, wood | MIO |
| | *Oncosperma tigillera* (Jack). Ridl. | I,C | Fiber, wood | MIO |
| | *Phoenix dactylifera* L. (Date palm) | I,C | Fiber, fruit, sugar | MES? |
| | *Phoenix reclinata* Jacq. | I,C | Fiber, ornamental | MIO |
| Plantaginaceae | *Plantago coronopus* L. (Buck's horn) | I,C | Vegetable | MIO? |
| | *Plantago maritima* L. (Sea plantain) | I,C | Vegetable | MES? |
| Plumbaginaceae | *Armeria maritima* (Mill.) Willd. (Sea pink/thrift) | I | Ornamental | MIO |
| | *Limonium bellidifolium* (Gouan) Dum. (Sea lavender) | C | Ornamental | EU/MES |
| | *Limonium gmelini* Kuntze | I | Ornamental | MIO? |
| | *Limonium humile* Mill. | C | Ornamental | MES |
| | *Limonium sinuatum* (L.) Mill. (Statice) | I,C | Ornamental | MES? |
| | *Limonium spicatum* Kuntze | I | Ornamental | MIO? |
| | *Limonium vulgare* Mill. | C | Ornamental | EU/MES |
| Polygonaceae | *Coccoloba uvifera* (L.) L. (Sea grape) | I,C | Ornamental, fruit | MIO |
| Portulacaceae | *Sesuvium portulacastrum* L. | I,C | Potherb, ornamental | EU/MES |
| | *Sesuvium verrucosum* Raf. | I,C | Potherb, ornamental | MES? |
| | *Trianthema portulacastrum* L. | I | Potherb, ornamental | MIO? |
| Rhizophoraceae | *Bruguiera gymnorhiza* (L.) Lam. | C | Timber, fuel | EU |
| | *Bruguiera sexangula* Poir (= *Bruguiera eriopetala* W. and A.) | C | Tannin, timber, edible seedlings | EU |
| | *Ceriops tagal* (Per.) C. B. Robinson | C | Tannin, fuel, adhesive | EU? |
| | *Rhizophora apiculata* Blume | C | Timber, tannin | MES? |
| | *Rhizophora harrisoni* (= *Rhizophora brevistyla*) | C | Tannin, charcoal | MES? |
| | *Rhizophora mangle* Roxb. (= *Rhizophora racemosa*) (Red mangrove) | C | Tannin, medicine | EU |
| | *Rhizophora mucronata* Lam. | C | Timber, tannin | EU |
| Rosaceae | *Potentilla anserina* L. (Silverweed) | I,C | Ornamental, medicine, vegetable | MIO? |

Table 2 (continued)
## REPRESENTATIVE LIST OF HALOPHYTIC SPECIES[5]

Partial List of Halophytic Species

| Family | Species and common name | Habitat[a] | Use | Salt tolerance[a] |
|---|---|---|---|---|
| Sonneratiaceae | *Sonneratia acida* L. | C | Paper | MES? |
| | *Sonneratia alba* G. Smith | C | Forage, timber | MES |
| Sterculiaceae | *Heritiera littoralis* (L.) Dyrand. | C | Timber, fuel | MIO? |
| | *Heritiera minor* (Sundri) | C | Timber, rayon | MIO? |
| Tamaricaceae | *Tamarix africana* Poir. (Tamarisk) | I | Ornamental, windbreak | MIO |
| | *Tamarix gallica* L. (Tamarisk) | I | Ornamental, windbreak | MIO |
| | *Tamarix pentandra* Pall. (Tamarisk) | I | Ornamental, windbreak | MIO |
| Umbelliferae | *Crithmum maritimum* L. (Rock samphire) | I,C | Spice, vegetable | MES/MIO |
| Zosteraceae | *Zostera marina* L. (Ell/surf grass) | C | Vegetable and food flavoring (North American Indian) | EU |

[a]  The habitat classifications include Coastal Salt Marsh (C), indicating the marsh habitat itself, and Inland Areas (I), indicating all other saline regions, including beaches, seacoast bluffs, inland salt marshes, deserts, and other inland saline/alkali areas. The salt tolerance classification in the final column uses abbreviations as follows: MIO = Miohalophyte, denoting tolerance to soil water NaCl levels of less than 0.5%; MES = Mesohalophyte, with tolerance to 0.5 to 1.0% NaCl in the soil water; and EU = Euhalophyte, indicating tolerance to soil water NaCl levels of over 1%. A question mark in this column indicates considerable uncertainty regarding the proper classification of that entry.

## Table 3
## BIOMASS PRODUCTIVITY OF SEVERAL HALOPHYTIC SPECIES[8,9]

| Species | Biomass of aerial parts (g dry weight/m²) | Net aboveground production (g dry weight/m²/year) | Locale | Ref. |
|---|---|---|---|---|
| *Atriplex hastata* | 700—800 | — | Southern England | 10 |
| *Distichlis spicata* | 360 | — | Virginia | 11 |
| | 885 | — | Connecticut | 12 |
| | 647 | — | New York | 13 |
| | 985 | — | New Jersey | 14 |
| | 603 | — | Georgia | 15 |
| *Jancus effusus* | 800 | — | England | 16 |
| *Juncus roemerianus* | 232 | 849 | Florida | 17 |
| | 1173 | 796 | North Carolina | 18 |
| | 786 | 1360 | North Carolina | 19 |
| | 340 | 850 | North Carolina | 20 |
| *Juncus squarrosus* | 690 | — | England | 16 |
| *Scirpus americanus* | 150 | 150 | South Carolina | 21 |
| *Spartina alterniflora* | — | 2000—3300 | Georgia | 22,23 |
| | 259—1320 | 329—1296 | North Carolina | 18 |
| | 545 | 650 | North Carolina | 24 |
| | 250—2100 | 1000 | North Carolina | 25 |
| | 413 | 445 | Delaware | 14 |
| *Spartina cynosuroides* | 1456 | — | Virginia | 11 |
| *Spartina patens* | 640 | 1296 | North Carolina | 19 |
| | 805 | — | Virginia | 11 |
| *Spartina townsendii* | 700—1060 | — | Southern England | 10,26 |

Table 4
## SALT TOLERANCE OF AGRICULTURAL PLANTS

| $EC_e \times 10^3$ | Field crops | Vegetable crops | Forage crops |
|---|---|---|---|
| 2—4 | Field beans | Radish | White Dutch clover |
| | | Celery | Meadow foxtail |
| | | Green beans | Alsike clover |
| | | | Red Clover |
| | | | Ladino Clover |
| | | | Burnet |
| 4—10 | Rye (grain) | Tomato | Alfalfa (California common) |
| | Wheat (grain) | Broccoli | Tall fescue |
| | Oats (grain) | Cabbage | Rye (hay) |
| | Rice | Bell pepper | Wheat (hay) |
| | Sorghum (grain) | Cauliflower | Oats (hay) |
| | Flax | Lettuce | Orchardgrass |
| | Sunflowers | Sweet corn | Blue grama |
| | Castorbeans | Potatoes (white rose) | Meadow fescue |
| | | Carrot | Reed canary |
| | | Onion | Big trefoil |
| | | Peas | Smooth brome |
| | | Squash | Tall meadow oatgrass |
| | | Cucumber | Cicer milkvetch |
| | | | Sourclover |
| | | | Sickle milkvetch |
| 10—12 | Rape | Garden beets | White sweetclover |
| | Cotton | Kale | Yellow sweetclover |
| | | Asparagus | Perennial ryegrass |
| | | Spinach | Mountain brome |
| | | | Strawberry clover |
| | | | Dallis grass |
| | | | Sudan grass |
| | | | Hubam clover |
| 12—18 | Barley (grain) | | Alkali sacaton |
| | Sugar beet | | Saltgrass |
| | | | Nuttall alkaligrass |
| | | | Bermuda grass |
| | | | Rescue grass |
| | | | Canada wildrye |
| | | | Western wheatgrass |
| | | | Barley (hay) |
| | | | Birdsfoot trefoil |

*Note:* The $EC_e \times 10^3$ numbers in the left column are the electrical conductivity values of the saturation extract in millimhos per centimeter at 25°C associated with a 50% decrease in yield. Plants within each group are in order of decreasing tolerance, with the most tolerant listed first and the least tolerant last.[27]

# REFERENCES

1. Flowers, T. J., Troke, P. F., and Yeo, A. R., The mechanism of salt tolerance in halophytes, *Annu. Rev. Plant Physiol.*, 28, 89, 1977.
2. Israelson, O. W. and Hansen, V. E., *Irrigation Principles and Practices*, 3rd ed., John Wiley & Sons, New York, 1962.
3. van Aart, R., Regional training course on drainage and land reclamation, *Nat. Res.*, 7, 22, 1971.
4. Boyko, H., Ed., *Salinity and Aridity — New Approaches to Old Problems, Monographiae Bibliogicae*, Vol. 16, W. Junk, The Hague, 1966.
5. Mudie, P. J., The potential economic uses of halophytes, in *Ecology of Halophytes*, Reimold, R. J. and Queen, W. H., Eds., Academic Press, New York, 1974.
6. Epstein, E. and Norlyn, J. D., Seawater based crop production: a feasibility study, *Science*, 197, 249, 1977.
7. Turner, R. E. and Gooselink, J. G., A note on standing crops of *Spartina alterniflora* in Texas and Florida, *Contrib. Mar. Sci.*, 19, 116, 1975.
8. Keefe, C. W., Marsh production: a summary of the literature, *Contrib. Mar. Sci.*, 16, 163, 1972.
9. Turner, R. E., Geographic variations in salt marsh macrophyte production: a review, *Contrib. Mar. Sci.*, 20, 47, 1976.
10. Ranwell, D. S., *Spartina* salt marshes in southern England. I. The effects of sheep grazing at the upper limits of *Spartina* marsh in Bridgwater Bay, *J. Ecol.*, 49, 325, 1961.
11. Wass, M. L. and Wright, T. D., Coastal Wetlands of Virginia, Interim Report to the Governor and General Assembly, Virginia Institute of Marine Science, Special Rep. in Applied Marine Science and Ocean Engineering #10, 1969.
12. Steever, E. Z., Productivity and Vegetation Studies of a Tidal Salt Marsh in Stonington, Connecticut; Cottrell Marsh, M.A. thesis, Connecticut College, New London, 1972.
13. Udell, A. F., Zarudsky, J., and Doheny, T. E., Productivity and nutrient values of plants growing in the salt marshes of the town of Hempstead, Long Island, *Bull. Torrey Bot. Club*, 96, 42, 1969.
14. Morgan, M. H., Annual Angiosperm Production on a Salt Marsh, M.S. thesis, University of Delaware, Newark, 1961.
15. Gallagher, J. L., Reimold, R. J., and Thompson, D. E., Remote sensing and salt marsh productivity, Proc. 38th Annual Meeting American Society of Photogram., Washington, 1972.
16. Pearsall, W. H. and Gorham, E., Production ecology. I. Standing crops of natural vegetation, *Oikos*, 7, 193, 1956.
17. Heald, E. J., The Production of Organic Detritus in a South Florida Estuary, Ph.D. thesis, University of Miami, Miami, 1969.
18. Stroud, L. M. and Cooper, A. W., Color-Infrared Aerial Photographic Interpretation and Net Primary Productivity of a Regularly-Flooded North Carolina Salt Marsh, Water Resources Research Institute, Rep. 14, University of North Carolina, Chapel Hill, 1969.
19. Waits, E. D., Net Primary Productivity of an Irregularly-Flooded North Carolina Salt Marsh, Ph.D. dissertation, North Carolina State University, Raleigh, 1967.
20. Williams, R. B. and Murdoch, M. B., Compartmental analysis of production and decay of *Juncus roemerianus*, *ASB Bull.*, 15, 59, 1968.
21. Boyd, C. E., Production, mineral accumulation and pigment concentrations in *Typha latifolia* and *Scirpus americanus*, *Ecology*, 51(2), 285, 1970.
22. Odum, E. P., *Fundamentals of Ecology*, W. B. Saunders, Philadelphia, 1959.
23. Odum, E. P., The role of tidal marshes in estuarin production, *N.Y. State Conserv.*, 15(6), 12, 35, 1961.
24. Williams, R. B. and Murdoch, M. B., The potential importance of *Spartina alterniflora* in conveying zinc, manganese, and iron into estuarine food chains, in Proc. 2nd Natl. Symp. on Radioecology, Nelson, D. J. and Evans, F. C., Eds., USAEC, CONF-670503, 1969.
25. Williams, R. B. and Murdoch, M. B., Annual production of *Spartina alterniflora* and *Juncus roemerianus* in salt marshes near Beaufort, North Carolina, *ASB Bull.*, 13, 49, 1966.
26. Ranwell, D. S., *Spartina* salt marshes in southern England. III. Rates of establishment, succession and nutrient supply at Bridgwater Bay, Somerset, *J. Ecol.*, 52, 95, 1964.
27. U.S. Salinity Laboratory, Diagnosis and Improvement of Saline and Alkali Soils, U.S. Department of Agriculture Handbook No. 60, U.S. Government Printing Office, Washington, D.C., 1954.

# LIGHT INTENSITY PREFERENCE AND TOLERANCE OF AQUATIC PHOTOSYNTHETIC MICROORGANISMS

## Edward J. Phlips and Akira Mitsui

## INTRODUCTION

The growth of aquatic photosynthetic organisms is limited to a narrow zone of surface illumination known as the euphotic zone. The exact depth of this region depends on incident light intensity and a myriad of factors related to the transparency of water.[1] In the tropics, afternoon surface irradiance normally exceeds 100 klx throughout the year, and the euphotic zone (i.e., depth at which 1% of incident light penetrates) can extend down to depths of greater than 100 m. On the other hand, surface illumination in polar waters seldom exceeds 20 klx during the summer, while winter intensities are very low and daylengths are short. Photosynthetic organisms have been found to successfully inhabit this entire range of light environments.

This section reviews the light intensity preference and tolerance of aquatic photosynthetic microorganisms (see References 2 to 26 for other reviews on this subject). In addition, some of the strategies used by organisms to deal with extreme light environments are discussed in terms of the concepts of "shade" (low light intensity) and "sun" (high light intensity) adaptation.[5,9,13,18,25]

## MECHANISMS OF LOW LIGHT INTENSITY ADAPTATION

The overriding problem faced by obligate photoautotrophic organisms living in low light intensity (or shade) habitats is the photochemical production of enough reducing power and chemical energy to sustain growth. The ultimate goal is to fix more carbon than is used up in respiration. Many terrestrial plants and some aquatic organisms, including the diatoms *Skeletonema costatum, Cyclotella meneghiniana, Nitzschia palea, Nitzschia closterium*, and the green alga *Scenedesmus quaricavda*,[4,9,18,26] appear to solve this problem by reducing their rate of dark respiration, thereby lowering the compensation light intensity (i.e., light intensity at which respiration equals photosynthesis). In the study of aquatic organisms, this strategy has been referred to as "*Cyclotella* type", based on the work done by Jorgensen[9,26] with this freshwater diatom.

In contrast, other studies have shown that many aquatic species and some terrestrial plants are capable of regulating their pigment content in response to changes in light intensity. This strategy has been referred to as "*Chlorella* type", after the work done by Steemann-Nielsen on *Chlorella vulgaris*.[27,28] In most cases, there is an inverse relationship between pigment content and light intensity (as shown in Figure 1). A number of different pigment groups have been shown to contain examples of this type of behavior, including some chlorophylls,[3,9,11,18,26-36] phycobilins,[30,33,37-39] carotenoids,[3,36,40] and bacteriochlorophylls.[41-43] Such increases in pigment content can result in both a significant enhancement of efficiency of photosynthesis (on a per cell basis) at low light intensities and a concomitant decrease in compensation light intensity. This is illustrated by the work done by Steemann-Nielsen and co-workers on *C. vulgaris* and *Chorella pyrenoidosa* (see Figure 2).[18,28]

In addition to the quantitative regulation of pigment content, some algae have been shown to contain special auxiliary pigments which assist in light absorption.[44-46] One example of this is the carotenoid *Siphonaxanthin*. Kageyama and Yokohama[44] found this pigment in a number of deep water and shade algae, but not in their intertidal counterparts. Apparently this pigment which strongly absorbs deep penetrating green light (550 nm) is of adaptive value within dimly lit deep water habitats.

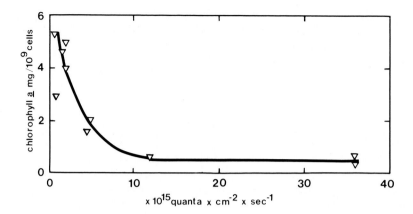

FIGURE 1.    The content of chlorophyll *a* per cell in *G. pyrenoidosa* (211/8b) as a function of the irradiance at which the algae were grown. Continuous incandescent light, 20°C.[17]

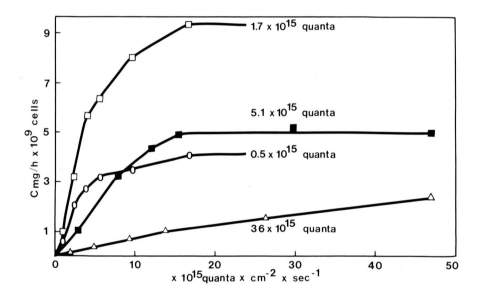

FIGURE 2.    The rate of photosynthesis per cell as a function of irradiance in *C. pyrenoidosa* (211/8b) grown in continuous incandescent light (different irradiance), 20°C,[17]

Along with these primary aspects of "*Cyclotella-*" and "*Chlorella-*type" adaptations, many shade-adapted organisms may exhibit a number of related metabolic and structural attributes including

1.    Relatively reduced concentrations of enzymes related to photosynthesis[3,47]
2.    Increased cell size[13,48]
3.    Diminished rates of cell division
4.    Enhanced cell bouyancy[49,50]
5.    Altered position and/or structure of chloroplasts and chromatophores[51,52]
6.    Generally reduced metabolic activity[3,9,13]

Taken together, these adaptations enable shade species to maximize the ratio between photosynthesis and respiration, thereby facilitating growth in low light environments. It is also important to note, however, that many photosynthetic species are not obligate autotrophs and may resort to a heterotrophic life style when light becomes limiting.

## HIGH LIGHT INTENSITY ADAPTATION

In high light intensity environments, the irradiance normally exceeds saturation levels for photosynthesis and growth. This surplus represents a potential source of photoinhibition. The two principal causes of photoinhibition are photooxidation and ultraviolet (UV) light damage. In addition, several researchers have suggested that photorespiration may play an important role in apparent photoinhibition[53,54] in certain species.

### Photooxidation

Photooxidative inhibition of photosynthesis appears to be primarily related to the inactivation and eventual destruction or bleaching of pigments, enzymes, and electron carriers active in photosynthesis.[55-70] The major active sites for photooxidation are at, or near, the reaction center pigments of the photosystems.[59,60] Recent studies indicate that inhibition of photosystem I is oxygen dependent and related to P700 activity, while photosystem II inhibition is centered at the linkage between the trapping centers and the electron acceptors.[58] There are other proteins, membrane lipids, and elements of the photosynthetic process which can become involved in photooxidation.[58,61,67,68,71] High light intensities can also inhibit respiration.[57,72]

The actual agents involved in the damaging effects of photooxidation are not yet completely understood, but could involve the production of harmful states of oxygen, especially singlet oxygen.[73-75]

The effects of short exposures to inhibiting intensities are normally reversible. Longer exposures can result in photobleaching and destruction of pigments which requires extensive repair time or is irreversible and leads to cell death.[65-69] The rate and extent of pigment destruction depends not only on light intensity, but also on the interaction of other environmental and biological factors. For example, low $CO_2$ and high $O_2$ concentration both enhance photooxidation.[55,58,63]

### UV Light

The other major source of photoinhibition is UV light.[76-91] UV light is generally divided into two categories: near UV (ranging from 300 to 380 nm) and far UV (ranging from 220 to 300 nm). The most serious effects identified with UV exposure are generally linked to the far UV range, although near UV effects have been observed.[82] Even moderate doses of far UV light can result in the inhibition of growth, photosynthesis, and morphogenesis. There appears to be an exponential relationship between dose and inhibition of growth[80,83,84] and photosynthesis.[85]

In terms of growth, the most serious effect is DNA damage.[77,86,87] The damage manifests itself during stages of DNA replication and cell division. Consequently the effects of UV light are most apparent during the exponential phase of culture growth.[80,84,88] The severity of far UV effects depends on a number of factors, including the dose, intensity, developmental state, test conditions (i.e., temperature, pH, etc.), and the species in question. For example, Sasa[84] inhibited the growth of *Chlorella* by exposure to 1 min of far UV light at an intensity of 30 nW/cm².

Far UV light also has a strong inhibitory effect on photosynthetic reactions.[59,60,89] For example, photosynthesis in *Chlorella* cultures is totally shut down after 5-min exposure to far UV light at an intensity of 35 erg/m²/sec.

There is still considerable controversy over the exact site of far UV inhibition. It has been proposed that the electron carrier plastoquinone is the site of action.[59,60] On the other hand, it has also been suggested that far UV light attacks the macromolecules which help to maintain the structural integrity of lamelar membrances and thereby disrupts the function of the photosystems.[89]

From an ecological standpoint, the importance of UV light depends strongly on the nature of the water in question.[77] Clear oceanic and oligotrophic freshwater are excellent transmitters of UV light, i.e., 15% absorbance per meter. On the other hand, the "yellow substances" in coastal water can diminish UV light transmission to the point where less than 1% of the incident wavelengths remain at 1 m. This means that the effects of solar UV light which is primarily composed of wavelengths from 290 to 380 nm are restricted to the top few meters of the open ocean, lakes, and the intertidal zone.

### Photorespiration

Recent studies have shown that photoinhibition in certain species of photosynthetic organisms may be partially attributed to the effects of photorespiraton.[54] Hence, there may be an apparent inhibition of photosynthesis resulting from the breakdown of glycolate and release of $CO_2$ via photorespiration. Even if this hypothesis is correct, it still remains to be seen how widespread this phenomenon is among various photosynthetic phyla. There are indications that high light intensity photorespiration is more significant in diatoms and some blue-green algae than in green algae,[53,54] but the evidence is still too sparse to draw definite conclusions.[56]

### Mechanisms of High Light Intensity Adaptation (Sun Adaptation)

Total inactivation of photosynthesis in whole cells normally occurs at light intensities in excess of 50 klx, although partial inhibition may be observed at considerably lower levels (see Figures 3 and 4).

The sensitivity of organisms to photooxidation differs according to conditions of preadaptation.[17,67] Shade-adapted organisms often exhibit photoinhibition at lower levels of irradiance than organisms adapted to high light intensities (sun adapted). There are several basic attributes of sun-adapted organisms which could account for this resistance to high irradiances, including the following:

1.  Aquatic organisms from high light intensity environments generally contain less chlorophyll and other light-harvesting pigments than shade-adapted organisms, even though the number of light-harvesting pigment molecules per trapping center may remain constant.[59,60] This means that sun cells tend to absorb a lower percentage of incident light.
2.  Conversely, some sun-adapted cells contain higher concentrations of carotenoids believed to provide protection from high light intensity photooxidation.[70,72,73,75,92-94]
3.  Sun cells frequently contain higher concentrations of the enzymes and electron carriers involved in the nonphotochemical reactions of photosynthesis.[3,47] This means that sun cells which absorb less incident radiation than shade cells are better equipped to process the reducing equivalents produced by the photosystems. This increases the light intensity threshold for both the saturation of photosynthesis and the onset of photooxidation.

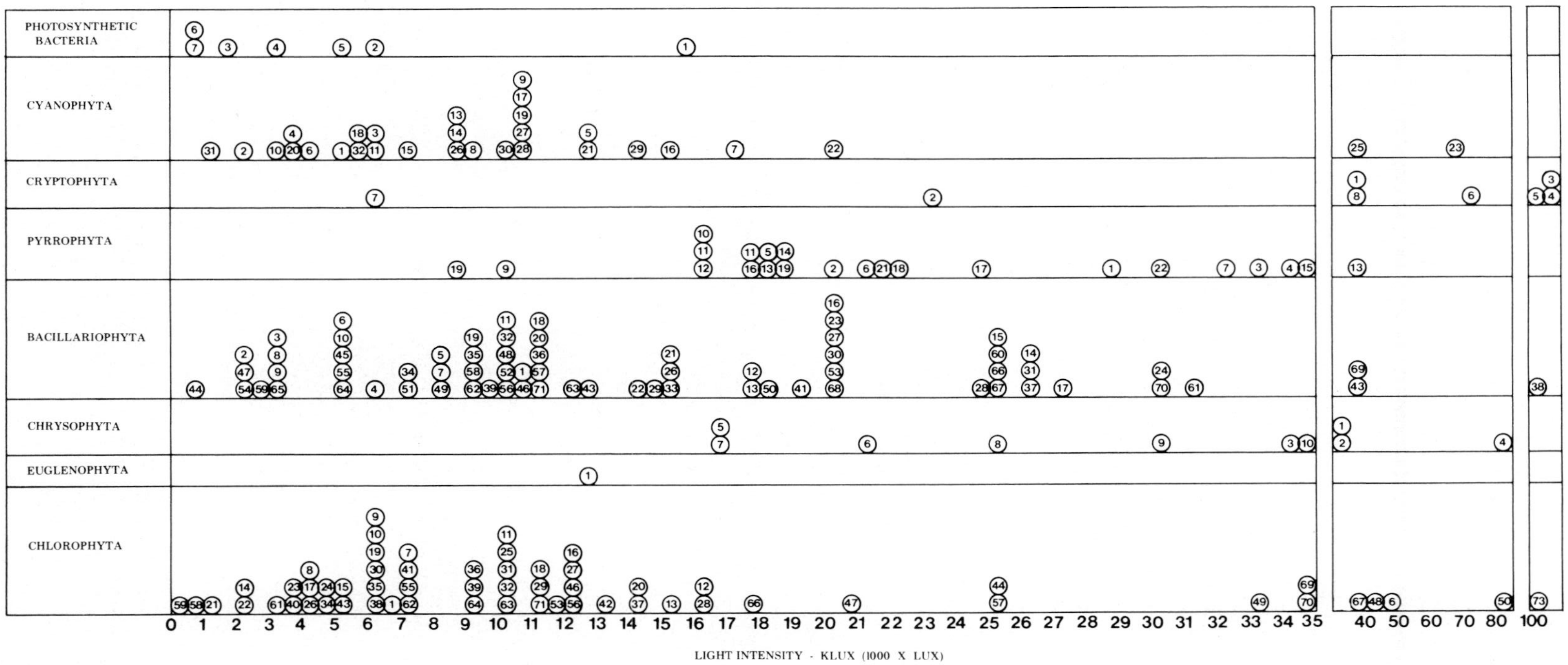

LIGHT INTENSITY - KLUX (1000 × LUX)

A

FIGURE 3. Species differences in adaptation of photosynthesis to light intensity. Figures 3(A) to (E) contain phylogenetic comparisons of light saturation, one half saturation light intensities, compensation intensities, $I_k$, and levels of photoinhibition for the experiments described in Table 1. The results in Figure 3 have been numbered within each phylum for reference purposes. These numbers correspond to those used in Table 1 (see right column of Table 1). (A) Light intensity at which photosynthesis saturates, (B) light intensity at which photosynthesis reaches one half saturation level, (C) compensation light intensity, (D) $I_k$ (for explanation see text), (E) highest light intensity at which less than 5% photoinhibition is observed. Note: "○" indicates that no inhibition was observed at the highest intensity tested. "□" indicates that photoinhibition was observed beyond the light intensity indicated.

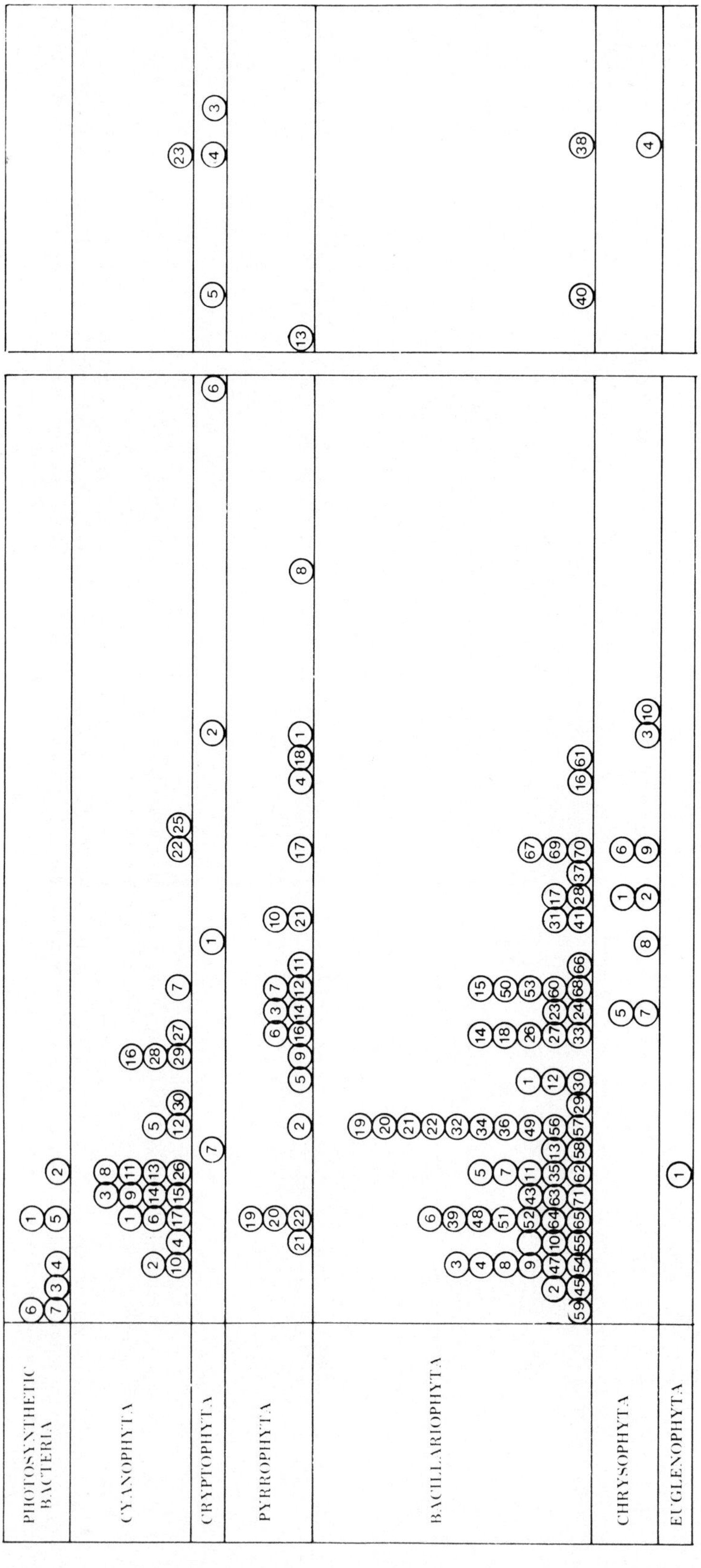
PHOTOSYNTHETIC BACTERIA
CYANOPHYTA
CRYPTOPHYTA
PYRROPHYTA
BACILLARIOPHYTA
CHRYSOPHYTA
EUGLENOPHYTA

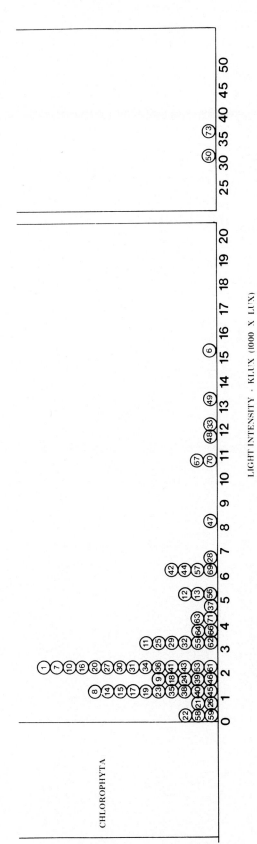

CHLOROPHYTA

LIGHT INTENSITY - KLUX (1000 X LUX)

FIGURE 3B

LIGHT INTENSITY - KLUX (1000 X LUX)

3C

3D

LIGHT INTENSITY - KLUX (1000 × LUX)

3E

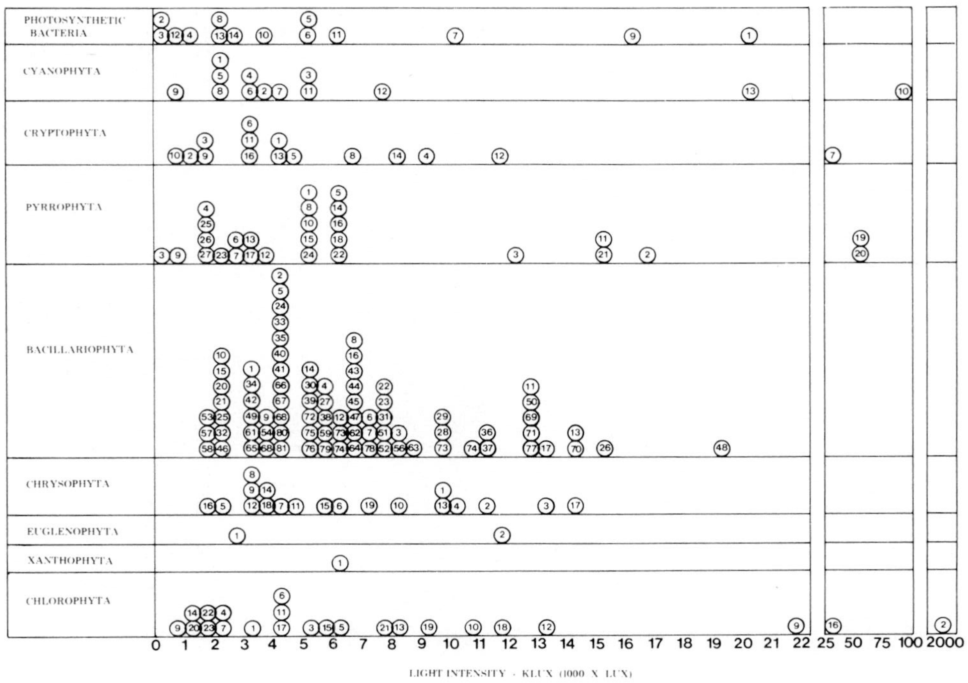

FIGURE 4. Species differences in adaptation of growth to light intensity. Figure 4(A) to (E) contain phylogenetic comparisons of light saturation, one half saturation light intensities, I$_k$, and levels of photoinhibition for the experiments described in Table 2. The results in Figure 4 have been numbered within each phylum for reference purposes. These numbers correspond to those used in Table 2 (see right column of Table 2). (A) Light intensity at which growth saturates, (B) light intensity at which growth reaches one half saturation level, (C) compensation light intensity, (D) I$_k$ for growth (for explanation see text), (E) highest light intensity at which less than 5% photoinhibition is observed. Note: O indicates that no inhibition was observed at the highest intensity tested. □ indicates that photoinhibition was observed beyond the light intensity indicated.

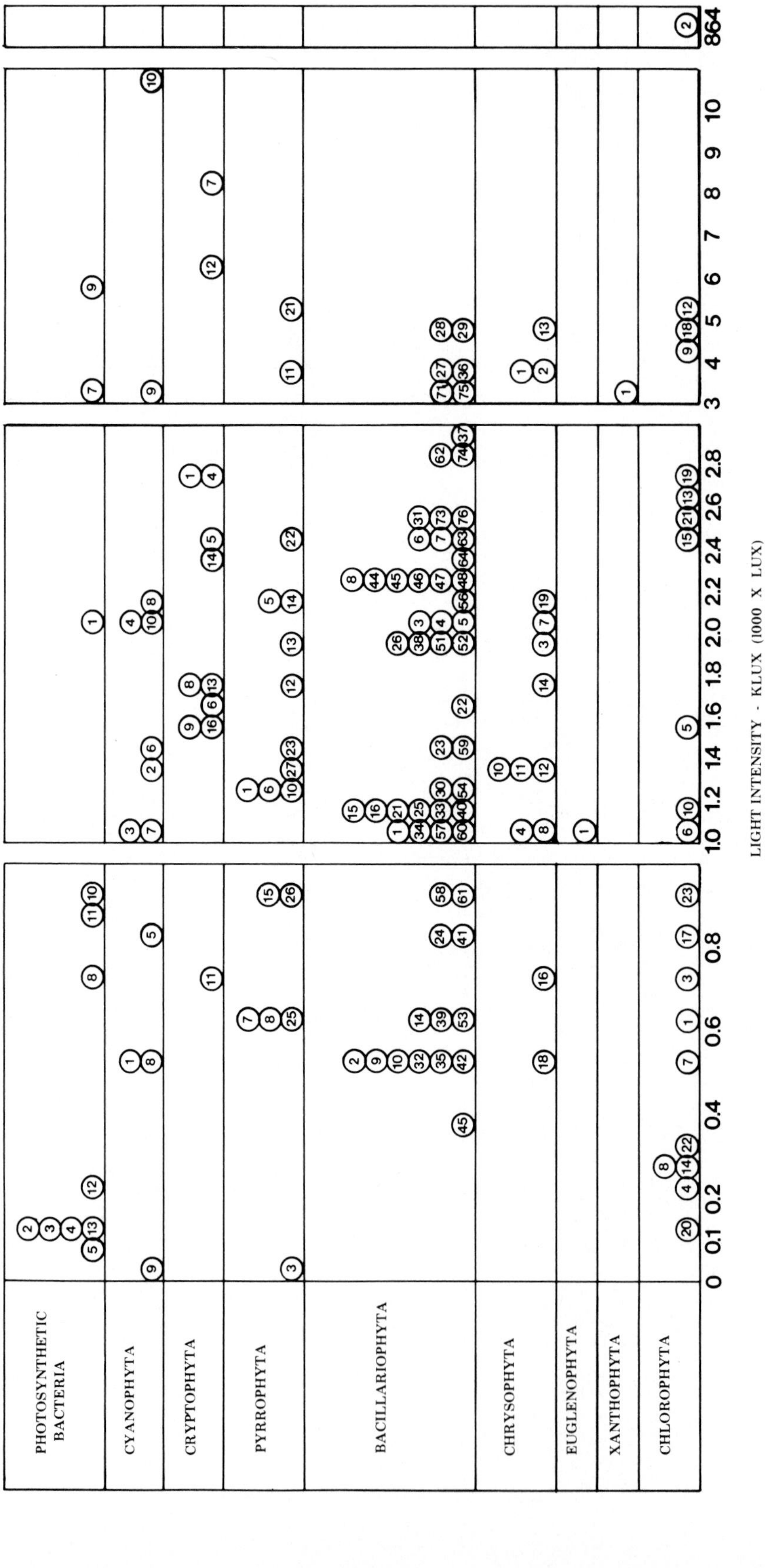
PHOTOSYNTHETIC BACTERIA
CYANOPHYTA
CRYPTOPHYTA
PYRROPHYTA
BACILLARIOPHYTA
CHRYSOPHYTA
EUGLENOPHYTA
XANTHOPHYTA
CHLOROPHYTA
0 0.1 0.2 0.4 0.6 0.8 1.0 1.2 1.4 1.6 1.8 2.0 2.2 2.4 2.6 2.8 3 4 5 6 7 8 9 10 86.4
LIGHT INTENSITY - KLUX (1000 X LUX)

FIGURE 4B

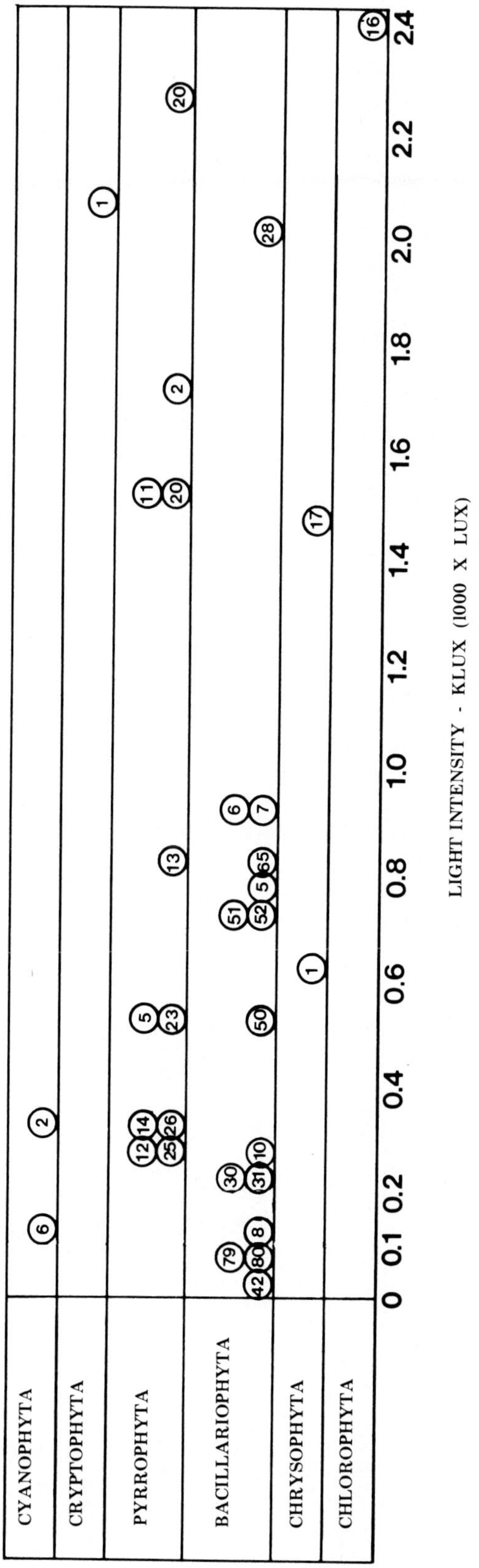

LIGHT INTENSITY - KLUX (1000 X LUX)

FIGURE 4C

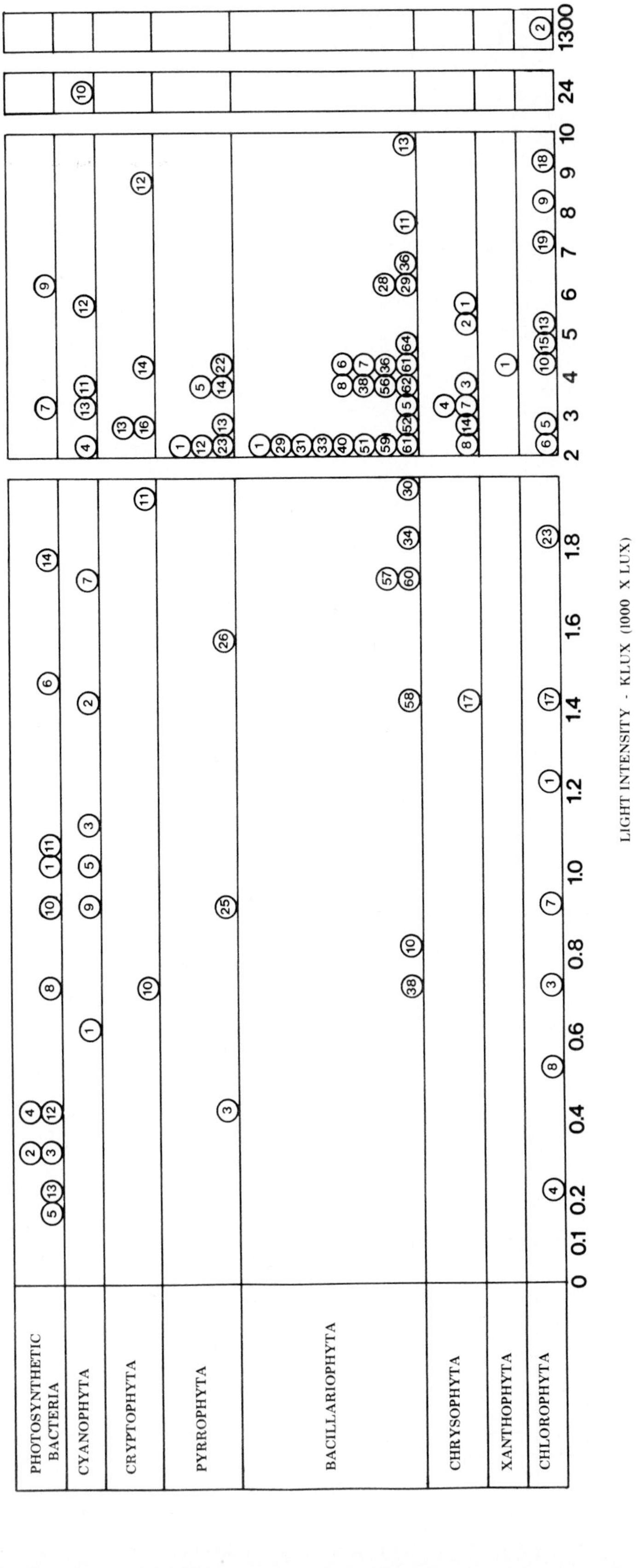

LIGHT INTENSITY - KLUX (1000 X LUX)

FIGURE 4D

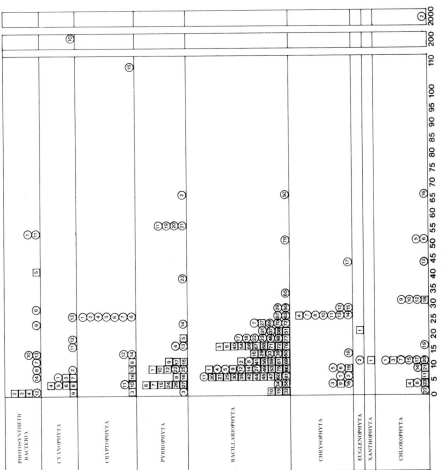

LIGHT INTENSITY - KLUX (1000 X LUX)

FIGURE 4E.

4.   Sun organisms may exhibit significant structural alterations in adapting to high light intensity. For example, several algae have been shown to respond to high irradiances by altering their arrangement of chloroplasts or chromatophores into a more light resistant configuration.[51,52]

## Mechanism of UV Light Resistance

All photosynthetic organisms appear to be susceptible to the inhibitory effects of far UV radiation, but there are species differences in maximum tolerance.[95] Relatively few studies have been made of the comparative resistance of photosynthetic organisms to UV light damage. In a study by McLeod and McLachlan,[80] six species of green algae were shown to have greater tolerance to 254-nm light than two species of diatoms. Other studies indicate that deep-water species are less resistant to UV light than shallow-water algae. For example, the resistance of *C. vulgaris* (green alga) is greater than the green algae *Ankistrodesmus* or *S. quadricavoa*. Some strains of blue-green algae appear to be exceptionally resistant to far UV irradiation, including *Agmenellum quadruplicatum* and *Anacystis nidulans*.[83,96-98]

One of the most important factors contributing to the resistance of organisms to UV light is the process of reactiviation or repair of UV damage.[81,99-102] Numerous species of photosynthetic and nonphotosynthetic organisms are capable of repairing damage to DNA, RNA, and elements of the photosynthetic apparatus.[99,100] Depending on the species, reactivation may require a period of dark incubation[77,88,101] and/or exposure to light of visible or near UV wavelengths. Light-dependent reactivation, which is enzyme mediated, has been observed in numerous species.[81,99-101]

## TIME COURSE OF ADAPTATION

The time course for nongenetic adaptation to new light intensity environments varies considerably between species. In the case of *C. vulgaris*, only 40 hr is necessary to complete adaptation from 3 to 30 klx (or vice versa).[28] *Oscillatoria rubens*, on the other hand, requires 1 to 2 weeks for full adaptation to 1.4 klx, after growth at 0.4 klx.[16] In a study of mixed phytoplankton from the North Pacific, Steemann-Nielsen and Park[103] noted that 102 hr were required for shade adaptation to take place. The studies of Steemann-Nielsen indicate that the time required for adaptation is closely related to cell division, growth, and doubling time.[28]

## FACTORS EFFECTING THE NATURE AND MAGNITUDE OF LIGHT INTENSITY ADAPTATION

The general characteristics of sun and shade adaptation were briefly reviewed in the previous sections. It must be kept in mind, however, that the exact response of organisms to changes in light intensity conditions may be dependent on a number of other environmental, physiological, and evolutionary factors.

## Genetic Vs. Nongenetic Adaptation

In discussing adaptation, it is important to make the distinction between genetic and nongenetic mechanisms. Studies of terrestrial plants have shown that certain species or "ecotypes" are specifically adapted to either shade or sun environments and are incapable of successfully adjusting to life in significantly different light intensity.[5] This lack of adaptability appears to stem from genetically determined limitations. Analogous examples of such phenomena in the aquatic environment are not as well documented. However, Jorgensen's[9,26] work with the diatom *C. meneghiniana* (discussed above) indicate that the degree of light adaptation of some aquatic species can be quite limited, especially when compared to the great flexibility of the green algae *C. vulgaris* and *Chlorella pyrenoidosa*.[18,28] In general, aquatic photosynthetic organisms appear to be characterized by considerable flexibility to changing light conditions. From an ecological standpoint, the purpose of this flexibility is to maximize productivity,[104] competitive ability, and in extreme cases to enhance survival. Nevertheless, a number of researchers have hypothesized that the genetic limitations in the response of organisms to light intensity can play a key role in spatial and temporal (e.g., seasonal) variations in species composition.[105,106]

### Environmental Effects

In natural habitats, the process of light intensity adaptation can be affected by a myriad of other environmental, physiological, and biological factors. The tolerance of organisms to extreme light environments can be strongly dependent on the existence of optimal environmental conditions.[8,17] One of the best-researched aspects of this issue is the interaction between temperature and light intensity.[12,25,32,38,104,107-122] It is well known that temperature plays a key role in the regulation of cell metabolism, primarily through its effect on enzyme activity. In terms of photosynthesis, increasing temperature raises the rate of light-saturated photosynthesis by enhancing the rate of the dark reactions. Raising the temperature can also increase the tolerance of organisms to high light intensities.[120]

The nature of the interrelationship between light intensity and other environmental factors has been investigated for a variety of parameters, including salinity,[112-114,123,124] pH,[12,125] $CO_2$ concentration,[12,126-130] $O_2$ concentration,[12,56,126,131,132] nutrient limitations,[12,113,133] day length,[16-32,134-137] light quality,[10,12,138-140] and diurnal rhythms.[141-143]

## PATTERNS OF LIGHT INTENSITY ADAPTATION

### Individual Preferences and Tolerances of Aquatic Photosynthetic Species

Over the last 100 years, the light intensity preferences and tolerances of a wide variety of aquatic photosynthetic species and populations have been studied. Tables 1 and 2 describe the relationships between light intensity, growth, and photosynthesis for a representative variety of species of microorganisms. It should be pointed out that the units used in the measurement of light intensity vary considerably from one publication to the next. Most of the literature reviewed represents light intensity in klux (klx = 1000 klx). However, recent research has switched over to the use of energy or quantum light units. An attempt has been made to include a conversion of these latter units to klx (shown in parentheses) for the purpose of comparison. However, these are only rough approximations, since accurate interconversion requires a knowledge of the exact spectral composition of the incident light utilized in the experiments. For further comparative examination of light intensity preferences and tolerances, the data in Tables 1 and 2 have been reorganized into a more suitable format in Figures 3 and 4.

The data compiled and compared in Figures 3 and 4 indicate that the unqualified separation of aquatic phyla according to light intensity preferences would be difficult to substantiate. However, certain patterns surface. Given the species included in this review, it is possible to discern the following phylogenetic tendencies in light preference:

| | |
|---|---|
| Groups with a general preference for low light intensity | Photosynthetic bacteria, blue-green algae (*Cyanophyta*) |
| Groups with a general preference for high light intensity | Dinoflagellates (*Pyrrophyta*), *Cryptophyta*, and *Chrysophyta* |
| Groups with a broad range of preferences | Diatoms (*Bacillariophyta*) and green algae (*Chlorophyta*) |

## Table 1
## LIGHT INTENSITY VS. PHOTOSYNTHESIS IN AQUATIC PHOTOSYNTHETIC MICROORGANISMS[a]

| Species (reference) | Environment[b] | Growth Conditions (klx)[c] | Exp. T° °C[d] | Type of Light[c] | Assay Method | Light Intensity at Psat[f] (klx) | Pmax[g] | Light Intensity at "Psat[h] one half (klx) | Compensation Light Intensity[i] (klx) | I_i' (klx) | Light Intensity for Initial Photoinhibition[k] (klx) | Maximum Light Intensity Tested (klx) | Inhibition[l] % | Case #[m] |
|---|---|---|---|---|---|---|---|---|---|---|---|---|---|---|
| **Photosynthetic Bacteria** | | | | | | | | | | | | | | |
| *Chloroflexus Aurantiacus*(144) | Yellowstone Hot Spring Ravine spring summer | Natural sunlight conditions | 58 | S | — | 15.9 | — | 2.1 | — | 4.2 | None — 84.8 | — | — | 1 |
| | | 14 days at 73% light reduction | 58 | S | — | 6.4 | — | 3.2 | — | 6.4 | 6.4 | 84.8 | 80 | 2 |
| *Chromatium D*(145) | — | — | 16 | N | — | 7 Kerg/cm²/sec (1.7) | — | 2 Kerg/cm²/sec (0.47) | — | 5 Kerg/cm²/sec (1.2) | None-35 kerg/cm²/sec (8.2) | — | — | 3 |
| | — | — | 22 | N | — | 15 Kerg/cm²/sec (3.5) | — | 6 Kerg/cm²/sec (1.4) | — | 10 Kerg/cm²/sec (2.3) | None-35 kerg/cm/sec (8.2) | — | — | 4 |
| | — | — | 28 | N | — | 22 Kerg/cm²/sec (5.2) | — | 9 Kerg/cm²/sec (2.1) | — | 15 Kerg/cm²/sec (3.5) | None-35 Kerg/cm²/sec (8.2) | — | — | 5 |
| *Chromatium D*(146) | — | 1 | 30 | T | — | 0.60 | 70.8 μmoles C/mgBchl/hr | 0.15 | 0.05 | 0.20 | None-30 | — | — | 6 |
| | — | 20 | 30 | T | — | 0.60 | 58.3 μmoles C/mgBchl/hr | 0.25 | 0.10 | 0.45 | None-30 | — | — | 7 |
| **Blue-Green Algae — Cyanophyta** | | | | | | | | | | | | | | |
| *Anacystis Nidulans*(116) | — | 30 watts of lamps 39°C | 39 | T | O₂ | 5.4 | 9.82 μmoles O₂/mg dry wt/hr | 2.2 | — | 3.2 | None-16.2 | — | — | 1 |
| | — | | 25 | T | O₂ | 2.4 | 3.79 μmoles O₂/mg dry wt/hr | 1.1 | — | 2.2 | None-16.2 | — | — | 2 |
| | — | 320 watts of lamps 39°C | 39 | T | O₂ | 6.6 | 14.06 μmoles O₂/mg dry wt/hr | 2.7 | — | 5.6 | None-16.2 | — | — | 3 |
| | — | | 25 | T | O₂ | 3.8 | 6.03 μmoles O₂/mg dry wt/hr | 1.6 | — | 2.7 | None-16.2 | — | — | 4 |
| | — | 960 watts of lamps 39°C | 39 | T | O₂ | 13.0 | 12.05 μmoles O₂/mg dry wt/hr | 4.3 | — | 9.2 | None-16.2 | — | — | 5 |
| | — | | 25 | T | O₂ | 4.3 | 4.29 μmoles O₂/mg dry wt/hr | 2.2 | — | 3.2 | None-16.2 | — | — | 6 |

| Organism | Habitat | Conditions | Temp | Type | Gas | | Rate | | Light | | None/Dark | | | No. |
|---|---|---|---|---|---|---|---|---|---|---|---|---|---|---|
| Anabaena Cylindrica (147) | — | 1320 watts of lamps 25°C | 39 | T | O₂ | 17.3 | 16.52 µmoles O₂/mg dry wt/hr | 7.0 | — | 13.0 | None- 16.2 | — | — | 7 |
| | — | — | 25 | T | O₂ | 9.2 | 7.59 µmoles O₂/mg dry wt/hr | 3.2 | — | 6.0 | None- 16.2 | — | — | 8 |
| | — | — | 25 | T | O₂ | 10.8 | 7.81 µmole O₂/mg dry wt/hr | 2.8 | — | 5.9 | None- 18.9 | — | — | 9 |
| | Freshwater | Sunlight 20°C | 10 | T | CO₂ | 3.0 | 20.83 µmoles C/mg Chl/hr | 1.5 | — | 1.5 | None- 100 | — | — | 10 |
| | | | 20 | T | CO₂ | 6.0 | 62.50 µmoles C/mg Chl/hr | 3.0 | — | 3.0 | None- 100 | — | — | 11 |
| | | | 30 | T | CO₂ | 10.0 | 133.3 µmoles C/mg Chl/hr | 4.0 | — | 6.0 | 20.0 | 100 | 50 | 12 |
| Anabaena Flos-Aquae (148) | — | 100 µE/m²/sec (5) | 25 | — | CO₂ | 170 µE/m²/sec (8.5) | 180 µmoles C/mg Chl/hr | 70 µE/m²/sec (3.5) | 20 µE/m²/sec (1.0) | 100 µE/m²/sec (5) | None- 970 µE/m²/sec (48.5) | — | — | 13 |
| Anabaena Variabilis (149) | — | — | 25 | T | O₂ | 8.6 | 6.90 µmoles O₂/mg dry wt/hr | 2.7 | — | 3.8 | — | — | — | 14 |
| Nostoc Muscorum (149) | — | — | 25 | T | O₂ | 7.0 | 3.55 µmoles O₂/mg dry wt/hr | 2.7 | — | 5.9 | None- 10.8 | — | — | 15 |
| Nostoc Muscorum (150) | Lake Itasca Minnesota | — | 25 | T | — | 15.1 | — | 5.9 | — | 11.9 | None- 25.9 | — | — | 16 |
| Oscillatoria Agardhii (151) | — | — | 20 | S | O₂ | 10.8 | — | 2.2 | — | — | 27.0 | 54.0 | 50 | 17 |
| Oscillatoria Limnetica (152) | — | — | 12 | F | CO₂ | 5.6 | — | — | — | — | None- 10.8 | — | — | 18 |
| Oscillatoria Terebriformis (153) | — | — | 20 | F | CO₂ | 10.8 | — | — | — | — | None- 10.8 | — | — | 19 |
| | — | — | 31 | — | — | 3.8 | — | — | — | — | None- 16.2 | — | — | 20 |
| Oscillatoria Thiebauth (154) | Subtropical Atlantic Ocean | — | 45 | — | — | 13.0 | — | — | — | — | None- 43.2 | — | — | 21 |
| | — | — | — | S | CO₂ | 400 µE/m²/sec (20.0) | 200µ E/m²/sec (10.0) | — | — | 300µE/m²/sec (15.0) | 600 µe/m²/sec (30.0) | 2000µE/ sec m²/sec (100 Klux) | 60 | 22 |
| Phormidium Sp (155) | Yellowstone hot spring | 42-47°C | 37 | — | CO₂ | 68.9 | — | 40.2 | 5.3 Klux | 58.3 | 70.0 | 95.4 | 33 | 23 |
| Synechococcus (156) | Yellowstone hot spring | — | — | S | CO₂ | — | — | — | — | — | None- 1.3 cal/cm²/min (100) | — | — | 24 |
| Synechococcus Lividus (144) | Yellowstone hot spring ravine spring summer | Natural sunlight conditions | 58 | S | CO₂ | 43.2 | — | 10.8 | — | 22.7 | 64.8 | 86.4 | 50 | 25 |
| | — | After 14 days of 73% light reduction | 58 | S | CO₂ | 8.6 | — | 3.2 | — | 5.4 | 8.6 | 86.4 | 80 | 26 |

## Table 1 (continued)
## LIGHT INTENSITY VS. PHOTOSYNTHESIS IN AQUATIC PHOTOSYNTHETIC MICROORGANISMS[a]

| Species (reference) | Environment[b] | Growth Conditions (klx)[c] | Exp. T° (°C)[d] | Type of Light[e] | Assay Method | Light Intensity at Psat[f] (klx) | Pmax[g] | Light Intensity at "Psat" one half (klx) | Compensation Light Intensity[i] (klx) | $I_c'$ (klx) | Light Intensity for Initial Photoinhibition[k] (klx) | Maximum Light Intensity Tested (klx) | Inhibition[l] % | Case #[m] |
|---|---|---|---|---|---|---|---|---|---|---|---|---|---|---|
| *Synechococcus Lividus* (157) | Yellowstone hot spring. Octopus spring summer | Natural sunlight conditions | 58 | S | $CO_2$ | 10.8 | — | 6.5 | — | 10.8 | 64.8 | 86.4 | 30 | 27 |
|  |  | After 16 days of 73% light reduction | 58 | S | $CO_2$ | 10.8 | — | 5.9 | — | 10.8 | 10.8 | 86.4 | 90 | 28 |
|  | Hot spring | 16.20 | 35 | T | $CO_2$ | 14.2 | — | 5.9 | — | 13.2 | None- 25.9 | — | — | 29 |
| *Synechocystis Parvula* (152) |  | 16.20 | 57 | T | $CO_2$ | 10.5 | — | 4.9 | — | 9.6 | None- 25.9 | — | — | 30 |
|  | — | — | 12 | F | $CO_2$ | 1.4 | — | — | — | — | None- 10.8 | — | — | 31 |
|  | — | — | 20 | F | $CO_2$ | 5.6 | — | — | — | — | 2.8 | 10.8 | 10 | 32 |
| **Cryptophytes — Cryptophyta** |  |  |  |  |  |  |  |  |  |  |  |  |  |  |
| *Cyanidium Caladarium* (158) | Yellowstone hot spring | 5.9 | 49 | F | $CO_2$ | 42.4 | — | 8.5 | — | 12.7 | 47.7 | 63.6 | 41 | 1 |
| *Cyanidium Caldarium* (125) | Hot spring acidophile | 74.2 | 49 | F | $CO_2$ | 23.3 | — | 12.7 | — | 21.2 | None- 63.6 | — | — | 2 |
|  |  | pH 1 | 40 | T | $CO_2$ | 150 | 138 μmoles C/mg Chl/hr | 45 | — | 95 | None- 300 | — | — | 3 |
|  |  | pH 3 | 40 | T | $CO_2$ | 200 | 168 μmoles C/mg Chl/hr | 40 | — | 80 | None- 300 | — | — | 4 |
|  |  | pH 7 | 40 | T | $CO_2$ | 100 | 150 μmoles C/mg Chl/hr | 25 | — | 60 | 150 | 300 | 16 | 5 |
|  |  | pH 9 | 40 | T | $CO_2$ | 80 | 54 μmoles C/mg Chl/hr | 20 | — | 40 | None- 300 | — | — | 6 |
| *Cyanophora Paradoxa* (159) | — | — | 17 | T | $CO_2$ | 142 μE/$m^2$/sec (6.2) |  | 75 μE/$m^2$/sec (3.3) | — | 142 μE/$m^2$/sec (6.2) | None- 290 μE/$m^2$/sec (12.6) | — | — | 7 |
| **Dinoflagellates — Pyrrophyta** |  |  |  |  |  |  |  |  |  |  |  |  |  |  |
| 4 Species (Averaged) *Exuviella, Gyrodinium, Gymnodinium Sp., Amphidinium Klebsi* (14) | New England marine planktonic | 10.6 20°C | 20 | S | $CO_2$ | 28.6 | — | 12.7 | — | 26.5 | 47.7 | 100 | 80 | 1 |
| *Amphidinium Carteri* (11) | Marine planktonic | 24°C | 24 | T | $O_2$ | 0.125 Ly/min (20.9) | — | 0.025 Ly/min (4.2) | 0.01 Ly/min (1.67) | 0.05 Ly/min (8.4) | 0.25 Ly/min (41.8) | 0.35 Ly/min (58.5) | 5 | 2 |

| No. | Species (ref) | Habitat | Light intensity | Temp (°C) | Method | Gas | 0.20 Ly/min (33.4) | Productivity | 0.04 Ly/min (6.7) | 0.01 Ly/min (1.67) | 0.07 Ly/min (11.7) | 0.25 Ly/min (41.8) | 0.35 Ly/min (58.5) | 5 |
|---|---|---|---|---|---|---|---|---|---|---|---|---|---|---|
| 3 | *Amphidinium Carteri* (160) | Marine planktonic | 3700 µW/cm² (7) 20°C | 24 | T | CO₂ | 14.5 mW/cm² (34.5) | 900 µmoles O₂/mg Chl²/hr | 5.0 mW/cm² (11.9) | 1.0 mW/cm² (2.4) | 6.3 mW/cm² (15.0) | None-20.0 mW/cm² (47.6) | — | 5 |
| 4 | *Ceratium Furca* (140) | Tropical marine planktonic | — | 20 | T | O₂ | 18.0 | — | 5.0 | — | 13.0 | 30.0 | — | — |
| 5 | *Cachonina Niei* (161) | Salton Sea California | — | — | S | CO₂ | 21.2 | 670 µmoles O₂/mg Chl²/hr | 6.4 | 2.12 | 11.1 | None-3.2 | 40.0 | 25 |
| 6 | *Dichtyocha Sp* (140) | | — | 23 | F | CO₂ | 32.0 | — | 7.0 | — | 15.0 | None-40.0 | — | — |
| 7 | *Dinophysis Miles* (140) | | — | — | S | CO₂ | 64.0 | — | 16.0 | — | 34.0 | None-40.0 | — | — |
| 8 | *Gonyaulax Polyedra* (35) | Marine planktonic | 400 µW/cm² (1.43) | 20 | F | O₂ | 3.8 mW/cm² (10.6) | 198 µmoles O₂/mg Chl²/hr | 1.7 mW/cm² (5.5) | — | 2.8 mW/cm² (9.0) | None-8.0 mW/cm² (25.8) | — | — |
| 9 | | | 1250 µW/cm² (4.48) | 20 | F | O₂ | 5.0 mW/cm² (16.2) | 219 µmoles O₂/mg Chl²/hr | 2.5 mW/cm² (8.1) | — | 4.0 mW/cm² (13.0) | None-8.0 mW/cm² (25.8) | — | — |
| 10 | | | 2500 µW/cm² (8.95) | 20 | F | O₂ | 5.0 mW/cm² (16.2) | 338 µmoles O₂/mg Chl²/hr | 2.4 mW/cm² (7.7) | — | 4.4 mW/cm² (14.2) | None-8.0 mW/cm² (25.8) | — | — |
| 11 | | | 4000 µW/cm² (14.32) | 20 | F | O₂ | 5.0 mW/cm² (16.2) | 222 µmoles O₂/mg Chl²/hr | 2.1 mW/cm² (6.8) | — | 4.1 mW/cm² (13.2) | None-8.0 mW/cm² (25.8) | — | — |
| 13 | *Gonyaulax Tamarensis* (162) | Marine planktonic | 250 µE/m²/sec (12.5) 15°C | 15 | T | O₂ | 1000 µE/m²/sec (43.5) | | 500 µE/m²/sec (21.7) | 140 µE/m²/sec (6.1) | 800 µE/m²/sec (34.8) | 3000 µE/m²/sec (130) | 4500 µE/m²/sec (196) | 30 |
| 14 | *Gymnodinium Kovalevskii* (4) | Black Sea planktonic | 0.03 cal/cm²/min (2.3) | 20 | — | CO₂ | 0.24 cal/cm²/min (18.5) | 67 µmoles C/10⁸ cells/hr | 0.09 cal/cm²/min (6.9) | 0.04 cal/cm²/min (3.1) | 0.19 cal/cm²/min (14.6) | None-0.45 cal/cm²/min (34.7) | — | — |
| 15 | | | 0.16 cal/cm²/min (12.3) | 20 | — | CO₂ | ≥0.45 cal/cm²/min (34.7) | 134 µmoles C/10⁸ cells/hr | — | 0.07 cal/cm²/min (5.4) | 0.27 cal/cm²/min (20.8) | None-0.45 cal/cm²/min (34.7) | — | — |
| 16 | *Peridinium Trichoideum* (4) | Black Sea planktonic | 0.03 cal/cm²/min (2.3) | 20 | — | CO₂ | 0.23 cal/cm²/min (17.7) | 916 µmoles C/10⁸ cells/hr | 0.08 cal/cm²/min (6.2) | 0.06 cal/cm²/min (4.6) | 0.16 cal/cm²/min (12.3) | None-0.45 cal/cm²/min (34.7) | — | — |
| 17 | | | 0.16 cal/cm²/min (12.3) | 20 | — | CO₂ | 0.32 cal/cm²/min (24.6) | 750 µmoles C/10⁸ cells/hr | 0.13 cal/cm²/min (10.0) | 0.07 cal/cm²/min (5.4) | 0.23 cal/cm²/min (17.7) | None-0.45 cal/cm²/min (34.7) | — | — |
| 18 | *Prorocentrum Micans* (160) | Marine planktonic | 3700 µW/cm² (7) 20°C | 20 | T | O₂ | 11.0 mW/cm² (26.2) | 600 µmoles O₂/mg Chl²/hr | 5.2 mW/cm² (12.4) | 1.6 mW/cm² (3.8) | 6.2 mW/cm² (14.8) | None-20.0 mW/cm² (47.6) | — | — |
| 19 | *Zooxanthellae (Symbiodinium Microadriaticum)* (163) | Indo-Pacific from massive coral Favia | 75 Kerg/cm²/sec (18.8) | — | — | O₂ | 75 Kerg/cm²/sec (18.8) | | 10 Kerg/cm²/sec (2.5) | — | 20 Kerg/cm²/sec (5.0) | None-225 Kerg/cm²/sec (56.3) | — | — |

## Table 1 (continued)
## LIGHT INTENSITY VS. PHOTOSYNTHESIS IN AQUATIC PHOTOSYNTHETIC MICROORGANISMS[a]

| Species (reference) | Environment[b] | Growth Conditions (klx)[c] | Exp. T° (°C)[d] | Type of Light[e] | Assay Method | Light Intensity at Psat[f] (klx) | Pmax[g] | Light Intensity at "Psat" one half (klx) | Compensation Light Intensity[i] (klx) | I_k[j] (klx) | Light Intensity for Initial Photoinhibition[k] (klx) | Maximum Light Intensity Tested (klx) | % Inhibition[l] | Case #[m] |
|---|---|---|---|---|---|---|---|---|---|---|---|---|---|---|
| Zooxanthellae From Tridacna Maxima (164) | Great Barrier Reef | 3m Depth | 25 | T | $CO_2$ | $120\times10^{18}Q/m^2/sec$ (8.7) | — | $30\times10^{18}Q/m^2/sec$ (2.2) | — | $40\times10^{18}Q/m^2/sec$ (2.9) | None-$320\times10^{18}Q/m^2/sec$ (23.2) | — | — | 20 |
| Zooxanthellae from Tridacna Maxima (164) | Great Barrier Reef | 3m Depth | 25 | T | $CO_2$ | $120\times10^{18}Q/m^2/sec$ (8.7) | — | $30\times10^{18}Q/m^2/sec$ (2.2) | — | $40\times10^{18}Q/m^2/sec$ (2.9) | None-$320\times10^{18}Q/m^2/sec$ (23.2) | — | — | 20 |
| Zooxanthellae from Bocillopora Damkornis (164) | Great Barrier Reef | 3m Depth | 25 | T | $CO_2$ | $>300\times10^{18}Q/m^2/sec$ (21.7) | — | $\geq120\times10^{18}Q/m^2/sec$ (8.7) | — | $\geq150\times10^{18}Q/m^2/sec2/sec$ (10.9) | None-$320\times10^{18}Q/m^2/sec$ (23.2) | — | — | 21 |
| Zooxanthellae from Blesiastria Urvill (165) | Temperate coral symbiosis | — | 15 | T | $O_2$ | 700 µE/m²/sec (30.4) | — | 45 µE/m²/sec (2.0) | — | 70 µE/m²/sec (3.0) | None-2300 µE/m²/sec (100) | — | — | 22 |
| **Diatoms — Bacillariophyta** | | | | | | | | | | | | | | |
| 3 Species Averaged Skeletonema Costatum Nitzschia Closterium Navicula (14) | New England marine planktonic | 10.8 / 20°C | 20 | S | $CO_2$ | 10.8 | — | 5.4 | — | 8.6 | 32.4 | 86 | 90 | 1 |
| Achnanthes Exigua (166) | Montana hot spring | 6.9 | 15 | T | $O_2$ | 2.1 | 714 µmoles $O_2$/mgChl/hr | 1.0 | — | 1.3 | None-11.7 | — | — | 2 |
| | | | 30 | T | $O_2$ | 3.2 | 1964 µmoles $O_2$/mg Chl/hr | 1.1 | — | 1.9 | None-11.7 | — | — | 3 |
| | | | 42 | T | $O_2$ | 6.4 | 2321 µmoles $O_2$/mg Chl/hr | 1.3 | — | 2.7 | None-11.7 | — | — | 4 |
| Amphiprora Alata (109) | Holland Benthic | 3 | 12 | — | $CO_2$ | 8 | — | 3 | — | 5 | 10 | 60 | 70 | 5 |
| Amphiprora Baludosa (167) | Holland Benthic | 4 | 12 | T | $CO_2$ | 5 | — | 2 | — | 3 | 6 | 60 | 65 | 6 |
| | estuarine | 20°C / 7 / 20°C | 12 | T | $CO_2$ | 8 | — | 3 | — | 5 | 10 | 60 | 70 | 7 |
| Asteroinella SP. (25) | Freshwater lake March | — | 5 | S | $O_2$ | 12 kerg/cm²/sec (3.0) | 28.1 µmoles $O_2$/10⁶ cells/hr | 4 kerg/cm²/sec (1.0) | 0.6 Kerg/cm²/sec (1.50) | 6 Kerg/cm²/sec (1.5) | None-90 Kerg/cm²/sec (22.5) | — | — | 8 |
| | Freshwater lake April | — | 5 | S | $O_2$ | 12 Kerg/cm²/sec (3.0) | 28.1 µmoles $O_2$/10⁶ cells/hr | 4 kerg/cm²/sec (1.0) | 0.6 Kerg/cm²/sec (0.15) | 6 Kerg/cm²/sec (1.5) | None-90 Kerg/cm²/sec (22.5) | — | — | 9 |

| | | | | | | | | | | | | | |
|---|---|---|---|---|---|---|---|---|---|---|---|---|---|
| Freshwater lake May | | 10 | S | 21 Kerg/$cm^2$ sec (5.3) | $O_2$ | 28.1 μmoles $O_2$/ $10^8$ cells/hr | 6 Kerg/$cm^2$ sec (1.5) | 0.7 Kerg/$cm^2$ sec (0.18) | 11 Kerg/$cm^2$ sec (2.8) | None-100 Kerg/$cm^2$ sec (25.0) | — | — | 10 |
| Freshwater lake June-July | | 15 | S | 40 Kerg/$cm^2$ sec (10.0) | $O_2$ | 68.8 - 137.5 μmoles $O_2$/$10^8$ cells/hr | 12 Kerg/$cm^2$ sec (3.0) | 0.18 Kerg/ $cm^2$/sec (0.20) | 21 Kerg/$cm^2$ sec (5.3) | 130 Kerg/ $cm^2$/sec (32.5) | 220 Kerg/ $cm^2$/sec (55.0) | 90 | 11 |
| Freshwater lake July | | 16 | S | 70 Kerg/$cm^2$/ sec (17.5) | $O_2$ | 75.0 μmoles $O_2$/ $10^8$ cells/hr | 21 Kerg/$cm^2$/ sec (5.3) | — | 31 Kerg/$cm^2$/ sec (7.8) | None-150 kerg.$m^2$/sec (37.5) | — | — | 12 |
| Asterionella Formosa (56) Freshwater planktonic | 40 μE/$m^2$ sec (1.7) | 18 | T | 400 μE/$m^2$ sec (17.4) | $CO_2$ | — | 90 μE/$m^2$/sec (3.9) | | 160 μE/$m^2$ sec (7.0) | 400 μE/$m^2$ sec (17.9) | 2000 μE/ $m^2$/sec (87) | 70 | 13 |
| | 200 μE/ $m^2$/sec (8.7) | 18 | T | 600 μE/$m^2$ sec (26.1) | $CO_2$ | — | 140 μE/$m^2$ sec (6.1) | | 240 μE/$m^2$ sec (10.4) | 600 μE/$m^2$ sec (26.1) | 2000 μE/ $m^2$/sec (87) | 25 | 14 |
| Asterionella Japonica (140) Tropical marine planktonic | | — | S | 25.0 | $CO_2$ | — | 7.0 | | 14.5 | None-30.0 | | — | 15 |
| Asterionella Socialis (168) Washington State surf zone summer | 3 / 15°C | 13 | S | 0.125 cal/ $cm^2$/min (20.8) | $O_2$ | — | 0.07 cal/$cm^2$/ min (11.7) | 0.02 cal/$cm^2$/ min (3.31) | 0.09 cal/$cm^2$/ min (15.0) | 0.15 cal/$cm^2$/ min (25.0) | 0.80 cal/ $cm^2$/min (133) | 100 | 16 |
| Biddulphia Sinensis (140) Tropical marine planktonic | | — | S | 27.0 | $CO_2$ | — | 9.0 | | 17.8 | None-30.0 | | — | 17 |
| Chaetoceros SP. (124) Brackish | 8 | 8 | T | 11 | $O_2$ | 219 μmoles $O_2$/ mg Chl/hr | 6 | | 6 | None-20 | | — | 18 |
| Chaetoceros Affinis (169) Pacific marine planktonic | 0.97 / 16°C / 54 / 16°C | 15 | F | 9 | $O_2$ | 625 μmoles $O_2$/ mg Chl/hr | 4 | | 6 | None-20 | | — | 19 |
| | | 15 | F | 11 | $O_2$ | 1375 μmoles $O_2$/mg Chl/hr | 4 | | 8 | None-20 | | — | 20 |
| | August | 20.5 | S | 15 | $O_2$ | — | 4 | | 10 | 42 | | 68 | 21 |
| | October | 18 | S | 14 | $O_2$ | — | 4 | | 8 | 25 | | 68 | 22 |
| Chaetoceros Armatum (168) Washington State surf zone summer | 3 / 15°C | 13 | S | 0.125 cal/ $cm^2$/min (20.8) | $O_2$ | — | 0.04 cal/$cm^2$/ min (6.7) | 0.01 cal/$cm^2$/ min (1.67) | 0.06 cal/$cm^2$/ min (10.0) | 0.5 cal/$cm^2$/ min (83.3) | 0.95 cal/ $cm^2$/min (160) | 15 | 23 |
| Chaetoceros Curvisetus (4) Black Sea planktonic | 0.03 cal/ $cm^2$/min (2.3) | 20 | — | 0.39 cal/$cm^2$/ min (30.0) | $CO_2$ | 666 μmoles C/ $10^8$ cells/hr | 0.09 cal/$cm^2$/ min (6.9) | | 0.18 cal/$cm^2$/ min (13.9) | None-0.39 cal/$cm^2$/min (30.0) | | — | 24 |
| | 0.16 cal/ $cm^2$/min (12.3) | 20 | — | — | $CO_2$ | 683 μmoles C/ $10^8$ cells/hr | — | | 0.26 cal/$cm^2$/ min (20.0) | | | — | 25 |
| Chaetoceros Lorenzianus (140) Tropical marine planktonic | | — | S | 15.0 | $CO_2$ | — | 6.0 | | 11.8 | 26.0 | 32.0 | 10 | 26 |
| Chaetoceros Socialis(4) Black Sea planktonic | 0.03 cal/ $cm^2$/min (2.3) | 20 | — | 0.27 cal/$cm^2$/ min (20.8) | $CO_2$ | 58.3 μmoles C/ $10^8$ cells/hr | 0.08 cal/$cm^2$/ min (6.2) | 0.028 cal/$cm^2$/ min (2.2) | 0.14 cal/ $cm^2$/min (10.8) | None-0.45 cal/$cm^2$/min (34.7) | | — | 27 |
| | 0.16 cal/ $cm^2$/min (12.3) | 20 | — | 0.32 cal/$cm^2$/ min (24.6) | $CO_2$ | 50.0 μmoles C/ $10^8$ cells/hr | 0.12 cal/$cm^2$/ min (9.2) | 0.07 cal/$cm^2$/ min (5.4) | 0.23 cal/$cm^2$/ min (17.7) | None-0.45 cal/$cm^2$/min (34.7) | | — | 28 |

## Table 1 (continued)
## LIGHT INTENSITY VS. PHOTOSYNTHESIS IN AQUATIC PHOTOSYNTHETIC MICROORGANISMS[a]

| Species (reference) | Environment[b] | Growth Conditions (klx)[c] | Exp. T °C[d] | Type of Light[e] | Assay Method | Light Intensity at Psat[f] (klx) | Pmax[g] | Light Intensity at "Psat" one half (klx) | Compensation Light Intensity[i] (klx) | L[j] (klx) | Light Intensity for Initial Photo-inhibition[k] (klx) | Maximum Light Intensity Tested (klx) | Inhibition, % | Case #[m] |
|---|---|---|---|---|---|---|---|---|---|---|---|---|---|---|
| Coscinodiscus Excentricus(91) | English Channel marine planktonic | — | 13-17 | S | $O_2$ | 20 joules/cm²/hr (14.9) | 3.4 µmoles $O_2$/10⁶ cells/hr | 6 joules/cm²/hr (4.5) | — | 11 joules/cm²/hr (7.2) | 42 joules/cm²/hr (40.0) | 80 joules/cm²/hr (75.0) | 50 | 29 |
| Coscinodiscus Radiatus(140) | Tropical marine planktonic | — | — | S | $CO_2$ | 20.0 | — | 5.0 | — | 11.5 | None-30.0 | — | — | 30 |
| Cyclotella Huxleyi(160) | Marine planktonic | 3.7 mW/cm² (7.6) 20°C | 20 | T | $O_2$ | 11.0 mW/cm² (26.2) | 1050 µmoles $O_2$/mgChl/hr | 3.7 mW/cm² (8.8) | 0.6 mW/cm² (1.4) | 6.5 mW/cm² (15.5) | None-20.0 mW/cm² (47.0) | — | — | 31 |
| Cyclotella Meneghiniana(26) | Freshwater | 3 21°C | 21 | T | $CO_2$ | 10 | 29.2 µmoles C/10⁶ cells/hr | 4 | — | 9 | None-30 | — | — | 32 |
| | | 30 21°C | 21 | T | $CO_2$ | 15 | 41.7 µmoles C/10⁶ cells/hr | 6 | — | 12 | None-30 | — | — | 33 |
| Cyclotella Nana(21) | Marine planktonic | — | 5 | — | $CO_2$ | 7 | 108 µmoles C/mg Chl/hr | 4 | — | 4.5 | None-25 | — | — | 34 |
| | | — | 10 | — | $CO_2$ | 9 | 266 µmoles C/mg Chl/hr | 3 | — | 5.5 | None-25 | — | — | 35 |
| | | — | 20 | — | $CO_2$ | 11 | 400 µmoles C/mg Chl/hr | 4 | — | 6 | None-25 | — | — | 36 |
| Cyclotella Nana(160) | Marine planktonic | 0.5 mW/cm² (7.6) | 20 | T | $O_2$ | 11.0 mW/cm² (26.2) | 750 µmoles $O_2$/mg Chl/hr | 4.0 mW/cm² (9.5) | 0.75 mW/cm² (1.8) | 6.0 mW/cm² (14.2) | None-20.0 mW/cm² (47.0) | — | — | 37 |
| Cyclotella Nana(162ª) | Marine planktonic | 250 µE/m²/sec (13) | 15 | T | $O_2$ | 2,800 µE/m²/sec (121) | — | 1000 µE/m²/sec (43.5) | 7.0 µE/m²/sec (0.30) | 850 µE/m²/sec (37.0) | 3,000 µE/m²/sec (130) | 4,500 µE/m²/sec (196) | 20 | 38 |
| Cylindrotheca Closterium(170) | Marine planktonic | 4.2 | 20 | T | $O_2$ | 9.5 | 8.0 µmoles $O_2$/10⁶ cells/hr | 2.1 | 0.5 | 2.6 | 21.2 | 47.7 | 10 | 39 |
| Dictylum Brightwelli(162ª) | Marine plankton | 250 µE/m²/sec (13) | 15 | T | $O_2$ | 1000 µE/m²/sec (43.5) | — | 600 µE/m²/sec (26.0) | 1.2 µE/m²/sec (0.05) | 1000 µE/m²/sec (43.5) | 3000 µE/m²/sec (130) | 4500 µE/m²/sec (196) | 30 | 40 |
| Ditylum Brightwellii(4) | Black Sea planktonic | 0.03 cal/cm²/min (2.3) | 20 | — | $CO_2$ | 0.25 cal/cm²/min (19.3) | 4666 µmoles C/10⁶ cells/hr | 0.11 cal/cm²/min (8.5) | 0.014 cal/cm²/min (1.10) | 0.23 cal/cm²/min (17.7) | 0.31 cal/cm²/min (23.9) | 0.42 cal/cm²/min (32.3) | 7 | 41 |
| | | 0.16 cal/cm²/min (12.3) | 20 | — | $CO_2$ | — | 8333 µmoles C/10⁶ cells/hr | — | 0.057 cal/cm²/min 4.40 | 0.27 cal/cm²/min (20.8) | — | — | — | 42 |
| Fragilaria Crotonensis(25) | Freshwater | — | 16 | S | $O_2$ | 50 Kerg/cm²/sec (12.5) | 100 µmoles $O_2$/10⁶ cells/hr | 10 Kerg/cm²/sec (2.5) | 1.8 Kerg/cm²/sec (0.45) | 18 Kerg/cm²/sec (4.5) | None-110 Kerg/cm²/sec (27.5) | — | — | 43 |

| Organism | Habitat | Col3 | Temp (°C) | T/S | Gas | Col7 | Rate | Col9 | Col10 | Col11 | Col12 | Col13 | Col14 | Ref. |
|---|---|---|---|---|---|---|---|---|---|---|---|---|---|---|
| *Fragilaria Sublinearis* (171) | Antarctic Sea ice | 0.02 — 7°C | — | — | CO₂ | 0.9 | — | — | — | — | 1.0 | 6.0 | 43 | 44 |
| | | 3.1 — 7°C | — | — | CO₂ | 5.3 | — | 0.9 | — | — | None- 5.3 | — | — | 45 |
| | | 10.6 | — | — | CO₂ | 10.6 | — | — | — | — | None- 3.0 | — | — | 46 |
| *Melosira Italica* (25) | Freshwater | 5°C | 5 | S | O₂ | 8 Kerg/cm²/sec (2) | 16.8 µmoles O₂/10⁸ cells/hr | 4 Kerg/cm²/sec (1) | — | 4 Kerg/cm²/sec (1) | None- 25 Kerg/cm²/sec (6.3) | — | — | 47 |
| *Navicula Pelliculosa* (148) | — | 100 µE/m²/sec (5) | 25 | T | CO₂ | 200 µE/m²/sec (10.0) | 210 µmoles C/mg Chl/hr | 45 µE/m²/sec (2.3) | 10 µE/m²/sec (0.5) | 65 µE/m²/sec (3.3) | 270 µE/m²/sec (13.5) | — | — | 48 |
| *Nitzschia Palea* (9) | Freshwater | 3 | 20 | T | CO₂ | 8 | 266 µmoles C/mg Chl/hr | 4 | — | 8 | 10 | 30 | 17 | 49 |
| | | 30 | 20 | T | CO₂ | 18 | 266 µmoles C/mgChl²/hr | 7 | — | 13 | None- 30 | — | — | 50 |
| *Phaeodactylum Tricornutum* (172) | Marine planktonic | — | 18 | T | O₂ | 3.0 mW/cm² (7.1) | — | 0.9 mW/cm² (2.1) | 0.11 mW/cm² (0.20) | 1.7 mW/cm² (4.0) | None- 6.0 mW/cm² (14.2) | 70 | — | 51 |
| *Phaeodactylum Tricornutum* (69) | Marine planktonic | 15 | 20 | T | CO₂ | 10.0 | — | 2.0 | — | 3.0 | 40 | — | 7 | 52 |
| | | 15 | 20 | T | O₂ | 20.0 | 343 µmoles O₂/mg Chl/hr | 7.0 | 1.5 | 12.0 | None- 45 | 60 | — | 53 |
| | | — | 20 | T | CO₂ | 2.5 | — | 1.0 | — | 1.5 | 3.0 | — | 95 | 54 |
| | | — | 20 | T | O₂ | 5.0 | 262 µmoles O₂/mg Chl/hr | 1.5 | 0.5 | 2.5 | None- 45 | — | — | 55 |
| *Phaeodactylum Tricornutum* (131) | Marine planktonic | 12 | 22 | T | CO₂ | 10 | — | 4 | — | 7 | None- 23 | — | — | 56 |
| *Phaeodactylum Tricornutum* (3) | Marine planktonic | — | — | — | CO₂ | 11.0 | — | 4.2 | — | 9.0 | None- 24 | — | — | 57 |
| | | 12 | — | — | CO₂ | 9.0 | — | 3.8 | — | 7.5 | None- 24 | — | — | 58 |
| | | 5 | — | — | CO₂ | 2.8 | — | 0.2 | — | 0.8 | None- 24 | — | — | 59 |
| | | 0.7 | — | S | O₂ | 25.0 | — | 7.0 | — | 14.5 | 32.0 | 40.0 | 30 | 60 |
| *Planktoniella Sol* (140) | Tropical marine planktonic | — | — | S | CO₂ | 31.0 | — | 12.0 | — | 23.5 | 28.0 | 45.0 | 20 | 61 |
| *Rhizosolenia Styliformis* (140) | Tropical marine planktonic | — | — | S | CO₂ | 9 | — | 3 | 1 | 5 | 15 | 36 | 60 | 62 |
| *Sketonema Costatum* (173) | N. Atlantic Marine planktonic | 10 — 20°C | 10 | T | CO₂ | 12 | — | 2.5 | 1 | 4 | 15 | 36 | 40 | 63 |
| | | — | 15 | T | CO₂ | 5 | — | 2 | 1 | 2.5 | 7 | 12 | 20 | 64 |
| | | — | 20 | T | CO₂ | 3 | — | 2 | 1 | 2.5 | 5 | 12 | 70 | 65 |
| *Skeletonema Costatum* (11) | Marine planktonic | 0.05 Ly/min (6.3) — 20°C | 20 | T | CO₂ | 0.15 Ly/min (25.1) | — | 0.045 Ly/min (7.5) | 0.006 Ly/min (1.0) | 0.090 Ly/min (15.0) | None- 0.36 Ly/min (60.1) | — | — | 66 |
| | | — | 20 | T | CO₂ | 0.15 Ly/min (25.1) | — | 0.06 Ly/min (10.0) | 0.006 Ly/min (1.0) | 0.105 Ly/min (17.5) | None- 0.36 Ly/min (60.1) | — | — | 67 |

Table 1 (continued)
## LIGHT INTENSITY VS. PHOTOSYNTHESIS IN AQUATIC PHOTOSYNTHETIC MICROORGANISMS[a]

| Species (reference)[a] | Environment[b] | Growth Conditions (klx)[c] | Exp. T° (°C)[d] | Type of Light[t] | Assay Method | Light Intensity at Psat[f] (klx) | Pmax[e] | Light Intensity at "Psat" one half (klx) | Compensation Light Intensity[f] (klx) | I[f]' (klx) | Light Intensity for Initial Photoinhibition[a] (klx) | Maximum Light Intensity Tested (klx) | Inhibition[a] % | Case #[a] |
|---|---|---|---|---|---|---|---|---|---|---|---|---|---|---|
| Skeletonema Costatum (128) | Brackish | 8 | 20 | T | O₂ | 20 | 937 μmoles O₂/mg Chl/hr | 7 | — | 12 | None- 20 | — | — | 68 |
| Skeletonema Costatum (162ª) | Marine planktonic | 250 μE/m²/sec (13) | 15 | T | O₂ | 1000 μE/m²/sec (43.5) | — | 250 μE/m²/sec (10.9) | 0.13 μE/m²/sec (0.001) | 150 μE/m²/sec (6.5) | None- 3000 μE/m²/sec (130) | — | — | 69 |
| Triceratium Favus (140) | Tropical marine planktonic | — | — | S | CO₂ | 30 | — | 10 | — | 17 | None- 40 | — | — | 70 |
| Tropidoneis SP 85% + Hantzschia amphioxys (174) | Massachusetts intertidal Benthic | — | 27 | S | CO₂ | 0.13 cal/cm²/min (11.1) | — | 0.03 cal/cm²/min (2.8) | — | 0.05 cal/cm²/min (4.2) | 0.25 cal/cm²/min (20.8) | 1.25 cal/cm²/min (108) | 10 | 71 |
| **Coccolithophores — Chrysophyta** | | | | | | | | | | | | | | |
| Coccolithus Hexleyi (175) | Temperate-tropical marine planktonic | 5.3 15°C | 14 | T | CO₂ | 37.1 | 170 μmoles C/mg Chl/hr | 9.3 | — | 15.9 | None- 37.1 | | — | 1 |
| Hymenomonas Sp. (175) | Marine planktonic | 5.3 15°C | 14 | T | CO₂ | 37.1 | 150 μmoles C/mg Chl/hr | 9.3 | — | 15.9 | None- 37.1 | | — | 2 |
| Isochrysis Galbana (160) | Marine planktonic | 3.7 mW/cm² (7.6) 20°C | 20 | T | O₂ | 14.5 mW/cm² (34.5) | 1100 μmoles O₂/mg Chl/hr | 5.4 mW/cm² (12.9) | 0.9 mW/cm² (2.1) | 7.1 mW/cm² (16.9) | None- 20.0 mW/cm² (47.6) | | — | 3 |
| Isochrysis Galbana (162ª) | Marine planktonic | 250 μE/m²/sec (13) | 15 | T | O₂ | 2,000 μE/m²/sec (86.0) | — | 1000 μE/m²/sec (43.5) | 10 μE/m²/sec (0.43) | 800 μE/m²/sec (34.8) | 4500 μE/m²/sec (196) | 6300 μE/m²/sec (274) | 10 | 4 |
| Monochrysis Lutheri (11) | Marine planktonic | 0.05 Ly/min 20°C (7.15) | 20 | T | O₂ | 0.10 Ly/min (16.7) | — | 0.04 Ly/min (6.7) | 0.01 Ly/min (1.7) | 0.065 Ly/min (10.9) | None- 0.36 Ly/min (60.1) | — | — | 5 |
| | | | 20 | T | O₂ | 0.13 Ly/min (21.7) | — | 0.06 Ly/min (10.0) | 0.01 Ly/min (1.7) | 0.11 Ly/min (18.4) | None- 0.36 Ly/min (60.1) | — | — | 6 |
| Syracosphaera Carterae (176) | Marine planktonic | 0.06 Ly/min (10.02) | — | S | O₂ | 0.10 Ly/min (16.7) | — | 0.04 Ly/min (6.7) | — | 0.08 Ly/min (13.4) | 0.15 Ly/min (25.1) | 0.37 Ly/min (62) | 100 | 7 |
| | | | — | S | CO₂ | 0.15 Ly/min (25.1) | — | 0.05 Ly/min (8.4) | — | 0.10 Ly/min (16.7) | 0.20 Ly/min (33.4) | 0.68 Ly/min (113) | 45 | 8 |
| | | 0.25 Ly/min (41.75) | — | S | O₂ | 0.18 Ly/min (30.1) | — | 0.06 Ly/min (10.0) | — | 0.12 Ly/min (20.0) | 0.25 Ly/min (41.8) | 0.65 Ly/min (108) | 30 | 9 |

| Species | | Habitat | °C | | | 0.21 Ly/min (35.1) | | 0.08 Ly/min (13.4) | | 0.15 Ly/min (25.1) | 0.35 Ly/min (58.5) | 0.65 Ly/min (108) | 20 | 10 |
|---|---|---|---|---|---|---|---|---|---|---|---|---|---|---|
| **Euglenophyta** | | | | | | | | | | | | | | |
| *Euglena Gracilis*(177) | — | — | 25 | S | $CO_2$ | 12.7 | — | 3.4 | — | 5.8 | None-31.8 | — | — | 1 |
| **Green Algae — Chlorophyta** | | | | | | | | | | | | | | |
| 7 Species (Averaged) *Duniella Buchlora, Chlamydomonas, Platymonas, Carteria, Mischococcus, Stichococcus, Nannochloris*(14) | 10.8 20°C | New England marine planktonic | 20 | S | $CO_2$ | 6.5 | — | 2.2 | — | 5.4 | 21.6 | 75.6 | 90 | 1 |
| *Actinastrum Hantzschii* (90) | 9.1 | Freshwater | 20 | T | $CO_2$ | — | | | | — | — | 75.3 | 9 | 2 |
| *Ankistrodesmus Falcatus* (90) | 9.1 | Freshwater | 20 | T | $CO_2$ | — | | | | — | — | 75.3 | 12 | 3 |
| *Asterococcus Superbus* (90) | 9.1 | Freshwater | 20 | T | $CO_2$ | — | | | | — | — | 75.3 | 48 | 4 |
| *Chlamydomonas Angulosa*(90) | 9.1 | Freshwater | 20 | T | $CO_2$ | — | | | | — | — | 75.3 | 69 | 5 |
| *Chlamydomonas Nivalis* (178) | 60.9 | Beartooth Mts. Montana Snow Alga summer | 10 | S | $CO_2$ | 54 | — | 15 | — | 30 | None-86 | — | — | 6 |
| *Chlorella 211* (160) | 3.7 mW/cm² (7.6) 20°C | Freshwater | 20 | T | $O_2$ | 3.0 mW/cm² (7.0) | 300 $\mu$moles $O_2$/mg Chl/hr | 1.0 mW/cm² (2.3) | 0.2 mW/cm² (0.47) | 1.0 mW/cm² (2.33) | None-20.0 mW/cm² (46.5) | — | — | 7 |
| *Chlorella Ellipsoidea* (179) | 0.7 | Freshwater planktonic | 22 | — | $O_2$ | 4.0 | 2.86 $\mu$moles $O_2$/10⁶ cells/hr | 1.0 | — | 1.9 | 6.0 | 18.4 | 18 | 8 |
| | 6 | Freshwater planktonic | 22 | — | $O_2$ | 6.0 | 11.00 $\mu$moles $O_2$/10⁶ cells/hr | 1.8 | — | 3.2 | None-27.0 | — | — | 9 |
| *Chlorella Ellipsoidea* (147) | — | Freshwater planktonic | 10 | T | $CO_2$ | 6 | 141 $\mu$moles C/mg Chl/hr | 2 | — | 3 | None-100 | — | — | 10 |
| | — | Freshwater planktonic | 20 | T | $CO_2$ | 10 | 250 $\mu$moles C/mg Chl/hr | 3 | — | 6 | None-100 | — | — | 11 |
| | — | Freshwater planktonic | 30 | T | $CO_2$ | 16 | 458 $\mu$moles C/mg Chl/hr | 5 | — | 11 | None-100 | — | — | 12 |
| *Chlorella Ellipsoidea* (124) | 8 | Freshwater planktonic | 22 | T | $O_2$ | 15 | 1250 $\mu$moles $O_2$/mg Chl/hr | 5 | — | 10 | None-20 | — | — | 13 |
| *Chlorella Fusca*(180) | — | Freshwater planktonic | 25 | T | $O_2$ | 10 Kergs/cm² sec (2.4) | 2100 $\mu$moles $O_2$/g dry wt/hr | 4 Kergs/cm²/sec (1.0) | — | 8 Kergs/cm² sec (2.0) | None-25 Kergs/cm²/sec (6.0) | — | — | 14 |
| *Chlorella Luteoviridis* (179) | 0.7 | — | 22 | — | $O_2$ | 5.0 | 1.79 $\mu$moles $O_2$/10⁶ cells/hr | 1.2 | — | 2.0 | 6.0 | 18.4 | 7 | 15 |
| | 6 | — | 22 | — | $O_2$ | 12.0 | 7.5 $\mu$moles $O_2$/10⁶ cells/hr | 2.0 | — | 3.7 | None-27.0 | — | — | 16 |
| *Chlorella Luteoviridis* | 0.7 | — | 22 | — | $O_2$ | 4.0 | 2.10 $\mu$moles $O_2$/10⁶ cells/hr | 1.0 | — | 1.8 | 6.0 | 18.4 | 15 | 17 |

## Table 1 (continued)
## LIGHT INTENSITY VS. PHOTOSYNTHESIS IN AQUATIC PHOTOSYNTHETIC MICROORGANISMS[a]

| Species (reference) | Environment[b] | Growth Conditions (klx)[c] | Exp. T° (°C)[d] | Type of Light[e] | Assay Method | Light Intensity at Psat[f] (klx) | Pmax[g] | Light Intensity at "Psat" one half (klx) | Compensation Light Intensity[i] (klx) | $I_c$[j] (klx) | Light Intensity for Initial Photoinhibition[k] (klx) | Maximum Light Intensity Tested (klx) | Inhibition[l] % | Case #[m] |
|---|---|---|---|---|---|---|---|---|---|---|---|---|---|---|
| Var. Aureoviridis (179) | — | 6 | 22 | — | $O_2$ | 11.0 | 8.53 μmoles $O_2$/ 10⁸ cells/hr | 1.7 | — | 3.0 | None-27.0 | — | — | 18 |
| Chlorella Luteoviridis Var. Lutescens (179) | — | 0.7 | 22 | — | $O_2$ | 6.0 | 1.96 μmoles $O_2$/ 10⁸ cells/hr | 1.2 | — | 2.0 | 6.0 | 18.4 | 16 | 19 |
| Chlorella Pyrenoidosa (24) | Freshwater | 6 | 22 | — | $O_2$ | 4.0 | 7.99 μmoles $O_2$/ 10⁸ cells/hr | 2.3 | — | 3.9 | None-27.0 | — | — | 20 |
| Chlorella Pyrenoidosa (24) | | 0.06 25°C | 25 | T | $O_2$ | 1.1 | 1.33 μmoles $O_2$/ μℓ pcv/hr | 0.5 | — | 0.8 | None-6.5 | — | — | 21 |
| | | 0.11 25°C | 25 | T | $O_2$ | 2.2 | 1.56 μmoles $O_2$/ μℓ pcv/hr | 0.4 | — | 0.7 | None-6.5 | — | — | 22 |
| | | 0.38 25°C | 25 | T | $O_2$ | 3.8 | 2.01 μmoles $O_2$/ μℓ pcv/hr | 1.1 | — | 1.5 | None-6.5 | — | — | 23 |
| | | 1 25°C | 25 | T | $O_2$ | 4.9 | 1.56 μmoles $O_2$/ μℓ pcv/hr | 1.6 | — | 1.4 | None-6.5 | — | — | 24 |
| | | 3.9 25°C | 25 | T | $O_2$ | 10.3 | 1.33 μmoles $O_2$/ μℓ pcv/hr | 3.2 | — | 2.8 | None-6.5 | — | — | 25 |
| Chlorella Pyrenoidosa (179) | Freshwater | 0.7 | 22 | — | $O_2$ | 4.0 | 2.46 μmoles $O_2$/ 10⁸ cells/hr | 0.9 | — | 1.9 | None-18.4 | — | — | 26 |
| | | 6 | 22 | — | $O_2$ | 12.0 | 9.64 μmoles $O_2$/ 10⁸ cells/hr | 2.0 | — | 3.7 | None-27.0 | — | — | 27 |
| Chlorella Pyrenoidosa (150) | Freshwater | 25°C | 25 | T | — | 16.2 | | 6.5 | — | 13.0 | None-25.9 | — | — | 28 |
| Chlorella Pyrenoidosa (90) | Freshwater | 9.1 | 20 | T | $CO_2$ | 11.0 | | 3.0 | — | 6.0 | None-75.3 | — | — | 29 |
| Chlorella Pyrenoidosa (18) | Freshwater | 0.32 20°C | 20 | T | $CO_2$ | 6 | 33.3 μmoles C/ 10⁸ cells/hr | 2 | — | 3 | None-15 | — | — | 30 |
| | | 1 20°C | 20 | T | $CO_2$ | 10 | 75.0 μmoles C/ 10⁸ cells/hr | 2 | 0.3 | 4 | None-15 | — | — | 31 |
| | | 3 20°C | 20 | T | $CO_2$ | 10 | 41.7 μmoles C/ 10⁸ cells/hr | 3 | — | 6 | None-30 | — | — | 32 |
| | | 21 20°C | 20 | T | $CO_2$ | 30 | 16.7 μmoles C/ 10⁸ cells/hr | 12 | 1.0 | 22 | None-30 | — | — | 33 |
| Chlorella Pyrenoidosa (148) | Freshwater | 100 μE/ m²/sec (4.4) | 25 | — | $CO_2$ | 105 μE/m²/ sec (4.6) | 100 μmoles C/ mg Chl/hr | 55 μE/m²/sec (2.4) | 20 μE/m²/sec (0.87) | 80 μE/m²/sec (3.5) | None-270 μE/m²/sec (11.7) | — | — | 34 |

| No. | Organism | Habitat | Growth | Temp (°C) | T | Gas | | Rate | | | | | | |
|---|---|---|---|---|---|---|---|---|---|---|---|---|---|---|
| 35 | Chlorella Saccharophila(179) | — | 0.7 | 22 | — | O₂ | 6.0 | 6.65 μmoles O₂/10⁸ cells/hr | 1.2 | — | 2.2 | 6.0 | 18.4 | 15 |
| 36 | | — | 6 | 22 | — | O₂ | 9.0 | 20.09 μmoles O₂/10⁸ cells/hr | 2.3 | — | 3.7 | None-27.0 | — | — |
| 37 | Chlorella Sorokinana (123) | Freshwater | 2.7 mW/cm²(6.4) | 39 | T | O₂ | 6.0 mW/cm²(14.3) | — | 2.0 mW/cm²(4.8) | — | 3.0 mW/cm²(7.1) | None-14 mW/cm²(33.3) | — | — |
| 38 | Chlorella Vulgaris(Columbia)(179) | — | 0.7 | 22 | — | O₂ | 6.0 | 4.06 μmoles O₂/10⁸ cells/hr | 1.0 | — | 2.1 | 11.6 | 18.4 | 15 |
| 39 | | — | 6 | 22 | — | O₂ | 9.0 | 15.85 μmoles O₂/10⁸ cells/hr | 1.8 | — | 3.3 | 27.0 | 82 | 18 |
| 40 | Chlorella Vulgaris(113) | — | — | 30 | T | O₂ | 3.7 | 0.48 μmoles C/μl pcv/hr | 1.3 | — | 2.1 | 4.4 | — | — |
| 41 | Chlorella Vulgaris(68) | — | 3  21°C | 20 | T | CO₂ | 7 | 0.68 μmoles C/μl pcv/hr | 2 | — | 5 | 16 | 28 | 15 |
| 42 | | — | 30  21°C | 20 | T | CO₂ | 13 | — | 6 | — | 11 | None-40 | — | — |
| 43 | Chlorella Vulgaris(28) | — | 3  21°C | 21 | T | CO₂ | 5 | 11.7 μmoles C/10⁸ cells/hr | 2 | — | 4 | None-16 | — | — |
| 44 | | — | 30  21°C | 21 | T | CO₂ | 25 | 13.3 μmoles C/10⁸ cells/hr | 6 | — | 13 | None-35 | — | — |
| 45 | Chlorella Vulgaris Var. Viridis(179) | — | 0.7 | 22 | — | O₂ | 4.0 | 2.46 μmoles O₂/10⁸ cells/hr | 1.1 | — | 2.2 | None-18.4 | — | — |
| 46 | | — | 6 | 22 | — | O₂ | 12.0 | 8.66 μmoles O₂/10⁸ cells/hr | 1.9 | — | 3.7 | 50 | 82 | 15 |
| 47 | Dunaliella Tertiolecta (11) | Marine planktonic | 0.05 Ly/min 25°C (13) | 25 | T | O₂ | 0.125 Ly/min(20.9) | — | 0.5 Ly/min(8.4) | 0.013 Ly/min(2.2) | 0.10 Ly/min(16.7) | None-0.35 Ly/min(58.5) | — | — |
| 48 | | | 0.15 Ly/min 25°C | 25 | T | O₂ | 0.27 Ly/min(45.1) | — | 0.07 Ly/min(11.7) | 0.013 Ly/min(2.2) | 0.13 Ly/min(21.7) | None-0.35 Ly/min(58.5) | — | — |
| 49 | | | 0.05 Ly/min 25°C | 25 | T | O₂ | 0.20 Ly/min(33.4) | — | 0.08 Ly/min(13.4) | 0.013 Ly/min(2.2) | 0.16 Ly/min(26.7) | None-0.35 Ly/min(58.5) | — | — |
| 50 | Dunaliella Tertiolecta (162)ᵃ | Marine planktonic | 250 μE/m²/sec (13) | 15 | T | O₂ | 2000 μE/m²/sec(87.0) | — | 700 μE/m²/sec(30.4) | 5.6 μE/m²/sec(0.24) | 600 μE/m²/sec(26.1) | 6300 μE/m²/sec(27.4) | 8000 μE/m²/sec(348) | 30 |
| 51 | Eudorina Elegans(90) | Freshwater | 9.1 | 20 | T | CO₂ | — | — | — | — | — | — | 75.3 | 44 |
| 52 | Gonium Sociale(90) | Freshwater | 9.1 | 20 | T | CO₂ | — | — | — | — | — | — | 75.3 | 88 |
| 53 | Hydrodictyon Africanum (13) | Freshwater shade alga | — | 15 | T | O₂ | 5.0 mW/cm²(11.6) | 20 μmoles O₂/mg Chl/hr | 1.0 mW/cm²(2.3) | 0.05 mW/cm²(0.12) | 1.5 mW/cm²(3.0) | 70.0 mW/cm²(46.5) | 50.0 mW/cm²(116) | 10 |
| 54 | Micractinium Pusillum (90) | Freshwater | 9.1 | 20 | T | CO₂ | — | — | — | — | — | None-75.3 | — | — |
| 55 | Nannochloris Atomus (21) | Marine planktonic | — | 5 | — | CO₂ | 7 | 83.3 μmoles C/mg Chl/hr | 3 | — | 5 | None-25 | 25 | 30 |
| 56 | | — | — | 10 | — | CO₂ | 12 | 250 μmoles C/mg Chl/hr | 5 | — | 9 | 17 | — | — |
| 57 | | — | — | 20 | — | CO₂ | 25 | 383 μmoles C/mg Chl/hr | 6 | — | 11 | None-25 | — | — |

Table 1 (continued)
LIGHT INTENSITY VS. PHOTOSYNTHESIS IN AQUATIC PHOTOSYNTHETIC MICROORGANISMS[a]

| Species (reference) | Environment[b] | Growth Conditions (klx)[c] | Exp. T °C[d] | Type of Light[e] | Assay Method | Light Intensity at Psat[f] (klx) | Pmax[g] | Light Intensity at "Psat" one half (klx) | Compensation Light Intensity[i] (klx) | $I_c$[j] (klx) | Light Intensity for Initial Photo-inhibition[k] (klx) | Maximum Light Intensity Tested (klx) | Inhibition[l] % | Case #[m] |
|---|---|---|---|---|---|---|---|---|---|---|---|---|---|---|
| Ostreobium (Siphonales) (163) | Coral endozoic associate Favia | Top green layer | — | — | $O_2$ | 1.0 Kerg/cm²/sec (0.48) | — | 0.4 Kerg/cm²/sec (0.19) | — | 0.8 Kerg/cm²/sec (0.38) | None- 2.0 Kerg/cm²/sec (0.96) | — | — | 58 |
|  |  | Middle green layer | — | — | $O_2$ | 0.1 Kerg/cm²/sec (0.05) | — | 0.025 Kerg/cm²/sec (0.01) | — | 0.05 Kerg/cm²/sec (0.02) | 0.6 Kerg/cm²/sec (0.29) | 1.8 Kerg/cm²/sec (0.86) | 100 | 59 |
| Pediastrum Duplex (90) | Freshwater | 9.1 | 20 | T | $CO_2$ | — | — | | | | — | 75.3 | 37 | 60 |
| Scenedesmus Sp (147) | Freshwater planktonic | — | 10 | T | $CO_2$ | 3 | 83 μmoles C/mg Chl/hr | 2 | — | 2 | None-100 | 75.3 | — | 61 |
|  |  |  | 20 | T | $CO_2$ | 7 | 229 μmoles C/mg Chl/hr | 3 | — | 4 | None-100 | — | — | 62 |
|  |  |  | 30 | T | $CO_2$ | 10 | 416 μmoles C/mg Chl/hr | 4 | — | 7.5 | None-100 | — | — | 63 |
|  |  |  | 40 | T | $CO_2$ | 9 | 291 μmoles C/mg Chl/hr | 3.5 | — | 5 | 20 | 100 | 50 | 64 |
| Scenedesmus Dimorphus (90) | Freshwater planktonic | 9.1 | 20 | T | $CO_2$ | — | — | — | — | — | None-75.3 | — | — | 65 |
| Scenedesmus Obliquus (181) | — | 1 | 30 | — | $O_2$ | 10.0 mW/cm² (17.9) | 225 μmoles $O_2$/mg Chl/hr | 2.0 mW/cm² (3.6) | 0.75 mW/cm² (1.3) | 3.0 mW/cm² (5.4) | None-62 mW/cm² (111) | — | — | 66 |
|  | — | 5 | 30 | — | $O_2$ | 25.0 mW/cm² (44.8) | 800 μmoles $O_2$/mg Chl/hr | 6.0 mW/cm² (10.7) | 2.15 mW/cm² (3.8) | 10.0 mW/cm² (17.9) | None-62 mW/cm² (111) | — | — | 67 |
| Scenedesmus Quadricauda (90) | Freshwater planktonic | 9.1 | 20 | T | $CO_2$ | — | — | — | — | — | — | 75.3 | 24 | 68 |
| Scenedesmus Quadricauda (4) | Black Sea planktonic | 0.03 cal/cm²/min (2.3) | 20 | — | $CO_2$ | 0.45 cal/cm²/min (34.7) | 266 μmoles C/10⁶ cells/hr | 0.08 cal/cm²/min (6.2) | 0.02 cal/cm²/min (1.5) | 0.18 cal/cm²/min (13.9) | None-0.45 cal/cm²/min (34.7) | — | — | 69 |
|  |  | 0.16 cal/cm²/min (12.3) | 20 | — | $CO_2$ | 0.45 cal/cm²/min (34.7) | 150 μmoles C/10⁶ cells/hr | 0.14 cal/cm²/min (10.8) | 0.06 cal/cm²/min (4.6) | 0.27 cal/cm²/min (20.8) | None-0.45 cal/cm²/min (34.7) | — | — | 70 |
| Staurastrum Sp (90) | — | 9.1 | 20 | T | $CO_2$ | 11 | — | 4 | — | 8 | 12 | 75.3 | 15 | 71 |
| Stichococcus Bacillaris (90) | — | 9.1 | 20 | T | $CO_2$ | — | — | — | — | — | — | 75.3 | 84 | 72 |
| Zygogonium Ericetorum (182) | Yellowstone algal mat | — | 25 | S | $CO_2$ | 1.3 cal/cm²/min (100) | — | 0.46 cal/cm²/min (35.4) | — | — | None- 1.3 cal/cm²/min (100) | — | — | 73 |

a    Abbreviations used: cal/cm²/min, kerg/cm²/sec, klx, Ly/min, Q/m²/sec, $\mu$E/m²/sec, mW/cm, BChl, C, Chl, N, pcv, $\mu$g at, calories centimeter$^{-2}$ minute$^{-1}$, kilo ergs centimeter$^{-2}$ second$^{-1}$, lux × 1000; langleys minute$^{-1}$; quanta meter$^{-2}$ second$^{-1}$; microEinsteins meter$^{-2}$ second$^{-1}$; milliwatts centimeter$^{-2}$; bacteriochlorophyll; carbon; chlorophyll; nitrogen; packed cell volume; microgram atoms.

b    Environment — general habitat or location of sampling.

c    Growth conditions — environmental conditions under which test organisms were grown, prior to photosynthetic experiments.

d    Exp. T° — experimental temperature in degrees centigrade.

e    Type of light — lighting used in experiments: S, sunlight; T, tungsten; F, fluorescent; N, sodium lamp. It should be noted that these are only rough characterizations. There are numerous different types of lamps in each category, e.g., cool-white vs. daylight fluorescent.

f    Light intensity at Psat — intensity at which photosynthesis reaches saturation.

g    Pmax — maximum rate of photosynthesis observed.

h    Light intensity at ½ Pmax — i.e. ½ of the maximum rate of photosynthesis.

i    Compensation light intensity — intensity at which photosynthesis equals respiration.

j    $I_k$ — light intensity at which a line through the initial slope of the light intensity vs. photosynthesis curve intersects a line parallel to the abscissa at Pmax. It is commonly used as a measure of "shade" and "sun" adaptation.[25]

k    Light intensity for initial photoinhibition — intensity at which there is 5% inhibition of photosynthesis. If there is no photoinhibition "none" is used followed by maximum intensity tested.

l    % inhibition — percent inhibition at maximum light intensity tested.

m    Case # — numbers used to refer to individual experimental results in Figures 3 and 4.

n    Some of the light intensity values reported in this paper appear to be uncharacteristically high.

## Table 2
## LIGHT INTENSITY VS. GROWTH IN AQUATIC PHOTOSYNTHETIC MICROORGANISMS. [a]

| Species (reference) | Environment[b] | Growth Conditions[c] | Exp. T° °C[d] | Type of Light[e] | Light Intensity at Growth Optima[f] (klx) | Maximum Growth[g] | Light Intensity at ½ Maximum Growth[h] (klx) | Compensation Light Intensity[i] (klx) | $I_s'$ (klx) | Light Intensity for Initial Photoinhibition[k] (klx) | Maximum Light Intensity Tested (klx) | Inhibition[l] (%) | Case # |
|---|---|---|---|---|---|---|---|---|---|---|---|---|---|
| **Photosynthetic Bacteria** | | | | | | | | | | | | | |
| Chloroflexus Aurantiacus (183) | Hot spring | — | 55 | T | 20 | 0.3db/day | 2 | — | 1 | None-54 | — | — | 1 |
| Chloropseudomonas Ethylicum(184) | — | Acetate substrate | 30 | T | 0.4 | — | 0.1 | — | 0.3 | 1.0 | 7.0 | 30 | 2 |
| | — | Ethanol substrate | 30 | T | 0.3 | — | 0.1 | — | 0.3 | 0.5 | 7.0 | 35 | 3 |
| | — | Ethanol substrate | 30 | T | 1.0 | — | 0.1 | — | 0.4 | 2.0 | 7.0 | 29 | 4 |
| Chloropseudomonas Ethylicum(41) | — | | 30 | T | 5.30 | — | 0.07 | — | 0.16 | 42. 4 | 106 | 12 | 5 |
| Rhodomicrobium Vannielii (43) | San Francisco Bay black mud | — | 34 | T | 5.30 | 13.6 dbl/day | — | — | 1.48 | None-27.6 | — | — | 6 |
| Rhodopseudomonas Acidophila(185) | — | | — | 522nm Light | 180 µE/m²/sec (10.0) | — | 60 µE/m²/sec (3.3) | — | 100 µE/m²/sec (5.5) | None-200 µE/m²/sec (11.0) | — | — | 7 |
| | — | | — | 860nm Light | 60 µE/m²/sec (2.2) | — | 20 µE/m²/sec (0.7) | — | 250 µE/m²/sec (0.9) | None-200 µE/m²/sec (7.2) | — | — | 8 |
| Rhodopseudomonas Capsulata(185) | — | | — | 522nm Light | 300 µE/m²/sec (16.5) | — | 100 µE/m²/sec (5.5) | — | 200 µE/m²/sec (11.0) | None-400 µE/m²/sec (22.6) | — | — | 9 |
| | — | | — | 860nm Light | 100 µE/m²/sec (3.6) | — | 250 µE/m²/sec (0.9) | — | 50 µE/m²/sec (1.8) | None-400 µE/m²/sec (14.4) | — | — | 10 |
| Rhodopseudomonas Spheroides(186) | — | | 34 | T | 6.36 | 9.6 dbl/day | 0.85 | — | 1.06 | None-53.0 | — | — | 11 |
| Rhodospirillum Rubrum(187) | — | | — | — | 0.74 | — | 0.21 | — | 0.42 | None-2.12 | — | — | 12 |
| Rhodospirillum Rubrum(42) | — | | 30 | T | 2.12 | — | 0.11 | — | 0.21 | None-12.7 | — | — | 13 |
| Rhodospirillum Rubrum(185) | — | | — | 885nm Light | 70 µE/m²/sec (2.5) | 11.4 dbl/day | 30 µE/m²/sec (1.1) | — | 50 µE/m²/sec (1.8) | None -200 µE/m²/sec (7.1) | — | — | 14 |
| **Blue-Green Algae — Cyanophyta** | | | | | | | | | | | | | |
| Anabaena(188) | New Zealand oxidation pond | — | 28 | F | 2.0 | 0.7 dbl/day | 0.5 | — | 0.6 | 4.0 | 6.0 | 5 | 1 |
| Anacystis Nidulans(189) | — | | 30 | T | 1.5 mW/cm² (3.5) | 4.8 dbl/day | 0.5 mW/cm² (1.3) | 0.13 mW/cm² (0.3) | 0.6 mW/cm² (1.4) | None-3.5 mW/cm² (8.0) | — | — | 2 |

| Organism | Source | L/D | Temp | F/T | | | | | | | | | | Ref |
|---|---|---|---|---|---|---|---|---|---|---|---|---|---|---|
| Aphanothece Halophytica (190) | Solar evaporation ponds | — | 43 | — | 5.3 | 1.8 dbl/day | 1.1 | — | 1.1 | 5.3 | 6.4 | 18 | | 3 |
| Gleocapsa Alpicola (30) | | — | 25 | F | 3.2 | 0.67 μfpcv/mfculture/day | 2.1 | — | 2.5 | 4.3 | 10.8 | 60 | | 4 |
| Microcystis Aeruginosa (188) | New Zealand oxidation pond | — | 28 | F | 2.0 | 1.0 dbl/day | 0.8 | — | 1.0 | None-3.0 | — | — | | 5 |
| Nostoc Muscorum (191) | | — | 24 | F | 3.1 | — | 1.4 | 0.11 | 1.7 | 3.1 | 4.9 | 10 | | 6 |
| Phormidium Luridum (30) | | — | 25 | F | 4.3 | 0.9 μfpcv/mfculture/day | 1.1 | — | — | 5.4 | 14.0 | 17 | | 7 |
| Phormidium Persicinum (30) | Marine | — | 25 | F | 2.2 | 0.5 μfpcv/mfculture/day | 0.54 | — | — | 3.2 | 10.8 | 90 | | 8 |
| Oscillatoria Agardhii (105) | Marine | — | 20 | — | 0.3 mW/cm² (0.54) | | 0.02 mW/cm² (0.04) | | 0.5 mW/cm² (0.09) | None-0.9 mW/cm² (1.67) | 1.2 mW/cm² (2.20) | 70 | | 9 |
| Synechococcus Elongatus F. Thermalis (192) | Kuril Islands hot springs | — | 57 | T | 40 mW/cm² (95.2) | | 5.0 mW/cm² (11.9) | | 10.0 mW/cm² (23.8) | 85 mW/cm² (2.02) | 140 mW/cm² (3.33) | 32 | | 10 |
| Synechococcus Lividus (115) | Oregon hot spring | — | 55 | — | 5.4 | 1.1 dbl/day | 2.2 | — | 3.8 | None-15.1 | — | — | | 11 |
| | | — | 65 | — | 7.6 | 1.9 dbl/day | 3.2 | — | 6.0 | None-18.4 | — | — | | 12 |
| Tolypothrix Tenuis (122) | Borneo rice paddy | — | 32 | — | 20 | | 2 | — | 3 | None-25 | — | — | | 13 |
| **Cryptophytes — Cryptophyta** | | | | | | | | | | | | | | |
| Chroomonas Saline (11) | Marine planktonic | 6L/18D | 10 | F | 6 cal/cm²/day (4.2) | 0.6 dbl/day | 4 cal/cm²/day (2.8) | 3 cal/cm²/day (2.1) | — | None-38 cal/cm²/day (26.4) | — | — | | 1 |
| | | 12L/12D | 10 | F | 4 cal/cm²/day (1.4) | 1.0 dbl/day | | — | — | None-75 cal/cm²/day (26.0) | — | — | | 2 |
| | | 18L/6D | 10 | F | 7 cal/cm²/day (1.6) | 1.0 dbl/day | | — | — | 7 cal/cm²/day (1.6) | 125 cal/cm²/day (28.9) | 50 | | 3 |
| | | 6L/18D | 20 | F | 13 cal/cm²/day (9.0) | 0.9 dbl/day | 4 cal/cm²/day (2.8) | — | — | None-38 cal/cm²/day (26.0) | — | — | | 4 |
| | | 12L/12D | 20 | F | 14 cal/cm²/day (4.9) | 1.8 dbl/day | 7 cal/cm²/day (2.4) | — | — | None-75 cal/cm²/day (26.0) | — | — | | 5 |
| | | 18L/6D | 20 | F | 14 cal/cm²/day (3.2) | 1.8 dbl/day | 7 cal/cm²/day (1.6) | — | — | None-125 cal/cm²/day (28.9) | — | — | | 6 |
| | | 6L/18D | 25 | F | 38 cal/cm²/day (26.4) | 1.4 dbl/day | 12 cal/cm²/day (8.3) | — | — | None-38 cal/cm²/day (26.4) | — | — | | 7 |
| | | 12L/12D | 25 | F | 20 cal/cm²/day (6.9) | 2.3 dbl/day | 5 cal/cm²/day (1.7) | — | — | None-75 cal/cm²/day (26.0) | — | — | | 8 |
| | | 18L/6D | 25 | F | 7 cal/cm²/day (1.6) | 2.5 dbl/day | 5 cal/cm²/day (1.5) | — | — | 50 cal/cm²/day (11.6) | 125 cal/cm²/day (28.9) | 8 | | 9 |

## Table 2 (continued)
## LIGHT INTENSITY VS. GROWTH IN AQUATIC PHOTOSYNTHETIC MICROORGANISMS.[a]

| Species (reference) | Environment[b] | Growth Conditions[c] | Exp. T° (°C)[d] | Type of Light[e] | Light Intensity at Growth Optima[f] (klx) | Maximum Growth[g] | Light Intensity at 1/2 Maximum Growth[h] (klx) | Compensation Light Intensity[i] (klx) | $I_k'$ (klx) | Light Intensity for Initial Photo-inhibition[k] (klx) | Maximum Light Intensity Tested (klx) | Inhibition[l] (%) | Case #[m] |
|---|---|---|---|---|---|---|---|---|---|---|---|---|---|
| Cryptomonas Ovata(108) | Freshwater | — | 8 | F | 0.003 Ly/min (0.7) | 0.1 dbl/day | — | — | 0.003 Ly/min (0.7) | 0.012 Ly/min (2.9) | 0.020 Ly/min (4.8) | 50 | 10 |
| | | | 14 | F | 0.013 Ly/min (3.1) | 0.3 dbl/day | 0.003 Ly/min (0.7) | — | 0.008 Ly/min (1.9) | None-0.020 Ly/min (4.8) | — | — | 11 |
| | | | 20 | F | 0.050 Ly/min (11.9) | 0.8 dbl/day | 0.027 Ly/min (6.4) | — | 0.037 Ly/min (8.8) | None-0.050 Ly/min (11.9) | — | — | 12 |
| | | | 26 | F | 0.017 Ly/min (4.1) | 0.5 dbl/day | 0.007 Ly/min (1.7) | — | 0.012 Ly/min (2.9) | 0.033 Ly/min (7.9) | 0.050 Ly/min (11.9) | 20 | 13 |
| Cryptomonas Ovata(30) | Freshwater | — | 25 | F | 8.5 | 1.8 μl pcv/ml culture/day | 2.3 | — | 4.5 | None-12.7 | — | — | 14 |
| Cyanidium Caldarium (193) | — | — | 50—55 | — | — | — | — | — | — | None-108 | — | — | 15 |
| Cyanidium Caldarium (30) | Hot springs | — | 25 | F | 3.7 | 0.30 μl pcv/ml culture/day | 1.6 | — | 2.7 | 5.3 | 10.6 | 70 | 16 |
| **Dinoflagellates—Pyrrophyta** | | | | | | | | | | | | | |
| Amphidinium Sp (30) | Marine planktonic | — | 25 | F | 5.3 | 0.1 μl pcv/ml culture/day | 1.3 | — | 2.3 | 8.5 | 12.7 | 20 | 1 |
| Amphidinium Carteri(107) | Woods Hole rock pools | 16L/8D | 23—26 | T | 0.10 Ly/min (16.7) | 0.96 dbl/day | — | 0.01 Ly/min (1.7) | — | None-0.40 Ly/min (66.8) | — | — | 2 |
| Amphidinium Carteri(194) | Mass., USA tidal pool | 12L/12D | 20 | F | 0.30 kerg/cm²/sec (0.1) | 5.0 dbl/day | 0.05 kerg/cm²/sec (0.02) | — | 0.10 kerg/cm²/sec (0.04) | None-1.0 kerg/cm²/sec (0.3) | — | — | 3 |
| Amphidinium Carteri(31) | Marine planktonic | 24L | 21 | F | 30 μE/m²/sec (1.9) | 1.0 dbl/day | — | — | — | None-256 μE/m²/sec (16.0) | — | — | 4 |
| Cachonina Nie(161) | Salton Sea California | — | 23 | F | 6.4 | 0.95 dbl/day | 2.1 | 0.53 | 3.7 | 17.0 | 21.2 | 5 | 5 |
| Cerarium Furca(195) | North Sea planktonic | — | — | F | 2.5 | 0.37 dbl/day | 1.2 | — | — | 2.5 | 10.0 | 100 | 6 |
| Cerarium Fusus(195) | North Sea planktonic | — | — | F | 2.5 | 0.35 dbl/day | 0.6 | — | — | 2.5 | 10.0 | 28 | 7 |
| Cerarium Lineatum(195) | North Sea planktonic | — | — | F | 5.0 | 0.46 dbl/day | 0.6 | — | — | 5.0 | 10.0 | 3 | 8 |

| Organism | Habitat | Light regime | Temp (°C) | Type | | | | | | | | | Ref |
|---|---|---|---|---|---|---|---|---|---|---|---|---|---|
| Ceratium Lineatum(31) | N.E. Pacific marine plaktonic | 24L | 21 | F | 10 µE/m²/sec (0.63) 5.0 | — | 0.4 dbl/day | — | — | 160 µE/m²/sec (10.0) 5.0 | 256 µE/m²/sec (16.0) 10.0 | 100 | 9 |
| Ceratium Tripos(195) | North Sea planktonic summer | — | — | F | 5.0 | 1.3 | 0.2 dbl/day | — | — | 10.0 | — | 5 | 10 |
| Cryptomonas Sp(196) | Marine planktonic | 24L | 18 | T | 60 kerg/cm²/sec (15.0) | 15 kerg/cm²/sec (3.7) | 0.5 dbl/day | 6 kerg/cm²/sec (1.5) | — | None-200 kerg/cm²/sec (50.0) | — | — | 11 |
| Dissodinium Lunula(197) | Tropical marine deep water | 12L/12D | 20 | F | 60 µE/m²/sec (3.7) | 28 µE/m²/sec (1.7) | 0.3 dbl/day | 4 µE/m²/sec (0.25) | 40 µE/m²/sec (2.5) | None-120 µE/m²/sec (7.5) | — | — | 12 |
| Gonyaulax Polyedra(35) | Marine planktonic | 12L/12D | 21 | F | 1.0 mW/cm² (3.2) | 0.6 mW/cm² (1.9) | 0.3 dbl/day | 0.25 mW/cm² (0.81) | 0.9 mW/cm² (2.9) | None-4.0 m W/cm² (13.0) | — | — | 13 |
| Gymnodinium Sp(198) | Tropical marine planktonic | 24L | 23—26 | F | 6.4 | 2.1 | 1.2 dbl/day | 0.32 | 4.0 | None-23.3 | — | — | 14 |
| Gymnodinium Simplex(31) | Marine planktonic | 24L | 21 | F | 80 µE/m²/sec (5.0) | 15 µE/m²/sec (0.93) | 1.3 dbl/day | 15 µE/m²/sec (0.93) | — | 80 µE/m²/sec (5.0) | 256 µE/m²/sec (16.0) | 15 | 15 |
| Gymnodinium Splendens (121) | La Jolla California planktonic | 24L | 15 | F | 6.0 | 6.0 | 0.3 dbl/day | — | — | 6.0 | 16.0 | 15 | 16 |
| Peridinium Sp(196) | Marine planktonic | 24L | 20 | F | 3.0 | — | 0.4 dbl/day | — | — | 10.0 | 16.0 | 10 | 17 |
| Peridinium Sp(196) | Marine planktonic | 24L | 27 | F | 6.0 | — | 0.4 dbl/day | — | — | 10.0 | 16.0 | 25 | 18 |
| Peridinium Sp(196) | Marine planktonic | 24L | 18 | T | >200 kerg/cm²/sec (50.0) | — | >1.0 dbl/day | — | — | None-200 kerg/cm²/sec (50.0) | — | — | 19 |
| Prorocentrum Gracile(196) | Marine planktonic | 24L | 18 | T | >200 kerg/cm²/sec (50.0) | — | >0.8 dbl/day | 6 kerg/cm²/sec (1.5) | — | None-200 kerg/cm²/sec (50.0) | — | — | 20 |
| Prorocentrum Micans(196) | Marine planktonic | 24L | 18 | T | 60 kerg/cm²/sec (15.0) | 20 kerg/cm²/sec (5.0) | 0.6 dbl/day | — | — | None-200 kerg/cm²/sec (50.0) | — | — | 21 |
| Prorocentrum Micans(199) | Marine planktonic | 24L | 20 | F | 0.025 cal/cm²/min (60) | 0.010 cal/cm²/min (2.4) | 0.65 dbl/day | — | 0.017 cal/cm²/min (4.0) | None-8.5 | — | — | 22 |
| Prorocentrum Micans(199) | Marine planktonic | 24L | 20 | T | 0.014 cal/cm²/min (2.3) | 0.008 cal/cm²/min (1.4) | 0.51 dbl/day | 0.003 cal/cm²/min (0.5) | 0.013 cal/cm²/min (2.2) | None-39 | — | — | 23 |
| Prorocentrum Micans(31) | Marine planktonic | 24L | 21 | F | 80 µE/m²/sec (5.0) | — | 0.6 dbl/day | — | — | 80 µE/m²/sec (5.0) | 256 µE/m²/sec (16.0) | 67 | 24 |
| Pyrocystis Fusiformis(197) | Tropical marine deep water | 12L/12D | 20 | F | 25 µE/m²/sec (1.6) | 10 µE/m²/sec (0.63) | 0.15 dbl/day | 4 µE/m²/sec (0.25) | 15 µE/m²/sec (0.93) | None-125 µE/m²/sec (7.5) | — | — | 25 |
| Pyrocystis Noctiluca(197) | Tropical marine deep water | 12L/12D | 25 | F | 30 µE/ms²/sec (1.9) | 15 µE/m²/sec (0.93) | 0.12 dbl/day | 5 µE/m²/sec (0.31) | 25 µE/m²/sec (1.6) | 70 µE/m²/sec (4.4) | 120 µE/m²/sec (7.5) | 10 | 26 |
| Scrippsiella Sweeneyae(31) | S. California marine planktonic | 24L | 21 | F | 30 µE/m²/sec (1.9) | 20 µE/m²/sec (1.3) | 0.8 dbl/day | — | — | 80 µE/m²/sec (5.0) | 160 µE/m²/sec (10.0) | 100 | 27 |

## Table 2 (continued)
## LIGHT INTENSITY VS. GROWTH IN AQUATIC PHOTOSYNTHETIC MICROORGANISMS.[a]

| Species (reference) | Environment[a] | Growth Conditions[c] | Exp. T° °C[d] | Type of Light[e] | Light Intensity at Growth Optima[f] (klx) | Maximum Growth[g] | Light Intensity at ½ Maximum Growth[h] (klx) | Compensation Light Intensity[i] (klx) | $I_c$[j] (klx) | Light Intensity for Initial Photo-inhibition[k] (klx) | Maximum Light Intensity Tested (klx) | Inhibition[l] (%) | Case #[m] |
|---|---|---|---|---|---|---|---|---|---|---|---|---|---|
| **Diatoms — Bacillariophyta** | | | | | | | | | | | | | |
| Achnanthes Exigua(166) | Montana hot spring | 24L | 40 | F | 3.2 | 2.0 dbl/day | 1.1 | — | 2.1 | 8.5 | 11.7 | 63 | 1 |
| Amphiprora Sp. (167) | Holland benthic estuarine | 16L/8D | 20 | T | 57.8 µE/m²/sec (4.3) | 2.0 dbl/day | 6.9 µE/m²/sec (0.5) | — | — | 1.39 µE/m²/sec (10.4) | 255 µE/m²/sec (19.0) | 10 | 2 |
| Amphiprora Paludosa(167) | Holland Benthic estuarine | 16L/8D | 20 | T | 8 | 2.0 dbl/day | 2 | — | — | 15 | 25 | 15 | 3 |
| Asterionella Japonica(200) | Marine planktonic | — | 18 | F | 5.5 | 1.4 dbl/day | 2.0 | — | — | None-9.0 | — | — | 4 |
| Asterionella Socialis(168) | Washington State surf zone | — | 13 | F | 4.0 | 1.1 dbl/day | 2.0 | 0.75 | 30 | None-7.5 | — | — | 5 |
| Biddulphia Aurita(134) | Coos Bay Oregon winter | 15L/9D | 11 | F | 0.03 cal/cm²/min (7.1) | 1.3 dbl/day | 0.01 cal/cm²/min (2.4) | 0.004 cal/cm²/min (2.95) | 0.018 cal/cm²/min (4.3) | 0.07 cal/cm²/min (16.7) | 0.14 cal/cm²/min (33.3) | 20 | 6 |
| | | 9L/15D | 11 | F | 0.03 cal/cm²/min (7.1) | 1.1 dbl/day | 0.01 cal/cm²/min (2.4) | 0.004 cal/cm²/min (0.95) | 0.017 cal/cm²/min (4.1) | None-0.1 cal/cm²/min (23.8) | — | — | 7 |
| Chaetoceros Sp. (198) | Tropical marine planktonic | — | 23—26 | F | 6.5 | 6.0 dbl/day | 2.2 | 0.11 | 3.5 | None-10.8 | — | — | 8 |
| Chaetoceros Sp. (31) | Tropical-subtropical marine | 24L | 20 | F | 75 µE/m²/sec (4.7) | 2.7 dbl/day | 10 µE/m²/sec (0.63) | — | — | 160 µE/m²/sec (10.0) | 256 µE/m²/sec (16.0) | 5 | 9 |
| Chaetoceros Armatum(168) | Washington State surf zone | — | 13 | F | 2.0 | 0.6 dbl/day | 0.5 | 0.25 | 0.8 | 2.0 | 7.5 | 8 | 10 |
| Coscinodiscus Sp. (201) | Marine planktonic | 14L/10D | — | F | 4.0 mW/cm² (12.9) | 1.0 dbl/day | 1.1 mW/cm² (3.6) | — | 2.1 mW/cm² (6.8) | None-6mW/cm² (19.4) | — | — | 11 |
| Coscinodiscus Pavillardii (202) | Marine planktonic | — | — | — | 6.0 | 1.3 dbl/day | 2.2 | 1.0 | 3.5 | None-15 | — | — | 12 |
| Cyclotella Cryptica(201) | Marine planktonic | 14L/10D | — | F | 4.0 mW/cm² (12.9) | 1.8 dbl/day | 1.5 mW/cm² (4.9) | — | 3.0 mW/cm² (9.7) | None-6 mW/cm² (19.4) | — | — | 13 |
| Cylindrotheca Fusiformis(31) | Tropical-subtropical marine | 24L | 20 | F | 100 µE/m²/sec (6.3) | 2.8 dbl/day | 10 µE/m²/sec (0.63) | — | — | 160 µE/m²/sec (10.0) | 256 E/m²/sec (16.0) | 5 | 14 |
| Detonula Confervacea(118) | Narragansett Bay | 24L | 2 | F | 2.2 | 1.2 dbl/day | 1.1 | — | — | 2.2 | 19.4 | 65 | 15 |
| | | | 7 | F | 6.5 | 1.5 dbl/day | 1.1 | — | — | 13.0 | 19.4 | 5 | 16 |
| | | | 12 | F | 13.0 | 1.5 dbl/day | — | — | — | None-19.4 | — | — | 17 |
| Detonula Confervacea(135) | Narragansett Bay | 8L/16D | 2 | F | — | 0.6 dbl/day | — | — | — | None-19.4 | — | — | 18 |
| | | 12L/12D | 2 | F | — | 0.6 dbl/day | — | — | — | None-19.4 | — | — | 19 |

| Species | Habitat | L/D | Temp | Type | (1) | Growth | (2) | (3) | (4) | (5) | (6) | Ref |
|---|---|---|---|---|---|---|---|---|---|---|---|---|
| *Ditylum Brightwellii* (137) | Marine planktonic | 15L/9D | 2 | F | 2.2 | 0.9 dbl/day | — | — | 6.5 | 19.4 | 78 | 20 |
| | | 24L | 2 | F | 2.2 | 1.4 dbl/day | 1.1 | — | 6.5 | 19.4 | 80 | 21 |
| | | 8L/16D | 7 | F | 7.6 | 0.9 dbl/day | 1.6 | — | None-19.4 | — | — | 22 |
| | | 12L/12D | 7 | F | 7.6 | 1.1 dbl/day | 1.4 | — | None-19.4 | — | — | 23 |
| | | 15L/9D | 7 | F | 4.4 | 1.3 dbl/day | 0.8 | — | 13.0 | 19.4 | 15 | 24 |
| | | 24L | 7 | F | 2.2 | 1.3 dbl/day | 1.1 | — | 6.5 | 19.4 | 23 | 25 |
| | | 16L/8D | 20 | T | 0.08 cal/cm²/min (15.3) | 2 dbl/day | 0.01 cal/cm² min (1.9) | — | None-0.12 cal/cm²/min (22.9) | — | — | 26 |
| *Fragilaria Striatula* (134) | Coos Bay Oregon summer | 24L | 20 | T | 0.03 cal/cm²/min (5.7) | 1.4 dbl/day | 0.02 cal/cm² min (3.8) | 0.025 cal/cm² min (6.0) | 0.05 cal/cm²/min (9.6) | 0.12 cal/cm²/min (22.9) | 43 | 27 |
| | | 15L/9D | 11 | F | 0.04 cal/cm²/min (9.5) | 1.4 dbl/day | 0.02 cal/cm² min (4.8) | 0.025 cal/cm² min (6.0) | 0.10 cal/cm²/min (23.8) | 0.13 cal/cm²/min (30.9) | 7 | 28 |
| *Melosira Moniliformis* (134) | Coos Bay Oregon | 9L/15D | 11 | F | 0.04 cal/cm²/min (9.5) | 0.9 dbl/day | 0.02 cal/cm² min (4.8) | 0.008 cal/cm²/min (1.9) | None-0.12 cal/cm²/min (28.6) | — | — | 29 |
| | | 15L/9D | 11 | F | 0.02 cal/cm²/min (4.8) | 0.9 dbl/day | 0.005 cal/cm² min (1.2) | 0.001 cal/cm²/min (0.24) | 0.02 cal/cm²/min (4.8) | 0.15 cal/cm²/min (35.7) | 33 | 30 |
| | | 9L/15D | 11 | F | 0.03 cal/cm²/min (7.1) | 0.6 dbl/day | 0.01 cal/cm² min (2.4) | 0.001 cal/cm²/min (0.24) | 0.08 cal/cm²/min (19.0) | 0.12 cal/cm²/min (28.6) | 10 | 31 |
| *Navicula Arenaria* (167) | Holland Benthic estuarine | 16L/8D | 20 | T | 28.9 µE/m²/sec (2.2) | 1.3 dbl/day | 6.9 µE/m²/sec (0.5) | — | 34.7 µE/m²/sec (2.5) | 1.39 µE/m²/sec (10.4) | 20 | 32 |
| *Nitzschia Closterium* (30) | Marine | — | 25 | F | 4.3 | 0.4 µl pcv/ml culture/day | 1.1 | 2.2 | None-13 | — | — | 33 |
| *Nitzschia Closterium* (203) | Sargasso Sea | — | 16 | F | 3.1 | 0.8 dbl/day | 1 | 1.8 | None-4.2 | — | — | 34 |
| *Nitzschia Dissipata* (167) | Holland Benthic estuarine | 16L/8D | 20 | T | 57.9 µE/m²/sec (4.3) | 2.2 dbl/day | 6.9 µE/m²/sec (0.5) | — | 6.4 µE/m²/sec (5.0) | 139 µE/m²/sec (10.4) | 25 | 35 |
| *Nitzschia Turgioula* (137) | Marine planktonic | 6L/18D | 20 | T | 0.06 cal/cm²/min (11.5) | 1.1 dbl/day | 0.02 cal/cm² min (3.8) | 0.035 cal/cm² min (6.7) | None-0.12 cal/cm²/min (22.9) | — | — | 36 |
| | | 10L/14D | 20 | T | 0.06 cal/cm²/min (11.5) | 2.0 dbl/day | 0.015 cal/cm² min (2.9) | 0.028 cal/cm²/min (5.4) | None-0.12 cal/cm²/min (22.9) | — | — | 37 |
| *Phaeodactylum Tricornutum* (204) | Marine planktonic | 24L | 20 | T | 0.04 cal/cm²/min (7.6) | 2.4 dbl/day | 0.01 cal/cm² min (1.9) | 0.02 cal/cm²/min (3.8) | None-0.12 cal/cm²/min (22.9) | — | — | 38 |
| | | — | 15 | T | 0.021 cal/cm² (3.5) | 1.7 dbl/day | 0.0036 cal/cm²/ (0.6) | 0.0042 cal/cm²/ (0.7) | None-0.079 cal/cm²/ (13.2) | — | — | 39 |

## Table 2 (continued)
## LIGHT INTENSITY VS. GROWTH IN AQUATIC PHOTOSYNTHETIC MICROORGANISMS. [a]

| Species (reference) | Environment[a] | Growth Conditions[c] | Exp. T° °C | Type of Light[d] | Light Intensity at Growth Optima[f] (klx) | Maximum Growth[b] | Light Intensity at ½ Maximum Growth[b] (klx) | Compensation Light Intensity[f] (klx) | Ik' (klx) | Light Intensity for Initial Photo-inhibition[a] (klx) | Maximum Light Intensity Tested (klx) | Inhibition[f] (%) | Case #[a] |
|---|---|---|---|---|---|---|---|---|---|---|---|---|---|
| Phaedactylum Tricornutum (3) | Marie planktonic | 24L | — | T | 4.0 | 1.4 dbl/day | 1.1 | — | 2.2 | None-12 (17.1) | — | — | 40 |
| Phaeodactylum Tricornutum (106) | Marine planktonic | 14L/10D | 17 | F&T | 0.026Ly/mi (4.3) | 1.8 dbl/day | 0.005 Ly/min (0.8) | — | — | 0.7 Ly/min (11) | 0.3 Ly/min (50.1) | 22 | 41 |
| | | 24L | 17 | F&T | 0.02 Ly/min (3.3) | 2.0 dbl/day | 0.0032 Ly/min (0.5) | 1.53 10⁻⁴ Ly/min (0.026) | — | 0.03 Ly/min (5.0) | 0.19 Ly/min (31.7) | 80 | 42 |
| Rhizosolenia Fragilissima (112) | Narragansett Bay | 24L 10°/₀₀ salinity 25 | 12 | F | 6.5 | 1.0 dbl/day | 2.2 | — | — | 6.5 | 19.4 | 20 | 43 |
| | | 24 L 25°/₀₀ salinity | 12 | F | 6.5 | 0.8 dbl/day | 2.2 | — | — | 6.5 | 19.2 | 25 | 44 |
| | | | | F | 6.5 | 1.2 dbl/day | 2.2 | — | — | 6.5 | 19.2 | 26 | 45 |
| | | 24L 35°/₀₀ Salinity | 25 | F | 2.2 | 1.5 dbl/day | 2.2 | — | — | None-19.4 | — | 35 | 46 |
| | | | 12 | F | 6.5 | 1.1 dbl/day | 2.2 | — | — | 6.5 | 19.4 | — | 47 |
| Skeletonema Costatum (31) | Tropical-subtropical marine planktonic | 24L | 25 | F | 19.4 | 1.2 dbl/day | 2.2 | — | — | None-19.4 | — | — | 48 |
| | | 24L | 20 | F | 50 µE/m²/sec (3.1) | 2.5 dbl/day | 7 µE/m²/sec (0.44) | — | — | 160 µE/m²/sec (10.0) | 256 µE/m²/sec (16.0) | 10 | 49 |
| Skeletonema Costatum (107) | Long Island Sound | 16L/8D | 20 | T | 0.075 Ly/min (12.5) | 2.4 dbl/day | — | 0.003 Ly/min (0.50) | — | None-0.4 Ly/min (66.8) | — | — | 50 |
| Synedra Tabulata (134) | Coos Bay Oregon | 15L/9D | 11 | F | 0.03 cal/cm²/min (7.1) | 1.2 dbl/day | 0.008 cal/cm²/min (1.9) | 0.003 cal/cm²/min (0.71) | 0.01 cal/cm²/min (2.4) | 0.08 cal/cm²/min (19.0) | 0.15 cal/cm²/min (35.7) | 15 | 51 |
| | | 9L/15D | 11 | F | 0.03 cal/cm²/min (7.1) | 0.8 dbl/day | 0.008 cal/cm²/min (1.9) | 0.003 cal/cm²/min (0.71) | 0.01 cal/cm²/min (2.4) | None-0.12 cal/cm²/min (28.6) | — | — | 52 |
| Thalassiosira Eccentrica (31) | Tropical-subtropical marine planktonic | 24L | 20 | F | 30 µE/m²/sec (1.9) | 2.1 dbl/day | 10 µE/m²/sec (0.63) | — | — | 160 µE/m²/sec (10.0) | 256 E/m²/sec (16.0) | 75 | 53 |
| Thalassiosira Floridana (31) | Tropical sub tropical marine planktonic | 24L | 20 | F | 75 µE/m²/sec (4.7) | 3.0 dbl/day | 20 µE/m²/sec (1.3) | — | — | None-256 µE/m²/sec (16.0) | — | — | 54 |
| Thalassiosira Fluviatilis (111) | Marine planktonic | 12L/12D | 10 | F | 12 cal/cm²/day (8.3) | 0.0 dbl/day | — | — | — | — | — | — | 55 |
| | | 6L/18D | 15 | F | — | 0.7 dbl/day | 3 cal/cm²/day (2.1) | — | 5 cal/cm²/da (3.5) | 15 cal/cm²/day (10.4) | 38 cal/cm²/day (26.4) | 12 | 56 |

| | | | | | | | | | | | | | |
|---|---|---|---|---|---|---|---|---|---|---|---|---|---|
| | | 12L/12D | 15 | F | 5 cal/cm²/day (1.7) | 0.8 dbl/day | 3 cal/cm²/day (1.0) | — | 5 cal/cm²/day (1.7) | None-75 cal/cm²/day (26.0) | — | — | 57 |
| | | 18L/6D | 15 | F | 8 cal/cm²/day (1.9) | 1.2 dbl/day | 4 cal/cm²/day (0.9) | — | 6 cal/cm²/day (1.4) | 15 cal/cm²/day (3.5) | 120 cal/cm²/day (27.8) | 25 | 58 |
| | | 6L/18D | 20 | F | 8 cal/cm²/day (5.5) | 1.1 dbl/day | 2 cal/cm²/day (1.4) | — | 3 cal/cm²/day (2.1) | 15 cal/cm²/day (10.4) | 38 cal/cm²/day (26.4) | 10 | 59 |
| | | 12L/12D | 20 | F | 10 cal/cm²/day (3.5) | 1.4 dbl/day | 3 cal/cm²/day (1) | — | 5 cal/cm²/day (1.7) | 15 cal/cm²/day (5.2) | 75 cal/cm²/day (26.0) | 2 | 60 |
| | | 18L/6D | 20 | F | 14 cal/cm²/day (3.2) | 1.8 dbl/day | 4 cal/cm²/day (0.9) | — | 9 cal/cm²/day (2.1) | 50 cal/cm²/day (11.6) | 120 cal/cm²/day (27.8) | 10 | 61 |
| | | 6L/18D | 25 | F | 10 cal/cm²/day (6.9) | 1.4 dbl/day | 4 cal/cm²/day (2.8) | — | 5 cal/cm²/day (3.5) | 20 cal/cm²/day (13.8) | 38 cal/cm²/day (26.4) | 29 | 62 |
| | | 12L/12D | 25 | F | 25 cal/cm²/day (8.7) | 2.3 dbl/day | 7 cal/cm²/day (2.4) | — | 12 cal/cm²/day (4.2) | None-75 cal/cm²/day (26.0) | — | — | 63 |
| | | 18L/6D | 25 | F | 30 cal/cm²/day (6.9) | 4.1 dbl/day | 10 cal/cm²/day (2.3) | — | 20 cal/cm²/day (4.6) | 50 cal/cm²/day (11.5) | 120 cal/cm²/day (27.8) | 27 | 64 |
| *Thalassiosira Nordenskioloii* (107) | Narragansett Bay planktonic | 16L/8D | 13 | T | 0.020 Ly/min (3.3) | 1.6 dbl/day | — | 0.005 Ly/min (0.83) | — | None-0.2 Ly/min (33.4) | — | — | 65 |
| *Thalassiosira Nordenskioloii* (32) | Narrangansett Bay planktonic | 15L/9D | 0 | F | 0.020 Ly/min (4.3) | 0.7 dbl/day | — | — | — | 0.07 Ly/min (15.1) | 0.11 Ly/min (23.7) | 43 | 66 |
| | | 12L/12D | 0 | F | 0.020 Ly/min (4.3) | 0.6 dbl/day | — | — | — | 0.07 Ly/min (15.1) | 0.11 Ly/min (23.7) | 5 | 67 |
| | | 9L/15D | 0 | F | 0.020 Ly/min (4.3) | 0.6 dbl/day | — | — | — | 0.07 Ly/min (15.1) | 0.11 Ly/min (23.7) | 5 | 68 |
| | | 15L/9D | 5 | F | 0.060 Ly/min (12.9) | 1.3 dbl/day | — | — | — | 0.07 Ly/min (15.1) | 0.11 Ly/min (23.7) | 12 | 69 |
| | | 12L/12D | 5 | F | 0.065 Ly/min (14.0) | 1.3 dbl/day | — | — | — | None-0.11 Ly/min (23.7) | — | — | 70 |
| | | 9L/15D | | F | 0.065 Ly/min (14.0) | 1.1 dbl/day | 0.015 Ly/min (3.2) | — | — | 0.11 Ly/min (23.7) | — | — | 71 |
| | | 15L/9D | 10 | F | 0.025 Ly/min (5.4) | 1.8 dbl/day | — | — | — | 0.07 Ly/min (15.1) | 0.11 Ly/min (23.7) | 8 | 72 |
| | | 12L/12D | 10 | F | 0.045 Ly/min (9.7) | 1.6 dbl/day | 0.010 Ly/min (2.2) | — | — | 0.07 Ly/min (15.1) | 0.11 Ly/min (23.7) | 10 | 73 |
| | | 9L/15D | 10 | F | 0.050 Ly/min (10.8) | 1.4 dbl/day | 0.013 Ly/min (2.8) | — | — | None-0.11 Ly/min (23.7) | — | — | 74 |
| | | 15L/9D | 15 | F | 0.025 Ly/min (5.4) | 1.5 dbl/day | — | — | — | 0.04 Ly/min (8.6) | 0.11 Ly/min (23.7) | 100 | 75 |

Table 2 (continued)
## LIGHT INTENSITY VS. GROWTH IN AQUATIC PHOTOSYNTHETIC MICROORGANISMS.[a]

| Species (reference) | Environment | Growth Conditions | Exp. T° °C | Type of Light | Light Intensity at Growth Optima (klx) | Maximum Growth | Light Intensity at ½ Maximum Growth (klx) | Compensation Light Intensity (klx) | $I_k$ (klx) | Light Intensity for Initial Photoinhibition (klx) | Maximum Light Intensity Tested (klx) | | |
|---|---|---|---|---|---|---|---|---|---|---|---|---|---|
| | | 12L/12D | 15 | F | 0.025 Ly/min (5.4) | 1.7 dbl/day | 0.010 Ly/min (2.2) | — | — | 0.07 Ly/min (15.1) | 0.11 Ly/min (23.7) | 100 | 76 |
| | | 9L/15D | 15 | F | 0.060 Ly/min (12.9) | 1.7 dbl/day | 0.015 Ly/min (3.2) | — | — | 0.08 Ly/min (17.2) | 0.11 Ly/min (23.7) | 100 | 77 |
| Thalassiosira Pseudonana (106) | Marine planktonic | 14L/10D | 17 | F&T | 0.034 Ly/min (5.68) | 2.5 dbl/day | 0.007 Ly/min (1.17) | $3.4 \times 10^{-4}$ Ly/min (0.057) | — | None-0.3 Ly/min (50.1) | — | — | 79 |
| | | 24L | 17 | F&T | 0.024 Ly/min (4.01) | 2.5 dbl/day | 0.004 Ly/min (0.67) | $5.3 \times 10^{-4}$ Ly/min (0.089) | — | 0.03 Ly/min (5.01) | 0.19 Ly/min (31.7) | 16 | 80 |
| Thalassiosira Rotula (117) | North Sea planktonic | — | 12 | — | 4 | 2.7 dbl/day | — | — | — | None-5.5 | — | — | 81 |
| **Coccolithophores — Chrysophyta** | | | | | | | | | | | | | |
| Coccolithus Huxleyi (136) | Marine planktonic | 6L/18D | 2 | T | 0.05 cal/cm²/min (9.6) | 0.8 dbl/day | 0.02 cal/cm²/min (3.8) | 0.004 cal/cm²/min (0.6) | 0.03 cal/cm²/min (5.7) | None-0.12 cal/cm²/min (22.9) | — | — | 1 |
| | | 10L/14D | 21 | T | 0.06 cal/cm²/min (11.5) | 1.3 dbl/day | 0.02 cal/cm²/min (3.8) | — | 0.027 cal/cm²/min (5.2) | — | — | — | 2 |
| | | 24L | 21 | T | 0.07 cal/cm²/min (13.4) | 1.8 dbl/day | 0.01 cal/cm²/min (1.9) | — | 0.02 cal/cm²/min (3.8) | — | — | — | 3 |
| Coccolithus Huxleyi (206) | Marine planktonic | — | 18.5 | T | 10 | 1.8 dbl/day | 1 | — | 3 | 26 | 50 | 20 | 4 |
| Coccolithus Huxleyi (175) | Temperate-tropical marine planktonic | — | 14 | F | 2.12 | — | — | — | — | None-8.48 | — | — | 5 |
| Hymenomonas Sp. (175) | Temperate-tropical marine planktonic | — | 14 | F | 6.36 | — | — | — | — | None-8.4 | — | — | 6 |
| Isochrysis Sp. (111) | Marine planktonic | 6L/18D | 10 | F | 6 cal/cm²/day (4.17) | 0.6 dbl/day | 3 cal/cm²/day (2.08) | — | 5 cal/cm²/day (3.47) | None 38 cal/cm²/day (26.39) | — | — | 7 |
| | | 12L/12D | 10 | F | 10 cal/cm²/day (3.47) | 0.9 dbl/day | 3 cal/cm²/day (1.04) | — | 6 cal/cm²/day (2.08) | None 75 cal/cm²/day (26.04) | — | — | 8 |
| | | 18L/6D | 10 | F | 14 cal/cm²/day (3.24) | 0.8 dbl/day | — | — | — | 14 cal/cm²/day (3.24) | 125 cal/cm²/day (28.93) | 38 | 9 |

| Organism | Source/habitat | L:D | Temp | Type | | | | | | | Ref. |
|---|---|---|---|---|---|---|---|---|---|---|---|
| | — | 6L/18D | 20 | F | 12 cal/cm²/day (8.33) | 1.7 dbl/day | 2 cal/cm²/day (1.39) | — | None-38 cal/cm²/day (26.39) | — | 10 |
| | | 12L/12D | 20 | F | 13 cal/cm²/day (4.51) | 1.8 dbl/day | 4 cal/cm²/day (1.39) | — | None-75 cal/cm²/day (26.04) | — | 11 |
| | | 18L/6D | 20 | F | 14 cal/cm²/day (3.24) | 2.0 dbl/day | 6 cal/cm²/day (1.39) | — | 125 cal/cm²/day (28.93) | — | 12 |
| | | 16L/18D | 25 | F | 14 cal/cm²/day (9.72) | 1.6 dbl/day | 7 cal/cm²/day (4.86) | — | 38 cal/cm²/day (26.39) | — | 13 |
| | | 12L/12D | 25 | F | 10 cal/cm²/day (3.47) | 3.5 dbl/day | 5 cal/cm²/day (1.74) | 8 cal/cm²/day (2.78) | None-75 cal/cm²/day (26.04) | — | 14 |
| | | 18L/6D | 25 | F | 25 cal/cm²/day (5.79) | 1.2 dbl/day | — | — | None-125 cal/cm²/day (28.93) | — | 15 |
| *Isochrysis Galbana*(207) | Port Erin marine fish pond | — | 25 | F | 1.5 | 1.3 dbl/day | 0.7 | 1.1 | None-3.2 | — | 16 |
| *Monochrysis Lutheri*(107) | Marine planktonic | 16L/8D | 19 | T | 0.1 Ly/min (14.3) | 1.2 dbl/day | 0.01 Ly/min (1.43) | — | None-0.3 Ly/min (42.9) | — | 17 |
| *Monochrysis Lutheri*(204) | Marine planktonic | — | 15 | T | 0.021 cal/cm²/min (3.5) | 1.2 dbl/day | 0.003 cal/cm²/min (0.5) | 0.006 cal/cm²/min (1.0) | None-0.079 cal/cm²/min (13.2) | — | 18 |
| *Ochromonas Danica*(30) | — | — | 25 | F | 7.4 | 3 µl pcv/ml culture/day | 2.1 | 7.4 | 13.8 / 9.5 | 15 | 19 |
| **Euglenophyta** | | | | | | | | | | | |
| *Euglena Gracilis*(177) | — | — | 25 | T | 2.7 | — | 1.1 | — | 31.8 / 21.2 | 9 | 1 |
| *Euglena Gracilis*(30) | — | — | 25 | F | 11.7 | 1.3 µl pcv/ml culture/day | — | — | None-11.7 | — | 2 |
| **Xanthophyta** | | | | | | | | | | | |
| *Tribonema Aequale*(30) | — | — | 25 | F | 6.4 | 4.7 ≫ l pcv/ml culture/day | 3.2 | 4.2 | 12.7 / 10.6 | 17 | 1 |
| **Green Algae — Chlorophyta** | | | | | | | | | | | |
| *Brachiomonas Submarina*(204) | Marine planktonic | — | 15 | T | 3.1 | 1.2 dbl/day | 0.6 | 1.2 | None-11.7 | — | 1 |
| *Chlorella 71105*(208) | — | — | 35 | T | 190,000 lumens ft² (2052) | — | 80,000 lumen ft² (864) | 120,000 lumen ft² (1296) | None 190,000 lumens/ft² (2052) | — | 2 |
| *Chlorella Ellipsoidea*(120) | Freshwater | 12L/12D | 7 | — | 5.0 | 0.3 dbl/day | 0.7 | 0.7 | 50 / 10.0 | 30 | 3 |
| | | 24L | 7 | — | 2.0 | 0.3 dbl/day | 0.2 | 0.2 | 50 / 4.0 | 100 | 4 |
| | | 12L/12D | 25 | — | 6.0 | 2.2 dbl/day | 1.5 | 2.5 | None-50 | — | 5 |
| | | 24L | 25 | — | 4.0 | 3.3 dbl/day | 1.0 | 2.1 | None-50 | — | 6 |
| *Chlorella Ovalis*(204) | Marine planktonic | — | 15 | T | 20 | 1.1 dbl/day | 0.5 | 0.9 | None-11.7 | — | 7 |

## Table 2 (continued)
## LIGHT INTENSITY VS. GROWTH IN AQUATIC PHOTOSYNTHETIC MICROORGANISMS.[a]

| Species (reference) | Environment[b] | Growth Conditions[c] | Exp. T° °C[d] | Type of Light[e] | Light Intensity at Growth Optima (klx) | Maximum Growth[g] | Light Intensity at ½ Maximum Growth[f] (klx) | Compensation Light Intensity (klx) | $I_k'$ (klx) | Light Intensity for Initial Photoinhibition (klx) | Maximum Light Intensity Tested (klx) | |
|---|---|---|---|---|---|---|---|---|---|---|---|---|
| Chlorella Pyrenoidosa(23) | — | | 25 | T | 0.97 | 2.8 dbl/day | 0.27 | — | 0.50 | None-3.90 | — | 8 |
| Chlorella Pyrendidosa(119) | — | | 40 | F | 21.6 | 13 dbl/day | 4.3 | — | 8.1 | None-32.4 | — | 9 |
| Chlorella Pyrenoidosa | — | | 25 | F | 10.8 | 5.5 dbl/day | 1.1 | — | 4.3 | None-32.4 | — | 10 |
| Chlorella Pyrenoidosa(30) | — | | 15 | F | 4.3 | 1.5 dbl/day | — | — | — | 5.4 | 20.5 / 100 | 11 |
| | — | | 25 | F | 15.0 | 1.2 μl pcv/ml culture/day | 5.4 | — | — | None-30.2 | — | 12 |
| Chlorella Sorokiniana(123) | Freshwater | | 39 | F | 2.5 mW/cm² (8.1) | 9.5 dbl/day | 0.8 mW/cm² (2.6) | — | 1.6 mW/cm² (5.2) | None-14 mW/cm² (45) | — | 13 |
| Chlorella Vulgaris(209) | — | | 22 | T | 1.0 | | 0.25 | — | — | None-8.0 | — | 14 |
| Chlorococcum Wimmeri(30) | — | | 25 | F | 5.9 | 0.4 μl pcv/ml culture/day | 2.4 | — | 4.9 | None-10.8 | — | 15 |
| Dunaliella Tertiolecta(107) | — | 16L/8D | 25 | T | 0.15 Ly/min (25.1) | 2.4 dbl/day | — | 0.015 Ly/min (2.5) | — | None-0.4 Ly/min (66.8) | — | 16 |
| Dunaliella Primolecta(204) | Marine planktonic | — | 15 | T | 4.0 | 1.1 dbl/day | 0.8 | — | 1.4 | None-11.7 | — | 17 |
| Mychonastes Ruminatus(210) | Chesapeake Bay | — | 25 | F | 11.75 | 1.6 dbl/day | 4.67 | — | 9.34 | 30.0 | 46.7 / 12 | 18 |
| Nannochloris(198) | Tropical marine plantonic | — | 28 | F | 8.6 | 4.5 dbl/day | 2.7 | — | 7.0 | None-15 | — | 19 |
| Scenedesmus Obliquus(209) | — | — | 22 | T | 1 | | 0.1 | — | — | 10 | — | 20 |
| Scenedesmus Obliquus(181) | — | 24L | 30 | F | 2.7 mW/cm² (7.6) | | 0.9 mW/cm² (2.5) | — | — | 3.0 mW/cm² (8.3) | 3.5 mW/cm² (9.7) | 21 |
| Scenedesmus Protuberans (105) | Freshwater plank-tonic | — | 20 | — | 0.6 mW/cm² (1.9) | | 0.1 mW/cm² (0.3) | — | — | None-0.6 mW/cm² (1.9) | 22 | 22 |
| Tetraselmis Sp. (133) | Sargasso Sea | — | 16 | F | 1.9 | 0.45 dbl/day | 0.9 | — | 1.8 | 3.1 | 4.2 / 20 | 23 |

[a] Abbreviations used: cal/cm²/min, calories centimeter⁻² minute⁻¹; kerg/cm²/sec, kiloergs centimeter⁻² second⁻¹; klx, lux × 1000; Ly/min, langleys minute⁻¹; Q/m²/sec, quanta meter⁻² second⁻¹; μE/m²/sec, micro Einsteins meter⁻² second⁻¹; μW/cm², microwatts centimeter⁻²; pcv, packed cell volume; L/D, hours light/hours dark.

[b] Environment — general habitat or location of sampling.

[c] Growth conditions — environmental conditions under which test organisms were grown.

[d] Exp. T° — experimental temperature in degree centigrade.

[e] Type of light — lighting used in experiments: T, tungsten; F, fluorescent.

[f] Light intensity at growth maximum — intensity at which growth reaches saturation.

[g] Maximum growth — maximum rate of growth observed.

<sup>a</sup>    Light intensity at ½ growth maximum — i.e., ½ maximum growth rate.

<sup>b</sup>    Compensation light intensity — intensity at which photosynthesis equals respiration.

<sup>c</sup>    $I_s$ — light intensity at which a line through the initial slope of the light intensity vs. growth curve intersects a line parallel to the abscissa at growth max$_{25}$. It is commonly used as a measure of "shade" and "sun" adaptation.

<sup>d</sup>    Light intensity for initial photoinhibition — intensity at which there is 5% inhibition of growth. If there is no photoinhibition, "none" is used followed by maximum intensity tested.

<sup>e</sup>    % inhibition — percent inhibition at maximum light intensity tested.

<sup>f</sup>    Case # — numbers used to refer to individual experimental results in Figures 3 and 4.

<sup>g</sup>    Some of the light intensity values reported in this paper appear to be uncharacteristically high.

These "tendencies" are based on a composite impression from Figures 3 and 4 and are marked by numerous exceptions. One of the problems in interpreting these data is the unevenness in the number of species tested in the different phyla. For example, relatively few studies have been done on blue-green algae. On the basis of those included, this group appears to have a preference for low light intensity. However, ecological records of blue-green algal distribution indicate the existence of many species with tolerance to high light intensity environments.

### Contrasting Strategies of Adaptation in Benthic Vs. Planktonic Organisms

Because of their contrasting life styles, the light intensity environment of planktonic organisms can be very different from that of benthic organisms inhabiting the same geographical location. For planktonic organisms, the light environment is not only a product of incident solar radiation and water transparency, but is, to a large extent, related to the existing hydrographic conditions, e.g., the rate and depth of vertical mixing. In well-mixed water columns (spring and summer), phytoplankton generally exhibit a mild form of sun adaptation. Conversely, in a stratified water column organisms inhabiting the deep water layers are shade adapted, and those at the surface are sun adapted.[15,25,104,204,211-221]

In the benthic environment, organisms are restricted to a certain depth and must be prepared to withstand longer periods of time at a given intensity as compared to the circulating planktonic organisms. It has been suggested that benthic species are more light tolerant than planktonic forms.[174] While proof for this hypothesis is not conclusive, it has been shown that circulation patterns, such as Langmuir cells, significantly reduce the net residence time of phytoplankton in zones of inhibiting light intensities.[54,211-214] This means that high surface light intensity need not lead to significant levels of photoinhibition in phytoplankton and may reduce the necessity for strong sun adaptation.

In contrast, shallow water and intertidal benthic species must cope with extended periods of exposure to such intensities. These organisms either revert to settling in shaded areas or develop sun-adapted characteristics.

### Topographical and Geographical Distribution of Sun and Shade Species

The geographical distribution of aquatic sun and shade organisms is, of course, a product of the pattern of solar insolation which, in turn, is related to seasonal and climatic changes. In the low latitudes of the tropics, light intensity is high year round. Therefore, adaptation in surface organisms is dominated by the sun strategy. In midlatitudes light intensity is seasonal, being high in the summer and low in the winter. The species composition and adaptational pattern in temperate waters are seasonally variable in response to this and other environmental changes. Polar regions are subjected to even sharper seasonal contrasts in light environment. The summer is characterized by long daylengths, and the winter is characterized by extended periods of near darkness. The exceptionally low levels of winter insolation in polar environments pose a serious problem for the survival of photosynthetic organisms.[215] Some researchers have hypothesized that Arctic species resort to a facultative heterotrophic life style for winter survival. In contrast, others have proposed that winter survival is based on the strong reduction of respiration made possible by the low polar temperatures. At the same time, the shade adaptation of polar species permits efficient utilization of the small amount of light available.

Zones of low light intensity in surface waters are not restricted to polar regions, but can be found throughout the aquatic environment where topographic features and biological overgrowth provide shade. The distribution and function of organisms within some of these special habitats have been studied, such as caves[19,216,217] and kelp forests.[218]

# REFERENCES

1. Jerlov, N. G., Light. General introduction, in *Marine Ecology*, Kinne, O., Ed., Wiley-Interscience, London, 1970, 95.
2. Bainbridge, R., Evans, G. C., and Rackham, O., *Light as an Ecological Factor*, Symp. Br. Ecology Soc., Blackwell Scientific, London, 1965.
3. Beardall, J. and Morris, I., The concept of light intensity adaptation in marine phytoplankton: some experiments with *Phaeopactylum tricornutum, Mar. Biol.*, 37, 377, 1976.
4. Berseneva, G. P., Sergeyeva, L. M., and Finenko, Z. Z., Adaptation of Marine Planktonic Algae to Light, *Oceanology*, 18, 197, 1978.
5. Boardman, N. K., Comparative photosynthesis of sun and shade plants, *Annu. Rev. Plant Physiol.*, 28, 355, 1977.
6. Evans, G. C., Bainbridge, R., and Rackham, D., *Light as an Ecological Factor*, Vol. 2, Blackwell Scientific, London, 1974.
7. Halldal, P., Light and photosynthesis of different marine algal groups, in *Optical Aspects of Oceanography*, Jerlov, N. G. and Steemann Nielsen, E., Eds., Academic Press, London, 1974, 345.
8. Hellebust, J. A., Light: plants, in *Marine Ecology*, Vol. 1 (Part 1), Kinne, O., Ed., Wiley-Interscience, 1970, 125.
9. Jørgensen, E. G., The adaptation of plankton algae. IV. Light adaptation in different algal species, *Physiol. Plant.*, 22, 1307, 1969.
10. Levring, T., Submarine daylight and the photosynthesis in marine algae, *Goteborgs K. Vet. o. Vitt. Samh. Handl.*, IV. (Ser. 5), 6, 1, 1947.
11. McAllister, C. D., Shah, N., and Strickland, J. D. H., Marine phytoplankton photosynthesis as a function of light intensity: a comparison of methods, *J. Fish. Res. Board Can.*, 21, 159, 1964.
12. Rabinowitch, E. I., *Photosynthesis and Related Processes*, Interscience, New York, 1951.
13. Raven, J. A. and Smith, F. A., 'Sun' and 'shade' species of green algae: relation to cell size and environment, *Photosynthetica*, 11, 48, 1977.
14. Ryther, J. H., Photosynthesis in the ocean as a function of light intensity, *Limnol. Oceanogr.*, 1, 61, 1956.
15. Ryther, J. H. and Menzel, D. W., Light adaptation by marine phytoplankton, *Limnol. Oceanogr.*, 4, 492, 1959.
16. Soeder, C. and Stengel, E., Physico-chemical factors affecting metabolism and growth rate, in *Algal Physiology and Biochemistry*, Stewart, W. D. P., Ed., University of California Press, Berkeley, 1975, 714.
17. Steemann-Nielsen, E., *Marine Photosynthesis*, Series 13, Elsevier Oceanography, 1975, 141.
18. Steemann-Nielsen, E. and Jørgensen, E. G., The adaptation of plankton algae. I. General part, *Physiol. Plant.*, 21, 401, 1968.
19. Titlyanov, E. A., Adaptation of benthic plants to light. I. Role of Light in distribution of attached marine algae, *Sov. J. Mar. Biol.*, 2, 1, 1976.
20. Weinberg, S., Submarine daylight and ecology, *Mar. Biol.*, 37, 291, 1976.
21. Yentsch, C. S., Some aspects of the environmental physiology of marine phytoplankton: a second look, *Oceanogr. Mar. Biol.*, 12, 41, 1974.
22. Yentsch, C. S. and Lee, R. W., A study of photosynthetic light reactions, and a new interpretation of sun and shade phytoplankton, *J. Mar. Res.*, 24, 319, 1974.
23. Myers, J., Culture conditions and the development of the photosynthetic mechanism, III. Influence of light intensity on cellular characteristics of *Chlorella, J. Gen. Physiol.*, 29, 419, 1946.
24. Myers, J., Culture conditions and the development of the photosynthetic mechanism. IV. Influence of light intensity on photosynthetic characteristics of *Chlorella, J. Gen. Physiol.*, 29, 429, 1946.
25. Talling, J. F., Photosynthetic characteristics of some freshwater plankton diatoms in relation to underwater radiation, *New Phytol.*, 56, 29, 1957.
26. Jørgensen, E. G., Adaptation to different light intensities in the diatom *Cyclotella meneghiniana* Kutz, *Physiol. Plant.*, 17, 136, 1964.
27. Steemann Nielsen, E., Chlorophyll concentration and rate of photosynthesis in *Chlorella vulgaris, Physiol. Plant.*, 14, 868, 1961.
28. Steemann Nielsen, E., Hansen, V. K., and Jørgensen, E. G., The adaptation to different light intensities in *Chlorella vulgaris* and the time dependence on transfer to a new light intensity, *Physiol. Plant.*, 15, 505, 1962.
29. Beale, S. I. and Appleman, D., Chlorophyll synthesis in *Chlorella*. Regulation by degree of light limitation of growth, *Plant Physiol.*, 47, 230, 1971.
30. Brown, T. E. and Richardson, F. T., The effect of growth environment on the physiology of algae: light intensity, *J. Physiol.*, 4, 38, 1968.
31. Chan, A. T., Comparative physiological study of marine diatoms and dinoflagellates in relation to irradiance and cell size. I. Growth under continuous light, *J. Phycol.*, 14, 396, 1978.

32. **Durbin, E. G.,** Studies on the autecology of the marine diatom *Thalassiosira nordenskoldii* Cleve. I. The influence of daylength, light intensity, and temperature on growth, *J. Phycol.,* 10, 220, 1974.

33. **Jones, L. W. and Myers, J.,** Pigment variations in *Anacystis nidulans* induced by light of selected wavelengths, *J. Phycol.,* 1, 7, 1965.

34. **Prezelin, B. B.,** The role of peridinin-chlorophyll *a*-proteins in the photosynthetic light adaption of the marine dinoflagellate, *Glenodinium* sp., *Planta,* 130, 225, 1976.

35. **Prezelin, B. B. and Sweeney, B. M.,** Photoadaptation of photosynthesis in *Gonyaulax polyedra,* *Mar. Biol.,* 48, 27, 1978.

36. **Ramus, J., Lemons, F., and Zimmerman, C.,** Adaptation of light-harvesting pigments to downwelling light and the consequent photosynthetic performance of the eulittoral rockweeds *Ascophyllum nodosum* and *Fucus vesiculosus, Mar. Biol.,* 42, 293, 1977.

37. **Brody, M. and Emerson, R.,** Effect of wavelength and intensity of light on the proportion of pigments in *Porphyridium cruentum, Am. J. Bot.,* 46, 433, 1959.

38. **Halldal, P.,** Pigment formation and growth in blue-green algae in crossed gradients of light intensity and temperature, *Physiol. Plant.,* 11, 401, 1958.

39. **Yocum, C. S. and Blinks, L. R.,** Light-induced efficiency and pigment alterations in red algae, *J. Gen. Physiol.,* 41, 1113, 1958.

40. **Brown, T. E., Richardson, F. L., and Vaughn, M. L.,** Development of red pigmentation in *Chlorococcum wimmeri* (Chlorophyta:Chlorococcales), *Phycologia,* 6, 167, 1967.

41. **Holt, S. C., Conti, S. F., and Fuller, R. C.,** Effect of light intensity on the formation of the photochemical apparatus in the green bacterium *Chloropseudomonas ethylicum, J. Bacteriol.,* 91, 349, 1966.

42. **Holt, S. C. and Marr, A. G.,** Effect of light intensity on the formation of intracytoplasmic membrane in *Rhodospirillum rubrum, J. Bacteriol.,* 89, 1421, 1965.

43. **Trentini, W. C. and Starr, M. P.,** Growth and ultrastructure of *Rhodomicrobium vannielii* as a function of light intensity, *J. Bacteriol.,* 93, 1699, 1967.

44. **Kageyama, A. and Yokohama, Y.,** Pigments and photosynthesis of deep-water green algae, *Bull. Jpn. Soc. Phycol.,* 25, 168, 1977.

45. **Shimura, S. and Fujita, Y.,** Changes in the activity of fucoxanthin-excited photosynthesis in the marine diatom *Phaeodactylum tricornutum* grown under different culture conditions, *Mar. Biol.,* 33, 185, 1975.

46. **Tanada, T.,** The photosynthetic efficiency of carotenoid pigments in *Navicula minima, Am. J. Bot.,* 38, 276, 1951.

47. **Sheridan, R. P.,** Sun-shade ecotypes of a blue-green alga in a hot spring, *J. Phycol.,* 12, 279, 1976.

48. **Raven, J. A. and Glidewell, S. M.,** Photosynthesis, respiration and growth in the shade alga *Hydrodictyon africanum, Photosynthetica,* 9, 361, 1975.

49. **Kahn, N. and Swift, E.,** Positive buoyancy through ionic control in the nonmotile marine dinoflagellate *Pyrocystis noctiluca* Murray ex Schuett, *Limnol. Oceanogr.,* 23, 649, 1978.

50. **Kondpka, A.,** Physiological ecology of planktonic cyanobacteria, in *Proc. 78th Annu. Meeting of the American Society for Microbiology,* Session 192, 1978.

51. **Jeffrey, S. W. and Vesk, M.,** Chloroplast structural changes induced by white light in the marine diatom *Stephanopyxis turris, J. Phycol.,* 14, 238, 1978.

52. **Herman, E. M. and Sweeney, B. M.,** Circadian rhythm of chloroplast ultrastructure in *Gonyaulax polyedra,* concentric organization around a central cluster of ribosomes, *J. Ultrastr. Res.,* 50, 347, 1975.

53. **Harris, G. P. and Lott, J. N. A.,** Light intensity and photosynthetic rates in phytoplankton, *J. Fish. Res. Board Can.,* 30, 1771, 1973.

54. **Harris, G. P. and Piccinin, B. B.,** Photosynthesis by natural phytoplankton populations, *Arch. Hydrobiol.,* 80, 405, 1977.

55. **Abeliovich, A. and Shilo, M.,** Photooxidative death in blue-green algae, *J. Bacteriol.,* 111, 682, 1972.

56. **Belay, A. and Fogg, F. E.,** Photoinhibition of photosynthesis in *Asterionella formosa* (Bacillariophyceae), *J. Phycol.,* 14, 341, 1978.

57. **Brown, D. L. and Tregunna, E. B.,** Inhibition of respiration during photosynthesis by some algae, *Can. J. Bot.,* 45, 1135, 1967.

58. **Harvey, G. W. and Bishop, N. I.,** Photolability of photosynthesis in two separate mutants of *Scenedesmus obliquus, Plant Physiol.,* 62, 330, 1978.

59. **Jones, L. W. and Kok, B.,** Photoinhibition of chloroplast reactions. I. Kinetics and action spectra, *Plant Physiol.,* 41, 1037, 1966.

60. **Jones, L. W. and Kok, B.,** Photoinhibition of chloroplast reactions. II. Multiple effects, *Plant Physiol.,* 41, 1044, 1966.

61. **Kandler, D. and Sironval, C.,** Phtooxidation processes in normal green *Chlorella* cells, II. Effects on metabolism, *Biochim. Biophys. Acta,* 33, 207, 1969.

62. **Kok, B. and Bongers, L. H.**, Radiation tolerances in photosynthesis and consequences of excesses, in *Medical and Biological Aspects of the Energies of Space,* Columbia University Press, New York, 1961, 299.

63. **Satoh, K.**, Mechanism of photoinactivation in photosynthetic systems II. The occurrence and properties of two different types of photoinactivation, *Plant Cell Physiol.,* 11, 29, 1970.

64. **Satoh, K.**, Mechanism of photoinactivation in photosynthetic systems III. Site and mode of photoinactivation in photosystem I, *Plant Cell Physiol.,* 11, 187, 1970.

65. **Sironval, C. and Kandler, O.**, Photooxidation processes in normal green *Chlorella* cells. I. The bleaching process, *Biochim. Biophys. Acta,* 29, 359, 1958.

66. **Sorokin, C.**, Injury and recovery of photosynthetic activity, *Physiol. Plant.,* 13, 20, 1960.

67. **Steemann Nielsen, E.**, On detrimental effects of high light intensities on the photosynthetic mechanism, *Physiol. Plant.,* 5, 334, 1952.

68. **Steemann Nielsen, E.**, Inactivation of the photochemical mechanism in photosynthesis as a means to protect cells against too high light intensities, *Physiol. Plant.,* 15, 161, 1962.

69. **Takahashi, M., Shimura, S., Yamaguchi, Y., and Fujita, Y.**, Photoinhibition of phytoplankton photosynthesis as a function of exposure time, *J. Oceanogr. Soc. Jpn.,* 27, 43, 1971.

70. **Forti, G. and Jagendorf, A. T.**, Inactivation by light of the phosphorylative activity of chloroplasts, *Biochim. Biophys. Acta,* 44, 34, 1960.

71. **Anderson, S. M., Krinsky, N. I., Stone, M. J., and Clagett, D. C.**, Effect of singlet oxygen quenchers on oxidative damage to liposomes initiated by photosensitization or by radiofrequency discharge, *Photochem. Photobiol.,* 20, 65, 1974.

72. **Anwar, M. and Prebble, J.**, The photoinactivation of the respiratory chain in *Sarcina lutea* (*Micrococcus luteus*) and protection by endogenous carotenoid, *Photochem. Photobiol.,* 26, 475, 1977.

73. **Anderson, S. M. and Kirnsky, N. I.**, Protective action of carotenoid pigments against photodynamic damage to liposomes, *Photochem. Photobiol.,* 18, 403, 1973.

74. **Foote, C. S.**, Mechanisms of photosensitized oxidation, *Science,* 162, 963, 1968.

75. **Foote, C. S., Chang, Y. C., and Denny, R. W.**, Chemistry of singlet oxygen. X. Carotenoid quenching parallels biological protection. *J. Am. Chem. Soc.,* 92, 5216, 1970.

76. **Garrard, L. A. and Brandle, J. R.**, Effects of UV-B radiation on photosynthesis of higher plants, in *Impacts of Climatic Changes on the Biosphere. 1. Ultraviolet Radiation Effects,* Climatic Impact Assessment Program Monogr., 5, 20, 1975.

77. **Halldal, P. and Taube, Ö.**, Ultraviolet action and photoreactivation in algae, in *Photophysiology,* Vol. 7, Giese, A., Ed., Academic Press, New York, 1972, 163.

78. **Jagger, J.**, *Introduction to Research in Ultraviolet Photobiology,* Prentice-Hall, Englewood Cliffs, N.J., 1967.

79. **Kumar, H. D.**, Ultraviolet lethality and spore survival in blue-green algae, *Phykos,* 9, 104, 1970.

80. **McLeod, G. C. and McLachlan, J.**, The quantum efficiency of photosynthesis in ultraviolet radiation of 2537 A, *Physiol. Plant.,* 12, 306, 1959.

81. **Rupert, C. S.**, Photoreactivation of ultraviolet damage, in *Photophysiology,* Vol. 2, Giese, A. C., Ed., Academic Press, New York, 1964, 283.

82. **Thomas, G.**, Effects of near ultraviolet light on microorganisms, *Photochem. Photobiol.,* 26, 699, 1977.

83. **Van Baalen, C.**, The effect of ultraviolet irradiation on a coccoid blue-green alga: survival, photosynthesis and photoreactivation, *Plant Physiol.,* 43, 1689, 1968.

84. **Sasa, T.**, Effect of ultraviolet light upon various physiological activities of *Chlorella* cells at different stages in their life cycle, *Plant Cell Physiol.,* 2, 253, 1961.

85. **Bell, L., Merinova, N., and Merinova, G. L.**, The effect of dose, wave length of UV rays on *Chlorella* photosynthesis, *Biofizika,* (Transl.), 6, 176, 1961.

86. **Dodge, J. D.**, Effects of ultraviolet light on the survival and nuclear division of a dinoflagellate, *Protoplasma,* 59, 485, 1965.

87. **Rahn, R. O.**, Denaturation in ultraviolet-irradiated DNA, in *Photophysiology,* Vol. 8, Giese, A., Ed., Academic Press, New York, 1973, 231.

88. **Davies, D. R.**, Repair mechanisms and variations in UV sensitivity within the cell cycle, *Mutat. Res.,* 2, 477, 1965.

89. **Mantai, K. E., Wond, J., and Bishop, N. I.**, Comparison studies of the effects of ultraviolet irradiation on photosynthesis, *Biochim. Biophys. Acta,* 197, 257, 1970.

90. **Nalewajko, C.**, Photosynthesis and excretion in various planktonic algae, *Limnol. Oceanogr.,* 11, 1, 1966.

91. **Jenkin, P. M.**, Oxygen production by the diatom *Coscinodiscus excentricus* EHR. in relation to submarine illumination in the English Channel, *J. Mar. Biol. Assoc. U.K.,* 22, 301, 1937.

92. **Anderson, I. C. and Robertson, D. S.**, Role of carotenoids in protecting chlorophyll from photodestruction, *Plant Physiol.,* 35, 531, 1960.

93. **Krinsky, N. I.,** The protective function of carotenoid pigments, in *Photophysiology,* Vol. 3, Giese, A., Ed., Academic Press, New York, 1968, 123.

94. **Mathews-Roth, M. M., Wilson, T., Fjuimori, E., and Krinsky, N. I.,** Carotenoid chromophore length and protection against photosensitization, *Photochem. Photobiol.,* 19, 217, 1974.

95. **Godward, M. B. E.,** Invisible radiations, in *Physiology and Biochemistry of Algae,* Lewin, R. A., Academic Press, New York, 1962, 551.

96. **Srivastava, B. S.,** A simple method of isolating ultraviolet resistant and sensitive strains of blue-green algae, *Phycologia,* 9, 205, 1970.

97. **Bhattacharjee, S. K. and David, K. A. V.,** Unusual resistance to ultraviolet light in dark phase of blue-green bacterium *Anacystis nidulans, Nature (London),* 265, 183, 1977.

98. **Kumar, H. D.,** Effects of radiations on blue-green algae. I. The production and characterization of a strain of *Anacystis nidulans* resistant to ultraviolet radiation, *Ann. Bot. (London),* 27, 723, 1963.

99. **Harm, W., Rupert, C. S., and Harm, H.,** The study of photoenzymatic repair of UV lesions in DNA by flash photolysis, in *Photophysiology,* Vol. 6, Giese, A., Ed., Academic Press, New York, 1971, 279.

100. **Howard-Flanders, P.,** DNA repair, *Annu. Rev. Biochem.,* 37, 175, 1968.

101. **Williams, E., Lambert, J., O'Brien, P., and Houghton, J. A.,** Evidence for dark repair of far ultraviolet light damage in the blue-green alga, *Gloeocapsa alpicola., Photochem. Photobiol.,* 29, 543, 1979.

102. **Nasim, A. and James, A. P.,** Life under conditions of high irradiation, in *Microbial Life in Extreme Environments,* Kushner, D. J., Ed., Academic Press, New York, 1978.

103. **Steemann Nielsen, E. and Park, T. S.,** On the time course in adapting to low light intensities in marine phytoplankton, J. Cons. Cons. Int. Explor. Mer., 29, 19, 1964.

104. **Jones, R. I.,** Adaptations to fluctuating irradiance by natural phytoplankton communities, *Limnol. Oceanogr.,* 23, 920, 1978.

105. **Mur, L. R., Gons, H. J., and van Liere, L.,** Some experiments on the competition between green algae and blue-green bacteria in light-limited environments, *FEMS Microbiol. Lett.,* 1, 335, 1977.

106. **Nelson, D. M., D'Elia, C. F., and Guillard, R. R. L.,** Growth and competition of the marine diatoms *Phaeodactylum tricornutum* and *Thalassiosira pseudonana.* II. Light Limitation, *Mar. Biol.,* 50, 313, 1979.

107. **Jitts, H. R., McAllister, C. D., Stephans, K., and Strickland, J. D. H.,** The cell division rates of some marine phytoplankton as a function of light and temperature, *J. Fish. Res. Board Can.,* 21, 139, 1964.

108. **Cloern, J. E.,** Effects of light intensity and temperature on *Cryptomonas ovata* (Cryptophyceae) growth and nutrient uptake rates, *J. Phycol.,* 13, 389, 1977.

109. **Colijn, F. and van Buurt, G.,** Influence of light and temperature on the photosynthetic rate of marine benthic diatoms, *Mar. Biol.,* 31, 209, 1975.

110. **Halldal, P. and French, C. S.,** Algal growth in crossed gradients of light intensity and temperature, *Plant Physiol.,* 33, 249, 1958.

111. **Hobson, L. A.,** Effects of interactions of irradiance, daylength and temperature on division rates of three species of marine unicellular algae, *J. Fish. Res. Board Can.,* 31, 391, 1974.

112. **Ignatiades, L. and Smayda, T. J.,** Autecological studies on the marine diatom *Rhizosolenia fragilissima* Bergon. I. The influence of light temperature and salinity, *J. Phycol.,* 6, 332, 1970.

113. **McCombie, A. M.,** Actions and interactions of temperature, light intensity and nutrient concentration of the green alga, *Chlamydomonas reinhardi, J. Fish. Res. Board Can.,* 17, 871, 1960.

114. **McIntire, C. D.,** The distribution of estuarine diatoms along environmental gradients: a canonical correlation, *Estuarine Coastal Mar. Sci.,* 6, 447, 1978.

115. **Meeks, J. C. and Castenholz, R. W.,** Growth and photosynthesis in an extreme thermophile *Synechococcus lividus* (Cyanophyta), *Arch. Mikrobiol.,* 78, 25, 1971.

116. **Myers, J. and Kratz, W. A.,** Relations between pigment content and photosynthetic characteristic in a blue-green alga, *J. Gen. Physiol.,* 29, 11, 1955.

117. **Schone, H. K.,** Experimentelle untersuchungen zur okologie der marinen kieselalge *Thalassiosira rotula.* I. Temperatur und Licht, *Mar. Biol.,* 13, 284, 1972.

118. **Smayda, T. J.,** Experimental observations on the influence of temperature, light, and salinity on cell division of the marine diatom *Detonula confervacea* (Cleve) Gran, *J. Phycol.,* 5, 150, 1969.

119. **Sorokin, C. and Krauss, R. W.,** Effects of temperature and illuminance on *Chlorella* growth uncoupled from cell division, *Plant Physiol.,* 37, 37, 1962.

120. **Tamiya, H., Sasa, T., Nihei, T., and Ishibasi, S.,** Effect of variation of day-length and day and night temperatures and intensity of daylight upon the growth of *Chlorella, J. Gen. Appl. Microbiol.,* 4, 298, 1955.

121. **Thomas, W. H., Dodson, A. N., and Linden, C. A.,** Optimum light and temperature requirements for *Gymnodinium splendens, a larval fish food organism,* Fish. Bull., 71, 599, 1973.

122. Ukai, Y., Fucjita, Y., Morimura, Y., and Watanabe, A., Studies on growth of blue green alga *Tolypothrix tenuis*, *J. Gen. Appl. Microbiol.*, 4, 163, 1958.

123. Chimiklis, P. and Karlander, E., Light and calcium interactions in *Chlorella* inhibited by sodium chloride, *Plant Physiol.*, 51, 48, 1973.

124. Nakanishi, M. and Monsi, M., Effect of variation in salinity on photosynthesis of phytoplankton growing in estuaries, *J. Fac. Sci. Univ. Tokyo Sect. 3*, 9, 19, 1965.

125. Enami, I. and Fukuda, I., Mechanisms of the acido- and thermophily of *Cyanidium caldarium* Geitler. I. Effects of temperature, pH, and light intensity on the photosynthetic oxygen evolution of intact and treated cells, *Plant Cell Physiol.*, 16, 211, 1975.

126. Fock, H., Canvin, D. T., and Grant, B. R., Effects of oxygen and carbon dioxide on photosynthetic $O_2$ evolution and $CO_2$ uptake in sunflower and *Chlorella*, *Photosynthetica*, 5, 389, 1971.

127. Whittingham, C. P., Rate of photosynthesis, and concentration of carbon dioxide in *Chlorella*, *Nature (London)*, 170, 1017, 1952.

128. Briggs, G. E. and Whittingham, C. P., Factors affecting the rate of photosynthesis of *Chlorella* at low concentrations of carbon dioxide and in high illumination, *New Phytol.*, 51, 236, 1952.

129. Hannan, P. J. and Petouillet, C., Gas exchange with mass cultures of algae. I. Effects of light intensity and rate of carbon dioxide input on oxygen production, *Appl. Microbiol.*, 11, 446, 1963.

130. Ogata, E. and Matsui, T., Photosynthesis in several marine plants of Japan in relation to carbon dioxide supply, light and inhibitors, *Jpn. J. Bot.*, 19, 83, 1965.

131. Beardall, J. and Morris, I., Effects of environmental factors on photosynthesis patterns in *Phaeodactylum tricornutum* (Bacillariophyceae). II. Effect of oxygen, *J. Phycol.*, 11, 430, 1975.

132. Turner, J. S. and Brittain, N., Oxygen as a factor in photosynthesis, *Biol. Rev.*, 37, 130, 1962.

133. Maddux, W. S. and Jones, R. F., Some interactions of temperature, light intensity, and nutrient concentration during continuous culture of *Nitzschia closterium* and *Tetraselmis* sp., *Limnol. Oceanogr.*, 9, 79, 1964.

134. Castenholz, R. W., The effect of daylength and light intensity on the growth of littoral marine diatoms in culture, *Physiol. Plant.*, 17, 951, 1964.

135. Holt, M. G. and Smayda, T. J., The effect of daylength and light intensity on the growth rate of the marine diatom *Detonula confervacea* (Cleve), Gran. *J. Phycol.*, 10, 231, 1974.

136. Paasche, E., Marine plankton algae grown with light-dark cycles. I. *Coccolithus huxleyi*, *Physiol. Plant.*, 20, 946, 1967.

137. Paasche, E., Marine plankton algae grown with light-dark cycles. II. *Ditylum brightwellii* and *Nitzschia turgidula*, *Physiol. Plant.*, 21, 66, 1968.

138. Crossett, R. N., Drew, E. A., and Larkum, A. W. D., Chromatic adaptation in benthic marine algae, *Nature (London)*, 207, 547, 1965.

139. Dring, M. J., Chromatic adaptation in marine algae: an examination of its significance using a computer model, *Br. Phycol. J.*, 12, 118, 1977.

140. Qasim, S. Z., Bhattathiri, P. M. A., and Devassy, V. P., The effect of intensity and quality of illumination on the photosynthesis of some tropical marine phytoplankton, *Mar. Biol.*, 16, 22, 1972.

141. Doty, M. S. and Oguri, M., Evidence for a photosynthetic daily periodicity, *Limnol. Oceanogr.*, 2, 37, 1957.

142. Peterson, R. B., Friberg, E. E., and Burris, R. H., Diurnal variation in $N_2$ fixation and photosynthesis by aquatic blue-green algae, *Plant Physiol.*, 59, 74, 1977.

143. Prezelin, B. B. and Sweeney, B. M., Characterization of photosynthetic rhythms in marine dinoflagellates. II. Photosynthesis-irradiance curves and *in vivo* chlorophyll *a* fluorescence, *Plant Physiol.*, 60, 388, 1977.

144. Madigan, M. T. and Brock, T. D., Adaptation by hot spring phototrophs to reduced light intensities, *Arch. Microbiol.*, 113, 111, 1977.

145. Wassink, E. C., Katz, E., and Dorrestein, R., On photosynthesis and fluorescence of bacteriochlorophyll in Thiorhodaceae, *Enzymologia*, 10, 285, 1942.

146. Shiokawa, K., Takahashi, M., and Ichimura, S., Physiological adaptation of photosynthetic bacteria to low light and its ecological meaning, *Jpn. J. Limnol.*, 34, 1, 1973.

147. Aruga, Y., Ecological studies of photosynthesis and matter production of phytoplankton. II. Photosynthesis of algae in relation to light intensity and temperature, *Bot. Mag.*, 78, 360, 1965.

148. Lloyd, N. D., Canvin, D., and Culver, D., Photosynthesis and photorespiration in alge, *Plant. Physiol.*, 59, 936, 1977.

149. Kratz, W. A. and Myers, J., Photosynthesis and respiration of three blue-green algae, *Plant. Physiol.*, 30, 275, 1955.

150. Clendenning, K. A., Brown, T. E., and Eyster, H. C., Comparative studies of photosynthesis in *Nostoc muscorum* and *Chlorella pyrenoidosa*, *Can. J. Bot.*, 34, 943, 1956.

151. Baker, A. L., Brock, A. J., and Klemer, A. R., Some photosynthetic characteristics of a naturally occurring population of *Oscillatoria agardhii* Gomont, *Limnol. Oceanogr.*, 14, 327, 1969.

152. Eberly, W. R., Problems in the laboratory culture of planktonic blue green algae, in *Environmental Requirements of Blue Green Algae,* Proceedings of a Symposium, Corvallis, Oregon, 1966, 7.

153. Castenholz, R. W., Environmental requirements of therophilic blue-green algae, in *Environmental Requirements of Blue-Green Algae,* Proceedings of a Symposium, Corvallis, Oregon, 1966, 55.

154. McCarthy, J. and Carpenter, E. J., *Oscillatoria (Trichodesmium) thiebauth (cyanophyta)* in the central north Atlantic Ocean, *J. Phycol.,* 15, 75, 1979.

155. Weller, D., Doemel, W., and Brock, T. D., Requirement of low oxidation-reduction potential for photosynthesis in a blue-green alga (*Phormidium* sp.), *Arch. Microbiol.,* 104, 7, 1975.

156. Brock, T. D. and Brock, M. L., Effect of light intensity on photosynthesis by thermal algae adapted to natural and reduced sunlight, *Limnol. Oceanogr.,* 14, 334, 1969.

157. Sheridan, R. P. and Ulik, T., Adaptive photosynthesis responses to temperature extremes by the thermophilic cyanophyte *Synechococcus lividus, J. Phycol.,* 12, 255, 1976.

158. Doemel, W. N. and Brock, T. D., The physiological ecology of *Cyanidium caldarium, J. Gen. Microbiol.,* 67, 17, 1971.

159. Trench, R. K., Pool, R. R., Jr., Logan, M., and Engelland, A., Aspects of the relation between *Cyanophora paradoxa* (Korschikoff) and its endosymbiotic cyanelles *Cyanocyta Korschikoffiana* (Hall & Claus). I. Growth, ultrastructure, photosynthesis and the obligate nature of the association, *Proc. R. Soc. London Ser. B,* 202, 423, 1978.

160. Dunstan, W. M., A comparison of the photosynthesis-light intensity relationship in phylogenetically different marine microalgae, *J. Exp. Mar. Biol. Ecol.,* 13, 181, 1973.

161. Loeblich, A. R., III., A seawater medium for dinoflagellates and the nutrition of *Cachonina niei, J. Phycol.,* 11, 80, 1975.

162. Falkowski, P. G. and Owens, T. G., Effects of light intensity of photosynthesis and dark respiration of six species of marine phytoplankton, *Mar. Biol.,* 45, 289, 1978.

163. Halldal, P., Photosynthetic capacities and hypotosynthetic action spectra of endozoic algae of the massive coral *Favia, Biol. Bull.,* 134, 411, 1968.

164. Scott, B. D. and Jitts, H. R., Photosynthesis of phytoplankton and zooxanthellae on a coral reef, *Mar. Biol.,* 41, 307, 1977.

165. Kevin, K. M. and Hudson, R. C. L., The role of zooxanthellae in the hermatypic coral *Plesiastrea urvillei* from cold waters, *J. Exp. Mar. Biol. Ecol.,* 36, 157, 1979.

166. Fairchild, E. and Sheridan, R. P., A physiological investigation of the hot spring diatom, *Achnanthes exigua* Grun, *J. Phycol.,* 10, 1, 1974.

167. Admiraal, W., Influence of light and temperature on the growth rate of estuarine benthic diatoms in culture, *Mar. Biol.,* 39, 1, 1977.

168. Lewin, J. and Mackas, D., Blooms of surf-zone diatoms along the coast of the Olympic Peninsula, Washington. I. Physiological investigations of *Chaetoceros armatum* and *Asterionella socialis* in laboratory cultures, *Mar. Biol.,* 16, 171, 1972.

169. Talling, J. F., Comparative laboratory and field studies of photosynthesis by a marine planktonic diatom, *Limnol. Oceanogr.,* 5, 62, 1960.

170. Humphrey, G. F. and Subba Rao, D. V., Photosynthetic rate of the marine diatom *Cylindrotheca closterium, Aust. J. Mar. Freshwater Res.,* 18, 123, 1967.

171. Bunt, J. S., Some characteristics of microalgae isolated from Antarctic sea ice, *Antarct. Res. Ser.,* 2, 1, 1968.

172. Mann, J. E. and Myers, J., On pigments, growth, and photosynthesis of *Phaeodactylum tricornutum, J. Phycol.,* 4, 349, 1968.

173. Curl, H. and McLeod, G. C., The physiological ecology of a marine diatom, *Skeletomema costatum* (Grev.) Cleve, *J. Mar. Res.,* 19, 70, 1961.

174. Taylor, W. R., Light and photosynthesis in intertidal benthic diatoms, *Helgol. Wiss. Meeresunters.,* 10, 29, 1964.

175. Jeffrey, S. W. and Allen, M. B., Pigments, growth and photosynthesis in cultures of two chrysomonads, *Coccolithus huxleyi* and *Hymenomonas* sp., *J. Gen. Microbiol.,* 36, 277, 1964.

176. McAllister, C. D., Observations on the variation of planktonic photosynthesis with light intensity, using both the $O_2$ and $C^{14}$ methods, *Limnol. Oceanogr.,* 6, 483, 1961.

177. Cook, J. R., Adaptations in growth and division in *Euglena* effected by energy supply, *J. Protozool.,* 10, 436, 1963.

178. Mosser, J. L., Mosser, A. G., and Brock, T. D., Photosynthesis in the snow; the alga *Chlamydomonas nivalis* (Chlorophyceae), *J. Phycol.,* 13, 22, 1977.

179. Winokur, M., Photosynthesis relationship of *Chlorella* species, *Am. J. Bot.,* 35, 207, 1948.

180. Lewenstein, A. and Bachofen, R., $CO_2$-fixation and ATP synthesis in continuous cultures of *Chlorella fusca, Arch. Microbiol.,* 116, 169, 1978.

181. Senger, H. and Fleischhacker, P., Adaptation of the photosynthetic apparatus of *Scenedesmus obliquus* to strong and weak light conditions. I. Differences in pigments, photosynthetic capacity, quantum yield and dark reactions, *Physiol. Plant.,* 43, 35, 1978.

182. **Lynn, R. and Broack, T. D.,** Notes on the ecology of a species of *Zygogonium* (Kutz.) in Yellowstone National Park, *J. Phycol.,* 5, 181, 1969.

183. **Pierson, B. K. and Castenholz, R. W.,** Studies of pigments and growth in *Chloroflexus aurantiacus,* a phototrophic filamentous bacterium, *Arch. Microbiol.,* 100, 283, 1974.

184. **Balitskaya, R. M.,** Development of green sulfur bacteria *Chloropseudomonas ethylicum* at different light intensities, *Mikrobiologiya,* 31, 961, 1962.

185. **Göbel, F.,** Quantum efficiencies of growth, in *The Photosynthetic Bacteria,* Clayton, R. K. and Sistrom, W. R., Eds., Plenum Press, New York, 1978, 907.

186. **Sistrom, W. R.,** The kinetics of the synthesis of photopigments in *Rhodopseudomonas spheroides,* *J. Gen. Microbiol.,* 28, 607, 1962.

187. **Clayton, R. K.,** Studies in the phototaxis of *Rhodospirillum rubrum.* II. The relation between phototaxis and photosynthesis, *Arch. Mikrobiol.,* 19, 125, 1953.

188. **Vincent, W. F. and Silvester, W. B.,** Growth of blue-green algae in the manukau (New Zealand oxidation ponds.) I. Growth potential of oxidation pond water and comparative optima for blue-green and green algal growth, *Water Res.,* 13, 711, 1979.

189. **Ihlenfeldt, M. J. A. and Gibson, J.,** CO₂ fixation and its regulation in *Anacystis nidulans* (*Synechococcus*), *Arch. Microbiol.,* 102, 13, 1975.

190. **Tindall, D. R., Yopp, J. H., Miller, D. M., and Schmid, W. E.,** Physiochemical parameters governing the growth of *Aphanothece halophytica* (Chroococcales) in hypersaline media, *Phycologia,* 17, 179, 1978.

191. **Lazaroff, N. and Vishniac, W.,** The effect of light on the developmental cycle of *Nostoc muscorum,* a filamentous blue-green alga, *J. Gen. Microbiol.,* 25, 365, 1961.

192. **Beljanin, V. N. and Trenkensu, A. P.,** Growth and spectrophotometric characteristics of the blue green alga *Synechococcus elongatus* under different temperature and light conditions, *Arch. Hydrobiol.,* 18 (Suppl. 51), 46, 1977.

193. **Fukuda, I.,** Physiological studies on a thermophilic blue-green algae alga *Cyanidium caldarium* Geitler, *Bot. Mag.,* 71, 79, 1958.

194. **Hersey, R. L. and Swift, E.,** Nitrate reductase activity of *Amphidinium carteri* and *Cachonina niel* (Dinophyceae) in batch culture: diel periodicity and effects of light intensity and ammonia, *J. Phycol.,* 12, 36, 1976.

195. **Nordi, E.,** Experimental studies on the ecology of ceratia, *Oikos,* 8, 201, 1957.

196. **Barker, H. A.,** The culture and physiology of the marine dinoflagellates, *Arch. Mikrobiol.,* 6, 157, 1935.

197. **Swift, E. and Meunier, V.,** Effects of light intensity on division rate, stimulable bioluminescence and cell size of the oceanic dinoflagellates *Dissodinium lunula,* *Pyrocystis fusiformis* and *P. noctiluca,* *J. Phycol.,* 12, 14, 1976.

198. **Thomas, W. H.,** Effects of temperature and illuminance on cell division rates of three species of tropical oceanic phytoplankton, *J. Phycol.,* 2, 17, 1966.

199. **Kain, J. M. and Fogg, G. E.,** Studies on the growth of marine phytoplankton. III. *Prorocentrum micans* Ehrenberg, *J. Mar. Biol. Assoc. U.K.,* 39, 33, 1960.

200. **Kain, J. M. and Fogg, G. E.,** Studies on the growth of marine phytoplankton. I. *Aesterionella japonica,* *J. Mar. Biol. Assoc. U.K.,* 37, 397, 1958.

201. **White, A. W.,** Growth of two facultatively heterotrophic marine centric diatoms, *J. Phycol.,* 10, 292, 1974.

202. **Findlay, I. W. O.,** Effects of external factors and cell size on the cell division rate of a marine diatom, *Coscinodiscus pavillardii* Forti, *Int. Rev. Ges. Hydrobiol.,* 57, 523, 1972.

203. **Stanbury, F. A.,** The effect of light of different intensities, reduced selectively and non-selectively, upon the rate of growth of *Nitzschia closterium,* *J. Mar. Biol. Assoc. U.K.,* 17, 633, 1931.

204. **Quraishi, F. D. and Spencer, C. P.,** Studies on the growth of some marine unicellular algae under different artificial light sources, *Mar. Biol.,* 8, 60, 1971.

205. **Ferguson, R. L., Collier, A., and Meeter, D. A.,** Growth response to *Thalassiosira pseudonana* Hasle and Heimdal clone 3H to illumination, temperature and nitrogen soruce, *Chesapeake Sci.,* 17, 148, 1976.

206. **Mjaaland, G.,** Some laboratory experiments on the coccolithophorid *Coccolithus huxleyi* (Lohm) Kamptner, *Oikos,* 7, 251, 1956.

207. **Kain, J. M. and Fogg, G. E.,** Studies on the growth of marine phytoplankton. II. *Isochrysis galbana,* *J. Mar. Biol. Assoc. U.K.,* 37, 781, 1958.

208. **Matthern, R. O., Kostick, J. A., and Okada, I.,** Effect of total illumination upon continuous *Chlorella* production in a high intensity light system, *Biotechnol. Bioeng.,* 11, 863, 1969.

209. **Mineeva, L. A.,** The effect of light intensity upon autotrophic and heterotrophic nutrition of *Chlorella vulgaris* and *Scenedesmus obliquus,* *Mikrobiologiya,* 31, 411, 1962.

210. **Simpson, P. D.,** The growth rate of *Mychonastes ruminatus* Simpson et Van Valkenburg under various light, temperature and salinity regimes, *Br. Phycol. J.,* 13, 291, 1978.

211. Jewson, D. H. and Wood, R. B., Some effects on integral photosynthesis of artificial circulation of phytoplankton through light gradients, *Verh. Int. Verein. Limnol.*, 19, 1037, 1975.
212. Quraishi, F. O. and Spencer, C. P., Studies on the responses of marine phytoplankton to light fields of varying intensity, in *4th European Marine Biology Symposium*, Cambridge University Press, Cambridge, 1969, 393.
213. Marra, J., Phytoplankton photosynthetic response to vertical movement in a mixed layer, *Mar. Biol.*, 46, 203, 1978.
214. Platt, T. and Jassby, A. D., The relationship between photosynthesis and light for natural assemblages of coastal marine phytoplankton, *J. Phycol.*, 12, 421, 1976.
215. Bunt, J. S., Owens, O. Van H., and Hoch, G., Exploratory studies on the physiology and ecology of a psychrophilic marine diatom, *J. Phycol.*, 2, 96, 1966.
216. Dellow, V. and Cassie, R. M., Littoral zonation in two caves in the Aukland district, *Trans. R. Soc. N.Z.*, 83, 321, 1955.
217. Norton, T. A., Ebling, F. J., and Kitching, J. A., Light and the distribution of organisms in a sea cave, in *4th European Marine Biology Symposium*, Cambridge University Press, Cambridge, 1969, 409.
218. Foster, M. S., Regulation of algal community development in a *Macrocystis pyrifera* forest, *Mar. Biol.*, 32, 331, 1975.

# LIGHT INTENSITY PREFERENCE AND TOLERANCE OF AQUATIC PHOTOSYNTHETIC MACROALGAE

Edward J. Phlips and Akira Mitsui

Light intensity is one of the most important factors governing the growth and productivity of macrophyte communities.[1-9] Macroalgae have been found to flourish in light environments ranging from dimly lit caves to intertidal zones, where the light intensity exceeds 100 klx (klux = 1000 klux), an indication of considerable adaptibility. Tables 1 and 2 present the relationships between photosynthesis, growth, and light intensity of a variety of macroalgal species. Most of the studies described deal with the measurement of photosynthetic oxygen evolution under controlled conditions, using algal samples collected in the field. Based on this work and a limited number of $CO_2$ uptake experiments, the light intensity preference and tolerance of most macrophytes appear to depend on conditions of growth as well as inherent genetic flexibility of the species in question. Organisms from low light environments (e.g., deep water or shade) normally exhibit low compensation light intensities (i.e., intensity where photosynthesis equals respiration), reduced saturation light intensities (i.e., intensity at which the rate of photosynthesis reaches saturation), and in some cases, susceptibility to photoinhibition at high light intensities. The adaptations associated with low light environments result in more efficient use of available light. This is particularly pronounced in extreme habitats such as caves. For example, photosynthesis in the red alga *Plumaria elegans*, collected from caves, reaches saturation at 2.3 klx and has a compensation light intensity of 0.05 klx.[22] These values are characteristic of shade adaptation, similar to that observed in microorganisms and terrestrial plants.[27,49] In an analogous light-limiting situation, some species of macroalgae such as the brown alga *Laminaria hyperborea* are able to grow actively during the winter in high latitudes.[7,34] Since light intensity during this time is believed to be too low to sustain such growth, it has been suggested that organic reserves built up during the summer serve as the basis for growth.[7]

At the other end of the light intensity scale, high light-tolerant or sun-adapted macrophyte species are usually found in summer intertidal habitats and shallow water regions of tropical latitudes. Photosynthesis in such organisms usually saturates at light intensities one tenth to one fifth of the maximum incident levels, yet photoinhibition is not observed at high light intensities, i.e., greater than 80 klx. For example, photosynthesis in the brown alga *Fucus serratus* saturates at 14 klx, and no inhibition is observed at 151 klx.[12] This pattern of adaptation is at least partially related to two important environmental considerations:

1.  Even when incident light intensities are high, self-shading, often experienced in benthic macrophyte communities, can reduce the level of penetrating radiation, thereby favoring a reduction in light intensity required for saturation of photosynthesis.
2.  At the same time, portions of intertidal algal populations are periodically subject to direct exposures to full sunlight, which during the summer can exceed 100 klx in intensity. This mandates the development of some resistence to photoinhibition and photodestruction.

From a phylogenetic viewpoint, several early investigators proposed that red algae dominate the dimly lit zones of deep water habitats and caves because of their light intensity and quality preferences, while green and brown algae dominate shallow and

## Table 1
## RELATIONSHIP BETWEEN LIGHT INTENSITY AND PHOTOSYNTHESIS IN MACROALGAE[a]

| Species (references) | Environment[b] | Conditions at sampling site[c] | Exp. T (°C)[d] | Type of light[e] | Assay method | Light intensity at Psat[f] (klx) | Pmax[g] | Light intensity at "Psat" one half (klx) | Compensation light intensity[j] (klx) | $I_c$[k] (klx) | Light intensity for initial photoinhibition[k] (klx) | Maximum light intensity tested (klx) | Inhibition[l] (%) | Case #[m] |
|---|---|---|---|---|---|---|---|---|---|---|---|---|---|---|
| **Rhodophyta — red algae** | | | | | | | | | | | | | | |
| Beckerella subcostatum(10) | Japan, marine | 12-m depth | 20 | T | $O_2$ | 5.0 | 1.3 μmol $O_2$/cm²/hr | 2.0 | — | 3.5 | None—40 | — | — | 1 |
| Calcareous reef-forming alga (11) | Marshall Islands | — | — | — | $O_2$ | 10.8 | 2.5 μmol $O_2$/cm²/hr | 3.2 | 0.6 | 10.8 | None—108 | — | — | 2 |
| Ceramium strictum (12) | West Baltic Sea, sublittoral | 15°C, autumn | 15 | T | $O_2$ | 20 mW/cm² (46.6) | 1000 μmol $O_2$/g dry wt/hr | 4 mW/cm² (9.3) | 0.85 mW/cm² (2.0) | 9 mW/cm² (20.9) | None—20 mW/cm² (46.6) | — | — | 3 |
| Ceramium tenuicorne (13) | — | — | 10.8 | F | $CO_2$ | 120 μE/m²/sec (6.3) | 266 μmol C/g dry wt/hr | 40 μE/m²/sec (2.1) | — | 80 μE/m²/sec (4.2) | None—700 μE/m²/sec (36.8) | — | — | 4 |
| C. tenuicorne(13) | Baltic Sea, sublittoral | — | 14.5 | F | $CO_2$ | 180 μE/m²/sec (9.5) | 667 μmol C/g dry wt/hr | 80 μE/m²/sec (4.2) | — | 150 μE/m²/sec (7.9) | None—600 μE/m²/sec (31.6) | — | — | 5 |
| | | — | 5.6 | F | $CO_2$ | 130 μE/m²/sec (6.8) | 167 μmol C/g dry wt/hr | 30 μE/m²/sec (1.6) | — | 80 μE/m²/sec (4.2) | 200 μE/m²/sec (10.5) | 350 μE/m²/sec (18.4) | 10 | 6 |
| Chondrococcus hornemannii(10) | Japan, marine | 5-m depth | 20 | T | $O_2$ | 5.0 | 0.9 μmol $O_2$/cm²/hr | 2.0 | — | 3.0 | None—40 | — | — | 7 |
| Chondrus crispus(4) | Marine | — | 2 | T,F | $O_2$ | 10.8 | 201 μmol $O_2$/g dry wt/hr | 4.3 | 0.27 | 9.7 | None—54 | — | — | 8 |
| | | — | 20 | T,F | $O_2$ | 16.2 | 446 μmol $O_2$/g dry wt/hr | 5.4 | — | 10.3 | None—54 | — | — | 9 |
| C. crispus(14) | New Hampsire, marine, lower littoral | — | 15 | T | $O_2$ | 10.8 | 147 μmol $O_2$/g dry wt/hr | 2.0 | — | 4.3 | 10.8 | 54 | 20 | 10 |
| | | — | 5 | T | $O_2$ | 10.8 | 94 μmol $O_2$/g dry wt/hr | 3.3 | — | 6.5 | 10.8 | 54 | 50 | 11 |
| C. crispus(15) | Northwest Atlantic intertidal | 0-m depth, October | 10 | T | $O_2$ | 7.9 | 136 μmol $O_2$/g dry wt/hr | 1.7 | 0.3 | 2.4 | 8.0 | 29.9 | 37 | 12 |
| | Northwest Atlantic, subtidal | 12-m depth, October | 10 | T | $O_2$ | 7.9 | 102 μmol $O_2$/g dry wt/hr | 1.3 | 0.2 | 1.9 | 8.0 | 29.9 | 47 | 13 |
| | Northwest Atlantic, subtidal | 12-m depth, February | 5 | T | $O_2$ | 4.9 | 131 μmol $O_2$/g dry wt/hr | 1.3 | — | 3.1 | 5.0 | 29.9 | 76 | 14 |
| | Northwest Atlantic, subtidal | 12-m depth, February | 15 | T | $O_2$ | 7.9 | 155 μmol $O_2$/g dry wt/hr | 1.4 | — | 4.2 | 8.0 | 29.9 | 45 | 15 |

| Species | Habitat | Condition | | T | | $P_{max}$ | | | | | | | | |
|---|---|---|---|---|---|---|---|---|---|---|---|---|---|---|
| *C. crispus*(16) | Marine | 10-m depth | 26 | T | O₂ | 3.0 mW/cm² (7.0) | 150 μmol O₂/μmol Chl/hr | 0.8 mW/cm² (1.9) | — | 1.2 mW/cm² (2.8) | 5.0 mW/cm² (11.6) | 6.3 mW/cm² (14.7) | 23 | 16 |
| | Marine | 0.5-m depth | 26 | T | O₂ | 3.0 mW/cm² (7.0) | 145 μmol O₂/μmol Chl/hr | 0.7 mW/cm² (1.6) | — | 1.0 mW/cm² (2.3) | 5.0 mW/cm² (11.6) | 6.3 mW/cm² (14.7) | 16 | 17 |
| *Chondrus verrucosa* (10) | Japan, marine, intertidal | Lower, intertidal | 20 | T | O₂ | 40 kerg/cm² sec (9.5) | — | 12 Kerg/cm² sec (2.9) | — | 21 kerg/cm²/sec (5.0) | None—60 kerg/cm² sec (14.3) | — | — | 18 |
| *Delesseria decipiens* (17) | Marine, shaded rock faces | — | — | T | O₂ | 3.5 kerg/cm²/sec (0.8) | — | 1.0 kerg/cm²/sec (0.2) | — | 2.0 kerg/cm²/sec (0.4) | None—10.0 kerg/cm² sec (2.3) | — | — | 19 |
| *Delesseria sanguinea* (12) | West Baltic Sea, sublittoral | 10°C, spring | 10 | T | O₂ | 6.0 mW/cm² (14.0) | 563 μmol O₂/g dry wt/hr | 1.0 mW/cm² (2.3) | 0.15 mW/cm² (0.35) | 1.7 mW/cm² (4.0) | None—65 mW/cm² (151) | — | — | 20 |
| | West Baltic Sea, sublittoral | 20°C, summer | 20 | T | O₂ | 5.0 mW/cm² (11.6) | 250 μmol O₂/g dry wt/hr | 1.1 mW/cm² (2.6) | 0.30 mW/cm² (0.70) | 2.0 mW/cm² (4.7) | None—10 mW/cm² (23.3) | — | — | 21 |
| | West Baltic Sea, sublittoral | 15°C, autumn | 15 | T | O₂ | 5.0 mW/cm² (11.6) | 186 μmol O₂/g dry wt/hr | 1.2 mW/cm² (2.8) | 0.35 mW/cm² (0.81) | 2.5 mW/cm² (5.8) | None—20 mW/cm² (46.6) | — | — | 22 |
| | West Baltic Sea, sublittoral | 5°C, winter | 5 | T | O₂ | 6.0 mW/cm² (14.0) | 156 μmol O₂/g dry wt/hr | 1.9 mW/cm² (4.4) | 0.13 mW/cm² (0.30) | 3.2 mW/cm² (7.4) | None—25 mW/cm² (58.1) | — | — | 23 |
| *Dumontia incrassata* (12) | West Baltic Sea, Eulittoral | 10°C, spring | 10 | T | O₂ | 12.0 mW/cm² (28.0) | 1063 μmol O₂/g dry wt/hr | 2.0 mW/cm² (4.7) | 0.25 mW/cm² (0.58) | 5.1 mW/cm² (11.9) | None—65 mW/cm² (151) | — | — | 24 |
| | | 5°C, winter | 5 | T | O₂ | 3.0 mW/cm² (7.0) | 625 μmol O₂/g dry wt/hr | 1.1 mW/cm² (2.6) | 0.15 mW/cm² (0.35) | 4.2 mW/cm² (9.8) | None—65 mW/cm² (151) | — | — | 25 |
| *Eucheuma acanthocladum* (18) | Florida Keys, marine | — | 20 | T | O₂ | 8.5 | 1.3 μmol O₂/g dry wt/hr | 3.2 | — | 6.4 | 8.5 | 13.8 | 44 | 26 |
| *Eucheuma gelidium* (18) | Florida Keys, marine | — | 18 | T | O₂ | 6.4 | 0.2 μmol O₂/g dry wt/hr | 0.7 | — | 1.4 | 6.5 | 13.8 | 27 | 27 |
| | Florida Keys, marine | — | 28 | T | O₂ | 3.8 | 0.3 μmol O₂/g dry wt/hr | 0.3 | — | 0.5 | 4.0 | 13.8 | 35 | 28 |
| *Eucheuma isiforme* (18) | Florida Keys, marine | — | 18 | T | O₂ | 6.4 | 0.2 μmol O₂/g dry wt/hr | 0.5 | — | 0.6 | 6.5 | 13.8 | 35 | 29 |
| | | — | 28 | T | O₂ | 9.0 | 0.3 μmol O₂/g dry wt/hr | 2.6 | — | 2.6 | 9.0 | 13.8 | 65 | 30 |
| *Eucheuma nudum* (18) | Florida Keys, marine | — | 18 | T | O₂ | 3.2 | 0.8 μmol O₂/g dry wt/hr | 0.1 | — | 0.3 | 3.2 | 13.8 | 90 | 31 |
| | | — | 28 | T | O₂ | 6.4 | 0.9 μmol O₂/g dry wt/hr | 3.2 | — | 1.2 | 6.5 | 13.8 | 50 | 32 |
| *Furcellaria fastigiata* (12) | West Baltic Sea, sublittoral | 10°C, spring | 10 | T | O₂ | 6.0 mW/cm² (14.0) | 203 μmol O₂/g dry wt/hr | 1.2 mW/cm² (2.8) | 0.13 mW/cm² (0.30) | 2.2 mW/cm² (5.1) | None—25 mW/cm² (58.1) | — | — | 33 |

## Table 1 (continued)
## RELATIONSHIP BETWEEN LIGHT INTENSITY AND PHOTOSYNTHESIS IN MACROALGAE[a]

| Species (references)[a] | Environment[a] | Conditions at sampling site[c] | Exp. T (°C)[d] | Type of light[e] | Assay method | Light intensity at Psat[f] (klx) | Pmax[g] | Light intensity at [h]Psat[h] one half (klx) | Compensation light intensity[i] (klx) | I_x[j] (klx) | Light intensity for initial photoinhibition[k] (klx) | Maximum light intensity tested (klx) | Inhibition (%) | Case #[m] |
|---|---|---|---|---|---|---|---|---|---|---|---|---|---|---|
| | West Baltic Sea, sublittoral | 15°C, autumn | 15 | T | $O_2$ | 5.5 mW/cm² (13.0) | 125 μmol $O_2$/g dry wt/hr | 1.5 mW/cm² (3.5) | 0.38 mW/cm² (0.88) | 4.2 mW/cm² (9.8) | None—20 mW/cm² (46.6) | — | — | 34 |
| | West Baltic Sea, sublittoral | 5°C, winter | 5 | T | $O_2$ | 6.0 mW/cm² (14.0) | 156 μmol $O_2$/g dry wt/hr | 1.2 mW/cm² (2.8) | 0.16 mW/cm² (0.37) | 2.1 mW/cm² (4.9) | None—25 mW/cm² (58.1) | — | — | 35 |
| F. fastigiata(13) | Baltic Sea, sublittoral | — | 14.5 | F | $CO_2$ | 180 μE/m²/sec (9.5) | 83 μmol C/g dry wt/hr | 95 μE/m²/sec (5.0) | — | 120 μE/m²/sec (6.3) | None—600 μE/m²/sec (31.6) | — | — | 36 |
| | — | — | 4.3 | F | $CO_2$ | 60 μE/m²/sec (3.2) | 42 μmol C/g dry wt/hr | 20 μE/m²/sec (1.1) | — | 50 μE/m²/sec (2.6) | None—650 μE/m²/sec (34.2) | — | — | 37 |
| Gelidium amansii (10) | Japan, marine | 1-m depth | 20 | T | $O_2$ | 15 | 3.6 μmol $O_2$/g dry wt/hr | 5 | — | 10 | None—40 | — | — | 38 |
| | — | 10-m depth | 20 | T | $O_2$ | 8 | 2.7 μmol $O_2$/g dry wt/hr | 2.5 | — | 4 | None—40 | — | — | 39 |
| G. amansii(19) | Japan, marine | — | 30 | T | $O_2$ | 15 | 491 μmol $O_2$/g dry wt/hr | 4 | 0.58 | 6 | 15 | 20 | 5 | 40 |
| Gigartina mamillosa (10) | Japan, marine intertidal | — | 20 | T | $O_2$ | 10 | 1.8 μmol $O_2$/cm²/hr | 4 | — | 7 | None—40 | — | — | 41 |
| Gigartina stellata(14) | Lower littoral | — | 15 | T | $O_2$ | 21.6 | 147 μmol $O_2$/g dry wt/hr | 6.5 | — | 10.8 | None—54 | — | — | 42 |
| | — | — | 5 | T | $O_2$ | 7.5 | 59 μmol $O_2$/g dry wt/hr | 1.1 | — | 2.2 | — | — | — | 43 |
| Gracilaria textorii(10) | Japan, marine | 1-m depth | 20 | T | $O_2$ | 12 | 1.6 μmol $O_2$/cm²/hr | 3 | — | 6 | None—40 | — | — | 44 |
| Graeloupia elliptica (10) | Japan, marine | 1-m depth | 20 | T | $O_2$ | 10 | 1.9 μmol $O_2$/cm²/hr | 2 | — | 4 | None—40 | — | — | 45 |
| Halosaccion glandiforme(20) | Artic, (Capes Eblin and Corwin) | — | 0 | T | $O_2$ | — | — | — | 0.001 Ly/min (0.18) | — | — | — | — | 46 |
| | Subice | — | 25 | T | $O_2$ | — | 446 μmol $O_2$/g dry wt/hr | — | — | — | — | — | — | 47 |
| Lithophyllum interme-dium(21) | Curacao, reef | — | 25 | T | $O_2$ | 6 | 1.6 μmol $O_2$/cm²/hr | 4 | 0.9 | 6 | None—80 | — | — | 48 |
| Meristotheca papulosa (10) | Japan, marine | 10-m depth | 20 | T | $O_2$ | 3 | 0.9 μmol $O_2$/cm²/hr | 1.5 | — | 2 | None—40 | — | — | 49 |

| Organism | Location | Temperature | S | F, S | Gas | 6 | | 4 | 1.5 | 6 | | | | Ref |
|---|---|---|---|---|---|---|---|---|---|---|---|---|---|---|
| Neogoniolithon solubile(21) | — | — | — | | | | 1.9 µmol O₂/cm²/hr | | | | None—80 | — | — | 50 |
| Phycorys rubens(12) | Curacao, reef | 10°C, spring | 10 | T | O₂ | 3.0 mW/cm² (7.0) | 313 µmol O₂/g dry wt/hr | 0.9 mW/cm² (2.1) | 0.15 mW/cm² (0.35) | 1.7 mW/cm² (4.0) | None—25 mW/cm² (58) | — | — | 51 |
| | West Baltic Sea, sublittoral | 20°C, summer | 20 | T | O₂ | 3.0 mW/cm² (7.0) | 219 µmol O₂/g dry wt/hr | 1.2 mW/cm² (2.8) | 0.39 mW/cm² (0.91) | 2.2 mW/cm² (5.1) | None—10 mW/cm² (23.3) | — | — | 52 |
| | West Baltic Sea, sublittoral | 15°C, autumn | 15 | T | O₂ | 5.5 mWcm² (12.8) | 256 µmol O₂/g dry wt/hr | 1.5 mW/cm² (3.5) | 0.35 mW/cm² (0.81) | 2.6 mW/cm² (6.1) | None—20 mW/cm² (46.6) | — | — | 53 |
| | West Baltic Sea, sublittoral | 5°C, winter | 5 | T | O₂ | 2.0 mW/cm² (4.7) | 125 µmol O₂/g dry wt/hr | 0.8 mW/cm² (1.9) | 0.18 mW/cm² (0.42) | 1.8 mW/cm² (4.2) | None—6 mW/cm² (14.0) | — | — | 54 |
| Phyllophora brodiaei (12) | West Baltic Sea, sublittoral | 10°C, spring | 10 | T | O₂ | 6.0 mW/cm² (14.0) | 113 µmol O₂/g dry wt/hr | 1.0 mW/cm² (2.3) | 0.15 mW/cm² (0.35) | 1.8 mW/cm² (4.2) | None—65 mW/cm² (151) | — | — | 55 |
| | West Baltic Sea, sublittoral | 15°C, autumn | 15 | T | O₂ | 6.0 mW/cm² (14.0) | 69 µmol O₂/g dry wt/hr | 1.7 mW/cm² (4.0) | 0.40 mW/cm² (0.93) | 3.5 mW/cm² (8.1) | None—20 mW/cm² (46.6) | — | — | 56 |
| | West Baltic Sea, sublittoral | 5°C, winter | 5 | T | O₂ | 6.0 mW/cm² (14.0) | 125 µmol O₂/g dry wt/hr | 0.65 mW/cm² (1.5) | 0.13 mW/cm² (0.30) | 1.3 mW/cm² (3.0) | None—65 mW/cm² (151) | — | — | 57 |
| Phyllophora truncata (13) | Baltic Sea, sublittoral | — | 5.6 | F | CO₂ | 70 µE/m²/sec (3.7) | 50 µmol C/g dry wt/hr | 30 µE/m²/sec (1.6) | — | 40 µE/m²/sec (2.1) | 80 µE/m²/sec (4.2) | 350 µE/m²/sec (18.4) | 60 | 58 |
| | Baltic Sea, sublittoral | — | 3.8 | F | CO₂ | 50 µE/m²/sec (2.7) | 42 µmol C/g dry wt/hr | 20 µE/m²/sec (1.1) | — | 30 µE/m²/sec (1.6) | None—650 µE/m²/sec (34.2) | — | — | 59 |
| Plocamium telfairiae (10) | Japan, marine | 5-m, depth | 20 | T | O₂ | 3.0 | 0.5 µmol O₂/cm²/hr | 1.5 | — | 2.5 | None—40 | — | — | 60 |
| Plumaria elegans(22) | Church Reef, Wembury, shaded areas | 75 ergs/mm²/sec (2.3) | 16 | F | O₂ | 7.5 kerg/cm²/sec (2.3) | 298 µmol O₂/g dry wt/hr | 1.2 kerg/cm²/sec (0.38) | 0.16 kerg/cm²/sec (0.05) | 2.4 kerg/cm²/sec (0.75) | None—10.0 kerg/cm²/sec (3.13) | — | — | 61 |
| Polysiphonia elongata (23) | Northwest Atlantic, estuarine | — | 10 | T | O₂ | 1.9 | 321 µmol O₂/g dry wt/hr | 0.9 | — | 1.9 | 19.1 | 38.2 | 50 | 62 |
| Polysiphonia lanosa (23) | Northwest Atlantic, estuarine | — | 10 | T | O₂ | 8.5 | 509 µmol O₂/g dry wt/hr | 2.2 | — | 3.2 | 10.6 | 38.2 | 5 | 63 |
| Polysiphonia nigrescens(23) | Northwest Atlantic, estuarine | — | 10 | T | O₂ | 10.6 | 67 µmol O₂/g dry wt/hr | 5.3 | 1.8 | 9.5 | None—38.2 | — | — | 64 |
| P. nigrescens(12) | West Baltic Sea, sublittoral | 10°C, spring | 10 | T | O₂ | 6.0 mW/cm² (14.0) | 281 µmol O₂/g dry wt/hr | 1.5 mW/cm² (3.5) | 0.20 mW/cm² (0.47) | 2.8 mW/cm² (6.5) | None—65 mW/cm² (151) | — | — | 65 |
| | West Baltic Sea, sublittoral | 15°C, autumn | 15 | T | O₂ | 6.0 mW/cm² (14.0) | 344 µmol O₂/g dry wt/hr | 2.8 mW/cm² (6.5) | 0.70 mW/cm² (1.60) | 5.8 mW/cm² (13.5) | None—20 mW/cm² (46.6) | — | — | 66 |

Table 1 (continued)
## RELATIONSHIP BETWEEN LIGHT INTENSITY AND PHOTOSYNTHESIS IN MACROALGAE[a]

| Species (references) | Environment[b] | Conditions at sampling site[c] | Exp. T (°C)[d] | Type of light[e] | Assay method | Light intensity at Psat (klx) | Pmax[f] | Light intensity at [1/2]Psat[g] one half (klx) | Compensation light intensity[i] (klx) | $I_k$ (klx) | Light intensity for initial photoinhibition[k] (klx) | Maximum light intensity tested (klx) | Inhibition (%) | Case #[m] |
|---|---|---|---|---|---|---|---|---|---|---|---|---|---|---|
| | West Baltic Sea, sublittoral | 5°C, winter | 5 | T | $O_2$ | 3.0 mW/cm² (7.0) | 156 μmol $O_2$/g dry wt/hr | 0.9 mW/cm² (2.1) | 0.15 mW/cm² (0.35) | 1.7 mW/cm² (4.0) | None—65 mW/cm² (151) | — | — | 67 |
| Polysiphonia subtilissima(23) | Northwest Atlantic, estuarine | — | 15 | T | $O_2$ | 10.6 | 375 μmol $O_2$/g dry wt/hr | 3.2 | — | 3.3 | None—38.2 | — | — | 68 |
| Porolithon pachydermum(21) | Curacao, reef | — | — | T | $O_2$ | 6.0 | 113 μmol $O_2$/g dry wt/hr | 5.0 | 2.5 | 6.0 | None—80 | — | — | 69 |
| Porphyra leucosticta (12) | West Baltic Sea, eulittoral | 10°C, spring | 10 | T | $O_2$ | 6.0 mW/cm² (14.0) | 1688 μmol $O_2$/g dry wt/hr | 1.1 mW/cm² (2.6) | 0.14 mW/cm² (0.32) | 3.3 mW/cm² (7.7) | None—12 mW/cm² (27.9) | — | — | 70 |
| | West Baltic Sea, eulittoral | 5°C, winter | 5 | T | $O_2$ | 6.0 mW/cm² (14.0) | 2000 μmol $O_2$/g dry wt/hr | 1.1 mW/cm² (2.6) | 0.12 mW/cm² (0.28) | 2.1 mW/cm² (4.9) | None—65 mW/cm² (151) | — | — | 71 |
| Porphyra tenera(24) | Japan, marine | — | 15 | T | $O_2$ | 10 | 1116 μmol $O_2$/g dry wt/hr | 2 | 0.20 | 4 | None—20 | — | — | 72 |
| | — | — | 20 | T | $O_2$ | 15 | 1786 μmol $O_2$/g dry wt/hr | 3 | 0.26 | 6 | None—20 | — | — | 73 |
| | — | — | 25 | T | $O_2$ | 15 | 2098 μmol $O_2$/g dry wt/hr | 4 | 0.35 | 7 | None—20 | — | — | 74 |
| Porphyra umbilicalis (16) | Marine | 10-m depth | 26 | T | $O_2$ | 6.3 mW/cm² (14.7) | 200 μmol $O_2$/μmol Chl/hr | 1.6 mW/cm² (3.8) | — | 3.0 mW/cm² (7.0) | None—14.0 mW/cm² (32.6) | — | — | 75 |
| | Marine | 0.5-m, depth | 26 | T | $CO_2$ | 6.3 mW/cm² (14.7) | 200 μmol $O_2$/μmol Chl/hr | 1.4 mW/cm² (3.2) | — | 2.5 mW/cm² (5.8) | None—14.0 mW/cm² (32.6) | — | — | 76 |
| Rhodomela confervoides (13) | Baltic Sea, sublittoral | — | 10.3 | F | $CO_2$ | 70 μE/m²/sec (3.7) | 100 μmol C/g dry wt/hr | 25 μE/m²/sec (1.3) | — | 40 μE/m²/sec (2.1) | 200 μE/m²/sec (10.5) | 800 μE/m²/sec (42.2) | 10 | 77 |
| | — | — | 3.7 | F | $CO_2$ | 100 μE/m²/sec (5.3) | 117 μmol C/g dry wt/hr | 10 μE/m²/sec (0.5) | — | 35 μE/m²/sec (1.8) | None—500 μE/m²/sec (26.3) | — | — | 78 |
| **Phaeophyta — brown algae** | | | | | | | | | | | | | | |
| Ascophyllum nodosum (25) | Long Island, salt marsh | Summer | — | 5 | $CO_2$ | 0.3 cal/cm²/min (23.1) | 125 μmol C/g dry wt/hr | 0.12 cal/cm²/min (9.2) | — | | None—53.9 | — | — | 1 |
| Ascophyllum nodosum (26) | Northwest Atlantic, marine | Winter | 1.5 | T | $O_2$ | 8.5 | 80 μmol $O_2$/g dry wt/hr | 2.1 | — | 3.2 | None—53.0 | — | — | 2 |
| | | Summer | 15 | T | $O_2$ | 5.3 | 121 μmol $O_2$/g dry wt/hr | 3.2 | — | 4.1 | 18.0 | 53.0 | 40 | 3 |

| | | | | | | | | | | | | | | |
|---|---|---|---|---|---|---|---|---|---|---|---|---|---|---|
| Carpophyllum plumosum(27) | New Zealand, intertidal | — | 20 | T | $O_2$ | 15.9 | 2.1 | — | 5.3 | — | 15.9 | 29.7 | 15 | 4 |
| Carpophyllum plumosum(sporeling)(27) | New Zealand, intertidal | — | 20 | T | $O_2$ | 15.9 | 9.5 | 2.12 | 15.9 | — | 15.9 | 29.7 | 100 | 5 |
| Chordaria flagelliformis(28) | Greenland, upper sublittoral | 5°C | 5 | T | $O_2$ | >70 | 49 μmol $O_2$/mgChl/hr | 7 | 1 | 8 | None—70 | — | — | 6 |
| Dictyosiphon foeniculaceus(13) | Baltic Sea, littoral | — | 14 | F | $CO_2$ | 350 μE/m²/sec (18.4) | 542 μmol C/g dry wt/hr | 130 μE/m²/sec (6.8) | 10 μE/m²/sec (0.53) | 250 μE/m²/sec (13.1) | None—600 μE/m²/sec (31.8) | — | — | 7 |
| Ectocarpus confervoides(12) | West Baltic Sea, eulittoral | 10°C, spring | 10 | T | $O_2$ | 6.0 mW/cm² (14.0) | 875 μmol $O_2$/g dry wt/hr | 1.0 mW/cm² (2.3) | 0.16 mW/cm² (0.37) | 2.4 mW/cm² (5.6) | None—40 mW/cm² (93.2) | — | — | 8 |
| Egregia laevigata(29) | Marine | — | 15 | F | $O_2$ | 13.5 | 589 μmol $O_2$/g dry wt/hr | 4.0 | — | 8.0 | None—13.5 | — | — | 9 |
| Fucus sp. (20) | Arctic, subice | — | <0 | T | $O_2$ | — | 223 μmol $O_2$/g dry wt/hr | — | 2.0 Ly/24 hr (0.20) | — | — | — | — | 10 |
| | | — | 25 | T | $O_2$ | — | 29 μmol $O_2$/mgChl/hr | 3 | — | — | — | — | — | 11 |
| Fucus distichus(28) | Sublittoral | 5°C | 5 | T | $O_2$ | 30 | 3 | — | — | 8 | None—65 | — | — | 12 |
| Fucus serratus(30) | — | — | 25 | — | $O_2$ | 25 | 7 | — | — | — | — | — | — | 13 |
| F. serratus(12) | West Baltic Sea, eulittoral | 10°C, spring | 10 | T | $O_2$ | 6.0 mW/cm² (14.0) | 156 μmol $O_2$/g dry wt/hr | 1.2 mW/cm² (2.8) | 0.16 mW/cm² (0.37) | 2.0 mW/cm² (4.6) | None—64 mW/cm² (149) | — | — | 14 |
| | West Baltic Sea, eulittoral | 15°C, autumn | 15 | T | $O_2$ | 5.5 mW/cm² (12.8) | 200 μmol $O_2$/g dry wt/hr | 2.0 mW/cm² (4.6) | 0.35 mW/cm² (0.82) | 4.0 mW/cm² (9.3) | None—20 mW/cm² (46) | — | — | 15 |
| | West Baltic Sea, eulittoral | 5°C, winter | 5 | T | $O_2$ | 12.0 mW/cm² (27.9) | 344 μmol $O_2$/g dry wt/hr | 1.2 mW/cm² (2.8) | 0.16 mW/cm² (0.37) | 2.6 mW/cm² (6.0) | None—64 mW/cm² (149) | — | — | 16 |
| Fucus vesiculosus(25) | Long Island, salt marsh | summer | — | S | $CO_2$ | 0.1 cal/cm²/min (7.7) | 104 μmol C/g dry wt/hr | — | — | — | None—38.5 | — | — | 17 |
| F. vesiculosus, apical portion(31) | West Baltic Sea, eulittoral | 15°C, autumn | 15 | T | $O_2$ | 10.0 mW/cm² (23.3) | 594 μmol $O_2$/g dry wt/hr | 2.2 mW/cm² (5.1) | 0.40 mW/cm² (0.93) | 5.5 mW/cm² (12.8) | None—21.5 mW/cm² (50.1) | — | — | 18 |
| F. vesiculosus, basal portion(31) | West Baltic Sea, eulittoral | 15°C, autumn | 15 | T | $O_2$ | 5.5 mW/cm² (12.8) | 75 μmol $O_2$/g dry wt/hr | 3.0 mW/cm² (7.0) | 0.57 mW/cm² (1.33) | 4.0 mW/cm² (9.3) | None—21.5 mW/cm² (50.1) | — | — | 19 |
| F. vesiculosus(12) | West Baltic Sea, eulittoral | 10°C, spring | 10 | T | $O_2$ | 6.0 mW/cm² (14.0) | 375 μmol $O_2$/g dry wt/hr | 1.2 mW/cm² (2.8) | 0.15 mW/cm² (0.35) | 2.7 mW/cm² (6.3) | None—65 mW/cm² (151) | — | — | 20 |
| | West Baltic Sea, eulittoral | 15°C, autumn | 15 | T | $O_2$ | 20.0 mW/cm² (46.6) | 438 μmol $O_2$/g dry wt/hr | 3.0 mW/cm² (7.0) | 0.6 mW/cm² (1.40) | 5.5 mW/cm² (12.8) | None—20.0 mW/cm² (46.6) | — | — | 21 |
| | West Baltic Sea, eulittoral | 5°C, winter | 5 | T | $O_2$ | 6.0 mW/cm² (14.0) | 344 μmol $O_2$/g dry wt/hr | 1.5 mW/cm² (3.5) | 0.12 mW/cm² (0.28) | 2.6 mW/cm² (6.0) | None—65 mW/cm² (151) | — | — | 22 |

## Table 1 (continued)
## RELATIONSHIP BETWEEN LIGHT INTENSITY AND PHOTOSYNTHESIS IN MACROALGAE[a]

| Species (references) | Environment[a] | Conditions at sampling site[c] | Exp. T° (°C)[d] | Type of light[e] | Assay method | Light intensity at Psat[f] (klx) | Pmax[g] | Light intensity at "Psat" one half (klx) | Compensation light intensity[j] (klx) | $I_c$[l] (klx) | Light intensity for initial photo-inhibition[k] (klx) | Maximum light intensity tested (klx) | Inhibition (%)[m] | Case #[n] |
|---|---|---|---|---|---|---|---|---|---|---|---|---|---|---|
| F. vesiculosus(4) | Marine | — | 2 | T | $O_2$ | 21.6 | 179 μmol $O_2$/g dry wt/hr | 6.5 | — | 13.8 | None—54 | — | — | 23 |
|  |  | — | 20 | T | $O_2$ | 43.2 | 446 μmol $O_2$/g dry wt/hr | 16.2 | — | 31.8 | None—54 | — | — | 24 |
| Ishige sinicola(32) | Japan, marine intertidal | — | 20 | T | $O_2$ | 60 kerg/cm²/sec (14.3) | — | 18 kerg/cm²/sec (4.3) | 3.5 kerg/cm²/sec (0.83) | 23 kerg/cm²/sec (5.5) | None—80 kerg/cm²/sec (19.0) | — | — | 25 |
| Laminaria digitata, terminal end(31) | West Baltic Sea, sublittoral | 15°C, autumn | 15 | T | $O_2$ | 5.5 mW/cm² (12.8) | 250 μmol $O_2$/g dry wt/hr | 1.0 mW/cm² (2.3) | 0.30 mW/cm² (0.70) | 2.2 mW/cm² (5.1) | None—21.5 mW/cm² (50.1) | — | — | 26 |
| L. digitata, basal end (31) | West Baltic Sea, sublittoral | 15°C, autumn | 15 | T | $O_2$ | 5.5 mW/cm² (12.8) | 69 μmol $O_2$/g dry wt/hr | 1.9 mW/cm² (4.4) | 0.35 mW/cm² (0.82) | 4.6 mW/cm² (10.7) | None—21.5 mW/cm² (50.1) | — | — | 27 |
| L. digitata(12) | West Baltic Sea, sublittoral | 10°C, spring | 10 | T | $O_2$ | 13.0 mW/cm² (30.3) | 250 μmol $O_2$/g dry wt/hr | 1.0 mW/cm² (2.3) | 0.16 mW/cm² (0.37) | 2.7 mW/cm² (6.3) | None—65 mW/cm² (151) | — | — | 28 |
|  | West Baltic Sea, sublittoral | 15°C, autumn | 15 | T | $O_2$ | 5.5 mW/cm² (12.8) | 250 μmol $O_2$/g dry wt/hr | 0.9 mW/cm² (2.1) | 0.32 mW/cm² (0.75) | 1.9 mW/cm² (4.4) | None—20 mW/cm² (4.4) | — | — | 29 |
|  | West Baltic Sea, sublittoral | 5°C, winter | 5 | T | $O_2$ | 0.6 mW/cm² (1.5) | 94 μmol $O_2$/g dry wt/hr | 0.2 mW/cm² (0.5) | 0.12 mW/cm² (0.28) | 0.5 mW/cm² (1.2) | None—3 mW/cm² (7.0) | — | — | 30 |
| Laminaria hyperborea (33) | Helgoland, marine | March | 4 | — | $O_2$ | 0.018 cal/cm²/min (3.0) | 0.8 μmol $O_2$/cm²/hr | 0.006 cal/cm²/min (1.0) | — | 0.008 cal/cm²/min (1.4) | None—0.042 cal/cm²/min (7.0) | — | — | 31 |
|  | Helgoland, marine | August | 16 | — | $O_2$ | 0.024 cal/cm²/min (4.0) | 1.4 μmol $O_2$/cm²/hr | 0.008 cal/cm²/min (1.4) | — | 0.011 cal/cm²/min (1.8) | None—0.042 cal/cm²/min (7.0) | — | — | 32 |
|  | Helgoland, marine | November | 10 | — | $O_2$ | 0.013 cal/cm²/min (2.1) | 0.7 μmol $O_2$/cm²/hr | 0.005 cal/cm²/min (0.9) | — | 0.008 cal/cm²/min (1.3) | None—0.033 cal/cm²/min (5.5) | — | — | 33 |
| L. hyperborea(34) | Helgoland, marine | February | 8 | T | $CO_2$ | 10.0 mW/cm² (23.9) | 1.3 μmol C/cm²/hr | 3.0 mW/cm² (7.2) | 0.5 mW/cm² (1.15) | 2.5 mW/cm² (5.9) | None—30.0 mW/cm² (71.7) | — | — | 34 |
|  | Helgoland, marine | February | 18 | T | $CO_2$ | 10.0 mW/cm² (23.9) | 2.3 μmol C/cm²/hr | 5.0 mW/cm² (11.5) | 0.5 mW/cm² (1.15) | 2.8 mW/cm² (6.7) | None—30.0 mW/cm² (71.7) | — | — | 35 |

| Species | Location | Condition | | | | | | | | | | | Ref. |
|---|---|---|---|---|---|---|---|---|---|---|---|---|---|
| *Laminaria saccharina* (12) | West Baltic Sea, eulittoral | 5°C, winter | 5 | T | $O_2$ | 0.65 mW/cm² (1.5) | 125 μmol $O_2$/g dry wt/hr | 0.17 mW/cm² (0.4) | 0.12 mW/cm² (0.28) | 0.3 mW/cm² (0.7) | None—3 mW/cm² (7.0) | — | 36 |
| | West Baltic Sea, eulittoral | 15°C, autumn | 15 | T | $O_2$ | 5 mW/cm² (11.6) | 125 μmol $O_2$/g dry wt/hr | 2 mW/cm² (4.6) | 0.55 mW/cm² (1.28) | 3.7 mW/cm² (8.6) | None—20 mW/cm² (46.6) | — | 37 |
| *L. saccharina* (sporophyes) (35) | North Sea | 0 μM NO₃, 5 klx, 12°C | 12 | F | — | 60 μE/m²/sec (3.2) | — | 20 μE/m²/sec (1.0) | — | 38 μE/m²/sec (2.0) | None—165 μE/m²/sec (8.7) | — | 38 |
| | | 3 μM NO₃, 5 klx, 12°C | 12 | F | — | 60 μE/m²/sec (3.2) | — | 20 μE/m²/sec (1.0) | — | 40 μE/m²/sec (2.1) | None—165 μE/m²/sec (8.7) | — | 39 |
| | | 10 μM NO₃, 5 klx, 12°C | 12 | F | — | 50 μE/m²/sec (2.6) | — | 12 μE/m²/sec (0.6) | — | 30 μE/m²/sec (1.6) | None—165 μE/m²/sec (8.7) | — | 40 |
| | | 500 μM NO₃, 5 klx, 12°C | 12 | F | — | 50 μE/m²/sec (2.6) | — | 10 μE/m²/sec (0.5) | — | 18 μE/m²/sec (0.9) | None—165 μE/m²/sec (8.7) | — | 41 |
| *Macrocystis* sp., kelp — mature blade (36) | Marine | — | 15 | — | — | 14.8 | — | 3.18 | 0.16 | 5.1 | None—21.2 | — | 42 |
| *Petalonia fasciata* (12) | West Baltic Sea, eulittoral | 10°C, spring | 10 | T | $O_2$ | 13 mW/cm² (30.3) | 2063 μmol $O_2$/g dry wt/hr | 1.5 mW/cm² (3.5) | 0.19 mW/cm² (0.44) | 4.0 mW/cm² (9.3) | None—65 mW/cm² (151.4) | — | 43 |
| *Pilayella littoralis* (13) | Baltic Sea, littoral | — | 14.2 | F | $CO_2$ | 200 μE/m²/sec (10.5) | 208 μmol C/g dry wt/hr | 100 μE/m²/sec (5.2) | 10 μEm²/sec (0.53) | 150 μE/m²/sec (7.9) | 250 μE/m²/sec (13.2) | 600 μE/m²/sec (31.6) | 44 |
| | | — | 4.0 | F | $CO_2$ | 180 μE/m²/sec (9.5) | 500 μmol C/g dry wt/hr | 40 μE/m²/sec (2.1) | — | 90 μE/m²/sec (4.7) | None—600 μE/m²/sec (31.6) | — | 45 |
| | Baltic Sea, sublittoral | — | — | F | $CO_2$ | 300 μE/m²/sec (15.8) | 375 μmol C/g dry wt/hr | 100 μE/m²/sec (5.2) | — | 160 μE/m²/sec (8.4) | None—600 μE/m²/sec (31.6) | — | 46 |
| *P. littoralis* (28) | Greenland, marine | 5°C | 5 | T | $O_2$ | 20 | 26.8 μmol $O_2$/mgChl/hr | 4 | 1 | 6 | None—70 | — | 47 |
| *Sargassum ringgoldianum* (32) | Japan, marine | Low-water mark | 20 | T | $O_2$ | 30 kerg/cm²/sec (7.1) | — | 10 kerg/cm²/sec (2.4) | 2 kerg/cm²/sec (0.48) | 18 kerg/cm²/sec (4.3) | None—80 kerg/cm²/sec (19.0) | — | 48 |
| *Scytosiphon lomentaria* (12) | West Baltic Sea, eulittoral | 10°C, spring | 10 | T | $O_2$ | 13 mW/cm² (30.3) | 813 μmol $O_2$/g dry wt/hr | 2.5 mW/cm² (5.8) | 0.23 mW/cm² (0.5) | 4.6 mW/cm² (10.7) | None—65 mW/cm² (151.4) | — | 49 |
| *Undaria pinnatifida* (32) | Japan, marine | Low-water mark | 20 | T | $O_2$ | 30 kerg/cm²/sec (7.1) | — | 12 kerg/cm²/sec (2.9) | 1.0 kerg/erg/cm²/sec (0.24) | 23 kerg/cm²/sec (5.5) | None—80 kerg/cm²/sec (19.0) | — | 50 |
| *Undaria peterseniana* (32) | Japan, marine | 15-m depth | 20 | T | $O_2$ | 25 kerg/cm²/sec (6.0) | — | 10 kerg/cm²/sec (2.4) | 0.5 erg/cm²/sec (0.12) | 19 kerg/cm²/sec (4.5) | None—80 kerg/cm²/sec (19.0) | — | 51 |

## Table 1 (continued)
## RELATIONSHIP BETWEEN LIGHT INTENSITY AND PHOTOSYNTHESIS IN MACROALGAE[a]

| Species (references) | Environment[a] | Conditions at sampling site[c] | Exp. T° (°C)[c] | Type of light[c] | Assay method | Light intensity at Psat[c] (klx) | Pmax[e] | Light intensity at ''Psat'' one half (klx) | Compensation light intensity[f] (klx) | I_k[g] (klx) | Light intensity for initial photo-inhibition[h] (klx) | Maximum light intensity tested (klx) | Inhibition (%) | Case #[m] |
|---|---|---|---|---|---|---|---|---|---|---|---|---|---|---|
| **Chlorophyta — green algae** | | | | | | | | | | | | | | |
| Acrosiphonia arcta (28) | Greenland, marine | 5°C | 5 | T | O₂ | 8 | 14.3 µmol O₂/mgChl/hr | 2 | 0.5 | 4 | None—60 | — | — | 1 |
| Acrosiphonia centralis (12) | West Baltic Sea, eulittoral | 10°C, spring | 10 | T | O₂ | 6 mW/cm² (14.0) | 938 µmol O₂/g dry wt/hr | 1.3 mW/cm² (3.0) | 0.2 mW/cm² (0.47) | 3.2 mW/cm² (7.4) | None—25 mW/cm² (58.1) | — | — | 2 |
| A. centralis (13) | Baltic Sea, littoral | — | 4.5 | F | CO₂ | 150 µE/m²/sec (7.9) | 150 µmol C/g dry wt/hr | 70 µE/m²/sec (3.7) | — | 120 µE/m²/sec (6.3) | 500 µE/m²/sec (26.3) | 700 µE/m²/sec (36.8) | 5 | 3 |
| Chaetomorpha sp. (20) | Arctic, subice | — | <0 | T | O₂ | — | — | — | 1.4 Ly/24 hr (0.16) | — | — | — | — | 4 |
| Chaetomorpha linum (37) | Marine, littoral | 150 µE/m²/sec (7.5) | 25 | T | O₂ | — | — | — | — | — | — | — | — | 5 |
| | | | 15 | T | O₂ | 20.0 mW/cm² (46.5) | 65 µmol O₂/mgChl/hr | 4.0 mW/cm² (9.3) | 0.4 mW/cm² (0.93) | 6.0 mW/cm² (140) | None—50.0 mW/cm² (116) | — | — | 6 |
| Cladophora sp. (38) | Upper intertidal | — | 20 | T | O₂ | 10 | 0.7 µmol O₂/cm²/hr | 4 | — | 6 | None—40 | — | — | 7 |
| Cladophora glomerata (12) | West Baltic Sea, eulittoral | 20°C, summer | 20 | T | O₂ | 20.0 mW/cm² (46.5) | 1750 µmol O₂/g dry wt/hr | 3.5 mW/cm² (8.1) | 0.30 mW/cm² (0.70) | 6.8 mW/cm² (15.8) | None—20.0 mW/cm² (46.5) | — | — | 8 |
| C. glomerata (13) | Baltic Sea, littoral | — | 14.2 | F | CO₂ | 400 µE/m²/sec (21.1) | 33 µmol C/g dry wt/hr | 150 µE/m²/sec (7.9) | 40 µE/m²/sec (2.1) | 300 µE/m²/sec (15.8) | None—600 µE/m²/sec (31.6) | — | — | 9 |
| | Baltic Sea, littoral | — | 9.4 | F | CO₂ | 300 µE/m²/sec (15.8) | — | 50 µE/m²/sec (2.6) | — | 100 µE/m²/sec (5.3) | None—800 µE/m²/sec (42.0) | — | — | 10 |
| Cladophora wrightiana (38) | Marine | 5-m depth | 20 | F | O₂ | 5 | 0.1 µmol O₂/cm²/hr | 2 | — | 2 | None—40 | — | — | 11 |
| Codium fragile (16) | Marine | 10-m depth | 26 | T | O₂ | 3.5 mW/cm² (8.1) | 80 µmol O₂/µmolChl/hr | 0.8 mW/m² (1.8) | — | 1.0 mW/cm² (2.3) | None—14.0 mW/cm² (32.6) | — | — | 12 |
| | Marine | 0.5-m depth | 26 | T | O₂ | 6.3 mW/cm² (14.7) | 60 µmol O₂/µmolChl/hr | 1.5 mW/cm² (3.5) | — | 2.5 mW/cm² (5.8) | None—14.0 mW/cm² (32.6) | — | — | 13 |
| Enteromorpha intestinalis (12) | West Baltic Sea, eulittoral | 10°C, spring | 10 | T | O₂ | 12.0 mW/cm² (27.4) | 1875 µmol O₂/g dry wt/hr | 1.8 mW/cm² (4.1) | 0.19 mW/cm² (0.44) | 3.7 mW/cm² (8.4) | None—15.0 mW/cm² (34.2) | — | — | 14 |

| Species | Habitat | Condition | °C | Method | Gas | | | | | | | | | | No. |
|---|---|---|---|---|---|---|---|---|---|---|---|---|---|---|---|
| | West Baltic Sea, eulittoral | 20°C, summer | 20 | T | O₂ | — | — | — | 0.45 mW/cm² (1) | — | — | — | — | | 15 |
| | West Baltic Sea, eulittoral | 5°C, winter | 5 | T | O₂ | 12.0 mW/cm² (27.4) | 1750 µmol O₂/dry wt/hr | 1.7 mW/cm² (4.0) | 0.12 mW/cm² (0.27) | 3.5 mW/cm² (8.1) | None—28.0 mW/cm² (65.1) | — | — | | 16 |
| *E. intestinalis*(37) | Marine, littoral | 150 µE/m²/sec (7.5) | 15 | T | O₂ | 300 W/m² (69.8) | 90 µmol O₂/mgChl/hr | 40 W/m² (9.3) | 4 W/m² (0.93) | 60 W/m² (14.0) | None—500 W/m² (116) | — | — | | 17 |
| *Enteromorpha linza* (12) | West Baltic Sea, eulittoral | 15°C, autumn | 15 | T | O₂ | 6.0 mW/cm² (14.0) | 1375 µmol O₂/dry wt/hr | 2.0 mW/cm² (4.6) | 0.25 mW/cm² (0.58) | 4.1 mW/cm² (9.5) | None—20.0 mW/cm² (46.5) | — | — | | 18 |
| *Enteromorpha prolifera*(12) | West Baltic Sea, eulittoral | 15°C, autumn | 15 | T | O₂ | 20.0 mW/cm² (46.5) | 2563 µmol O₂/dry wt/hr | 4.0 mW/cm² (9.3) | 0.35 mW/cm² (0.81) | 9.0 mW/cm² (20.9) | None—20.0 mW/cm² (46.5) | — | — | | 19 |
| | West Baltic Sea, eulittoral | 20°C, summer | 20 | T | O₂ | 5.5 mW/cm² (12.8) | 875 µmol O₂/dry wt/hr | 3.5 mW/cm² (8.1) | 1.00 mW/cm² (2.30) | 5.0 mW/cm² (11.6) | None—10.0 mW/cm² (23.3) | — | — | | 20 |
| *Hormidium flaccidum* (39) | Sun adapted | — | 18 | T | — | 6.0 | — | 1.2 | — | — | — | — | — | | 21 |
| | Shade adapted | — | 18 | T | — | 2.4 | — | 0.2 | — | — | — | — | — | | 22 |
| *Monostroma grevillei* (13) | Baltic Sea, littoral | — | 4.8 | F | CO₂ | 200 µE/m²/sec (10.5) | 983 µmol C/g dry wt/hr | 60 µE/m²/sec (3.2) | — | 120 µE/m²/sec (6.3) | None—650 µE/m²/sec (34.2) | — | — | | 23 |
| *Monostroma nitidum* (38) | Upper intertidal | — | 20 | T | O₂ | 20 | 1.0 µmol O₂/cm³/hr | 6 | — | 11 | None—40 | — | — | | 24 |
| *Ulothrix speciosa*(28) | Greenland, marine | 5°C | 5 | T | O₂ | 25 | 17.9 µmol O₂/mgChl/hr | 5 | 2 | 9 | None—70 | — | — | | 25 |
| *U. speciosa*(12) | West Baltic Sea, eulittoral | 10°C, spring | 10 | T | O₂ | 25.0 mW/cm² (58) | 219 µmol O₂/dry wt/hr | 2.8 mW/cm² (6.5) | 0.18 mW/cm² (0.42) | 6.5 mW/cm² (15.1) | None—63 mW/cm² (151) | — | — | | 26 |
| *Ulva japonica*(40) | 20-m depth | — | 25 | — | — | 3 × 10¹⁹ quanta/m²/sec (2.5) | — | 1.5 × 10¹⁹ quanta/m²/sec (1.3) | — | 2.5 × 10¹⁹ quanta/m²/sec (2.1) | None—10 × 10¹⁹ quanta/m²/sec (8.5) | — | — | | 27 |
| *U. japonica*(38) | 10-m depth | — | 20 | T | O₂ | 6 | 1.2 µmol O₂/cm³/hr | 2 | — | 4 | 20 | 40 | 10 | | 28 |
| *Ulva lactuca*(16) | Marine | 6-m depth | 26 | T | O₂ | 3.5 mW/cm² (8.1) | 80 µmol O₂/µmolChl/hr | 1 mW/cm² (2.3) | — | 2 mW/cm² (4.7) | None—6.3 mW/m² (14.7) | — | — | | 29 |
| | Marine | 0.5-m depth | 26 | T | O₂ | 5.5 mW/cm² (12.8) | 275 µmol O₂/µmolChl/hr | 2 mW/cm² (4.7) | — | 3.5 mW/cm² (8.1) | None—6.3 mW/m² (14.7) | — | — | | 30 |
| *U. lactuca*(41) | Marine | — | 28 | — | CO₂ | 9.5 | — | 3.3 | — | 6.9 | None—18.0 klux | — | — | | 31 |
| *U. pertusa*(24) | Japan, marine | — | 20 | T | O₂ | 15 | 1100 µmol O₂/g dry wt/hr | 5 | 0.30 | 9 | None—20 klux | — | — | | 32 |
| *U. pertusa*(40) | Japan, middle intertidal | — | 25 | — | — | 10 × 10¹⁹ quanta/m²/sec (8.5) | — | 3 × 10¹⁹ quanta/m²/sec (2.5) | — | 5.5 × 10¹⁹ quanta/m²/sec (4.6) | None—15 × 10¹⁹ quanta/m²/sec (12.5) | — | — | | 33 |

## Table 1 (continued)
## RELATIONSHIP BETWEEN LIGHT INTENSITY AND PHOTOSYNTHESIS IN MACROALGAE[a]

| Species (references)[b] | Environment[b] | Conditions at sampling site[c] | Exp. T° (°C)[d] | Type of light[e] | Assay method | Light intensity at Psat[f] (klx) | Pmax[g] | Light intensity at ¹/₂Psat[h] one half (klx) | Compensation light intensity[i] (klx) | I_k[j] (klx) | Light intensity for initial photo-inhibition[k] (klx) | Maximum light intensity tested (klx) | Inhibition (%)[l] | Case #[m] |
|---|---|---|---|---|---|---|---|---|---|---|---|---|---|---|
| *U. pertusa*(38) | Middle interridal | — | 20 | T | O₂ | 10 | 1.3 µmol O₂/cm²/hr | 5 | — | 10 | None—40 | — | — | 34 |
| *Ulva taeniata*(17) | Marine | — | — | — | O₂ | 10 kerg/cm²/sec (2.4) | | 4.2 kerg/cm²/sec (1.0) | — | 8.0 kerg/cm²/sec (1.9) | None—10.0 kerg/cm²/sec (2.4) | — | — | 35 |
| *Ulvopsis grevillei*(12) | West Baltic Sea, eulittoral | 10°C, spring | 10 | T | O₂ | 12 mW/cm² (27.4) | 3000 µmol O₂/g dry wt/hr | 2 mW/cm² (4.7) | 0.18 mW/cm² (0.42) | 4.2 mW/cm² (9.8) | None—65 mW/cm² (151) | — | — | 36 |
| | | 10°C | 5 | T | O₂ | 7 mW/cm² (16.3) | 875 µmol O₂/g dry wt/hr | 1.5 mW/cm² (3.5) | 0.12 mW/cm² (0.28) | 3.7 mW/cm² (8.6) | None—40 mW/cm² (93) | — | — | 37 |
| *Urospora penicilliformis*(28) | — | 5°C | 5 | T | O₂ | 60 | 15.6 µmol O₂/mgChl/hr | 9 | 3 | 12 | None—60 | — | — | 38 |

[a] Abbreviations used: cal/cm²/min, calories centimeter⁻² minute⁻¹; kerg/cm²/sec, kiloerg centimeter⁻² second⁻¹; klx, 1000 × Lux; Ly/min, langleys minute⁻¹; Q/m²/sec, quanta meter⁻² second⁻¹; µE/m²/sec, microeinsteins meter⁻² second⁻¹; mW/cm², milliwatts centimeter⁻²; BChl, bacteriochlorophyll; C, carbon; cpm, counts per minute; N, nitrogen; Prel, photosynthetic rate in relative unites; µg at, microgram atoms; µmolO₂/cm²/hr, micromole O₂/cm² of tissue/hour; µmolO₂/g dry wt/hr, micromole O₂/gram dry weight/hour.

[b] Environment — general habitat or location of sampling.

[c] Conditions at sampling site — specific environmental conditions at the site where macroalgal samples were collected.

[d] Exp. T° — experimental temperature in degrees centigrade.

[e] Type of light — lighting used in experiments: S, sunlight; T, tungsten; F, fluorescent. It should be noted that these are only rough categorizations of the light sources used. Each category contains a wide diversity of lamp types, e.g., cool-white vs. daylight fluorescent.

[f] Light intensity at Psat — intensity at which photosynthesis reaches saturation.

[g] Pmax — maximum rate of photosynthesis observed.

[h] Light intensity at one half Pmax — i.e., one half of the maximum photosynthetic rate.

[i] Compensation light intensity — intensity at which photosynthesis equals respiration.

[j] I_k — light intensity at which a line through the initial slope of the light intensity vs. photosynthesis curve intersects a line parallel to the abscissa at Pmax. It is commonly used as a measure of "shade" and "sun" adaptation.

[k] Light intensity for initial photoinhibition — intensity at which there is 5% inhibition of photosynthesis. If there is no photoinhibition, "none" is used followed by maximum intensity tested.

[l] Inhibition (%) — percent inhibition at maximum light intensity tested.

[m] Case # — numbers used to refer to individual experimental results in Figures 1.

# Table 2
## RELATIONSHIP BETWEEN LIGHT INTENSITY AND GROWTH IN MACROALGAE[a]

| Species (Reference) | Environment[b] | Growth conditions[c] | Exp. T[od] (°C) | Type of light[e] | Light intensity at growth optima[f] (klx) | Maximum growth rate[g] | Light intensity at one half maximum growth rate[h] (klx) | Compensation light intensity[i] (klx) | $I_k$(klx) | Light intensity for initial photoinhibition[k] (klx) | Maximum light intensity tested (klx) | Inhibition[l] (%) |
|---|---|---|---|---|---|---|---|---|---|---|---|---|
| **Rhodophyta — red algae** | | | | | | | | | | | | |
| *Antithamnion plumula* (sporelings) (42) | — | — | 16 | F | 1.5 kerg/cm²/sec (0.47) | — | 0.2 kerg/cm²/sec (0.06) | — | — | None—2.2 kerg/cm²/sec (0.69) | — | — |
| *Bonnemaisonia asparagoides* (sporelings) (43) | Sublittoral | — | 17—20 | F | 0.6 | — | 0.2 | — | — | 0.6 | 6.5 | 100 |
| *Brongniatrella byssoides* (sporelings) (42) | — | — | 16 | F | 0.25 kerg/cm² sec (0.08) | — | 0.05 kerg/cm² (0.02) | — | — | 1.0 kerg/cm²/sec (0.31) | 2.0 kerg/cm¹/skc (0.63) | 0 |
| *Clathromorphum circumscriptum* (44) | Subarctic marine | — | 15 | F | 0.4 | 12 μm/day | — | — | — | — | — | — |
| *Dilsea carnosa* (sporelings) (43) | Littoral | — | 8—9 | F | 0.1 | — | 0.03 | — | — | 2.0 | 9.2 | 100 |
| *Dumontia incrassata* (sporelings) (43) | Intertidal | — | 16—20 | F | 11.0 | — | 2.1 | — | — | None—11.0 | — | — |
| *Halarachnion ligulatum* (sporelings) (43) | Sublittoral | — | 15—19 | F | 0.5 | — | 0.2 | — | — | 0.5 | 10.6 | 100 |
| *Leptophytum laeve* (44) | Subarctic marine | — | 10 | F | 0.35 | 10 μm/day | — | — | — | — | — | — |
| *Leptophytum foecundum* (44) | Subarctic marine | — | 10 | F | 0.050 | 10 μm/day | — | — | — | — | — | — |
| *Lithophyllum orbiculatum* (44) | Subarctic marine | — | 8 | F | 0.36 | 4 μm/day | — | — | — | — | — | — |
| *Lithothamnion glaciale* (44) | Subarctic marine | — | 13 | F | 0.36 | 14 μm/day | — | — | — | — | — | — |
| *Naccaria wiggii* (sporelings) (43) | Sublittoral | — | 15—19 | F | 1 | — | 0.2 | — | — | 2 | 11 | 100 |

## Table 2 (continued)
## RELATIONSHIP BETWEEN LIGHT INTENSITY AND GROWTH IN MACROALGAE[a]

| Species (Reference) | Environment[b] | Growth conditions[c] | Exp. T°[d] (°C) | Type of light[e] | Light intensity at growth optima[f] (klx) | Maximum growth rate[g] | Light intensity at one half maximum growth rate[h] (klx) | Compensation light intensity[i] (klx) | $I_k$ (klx) | Light intensity for initial photoinhibition[k] (klx) | Maximum light intensity tested (klx) | Inhibition[l] (%) |
|---|---|---|---|---|---|---|---|---|---|---|---|---|
| Phymatolithon polymorphum (44) | — | — | 10 | F | 0.65 | 15 μm/day | — | — | — | — | — | — |
| Plumaria elegans (sporelings) (22) | — | — | 16 | F | 0.5 kerg/cm²/sec (0.16) | 60 cells/100 spore/day | 0.4 kerg/cm²/sec (0.13) | — | 0.5 kerg cm² 0.5 kerg/cm²/sec (0.16) | 1.1 kerg/cm²/sec (0.34) | 4.2 kerg/cm²/sec (1.30) | 83 |
| | | | 16 | | 1.9 kerg/cm²/sec (0.59) | 42 cells/100 spore/day | 0.4 kerg/cm²/sec (0.13) | — | 0.8 kerg/cm²/sec (0.25) | None—5.0 kerg/cm² sec (1.60) | — | 50 |
| Rhodymenia palmata (sporelings) (43) | Intertidal | — | 8—11 | F | 6.5 | — | 0.7 | — | — | 6.7 | 11 | — |
| **Phaeophyta-Brown Algae** | | | | | | | | | | | | |
| Alaria marginata (gametophyte) (45) | North Pacific, lower intertidal | — | 12 | F | 40 μE/m²/sec (2.1) | Increase of 14 μm diameter/1 week | 8 μE/m²/sec (0.4) | — | 20 μE/m²/sec (1.0) | None—60 μE/m²/sec (3.1 klux) | — | — |
| | | — | 12 | F | 2.1 | 30% fertile females/2 weeks | 1.5 | 0.5 | — | None—3.1 | — | — |
| Egregia menziesii (gametophyte) (45) | Pacific, subtidal | — | 17 | F | 40 μE/m²/sec (2.1) | Increase of 14 μm dia meter/1 week | 9 μE/m²/sec (0.45) | — | 22 μE/m²/sec (1.1) | None—60 μE/m²/sec (3.1) | — | — |
| | | — | 12 | F | 1.0 | 90% fertile females/2 weeks | 0.4 | 0.2 | — | None—3.1 | — | — |
| Hedophyllum sessile (gametophyte) (45) | North Pacific, subtidal | — | 12 | F | 40 μE/m²/sec (2.1) | Increase of 13 μm diameter/1 week | 5 μE/m²/sec (0.2) | — | 18 μE/m²/sec (0.8) | None—60 μE/m²/sec (3.1) | — | — |
| | | — | 12 | F | 2.1 | 90% fertile females/2 weeks | 0.9 | 0.5 | — | 2.1 | 3.1 | 5 |

| Species | Location | | | | | | | | | | | | |
|---|---|---|---|---|---|---|---|---|---|---|---|---|---|
| *Laminaria dentigera* (gametophyte) (45) | North Pacific, lower intertidal | — | — | 12 | F | 40 µE/m² sec (2.1) | Increase of 1.55µm diameter/1 week | 7 µE/m² sec (0.3) | — | 17 µE/ m²/sec (0.7) | 40 µE/m²/sec (2.1) | 60 µE/m²/ sec (3.1) | 5 |
| | | — | — | 12 | F | 3.1 | 35% fertile females/1 week | 1.7 | 0.5 | — | None—3.1 | — | — |
| *Laminaria farlowii* (gametophyte) (45) | California, lower intertidal | — | — | 17 | F | 40 µE/m² (2.1) | Increase of 12.5 µm diameter/1 week | 6 µE/m² sec (0.3) | — | 18 µE/ m²/sec (0.8) | 40 µE/m²/sec (2.1) | 60 µE/m²/ sec (3.1) | 5 |
| | | — | — | 12 | F | 3.1 | 60% fertile females/2 weeks | 2.2 | 0.05 | — | None—3.1 | — | — |
| *Laminaria hyperborea* (sporophyte) (46) | North Atlantic | — | — | 10 | F | 0.003 cal/ cm²/min (0.73) | 5.5 dbl/day | 0.001 cal/ cm²/min (0.29) | 0.00005 cal/ cm²/min (0.01) | 0.0024 cal/ cm²/min (0.58) | None—0.0084 cal/cm²/min (2.03) | — | — |
| *Laminaria digitata* (sporophyte) (46) | North Atlantic | — | — | 10 | F | 0.003 cal/ cm²/min (0.73) | 5.5 dbl/day | 0.001 cal/ cm²/min. (0.29) | 0.00005 cal/ cm²/min (0.01) | 0.0024 cal/ cm²/min (0.58) | None—0.0084 cal/cm²/min (2.03) | — | — |
| *Laminaria saccharina* (sporophyte) (46) | North Atlantic | — | — | 10 | F | 0.003 cal/ cm²/min (0.73) | 5.5 dbl/day | 0.001 cal/ cm²/min (0.29) | 0.00013 cal/ cm²/min (0.03) | 0.0024 cal/ cm²/min (0.58) | None—0.0084 cal/cm²/min (2.03) | — | — |
| *Laminaria sinclairii* (gametophyte) (45) | North Pacific, lower intertidal | — | — | 12 | F | 40 µE/m² sec (2.1) | Increase of 1.55 µm diameter/1 week | 7 µE/m² sec (0.35) | — | 12 µE/ m²/sec (0.6) | None—60 µE/ m²/sec (3.1) | — | — |
| | | — | — | 12 | F | 3.1 | 60% fertile females/2 weeks | 1.0 | 0.5 | — | None—3.1 | — | — |
| *Macrocystis integrifolia* (gametophyte) (45) | North Pacific, lower intertidal | — | — | 17 | F | 40 µE/m² sec (2.1) | Increase of 20 µm diameter/1 week | 8 µE/m² (0.4) | — | 19 µE/ m²/sec (0.8) | 40 µE/m²/sec (2.1) | 60 µE/m²/ sec (3.1) | 5 |
| | | — | — | 12 | F | 3.1 | 70% fertile females/2 weeks | 2.1 | 0.5 | — | None—3.1 | — | — |
| *Macrocystis pyrifera* (gametophyte) (45) | Pacific, subtidal | — | — | 17 | F | 40 µE/m² sec (2.1) | Increase of 20 µm diameter/1 week | 8 µE/m² sec (0.4) | — | 18 µE/ m²/sec (0.8) | None—60 µE/ m²/sec (3.1) | — | — |
| | | — | — | 12 | F | 3.1 | 60% fertile females/2 weeks | 2.1 | 0.5 | — | None—3.1 | — | — |
| *Pterygophora californica* (gametophyte) (45) | Pacific subtidal | — | — | 12 | F | 1.5 | 40% fertile females/2 weeks | 0. | 0.5 | — | None—2.1 | — | — |

## Table 2 (continued)
## RELATIONSHIP BETWEEN LIGHT INTENSITY AND GROWTH IN MACROALGAE[a]

| Species (Reference) | Environment[b] | Growth conditions[c] | Exp. T[o,d] (°C) | Type of light[e] | Light intensity at growth optima[f] (klx) | Maximum growth rate[g] | Light intensity at one half maximum growth rate[g] (klx) | Compensation light intensity[i] (klx) | $I_k$(klx) | Light intensity for initial photoinhibition[k] (klx) | Maximum light intensity tested (klx) | Inhibition[l] (%) |
|---|---|---|---|---|---|---|---|---|---|---|---|---|
| *Saccorhiza* sp. (sporophyte) (46) | North Atlantic, marine | — | 10 | F | 0.003 cal/cm²/min (0.73) | 5.5 dbl/day | 0.001 cal/cm²/min (0.29) | 0.0001 cal/cm²/min (0.03) | 0.0004 cal/cm²/min. (0.58) | None—0.0084 cal/cm²/min (2.03) | — | — |
| **Chlorophyta — green algae** | | | | | | | | | | | | |
| *Acetabularia orenulata*(47) | Marine | 5klx, 23°C | 25 | F | 0.76 | 0.055 dbl in length/day | 0.38 | 0.09 | 0.76 | None—3.24 | — | — |
| | | 12L/12D | 24 | F | 80 µE/m²/sec (4.2) | — | 10 µE/m²/sec (0.5) | — | 15 µE/m²/sec (0.8) | 80 µE/m²/sec (4.2) | 140 µE/m²/sec (7.3) | 25 |
| *Codium fragile*(sporelings) (48) | Marine | 24L/0D | 24 | F | 30 µE/m²/sec (1.6) | — | 5 µE/m²/sec (0.2) | — | 10 µE/m²/sec (0.5) | 30 µE/m²/sec (1.6) | 140 µE/m²/sec (7.3) | 63 |
| *C. fragile*(48) | Northwest Atlantic, Narragansett Bay | 12L/12D | 12 | F | 80 µE/m²/sec (4.2) | 12 mg dry wt/21 days | 15 µE/m²/sec (0.8) | — | 35 µE/m²/sec (1.8) | None—140 µE/m²/sec (7.3) | — | — |
| | | 12L/12D | 24 | F | 80 µE/m²/sec (4.2) | 25 mg dry wt/21 days | 15 µE/m²/sec (0.8) | — | 32 µE/m²/sec (1.7) | 80 µE/m²/sec (4.2) | 140 µE/m²/sec (7.3) | 10 |
| | | 12L/12D | 30 | F | 30 µE/m²/sec (1.6) | 10 mg dry wt/21 days | 15 µE/m²/sec (0.8) | — | 30 µE/m²/sec (1.6) | 60 µE/m²/sec (3.1) | 140 µE/m²/sec (7.3) | 50 |
| | | 24L/0D | 12 | F | 80 µE/m²/sec (4.2) | 20 mg dry wt/21 days | 20 µE/m²/sec (1.0) | — | 30 µE/m²/sec (1.6) | None—140 µE/m²/sec(7.3) | — | — |
| | | 24L/0D | 24 | F | 80 µE/m²/sec (4.2) | 52 mg dry wt/21 days | 15 µE/m²/sec (0.8) | — | 35 µE/m²/sec (1.8) | 80 µE/m²/sec (4.2) | 140 µE/m²/sec (7.3) | 12 |

| 24/0D | 30 | F | 30 μE/m²/sec (1.6) | 28 mg dry wt/21 days | 10 μE/m²/sec (0.5) | — | 30 μE/m²/sec (1.6) | 30 μE/m²/sec (1.6) | 140 μE/m²/sec (7.3) | 100 |

a  Abbreviations used: cal/cm²/min, calories centimeter⁻² minute⁻¹; kerg/cm²/sec, kiloerg centimeter⁻² minute⁻¹; klx, 1000 × lx; Ly/min, langleys minute⁻¹; Q/m²/sec, quanta meter⁻² second⁻¹; μE/m²/sec, microeinsteins meter⁻² second⁻¹; mW/cm², milliwatts centimeter⁻²; dbl, doubling; pcv, packed cell volume; L/D, hours light/hours dark.

b  Environment — general habitat or location of sampling.

c  Growth conditions — environmental conditions under which test organisms were grown.

d  Exp. T° — experimental temperature in degrees centigrade.

e  Type of light — lighting used in experiments: T, tungsten; F, fluorescent. It should be noted that these are only rough categorizations of the light sources used. Each category contains a wide diversity of lamp types, e.g., cool white vs. daylight fluorescent.

f  Light intensity at growth maximum—intensity at which the growth rate reaches saturation.

g  Maximum growth — maximum rate of growth observed.

h  Light intensity at one half growth maximum — i.e., one half of the peak rate of growth.

i  Compensation light intensity — intensity at which photosynthesis equals respiration.

j  $I_k$ — light intensity at which a line through the initial slope of the light intensity versus growth rate curve intersects a line parallel to the abscissa at 6 max. It is commonly used as a measure of "shade" and "sun" adaptation.

k  Light intensity for initial photoinhibition — intensity at which there is 5% inhibition of growth. If there is no photoinhibition "none" is used followed by maximum intensity tested.

l  Inhibition (%) — percent inhibition at maximum light intensity tested.

intermediate zones.[9,50-54] While data on the vertical distribution of algae does lend some weight to this concept, there are many noteworthy exceptions to the rule, as can be seen in the tables and Figure 1, which contains a comparative look at the light intensity preferences and tolerances of photosynthesis in red, brown, and green macroalgae.

Even within a given species and environment, there are a considerable number of biological and environmental factors which can change the light intensity tolerance or preference of organisms. Early stages in development, for example, can be more sensitive to light intensity than adult forms.[22,43,45,55-58] This is illustrated by the comparative inhibition of photosynthesis in adult and sporeling forms of the brown alga *Carpophyllum plumosum*.[27] At 30 klx, photosynthesis is inhibited by 15% in the former and by 100% in the latter.

It is also well recognized that the existence of suboptimal conditions can lead to greater susceptibility to light damage or at least to a reduction of efficiency in light utilization. For example, a light intensity of 54 klx inhibits photosynthesis in the red alga *Chondrus crispus* by 20% at 15°C, but by 50% at 5°C.[14] Hence, it is important to determine the optimal conditions for growth or photosynthesis before establishing the light intensity preference and tolerance of given organisms.

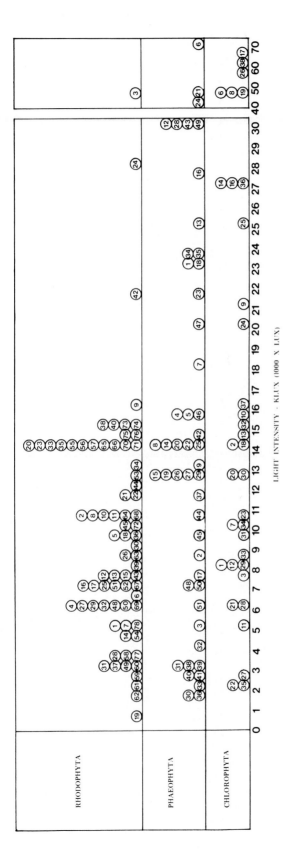

LIGHT INTENSITY - KLUX (1000 X LUX)

FIGURE 1. Species differences in adaptation of photosynthesis to light intensity. Figures 1A to 1E contain phylogenetic comparisons of light saturation, one half saturation light intensities, compensation intensities, $I_k$, and levels of photoinhibition for the experiments described in Table 1. The results in Figure 1 have been numbered within each phylum for reference purposes. These numbers correspond to those used in Table 1 (see right column of Table 1, i.e., case #). (A) Light intensity at which photosynthesis saturates, (B) light intensity at which photosynthesis reaches one half saturation level, (C) compensation light intensity, (D) $I_k$ (for explanation see text), (E) highest light intensity at which less than 5% photoinhibition is observed. Note: "○" indicates that no inhibition was observed at the highest intensity tested and "□" indicates that photoinhibition was observed beyond the light intensity indicated.

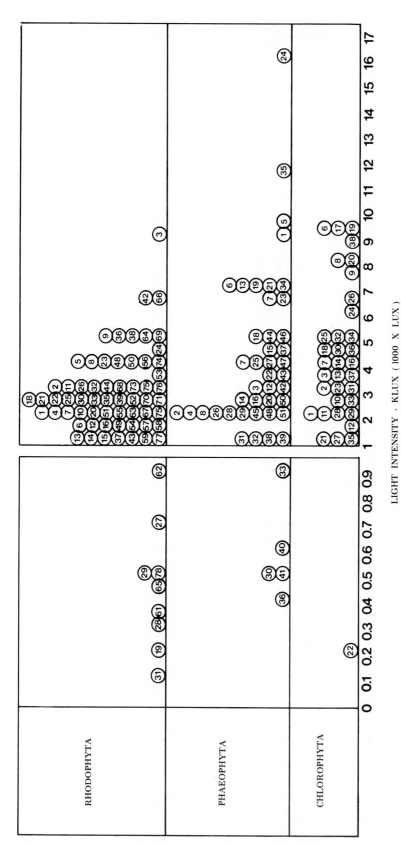

LIGHT INTENSITY - KLUX (1000 X LUX)

FIGURE 1B.

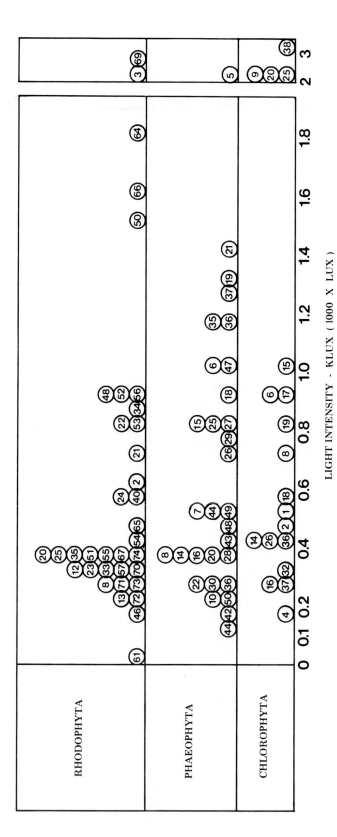

LIGHT INTENSITY - KLUX ( 1000 X LUX )

FIGURE 1C.

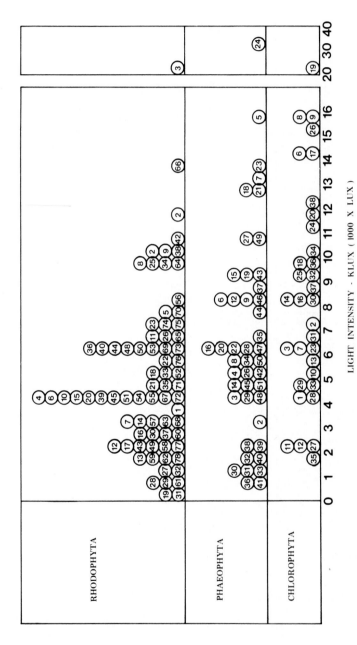

LIGHT INTENSITY - KLUX ( 1000 X LUX )

FIGURE 1D.

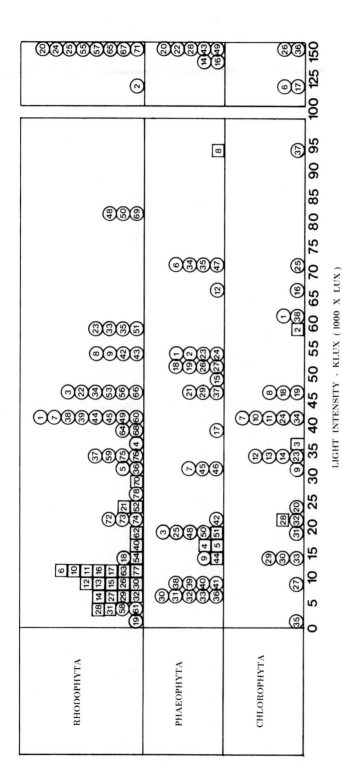

LIGHT INTENSITY - KLUX (1000 X LUX )

FIGURE 1E.

# REFERENCES

1. Bainbridge, R., Evans, G. C., and Rackham, O., *Light as an Ecological Factor,* Symp. Brit. Ecol. Soc., Blackwell Scientific, London, 1965.
2. Evans, G. C., Bainbridge, R., and Rackham, D., *Light as an Ecological Factor,* Vol. II, Blackwell Scientific, London, 1974.
3. Hellebust, J. A., Light: plants, in *Marine Ecology,* Vol. 1 (Part 1), Kinne, O., Ed., Wiley-Interscience, 1970, 125.
4. Kanwisher, J. W., Photosynthesis and respiration in some seaweeds, in *Some Contemporary Studies in Marine Science,* Barnes, H., Ed., George Allen and Unwin, Ltd., London, 1966, 407.
5. Levring, T., Submarine daylight and the photosynthesis in marine algae, *Göteborgs K. Vet. o. Vitt. Samh. Handl.,* IV., Ser. 5, 5A, 6, 1, 1947.
6. Levring, T., Light conditions, photosynthesis, and growth of marine algae in coastal and clear oceanic water, in *Proc. 6th Int. Seaweed Symp.,* Margalef, R., Ed., Subsecretaria De La Marina Merante, Madrid, 1969, 235.
7. Mann, K. H., Seaweeds: their productivity and strategy for growth, *Science,* 182, 975, 1973.
8. Parsons, T. R., Takahashi, M., and Hargrave, B., *Biological Oceanographic Processes,* 2nd ed., 1977.
9. Soeder, C. and Stengel, E., Physico-chemical factors affecting metabolism and growth rate, in *Algal Physiology and Biochemistry,* Stewart, W. D. P., Ed., University of California Press, 1975, 714.
10. Yokohama, Y., Photosynthetic properties of marine benthic red algae from different depths in coastal area, *Bull. Jpn. Soc. Phycol.,* 21, 119, 1973.
11. Marsh, J. A., Jr., Primary productivity of reef-building calcareous red algae, *Ecology,* 51, 255, 1970.
12. King, R. J. and Schramm, W., Photosynthetic rates of benthic marine algae in relation to light intensity and seasonal variations, *Mar. Biol.,* 37, 215, 1976.
13. Wallentinus, I., Productivity studies on Baltic macroalgae, *Bot. Mar.,* 21, 365, 1978.
14. Mathieson, A. E. and Burns, R. L., Ecological studies of economic red algae. I. Photosynthesis and respiration of *Chondrus crispus* Stackhouse and *Gigartina stellata* (Stackhouse) Batters, *J. Exp. Mar. Biol. Ecol.,* 7, 197, 1971.
15. Mathieson, A. C. and Norall, T. L., Photosynthetic studies of *Chrondrus crispus,* *Mar. Biol.,* 33, 207, 1975.
16. Ramus, J., Beale, S. I., and Mauzerall, D., Correlation of changes in pigment content with photosynthetic capacity of seaweeds as a function of water depth, *Mar. Biol.,* 37, 231, 1976.
17. Haxo, F. T. and Blinks, L. R., Photosynthetic action spectra of marine algae, *J. Gen. Physiol.,* 33, 389, 1950.
18. Mathieson, A. C. and Dawes, C. J., Ecological studies of Floridian *Eucheuma.* II. Photosynthesis and respiration, *Bull. Mar. Sci.,* 24, 274, 1974.
19. Ogata, E. and Matsui, T., Photosynthesis in several marine plants of Japan in relation to carbon dioxide supply, light and inhibitors, *Jpn. J. Bot.,* 19, 83, 1965.
20. Healey, F. P., Photosynthesis and respiration of some Arctic seaweeds, *Phycologia,* 11, 267, 1972.
21. Wanders, J. B. W., The role of benthic algae in the shallow reef of Curacao. I. Primary productivity in the coral reef, *Aquat. Bot.,* 2, 235, 1976.
22. Boney, A. D. and Corner, E. D. S., The effect of light on the growth of sporelings of the intertidal red alga *Plumaria elegans* SCHM, *J. Mar. Biol. Assoc. U.K.,* 42, 65, 1962.
23. Fralick, R. A. and Mathieson, A. C., Physiological ecology of four *Polvsiphonia* species (Rhodophyta, Ceramiales), *Mar. Biol.,* 29, 29, 1975.
24. Ogata, E. and Matsui, T., Photosynthesis in several marine plants of Japan as affected by salinity, drying and pH, with attention to their growth habitats, *Bot. Mar.,* 8, 199, 1965.
25. Brinkhuis, H., Tempel, N. R., and Jones, R. F., Photosynthesis and respiration of exposed salt-marsh fucoids, *Mar. Biol.,* 34, 349, 1976.
26. Chock, J. S. and Mathieson, A. C., Physiological ecology of *Ascophyllum noposum* (L.) Le Jolis and its detached escad *Scorpioides* (Hornemann) Hauck (Facales, Phaeophyta), *Bot. Mar.,* 22, 21, 1979.
27. Chapman, V. J., The physiological ecology of some New Zealand seaweeds, in *Proc. 5th Int. Seaweed Symp.,* Young, E. G. and McLachlan, J. L., Eds., Pergamon Press, New York, 1965, 29.
28. Prahl, C., Photosynthesis and respiration of some littoral marine algae from Greenland, *Phycologia,* 18, 166, 1979.
29. Chapman, V. J., A contribution to the ecology of *Egregia laevigata* Setchell. III. Photosynthesis and respiration; conclusions, *Bot. Mar.,* 3, 101, 1962.
30. Steemann Nielsen, E., Der mechanismus der photosynthese, *Dan. Bot. Ark.,* 11, 1, 1942.
31. King, R. J. and Schramm, W., Determination of photosynthetic rates for the marine algae *Fucus vesiculosus* and *Laminaria digitata,* *Mar. Biol.,* 37, 209, 1976.

32. Kageyama, A. and Yokohama, Y., Photosynthetic properties of marine benthic brown algae from different depths in coastal area, *Bull. Jpn. Soc. Phycol.,* 22, 119, 1974.

33. Lüning, K., Seasonal growth of *Laminaria hyperborea* under recorded underwater light conditions near Helgoland, in *4th Eur. Marine Biology Symp.,* Crisp, D. J., Ed., Cambridge University Press, Cambridge, 1969, 347.

34. Kain, J. M., Drew, E. A., and Jupp, B. P., Light and the ecology of *Laminaria hyperborea* II, in *Light as an Ecological Factor* Vol. 2, Evans, G. C., Bainbridge, R., and Rackham, D., Eds., 1975, 63.

35. Chapman, A. R. O., Markham, J. W., and Lüning, K., Effects of nitrate concentration on the growth and physiology of *Laminaria saccharina* (Phaeophyta) in culture, *J. Phycol.,* 14, 195, 1978.

36. Clendenning, K. A., Photosynthesis and general development in macrocystis, *Nova Hedwigia Z. Kryptogamen Kd.,* 32, 169, 1971.

37. Raven, J. A. and Smith, F. A., 'Sun' and 'shade' species of green algae: relation to cell size and environment, *Photosynthetica,* 11, 48, 1977.

38. Yokohama, Y., Photosynthetic properties of marine benthic green algae from different depths in the coastal area, *Bull. Jpn. Soc. Phycol.,* 21, 70, 1973.

39. Rabinowitch, E. I., *Photosynthesis,* John Wiley & Sons, New York, 1951, 964.

40. Kageyama, A. and Yokohama, Y., Pigments and photosynthesis of deep-water green algae, *Bull. Jpn. Soc. Phycol.,* 25, 168, 1977.

41. Gemmill, E. R. and Galloway, R. A., Photoassimilation of $^{14}$C-acetate by *Ulva lactuca, J. Phycol.,* 10, 359, 1974.

42. Boney, A. D. and Corner, E. D. S., The effect of light on the growth of the red algae *Antithamnion plumula* and *Brongniartella byssoides, J. Mar. Biol. Assoc. U.K.,* 43, 319, 1963.

43. Jones, W. E. and Dent, E. W., The effect of light on the growth of algal spores, in *4th Eur. Marion Biology Symp.,* Crisp, D. J., Ed., Cambridge University Press, Cambridge, 1969, 363.

44. Adey, W. H., The effects of light and temperature on growth rates in Boreal-subarctic crustose corrallines, *J. Phycol.,* 6, 269, 1970.

45. Lüning, K. and Neushul, M., Light and temperature demands for growth and reproduction of laminarian gametophytes in southern and central California, *Mar. Biol.,* 45, 297, 1978.

46. Kain, J. M., The biology of *Laminaria hyperborea.* V. Comparison of early stages of competitors, *J. Mar. Biol. Assoc. U.K.,* 49, 455, 1969.

47. Terborgh, J. and Thimann, K. V., Interaction between daylength and light intensity in the growth and chlorophyll content of *Acetabularia crenulata, Planta,* 63, 83, 1964.

48. Hanisak, M. D., Growth patterns of *Codium fragile* sp. *Tomentosoides* in response to temperature, irradiance, salinity and nitrogen source, *Mar. Biol.,* 50, 319, 1979.

49. Boardman, N. K., Comparative photosynthesis of sun and shade plants, *Annu. Rev. Plant Physiol.,* 28, 355, 1977.

50. Crossett, R. N., Drew, E. A., and Larkum, A. W. D., Chromatic adaptation in benthic marine algae, *Nature (London),* 207, 547, 1965.

51. Dring, M. J., Chromatic adaptation in marine algae: an examination of its significance using a computer model, *Br. Phycol. J.,* 12, 118, 1977.

52. Dellow, V. and Cassie, R. M., Littoral zonation in two caves in the Aukland district, *Trans. R. Soc. N.Z.,* 83, 321, 1955.

53. Norton, T. A., Ebling, F. J., and Kitching, J. A., Light and the distribution of organisms in a sea cave, in *4th Eur. Marine Biology Symp.,* Crisp, D. J., Ed., Cambridge University Press, Cambridge, 1969, 409.

54. Zavodnik, D., Light conditions and shade seeking populations among algal settlements, in *4th Eur. Marine Biology Symp.,* Crisp, D. J., Ed., Cambridge University Press, Cambridge, 1969, 433.

55. Arasaki, S., An experimental note on the influence of light on the development of spores of algae, *Bull. Jpn. Soc. Sci. Fish.,* 19, 466, 1953.

56. Kain, J. M., Aspects of the biology of *Laminaria hyperborea.* III. Survival and growth of gametophytes, *J. Mar. Biol. Assoc. U.K.,* 44, 415, 1964.

57. Kain, J. M., Aspects of the biology of *Laminaria hyperborea.* IV. Growth of early sporophytes, *J. Mar. Biol. Assoc. U.K.,* 45, 129, 1965.

58. Segi, T. and Kida, W., Studies on the development of *Undaria undarioides* (Yendo) Okamura. I. On the development of gametophytes and influence of light intensity on it, *Rep. Fac. Fish. Prefect. Univ. Mie,* 2, 517, 1957.

# TEMPERATURE PREFERENCE AND TOLERANCE OF AQUATIC PHOTOSYNTHETIC MICROORGANISMS

## Edward J. Phlips and Akira Mitsui

## INTRODUCTION

Temperatures within the aquatic habitat range from well below freezing in the ice packs of polar latitudes to greater than 93°C in some hot springs.[1-5] A variety of different life forms have been found to grow throughout almost this entire range of environments. Life at the extreme ends of the temperature scale involves special hardships, as exemplified by the significant decrease of species diversity of plants and animals at both low and high temperatures.[4] This section deals with the tolerance of aquatic photosynthetic organisms to low and high temperatures (previously reviewed in References 1-27), particularly as it relates to photosynthesis, growth, and productivity.

The thermal preferences and tolerances of organisms have been traditionally subdivided into three principle categories: psychrophiles, organisms which prefer cold temperatures, mesophiles, organisms which prefer intermediate temperatures, and thermophiles, organisms which prefer high temperatures. The quantitative values associated with these divisions are dependent upon the group of organisms being discussed. This point is particularly clear when discussing maximum temperatures, since there is a phylogenetic pattern to high-temperature tolerance.[4] It appears that the maximum temperatures for survival and optimal growth are inversely proportional to cellular and morphological complexity. In other words, the two structurally simplest forms of life, bacteria and blue-green algae, are also the most temperature tolerant. Keeping this in mind, the following rough values can be given for the temperature categories of aquatic photosynthetic organisms:

1. Thermophiles, eucaryotic organisms with optimal growth at temperatures greater than 35°C, or blue-green algae with optima greater than 45°C[5]
2. Mesophiles, organisms with optimal growth temperatures between 20 and 45°C
3. Psychrophiles: (a) obligate, organisms which can grow at 0°C and have optimal growth temperatures below 20°C, and (b) facultative, organisms which can grow at 0°C, but which have optimal growth temperatures greater than or equal to 20°C[18]

The majority of photosynthetic species fit into the category of mesophiles. This point is exemplified in Tables 1 and 2 and Figures 1 and 2 which describe the results of temperature experiments with a wide range of species. There are, however, a significant number of species with excellent growth and photosynthetic capabilities at extreme temperatures. The characteristics of these psychrophiles and thermophiles are reviewed briefly below.

## LOW-TEMPERATURE TOLERANCE OF PHOTOSYNTHETIC MICROORGANISMS

Since a large portion of the aquatic environment is characterized by cold temperatures, the ability of certain organisms to function normally under these conditions is of considerable importance to world primary productivity. The problems which low temperature present to life can be looked at from two viewpoints:

Table 1

THE RELATIONSHIP BETWEEN TEMPERATURES AND PHOTOSYNTHESIS IN AQUATIC PHOTOSYNTHETIC MICROORGANISMS

| Species (reference) | Environment | Growth conditions | Light intensity (klx) | Optimum temperature (°C) | Pmax, maximum rate of photosynthesis | Maximum temperature for one half Pmax (°C) | Highest temperature tested[a] (°C) | Percent of Pmax at highest temperature tested | Minimum temperature for one half Pmax (°C) | Lowest temperature tested[a] (°C) | Percent of Pmax at lowest temperature tested |
|---|---|---|---|---|---|---|---|---|---|---|---|
| Photosynthetic bacteria | | | | | | | | | | | |
| Chloroflexis(28) | Twin Butte Vista, hot spring | 72°C | 65 | 66 | — | 71 | 74 | 20 | 56 | 54 | 35 |
| | | 60°C | 11—55 | 60 | — | 65 | 70 | 0 | 52 | 45 | 20 |
| | | 50°C | 12—43 | 54 | — | 64 | 65 | 40 | 45 | 35 | 20 |
| | | 45°C | 14—89 | 50 | — | 55 | 60 | 30 | 33 | 30 | 25 |
| Blue-green algae — Cyanophyta | | | | | | | | | | | |
| Anabaena cylindrica(29) | Freshwater | 20°C | Saturating | 28 | 133 μmol C/ mg Chl/hr | 33 | 36 | 40 | 19 | 5 | 17 |
| Nostoc muscorum(30) | — | — | Saturating | 35 | — | 41 | 45 | 9 | 26 | 0 | 3 |
| Oscillatoria(31) | Cold mountain stream | 0°C | 11.8 | 25 | — | 35 | 40 | 25 | 6 | 0 | 30 |
| Oscillatoria geminata(32) | — | — | — | 35—40 | — | 43 | 45 | 25 | 20 | — | — |
| Phormidium (31) | Cold mountain stream | 0°C | 11.8 | 25 | — | 35 | 40 | 25 | 6 | 0 | 30 |
| Synechococcus lividus(33) | Oregon, hot springs | 70°C | 6 | 66 | — | 72 | 75 | 13 | 48 | 35 | 7 |
| S. lividus(34) | Hot springs | 35°C | 16 | 50 | 1000 μmol O₂/mg Chl/hr | 60 | 62 | 23 | 41 | 35 | 28 |
| | | 45°C | 16 | 50 | 563 μmol O₂/ mg Chl/hr | 60 | 62 | 26 | 33 | 30 | 36 |
| | | 57°C | 16 | 55 | 214 μmol O₂/ mg Chl/hr | 61 | 62 | 38 | 46 | 35 | 31 |

| | | | | | | | | | | | |
|---|---|---|---|---|---|---|---|---|---|---|---|
| *S. lividus* (35) | Hot springs | 35°C | 25 | 45 | — | 57 | 65 | 0 | 36 | 30 | 32 |
| | | 45°C | 25 | 50 | — | 58 | 65 | 0 | 36 | 30 | 37 |
| | | 57°C | 25 | 55 | — | 60 | 65 | 0 | 48 | 30 | 0 |
| | | 65°C | Satu-rating | 65 | 296 $\mu$mol O$_2$/mg Chla/hr | — | 73 | 69 | — | 55 | 69 |
| | | 65°C | Satu-rating | 65 | 182 $\mu$mol O$_2$/mg Chla/hr | 71 | 73 | 31 | — | 55 | 57 |
| Cryptophyta | | | | | | | | | | | |
| *Cyanidium cal-darium* (36) | Yellowstone, hot springs | 56°C | 5 | 45 | — | 50 | 74 | 0 | 32 | 20 | 5 |
| Dinoflagellates — Pyrrophyta | | | | | | | | | | | |
| *Gonyaulax cate-nella* (37) | Northeast Pa-cific | — | 1.5—3.5 | 13—17 | — | — | 23 | — | — | 12 | — |
| Diatoms — Bacillariophyta | | | | | | | | | | | |
| *Achnanthes exi-gua* (38) | Montana, hot spring | 40°C | 6.89 | 42 | 2100 $\mu$mol O$_2$/mg Chl/hr | 46 | 48 | 0 | 23 | 15 | 25 |
| *Cymbella* (39) | Wisconsin, fro-zen lake | <4°C | 2.65 | 26 | — | 32 | 45 | 3 | 8 | 4 | 38 |
| *Fragilaria subli-nearis* (39) | Antarctic ma-rine | 1 klx, 2—3°C | Satu-rat-ing, 678 nm | 3—10 | — | — | 24 | 18 | — | 3 | 100 |
| *Nitzschia clos-terium* (40) | North Atlantic, planktonic | — | — | 26—30 | 7.8 $\mu$mol O$_2$/10 $\mu$L pcv/hr | 35 | 37 | 0 | 18 | 5 | 6 |
| *Nitzschia palea* (40) | Freshwater, planktonic | — | — | 33 | 17.9 $\mu$mol O$_2$/10 $\mu$L pcv/hr | — | 39 | 12 | 15 | 6 | 10 |
| *Phaeodactylum tricornutum* (41) | Marine, plank-tonic | 5°C | 8 | 10 | 6.8 $\mu$mol C/10$^8$ cells/hr | — | 20 | 58 | — | 5 | 78 |
| | | 10°C | 8 | 10 | 5.3 $\mu$mol C/10$^8$ cells/hr | — | 20 | 81 | — | 5 | 80 |

## Table 1 (continued)
## THE RELATIONSHIP BETWEEN TEMPERATURES AND PHOTOSYNTHESIS IN AQUATIC PHOTOSYNTHETIC MICROORGANISMS

| Species (reference) | Environment | Growth conditions | Light intensity (klx) | Optimum temperature (°C) | Pmax, maximum rate of photosynthesis | Maximum temperature for one half Pmax (°C) | Highest temperature tested[a] (°C) | Percent of Pmax at highest temperature tested | Minimum temperature for one half Pmax (°C) | Lowest temperature tested[a] (°C) | Percent of Pmax at lowest temperature tested |
|---|---|---|---|---|---|---|---|---|---|---|---|
| | | 20°C | 8 | 20 | 7.8 µmol C/10$^8$ cells/hr | — | 20 | 100 | — | 5 | 54 |
| Skeletonema costatum(42) | Marine, planktonic | 10 klx, 20°C, low nutrient | 13 | 20 | — | 26 | 30 | 25 | 14 | 5 | 0 |
| | | 10 klx, 20°C, normal nutrients | 13 | 30 | — | — | 30 | 100 | 16 | 5 | 0 |
| S. costatum(17) | Marine, planktonic | — | 10 | 20 | 375 µmol C/mg Chl/hr | — | 20 | 100 | — | 2 | 62 |
| Synedra(29) | Freshwater, planktonic | 20°C | Saturating | 23 | 167 µmol C/mg Chl/hr | 32 | 35 | 25 | 10 | 5 | 25 |
| Green algae — Chlorophyta | | | | | | | | | | | |
| Ankistrodesmus falcatus(43) | — | — | 5 | 25 | 2.14 µmol O$_2$/mg dry wt/hr | 32 | 40 | 0 | 14 | 7 | 0 |
| Chlamydomonas nivalis(44) | Snow alga | Station 78-3 | Saturating | 10 | — | 25 | 30 | 10 | — | 0 | 60 |
| | | Station 89-3 | Saturating | 10 | — | — | 30 | 65 | — | −3 | 85 |
| | | Station 90-3 | Saturating | 20 | — | 25 | 30 | 3 | — | −3 | 70 |
| | | Station 100-4 | Saturating | 0 | — | 8 | 30 | 3 | — | 0 | 100 |

| Organism | Habitat | | | | | | | | | | |
|---|---|---|---|---|---|---|---|---|---|---|---|
| *Chlamydomonas reinhardii* (45) | Snow alga | — | 0.75 | 30 | — | — | 35 | 86 | 18 | 5 | 0 |
| *Chlorella ellipsoidea* (29) | Freshwater, planktonic | 10—13° C | 1.5 | 35 | — | — | 35 | 100 | 19 | 5 | 0 |
| | | 20—22° C | Saturating | 20—30 | 458 $\mu$mol C/mg Chl/hr | 38 | 40 | 10 | 6 | 4 | 45 |
| | | 30—33° C | Saturating | 30 | — | 37 | 45 | 0 | 18 | 5 | 15 |
| *Chlorella fusca* (46) | Freshwater, planktonic | — | Saturating | 25 | — | 41 | 45 | 0 | 22 | 5 | 5 |
| *Chlorella vulgaris* (47) | Freshwater, planktonic | — | 0.80 | 21—28 | — | — | 35 | 85 | 16 | 15 | 45 |
| *Coelastrum microporum* (43) | — | — | 2 | 28 | — | — | 28 | 100 | 12 | 10 | 42 |
| | | — | 4.1 | ≥28 | — | — | 28 | 100 | 15 | 10 | 27 |
| | | — | 5 | 30 | 3.1 $\mu$mol O$_2$/mg dry wt/hr | 40 | 28 | 0 | 18 | 10 | 20 |
| *Hydrodictyon africanum* (48) | Freshwater | — | Saturating | 30 | 35 $\mu$mol O$_2$/mg Chl/hr | — | 44 | 93 | 20 | 8 | 0 |
| *Protoccus* (31) | Cold mountain stream | 0°C | 11.8 | 30 | — | 35 | 40 | 20 | 2 | 0 | 60 |
| *Scenedesmus* (29) | Freshwater | 20°C | Saturating | 35 | 417 $\mu$mol C/mg Chl/hr | 41 | 45 | 5 | 20 | 20 | 40 |
| *Scenedesmus obliquus* (49) | Freshwater | — | 28 W/m$^2$ (9) | 30 | 730 $\mu$mol O$_2$/mg Chl/hr | 35 | 35 | 50 | — | — | 20 |
| *Scenedesmus obtusiusculus* (43) | — | — | 5 | 35 | 3.5 $\mu$mol O$_2$/mg dry wt/hr | 39 | 44 | 0 | 18 | 18 | 62 |
| *Scenedesmus quadricauda* (50) | Freshwater | — | — | 30 | — | — | 35 | 95 | 9 | 9 | 50 |
| *Spirogyra* (51) | — | — | — | 20 | — | — | >25 | — | <5 | — | — |
| *Zygnema* (31) | Cold mountain stream | 0°C | 11.8 | 30 | — | 35 | 40 | 15 | 8 | 0 | 30 |
| *Zygogonium* (52) | Yellowstone, acidic ponds | 25°C | Saturating | 25 | — | 29 | 45 | 4 | 19 | 18 | 40 |

[a] Or the temperature at which no photosynthesis is first observed in experiments which span a sufficient temperature range.

Table 2

## THE RELATIONSHIP BETWEEN TEMPERATURE AND GROWTH IN AQUATIC PHOTOSYNTHETIC MICROORGANISMS

| Species (Reference) | Environment | Preculture conditions | Light intensity (klx) | Optimum temperature (°C) | Gmax, Maximum growth rate | Maximum temperature for one half Gmax (°C) | Highest temperature tested (°C) | Percent of Pmax at highest temperature tested | Minimum temperature for one half Gmax (°C) | Lowest temperature tested (°C) | Percent of Pmax at lowest temperature tested |
|---|---|---|---|---|---|---|---|---|---|---|---|
| **Blue-green algae — Cyanophyta** | | | | | | | | | | | |
| Anabaena (53) | Gulf Coast of U.S., marine | NH₄Cl added | Saturating | 45 | 5.5 dbl/day | 49 | 50 | 0 | — | 30 | 59 |
| Anabaena (54) | Freshwater, planktonic | Grown on N₂ | Saturating | 42 | 4.5 dbl/day | 49 | 50 | 0 | — | 30 | 64 |
|  |  |  | 4 | 36 |  |  |  |  |  |  |  |
| Anabaena variabilis(55) | Freshwater | — | 3.5 | 34 | 4 dbl/day | — | 38 | 67 | — | 25 | 58 |
| A. variabilis(56) | Egypt, rice fields | — | — | 36 | — | 40 | 45 | 10 | — | 20 | 70 |
| Anacystis nidulans(55) | Freshwater | — | 4.5 | 41 | 11.1 dbl/day | — | 44 | 71 | — | 25 | 26 |
| A. nidulans(54) | Freshwater, planktonic | — | Saturating | 40 | — | — | — | — | 30 | — | — |
| Calothrix(57) | Oregon, hot spring | 45—50°C | 4 | 45 | 2.3 dbl/day | — | 50 | 74 | 31 | 31 | 50 |
| Calothrix elenkinii(56) | Egypt, rice fields | 25°C | — | 36 | — | 45 | 45 | 50 | — | 20 | 70 |
| Gomphosphaeria aponina(58) | Florida, marine, planktonic | — | — | 26—30 | — | 35 | 38 | 40 | — | 22 | 60 |
| Hapalosiphon fontinalis(56) | Egypt, rice fields | — | — | 34 | — | 42 | 45 | 18 | 21 | 20 | 45 |
| Mastigocladus laminosus(59) | British Columbia, hot spring | — | Saturating | 50 | — | 56 | 60 | 0 | 30 | 25 | 5 |
| Nostoc muscorum(55) | Freshwater | — | 3.5 | 33 | 2.6 dbl/day | — | 35 | 88 | — | 25 | 63 |
| Oscillatoria geminata(32) | — | — | — | 40 | — | 52 | 55 | 0 | 31 | 22 | 25 |
| Oscillatoria terebriformis(60) | Hot spring | — | — | 48—52 | — | — | — | — | — | — | — |
| Pleurocapsa sp. (57) | Oregon, hot spring | 45—50°C | 4 | 45 | — | 53 | 55 | 38 | 35 | 35 | 50 |
| Synechococcus elongatus(61) | Kuril Islands, hot spring | — | 340 W/m² (100) | 57 | 1.3 dbl/day | 63 | 67 | 0 | 46 | 37 | 0 |
| Synechococcus lividus(62) | Yellowstone, hot springs | 50—55°C | 4.24 | 48 | 4.2 dbl/day | — | 55 | 76 | 36 | 35 | 43 |
| S. lividus(62) | Hot springs | 50—55°C | 15.9 | 52 | 6.6 dbl/day | 62 | 55 | 80 | 40 | 45 | 8 |
|  |  | 53°C |  | 45 | 10 dbl/day | 66 | 65 | 80 | 45 | 30 | 20 |
|  |  | 60°C |  | 50 | 4.5 dbl/day | 71 | 70 | 0 | 51 | 35 | 0 |
|  |  | 66°C |  | 55 | 5.5 dbl/day | 69 | 75 | 0 | 53 | 45 | 0 |
|  |  | 75°C |  | 65 | 2 dbl/day | — | — | 0 | 55 | 50 | 0 |
| S. lividus(33) | Oregon, hot springs | — | 6 | 66 | 1.9 dbl/day | — | 72 | 11 | — | 54 | 0 |
| Tolypothrix tenuis(63) | Rice paddies | — | 15 | 38 | 4 dbl/day | 41 | 43 | 33 | 28 | 26 | 42 |
| **Cryptophyta** | | | | | | | | | | | |
| Cryptomonas frigoris(64) | Snow | — | — | 5 | — | — | 10 | — | — | 2 | — |
| Cryptomonas ovata(65) | British Columbia, lake | — | Saturating | 23 | — | 26 | 27 | 0 | 18 | 8 | 10 |

| Organism | Habitat | | | | | | | | | |
|---|---|---|---|---|---|---|---|---|---|---|
| *Cyanidium caldarium*(54) | Hot springs | 3.5 | 43 | — | — | — | — | — | — | — |
| *C. caldarium*(36) | Yellowstone, hot springs | 5 | 45 | — | 52 | 58 | 0 | 35 | 25 | 10 |
| **Dinoflagellates — Pyrrophyta** | | | | | | | | | | |
| *Amphidinium carteri*(66) | Woods Hole, marine rock pool | 0.30 Ly/min (50) | 24 | 1 dbl/day | 26 | 33 | 0 | 16 | 18 | 0 |
| *Cachonina niei*(67) | Salton Sea, Calif. | 10.6 | 20 | 0.76 dbl/day | — | 27 | 63 | 13 | 15 | 13 |
| *Ceratium furca*(68) | Marine, planktonic | Saturating | 20 | 0.35 dbl/day | 25 | 25 | 50 | 9 | 5 | 0 |
| *Ceratium fusus*(68) | Marine, planktonic | Saturating | 15 | 0.30 dbl/day | — | 25 | 80 | 10 | 5 | 0 |
| *Ceratium lineatum*(68) | Marine, planktonic | Saturating | 20 | 0.35 dbl/day | — | 25 | 71 | 12 | 5 | 9 |
| *Ceratium tripos*(68) | Marine, planktonic | Saturating | 20 | 0.28 dbl/day | 30 | 25 | 50 | 20 | 5 | 82 |
| *Exuviella baltica*(69) | Tropical, marine, planktonic | — | 22—26 | 1.2 dbl/day | 30 | 30 | 0 | 18 | 7 | 20 |
| *Gymnodinium* strain 581 (70) | Tropical, marine, planktonic | 15 | 27 | 1.7 dbl/day | 31 | 32 | 0 | 13 | 16 | 0 |
| *Gymnodinium* strain 582 (70) | Tropical, marine, planktonic | 12.5 | 23—27 | — | 30 | 34 | 0 | 14 | 15 | 0 |
| *Gymnodinium breve*(58) | Florida, marine, planktonic | 25°C | 22—25 | — | 29 | 32.5 | 0 | 13 | 12 | 45 |
| *Gymnodinium splendens*(71) | Southern California, marine, planktonic | 6 | 20 | 0.44 dbl/day | 29 | 30 | 0 | 14 | 10 | 0 |
| *Gonyaulax polyedra*(69) | Marine, planktonic | — | 21 | — | 24 | 29 | 50 | 13 | 10 | 20 |
| *Peridinium*(72) | Marine, planktonic | 18°C | 18 | 1.0 dbl/day | 25 | 25 | 33 | 12 | 8 | 0 |
| *Peridinium trochoideum*(69) | Oslo Fjord, marine, planktonic | — | 19 | — | — | 25 | 50 | 10 | 7 | 20 |
| *P. trochoideum*(69) | Bay of Naples, marine, planktonic | — | 25 | — | — | 30 | 80 | 15 | 10 | 10 |
| *Prorocentrum gracile*(72) | Marine, planktonic | 18°C | 20 | 0.9 dbl/day | 21 | 25 | 55 | 11 | 7 | 10 |
| *Prorocentrum micans*(72) | Marine, planktonic | 18°C | 18 | 0.75 dbl/day | 28 | 25 | 0 | 13 | 8 | 0 |
| *P. micans*(69) | Oslo Fjord, marine, planktonic | — | 25 | — | — | 30 | 0 | 15 | 8 | 0 |
| *P. micans*(69) | Pacific, marine, planktonic | — | 20 | — | 29 | 27 | 55 | 11 | 9 | 30 |
| *P. micans*(69) | Marine, planktonic | — | 25 | — | — | 30 | 60 | 14 | 10 | 10 |
| **Diatoms — Bacillariophyta** | | | | | | | | | | |
| *Achnanthes brevipes*(73) | Marine, planktonic | 40°C | 18 | — | 28 | 26 | 0 | 8 | 10 | 40 |
| *Achnanthes exigua*(38) | Montana, hot spring | 6.89 | 40 | 2.0 dbl/day | 42 | 33 | 0 | 20 | 15 | — |
| *Actinocyclus*(74) | Marine, planktonic | — | 18—27 | — | — | 44 | — | — | — | 40 |
| *Amphiprora*(74) | Marine, planktonic | — | 8—30 | — | — | — | — | — | — | — |
| *Amphiprora*(75) | Holland, estuarine, benthic | 85 μE/m²/sec (4) | 25 | 2.5 dbl/day | 25 | 30 | 84 | 15 | 4 | 20 |
| *Asterionella japonica*(76) | Marine, planktonic | Saturating | 25 | — | 27 | 30 | 0 | 13 | 10 | 33 |
| *Asterionella socialis*(77) | Washington State, surf zone | 3 | 18 | 1.6 dbl/day | 21 | 24 | 0 | 8 | 5 | 25 |

Table 2 (continued)
## THE RELATIONSHIP BETWEEN TEMPERATURE AND GROWTH IN AQUATIC PHOTOSYNTHETIC MICROORGANISMS

| Species (Reference) | Environment | Preculture conditions | Light intensity (klx) | Optimum temperature (°C) | Gmax, Maximum growth rate | Maximum temperature for one half Gmax (°C) | Highest temperature tested* (°C) | Percent of Pmax at highest temperature tested | Minimum temperature for one half Gmax (°C) | Lowest temperature tested* (°C) | Percent of Pmax at lowest temperature tested |
|---|---|---|---|---|---|---|---|---|---|---|---|
| *Chaetoceros*(70) | Tropical, marine, planktonic | — | 14 | 29 | 6.2 dbl/day | 38 | 41 | 0 | 19 | 10 | 0 |
| *Chaetoceros armatum*(77) | Washington State, surf zone | — | 3 | 20 | 0.6 dbl/day | 23 | 24 | 65 | 9 | 5 | 0 |
| *Cyclotella nana*(Clone 3 H)(78) | Long Island, marine, planktonic | September | 4.5 | ≥25 | 2.7 dbl/day | — | 25 | <100 | 8 | 4 | 9 |
| *C. nana*(Clone 5 A)(78) | Long Island, marine, planktonic | September | 4.5 | 20 | 2.7 dbl/day | — | 25 | 93 | 11 | 4 | 9 |
| *C. nana*(Clone E.P.)(78) | Woods Hole, marine, planktonic | June | 4.5 | ≥25 | 2.8 dbl/day | — | 25 | ≤100 | 11 | 4 | 20 |
| *C. nana*(Clone 7-15)(78) | Marine, planktonic | December | 4.5 | 15—20 | 2.2 dbl/day | — | 25 | 59 | 6 | 4 | 27 |
| *C. nana*(Clone 13-1)(78) | Sargasso Sea, planktonic | — | 4.5 | 20—25 + | 1.8 dbl/day | — | 25 + | 100 | — | 15 | 55 |
| *Detonula confervacea*(78) | Marine, planktonic | January | 11 | 10 | 1.4 dbl/day | — | 15 | 55 | — | 4 | 55 |
| *D. confervacea*(78) | Marine, planktonic | — | — | 12 | 1.5 dbl/day | 15 | 16 | 0 | 3 | 2 | 40 |
| *D. confervacea*(79) | Arctic | — | 12 | 12 | 1.5 dbl/day | 15 | 16 | 0 | 3 | 2 | 41 |
| *Ditylum brightwelli*(80) | Marine, planktonic | — | 0.12 cal/cm²/min (20) | 26 | 2.5 dbl/day | 31 | 32 | 0 | 13 | 13 | 50 |
| *Fragilaria sublinearis*(22) | Antarctic ice alga | — | — | 5 | 1.0 dbl/day | — | — | — | -2 | 8—9 | — |
| *Melosira*(74) | Marine, planktonic | — | — | 15—24 | — | — | 27—30 | 0 | — | 4 | 13 |
| *Navicula arenaria*(75) | Holland, estuarine, benthic | — | 85 μE/m²/sec (4) | 15 | 1.5 dbl/day | 22 | 25 | 33 | 8 | 4 | 50 |
| *Nitzschia closterium (Phaeodactylum tricornatum)*(81) | Marine, planktonic | — | Saturating | 25 | 2.4 dbl/day | — | 25 | 100 | 10 | 10 | 50 |
| *Nitzschia closterium*(82) | Sargasso Sea, planktonic | 8.9 μg at N/1; 0.42 μg at P/1 | 0.86 | 16 | 0.3 dbl/day | 20 | 32 | 0 | 12 | 10 | 80 |
| | | 10 mg at N/1; 0.52 mg at P/1 | 1.90 | 23 | 0.8 dbl/day | — | 32 | 52 | — | 10 | 27 |
| *Nitzschia dissipata*(75) | Holland, estuarine, benthic | — | 85 μE/m²/sec (4) | 25 | 2.5 dbl/day | — | 25 | 100 | 13 | 4 | 16 |
| *Nitzschia laevis*(74) | Marine, planktonic | — | — | 15—24 | — | — | 30 | 0 | — | 4 | — |
| *Nitzschia sigma*(75) | Holland, estuarine, benthic | — | 85 μE/m²/sec (4) | 25 | — | — | 25 | 100 | 4 | 4 | 50 |

| Organism | Habitat | | | | | | | | | |
|---|---|---|---|---|---|---|---|---|---|---|
| *Phaeodactylum tricornutum*(74) | Marine, planktonic | 8—24 | — | — | — | 29—35 | 0 | — | — | — |
| *P. tricornutum*(73) | Marine, planktonic | 20 | — | — | 28 | 35 | 0 | 5 | — | — |
| *Rhizosolenia fragilissima*(83) | Near-shore marine | 18—25 | 12 | 1.2 dbl/day | — | 30 | 75 | 9 | 7 | 0 |
| *Skeletonema costatum*(66) | Long Island Sound, marine, planktonic | 20 | 0.25 Ly/min (40) | 2.4 dbl/day | — | >28 | 0 | 9 | 6 | 0 |
| *S. costatum*(84) | Marine, planktonic | 20 | 3 | 2.3 dbl/day | 33 | 20 | 100 | 17 | 7 | 40 |
| *Skeletonema tropicum*(Clone "*S. trop*")(85) | Tropical, marine, planktonic | 25—31 | 10.6 | 3.0 dbl/day | 32 | 36 | 0 | 19 | 13 | 0 |
| *S. tropicum*(Clone "21-L")(85) | Tropical, marine, planktonic | 31 | 5.3 | 2.4 dbl/day | — | 34 | 0 | 14 | 15 | 0 |
| *Thalassiosira fluviatilis*(86) | Woods Hole, marine, planktonic | 25 | Saturating | 1.2 dbl/day | 25 | 25 | 100 | 10 | 10 | 0 |
| *Thalassiosira nordenskioldii*(66) | Narragansett Bay, marine, planktonic | >25 | Saturating | 2.3 dbl/day | — | <100 | >17 | 10 | 0 | 0 |
| | | >25 | Saturating | 3.6 dbl/day | — | 25 | ≤100 | >19 | 10 | 0 |
| | | 13 | 0.075 Ly/min (13) | 1.6 dbl/day | — | 19 | 0 | — | <2 | 0 |
| *T. nordenskioldii*(87) | Marine, planktonic, cold water | 10 | 0.066 Ly/min (11) | 1.8 dbl/day | — | 15 | 83 | 2 | 0 | 39 |
| *Thalassiosira rotula*(88) | North Sea, marine, planktonic | >22 | 2.4 | 2.6 dbl/day | — | 22 | ≤100 | >10 | 4 | ≤25 |
| **Chrysophyta** | | | | | | | | | | |
| *Chromulina chionophila*(89) | Snow alga | 5—10 | — | — | — | 15 | — | — | 0 | 0 |
| *Coccolithus huxleyi*(90) | Plymouth, marine, planktonic | 21 | 0.55 | 0.8 dbl/day | 24.5 | 26 | 0 | 15 | 8 | 13 |
| | Plymouth, marine, planktonic | 19 | 0.55 | 0.9 dbl/day | 23 | 24 | 0 | 13 | 7 | 8 |
| | Norweigen Sea, marine, planktonic | 22 | 0.55 | 0.8 dbl/day | 26 | 27 | 0 | 15 | 7 | 0 |
| *C. huxleyi*(91) | Marine, planktonic | 21—23 | 0.07 cal/cm²/min (12) | 1.8 dbl/day | 27 | 29 | 0 | 16 | 12 | 28 |
| *Isochrysis*(86) | Marine, planktonic | >25 | Saturating | 1.6 dbl/day | — | 25 | ≤100 | >13 | 10 | ≤33 |
| | | >25 | Saturating | 3.5 dbl/day | — | 25 | ≤100 | >21 | 10 | ≤27 |
| | | 20 | Saturating | 1.9 dbl/day | — | 25 | 63 | 12 | 10 | 37 |
| *Isochrysis galbana*(92) | Marine, planktonic | 25 | Saturating | — | 28 | 30 | 0 | 12 | 10 | 40 |
| *I. galbana*(74) | Marine, planktonic | 14—22 | — | — | — | 27—35 | 0 | — | 8—9 | 0 |
| *Monochrysis lutheri*(74) | Marine, planktonic | 14—25 | 0.15 Ly/min (25) | 1.2 dbl/day | — | 29—35 | 0 | — | 8—9 | 0 |
| *M. lutheri*(66) | Marine, planktonic | 19 | — | — | — | >27 | 0 | — | 8 | 0 |
| *Ochromonas smithii*(93) | Snow alga | 5 | — | — | — | 15 | — | — | 4 | — |
| *Olisthodiscus luteus*(94) | Narragansett Bay, marine | 20 | Saturating | dbl/day | — | 30 | 70 | 11 | 5 | 20 |
| **Green algae — Chlorophyta** | | | | | | | | | | |
| *Chlainomonas kolii*(9) | Snow alga | ≤4 | — | — | — | — | — | — | — | — |
| *Chlainomonas rubra*(9) | Snow alga | ≤4 | — | — | — | 4 | — | — | 0 | — |
| *Chlamydomonas*(74) | Marine, planktonic | 14—28 | — | dbl/day | — | 32—35 | 0 | 0 | 8—9 | 0 |

Table 2 (continued)

## THE RELATIONSHIP BETWEEN TEMPERATURE AND GROWTH IN AQUATIC PHOTOSYNTHETIC MICROORGANISMS

| Species (Reference) | Environment | Preculture conditions | Light intensity (klx) | Optimum temperature (°C) | Gmax, Maximum growth rate | Maximum temperature for one half Gmax (°C) | Highest temperature tested* (°C) | Percent of Pmax at highest temperature tested | Minimum temperature for one half Gmax (°C) | Lowest temperature tested* (°C) | Percent of Pmax at lowest temperature tested |
|---|---|---|---|---|---|---|---|---|---|---|---|
| Chlamydomonas sp. (73) | Marine, planktonic | — | — | 20 | — | 33 | 39 | 0 | 5 | 5 | — |
| Chlamydomonas geitleri(95) | Freshwater | — | 30 W/m² (9) | 21 | — | 28 | 31 | 0 | 11 | 5 | 0 |
| Chlamydomonas nivalis(9) | Snow alga | — | — | 0—2 | — | — | 25 | — | — | 0 | — |
| Chlamydomonas nivalis(strain 464) (9) | Culture collection | — | — | 15—20 | — | — | — | — | — | — | — |
| Chlamydomonas reinhardii(96) | Canada, freshwater, planktonic | — | 0.75 | 28 | — | 34 | 34 | 50 | 15 | 6 | 0 |
|  |  | Double media conc. | 1.5 | 29 | — | 34.5 | 35 | 0 | 12 | 6 | 0 |
|  |  |  | 2 | 29 | — | 34.5 | 35 | 0 | 12 | 6 | 0 |
|  |  |  | 2 | 19 | — | 33 | 33 | 50 | 13 | 12 | 0 |
| Chlamydomonas yellowstonensis (97) | Snow alga | — | — | — | — | — | 25 | — | — | 0 | — |
| Chlorella(#580)(74) | Marine, planktonic | — | — | 14—35 | — | — | — | — | — | 8—9 | 0 |
| Chlorella(UHMC)(74) | Marine, planktonic | — | — | 14—29 | — | — | 32—35 | 0 | — | 8—9 | 0 |
| Chlorella(1-9-30 high T° strain) (98) | Freshwater | — | 17.6 | 38 | 12 dbl/day | 41 | 42 | 42 | 28 | 15 | 0 |
| Chlorella fusca(46) | Freshwater, planktonic | — | Saturating | 30 | 1.6 dbl/day | — | 35 | — | 20 | 15 | — |
| Chlorella pyrenoidosa(54) | Freshwater, planktonic | — | 2.5 | 29 | — | — | — | — | — | — | — |
| C. pyrenoidosa(high T° strain 7-11-05) (99) | Freshwater | — | 16 | 38—40 | 9 dbl/day | — | 41 | 89 | 27 | 25 | 37 |
| C. pyrenoidosa(Emerson strain) (99) | Freshwater | — | 16 | 25 | 2.9 dbl/day | 29 | 29 | 50 | 18 | 18 | 50 |
| C. pyrenoidosa(strain 7-11-05) (100) | Freshwater | — | Saturating | 38—40 | 13 dbl/day | 43 | 44 | 0 | 26 | 15 | 5 |
| Chlorella vulgaris(high T° strain) (101) | — | — | 4 | — | — | — | 42 | — | — | — | — |
| Chlorococcum(74) | Marine, planktonic | — | — | 15—30 | — | — | — | — | — | — | — |
| Chloromonas(9) | Snow alga | — | — | 5 | — | — | 10 | 0 | — | 0 | — |
| Chloromonas pichinchae(9) | Snow alga | — | — | 1 | — | — | 10 | — | — | 0 | — |
| Cylindrocystis brebissonii(9) | Snow alga | — | — | 10 | — | — | 20 | — | — | 1 | — |
| Dunaliella(102) | Brine Lake, Utah | — | 5.4 | 28—35 | 1.2 dbl/day | — | 40 | 0 | — | 5 | 0 |
| Dunaliella euchlora(74) | Marine, planktonic | — | — | 12—35 | — | — | 39 | 0 | — | 8—9 | 0 |

| | | | | | | | | | | | |
|---|---|---|---|---|---|---|---|---|---|---|---|
| *Duniella tertiolecta*(66) | Marine, planktonic | — | 0.25 Ly/min (42) | 25 | 2.4 dbl/day | — | 36 | 0 | — | 11 | 0 |
| *D. tertiolecta*(6) | Marine, planktonic | — | — | 34 | 5.5 dbl/day | 38 | 39 | 0 | 22 | 9 | 0 |
| *Mychonastes ruminatus*(103) | Chesapeake Bay | — | 28.6 | 25 | 1.5 dbl/day | — | 35 | 79 | 10 | 5 | 0 |
| *Nannochloris*(70) | Tropical, marine, planktonic | — | 14.5 | 33 | 4.5 dbl/day | 38 | 40 | 0 | 20 | 10 | 0 |
| *Nannochloris atomus*(73) | Marine, planktonic | — | — | 30 | — | 40 | — | — | <5 | — | — |
| *Platymonas*(74) | Marine, planktonic | — | — | 12—32 | — | — | 35 | 0 | — | — | — |
| *Protococcus*(74) | Marine, planktonic | — | — | 12—32 | — | — | 35 | 0 | — | 8—9 | 0 |
| *Protococcus* sp. (74) | Marine, planktonic | — | — | 12—25 | — | — | 26—32 | 0 | — | 8—9 | 0 |
| *Raphidonema nivale*(9) | Snow alga | — | — | 5 | — | 10 | 20 | — | 1 | 0 | — |
| *Raphidonema tatrae*(104) | Snow alga | — | — | — | — | — | — | — | — | 10 | — |
| *Scenedesmus quadricauda*(50) | Freshwater | — | 0.06 cal/cm²/min (10) | 35—36 | — | — | 36 | 100 | 18 | 11 | 25 |
| *Stichococcus bacillaris*(9) | Snow alga | — | — | 4 | — | — | 15 | — | — | 0 | — |
| *Tetraselmis (Platymonas)*(82) | Sargasso Sea, planktonic | 8.9 µg at N/L, 0.42 µg at P/L | 0.86 | 16—22 | 0.08 dbl/day | 27 | 32 | 0 | 10 | 10 | 50 |
| | | 10 mg at N/L, 0.52 mg at P/L | 1.9 | 22 | 0.48 dbl/day | — | 32 | 81 | 12 | 10 | 25 |

ᵃ  Or the temperature at which no growth is first observed in experiments which span a sufficient temperature range.

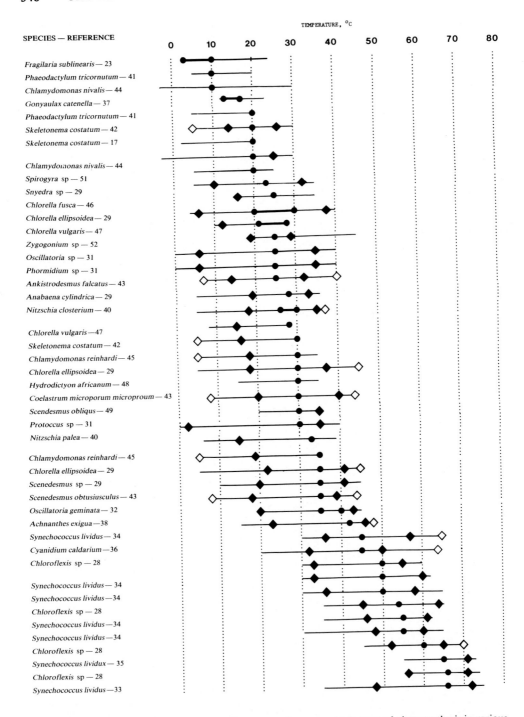

FIGURE 1.   Graphic comparison of the temperature preference and tolerance of photosynthesis in various aquatic photosynthetic microorganisms. Species from Table 1 are ranked according to their optimal temperature for photosynthesis and designated by "●"; "◆ " refers to the temperatures at which one half Pmax is achieved; "◇ " refers to the temperatures at which no photosynthetic activity occurs. Broad temperature optima are indicated by two filled circles linked by a thick line. Some species appear more than once because of disparity in temperature responses observed by different authors or different experimental conditions.

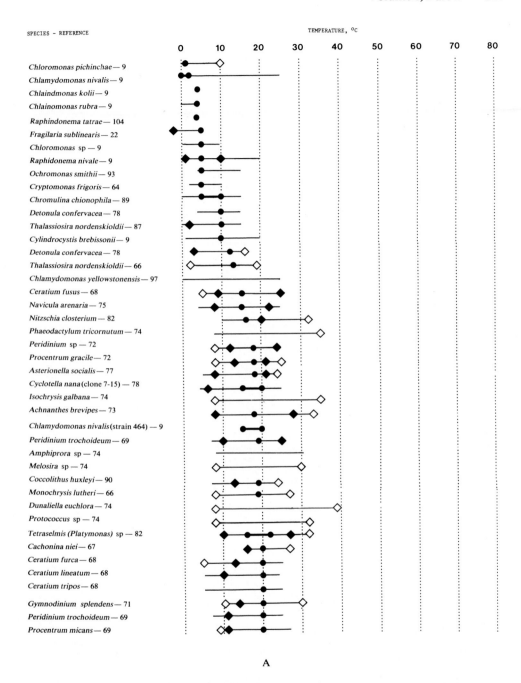

SPECIES - REFERENCE

TEMPERATURE, °C

A

FIGURE 2. (A, B, and C).    Graphic comparison of the temperature preference and tolerance of growth in various aquatic photosynthetic microorganisms. Species from Table 2 are ranked according to their optimal temperatures for growth and designated by "●" "◆" refers to the temperatures at which one half Pmax is achieved; "◇" refers to the temperatures at which no growth occurs. Broad temperature optima are indicated by two filled circles linked by a thick line. Some species appear more than once because of disparity in the temperature responses observed by different authors or different experimental conditions.

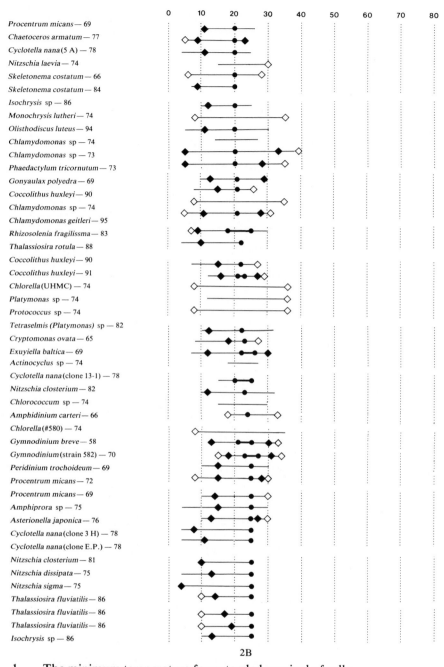

2B

1. The minimum temperature for extended survival of cells
2. The minimum temperature at which organisms can develop and grow efficiently enough to be active members of the photosynthetic community.[1,2,7,10-12,15,21]

## Low-Temperature Limit for Survival

For many aquatic species, the lower limit for survival is found at freezing temperatures.[2,8-11,26,104-107] From a physical standpoint, freezing involves the crystallization of intracellular water and the exclusion of salts.[11,26] This can result in the physical disruption of membranes and other structures, leading to death.[1,7,10,26,108] The importance of water content to freezing tolerance has been demonstrated in experiments with the red alga *Porphyra*. These experiments revealed that the reduction of cellular water content significantly increases resistance to freezing.[29] This is probably related to a concomi-

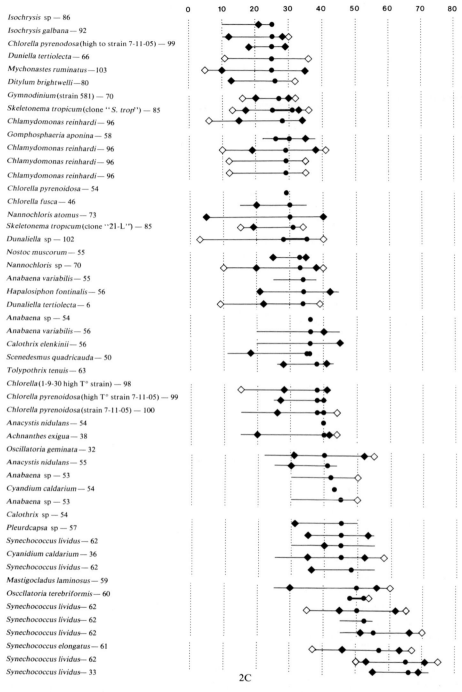

Isochrysis sp — 86
Isochrysis galbana — 92
Chlorella pyrenodosa (high to strain 7-11-05) — 99
Duniella tertiolecta — 66
Mychonastes ruminatus — 103
Ditylum brightwelli — 80
Gymnodinium (strain 581) — 70
Skeletonema tropicum (clone "S. trop") — 85
Chlamydomonas reinhardi — 96
Gomphosphaeria aponina — 58
Chlamydomonas reinhardi — 96
Chlamydomonas reinhardi — 96
Chlorella pyrenoidosa — 54
Chlorella fusca — 46
Nannochloris atomus — 73
Skeletonema tropicum (clone "21-L") — 85
Dunaliella sp — 102
Nostoc muscorum — 55
Nannochloris sp — 70
Anabaena variabilis — 55
Hapalosiphon fontinalis — 56
Dunaliella tertiolecta — 6
Anabaena sp — 54
Anabaena variabilis — 56
Calothrix elenkinii — 56
Scenedesmus quadricauda — 50
Tolypothrix tenuis — 63
Chlorella (1-9-30 high T° strain) — 98
Chlorella pyrenoidosa (high T° strain 7-11-05) — 99
Chlorella pyrenoidosa (strain 7-11-05) — 100
Anacystis nidulans — 54
Achnanthes exigua — 38
Oscillatoria geminata — 32
Anacystis nidulans — 55
Anabaena sp — 53
Cyandium caldarium — 54
Anabaena sp — 53
Calothrix sp — 54
Pleurdcapsa sp — 57
Synechococcus lividus — 62
Cyanidium caldarium — 36
Synechococcus lividus — 62
Mastigocladus laminosus — 59
Osccllatoria terebriformis — 60
Synechococcus lividus — 62
Synechococcus lividus — 62
Synechococcus lividus — 62
Synechococcus elongatus — 61
Synechococcus lividus — 62
Synechococcus lividus — 33

2C

tant decrease in the problem of water crystallization. Of course, water content is not the only important factor effecting freezing survival. The temperatures at which cells are grown[109,110] and the rates of freezing and thawing[109,111,112] also influence survival. In addition, there are molecular factors within and outside of cells which can enhance durability to freezing temperatures.[11,110] One of the best-studied examples of this is the lipid content of cell membranes. Recent studies indicate that there is a correlation between high unsaturated fatty acid content and resistance to freezing, as demonstrated in the green alga *Chlorella*.[110,113] A-p Apparently an increase of the former elements and an associated decrease of saturated fatty acid content results in a significant reduction in the temperature at which the transition from liquid to solid state takes place inside the membrane.[7,10] This alteration in membrane composition not only aids in the resistance to solidification and membrane damage at freezing temperatures, but also facilitates membrane transport and has been linked to enhanced rates of growth and metabolism.[7,10]

## Metabolism at Low Temperatures

Organisms which are adapted to life in cold water habitats often exhibit rates of growth and photosynthesis comparable to temperate water species. For example, cultures of the marine diatom *Skeletonema costatum* grown at 20 and 7°C exhibit very similar rates of photosynthesis.[84] Steemann-Nielsen and Jørgensen have attributed this similarity to the fact that cells grown at low temperatures contain twice as much protein as cells cultured at 20°C. This conclusion was supported by a study of the green alga *Duniella tertiolecta*.[84] However, contradictory evidence has been obtained in other studies;[14] thus, the issue remains unresolved.

Notwithstanding this controversy, there are many indications that the ability to grow rapidly at cold temperatures does involve special adaptations in key areas of cell[2,8,10,26] and enzyme structure and function.[1,7,8,10,11,26]

## Temperature Vs. Growth and Photosynthesis in Cold-Tolerant Species

By definition, the minimum temperature for the growth of a psychrophile is 0°C or less. A number of photosynthetic species have been shown to photosynthesize[9,23,31,39,44,45,114-119] and grow[9,64,89,93,97,104,120-125] at or below this limit, as shown in Tables 1 and 2 and Figures 1 and 2. In addition, a large number of species have been observed in field studies of sub 0°C habitats, but the specific growth characteristics of most of these have not been studied.

In the case of most mesophilic and thermophilic organisms, the optimum temperatures for growth and photosynthesis resembles that found in the natural environment. This is not necessarily the case for psychrophiles. The optimal growth temperatures for organisms from cold water habitats are often 10 to 15°C, or more, above the level at the sampling site. The psychrophile *Chlamydomonas nivalis* is a rather typical example of this phenomenon. In studies by Mosser and co-workers,[44] samples of this green alga were collected from snow with a temperature of less than 0°C. However, laboratory experiments indicated that the optimum temperature for photosynthesis in this alga could be as high as 10 to 20°C. The successful bloom of the alga below optimal temperatures is based on the fact that over 70% of the maximal photosynthetic activity remains at −3°C. Other species of algae have been shown to exhibit similar reactions to temperature.[9,44,89]

Not all psychrophiles exhibit a divergence between optimal and ambient temperatures. A number of snow algae and freshwater and marine psychrophiles exhibit maximum growth[9,22,93,97,104] or photosynthesis[22,23,44] maximally between 0 and 5°C. These organisms also tend to exhibit lower upper limits for growth. For instance, *Raphidonema tatrae* grows optimally at 4°C, but will not develop above 10°C.[9] In contrast, some cold-tolerant species will photosynthesize at temperatures greater than 40°C (e.g., the diatom *Cymbella*).

## Primary Productivity in Low-Temperature Environments

Populations of marine and freshwater phytoplankton can be found throughout the cold water regions of the oceans. Diatoms appear to dominate cold water phytoplankton communities, although dinoflagellates, green flagellates, and chrysophytes may also be well represented.[8,116,117] In the benthic community, a wide variety of microalgal species have been found in cold environments.[126]

Despite the low temperatures, many of the photosynthetic communities inhabiting these areas exhibit significant, and in some cases high, levels of primary productivity (see Table 3), as shown in studies of the Antarctic marine environment,[22,115-118,127-129,131,138-145] arctic and subarctic marine environments,[122,126,132-137,146,147] polar and high altitude lakes,[2,125,130,148-150,151] mountain streams,[31] and snow.[9,44,119,121,124,144] Primary productivity in these areas depends on the ability of phytoplankton to grow successfully at low temperatures. In a number of experimental studies with polar sea ice, phytoplankton, and snow algae, relatively high rates of productivity have been observed at temperatures ranging from −3 to 0°C.[22,44,115]

## HIGH-TEMPERATURE TOLERANCE OF PHOTOSYNTHETIC MICROORGANISMS

Unlike the psychrophiles of cold water regions, thermophilic species do not play a major role in the total productivity of the aquatic environment. They are, however, becoming ecologically important from the standpoint of their ability to grow in areas of thermal pollution, e.g., heat waste from nuclear reactors and other industrial sources.[3,5,20] In addition, they may become an important part of new applied biosolar energy technologies from the standpoint of their exceptional tolerance to the types of high-temperature situations frequently encountered in commercial-scale food, fuel, and fertilizer production systems.

In the natural aquatic environment, the distribution of thermophilic organisms is primarily restricted to areas where there are hot springs. The distribution of the major hot springs around the world has been reviewed by Castenholz[5] and Brock.[4] Some characteristically thermophilic species have also been found in tropical latitudes where solar radiation heats up shallow areas of water.[152,153]

### Upper Temperature Limit of Thermophiles

The upper temperature limit for growth of photosynthetic organisms is 74°C.[4,5,19] However, most thermophilic species exhibit optimal growth between 45 and 60°C.[3-5,7,19] Since the upper limits for the growth of some bacteria lie above 90°C, it has been suggested that the thermal vulnerability of the photosynthetic apparatus sets the upper limit for the growth of photoautotrophic organisms.[4] The ability of these thermophilic organisms to grow at such high temperatures depends on special structural and functional adaptations.[3,4,7,8,15]

The sensitivity of membranes to thermal disruption is one of the key elements in adaptation to thermal environments. A number of researchers have shown that the membranes of thermophilic organisms are more resistant to high temperatures than those of mesophilic species.[8] This stability appears to be related to the increased composition of high melting point, saturated fatty acids in the membranes of cells grown at high temperatures.[4,8,113,154]

Another key area for adaptation to thermal environments appears to be enzyme structure. The results of a considerable number of in vivo and in vitro studies with specific enzymes indicate that proteins of thermophilic origin are often more stable at high temperatures than mesophilic forms.[4,7,8,155]

### Temperature Preferences and Tolerances of Photosynthetic Thermophiles

As a reflection of the relative harshness of thermal environments, the list of photosynthetic thermophiles is quite short and consists primarily of blue-green algae.[4,5,19] There is a strong "fidelity" of thermophiles for their high-temperature environments.[156] In fact, only a handful of thermophilic species will tolerate temperatures less than 30°C.[4,5,156] One such species, *Phormidium tenuis*, has been isolated

## Table 3
## PRIMARY PRODUCTIVITY IN VARIOUS COLD WATER POPULATIONS

| Region or species | Habitat | Season/ month | Maximum productivity | Ref. |
|---|---|---|---|---|
| **Antarctic** | | | | |
| Antarctic | Planktonic | — | 2.3 mg C/mg Chl a/hr | 118 |
| | Ice flora | — | 2.6 mg C/mg Chla/hr | |
| McMurdo Sound | Planktonic, below ice | — | 0.08 mg C/mg Chla/hr | 127 |
| McMurdo Sound | Sea ice | — | 0.5 mg C/mg Chla/hr | 128 |
| South Wendall Sea | Planktonic | — | 3.9 mg C/mg Chl/hr | 129 |
| South Orkey and South Sandwich Island | Planktonic | — | 2.3 mg C/mg Chla/hr (10 mg C/m³/hr) | 129 |
| Antarctic Lake | Planktonic | — | 1900 mg C/m²/day | 130 |
| Drake Passage | Planktonic | — | 10.5 mg C/mg Chla/hr (14.1 mg C/m³/hr) | 131 |
| Bransfield Strait | Planktonic | — | 9.9 mg C/mg Chla/hr (18.0 mg C/m³/hr) | 131 |
| **Arctic and Subarctic** | | | | |
| Barrow, Alaska | Sea ice | May | 32.4 mg C/m³/hr | 132 |
| | Water under ice | May | 2.8 mg C/m³/hr | 132 |
| | Benthic, after ice break up | — | 56.0 mg C/m³/hr | 132 |
| Prudhoe Bay, Beaufort Sea | Planktonic | — | 21.2 mg C/m³/hr | 132 |
| Harrison Bay, Beaufort Sea | Planktonic | — | 35.5 mg C/m³/hr | 132 |
| Wainwright and Peard Bay, Beaufort Sea | Planktonic | — | 133.0 mg C/m³/hr | 132 |
| Northeast Pacific | Planktonic | Winter | 50 mg C/m²/day | 133 |
| | Planktonic | Summer | 350 mg C/m²/day | 133 |
| Mid-Subarctic Pacific | Planktonic | | 410 mg C/m²/day | 134 |
| Lake, Japan | Ice covered | February | 0.7 mg/mg Chl a/hr (204 mg C/m²/day | 135 |
| | Planktonic, no ice cover | December | 0.9 mg/mg Chl a/hr (215 mg C/m²/day) | 135 |
| Chuckchi Sea | Benthic, with sea ice | Winter | 0.5 mg C/m²/hr | 126 |
| | Benthic, no sea ice | Summer | 57.0 mg C/m²/hr | 126 |
| | Planktonic | August | 19.0 mg C/m²/hr | 126 |
| | Ice algae | May | 5 mg C/m²/hr | 126 |
| | Planktonic | May | 1 mg C/m²/hr | 126 |
| | Benthic | May | >0.5 mg C/m²/hr | 126 |
| Japan Sea | Planktonic | Winter | 12.5 mg C/m²/hr | 136 |
| Barrow, Alaska | Planktonic | May | 0.25 mg C/mg Chl a/hr (7.7 mg C/m²/hr) | 137 |
| Arctic | Planktonic | Summer | 1.5 mg C/mg Chl a/hr | 16 |
| Beaver Pond, Massachusetts | Planktonic, ice cover | February | 0.54 mg C/mg Chl a/hr (9.1 mg C/m³/hr) | 125 |
| **Snow** | | | | |
| *Chlamydomonas nivalis* | Snow | — | 0.58 mg C/mg Chl a/hr | 44 |
| *Raphdonema tatrae* and *Chlamydomonas* sp. | Snow | — | 0.26 mg C/mg Chl a/hr | 121 |
| South Orkney Islands | Snow | — | 0.86 μg C/mm³/hr | 119 |

from the Florida Everglades.[156] It grows well at temperatures of up to 50°C. However, it will not tolerate greater than 50°C and therefore may not truly qualify as an "obligate" thermophilic.

Another low temperature tolerant species is *Cyanidium caldarium* which will grow in the laboratory at 20°C, but in nature is usually out-competed by other species below

20°C.[157] In many cases, it is difficult to evaluate whether the lower limit of a thermophile is dictated by temperature, competition, or other factors such as grazing by animals.[4]

Brock[4] and Castenholz[5] have established lists of presently identified thermophilic species and their limits for growth and/or distribution (see Table 4). In addition, the temperature preferences and tolerances for growth and photosynthesis of a number of thermophilic species are described in Tables 1 and 2 and Figures 1 and 2. Several of these species exhibit special characteristics worth further consideration.

### Photosynthetic Bacteria

The most common photosynthetic bacteria of neutral and alkaline hot springs is *Chloroflexus*, a filamentous gliding form.[4,5,28,158,159] In most cases, it is found in association with blue-green algae, often the common unicellular form *Synechococcus*. In this association, it provides a supporting matrix or mat for the blue-green algae,[160] and, in return, can utilize substances excreted by the algae.[161] Photosynthetically *Chloroflexus* is able to fix carbon efficiently at low light intensities.[162] This adaptation permits photosynthesis to take place within deeper layers of the mat. In high sulfide springs, *Chloroflexus* grows photoautotrophically using sulfide as an electron donor.[163] Because high concentrations of sulfide can inhibit the growth of many thermophilic blue-green algae, *Chloroflexus* can be found growing by itself in such environments.[164] The maximum temperature for growth of *Chloroflexus* is similar to its common associate *Synechococcus*, i.e., 73°C.

The only other photosynthetic bacteria so far found in hot springs is a species of *Chromatium*.[5]

### Blue-green Algae

Blue-green algae are the most extensively represented photosynthetic organisms in thermal environments.[4,5,19] One of the best-studied and common species of this group is *Synechococcus lividus*, a rod-shaped single-celled form.[4,5,19,33-35,60,165-167] It is the most temperature-tolerant photosynthetic organism so far discovered. One strain of *S. lividus* grows at temperatures of up to 75°C and exhibits an optimum growth temperature of around 65°C. The experiments of Peary and Castenholz[60] indicate that at least four distinct temperature strains of *S. lividus* can be found, even within different locations of one hot spring. Laboratory experiments show that the minimum, optimum, and maximum temperatures of these strains are to some extent genetically fixed. The minimum temperatures for growth range from below 30 to 50°C. In one strain, however, the minimum temperature for photosynthesis (in terms of $CO_2$ fixation) is around 33°C, considerably lower than the minimum for growth. The maximum temperatures for growth range from 65 to 75°C which coincide closely to the maximum temperatures for photosynthesis.

Another interesting physiological feature of *S. lividus* is its ability to adapt to growth at low light intensities.[162] This capability is shared by other hot spring species, including *Chloroflexus* and *Mestigocladus*, and may play an important role in survival during periods of reduced winter irradiance.[162,168,169] At the other end of the light intensity spectrum, *Synechococcus* is also fairly resistant to photoinhibition, especially when grown at high light intensities.[162,168] This is probably an adaptation to the high summer intensities associated with many thermal ponds.

Perhaps the most widely distributed thermophilic blue-green alga is *Mastigocladus laminosus*, a filamentous branching form with heterocysts.[59,170,171] This species is particularly unique in being the most heat-tolerant nitrogen-fixing species found to date.[172] While it can be found throughout the world, it is the dominant species in Iceland and New Zealand where *S. lividus* is not found.

Table 4
## OPTIMUM AND MAXIMUM TEMPERATURES FOR THE GROWTH OF PHOTOSYNTHETIC THERMOPHILES

| Organisms | Optimal temperature (°C) | Maximum growth rate | Maximum temperature (°C) | Ref. |
|---|---|---|---|---|
| Photosynthetic bacteria | | | | |
| *Chloroflexus aurantiacus* | 55 | 7.2 dbl/day | 73 | 158 |
| *Chromatium* | — | — | 60 | 5 |
| | | | | |
| Cyanophyta (blue-green algae) | | | | |
| *Anabaena* sp. | 45 | — | 50 | 53 |
| *Aphanocapsa thermalis* | — | — | >55[a] | 4,5 |
| *Calothrix* sp. | 45 | 2.3 dbl/day | 52—54 | 57 |
| *Mastigocladus laminosus* | 40—50 | 1.5 dbl/day | 63—64 | 4,5,59 |
| *Oscillatoria amphibia* | — | — | 57[a] | 4,5 |
| *Oscillatoria animalis* | — | — | >55[a] | 4 |
| *Oscillatoria germinata* | 40 | — | 55 | 32 |
| *Oscillatoria okenii* | — | — | >60[a] | 4,5 |
| *Oscillatoria tenuis* | | | 46—47[a] | |
| *Oscillatoria terebriformis* | 48—52 | 4.8 dbl/day | 53 | 4,5 |
| *Phormidium lamindsum* | — | — | 57 | 4,5 |
| *Phormidium purpurasieins* | — | — | 46—47[a] | 4,5 |
| *Pleurocapsa* sp. | 45 | 1.3 dbl/day | >55 | 57 |
| *Spirulina* sp. | — | — | 55—60[a] | 4,5 |
| *Symploca thermalis* | — | — | 45—47[a] | 4,5 |
| *Synechocystis aquatilus* | — | — | 45—50[a] | 4,5 |
| *Synechococcus elongatus* | 57 | — | 67 | 61 |
| *Synechococcus lividus* | | | | 60 |
| Strain I | 45 | 10 dbl/day | — | |
| Strain II | 50 | 4.5 dbl/day | 65 | |
| Strain III | 55 | 5.5 dbl/day | 70 | |
| Strain IV | 65 | 2.0 dbl/day | 75 | |
| *Synechococcus minervae* | — | — | 60[a] | 4,5 |
| | | | | |
| Cryptophyta | | | | |
| *Cyanidium caldarium* | 45 | — | 58 | 36 |
| | | | | |
| Bacillariophyta (diatoms) | | | | |
| *Acanthes exigua* | 40 | 2 dbl/day | 44 | 38 |
| | | | | |
| *Chlorella pyrenoidosa* strain 7-11-05 | 40 | 13 dbl/day | 44 | 100 |
| *Chlorella vulgaris* high to strain | — | — | >42 | 101 |
| *Chlorella* sp. strain 1-9-30 | 38 | — | >42 | 98 |

[a]    Based on field observations.

Another interesting thermophilic species is *Oscillatoria terebriformis* which forms dense mats on the surface of hot springs in Oregon, within a temperature range of 35 to 54°C.[173-175] The interesting feature of these mats is the movement which they exhibit in response to environmental change. For example, under high light intensity conditions, the mats contract and aggregate for protection through self-shading. The mats also exhibit movement in response to changing temperature conditions, thereby ensuring optimal location.[174]

*Eucaryotic Algae*

The best-studied thermophilic eucaryote is *C. caldarium*.[4,157,176-178] While the taxonomic position of *Cyanidium* is still an issue of considerable debate, there is no doubt that it is unique among the photosynthetic phyla in its ability to grow in very acidic hot springs. *Cyanidium* clearly dominates hot springs where the pH drops below 4, and laboratory studies indicate that it grows well at pH 1. The upper temperature limit for photosynthesis and growth of *Cyanidium* is about 57°C.[176] Most populations of *Cyanidium* seem to grow optimally at about 45°C. In the laboratory, growth and photosynthesis have been obtained at temperatures as low as 20°C. As discussed above, natural populations of *Cyanidium* are limited below 40°C because of competitive stress from other species.

Beyond *Cyanidium*, there are no eucaryotic species which fit the strict definition of thermophiles proposed by Castenholz for blue-green algae, i.e., organisms which grow optimally at 45°C or higher.[4,5,19] If this temperature limit is redefined for eucaryotic cells to organisms which can grow at 38°C, as suggested by Kessler,[101,179] or to organisms with optimal temperatures greater than 35°C, then a number of other eucaryotic organisms can be designated as thermophilic. For example, a number of diatoms have been observed in hot spring environments, including *Achanthes exigua* which exhibits an optimal growth temperature of 40°C.[38]

Several green algae have been shown by Kessler to grow at 38°C, including two strains of *Chlorella* and two strains of *Scenedesmus*.[54,55] Studies by Sorokin[98,99] and Clendenning[30] have revealed two strains of *Chlorella pyrenoidosa* with high growth rates at 38 to 40°C.

## REFERENCES

1. **Gessner, F.,** Temperature: plants, in *Marine Ecology,* Kinne, O., Ed., Interscience, New York, 1970, 363.
2. **Baross, J. A. and Morita, R. Y.,** Microbial life at low temperature: ecological aspects, in *Microbial Life in Extreme Environments,* Kushner, D. J., Ed., Academic Press, New York, 1978, 9.
3. **Brock, T. D.,** High temperature systems, *Annu. Rev. Ecol. Syst.,* 1, 1970, 191.
4. **Brock, T. D.,** *Thermophilic Microorganisms and Life at High Temperatures,* Springer-Verlag, New York, 1978.
5. **Castenholz, R. W.,** Thermophilic blue-green algae and the thermal environment, *Bacteriol. Rev.,* 33, 476, 1969.
6. **Eppley, R. W.,** Temperature and phytoplankton growth in the sea, *Fish. Bull., U.S.,* 70(4), 1063, 1972.
7. **Farrel, J. and Rose, A. H.,** Temperature effects on microorganisms, in *Thermobiology,* Rose, A. H., Ed., Academic Press, New York, 1967, 147.
8. **Amelunxen, R. E. and Murdock, A. L.,** Microbial life at high temperatures: mechanisms and molecular aspects, in *Microbial Life in Extreme Environments,* Kushner, D. J., Ed., Academic Press, New York, 1978, 217.
9. **Hoham, R. W.,** Optimal temperatures and temperature ranges for growth of snow algae, *Arct. Alp. Res.,* 7, 13, 1975.
10. **Inniss, W. E. and Ingraham, J. L.,** Microbial life at low temperature: mechanisms and molecular aspects, in *Microbial Life in Extreme Environments,* Kushner, D. J., Ed., 1978, 73.
11. **Li, P. H. and Sakai, A.,** *Plant Cold Hardiness and Freezing Stress,* Academic Press, New York, 1978.
12. **Kol, E.,** Kryobiologie, Biologie und Limnologie des Schnees und des Eisec. I. Kryovegetation, in *Die Binnengewasser,* Vol. 24, Elster, H.-J. and Ohle, W., Eds., E. Schweizerbartsche, Stuttgart, 1968, 216.
13. **Morris, J. and Glover, H. E.,** Questions on the mechanism of temperature adaptation in marine phytoplankton, *Mar. Biol.,* 24, 147, 1974.

14. **Parsons, T. R., Takahashi, M., and Hargrave, B.,** *Biological Oceanographic Processes,* 2nd ed., Pergamon Press, New York, 1977.

15. **Soeder, C. and Stengel, E.,** Physico-chemical factors affecting metabolism and growth rate, in *Algal Physiology and Biochemistry,* Stewart, W. D. P., Ed., 1975, 714.

16. **Steeman-Nielsen, E. and Hansen, V. K.,** Light adaptation in marine phytoplankton populations and its interrelation with temperature, *Physiol. Plant.,* 12, 353, 1959.

17. **Steemann-Nielsen, E. and Jørgensen, E. G.,** The adaptation of plankton algae. I. General part, *Physiol. Plant.,* 21, 401, 1968.

18. **Stokes, J. L.,** General biology and nomenclature of psychrophile microorganisms, in *Recent Progress in Microbiology,* Gibbons, N. E., Ed., Symp. of 8th Int. Congr. Microbiology, Montreal, 1962, Toronto, University of Toronto Press, Toronto, 1963, 187.

19. **Tansey, M. R. and Brock, T. D.,** Microbial life at high temperatures: ecological aspects, in *Microbial Life in Extreme Environments,* Kushner, D. J., Ed., Academic Press, New York, 1978, 159.

20. **Brock, T. D.,** Predicting the ecological consequences of thermal pollution from observations on geothermal habitats, in *Environmental Effects of Cooling Systems at Nuclear Power Plants,* International Atomic Energy Agency, Vienna, 1975.

21. **Alexandrov, V. Y.,** Cytophysiological and cytoecological investigations of resistance of plant cells toward the action of high and low temperature, *Q. Rev. Biol.,* 39, 35, 1964.

22. **Bunt, J. S.,** Some characteristics of microalgae isolated from Antarctic sea ice, *Antarct. Res. Ser.,* 2, 1, 1968.

23. **Bunt, J. S., Owens, O. V. H., and Hoch, G.,** Exploratory studies on the physiology and ecology of a psychrophilic marine diatom, *J. Phycol.,* 2, 96, 1966.

24. **Fogg, G. E. and Horne, A. J.,** The physiology of Antarctic freshwater algae, *Antarct. Res.,* 11, 632, 1968.

25. **Lutova, M. I., Zavadskaya, I. G., Luknitskaya, A. F., and Feldman, N. L.,** Temperature adaptation of cells of marine and freshwater algae, in *The Cell and Environmental Temperature,* Troschin, A. S., Ed., Academic Press, New York, 1967, 166.

26. **Mazur, P.,** Freezing injury in plants, *Annu. Rev. Plant. Physiol.,* 20, 419, 1969.

27. **Morita, R. Y.,** Psychrophilic bacteria, *Bacteriol. Rev.,* 39, 144, 1975.

28. **Bauld, J. and Brock, T. D.,** Ecological studies of *Chloroflexis,* a gliding photosynthetic bacterium, *Arch. Mikrobiol.,* 92, 267, 1973.

29. **Aruga, Y.,** Ecological studies of photosynthesis and matter production of phytoplankton, II. Photosynthesis of algae in relation to light intensity and temperature, *Bot. Mag.,* 78, 360, 1965.

30. **Clendenning, K. A., Brown, T. E., and Eyster, H. C.,** Comparative studies of photosynthesis in *Nostoc muscorum* and *Chlorella pyrenoidosa, Can. J. Bot.,* 34, 943, 1956.

31. **Mosser, J. L. and Brock, T. D.,** Temperature optima for algae inhabiting cold mountain streams, *Arct. Alp. Res.,* 8, 111, 1976.

32. **Bünning, E. and Herdtle, H.,** Physiologische Untersuchungen und thermophilen Blaualgen, *Z. Naturforsch.,* 1, 93, 1946.

33. **Meeks, J. C. and Castenholz, R. W.,** Growth and photosynthesis in an extreme thermophile, *Synechococcus lividus* (Cyanophyta), *Arch. Microbiol.,* 78, 25, 1971.

34. **Sheridan, R. P. and Ulik, T.,** Adaptive photosynthesis responses to temperature extremes by the thermophilic cyanophyte *Synechococcus lividus, J. Phycol.,* 12, 255, 1976.

35. **Meeks, J. C. and Castenholz, R. W.,** Photosynthetic properties of the extreme thermophile *Synechococcus lividus, J. Thermal Biol.,* 3, 19, 1978.

36. **Doemel, W. N. and Brock, T. D.,** The physiological ecology of *Cyanidium caldarium, J. Gen. Microbiol.,* 67, 17, 1971.

37. **Norris, L. and Chew, K.,** Effect of environmental factors on growth of *Gonyaulax catenella,* in *The First International Conference on Toxic Dinoflagellates 1974-1975,* Massachusetts Science and Technology Foundation, Wakefield, 143.

38. **Fairchild, E. and Sheridan, R. P.,** A physiological investigation of the hot spring diatom, *Achanthes exigua* grün, *J. Phycol.,* 10, 1, 1974.

39. **Boylen, C. W. and Brock, T. D.,** A seasonal diatom in a frozen Wisconsin lake, *J. Phycol.,* 10, 210, 1974.

40. **Barker, A.,** Photosynthesis in diatoms, *Arch. Mikrobiol.,* 6, 141, 1935.

41. **Morris, I. and Farrell, K.,** Photosynthetic rates, gross patterns of carbon dioxide assimilation and activities of ribulose diphosphate carboxylase in marine algae grown at different temperatures, *Physiol. Plant.,* 25, 372, 1971.

42. **Curl, H. and McLeod, G. C.,** The physiological ecology of a marine diatom, *Skeletomema costatum* (Grev.) Cleve., *J. Mar. Res.,* 19, 70, 1961.

43. **Felföloy, L. J. M.,** Effect of temperature on photosynthesis in three unicellular green algal strains, *Acta Biol. Acad. Sci. Hung.,* 12, 153, 1961.

44. Mosser, J. L., Mosser, A. G., and Brock, T. D., Photosynthesis in the snow: the alga *Chlamydomonas nivalis* (Chlorophyceae), *J. Phycol.*, 13, 22, 1977.
45. Ehrke, G., Uber die wirkung der temperatur und des lichtes auf atmung und assimilation einiger meeres- und subwasseralgen, *Planta*, 13, 221, 1931.
46. Lewenstin, A. and Bachofen, R., $CO_2$-fixation and ATP synthesis in continuous cultures of *Chlorella fusca*, *Arch. Microbiol.*, 116, 169, 1978.
47. Wassink, E., Vermeulen, D., Reman, G., and Katz, E., *Enzymologia*, 5, 100, 1938.
48. Raven, J. A. and Smith, F. A., Sun and shade species of green algae: relation to cell size and environment, *Photosynthetica*, 11, 48, 1977.
49. Senger, H. and Fleischhacker, P., Adaptation of the photosynthetic apparatus of *Scenedesmus obliquus* to strong and weak light conditions. I. Differences in pigments, photosynthetic capacity, quantum yield and dark reactions, *Physiol. Plant.*, 43, 35, 1978.
50. Komárek, J. and Ruzicka, J., Effect of temperature on the growth of *Scenedesmus quadricauda* (Turpl) Bréb., in *Studies in Phycology*, Fott, B., Ed., Academia, Prague, 1969, 262.
51. Ehrke, G., Uber die wirkung der temperatur und des lichtes auf atmung und assimilation einiger meeres- und subwasseralgen, *Planta*, 13, 221, 1931.
52. Lynn, R. and Brock, T. D., Notes on the ecology of a species of *Zygogonium* (Kütz.) in Yellowstone Nat. Park, *J. Phycol.*, 5, 181, 1969.
53. Stacey, G., van Paalen, C., and Tabita, F. R., Isolation and characterization of a marine *Anabaena* sp. capable of rapid growth on molecular nitrogen, *Arch. Microbiol.*, 114, 197, 1977.
54. Halldal, P. and French, C. S., Algal growth in crossed gradients of light intensity and temperature, *Plant Physiol.*, 33, 249, 1958.
55. Kratz, W. A. and Myers, J., Nutrition and growth of several blue-green algae, *Am. J. Bot.*, 42, 282, 1955.
56. Taha, M. S., The effect of the hydrogen ion concentration in the medium and of temperature on the growth and nitrogen fixation by blue-green algae, *Mikrobiologyia*, 32, 968, 1963.
57. Wickstrom, C. E. and Castenholz, R. W., Association of *Pleurocapsa* and *Calothrix* (Cyanophyta) in thermal streams, *J. Phycol.*, 14, 84, 1978.
58. Eng-wilmot, D. L., Hitchcock, W. S., and Martin, D. F., Effect of temperature on the proliferation of *Gymnodinium breve* and *Gomphosphaeria aponina*, *Mar. Biol.*, 41, 71, 1977.
59. Holton, R. W., Isolation, growth, and respiration of a thermophilic blue-green alga, *Am. J. Bot.*, 49, 1, 1962.
60. Peary, J. A. and Castenholz, R. W., Temperature strains of a thermophilic blue-green alga, *Nature (London)*, 202, 720, 1964.
61. Beljanin, V. N. and Trenkensu, A. P., Growth and spectrophotometric characteristics of the blue-green alga *Synechococcus elongatus* under different temperature and light conditions, *Arch. Hydrobiol. Suppl.*, 51 (18), 46, 1977.
62. Dyer, D. L. and Gafford, R. D., Some characteristics of a thermophilic blue-green alga, *Science*, 134, 616, 1961.
63. Ukai, Y., Fugita, Y., Morimura, Y., and Watanabe, A., Studies on growth of blue-green alga *Tolypothrix tenuis*, *J. Gen. Appl. Microbiol.*, 4, 163, 1958.
64. Stien, J. R., A *Chromulina* (Chrysophyceae) from snow, *Can. J. Bot.*, 41, 1367, 1963.
65. Cloern, J. E., Effects of light intensity and temperature on *Cryptomonas ovata* (Cryptophyceae) growth and nutrient uptake rates, *J. Phycol.*, 13, 389, 1977.
66. Jitts, H. R., McAllister, C. D., Stephans, K., and Strickland, J. D. H., The cell division rates of some marine phytoplankters as a function of light and temperature, *J. Fish. Res. Board Can.*, 21, 139, 1964.
67. Loeblich, A. R., III, A seawater medium for dinoflagellates and the nutrition of *Cachonina niei*, *J. Phycol*, 11, 80, 1975.
68. Nordli, E., Experimental studies on the ecology of *Ceratia*, *Oikos*, 8, 201, 1957.
69. Braarud, T., Cultivation of marine organisms as a means of understanding environmental influences on populations, in *Oceanography*, Sears, M., Ed., American Association for the Advancement of Science, Washington, D.C., 271, 1961.
70. Thomas, W. H., Effects of temperature and illuminance on cell division rates of three species of tropical oceanic phytoplankton, *J. Phycol.*, 2, 17, 1966.
71. Thomas, W. H., Dodson, A. N., and Linden, C. A., Optimum light and temperature requirements for *Gymnodinium splendens*, a larval fish food organism, *Fish. Bull.*, 71, 599, 1973.
72. Barker, A., The culture and physiology of the marine dinoflagellates, *Arch. Mikrobiol.*, 6, 157, 1935.
73. Styron, C. E., Hagen, T. M., Campbell, D. R., Harvin, J., Whittenburg, N. K., Baughman, G. A., Bransford, M. E., Saunders, W. H., Williams, D. C., Woodle, C., Dixon, N. K., and McNeill, C. R., Effects of temperature and salinity on growth and uptake of $^{65}Zn$ and $^{137}Cs$ for six marine algae[1], *J. Mar. Biol. Assoc. U.K.*, 56, 13, 1976.

74. **Ukeles, R.,** The effect of temperature on the growth and survival of several marine algal species, *Biol. Bull.,* 120, 255, 1961.
75. **Admiraal, W.,** Influence of light and temperature on the growth rate of estuarine benthic diatoms in culture, *Mar. Biol.,* 39, 1, 1977.
76. **Kain, J. M. and Fogg, G. E.,** Studies on the growth of marine phytoplankton. I. *Asterionella japonica, J. Mar. Biol. Assoc. U.K.,* 37, 397, 1958.
77. **Lewin, J. and Mackas, D.,** Blooms of surf-zone diatoms along the coast of the Olympic Peninsula, Washington. I. Physiological investigations of *Chaetoceros armatum* and *Asterionella socialis* in laboratory cultures, *Mar. Biol.,* 16, 171, 1972.
78. **Guillard, R. R. L. and Ryther, J. H.,** Studies of marine planktonic diatoms. I. *Cyclotella nana* Hustedt and *Detonula confervacea* (Cleve) Gran, *Can. J. Microbiol.,* 8, 229, 1962.
79. **Smayda, T. J.,** Experimental observations on the influence of temperature, light, and salinity on cell division of the marine diatom *Detonula confervacea* (Cleve) Gran, *J. Phycol.,* 5, 150, 1969.
80. **Paasche, E.,** Marine plankton algae grown with light-dark cycles. II. *Ditylum brightwellii* and *Nitzschia turgidula, Physiol. Plant.,* 21, 66, 1968.
81. **Spencer, C. P.,** Studies on the culture of a marine diatom, *J. Mar. Biol. Assoc. U.K.,* 33, 265, 1954.
82. **Maddux, W. S. and Jones, R. F.,** Some interactions of temperature, light intensity, and nutrient concentration during the continuous culture of *Nitzschia closterium* and *Tetraselmis* sp., *Limnol. Oceanogr.,* 9, 79, 1964.
83. **Ignatiades, L. and Smayda, T. J.,** Autoecological studies on the marine diatom *Rhizosolenia fragilissima* Bergon. I. The influence of light, temperature and salinity, *J. Phycol.,* 6, 332, 1970.
84. **Jørgensen, E. G.,** The adaptation of plankton algae. II. Aspects of the temperature adaptation of *Skeletonema costatum, Physiol. Plant.,* 21, 423, 1968.
85. **Hulburt, E. and Guillard, R.,** The relationship of the distribution of the diatom *Skeletonema tropicum* to temperature, *Ecology,* 49, 337, 1968.
86. **Hobson, L. A.,** Effects of interactions of irradiance, daylength and temperature on division rates of three species of marine unicellular algae, *J. Fish. Res. Board Can.,* 31, 391, 1974.
87. **Durbin, E. G.,** Studies on the autoecology of the marine diatom *Thalassiosira nordenskoldii* Cleve. I. The influence of daylength, light intensity, and temperature on growth, *J. Phycol.,* 10, 220, 1974.
88. **Schöne, H. K.,** Experimentelle untersuchungen sur okologie der marinen kieselalge *Thalassiosira rotula.* I. Temperatur and licht, *Mar. Biol.,* 13, 284, 1972.
89. **Hardy, J. T.,** Identification, Culture and Physiological Ecology of Cryophilic Algae, M.S. thesis, Oregon State University, Corvallis, 1966.
90. **Mjaaland, G.,** Some laboratory experiments on the coccolithophorid *Coccolithus huxleyi* (Lohm) Kamptner., *Oikos,* 7, 251, 1956.
91. **Paasche, E.,** Marine plankton algae grown with light-dark cycles. I. *Coccolithus huxleyi, Physiol. Plant.,* 20, 946, 1967.
92. **Kain, J. M. and Fogg, G. E.,** Studies on the growth of marine phytoplankton. II. *Isochrysis galbana, J. Mar. Biol. Assoc. U.K.,* 37, 781, 1958.
93. **Fukushima, H.,** Studies on cryophytes in Japan, *J. Yokohama Munic. Univ., Ser. C, Nat. Sci.,* 43, 1, 1963.
94. **Tomas, C. R.,** *Olisthodiscus luteus* (Chrysophyceae). I. Effects of salinity and temperature on growth, motility and survival, *J. Phycol.,* 14, 309, 1978.
95. **Tetik, K. and Necas, J.,** Growth characteristics of *Chlamydomonas geitleri, Arch. Hydrobiol. Suppl.,* 51(19), 164, 1977.
96. **McCombie, A. M.,** Actions and interactions of temperature, light intensity and nutrient concentration of the green alga, *Chlamydomonas reinhardi, J. Fish. Res. Board Can.,* 17, 871, 1960.
97. **Sutton, E. A.,** The Physiology and Life Histories of Selected Cryophytes of the Pacific Northwest, Ph.D. thesis, Oregon State University, Corvallis, 1972.
98. **Sorokin, C.,** New high-temperature *Chlorella, Science,* 158, 1204, 1967.
99. **Sorokin, C.,** Tabular comparative data from the low- and high-temperature strains of *Chlorella, Nature (London),* 184, 613, 1959.
100. **Sorokin, C. and Krauss, R. W.,** Effects of temperature and illuminance on *Chlorella* growth uncoupled from cell division, *Plant Physiol.,* 37, 37, 1962.
101. **Kessler, E.,** Physiologische and biochemische Beitrage zur Taxonomie der Gattung *Chlorella.* VII. Die Thermophilie von *Chlorella vulgaris* f. *Tertia* Fott et Novakova, *Arch. Mikrobiol.,* 87, 243, 1972.
102. **van Auken, D. W. and McNulty, I. B.,** The effect of environmental factors on the growth of a halophylic species of algae, *Biol. Bull.,* 145, 210, 1973.
103. **Simpson, P. D.,** The growth rate of *Mychonastes ruminatus* Simpson et Van Valkenburg under various light, temperature and salinity regimes, *Br. Phycol. J.,* 13, 291, 1978.
104. **Hindak, F. and Komarek, J.,** Cultivation of the cryosestonic alga *Koliella tatrae* (Kol) Hind., *Biol. Plant.,* 10, 95, 1968.

105. Duthie, H. C., The survival of desmids in ice, *Br. Phycol. Bull.*, 2, 376, 1964.
106. Holm-Hansen, O., Viability of blue-green and green algae after freezing, *Physiol. Plant.*, 16, 530, 1963.
107. Stockner, J. G. and Lund, J. W. G., Live algae in postglacial lake deposits, *Limnol. Oceanogr.*, 15, 41, 1970.
108. Stein, J. R. and Bisalputra, T., Crystalline bodies in an algal chloroplast, *Can. J. Bot.*, 47, 233, 1969.
109. Morris, G. J., The cryopreservation of *Chlorella*. I. Interactions of rate of cooling, protective additive and warming rate, *Arch. Microbiol.*, 107, 57, 1976.
110. Morris, G. J., The cryopreservation of *Chlorella*. II. Effect of growth temperature on freezing tolerance, *Arch. Microbiol.*, 107, 309, 1976.
111. Migita, S., Freezing preservation of *Porphyra thalli* in viable state. II. Effects of cooling velocity and water content of thalli on frost resistance, *Bull. Fac. Fish. Nagasaki Univ.*, 21, 131, 1966.
112. Hwang, S. and Horneland, W., Survival of algal cultures after freezing by controlled and uncontrolled cooling, *Cryobiology*, 1, 305, 1965.
113. Raison, J. K. and Berry, J. A., The physical properties of membrane lipids in relation to the adaptation of higher plants and algae to contrasting thermal regimes, *Carnegie Inst. Washington Yearb.*, 77, 276, 1978.
114. Boylen, C. W. and Brock, T. D., A seasonal diatom in a frozen Wisconsin lake, *J. Phycol.*, 10, 210, 1974.
115. Bunt, J. S., Primary productivity under sea ice in antarctic waters. II. Influence of light and other factors on photosynthetic activities of antarctic marine microalgae, *Antarct. Res. Ser.*, 1, 27, 1964.
116. Bunt, J. S., Diatoms of antarctic sea ice as agents of primary production, *Nature (London)*, 199, 1255, 1963.
117. Bunt, J. S., Microalgae of the Antarctic pack ice zone, in *Proc. Symp. Antarctic Oceanography*, Currie, R. I., Ed., Scott Polar Research Institute, 1968, 198.
118. Burkholder, P. R. and Mandelli, E. F., Productivity of microalgae in Antarctic sea ice, *Science*, 149, 872, 1965.
119. Fogg, G. E., Observations on the snow algae of the South Orkney Islands, *Philos. Trans. R. Soc. London Ser. B*, 252, 279, 1967.
120. Hoham, R. W. and Mullet, J. E., The life history and ecology of the snow alga *Chloromonas cryophila* sp. nov. (Chlorophyta, Volvocales), *Phycologia*, 16, 53, 1977.
121. Komarek, J., Hindak, F., and Javornicky, P., Ecology of the green kryophilic algae from Belanske Tarrey Mountains (Czechoslovakia), *Arch. Hydrobiol. Suppl.*, 41, 427, 1973.
122. Meguro, H., Ito, K., and Fukushima, H., Diatoms and the ecological conditions of their growth in sea ice in the Arctic Ocean, *Science*, 152, 1089, 1966.
123. Meguro, H., Ito, K., and Fukushima, H., Ice flora (bottom type): a mechanism of primary production and growth of diatoms in the sea ice, *Arctic*, 20, 114, 1967.
124. Thomas, W. H., Observations on snow algae in California, *J. Phycol.*, 8, 1, 1972.
125. Wright, R. T., Dynamics of a phytoplankton community in an ice-covered lake, *Limnol. Oceanogr.*, 9, 163, 1964.
126. Matheke, G. E. M. and Horner, R., Primary productivity of benthic microalgae in the Chuckchi Sea near Barrow, Alaska, *J. Fish. Res. Board Can.*, 31, 1779, 1974.
127. Bunt, J. S., Primary productivity under sea ice in Antarctic waters. I. Concentrations and photosynthetic activities of microalgae in the waters of McMurdo Sound, Antarctica, *Antarct. Res. Ser.*, 1, 13, 1964.
128. Bunt, J. S. and Lee, C. C., Seasonal primary production in Antarctic sea ice at McMurdo Sound in 1967, *J. Mar. Res.*, 28, 304, 1970.
129. El-Sayed, S. Z. and Mandelli, E. F., Primary production and standing crop of phytoplankton in the Weddell Sea Drake Passage, *Antarct. Res. Ser.*, 5, 87, 1965.
130. Goldman, C. R., Mason, D. T., and Wood, B. J. B., Comparative study of the limnology of two small lakes on Ross Island, Antarctica, in *Antarctic Terrestrial Biology*, Llano, B. A., Ed., American Geophys. Union, Washington, D.C., *Antarct. Res. Ser.*, 20, 1, 1972.
131. El-Sayed, S. Z., Mandelli, E. F., and Sugimura, Y., Primary organic production in the Drake Passage and Bransfield Strait, *Antarct. Res. Ser.*, 1, 1, 1964.
132. Horner, R. A., Ecological Studies on Arctic Sea Ice Organisms, Progress report to ONR #R72-17, 1972.
133. McAllister, C. D., Aspects of estimating zooplankton production from phytoplankton production, *J. Fish. Res. Board Can.*, 26, 199, 1969.
134. Larrance, J. D., Primary production in the mid-subarctic Pacific region, *Fish. Bull.*, 69, 595, 1971.
135. Maeda, O. and Ichimura, S., On the high density of a phytoplankton population found in a lake under ice, *Int. Rev. Ges. Hydrobiol.*, 58, 673, 1973.

136. **Sorokin, Y. U. and Kondvalova, I. W.,** Production and decomposition of organic matter in a bay of the Japan Sea during the winter diatom bloom, *Limnol. Oceanogr.,* 18, 962, 1973.

137. **Clasby, R. C., Horner, R., and Alexanger, V.,** An *in situ* method for measuring primary productivity of Arctic sea ice algae, *J. Fish. Res. Board Can.,* 30, 835, 1973.

138. **Ackley, S. F., Buck, K. R., and Taguchi, S.,** Standing crop of algae in the sea ice of the Weddell Sea region, *Deep-Sea Res.,* 26, 269, 1979.

139. **Burkholder, P. R. and Mandelli, E. F.,** Carbon assimilation of marine phytoplankton in Antarctica, *Proc. Natl. Acad. Sci., U.S.A.,* 54, 437, 1965.

140. **El-Sayed, S. Z.,** On the productivity of the Southern Ocean (Atlantic and Pacific Sectors), in *Antarctic Ecology,* Vol. 1, Holdgate, M. W., Ed., Academic Press, New York, 1968, 119.

141. **Klyashtorin, L. G.,** Primary production in the Atlantic and Southern Oceans according to the data obtained during the fifth Antarctic voyage of the diesel-electric, *Ob. Kokl. Akad. Nauk USSR,* 141, 5, 1204, 1961.

142. **Mandelli, E. F. and Burkholder, P. R.,** Primary productivity in the Gerlache and Bransfield Straits of Antarctica, *J. Mar. Res.,* 24, 15, 1966.

143. **Saijo, Y. and Kawshima, T.,** Primary production in the Antarctic Ocean, *J. Oceanogr. Soc. Jpn.,* 19, 4, 22, 1964.

144. **Tsurikov, V. L. and Vedernikov, V. I.,** Photosynthesis of algae in ice, physiology of algae, *Sov. J. Mar. Biol.,* 3, 34, 1978.

145. **Whiteker, T. M.,** Sea ice habitats of Signy Island (South Orkneys) and their primary productivity, in *Adaptations Within Antarctic Ecosystems,* Llano, G. A., Ed., Gulf Publishing, Houston, 1977, 75.

146. **Hoshiai, T.,** Ecological observations of the colored layer of the sea ice at Syowa Station, *Antarct. Rec.,* 34, 60, 1969.

147. **Saito, K. and Taniguchi, A.,** Phytoplankton communities in the Bering Sea and adjacent seas. II. Spring and summer communities in seasonally ice-covered areas, *Astarte,* 11, 27, 1978.

148. **Goldman, C. R.,** Antarctic freshwater ecosystems, in *Antarctic Ecology,* Holdgate, M. W., Ed., Academic Press, New York, 1970, 609.

149. **Hobbie, J. E.,** Carbon-14 measurements of primary production in two Arctic Alaskan lakes, *Verh. Int. Verein. Theor. Angew. Limnol.,* 1, 360, 1964.

150. **Kalff, J.,** Arctic lake ecosystems, in *Antarctic Ecology,* Vol. 2, Holdgate, M. W., Ed., Academic Press, New York, 1970, 651.

151. **Tilzer, M.,** Dynamics and productivity of phytoplankton and pelagic bacteria in high mountain lakes, *Arch. Hydrobiol.,* 40, 201, 1972.

152. **Jackson, J. E., Jr. and Castenholz, R. W.,** Fidelity of thermophilic blue-green algae to hot springs habitats, *Limnol. Oceanogr.,* 20, 305, 1975.

153. **Por, F. D.,** Limnology of heliothermal solar lake on the coast of Sinai (Gulf of Eilat), *Verh. Int. Verein. Limnol.,* 17, 1031, 1969.

154. **Adams, B. L., McMahon, V., and Seckbach, J.,** Fatty acids in the thermophilic algae, *Cyanidium caldarium, Biochem. Biophys. Res. Commun.,* 42, 359, 1971.

155. **Ljungdahl, L. G. and Sherod, D.,** Proteins from thermophilic microorganisms, in *Extreme Environments: Mechanisms of Microbial Adaptation,* Heinrich, M. R., Ed., Academic Press, New York, 1976, 147.

156. **Jackson, J. E., Jr. and Castenholz, R. W.,** Fidelity of thermophilic blue-green algae to hot spring habitats, *Limnol. Oceanogr.,* 20, 305, 1975.

157. **Doemel, W. N. and Brock, T. D.,** The physiological ecology of *Cyanidium caldarium, J. Gen. Microbiol.,* 67, 17, 1971.

158. **Pierson, B. K. and Castenholz, R. W.,** Studies of pigments and growth in *Chloroflexus aurantiacus,* a phototrophic filamentous bacterium, *Arch. Microbiol.,* 100, 283, 1974.

159. **Pierson, B. K. and Castenholz, R. W.,** A phototrophic gliding filamentous bacterium of hot springs, *Chloroflexus aurantiacus,* gen. and sp. nov., *Arch. Microbiol.,* 100, 5, 1974.

160. **Brock, T. D.,** Vertical zonation in hot spring algal mats, *Phycologia,* 8, 201, 1969.

161. **Bauld, J. and Brock, T. D.,** Algal excretion and bacterial assimilation in hot spring algal mats, *J. Phycol.,* 10, 101, 1974.

162. **Madigan, M. T. and Brock, T. D.,** Adaptation by hot spring phototrophs to reduced light intensities, *Arch. Microbiol.,* 113, 111, 1977.

163. **Madigan, M. T., and Brock, T. D.,** Photosynthetic sulfide oxidation by *Chloroflexus aurantiacus,* a filamentous, photosynthetic, gliding bacterium, *J. Bacteriol.,* 122, 782, 1975.

164. **Castenholz, R. W.,** the Possible photosynthetic use of sulfide by the filamentous phototrophic bacteria of hot springs, *Limnol. Oceanogr.,* 18, 863, 1973.

165. **Brock, T. D. and Brock, M. L.,** The measurement of chlorophyll, primary productivity, photophosphorylation, and macromolecules in benthic algal mats, *Limnol. Oceanogr.,* 12, 600, 1967.

166. Brock, T. D. and Brock, M. L., Measurements of steady-state growth rates of a thermophilic alga directly in nature, *J. Bacteriol.*, 95, 811, 1968.

167. Meeks, J. C. and Castenholz, R. W., Photosynthetic properties of the extreme thermophile *Synechococcus lividus*. I. Effect of temperature on fluorescence and enchancement of $CO_2$ assimilation, *J. Thermal Biol.*, 3, 11, 1978.

168. Brock, T. D. and Brock, M. L., Effect of light intensity on photosynthesis by thermal alga adapted to natural and reduced sunlight, *Limnol. Oceanogr.*, 14, 334, 1969.

169. Sperling, J. A., Algal ecology of southern Icelandic hot springs in winter, *Ecology*, 56, 183, 1975.

170. Castenholz, R. W., The thermophilic cyanophytes of Iceland and the upper temperature limit, *J. Phycol.*, 5, 360, 1969.

171. Schwabe, G. H., Uber den thermobionten Kosmopolitan *Mastigocladus laminosus* Cohn. Blau-Algen und Lebenstraum V, *Schwiez. Z. Hydrol.*, 22, 757, 1960.

172. Stewart, W. D. P., Nitrogen fixation by blue-green algae in Yellowstone thermal areas, *Phycologia*, 9, 261, 1970.

173. Castenholz, R. W., Aggregation in a thermophilic *Oscillatoria*, *Nature (London)*, 215, 1285, 1967.

174. Castenholz, R. W., The behavior of *Oscillatoria terebriformis* in hot springs, *J. Phycol.*, 4, 132, 1968.

175. Castenholz, R. W., Low temperature acclimation and survival in thermophilic *Oscillatoria terebriformis*, in *Taxonomy and Biology of Blue-green Algae*, Desikachary, T. V., Ed., University of Madras, India, 1972, 406.

176. Doemel, W. N. and Brock, T. D., The upper temperature limit of *Cyanidium caldarium*, *Arch. Mikrobiol.*, 72, 326, 1970.

177. Enami, I. and Fukuda, I., Mechanisms of the acido- and thermophily of *Cyanidium caldarium* Geitler. I. Effects of temperature, pH, and light intensity on the photosynthetic oxygen evolution of intact and treated cells, *Plant Cell Physiol.*, 16, 211, 1975.

178. Fukuda, I., Physiological studies on a thermophilic blue-green alga, *Cyanidium caldarium* Geitler, *Bot. Mag.*, 71, 79, 1957.

179. Kessler, E., Physiological and biochemical contributions to the taxonomy of the genera *Ankistrodesmus* and *Scenedesmus*. IV. Salt tolerance and thermophily, *Arch. Microbiol.*, 113, 143, 1977.

# TEMPERATURE PREFERENCE AND TOLERANCE OF AQUATIC PHOTOSYNTHETIC MACROALGAE

Edward J. Phlips and Akira Mitsui

Marine macrophytes play an important role in the productivity of coastal environments, even in arctic and boreal latitudes.[1-3] This attests to their ability to attain high productivity and biomass even at low temperatures, i.e., ≃0°C. However, most seaweeds, even those from arctic latitudes, exhibit optimal temperatures for photosynthesis and growth in the 20 to 30°C range.[4-8] Tables 1 and 2 and Figure 1 describe the temperature preference and tolerance of a wide variety of macrophytes.

The success of cold-adapted organisms depends on their ability to retain a substantial percentage of their maximal growth and photosynthetic capability at low temperatures. One example of this is the marine red alga *Chondrus crispus*. A study by Mathieson and Norall[10] has shown that this alga maintains at least half of its maximal photosynthetic activity from near 0 to 32°C. The fact that many species of macrophytes have been shown to grow successfully over a wide range of temperatures indicates that they are physiologically adjusting to different habitats.

In terms of the lethal lower limit of macrophytes, many species withstand temperatures well below 0°C (see Figure 2).[21-37] This is true even for species indigenous to intertidal zones of warmer latitudes and indicates that adaptation to the intertidal zone involves mechanisms which directly or indirectly enhance freezing tolerance. At the same time, within a given vertical zone, there is a tendency for organisms native to cold waters to exhibit lower lethal temperature limits than those from warm waters.[21-23]

The tolerance of certain macrophytes to freezing seems to be related to several key factors. One of the most important factors appears to be water content. Experiments with the adult red alga *Porphyra* indicate that the reduction of cellular water content significantly increases resistance to freezing.[26-28] This is probably related to a concomitant reduction in water of crystallization.

As for the other end of the temperature scale, there are no macroalgae which fit the category of "thermophiles" as defined in the section on microalgae.[38] This is in accordance with the theory that resistance to heat is inversely proportional to the structural complexity of organisms. The studies of Biebl[21-24] on the upper lethal temperature limits of macroalgae indicate that the threshold for most species lies between 30 and 35°C.

Phylogenetically, the temperature tolerance of photosynthesis in macrophytes seems to be greatest among the green algae and lowest in brown algae, with red algae in an intermediate position. However, the general pattern proposed by Yokohama[6] is marked by numerous exceptions, including the eurythermal species.[4] In addition, some tropical macrophytes readily survive temperatures in excess of 35°C which is not surprising, since summer temperatures in shallow tropical waters can reach 35°C and above.

Table 1
## THE RELATIONSHIP BETWEEN TEMPERATURE AND PHOTOSYNTHESIS IN AQUATIC PHOTOSYNTHETIC MACROALGAE

| Species (reference) | Environment | Conditions at sampling location | Light intensity (klx) | Optimum temperature (°C) | Pmax, maximum rate of photosynthesis (μmol O$_2$/g dry wt/hr) | Maximum temperature for one half Pmax (°C) | Highest temperature tested* (°C) | Percent of Pmax at highest temperature tested | Minimum temperature for one half Pmax (°C) | Lowest temperature tested* (°C) | Percent of Pmax at lowest temperature tested |
|---|---|---|---|---|---|---|---|---|---|---|---|
| **Red algae — Rhodophyta** | | | | | | | | | | | |
| Chondrus crispus(9) | Northwest Atlantic, lower littoral | — | 21 | 20 | 362 | — | 32 | 60 | 13 | 2 | 24 |
| C. crispus(10) | Northwest Atlantic | September 0-m depth | 5.1 | 17 | 99 | 30 | 34 | 0 | — | 2 | 60 |
| | Northwest Atlantic | August, 12-m depth | 5.1 | 15 | 86 | 32 | 35 | 0 | 1 | 1 | 50 |
| | Northwest Atlantic | February, 12-m depth | 5.1 | 7 | 96 | 23 | 26 | 12 | — | 1 | 72 |
| Compsopogon hooker (11) | Freshwater, tropical, and heat-affluent areas | — | Saturating | 35 | — | 44 | 45 | 0 | 20 | 5 | 11 |
| Eucheuma acanthocladum(12) | Florida Keys, marine | — | 9.54 | 21 | 0.57 | 26 | 33 | 4 | 16 | 12 | 12 |
| Eucheuma gelidium (12) | Florida Keys, marine | — | 6.36 | 24 | 1.02 | 26 | 36 | 7 | 20 | 12 | 25 |
| Eucheuma nudum(12) | Florida Keys, marine | — | 3.18 | 21 | 0.94 | 25 | 33 | 0 | 16 | 12 | 9 |
| Gelidium amansii(6) | Japan, low tide mark | Winter | 30 | 26—30 | 558 | 34 | 35 | 0 | 12 | 5 | 8 |
| Gigartina stellata(9) | Northwest Atlantic, lower littoral | Summer | 30 | 30 | 603 | — | 35 | 75 | 15 | 10 | 22 |
| | | — | 21 | 20 | 126 | — | 32 | 70 | 9 | 2 | 35 |
| Gloiopeltis complanta (6) | Japan lower intertidal | Winter | 30 | 23—26 | 335 | — | 35 | 60 | 8 | 5 | 40 |
| G. complanata(6) | Japan, lower intertidal | Summer | 30 | 27—30 | 281 | — | 35 | 72 | 16 | 10 | 25 |

| Species | Location | Season | | | | | | | | | |
|---|---|---|---|---|---|---|---|---|---|---|---|
| *Gymnogongrus flabelliformis* (6) | Japan, upper intertidal | Winter | 30 | 30 | 424 | 34 | 35 | 0 | 12 | 5 | 8 |
| *Halosaccion glandiforme* (8) | Arctic marine | Summer | 30 | 35 | 759 | — | 35 | 100 | 20 | 10 | 8 |
| | | -1.5°C | 15 | 25 | 446 | 29 | 30 | 5 | 8 | -1.5 | 22 |
| *Lemanea annulata* (11) | Central Europe freshwater | — | Saturating | 25 | 500 | 34 | 40 | 0 | 14 | 5 | 30 |
| *Plocamium coccineum* (13) | — | — | — | 11 | — | 27 | 30 | 20 | 5 | 0 | 25 |
| *Polysiphonia elongata* (14) | Northwest Atlantic, estuarine | 15—20°C | 10.6 | 24 | 870 | 29 | 35 | 15 | 16 | 5 | 18 |
| *Polysiphonia lanosa* (14) | Northwest Atlantic, estuarine | 15—20°C | 10.6 | 22 | 536 | 31 | 35 | 10 | 8 | 5 | 30 |
| *Polysiphonia nigrescens* (14) | Northwest Atlantic, estuarine | 15—20°C | 10.6 | 27 | 214 | 32 | 35 | 0 | 17 | 5 | 0 |
| *Polysiphonia subtilissima* (14) | Northwest Atlantic, estuarine | 15—20°C | 10.6 | 27 | 536 | — | 35 | 55 | 20 | 5 | 0 |
| *Porphyra subarbiculata* (6) | Japan, high tide mark | Winter | 30 | 20—25 | 2232 | 33 | 35 | 36 | 5 | 5 | 50 |
| **Brown algae—Phaeophyta** | | | | | | | | | | | |
| *Ascophyllum nodosum* (15) | Northwest Atlantic | Winter | 10 | 10—25 | 51 | 30 | 35 | 0 | — | 2 | |
| *A. nodosum escad scorpioides* (15) | Northwest Atlantic | Summer | 10 | 18 | 241 | 27 | 35 | 0 | 6 | 5 | 40 |
| | | Winter | 10 | 10—25 | 51 | 32 | 35 | 0 | 2 | 2 | 50 |
| *Colpomenia bullosa* (6) | Japan, middle intertidal | Summer | 10 | 18—21 | 201 | 26 | 35 | 0 | 8 | 5 | 40 |
| | | Winter | 30 | 20 | 670 | 29 | 30 | 33 | 9 | 5 | 33 |
| *Colpomenia sinuosa* (6) | Japan, low tide mark | Winter | 30 | 22—25 | 714 | 29 | 30 | 31 | 11 | 5 | 25 |
| *Egregia laevigata* (16) | Shallow coastal | — | 13 | 20 | — | — | 25 | 75 | — | 10 | 77 |
| *Endarachne binghamiae* (6) | Japan, middle intertidal | Winter | 30 | 21—25 | 1607 | 31 | 33 | 0 | 5 | 5 | 50 |
| *Fucus* (8) | Arctic, marine | -1.5°C | 10 | 20—22 | 223 | — | 30 | 70 | 9 | -1.5 | 20 |
| *Fucus serratus* (13) | — | — | — | 21 | — | — | >30 | — | <0 | 5 | — |
| *Hizikia fusiforme* (6) | Japan, middle intertidal | Winter | 30 | 15—25 | 56 | — | 35 | 71 | 7 | 5 | 43 |
| *Ishige okamurai* (6) | Japan, middle intertidal | Summer | 30 | 30 | 281 | — | 35 | 97 | 12 | 10 | 40 |
| | | Winter | 30 | 20—25 | 232 | 33 | 35 | 40 | 8 | 5 | 35 |
| | Japan, middle intertidal | Summer | 30 | 30 | 246 | — | 35 | 95 | 13 | 10 | 36 |

Table 1 (continued)

## THE RELATIONSHIP BETWEEN TEMPERATURE AND PHOTOSYNTHESIS IN AQUATIC PHOTOSYNTHETIC MACROALGAE

| Species (reference) | Environment | Conditions at sampling location | Light intensity (klx) | Optimum temperature (°C) | $P_{max}$, maximum rate of photosynthesis (μmol $O_2$/g dry wt/hr) | Maximum temperature for one half $P_{max}$ (°C) | Highest temperature tested[a] (°C) | Percent of $P_{max}$ at highest temperature tested | Minimum temperature for one half $P_{max}$ (°C) | Lowest temperature tested[a] (°C) | Percent of $P_{max}$ at lowest temperature tested |
|---|---|---|---|---|---|---|---|---|---|---|---|
| Laminaria grodenlandica(17) | Vancouver Island, marine | — | 4.5 | ≤7 | 0.25 | 9 | 13 | 0 | — | 7 | ≤100 |
| Laminaria saccharine (17) | Vancouver Island, marine | — | 4.5 | 10 | 0.26 | 13 | 17 | 0 | — | 7 | 97 |
| | | — | 4.5 | ≥18 | 0.48 | — | 18 | 100 | — | 7 | 56 |
| | | — | 4.5 | ≥18 | 0.57 | — | 18 | 100 | — | 7 | 55 |
| Padina arborescens(6) | Japan, subtidal | Winter | 30 | 20—25 | 313 | 33 | 35 | 21 | 10 | 5 | 21 |
| | | Summer | 30 | 26—30 | 536 | — | 35 | 88 | 15 | 10 | 29 |
| Undaria pinnatifida (6) | Japan, subtidal | Winter | 30 | 21—25 | 804 | 30 | 30 | 50 | 12 | 10 | 35 |
| **Green Algae—Chlorophyta** | | | | | | | | | | | |
| Chaetomorpha(8) | Arctic, marine | -1.5°C | 15 | 25 | 223 | — | 30 | 60 | 10 | 2 | 26 |
| Cladophora fracta(13) | — | — | — | 6 | — | — | >25 | — | <—2 | — | — |
| Enteromorpha compressa(13) | — | — | — | 15 | — | — | >30 | — | <0 | — | — |
| E. compressa(6) | Japan, upper intertidal | Winter | 30 | 25 | 2455 | 33 | 35 | 45 | 10 | 5 | 36 |
| | | Summer | 30 | 30 | 1875 | — | 35 | 93 | 10 | 10 | 50 |
| Entermorpha linza(6) | Japan, lower intertidal | Winter | 30 | 25 | 1875 | — | 35 | 60 | — | 10 | 59 |
| Monostroma nitidum (6) | Japan, high tide mark | Winter | 30 | 30—35 | 848 | — | 35 | 100 | 13 | 10 | 37 |
| Ulva pertusa(6) | Japan, middle intertidal | Winter | 30 | 25—30 | 1250 | — | 35 | 79 | 10 | 5 | 29 |
| | | Summer | 30 | 30 | 804 | — | 35 | 72 | 16 | 10 | 28 |

[a] Or the temperature at which no photosynthesis is first observed in experiments which span a sufficient temperature range.

Table 2

## THE RELATIONSHIP BETWEEN TEMPERATURE AND GROWTH IN AQUATIC PHOTOSYNTHETIC MACROPHYTES

| Species (reference) | Environment | Preculture conditions | Light intensity (klx) | Optimum temperature (°C) | Gmax, maximum growth rate (μm/day) | Maximum temperature for one half Gmax (°C) | Highest temperature tested[a] (°C) | Percent of Pmax at highest temperature tested | Minimum temperature for one half Gmax (°C) | Lowest temperature tested[a] (°C) | Percent of Pmax at lowest temperature tested |
|---|---|---|---|---|---|---|---|---|---|---|---|
| **Red algae — Rhodophyta** | | | | | | | | | | | |
| Clathromorphum circumscriptum (18) | Arctic coralline | — | 0.7 | 5—16 | 5 | 18 | 19 | 0 | 1 | 0 | 30 |
| Corallina officinalis (19) | Northwest Atlantic, marine coralline | 6°C | — | 12 | 2.89 mm growth/6 weeks | 21 | 25 | 0 | 9 | 5 | 0 |
| Leptophytum foecundum (18) | Subarctic coralline, deepwater | — | 0.05 | 10 | 9 | 13 | 15 | 0 | 5 | 0 | 33 |
| Leptophytum laeve (18) | Subarctic coralline, deepwater | — | 0.05 | 8—10 | 10 | 14 | 15 | 0 | 1 | 0 | 30 |
| Lithophyllum oriculatum (18) | Boreal coralline | — | 0.7 | 10—12 | 4 | 17 | 20 | 0 | 0 | 0 | 50 |
| Lithothamnium glaciale (18) | Subarctic coralline | — | 0.7 | 10 | 12 | 17 | 19 | 0 | 4 | 0 | 8 |
| Phymatolithon polymorphum (18) | North Atlantic, coralline | — | 0.65 | 10 | 15 | 18 | 20 | 0 | 6 | 0 | 20 |
| **Green algae — Chlorophyta** | | | | | | | | | | | |
| Codium fragile (20) | Northwest Atlantic, Narragansett Bay | 12 hr light/12 hr dark | Saturating | 24 | 25 mg dry wt/21 days | 28 | 30 | 10 | 14 | 6 | 14 |
| | Northwest Atlantic, Narragansett Bay | Continuous light | Saturating | 24 | 52 mg dry wt/21 days | 28 | 30 | 35 | 13 | 6 | 15 |

[a]  Or the temperature at which no growth is first observed in experiments which span a sufficient temperature range.

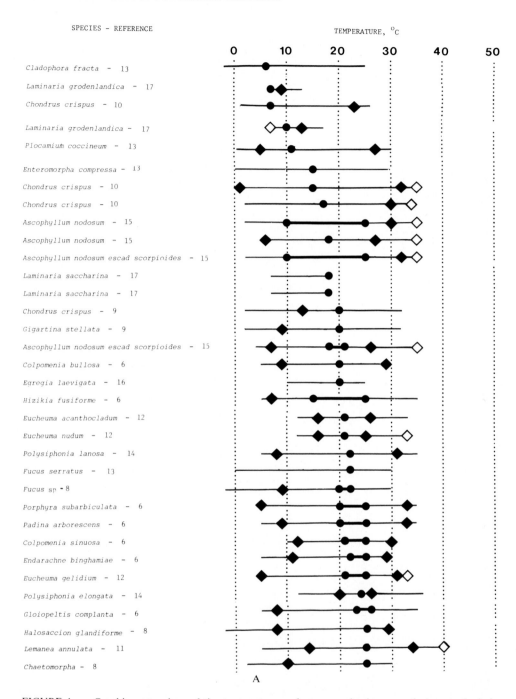

FIGURE 1. Graphic comparison of the temperature preferences and tolerances of photosynthesis in various aquatic photosynthetic macroorganisms. Species from Table 1 are ranked according to their optimal temperature for photosynthesis, designated by "●"; "◆" refers to the temperatures at which one half Pmax is achieved; "◇" refers to the temperatures at which no photosynthetic activity occurs. Broad temperature optima are indicated by two filled circles linked by a thick line. Some species appear more than once because of disparity in temperature responses observed by different authors or different experimental conditions.

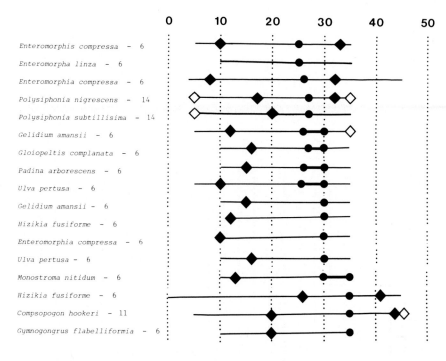

FIGURE 1B

GREEN ALGAE — CHLOROPHYTA

| SPECIES | ZONE | SAMPLING AREA |
|---|---|---|
| Anadyomene stellata | Sublittoral | Puerto Rico |
| Chaetomorpha cannabina | Littoral | Izembek Lagoon, Alaska |
| Chaetomorpha cannabina | Littoral | Douglas Island, Alaska |
| Chaetomorpha cannabina | Littoral | Alaska |
| Cladophora crispula | Lower Littoral | Puerto Rico |
| Cladophora fascicularis | Lower Littoral | Puerto Rico |
| Cladophora hutchinsiae | Sublittoral | Roscoff |
| Cladophora pellucida | Sublittoral | Roscoff |
| Cladophora rupestris | Littoral | Roscoff |
| Cladophora trichotoma | Littoral | Friday Harbor, Wash. |
| Cladophora trichotoma | Littoral | Pacific Grove, Calif. |
| Cladophoropsis membranacea | Littoral | Puerto Rico |
| Enteromorpha compressa | Littoral | Roscoff |
| Enteromorpha flexuosa | Littoral | Puerto Rico |
| Enteromorpha linza | Littoral | Friday Harbor, Wash. |
| Enteromorpha linza | Littoral | Cape Glazenap, Alaska |
| Enteromorpha torta | Littoral | Douglas Island, Alaska |
| Monostroma zostericola | Sublittoral | Friday Harbor, Wash. |
| Rhizoclonium hookeri | Littoral | Puerto Rico |
| Rhizoclonium riparium | Littoral | Friday Harbor, Wash. |
| Ulva sp | Lower Littoral | Izembek, Alaska |
| Ulva fasciata | Lower Littoral | Puerto Rico |
| Ulva lactuca | Littoral | Roscoff |
| Ulva lactuca | Littoral | Douglas Island, Alaska |
| Ulva lactuca | Littoral | Friday Harbor, Wash. |
| Ulva lactuca | Lower Littoral | Puerto Rico |
| Ulva olivacea | Sublittoral | Roscoff |

Temperature, °C

A

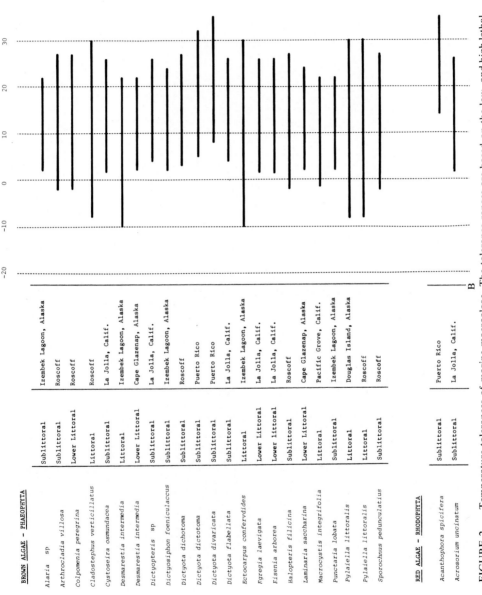

FIGURE 2.    Temperature tolerances of marine macroalgae. The tolerance ranges are based on the low  and high lethal temperatures determined by Biebl.[21-23]

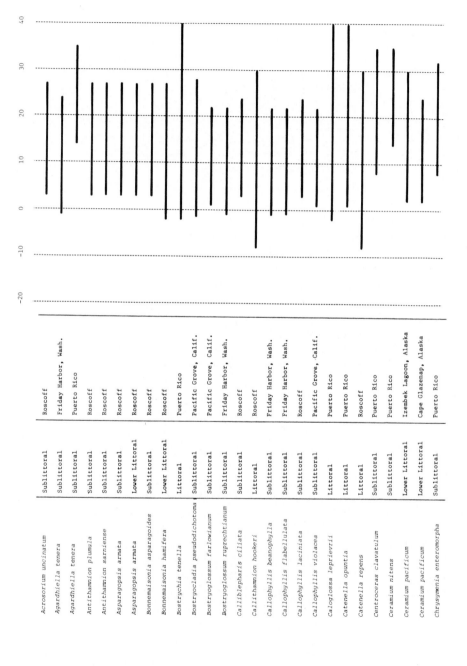

| | | |
|---|---|---|
| Acrosorium uncinatum | Sublittoral | Roscoff |
| Agardhiella tenera | Sublittoral | Friday Harbor, Wash. |
| Agardhiella tenera | Sublittoral | Puerto Rico |
| Antithamnion plumula | Sublittoral | Roscoff |
| Antithamnion sarniense | Sublittoral | Roscoff |
| Asparagopsis armata | Sublittoral | Roscoff |
| Asparagopsis armata | Lower Littoral | Roscoff |
| Bonnemaisonia asparagoides | Sublittoral | Roscoff |
| Bonnemaisonia hamifera | Lower Littoral | Roscoff |
| Bostrychia tenella | Littoral | Puerto Rico |
| Bostryocladia pseudodichotoma | Sublittoral | Pacific Grove, Calif. |
| Bostryoglossum farlowianum | Sublittoral | Pacific Grove, Calif. |
| Bostryoglossum ruprechtianum | Sublittoral | Friday Harbor, Wash. |
| Calliblepharis ciliata | Sublittoral | Roscoff |
| Callithamnion hookeri | Littoral | Roscoff |
| Callophyllis beanophylla | Sublittoral | Friday Harbor, Wash. |
| Callophyllis flabellulata | Sublittoral | Friday Harbor, Wash. |
| Callophyllis laciniata | Sublittoral | Roscoff |
| Callophyllis violacea | Sublittoral | Pacific Grove, Calif. |
| Caloglossa leprievrii | Littoral | Puerto Rico |
| Catenella opuntia | Littoral | Puerto Rico |
| Catenella repens | Littoral | Roscoff |
| Centroceras clavatulum | Sublittoral | Puerto Rico |
| Ceramium nitens | Sublittoral | Puerto Rico |
| Ceramium pacificum | Lower Littoral | Izembek Lagoon, Alaska |
| Ceramium pacificum | Lower Littoral | Cape Glazenap, Alaska |
| Chrysymenia enteromorpha | Sublittoral | Puerto Rico |

2C

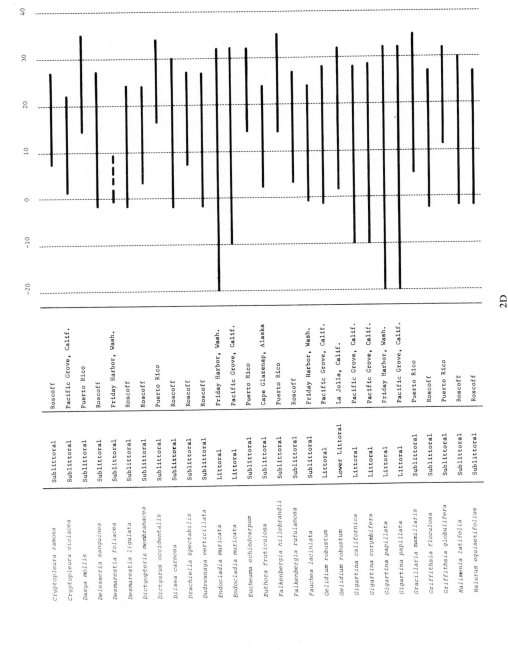

2D

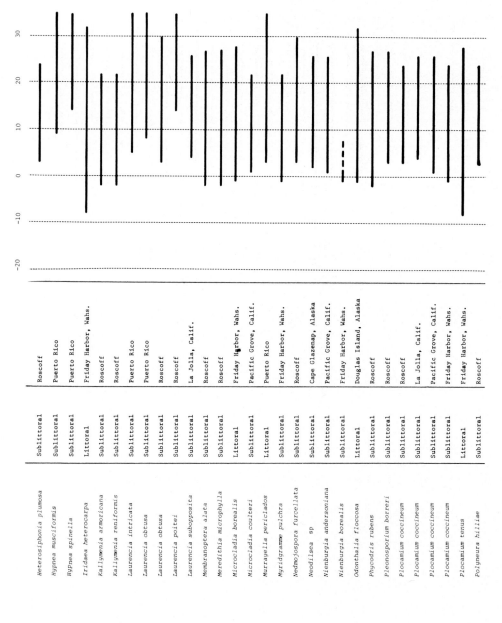

| | | |
|---|---|---|
| *Heterosiphonia plumosa* | Sublittoral | Roscoff |
| *Hypnea musciformis* | Sublittoral | Puerto Rico |
| *Hypnea spinella* | Sublittoral | Puerto Rico |
| *Iridaea heterocarpa* | Littoral | Friday Harbor, Wahs. |
| *Kallymenia armoricana* | Sublittoral | Roscoff |
| *Kallymenia reniformis* | Sublittoral | Roscoff |
| *Laurencia intricata* | Sublittoral | Puerto Rico |
| *Laurencia obtusa* | Sublittoral | Puerto Rico |
| *Laurencia obtusa* | Sublittoral | Roscoff |
| *Laurencia poitei* | Sublittoral | Roscoff |
| *Laurencia subopposita* | Sublittoral | La Jolla, Calif. |
| *Membranoptera alata* | Sublittoral | Roscoff |
| *Meredithia microphylla* | Sublittoral | Roscoff |
| *Microcladia borealis* | Littoral | Friday Harbor, Wahs. |
| *Microcladia coulteri* | Sublittoral | Pacific Grove, Calif. |
| *Murrayella periclados* | Littoral | Puerto Rico |
| *Myriogramme pulchra* | Sublittoral | Friday Harbor, Wahs. |
| *Nedmojospora furcellata* | Sublittoral | Roscoff |
| *Neodilsea* sp | Sublittoral | Cape Glazenap, Alaska |
| *Nienburgia andersoniana* | Sublittoral | Pacific Grove, Calif. |
| *Nienburgia borealis* | Sublittoral | Friday Harbor, Wahs. |
| *Odonthalia floccosa* | Littoral | Douglas Island, Alaska |
| *Phycodris rubens* | Sublittoral | Roscoff |
| *Pleonosporium borreri* | Sublittoral | Roscoff |
| *Plocamium coccineum* | Sublittoral | Roscoff |
| *Plocamium coccineum* | Sublittoral | La Jolla, Calif. |
| *Plocamium coccineum* | Sublittoral | Pacific Grove, Calif. |
| *Plocamium coccineum* | Sublittoral | Friday Harbor, Wahs. |
| *Plocamium tenus* | Littoral | Friday Harbor, Wahs. |
| *Polyneura hilliae* | Sublittoral | Roscoff |

2E

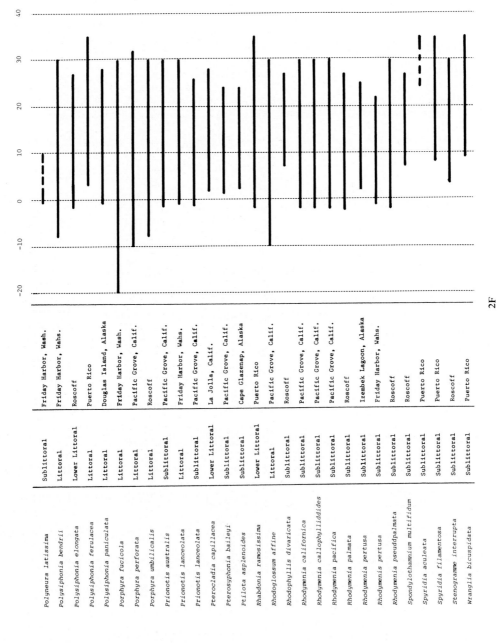

| | |
|---|---|
| *Polyneura latissima* | Sublittoral | Friday Harbor, Wash. |
| *Polysiphonia bendrii* | Littoral | Friday Harbor, Wahs. |
| *Polysiphonia elongata* | Lower Littoral | Roscoff |
| *Polysiphonia ferulacea* | Littoral | Puerto Rico |
| *Polysiphonia paniculata* | Littoral | Douglas Island, Alaska |
| *Porphyra fucicola* | Littoral | Friday Harbor, Wash. |
| *Porphyra perforata* | Littoral | Pacific Grove, Calif. |
| *Porphyra umbilicalis* | Littoral | Roscoff |
| *Prionotis australis* | Sublittoral | Pacific Grove, Calif. |
| *Prionotis lanceolata* | Littoral | Friday Harbor, Wahs. |
| *Prionotis lanceolata* | Sublittoral | Pacific Grove, Calif. |
| *Pterocladia capillacea* | Lower Littoral | La Jolla, Calif. |
| *Pterosiphonia baileyi* | Sublittoral | Pacific Grove, Calif. |
| *Ptilota asplenoides* | Sublittoral | Cape Glazenap, Alaska |
| *Rhabdonia ramosissima* | Lower Littoral | Puerto Rico |
| *Rhodoglossum affine* | Littoral | Pacific Grove, Calif. |
| *Rhodophyllis divaricata* | Sublittoral | Roscoff |
| *Rhodymenia californica* | Sublittoral | Pacific Grove, Calif. |
| *Rhodymenia callophylliddides* | Sublittoral | Pacific Grove, Calif. |
| *Rhodymenia pacifica* | Sublittoral | Pacific Grove, Calif. |
| *Rhodymenia palmata* | Sublittoral | Roscoff |
| *Rhodymenia pertusa* | Sublittoral | Izembek Lagoon, Alaska |
| *Rhodymenis pertusa* | Sublittoral | Friday Harbor, Wahs. |
| *Rhodymenia pseuddpalmata* | Sublittoral | Roscoff |
| *Spondylothamnium multifidum* | Sublittoral | Roscoff |
| *Spyridia aculeata* | Sublittoral | Puerto Rico |
| *Spyridia filamentosa* | Sublittoral | Puerto Rico |
| *Stenogramme interrupta* | Sublittoral | Roscoff |
| *Wrangia bicuspidata* | Sublittoral | Puerto Rico |

2F

# REFERENCES

1. **Mann, K. H.**, Seaweeds: their productivity and strategy for growth, *Science,* 182, 975, 1973.
2. **Kanwisher, J. W.**, Photosynthesis and respiration in some seaweeds, in *Some Contemporary Studies in Marine Science,* Barnes, H., Ed., G. Allen and Unwin Ltd., London, 1966, 407.
3. **Parsons, T. R., Takahashi, M., and Hargrave, B.**, *Biological Oceanographic Processes,* 2nd ed., Pergamon Press, New York, 1977.
4. **Gessner, F.**, Temperature: plants, in *Marine Ecology,* Kinne, O., Ed., Wiley Interscience, New York, 1970, 363.
5. **Soeder, C. and Stengel, E.**, Physico-chemical factors affecting metabolism and growth rate, in *Algal Physiology and Biochemistry,* Stewart, W. D. P., Ed., 1975, 714.
6. **Yokohama, Y.**, Photosynthesis-temperature relationships in several benthic marine algae, in *Proc. 7th Int. Seaweed Symp.,* John Wiley & Sons, New York, 1971, 286.
7. **Rabinowitch, E. I.**, *Photosynthesis and Related Processes,* Interscience, New York, 1951.
8. **Healey, F. P.**, Photosynthesis and respiration of some arctic seaweeds, *Phycologia,* 11, 267, 1972.
9. **Mathieson, A. E. and Burns, R. L.**, Ecological studies of economic red algae. I. Photosynthesis and respiration of *Chondrus crispus* Stackhouse and *Gigartina stellata* (Stackhouse) Batters, *J. Exp. Mar. Biol. Ecol.,* 7, 197, 1971.
10. **Mathieson, A. C. and Norall, T. L.**, Photosynthetic studies of *Chondrus crispus,* *Mar. Biol.,* 33, 207, 1975.
11. **Kremer, B. P.**, Aspects of $CO_2$-fixation in some freshwater Rhodophyceae, *Phycologia,* 17, 430, 1978.
12. **Mathieson, A. C. and Dawes, C. J.**, Ecological studies of Floridian *Eucheuma.* II. Photosynthesis and respiration, *Bull. Mar. Sci.,* 24, 274, 1974.
13. **Ehrke, G.**, Uber die wirkung der temperatur and des lichtes auf atmung und assimilation einiger meeres- und subwasseralgen, *Planta,* 13, 221, 1931.
14. **Fralick, R. A. and Mathieson, A. C.**, Physiological ecology of four *Polysiphonia* species (Rhodophyta, Ceramiales), *Mar. Biol.,* 29, 29, 1975.
15. **Chock, J. S. and Mathieson, A. C.**, Physiological ecology of *Ascophyllum noposum* (L.) le jolis and its detached escao *Scorpioides* (Hornemann) Hauck (Fucales, Phaeophyta), *Bot. Mar.,* 22, 21, 1979.
16. **Chapman, V. J.**, A contribution to the ecology of *Egregia laevigata* Setchell. III. Photosynthesis and respiration; conclusions, *Bot. Mar.,* 3, 101, 1962.
17. **Druehl, L. D.**, Distribution of two species of *Laminaria* as related to some environmental factors, *J. Phycol.,* 3, 103, 1967.
18. **Adey, W. H.**, The effects of light and temperature on growth rates in Boreal-subarctic crustose corrallines, *J. Phycol.,* 6, 269, 1970.
19. **Colthart, B. J. and Johansen, H. W.**, Growth rates of *Corralina officinalis* (Rhodophyta) at different temperatures, *Mar. Biol.,* 18, 46, 1973.
20. **Hanisak, M. D.**, Growth patterns of *Codium fragile* sp. *Tomentosoides* in response to temperature, irradiance, salinity and nitrogen source, *Mar. Biol.,* 50, 319, 1979.
21. **Biebl, R.**, Temperaturresistenz Tropischer Meeresalgen, *Bot. Mar.,* 4, 241, 1962.
22. **Biebl, R.**, Temperatur und osmotische Resistenz von Merresalgen der bretonischen Kueste, *Protoplasma,* 50, 217, 1958.
23. **Biebl, R.**, Protoplasmatische Oekologie der Pflanzenn. I. Wasser und Temperatur, *Handb. Protoplasmaforsch.,* 12, 1, 1962.
24. **Biebl, R.**, Vergleichende Untersuchungen zur Temperaturresistenz von Merresalgen entlang der pazifischen Kueste Nordamerikas, *Protoplasma,* 69, 61, 1970.
25. **Kanwisher, J.**, Freezing and drying, in intertidal algae, *Biol. Bull.,* 113, 275, 1957.
26. **Migita, S.**, Viability and spore liberation of Conchocelis phase, *Porphyra tenera,* freeze-preserved in sea water, *Bull. Fac. Fish. Nagasaki Univ.,* 22, 33, 1967.
27. **Migita, S.**, Freezing preservation of *Porphyra* thalli in viable state, II. Effects of cooling velocity and water content of thalli on frost resistance, *Bull. Fac. Fish. Nagasaki Univ.,* 21, 131, 1966.
28. **Terumoto, I.**, Freezing and drying in a red marine alga *Porphyra yezoensis,* *Low Temp. Sci.,* 23, 11, 1965.
29. **Hwang, S. and Horneland, W.**, Survival of algal cultures after freezing by controlled and uncontrolled cooling, *Cryobiology,* 1, 305, 1965.
30. **Bird, C. J. and McLachlan, J.**, Cold-hardiness of zygotes and embryos of *Fucus* (Phaeophyceae, Fucales), *Phycologia,* 13, 215, 1974.
31. **Kylin, H.**, Uber die Kaelteresistenz der Meeresalgen, *Ber. Dtsch. Bot. Ges.,* 35, 370, 1917.
32. **Parker, J.**, Seasonal changes in cold-hardiness of *Fucus vesiculosus,* *Biol. Bull.,* 119, 474, 1960.
33. **Scholm, H. E.**, Untersuchungen zur Hitze- und Frostresistenz einheimishcer Subwasseralgen, *Protoplasma,* 65, 97, 1968.

34. **Terumoto, J.,** Frost-resistance in a marine alga *Ulva pertusa kjellmann, Low Temp. Sci., Ser. B.,* 18, 35, 1960.
35. **Terumoto, I.,** Frost resistance in some marine algae *Enteromorpha intestinalis* (L) *Low Temp. Sci.,* 119, 23, 1961.
36. **Terumoto, I.,** Frost resistance in some marine algae from the winter intertidal zone, *Low Temp. Sci.,* 22, 19, 1964.
37. **Tsuru, S.,** Preservation of marine and fresh water algae by means of freezing and freeze-drying, *Cryobiology,* 10, 445, 1973.
38. **Brock, T. D.,** *Thermophilic Microorganisms and Life at High Temperatures,* Springer-Verlag, Basel, 1978.

# DROUGHT-TOLERANT PLANTS: DRY MATTER PRODUCTION OF CROP AND NATIVE SPECIES

Jack R. Mauney and Stan R. Szarek

## INTRODUCTION

Levitt[1,2] has pointed out that drought-resistant plants may exhibit varying degrees of drought avoidance (the ability to conserve water or find additional water so that tissue desiccation does not occur) and drought tolerance (the ability to survive a severe reduction in internal water content). Ephemerals exhibit an alternate form of drought avoidance, the ability to grow rapidly and set seed before soil moisture is depleted. Though native ephemerals provide considerable dry matter production in xeric climates during years with high rainfall,[3] they are regarded as having no drought-resisting adaptations and are not considered in this review. Rather, we will consider those plants which have mechanisms for survival during periods of water stress (drought resistant), and evaluate their water-use efficiency (WUE, gram dry matter production per kilogram of water evapotranspired, g/kg) under cultivated or natural conditions. Generally, these plants are from areas with less than 50 cm water in rainfall or irrigation per year.

Ideally, a drought-resistant plant would be one in which WUE increases as water becomes limited. In practice, few species exhibit this capability. In many instances,[4-7] the highest WUE for the species in question has been observed with the amount of water (optimum) which produces 75 to 90% of the maximum potential yield. However, in species or cultivars found to be drought tolerant, the reduction in growth and yield under water stress is less than in drought-sensitive types.[8,9] The question of the effect of drought on WUE is complicated in crop plants by the fact that most are grown for products other than total biomass. Development of flowers and fruits involves processes requiring both cell division and cell expansion, the latter being especially sensitive to water stress. Thus, yield of the agronomic product may be suppressed by water stress much more than yield of total biomass. On the other hand, yield of a differentiation product such as the rubber-containing latex of guayule may be increased by water stress at the same time that total plant biomass is decreased.[10]

Comparisons between drought-resistant and drought-sensitive crops and between crops and native species have been used to establish the growth characteristics which are typical for resistant species.[8,11-13] These studies and others[4,7,14] which characterize the response of WUE in desert and crop ecosystems to increasing water stress can be used to predict the productivity of the stressed ecosystem. This is important, since deserts are expanding worldwide, converting semiarid lands (40- to 80-cm annual rainfall) into arid lands with more restricted productivity.[13] Such areas can be used by employing runoff (e.g., Avdat Runoff Farm)[15] or drought control.[16] These measures could increase production with little increase in economic cost. However, the feasibility analysis and choice of crops must include species responses to increments of added water.[16]

## CROPS

In an excellent review of the effects of water deficit on crops, Begg and Turner[11] listed more than 30 books and review articles dealing with plant water relations since 1960. We will not attempt to duplicate this exhaustive review. Instead, we will document productivity of several crops which have traditionally been grown under dryland culture in areas with less than 50-cm annual rainfall. They are cotton (*Gossypium hirsutum* L.), grain sorghum (*Sorghum bicolor*), and wheat (*Triticum aestivum* L.). In recent years, several native species have been shown to be suitable for dryland cultivation in arid climates. Guayule (*Parthenium argentatum* Gray), tepary bean (*Phaseolus acutifolius* A. Gray), and jojoba (*Simmondsia chinensis* (Link) Schneider) also may have potential as dryland crops.

### Grain Sorghum

In dryland farming areas in Kansas, Oklahoma, and Texas, sorghum is the feed grain of choice. Though some workers have reported the WUE of sorghum to be somewhat higher than maize (see Table 1), WUE may not be the principal factor which makes sorghum better suited to the variable weather of the Great Plains. Martin[17] found that both cuticular and stomatal water loss under moisture stress were greater in maize than sorghum. Sorghum could also withstand longer periods of extreme stress. Thus, sorghum exhibits both drought avoidance through water conservation and drought tolerance under extreme stress. Arkley[18] applied a correction to WUE data for the humidity in which the data were taken and calculated that sorghum is 20% more water-use efficient than maize. Stewart et al.,[7] on the other hand, found that at maximum seasonal evapotranspiration (ET) (63 cm in Davis, California), maize was more productive and exhibited higher WUE (1.8 g/kg) than did sorghum (1.6 g/kg). However, if less than 65% of maximum ET was available, sorghum was the more productive crop. At 50% ET, sorghum had WUE of 1.6 g/kg and maize had 1.3 g/kg. Management of soil water availability when water was limited was much more critical for maize than for sorghum. This suggests that with uncertain rainfall, sorghum would be the better choice. Hsaio et al.[8] considered several characteristics of these two species which might contribute to drought resistance: rooting patterns, stomatal behavior, developmental morphology, and source-sink relationships. They concluded that only a dynamic mixture of traits could account for the observed differences in yielding ability under water stress. They also observed that differences between cultivars within each species are considerable so that there is overlap in the range of sensitivity to drought.

### Wheat

Although wheat is a $C_3$ plant that is less productive than sorghum (see Table 1), it is grown primarily in areas in which severe winters and short growing seasons limit sorghum production. Musick et al.[5] observed that wheat depletes soil moisture to a greater depth than does sorghum. In addition, it may have greater drought tolerance. Leaf water potentials ($\Psi$) of −50 bars have been measured[9,19] in wheat plants which recovered upon rewatering. Sullivan and Eastin[20] found that 50% kill of sorghum plants were associated with $\Psi$ values of −33 to −48 bars. When barley was compared to wheat, Fischer and Maurer[9] concluded that wheat was more drought resistant, but that barley was 9 days earlier than the average wheat cultivars. This earliness provided significant drought avoidance in some years.

### Cotton

The WUE of cotton is about half that of the $C_4$ grains. Because it is very sensitive to water stress during the flowering phase of growth,[21] the WUE can be increased by supplemental irrigation during peak flowering. Quisenberry and Roark[22] found that under dryland conditions, cultivars with a relatively indeterminate flowering habit had superior yield to the highly determinate cultivars. The indeterminate trait probably gives the cultivar an "opportunistic" capability for utilizing rainfall at any time during the season.

Table 1
WATER USE AND WUE OF BIOMASS PRODUCTION AND
HARVESTED YIELD OF SELECTED CROPS AND NATIVE
SPECIES

| | | | WUE | | |
| | | | Harvested | Total biomass (g/ | |
| Species | Location | Water use (cm) | (g/kg H$_2$O) | kg H$_2$O) | Ref. |
| --- | --- | --- | --- | --- | --- |
| Sorghum | Arizona | 63 | 0.7 | 1.4[a] | 44 |
| Sorghum | Australia | 60 | 1.1 | 2.2 | 45 |
| Sorghum | California | 34 | 1.6 | 3.2 | 7 |
| Sorghum | Colorado | | | 3.5 | 30 |
| Sorghum | Texas | 60 | 1.4 | 2.8[a] | 4 |
| Corn | Colorado | | | 2.7 | 30 |
| Corn | Ohio | 56 | 1.6 | 4.0 | 46 |
| Corn | Illinois | 38 | 3.0 | 7.5 | 47 |
| Cotton | Arizona | 100 | 0.33 | 1.2[a] | 41 |
| Cotton | Texas | 50 | 0.5 | 1.5[a] | 26 |
| Cotton | California | 25 | 0.7 | 1.7[a] | 24 |
| Soybean | Australia | 50 | 1.1 | 2.2 | 45 |
| Soybean | Illinois | 39 | 0.6 | 1.2 | 48 |
| Wheat | Arizona | 65 | 1.0 | 2.5[a] | 49 |
| Wheat | North Dakota | | 1.0 | 2.6[a] | 50 |
| Jojoba | Arizona | | | 0.6 | 29 |
| Jojoba | Arizona | | 0.08 | | 34 |
| Guayule | California | 3.5 | 0.09 | | 36 |
| Tepary bean | Arizona | 24 | 0.2 | 0.6 | 28 |
| Foot hill paloverde | | | | 1.3 | 29 |
| Russian thistle | Colorado | | | 3.0 | 30 |
| Russian thistle | North Dakota | | | 4.5 | 50 |

[a] Based on harvest indexes of Loomis and Gerakis.[51]

The planting configuration of the crop may also influence WUE. Namken et al.[23] compared rapidly maturing cultivars in close-row plantings with conventional cultivars in 1-m rows. The WUE of the unconventional system was 25% higher[11] than that of the conventional. El-Zik and Walhood[24] observed similar WUE shifts in plantings in California. Cotton possesses good heat and drought tolerance as shown by the observation that it retained stomatal conductance at $\Psi$ below −30 bars[25] at which sorghum and maize were closed. When grown on rainfall in areas with 50-cm annual and 15-cm crop season rainfall, cotton produces approximately 1000 kg/ha seed cotton and 3000 kg/ha total dry matter. The addition of as little as 10- to 15-cm irrigation will double these values.[22,26]

**Tepary Bean**

The very short productive season of tepary beans, 60 days,[27] made the cultivation of this species more reliable than other dry beans for native American Indians of the Southwest. In locations with as little as 20 cm of annual rainfall, a crop of 700 kg/ha has been reported.[28] Though the WUE for tepary is similar to that of other beans (see Table 1), its rapid maturity and low consumptive use made it a staple among the natives of the Southwest.

## NATIVE SPECIES

Less information is available on the WUE of drought-resistant native species growing in desert environments, since most are not economically important. The survey of McGinnis and Arnold[29] and Shantz and Piemeisel[30] concluded that most desert-adapted species have relatively low WUE. Only foothills paloverde (*Cercidium microphyllum* (TORR., Rose, and JTN.) had WUE approaching that of cultivated species (see Table 1). Desert ephemerals as represented by Russian thistle (*Salsola hali* L. var. *tenuifolia tausch*) may have WUE in the range of the most efficient crops (see Table 1).

Foliar WUE (the ratio of milligrams $CO_2$ uptake to grams water transpired per unit of leaf area) have been calculated for many desert species. Studies comparing native desert perennials (the $C_3$ jojoba, *S. chinensis*, and the $C_4$ saltbush, *Atriplex halimus* L.) to crop species within the same photosynthetic group ($C_3$ soybean and sunflower, $C_4$ sorghum) have documented the slightly higher foliar WUE of the native species.[12] The basis for relatively high foliar WUE by plants native to the desert ecosystem was recently reviewed by Fisher and Turner.[13] However, for comparative purposes, we will emphasize the former definition of WUE which considers the ratio of dry matter yield to ET.

### Creosote Bush (*Larrea tridentata* (DC.) Cov.)

This species is characteristic of the hot deserts in North America and is generally regarded as being drought tolerant. Adaptive strategies, including water use, have recently been reviewed by Barbour et al.[31] and are briefly reviewed from this source.

*Larrea* appears to be capable of obtaining water from the soil by distillation of lower soil horizons. This would necessitate maintaining low $\Psi$ values for the continuation of water uptake. Minimum values of $-60$ to $-70$ bars for active, living plants are frequently reported. Transpiration rates of *Larrea* are lower than those of other desert perennials, implying greater regulation of water loss. *Larrea* can maintain net photosynthesis at levels of $\Psi$ between $-60$ and $-78$ bars. Despite the species appearing to be drought tolerant, WUE decreased as soil and plant water stress increased.

Monocultures do not occur in native ecosystems, but in some locations *Larrea* clearly dominates the perennial vegetation. Within the Chihuahuan Desert, at a bajada site, the computed WUE for *Larrea* from 1971 to 1974 averaged 0.6 g/kg.[32]

### Jojoba

This species occurs in mediterranean and desert environments of the southwestern U.S. and Mexico, but because of its broadleaf, evergreen nature, it is considered to be a drought-tolerant species. Field studies of water use have been conducted by Halvorson and Patten,[33] and its interrelationship with photosynthesis has been studied by Al-Ani et al.[34] and Woodhouse[35] and are briefly reviewed from these sources.

Seasonal minima for plant $\Psi$ have been measured in the range of $-60$ to $-70$ bars, with net photosynthesis being measured at $\Psi$ of $-70$ bars and leaf temperatures being between 40 and 47°C. Carbon dioxide efflux in the light (negative carbon balance) and complete stomatal closure have not been observed during daytime studies. Stomatal regulation of transpirational water loss occurs at leaf $\Psi$ levels down to $-70$ bars. During periods with both high temperature and water stress conditions, day-long average foliar WUE is between 1.4 to 1.6 mg $CO_2$ assimilated per gram $H_2O$ transpired. Ehrler et al.[36] found that catchments which increased the available water by about 5 times increase the seed yield more than 7.5 times.

**Guayule**

This species has been gathered by natives from wild habitats for centuries and processed for rubber. For a time, 1942 to 1945,[37] there was considerable activity aimed at cultivation of the shrub as a crop. Recently the interest has been renewed, but production is presently confined to native stands. The WUE of guayule is low (see Table 1). Primary dry matter (biomass) production is improved by additional water,[10] but latex production is reduced by adequate water.[10,38]

## DESERT ECOSYSTEMS

Vegetation communities of desert ecosystems are composed of a variety of life forms and photosynthetic groups. Each species possesses a combination of morphological and physiological characteristics which contribute to survival in such environments. Intensive studies recently reviewed by Szarek[32] indicate that the dominant perennial vegetation type is a $C_3$, woody shrub. A "conservative strategy" for survival is generally applied to the Crassulacean acid metabolism (CAM) photosynthetic group which sacrifices carbon fixation potential for maintaining high seasonal levels of tissue $\Psi$. A more "opportunistic strategy" for survival is attributed to the $C_4$ photosynthetic group which maximizes carbon fixation during periods of available soil moisture, a form of drought avoidance. Thus, drought-tolerant species may be best exemplified by the $C_3$ group.

Water expenditure for dry matter production is large (low WUE) for desert ecosystems (see Table 2). At these sites, $C_3$ species clearly dominate the community composition and contribution to water use. Only at the Great Basin site does a $C_4$ species contribute 34% of the community importance. At the Chihuahuan Desert site, two CAM species contribute to 9% of the community importance. Webb et al.[14] reported that dry matter production at these study sites is linearly related to annual ET, as long as precipitation was greater than the minimum amount of water required to sustain the ecosystem. Secondly, the increase in production per unit of precipitation above the minimum requirement is low in comparison to crop species. The WUE for incremental increases in dry matter production averaged 0.5 g/kg. Desert ecosystems are typically nitrogen deficient.[39] Thus, when water stress is relieved, the plants may be limited in growth by nutrient availability. Certainly high annual productivity and high annual WUE would require a large nutrient input.[40] It also could be argued that the drought-resistant native species have evolved a "conservative strategy" for accumulation of dry matter in response to additional precipitation or irrigation. The year-to-year variation in timing and magnitude of precipitation is high for desert environments.[41] Thus, additional increments of water during wet years without a "conservative" growth strategy might contribute to a plant biomass which could be severely stressed in succeeding dry years. The response of desert vegetation to additional precipitation is less than that of the shortgrasses which also occur in arid environments.[14] The grasses survive periods of drought by becoming dormant, a strategy which places them very close to annual ephemerals with respect to drought tolerance.

## DISCUSSION

Although the species which are adapted to deserts have ratios of photosynthetic rate to transpiration rate (foliar WUE) which are in the same range as crops plants,[12,13] the annual productivity of desert species and ecosystems is low (see Tables 1 and 2). Growth of desert natives when water is abundant is limited by nutrient supply[39] and the "conservative" growth strategy of the species. Increasing the productivity of desert

Table 2
WUE FOR ABOVEGROUND
PRIMARY PRODUCTION BY
PERENNIAL VEGETATION IN
FOUR NORTH AMERICAN
DESERTS[a]

| Location | WUE[b] (g dry matter/kg $H_2O$ ET) |
|----------|------------------------------------|
| Chihuahuan | 0.83 |
| Great Basin | 1.26 |
| Mojave | 0.37 |
| Sonoran | 0.45 |

[a]   References in Szarek.[32]
[b]   Per growing-season basis.

areas through fertilization or alteration of the ecosystem by introduction of more actively growing species would probably be self-defeating because, as Evenari et al.[42] have observed, "the price of survival for desert plants and communities is the utter restriction of growth and production". Loomis[40] has aptly pointed out that if marginal lands could be as productive as some[43] have imagined, they would no longer be considered marginal.

The crop situation maximizes WUE through selection of adapted cultivars, optimum fertility, and rapidly developing ground cover which minimizes soil evaporation. Species adapted to dryland agriculture exhibit drought avoidance (because of stomatal or other reactions which limit water loss before severe stress occurs) or drought tolerance (able to survive severe stress in the same ways as desert perennials). In addition, drought-adapted crops have either an indeterminate flowering habit (cotton) or the ability to tiller (sorghum and wheat), enabling them to respond rapidly to uncertain rainfall patterns with added yield capability. Where water is available for supplemental irrigation, the most effective use for total biomass production would be to irrigate a $C_4$ species with optimum cultivation management.

## REFERENCES

1. **Levitt, J.,** Frost, drought and heat resistance, in *Protoplasmatologia, Bd VII. Physiologie des Protoplasmas,* Springer-Verlag, Wien, 1958, 1.
2. **Levitt, J.,** *Responses of Plants to Environmental Stresses,* Academic Press, New York, 1972.
3. **Patten, D. T.,** Productivity and production efficiency of an upper Sonoran Desert ephemeral community, *Am. J. Bot.,* 65, 891, 1978.
4. **Musick, J. T. and Dusek, D. A.,** Grain sorghum response to number, timing and size of irrigations in the Southern high plains, *Trans. Am. Soc. Agric. Eng.,* 14, 401, 1971.
5. **Musick, J. T., New, L. L., and Dusek, D. A.,** Soil water depletion — yield relationships of irrigated sorghum, wheat, and soybean, *Trans. Am. Soc. Agric. Eng.,* 19, 489, 1976.
6. **Sinclair, T. R., Bingham, G. E., Lemon, E. R., and Allen, L. H., Jr.,** Water-use efficiency of field-grown maize during moisture stress, *Plant Physiol.,* 57, 245, 1975.
7. **Stewart, J. I., Misra, R. D., Pruitt, W. D., and Hagan, R. M.,** Irrigating corn and grain sorghum with a deficient water supply, *Trans. Am. Soc. Agric. Eng.,* 18, 270, 1975.
8. **Hsiao, T., Ferres, C. E., Acevado, E., and Henderson, D. W.,** Water stress and dynamics of growth and yield of crop plants, in *Water and Plant Life: Problems and Modern Approaches,* Lange, D. L., Kappen, L., and Schulze, E. D., Eds., Springer-Verlag, Basel, 1976, 311.

9. **Fischer, R. A. and Maurer, R.,** Drought resistance in spring wheat cultivars. I. Grain yield responses, *Aust. J. Agric. Res.,* 29, 897, 1978.

10. **Wadleigh, C. H., Gauch, H. G., and Magistad, O. C.,** Growth and Rubber Accumulation in Guayule as Conditioned by Soil Salinity and Irrigation Regime, Technical Bulletin 925, U.S. Department of Agriculture, Washington, 1946, 1.

11. **Begg, J. E. and Turner, N. C.,** Crop water deficits, *Adv. Agron.,* 28, 161, 1976.

12. **Rawson, H. M., Begg, J. E., and Woodward, R. G.,** The effect of atmospheric humidity on photosynthesis, transpiration and water use efficiency of leaves of several plant species, *Planta,* 134, 5, 1977.

13. **Fischer, R. A. and Turner, N. C.,** Plant productivity in the arid and semiarid zones, *Annu. Rev. Plant Physiol.,* 29, 277, 1978.

14. **Webb, W. Szarek, S., Lauenroth, W., Kinerson, R., and Smith, M.,** Primary productivity and water use in native forest, grassland and desert ecosystems, *Ecology,* 59, 1239, 1978.

15. **Evenari, M., Lange, O. L., Schulze, E-D. Kappen, L., and Buschbom, U.,** New photosynthesis, dry matter production, and phenological development of apricot trees (*Prunus armeniaca* L.) cultivated in the Negev Highlands (Israel), *Flora,* 166, 383, 1977.

16. **Viet, F. G., Jr.,** Effective drought control for successful dryland agriculture, in *Drought Injury and Resistance in Crops,* Larson, K. L. and Eastin, J. D., Eds., Crops Science Society of America, Madison, 1971, 57.

17. **Martin, J. H.,** The comparative drought resistance of sorghum and corn, *J. Am. Soc. Agron.,* 22, 993, 1930.

18. **Arkley, R. J.,** Relationship between plant growth and transpiration, *Hilgardia,* 34, 559, 1963.

19. **Angus, J. F. and Moncur, M. W.,** Water stress and phenology of wheat, *Aust. J. Agric. Res.,* 28, 177, 1977.

20. **Sullivan, C. W. and Eastin, J. W.,** Plant physiological responses to water stress, *Agric. Meteorol.,* 14, 133, 1974.

21. **Grimes, D. W., Miller, R. J., and Dickens, L.,** Water stress during flowering of cotton, *Calif. Agric.,* 24(3), 4, 1970.

22. **Quisenberry, J. E. and Roark, B.,** Influence of indeterminate growth habit on yield and irrigation water-use efficiency in upland cotton, *Crop Sci.,* 16, 762, 1976.

23. **Namken, L. N., Wiegand, C. L., and Willis, W. O.,** Soil and air temperature as limitation to more efficient water use, *Agric. Meterol.,* 14, 169, 1974.

24. **El-Zik, K. M. and Walhood, V. T.,** Increasing efficiency of water for cotton, in Proc. Western Cotton Production Conf., Phoenix, Ariz., Feb. 21-23, 1978, 34.

25. **Turner, N. C.,** Stomatal response to light and water under field conditions, in *Mechanisms of Regulation of Plant Growth,* R. L. Bielenski, Ferguson, A. R., and Cresswell, M. M., Eds., Royal Society New Zealand, Wellington, 1974, 423.

26. **Bordovsky, D. G., Jordan, W. R., Hiler, E. A., and Howell, T. A.,** Choice of irrigation timing indicator for narrow row cotton, *Agron. J.,* 66, 88, 1974.

27. **Rachie, K. D. and Robert, L. M.,** Grain legumes of the lowland tropics, *Adv. Agron.,* 26, 1, 1974.

28. **Clothier, R. W.,** Dry farming in the arid southwest, *Ariz. Agric. Exp. St. Bull.,* 70, 725, 1913.

29. **McGinnis, W. G. and Arnold, J. F.,** Relative water requirements of Arizona range plants, *Ariz. Agric. Exp. St. Bull.,* 80, 167, 1939.

30. **Shantz, H. L. and Piemeisel, L. N.,** The water requirements of plant at Akron, Colorado, *J. Agric. Res.,* 34, 1093, 1927.

31. **Barbour, M. G., Cuningham, G. L., Oechel, W. C., and Bamber, S. A.,** Growth and development, form and function, in *Creosote Bush: Biology and Chemistry of Larrea in New World Deserts,* Vol. 6, US/IBP Synthesis Series, Mabry, T. J., Hunziker, J. H., and Difeo, D. R., Jr., Eds., Dowden, Hutchinson and Ross, Stroudsburg, Pa., 1977.

32. **Szarek, S. R.,** Primary production in four North American deserts: indices of efficiency, *J. Arid. Environ.,* 2, 187, 1979.

33. **Halvorson, W. L. and Patten, D. T.,** Seasonal water potential changes in Sonoran desert shrubs in relation to topography, *Ecology,* 55, 173, 1974.

34. **Al-Ani, H. A., Strain, B. R., and Mooney, H. A.,** The physiological ecology of diverse populations of the desert shrub *Simmondsia chinensis, J. Ecol.,* 60, 41, 1972.

35. **Woodhouse, R. M.,** The Physiological Ecology of the Desert Evergreen Shrub *Simmondsia chinensis,* Ph.D. dissertation, Arizona State Univrsity, Tempe, 1978.

36. **Ehrler, W. L., Fink, D. H., and Mitchell, S. T.,** Growth and yield of jojoba plants in native stands using runoff-collecting microcatchments, *Agron. J.,* 70, 1005, 1978.

37. **Taylor, K. W.,** Guayule — an American source of rubber, *Econ. Bot.,* 5, 255, 1951.

38. **Polhamus, L. G.,** *Rubber,* Leonard Hill, London, 1962, 449.

39. **Charley, J. L. and West, N. E.,** Micropatterns of nitrogen mineralization activity in soils of some shrub-dominated semidesert ecosystems of Utah, *Soil Biol. Biochem.,* 9, 357, 1977.
40. **Loomis, R. S.,** Agriculture, in *Symposium on Organic Raw Materials,* St. Pierre, L. E., Ed., Permagon Press, New York, in press, 1979.
41. **Noy-Meir, I.,** Desert ecosystems: environment and producers, *Annu. Rev. Ecol. Syst.,* 4, 25, 1974.
42. **Evenari, M., Bamberg, M. J., Schulze, E-D, Kappen, L., Lange, O. L., and Buschbom, U.,** The biomass production of some higher plants in near eastern and American deserts, in *Photosynthesis and Productivity in Different Environments,* Cooper, J. P., Ed., Cambridge University Press, Cambridge, 1975, 121.
43. **Calvin, M.,** Green factories, *Chem. Eng. News,* 56(12), 30, 1978.
44. **Erie, L. J., French, O. F., and Harris, K.,** Consumptive use of water by crops in Arizona, *Ariz. Exp. St. Tech. Bull.,* 169, 1, 1965.
45. **Burch, G. J., Smith, R. C. S., and Mason, W. K.,** Agronomic and physiological response of soybean and sorghum crops to water deficits. II. Crop evaporation, soil water depletion and root distribution, *Aust. J. Plant. Physiol.,* 5, 169, 1978.
46. **Harrold, L. L., Peters, D. B., Freibelbis, F. R., and McGuinnes, J. L.,** Transpiration evaluation of corn grown on a plastic covered lysimeter, *Soil Sci. Soc. Am. Proc.,* 23, 174, 1959.
47. **Peters, D. B. and Russell, M. B.,** Relative water losses by evaporation and transpiration in field corn, *Soil Sci. Soc. Am. Proc.,* 23, 170, 1959.
48. **Peters, D. B. and Johnson, L. C.,** Soil moisture use by soybeans, *Agron. J.,* 52, 687, 1960.
49. **Erie, L. J., Bucks, D. A., and French, O. F.,** Consumptive use and irrigation management for high-yielding wheats in central Arizona, *Prog. Agric. Ariz.,* 25, 14, 1973.
50. **Dillman, A. C.,** The water requirement of certain crop plants and weeds in the northern great plains, *J. Agric. Res.,* 42, 187, 1931.
51. **Loomis, R. S. and Gerakis, P. A.,** Productivity of agricultural ecosystems, in *Photosynthesis and Productivity in Different Environment,* Cooper, J. D., Ed., Cambridge University Press, Cambridge, 1975, 145.

*Section 7*
*Biological Resources: Primary Productivity*

# PRIMARY PRODUCTIVITY IN FRESHWATER ENVIRONMENTS

## Ferdinand Schanz and Kurt Wälti

## DEFINITIONS[1,2]

Primary production — weight of new organic material formed by photosynthesis (It is the increase in biomass of green plants observed over a period of time plus any losses during that period.)

Primary productivity — rate of production, expressed as production divided by the period of time

Gross productivity — observed change in biomass plus all losses including respiration, divided by the time interval

Net productivity — rate of accumulation or production of new organic matter or stored energy, less losses, divided by the time interval

## METHODS

Primary production can be measured as follows (for general discussion and further methods, see Hall et al.).[3]

### Increase in Biomass

The difference in biomass at the beginning and end of an investigated period has been determined by the following methods:

1. ATP level[4]
2. Species determination, enumeration, and calculation of the total volume[5,6]
3. Dry weight or ash-free dry weight[7]
4. The content of total nitrogen[8] or total organic carbon[9]
5. The level of photosynthetic pigments[10,11]

The biomass of macrophytes or periphyton is rather difficult to determine, and only a few methods are suitable.[12]

### Oxygen Production or Carbon Dioxide Fixation

*Oxygen Method*

After determining the $O_2$ content, the sample is re-exposed during a few hours at the same depth and the $O_2$ content is measured again. The difference in $O_2$ content is equal to the net production. The same procedure is made with a dark bottle, showing an $O_2$ loss because of respiration. The sum of the two results is equal to the gross production. The method is presented by Carpenter[13] and Tschumi et al.[14] and discussed by Friedli.[15]

*[14]C Method*

Instead of measuring the $O_2$ content, the total amount of $CO_2$ fixed is calculated, based on the incorporation of [14]$CO_2$. The [14]$CO_2$ is added in a small amount to the sample. Otherwise the procedure is the same as for the $O_2$ method. Dependent on the author, the interpretation of the relation between the dark and light bottles is very different. The [14]C method is described by Steeman Nielsen et al.,[16] Schegg,[17] and Müller.[18]

## RESULTS

### Lakes

*Phytoplankton*

The measured production (usually in milligrams C $m^{-3}$ 4 $hr^{-1}$) in various depths gives the possibility for calculating the daily productivity per unit area (in milligrams C $m^{-2}$ $day^{-1}$). A great number of measurements made over the year is prerequisite to calculate the annual productivity (in grams C $m^{-2}$ $year^{-1}$). For a reliable comparison of lakes of different trophic status, many data of daily productivity are needed because of the heterogeneous nature of the samples analyzed, even from the same lake and the same season. Table 1 gives the productivity data of several lakes around the world. The primary production was measured mostly with the $^{14}C$ technique.

*Macrophytes*

The most commonly used technique for estimating the seasonal changes in macrophyte biomass and productivity is the harvesting technique.[19] Publications concerning the production of macrophytes are relatively few when compared with phytoplankton. Some results are summarized by Wetzel,[1] Barko et al.,[19] and Søndergaard.[20]

*Periphyton (Aufwuchs)*

Several methods exist for determining the Aufwuchs production. Simulated substrata are often used with attached algae; they are exposed under natural conditions during several weeks (discussion by Tippett).[21] The production is followed by harvesting (biomass difference). Furthermore, the methods described for phytoplankton have been applied as well to substrata covered with periphyton entrapped in closed vessels.[18] Results of periphyton productivity studies are given by Wetzel,[1] Pieczynska,[22] and Søndergaard and Sand-Jensen.[20]

### Rivers and Streams

Depending on the local situation, macrophytes or attached algae are the dominant primary producers. The methods for determining their productivity correspond generally to the situation of the lakes. It can be assumed that the plankton population does not contribute much to the total primary production of a river. In large and slowly flowing streams, the phytoplankton production is an important part of the total primary production. Here, the situation is similar to lakes.[23] Results and special techniques are described by Backhaus,[24] Naiman and Gerking,[25] and Vollenweider.[12]

## Table 1
## PRODUCTIVITY DATA OF LAKES AROUND THE WORLD

| Lake | Latitude | Primary productivity | Remarks | Ref. |
|---|---|---|---|---|
| **America** | | | | |
| Bluehill South (Canada) | 49° N | 6.2 g C · m$^{-2}$ · year$^{-1}$, 1969/1970 | Oligotrophic | 26 |
| Eiffel (Canada) | 51° N | 25.6 mg C · m$^{-2}$ · day$^{-1}$, mean May—October | Oligotrophic | 27 |
| | | 5.5 g C · m$^{-2}$ · year$^{-1}$, 1965 | Alpine | |
| | | 95 mg C · m$^{-2}$ · day$^{-1}$, mean July—September | | |
| Castle Lake (U.S.) | 36° N | 98.3 mg C · m$^{-2}$ · day$^{-1}$, annual mean | Oligotrophic, alpine | 28 |
| Lawrence Lake (U.S.) | 45° N | 118.9 mg C · m$^{-2}$ · day$^{-1}$, annual mean | Oligotrophic, hard water | 29 |
| Clear Lake (U.S.) | 36° N | 437.8 mg C · m$^{-2}$ · day$^{-1}$, annual mean | Mesotrophic, shallow | 30 |
| Waco Reservoir (U.S.) | 31° N | 857 mg C · m$^{-2}$ · day$^{-1}$, mean | Eutrophic | 31 |
| Nr 1 (Alaska, U.S.) | 67° N | 2.51 mg C · m$^{-2}$ · hr$^{-1}$, July 13 | Shallow | 32 |
| Pyramid (Canada) | 52° N | 108.2 mg C · m$^{-2}$ · day$^{-1}$, August 5 | Oligotrophic, alpine | 33 |
| Atikonate (Canada) | 52° N | 120 mg C · m$^{-2}$ · day$^{-1}$, mean June—September | | 34 |
| Oliver (U.S.) | 40° N | 123 g C · m$^{-2}$ · year$^{-1}$ | Mesotrophic, large, deep | 35 |
| Lake Michigan (U.S.) | 42° N | ca.130 g C · m$^{-2}$ · year$^{-1}$ | Oligotrophic | 36 |
| Sylvan (U.S.) | 40° N | 570 g C · m$^{-2}$ · year$^{-1}$ | Eutrophic | 37 |
| | | 1564 mg C · m$^{-2}$ · day$^{-1}$, annual mean | | |
| Lake Wabamun (Canada) | 53° N | 170 g C · m$^{-2}$ · year$^{-1}$, 1971/1972 | Mesotrophic | 38 |
| | | 880 mg C · m$^{-2}$ · day$^{-1}$, annual mean | | |
| Lake Tahoe (U.S.) | 38° N | 58.5 g C · m$^{-2}$ · year$^{-1}$ 1971 | Oligotrophic | 39 |
| | | 158 mg C · m$^{-2}$ · day$^{-1}$, annual mean | | |
| Turtle Creek Reservoir (U.S.) | 38° N | 66.8 mg C · m$^{-2}$ · day$^{-1}$, annual mean | Shallow, high turbidity | 40 |
| **Europe** | | | | |
| Finstertaler See (Austria) | 46° N | 80—100 mg C · m$^{-2}$ · day$^{-1}$, mean ice-free period | Oligotrophic, alpine | 41 |
| Lake Ohrid (Yugoslavia) | 41° N | 212.7 mg C · m$^{-2}$ · day$^{-1}$, annual mean | Large, deep, hard water, oligotrophic | 42 |
| Furesø (Denmark) | 57° N | 462 mg C · m$^{-2}$ · day$^{-1}$, annual mean | Mesotrophic | 43 |
| Søllerod Sø (Denmark) | 57° N | 522 g C · m$^{-2}$ · year$^{-1}$ | Eutrophic, large amount of treated sewage | 44 |
| | | 1430 mg C · m$^{-2}$ · day$^{-1}$, annual mean | | |

## Table 1 (continued)
## PRODUCTIVITY DATA OF LAKES AROUND THE WORLD

| Lake | Latitude | Primary productivity | Remarks | Ref. |
|------|----------|---------------------|---------|------|
| Lake of Lucerne (Switzerland) | 47° N | 415 g C · m$^{-2}$ · year$^{-1}$, 1969/1970 | Oligotrophic | 45 |
| Rotsee (Switzerland) | 47° N | 511 g C · m$^{-2}$ · year$^{-1}$, 1969/1970 | Eutrophic | 45 |
| Mauensee (Switzerland) | 47° N | 303 g C · m$^{-2}$ · year$^{-1}$, 1974 | Eutrophic | 46 |
| Grössin-See (Germany) | 52° N | 2170 mg C · m$^{-2}$ · day$^{-1}$, mean May—November 1968 | Hypertrophic, shallow | 47 |
| Linzer Untersee (Austria) | 46° N | 45 g C · m$^{-2}$ · year$^{-1}$ | Oligotrophic, alpine, small | 48 |
| Erken (Sweden) | 58° N | 123 mg C · m$^{-2}$ · day$^{-1}$, annual mean<br>104 g C · m$^{-2}$ · year$^{-1}$, 1954 | Large, deep | 49 |
| Lake Trummen (Sweden) | 58° N | 285 mg C · m$^{-2}$ · day$^{-1}$, annual mean<br>225 g C · m$^{-2}$ · year$^{-1}$, 1973 | | 50 |
| Other parts of the world | | | | |
| Vanda (Antarctica) | 77° S | 14 mg C · m$^{-2}$ · day$^{-1}$, mean | Permanently frozen | 51 |
| Kasumigaura (Japan) | 35° N | 346 mg C · m$^{-2}$ · day$^{-1}$, mean September—April | Eutrophic, shallow | 52 |
| Lake Suwa (Japan) | 35° N | 531 mg C · m$^{-2}$ · day$^{-1}$, mean March—August | Eutrophic | 52 |
| Lake Kinneret (Israel) | 32° N | 790 mg C · m$^{-2}$ · day$^{-1}$, mean July—March | | 53 |
| Lake Waipori (New Zealand) | 44° S | 9.5 g C · m$^{-2}$ · year$^{-1}$ | | 54 |
| Lake Lanao (Philippines) | 8° N | 620 g C · m$^{-2}$ · year$^{-1}$, 1970/1971<br>1700 mg C · m$^{-2}$ · day$^{-1}$, annual mean | Eutrophic, large | 55 |
| Lake Victoria (Africa) | 2° S | 640 g C · m$^{-2}$ · year$^{-1}$<br>1750 mg C · m$^{-2}$ · day$^{-1}$, annual mean | Eutrophic, large, deep | 56 |
| Lake Sibaya (South Africa) | 27° S | 800 mg C · m$^{-2}$ · day$^{-1}$, annual mean, 1973/1974 | | 57 |

# REFERENCES

1. **Wetzel, R. G.,** *Limnology,* W. B. Saunders, Philadelphia, 1975.
2. **Westlake, D. F.,** Theoretical aspects of the comparability of productivity data, in *Primary Productivity in aquatic environments, Mem. Ist. Ital. Idrobiol.,* 18(Suppl.), 315, 1965.
3. **Hall, A. S. C. and Moll, R.,** Methods of assessing aquatic primary productivity, in *Primary Productivity of the Biosphere,* Lieth, H. and Whittacker, H. R., Eds., Springer-Verlag, 1975.
4. **Hofer-Siegrist, L.,** Eine verbesserte Methode zur Bestimmung von ATP in Seewasser, *Schweiz. Z. Hydrol.,* 38, 49, 1976.
5. **Nauwerk, A.,** Die Beziehungen zwischen Zooplankton und Phytoplankton im See Erken, *Symb. Bot. Ups.,* 17, 1, 1963.
6. **Kavanaugh, M. C., Zimmermann, U., and Vagenknecht, A.,** Determinations of particle size distributions in natural waters, use of Zeiss Micro-Videomat image analyzer, *Schweiz. Z. Hydrol.,* 39, 86, 1977.
7. **Strickland, J. D. H. and Parsons, T. R.,** A practical handbook of seawater analysis, 2nd ed., *Bull. Fish. Res. Board Can.,* 167, 1, 1972.
8. **Pavoni, M.,** Beziehung zwischen Biomasse und Stickstoffgehalt des Phytoplanktons und die daraus ableitbare Anwendung der Bestimmungsmethoden für die Praxis, *Schweiz. Z. Hydrol.,* 31, 110, 1969.
9. **Russell-Hunter, V. D., Meadows, R. T., Apley, M. L., and Burky, A. J.,** On the use of a "wet-oxidation" method for estimates of total organic carbon in mollusc growth studies, *Proc. Malacol. Soc. London,* 38, 1, 1968.
10. **Moss, B.,** A spectrophotometric method for the estimation of percentage degradation of chlorophylls to phaeopigments, *Limnol. Oceanogr.,* 12, 335, 1969.
11. **Slovacek, R. E. and Hannan, P. J.,** *In vivo* fluorescence determinations of phytoplankton chlorophyll a, *Limnol. Oceanogr.,* 22, 919, 1977.
12. **Vollenweider, R. A., Ed.,** A manual on methods for measuring primary production in aquatic environments, Int. Biol. Program Handbook 12, 2nd ed., Blackwell Scientific, Oxford, 1974.
13. **Carpenter, J. H.,** The accuracy of the Winkler method for dissolved oxygen analysis, *Limnol. Oceanogr.,* 10, 135, 1965.
14. **Tschumi, P. A., Zbären, D., and Zbären, J.,** An improved oxygen method for measuring primary production in lakes, *Schweiz. Z. Hydrol.,* 39, 306, 1978.
15. **Friedli, P.,** Die Tages- und Jahresprimarproduktion des Bielersees unter Berucksichtigung der Extrapolation von Kurzzeitmessungen, der Biomasse, des Chlorophylls und der Einstrahlung; Selbstverlag, Bern, 1978.
16. **Steeman Nielsen, E. and Hansen, V. K.,** Measurements with the C-14 technique of the respiration rates in natural populations of phytoplankton, *Deep-Sea Res.,* 5, 222, 1959.
17. **Schegg, E.,** Produktion und Destruktion in der trophogenen Schicht, Untersuchungen ökologischer Parameter im polytrophen Rotsee und in der mesotrophen Horwer Bucht (Vierwaldstättersee), *Schweiz. Z. Hydrol.,* 33, 425, 1971.
18. **Müller, P.,** Die Primärproduktion des epilithischen Aufwuchses und des Phytoplanktons am Ufer des eutrophen Greifensees, Zürich, 1976.
19. **Barko, J. W., Murphy, P. G., and Wetzel, R. G.,** An investigation of primary production and ecosystem metabolism in a Lake Michigan dune pond, *Arch. Hydrobiol.,* 81, 155, 1977.
20. **Søndergaard, M. and Sand-Jensen, K.,** Total autotrophic production in oligotrophic Lake Kalgaard, Denmark, *Verh. Internat. Verein. Limnol.,* 20, 667, 1978.
21. **Tippett, R.,** Artificial surfaces as a method of studying populations of benthic micro-algae in freshwater, *Br. Phycol. J.,* 5, 187, 1970.
22. **Pieczynska, E.,** Variations in the primary production of plankton and periphyton in the littoral zone of lakes, *Bull. Acad. Pol. Sci. Cl. 2,* 13, 219, 1965.
23. **Rosemarin, A. S. and Hart, J. S.,** Annual seasonal variation of phytoplankton primary productivity and biomass, correlated with physical parameters in the Ottawa River, Canada, *Verh. Internat. Verein. Limnol.,* 20, 1299, 1978.
24. **Backhaus, D.,** Oekologische und experimentelle Untersuchungen an den Aufwuchsalgen der Donauquellflüsse Breg und Brigach und der obersten Donau bis zur Versickerung bei Immendingen, Dissertation Universität Freiburg, 1965.
25. **Naiman, R. J. and Gerking, S. D.,** Interrelationships of light, chlorophyll and primary production in a thermal stream, *Verh. Internat. Verein. Limnol.,* 19, 1659, 1975.
26. **Kerkes, J.,** Factors relating to annual planktonic primary production in five small oligotrophic lakes in Terra Nova National Park, Newfoundland, *Int. Rev. Ges. Hydrobiol.,* 62, 345, 1977.
27. **Fabris, G. L. and Hammer, U. T.,** Primary production in four small lakes in the Canadian Rocky Mountains, *Verh. Internat. Verein. Limnol.,* 19, 530, 1975.

28. **Kimmel, L. R. and Goldman, C. R.,** Production, sedimentation and accumulation of particulate carbon and nitrogen in a sheltered subalpine lake, in *Interactions Between Sediments and Freshwater,* Golterman, H. L., Ed., 1977, 148.

29. **Wetzel, R. G., Rich, P. H., Miller, M. G., and Allen, H. L.,** Metabolism of dissolved and particulate detrital carbon in a temperate hardwater Lake, Mem. Ist. Ital. Idrobiol., 29 Suppl., 185, 1972.

30. **Goldman, C. R. and Wetzel, R. G.,** A study of the primary productivity of Clear Lake, Lake County, California, *Ecology,* 44, 283, 1963.

31. **Kimmel, L. R. and Lind, O. T.,** Factors affecting phytoplankton production in a eutrophic reservoir, *Arch. Hydrobiol.,* 71, 124, 1972.

32. **O'Brien, W. J., Huggins, D. G., and De Noyelles, F.,** Primary productivity and nutrient limiting factors in lakes and ponds of the Noatak River Valley, Alaska, *Arch. Hydrobiol.,* 75, 263, 1975.

33. **Anderson, R. S. and Dokulil, M.,** Assessments of primary and bacterial production in three large mountain lakes in Alberta, Western Canada, *Int. Rev. Ges. Hydrobiol.,* 62, 97, 1977.

34. **Ostrofsky, M. L. and Duthie, H. C.,** Primary productivity, phytoplankton and limiting nutrient factors in Labrador lakes, *Int. Rev. Ges. Hydrobiol.,* 60, 145, 1975.

35. **Wetzel, R. G.,** Productivity investigations of interconnected lakes. I. The eight lakes of the Oliver and Waters chains, northeastern Indiana, *Hydrobiol. Stud.,* 3, 91, 1973.

36. **Vollenweider, R. A., Munawar, M., and Stadelmann, P.,** A comparative review of phytoplankton and primary production in the Laurentian Great Lakes, *J. Fish. Res. Board Can.,* 31, 739, 1974.

37. **Wetzel, R. G.,** Variations in productivity of Goose and hypereutrophic Sylvan lakes, Indiana, *Invest. Indiana Lakes Streams,* 7, 147, 1966.

38. **Noton, L. R.,** The effect of thermal effluent on phytoplankton productivity in Lake Wabamun, Alberta, Verh. Internat. Verein. Limnol., 19, 542, 1975.

39. **Goldman, C. R. and De Amezaga, E.,** Spatial and temporal changes in the primary productivity of Lake Tahoe, California-Nevada, between 1959 and 1971, Verh. Internat. Verein. Limnol., 19, 812, 1975.

40. **Marzolf, G. R. and Osborne, J. A.,** Primary production in a Great Plains reservoir, Verh. Internat. Verein. Limnol., 18, 126, 1972.

41. **Pechlaner, R.,** Die Finstertaler Seen (Kühtai, Oesterreich). II. Das Phytoplankton, *Arch. Hydrobiol.,* 63, 145, 1967.

42. **Ocevski, T. B. and Allen, L. H.,** Limnological studies in a large, deep, oligotrophic lake (Lake Ohrid, Yugoslavia), *Arch. Hydrobiol.,* 79, 429, 1977.

43. **Jonasson, P. M. and Mathiesen, H.,** Measurements of primary production in two Danish eutrophic lakes, Esrom So and Fureso, *Oikos,* 10, 137, 1959.

44. **Steeman Nielsen, E.,** The production of organic matter by the phytoplankton in a Danish lake receiving extraordinarily great amounts of nutrient salts, *Hydrobiologia,* 7, 68, 1955.

45. **Bloesch, J., Stadelmann, P., and Bührer, H.,** Primary production, mineralization and sedimentation in the euphotic zone of two Swiss lakes, *Limnol. Oceanogr.,* 22, 511, 1977.

46. **Gächter, R.,** Die Tiefenwasserableitung, ein Weg zur Sanierung von Seen, *Schweiz. Z. Hydrol.,* 38, 1, 1976.

47. **Kalbe, L.,** Sauerstoff und Primärproduktion in hypertrophen Flachseen des Havelgebietes, *Int. Rev. Ges. Hydrobiol.,* 57, 825, 1972.

48. **Jonasson, P. M.,** Ecology and production of the profundal benthos in relation to phytoplankton in Lake Esrom, *Oikos,* Suppl. 14, 1, 1972.

49. **Rodhe, W.,** Primarproduktion und Seetypen, Verh. Internat. Verein. Limnol., 13, 121, 1958.

50. **Cronberg, G., Gelin, C., and Larsson, K.,** Lake Trummen restoration project. II. Bacteria, phytoplankton and phytoplankton productivity, Verh. Internat. Verein. Limnol., 19, 1088, 1975.

51. **Goldman, C. R., Mason, D. T. and Hobbie, J. E.,** Two antarctic desert lakes, *Limnol. Oceanogr.,* 12, 295, 1967.

52. **Sakamoto, M.,** Primary production by phytoplankton community in some Japanese lakes and its dependence on lake depth, *Arch. Hydrobiol.,* 62, 1, 1966.

53. **Rodhe, W.,** Evaluation of primary production parameters in Lake Kinneret (Israel), Verh. Internat. Verein. Limnol., 18, 93, 1972.

54. **Mitchell, S. F.,** Phytoplankton productivity in Tomhawk Lagoon, Lake Waipori and Lake Mahinerangi, *N. Z. Mar. Dep. Fish. Res. Bull.,* 3, 1, 1971.

55. **Lewis, W. M., Jr.,** Primary production in the plankton community of a tropical lake, *Ecol. Monogr.,* 44, 377, 1974.

56. **Talling, J. F.,** The photosynthetic activity of phytoplankton in East African lakes, *Int. Rev. Ges. Hydrobiol.,* 50, 1, 1965.

57. **Allanson, B. R. and Hart, R. C.,** The primary production of Lake Sibaya, Kwa Zulu, South Africa, Verh. Internat. Verein. Limnol., 19, 1426, 1975.

# GENERAL DESCRIPTION OF PHYTOPLANKTON PRODUCTIVITY IN MARINE AND ESTUARINE ENVIRONMENTS

## B. Zeitzschel

Marine plankton are the most abundant and widely distributed forms of life on earth. Plankton, defined by Hensen in 1887, is a comprehensive term which includes all organisms, plants, and animals, that are passively "drifting" along with water movements. The plant component of plankton — phytoplankton — is made up of unicellular (exceptionally multicellular) algae which are solitary or colonial. Phytoplankton components are diatoms, dinoflagellates, coccolithophores, and some other flagellates. Blue-green and green algae can be very abundant in estuaries, but are of lesser significance in the sea. Phytoplankton organisms are autotrophs, i.e., they fix solar energy by photosynthesis using carbon dioxide, nutrients, and trace metals. All these autotrophs contain photosynthetic pigments, such as chlorophylls and carotenoids. Some phytoplankton organisms, mainly species of the dinoflagellates, can be temporarily heterotrophic, i.e., they build up organic particulate matter from dissolved organic substances (osmotrophy) or even particulate organic matter (phagotrophy).

Plankton may be arbitrarily classified by size into nanoplankton, cells $<20$ $\mu$m, microplankton, organisms between 20 and 200 $\mu$m, and then mesoplankton, macroplankton, and megaplankton. Phytoplankton belong mainly to the nano- and microplankton fraction. The larger phytoplankton species may be concentrated by plankton nets. For most quantitative investigations, however, phytoplankton will be sampled by water bottles and concentrated by other methods, e.g., the Utermöhl technique.[1] Malone[2] compared the nanoplankton and netplankton primary productivity and standing stock in neritic (coastal) and oceanic waters and came to the conclusion that nanoplankton was the most important producer in all the environments studied, but that net productivity was significantly higher in neritic rather than in oceanic regions.

The tremendous diversity in the shape of planktonic organisms has attracted naturalists for over 100 years. It has been traditionally held that the diverse surfaces characterizing diatoms and dinoflagellates are factors directly related to their suspension. Smayda[3] reviewed the literature on suspension and sinking of phytoplankton in the sea and stated that three principal mechanisms can be recognized: morphological, physiological, and physical. Smayda formulated a phytoplankton suspension hypothesis that "the various morphological adaptations of the diatoms in particular are to be taken not as aids to suspension (flotation) per se, as commonly held, but as mechanisms to permit twisting and vertical movements within the water." The problem for phytoplankton is not to float, but to sink or rise and rotate. Smayda concluded that biological suspension mechanisms are of an ambiguous nature and that physical mechanisms appear to provide a satisfactory explanation for phytoplankton suspension.

Most phytoplankton organisms have a density greater than water. The higher density is in part caused by skeletons which consist of silica, calcium carbonate, and cellulose. Water turbulence combined with other factors, such as shape or physiological state, reduce the sinking rate of nonmotile organisms such as diatoms. There is recent evidence that the settling of phytoplankton to the bottom, at least in neritic waters, is not uniform, but occurs at irregular intervals. Motile phytoplankton, like most dinoflagellates, may actively swim to compensate for sinking.

Autotrophic algae are most abundant in the euphotic zone (see Figure 1). The latter is defined as a zone reaching from the surface of the sea to a depth where the light energy intensity is such that production of organic matter by photosynthesis in respect to an individual phytoplankton cell balances destruction by respiration. This depth is

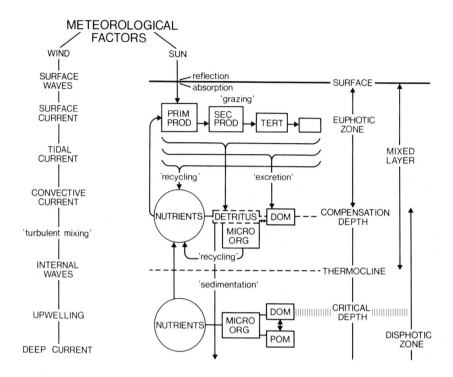

FIGURE 1.   Schematic diagram of the major elements of a pelagic ecosystem and the main environmental factors influencing the system.[4]

called the compensation depth, where light intensity generally ranges between 0.1 and 1% of the incident radiation reaching the surface of the water or very approximately 1 to 10 Ly day$^{-1}$. The zone beneath the compensation depth is called the disphotic zone. In the open sea, it may reach down to approximately 1000 m. The aphotic zone is defined as a region further down where no daylight penetrates. Gran and Braarud[5] introduced the concept of critical depth. According to Sverdrup,[6] the critical depth is defined as the depth to which phytoplankton may be mixed and at which the total production for the water column is equal to the respiration of primary producers for a period of 24 hr. It follows that a net increase in production can take place only if the critical depth is greater than the depth of mixing.

Thermoclines often occur in the oceans. Their occurrence is clearest in tropical and subtropical seas. A permanent shallow thermocline characterizes the inner tropical region, and a permanent deep thermocline characterizes the outer tropical region. The subtropical and the boreal zones, where the seasonal variations in surface temperature become appreciable, are characterized by the development of a shallow summer thermocline. The zone above the thermocline is referred to as the mixed layer.

A schematic diagram of the major organic elements of a pelagic ecosystem is given in Figure 1. It consists of a living portion, the primary producers (autotrophic phytoplankton), the secondary producers (herbivorous zooplankton), the tertiary producers (carnivorous zooplankton I), and several more carnivorous links if it is a long, open ocean food chain, degraders (e.g., bacteria), and the dead particulate fraction, the detritus. The living and dead particulate organic matter is abbreviated POM. The transfer of organic matter in the food chain (or food web) is indicated in Figure 1 by the different sizes of the boxes. There is a loss of about 80 to 90% from one link to the next, meaning that only about 10 to 20% of the organic matter is transferred and incorporated in the next higher level.[7]

The living organisms exude or excrete a considerable amount of dissolved organic matter (DOM) (phytoplankton between 10 to 30%). The pool of dissolved organic carbon in the sea is about ten times the particulate carbon. The dissolved organic substances can be used by heterotroph organisms like bacteria. By turbulent mixing processes, POM might be formed out of DOM.[8] The recycling of nutrients induced by the activity of bacteria (remineralization) takes place in the euphotic zone as well as in the disphotic zone. A considerable amount of particulate matter sinks out of the euphotic zone as indicated in the diagram by the term "sedimentation". The major physical factors affecting a pelagic ecosystem in the different water layers are indicated on the left side of the diagram.

Phytoplankton is not distributed evenly in the oceans. It is believed to occur in three dimensional patches of various size, caused by biological and physical phenomena. Steele[9] suggests that although variability of spatial heterogeneity occurs at all scales, there may be patches with, typically, dimensions of 10 to 100 km. According to Steele, many of these features can be explained by a combination of accumulation because of phytoplankton growth and dispersion because of turbulent diffusion.

The complex movements of water in the sea can be separated formally into transport (advection) and mixing processes. The large-scale transport of water in currents, such as the equatorial current system or the Gulf Stream, is the best-known form of lateral movement. The largest vertical motion of water is found in those regions mainly on the western sides of the continents where upwelling occurs. Vertical velocities in such areas are in the order of $10^{-3}$ cm $sec^{-1}$, several orders of magnitude smaller than average horizontal transport in the main currents. Turbulent mixing, both lateral and vertical, is always present, especially in the surface layers of the sea. Thus, although lateral movements disperse and intermix plankton populations, it is turbulent mixing which generally controls the production of phytoplankton.

Horizontal movements in the sea are most pronounced in surface currents, undercurrents and deep currents, tides, and internal waves; vertical movements of water occur in zones of mixing, upwelling regions, and zones of convergence and divergence. Special circulation patterns like the Langmuir circulation induced by wind and small-scale horizontal vortices caused by thermohaline driving forces influence the pattern of production and distribution of plankton organisms. The major physical factors which influence primary productivity and distribution of plankton are summarized in Table 1.

Phytoplankton is of great ecological significance because it comprises the major portion of primary producers in the sea. Like plants on land, phytoplankton is the basic food in the sea for all consumers, such as zooplankton and fish. It is a well-accepted fact that primary production by phytoplankton contributes to the energy requirements by benthic animals in shallow waters, e.g., estuaries. There is a controversy, however, as to whether or not there is a direct input of organic matter produced in the euphotic zone to the bottom of the deep sea. The main difference between primary production in the sea and on land is that phytoplankton in the open ocean is eaten almost entirely by zooplankton, whereas on land only about 10% of plant material is eaten by herbivores. On land, a large long-lived plant population is found in the form of grasses, bushes, and trees. In the sea, the population of primary producers has a lifetime of about 1 day.

Table 1

## MAJOR HYDROGRAPHICAL FACTORS INFLUENCING THE PRIMARY PRODUCTIVITY AND DISTRIBUTION OF PLANKTON[4]

| | | | Influence on | |
|---|---|---|---|---|
| Type of motion | Forms of water motion | Special phenomena | Productivity | Distribution |
| Motions with large-scale effects | Permanent (gradient) flow | Meanders | ( + ) | + + |
| | | | ( + ) | + + |
| | Wind currents | Eddies | | |
| | Divergence/ convergence | Fronts | + | ( + ) |
| | (Upwelling) | Patches | + + | ( + ) |
| | (Convective circulation) | — | ( + ) | ( + ) |
| Motions with small-scale effects | Tidal currents | Eddies, fronts | ( + ) | + |
| | Surface wave action | Foam, bubbles | ( + ) | − |
| | Wind mixing | Langmuir circulation | ( + ) | ( + ) |
| | Internal wave action | Slicks | ( + ) | ( + ) |
| | Upwelling | Patches | + + | ( + ) |
| | Divergence/ convergence | Fronts, eddies | + | ( + ) |
| | Molecular diffusion | Step structure | ( + ) | − |
| | Convective circulation | — | ( + ) | ( + ) |
| | Shear instabilities | Fronts | ( + ) | − |
| Motions caused by morphological features, small-scale effects | Island circulation | Eddies | + | + |
| | Land promontories | Eddies | + | ( + ) |
| | Topography (ridges, submarine canyons) | Meanders | + | ( + ) |
| | Runoff and land drainage | Plumes | + | + |

*Note:* The symbols in Columns four and five indicate different categories of influence. The parentheses around the + signs imply some degree of reservation.

# REFERENCES

1. **Sournia, A.,** Phytoplankton manual, in *Monographs on Oceanographic Methodology*, Sournia, A., Ed., UNESCO, Paris, 6, 1, 1978.
2. **Malone, T. C.,** The relative importance of nanoplankton and netplankton as primary producers in tropical oceanic and neritic phytoplankton communities, *Limnol. Oceanogr.*, 16, 633, 1971.
3. **Smayda, T. J.,** The suspension and sinking of phytoplankton in the sea, *Oceanogr. Mar. Biol. Annu. Rev.*, 8, 353, 1970.
4. **Zeitzschel, B.,** Oceanographic factors influencing the distribution of plankton in space and time, *Micropaleontology*, 24, 139, 1978.
5. **Gran, H. H. and Braarud, T.,** A quantitative study of the phytoplankton in the Bay of Funday and the Gulf of Maine including observations on hydrography, chemistry and turbidity, *J. Biol. Board Can.*, 1, 219, 1935.
6. **Sverdrup, H. U.,** On conditions for the vernal blooming of phytoplankton, *J. Cons. Int. Explor. Mer.*, 18, 287, 1953.
7. **Parsons, T. R., Takahashi, M., and Hargrave, B.,** *Biological Oceanographic Processes*, 2nd ed., Pergamon Press, 1977, chap. 4.
8. **Riley, G. A.,** Particulate and organic matter in sea water, in *Adv. Mar. Biol.*, 8, 1970.
9. **Steele, J.,** *Patchiness in the Ecology of the Seas*, Cushing, D. H. and Walsh, J. J., Eds., Blackwell Scientific, Oxford, 1976, chap. 5.

# PHYTOPLANKTON PRODUCTIVITY IN THE PACIFIC OCEAN

## Masayuki Takahashi and Shun-ei Ichimura

The Pacific Ocean, including adjacent seas surrounded by the American, Asian, Australian, and Antarctic continents, has a surface area of $180 \times 10^6$ km$^2$, which is about 50% of the all world oceans. Major ocean currents flow clockwise in the northern hemisphere and counterclockwise in the southern hemisphere (see Figure 1). Upwellings because of ocean currents and the prevailing wind are significant along California, Peru, and along the equator. Large-scale divergence (upwelling) is also known in the Antarctic Sea around 60 to 70°S.

Yearly total solar radiation reaching the sea surface at different areas in the Pacific Ocean varies by as much as a factor of three (see Figure 2). The total radiation is more than 140 kcal (cm$^2 \cdot$year)$^{-1}$, with no obvious seasonal change in the tropical Pacific, and is less than 80 kcal (cm$^2 \cdot$year)$^{-1}$, with remarkable seasonal changes at higher latitudes because of seasonal variations of day length and sun angle.[3] Daily average solar radiation is the highest in the polar region during summer. Photosynthetically active radiation (wavelength between 300 and 700 nm) is believed to be as much as 50% of the total incoming solar radiation at the sea surface.[4]

The photosynthetically productive (euphotic) zone depends on the transparency of the water column and is different at different areas and seasons (see Figure 3). Solar energy is absorbed mostly by algal cells in shallow euphotic zone and by water in deep euphotic zone. For example, 60% of the solar energy is absorbed by algal cells in the 10-m euphotic zone, but 65% is absorbed by water in the 80-m euphotic zone in the North Pacific.[8] The total algal biomass decreases with increasing depth of the euphotic zone. In the North Pacific, the following empirical relation is observed between the total chlorophyll in the euphotic zone and the euphotic zone depth: $C_{chl} = 260 \, \ell n$ ($-0.032$d), in which $C_{chl}$ is the total chlorophyll $a$ amount within the euphotic zone in mg chl $a$/m$^2$, and d is the depth of euphotic zone in meters.[8,9,10] Subsurface development of phytoplankton is also well documented in offshore waters.[11,12] Those algal cells developing in the subsurface zone utilize solar energy quite efficiently, particularly at low light levels.

Nutrients are another important factor for controlling algal productivity. Nutrient concentrations in the Pacific Ocean are about 35 $\mu$g-at/$\ell$ NO$_3$, 3 $\mu$g-at/$\ell$ PO$_4$, and 170 $\mu$g-at/$\ell$ SiO$_3$ in the aphotic zone, values which are about 50 to 200% greater than in the Atlantic Ocean.[1] Within the euphotic zone, most limiting nutrients are always undetectable in the tropical Pacific. Towards the polar region, the water column becomes nutrient depleted during only certain time periods of the year, including summer. Nutrients in a given euphotic zone are supplied by local regeneration processes and by diffusion or vertical water mixing from the subsurface zone. Photosynthetic rate at the optimum light intensity (assimilation number, mg C/mg chl $a$/hr) is known to be highly dependent on available nutrients: 0.3 to 0.7 and 2 to 6 mg C/mg chl $a$/hr for oligotrophic and mesotrophic samples, respectively.[13] Under the conditions of rich nutrients, the assimilation number becomes larger in proportion to temperature increases within a certain temperature range.[14,15]

In tropical waters photosynthetic productivity is maintained at low level because of limited nutrient supply, even though there is plenty of solar radiation. In the polar region, on the other hand, algal photosynthesis in the water column is highly dependent on available solar radiation. In temperate water, the condition is more or less intermediate; algal photosynthesis is limited by solar radiation during winter and by nutrients during summer.[16] Phytoplankton spring bloom first occurs at lower latitudes

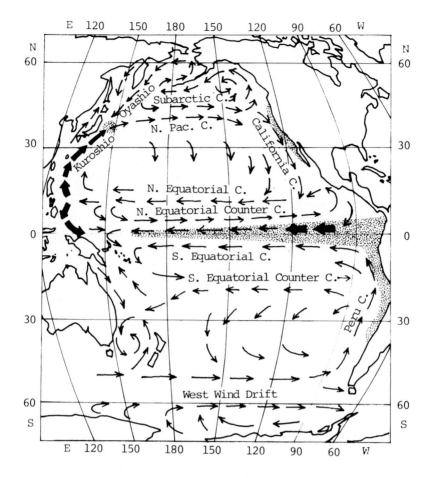

FIGURE 1.    Surface currents of the Pacific Ocean. Dotted area indicates large-scale upwelling area.[1]

of temporal region in February, and blooming area travels towards higher latitudes with delayed timing.[17] This is mainly because of the increase of available solar radiation and development of water column stability. Intense nutrient supply from subsurface zone in upwelling areas maintains high productivity (continuous blooming). The magnitude upwelling depends on currents and wind.

Primary productivity in the Pacific Ocean generally is low in the central portion of the ocean; high productivity is observed near shore, particularly at higher latitudes (see Figure 4). Along the equator, where there are equatorial upwellings, high productivity is also observed. Low productivity in tropical and temperate regions is mainly because of limited supply of nutrients and low incoming solar radiation. Possibly deep mixing also causes low productivity in the open sea of the subarctic region.

Photosynthetic efficiency is derived as follows:

$$\text{Photosynthetic efficiency} = \frac{11.4 \ (\text{g} \cdot \text{cal/mg C}) \times \text{photosynthetic productivity} \ [\text{mg C/} (\text{m}^2 \cdot \text{day})]}{\text{Incoming solar radiation} \ [\text{g} \cdot \text{cal/}(\text{m}^2 \cdot \text{day})]}$$

The results obtained varied between 0.00 to 0.37% (per total solar energy) in the Pacific (see Table 1). This range is doubled if photosynthetically active radiation is used. In the present estimate, extracellular production is not taken into account. Assuming

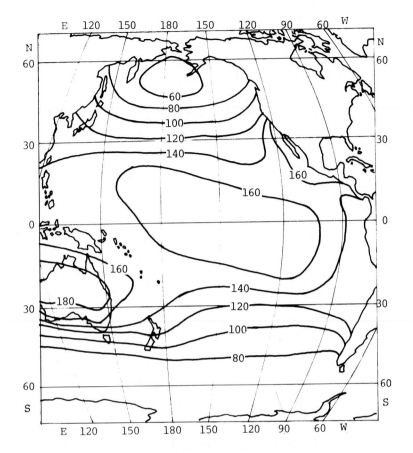

FIGURE 2.   Total solar radiation reaching the horizontal plate at the sea surface in the Pacific Ocean.[2] Radiation unit = k cal (cm² · year)⁻¹.

simply that extracellular production is 50% in the oceanic region and 10% in the coastal region, photosynthetic efficiency could increase 10 to 50% at different locations.

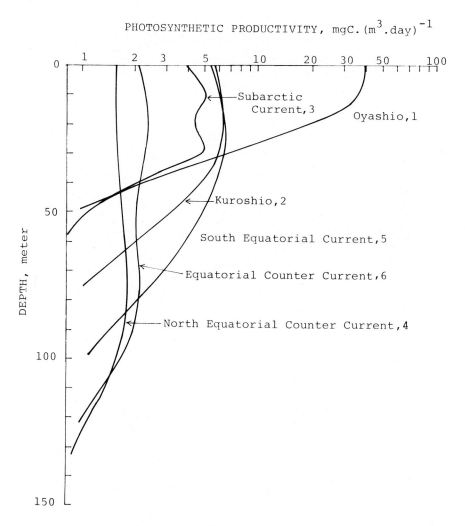

FIGURE 3.   Vertical profiles of daily photosynthetic productivity in different areas of the Pacific Ocean. 1 and 2 from Aruga and Monsi;[5] 3 and 6 from Takahashi et al.;[6] 4 and 5 from Taniguchi and Kawamura.[7]

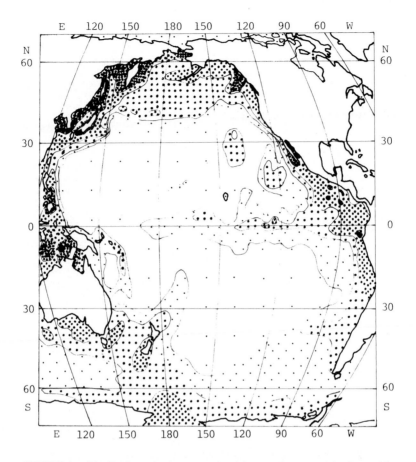

FIGURE 4.  Distribution of primary productivity (yearly average) in the Pacific Ocean. Units are in mg C/m²/day. [· · ·] less than 100; [· · ·], 100 to 150; [: : : :], 150 to 250; [: : : : :], 250 to 500; [▩▩▩], greater than 500.[18]

## Table 1
## PHOTOSYNTHETIC EFFICIENCY IN DIFFERENT PARTS OF THE PACIFIC OCEAN[a]

| Regions | Efficiency (%) | Regions | Efficiency (%) |
|---|---|---|---|
| Subarctic region | | Tropical and subtropical Pacific | |
| Strait of Georgia | 0.03—0.17 | North Equatorial Current | 0.01—0.04 |
| Washington and Oregon coast | | Equatorial Counter Current | 0.03—0.06 |
| Oceanic water | 0.03—0.13 | South Equatorial Current | 0.08—0.10 |
| Upwelling water | 0.04—0.24 | Tropical Eastern Pacific | 0.05—0.09 |
| Plume water | 0.05—0.18 | | |
| Mid Subarctic region | 0.03—0.08 | | |
| Bering Sea | 0.00—0.20 | | |
| Alaska Stream | 0.02—0.28 | | |
| Northwestern Pacific | | | |
| Oyashio Current | 0.08—0.37 | | |
| Kuroshio Current | 0.08—0.15 | | |
| Western Central Pacific water | 0.02—0.06 | | |

[a] Presented at the 13th Pacific Science Congress, held in Vancouver, B.C., 1975, by S. Ichimura under the title of "Energy Transformation in Different Parts of the Pacific Ocean".

# REFERENCES

1. **Sverdrup, H. U., Johnson, M. W., and Fleming, R. H.,** *The Oceans, their Physics, Chemistry and General Biology,* Prentice-Hall, Englewood Cliffs, N. J., 1942, 1087.
2. **Budyko, M. T.,** *Atlas of the Heat Balance,* Gidrometeorologicheskoe izdatel'stvo, Leningrad, 1955.
3. **Gates, D. M.,** *Energy Exchange in the Atmosphere,* Harper & Row, New York, 1962, 151.
4. **Parsons, T. R., Takahashi, M., and Hargrave, B.,** Biological Oceanographic Processes, *Pergamon*
5. **Aruga, Y. and Monsi, M.,** Primary production in the northwestern part of the Pacific Ocean off Honshu, Japan, *J. Oceanogr. Soc. Jpn.,* 18, 85, 1962.
6. **Takahashi, M., Satake, K., and Nakamoto, N.,** Chlorophyll distribution and photosynthetic activity in the north and equatorial Pacific Ocean along 155° W, *J. Oceanogr. Soc. Jpn.,* 28, 27, 1972.
7. **Taniguchi, A. and Kawamura, T.,** Primary production in the western tropical and subtropical Pacific Ocean, in *The Kuroshio II, Proc. of the 2nd CSK Symp., Tokyo, 1970,* Sugawara, K. Ed., Saikon, Tokyo, 159, 1972.
8. **Lorenzen, C.,** Primary production in the sea, in *Ecology of the Seas,* Cushing, D. H. and Walsh, J. J., Eds., Blackwell Scientific, Oxford, 1976, 467.
9. **Aruga, Y. and Ichimura, S.,** Characteristics of photosynthesis of phytoplankton and primary production in the Kuroshio, *Bull. Misaki Mar. Biol. Inst. Kyoto Univ.,* 21, 3, 1968.
10. **Takahashi, M. and Parsons, T. R.,** Maximization of the standing stock and primary productivity of marine phytoplankton under natural conditions, *Indian J. Mar. Sci.,* 1, 61, 1972.
11. **Anderson, G. C.,** Subsurface chlorophyll maximum in the northeast Pacific Ocean, *Limnol. Oceanogr.,* 14, 386, 1969.
12. **Hobson, L. A. and Ketcham, D. E.,** Observations on subsurface distributions of chlorophyll *a* and phytoplankton carbon in the northeast Pacific Ocean, *J. Fish. Res. Board Can.,* 31, 1919, 1974.
13. **Ichimura, S. and Aruga, Y.,** Photosynthetic natures of natural algal communities in Japanese waters, in *Recent Researches in the Fields of Hydrosphere and Nuclear Geochemistry,* Miyake, Y. and Koyama, T. Eds., Maruzen, Tokyo, 13, 1964.
14. **Aruga, Y.,** Ecological studies of photosynthesis and matter production of phytoplankton. II. Photosynthesis of algae in relation to light intensity and temperature, *Bot. Mag.,* Tokyo, 78, 360, 1965.
15. **Eppley, R. W.,** Temperature and phytoplankton growth in the sea, *Fish. Bull.,* 70, 1063, 1972.
16. **Takahashi, M., Fujii, K., and Parsons, T. R.,** Simulation study of phytoplankton photosynthesis and growth in the Fraser River estuary, *Mar. Biol.,* 19, 102, 1973.
17. **Parsons, T. R., Giovando, L. F., and LeBrasseur, R. J.,** The advent of the spring bloom in the eastern subarctic Pacific Ocean, *J. Fish. Res. Board Can.,* 23, 539, 1962.
18. **Koblenz-Mishke, O. J., Volkovinsky, V. V., and Kabanova, J. G.,** Plankton primary production of the world ocean, Scientific Exploration of the South Pacific, Standard Book No. 309-01755-6, National Academy of Sciences, Washington, D.C., 1970, 183.

# PHYTOPLANKTON PRODUCTIVITY IN THE BALTIC SEA, NORTH SEA, AND ATLANTIC OCEAN

## Ulrich Horstmann

## INTRODUCTION

The annual productivity pattern of the Baltic, North Sea, and North Atlantic follows the normal production cycle of temperate waters. There are spring and autumn blooms separated by a nutrient-limited low production in mid-summer in the lower latitudes. In higher latitudes (above the fifties), insufficient light conditions only allow one phytoplankton peak in the summer. Dependent on the limiting factor of light, mineral nutrients, or the mixing of the water column, productivity undergoes considerable changes in different areas and at different times. Average data of surface irradiance at 55° N (Strait of Denmark) are given in Figure 1. Extremely low light intensities limit productivity to insignificant values in the Baltic, North Sea, and the northern part of the North Atlantic from November to February. The amount of mineral nutrients in the surface layers of the three areas varies widely. The supply of biogenic elements into the euphotic zone occurs through different processes. In the North Atlantic, northern North Sea, and Central Baltic, thermal convection is the most efficient mechanism for nutrient transport into the surface layer. In the southern North Sea, wind- and tidal-induced mixing processes result in full convection almost throughout the year, resulting in a nutrient supply to the surface layer. In nearshore areas, especially in the land-locked waters of Baltic and North Sea, eutrophication processes, as the result of runoff, increase to a large extent the primary productivity. The effect of sewage is clearly evident near big cities, such as Helsinki, Leningrad, or Copenhagen, and to an even greater extent around the estuaries of the heavily polluted European rivers, e.g., the Wista and Oder draining into the Baltic, the Elbe, Weser, Rhein, and Thames flowing into the North Sea, and the Seine, Loire, and Garonne leading into the Atlantic. There are a number of other processes, such as interaction of warm and cold waters, inclination of thermocline, or divergence processes, especially in the North Atlantic, that lead to upwelling and contribute to increased productivity.

The primary producers are phytoplanktonic organisms which follow distinct patterns of succession in different areas, as described, for example, by Niemi[1] for the northern Baltic, Hobro[2] for stations in the Central Baltic, Smetacek[3] for Kiel Bight, Hagmeier[4] for a station near Heligoland in the North Sea, and Holligan and Harbour[5] for the western English Channel.

## THE BALTIC SEA

Systematic observations have been made on the primary productivity in the Baltic Sea and the Strait of Denmark. The latter can be considered as the most thoroughly investigated area with respect to phytoplankton productivity. Early investigations in 1933 by Steemann-Nielsen (Gargas)[6] in the Oeresund ($O_2$ method) showed productivity values between 45 and 50 g C m$^{-2}$ year$^{-1}$ and values of 125 g C m$^{-2}$ year$^{-1}$ in the Kattegat (see Figure 2A). Gargas[6] obtained 80 to 90 g C m$^{-2}$ year$^{-1}$ in the Great Belt. There is a fluctuation of primary productivity in the Great Belt to values above 150 g C m$^{-2}$ year$^{-1}$ in 1977[6] (see Figure 2B). Table 1 summarizes the various productivity investigations in the Baltic since 1970. The data show a considerable variability of phytoplankton productivity in the different areas of the Baltic in different years. The variability cannot be attributed to the methods employed, since a number of calibration exercises

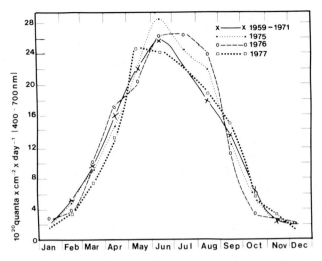

FIGURE 1. Surface irradiance at 55° 45′ N, 12° 30′ E in Strait of Denmark during the years 1975, 1976, and 1977 and as an average of the years 1959 to 1971. (Modified from Gargas.[6])

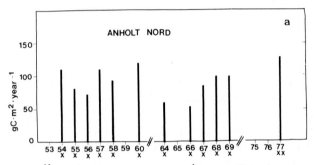

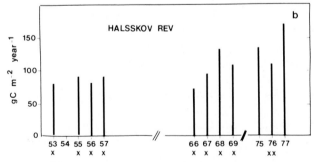

FIGURE 2. Changes in estimated yearly phytoplankton production at Anholt Nord since 1954.[6] (b) Changes in estimated yearly phytoplankton production at Halsskov Rev. (Modified from Gargas.[6])

Table 1
## CHANGES IN PHYTOPLANKTON PRODUCTION AND CHLOROPHYLL *a* IN DIFFERENT AREAS OF THE BALTIC SEA AND THE OPEN DANISH WATERS SINCE 1970[6]

| Locality | Year | g<br>$C \cdot m^{-2} \cdot year^{-1}$ | mg<br>$C \cdot m^{-2} \cdot day^{-1}$<br>(in August) | $\mu g\, chl \cdot a \cdot L^{-1}$<br>(in August) | Ref. |
|---|---|---|---|---|---|
| Nordbyskar (Northern | 1973 | 71 | 600 | 1.9 | 57 |
| Bothnian Bay) | 1974 | 70 | 490 | 2.5 | |
| Oregrund (Northern | 1973 | — | 380 | — | 57 |
| Aalandsea) | 1974 | — | 220 | — | |
| | 1975 | — 67 | 400 | 2.4 | |
| | 1976 | 65 | 500 | 1.9 | |
| Oregrund (Southern | 1973 | 94 | 450 | 1.5 | 57 |
| Aalandsea) | 1974 | — | 260 | 1.3 | |
| Bojen (Askö area, | 1970 | 203 | — | — | 58 |
| South of Stockholm) | 1972 | 162 | — | — | |
| | 1973 | 168 | — | — | |
| | 1974 | 182 | — | — | |
| | 1975 | 160 | — | — | |
| | 1976 | 143 | — | — | |
| | 1977 | 128 | 750 | 2.9 | |
| Horvik | 1973 | 91 | 600 | 0.7 | 57 |
| st. no. 2 | 1974 | 116 | 830 | 2.2 | |
| (Hano Baight) | 1975 | 96 | 533 | 1.4 | |
| | 1976 | 87 | 550 | 1.7 | |
| Horvik | 1973 | 105 | 540 | 0.8 | 57 |
| st. no. 1 | 1974 | 121 | 620 | 0.7 | |
| (Hano Baight) | 1975 | 132 | 845 | 0.4 | |
| | 1976 | 132 | 480 | 1.0 | |
| Danzig Bay | 1970 | 118 | 800 | 2.5 | 59 |
| | 1971 | 73 | 200 | 1.5 | |
| | 1972 | 69 | 480 | — | |
| | 1973 | 60 | 200 | — | |
| Bornholm Sea | 1972 | — | 155 | 0.9 | 60,61,62 |
| | 1973 | — | 885 | 0.7 | |
| | 1974 | — | 715 | 0.8 | |
| | 1975 | 77 | 737 | 0.8 | |
| | 1976 | — | 494 | 1.2 | |
| | 1977 | — | 469 | 1.0 | |
| Arkona Sea | 1972 | — | 440 | 1.0 | 60,61,62 |
| | 1973 | — | 1045 | 0.9 | |
| | 1974 | — | 685 | 0.8 | |
| | 1975 | — | 500 | 0.8 | |
| | 1976 | — | 397 | 0.9 | |
| | 1977 | 96 | 576 | 1.2 | |
| Belt Project | 1975 | — | 670 | 2.5 | 6 |
| (South of Lolland) | 1976 | 94 | 268 | 1.0 | |
| (Belt Sea) | 1977 | 134 | 820 | 3.0 | |
| Belt Project | 1975 | — | 1080 | 3.5 | 6 |
| (North of Langeland) | 1976 | 73 | 367 | 1.0 | |
| (Belt Sea) | 1977 | 92 | 378 | 1.5 | |
| Belt Project | 1975 | — | 478 | 2.1 | 6 |
| (North of Fyn) | 1976 | — | 320 | 1.7 | |
| (Belt Sea) | 1977 | — | 557 | 2.0 | |
| Belt Project | 1975 | 158 | 230 | 1.0 | 6 |
| (West of Läsko) | 1976 | 81 | 206 | 1.1 | |
| (Northern Kattegat) | 1977 | 174 | 390 | 1.0 | |

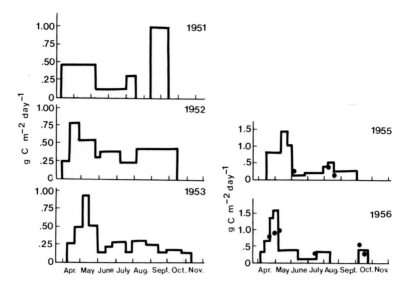

FIGURE 3.    Seasonal changes in primary productivity on Fladen Grounds (black dots: production rate measured by radio-carbon method. (Modified from Steele.[8,9])

and workshops have been undertaken among the investigators to avoid procedural errors. The workshops also led to the publication of a manual for phytoplankton primary production studies in the Baltic[7] which serves as a useful guide for future comparable primary productivity measurements.

## THE NORTH SEA

Seasonal series of productivity studies in the North Sea are rather few and almost all investigations are restricted to limited areas. Estimates of productivity over the years 1951 to 1956 by Steele[8] on Fladen Ground are based on phosphate determinations, but were shown to correlate well with values obtained by the radio-carbon method (see Steele)[9] (see Figure 3). Investigations by Cushing[10] were undertaken over a series of 3 months off the northeast coast of England. Expressed in mg C $m^{-3}$ $day^{-1}$, the data range from 3.6 during April 5 to 15 to 13.2 during the second half of May.

A few data exist from the Galathea-Expedition (Steemann-Nielsen et al.)[11] from which rough estimates were made for the annual production in the English Channel (between 78 and 202 g C $m^{-2}$ $year^{-1}$) and for the Fladen Ground (between 54 and 128 g C $m^{-2}$ $year^{-1}$). These values can be compared to 60 g C $m^{-2}$ $year^{-1}$ in the Danish waters. The available primary productivity data of the North Sea up to 1968 were summarized by Koblentz-Mischke et al.[12] The authors suggested a daily average production for the North Sea of 150 to 500 mg C $m^{-2}$ $day^{-1}$.

An excellent example for the interaction of environmental factors and primary productivity can be obtained from a monitoring station near Helgoland in the southern North Sea.[13] No direct primary production values, but significant links towards the phytoplankton productivity and even towards production cycles in the North Sea and North Atlantic, can be obtained from the results of the planktonic recorder network of the Oceanographic Laboratory of Edinburgh. Though only large phytoplankton organisms can be caught, the monthly recordings since 1948, on standardized routes, also provide useful material for phytoplankton studies.[14] Cushing[15] found three types of production cycles for the North Sea from the recorder data:

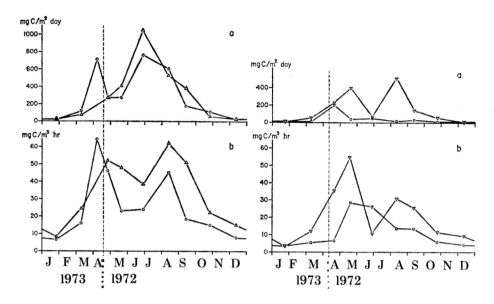

FIGURE 4.    Seasonal variation of (A) estimated *in situ* production and (B) potential production in the southern North Sea (O, western Wadden Sea; △, eastern Wadden Sea; ▽, Eems estuary; and ☐ Dollard). (Modified from Cadée and Hegeman.[18])

1.  In the southern North Sea, production rises in March and remains high until October.
2.  In the Central North Sea, the spring bloom starts also in March, peaks in April, and then dies away in June, to recover to another autumn bloom in September or October. In the western part of the North Sea, the autumn peak is considered more important than the spring peak.
3.  In the northern North Sea, the planktonic recorder data indicate a single peak of phytoplankton production in May.

Additional information as to phytoplankton succession or species distribution has been obtained from recorder data.[16] However, it must be emphasized that information is limited as to the phytoplankton that does not pass the 270-μm apertures of the silk of the plankton recorder.

Extensive studies of [14]C primary productivity in the North Sea, especially the southern parts, started in 1970. Postma and Rommets[17] investigated the potential productivity of the Wadden Sea. Cadee and Hegem[18] studied the seasonal variations in different areas of the Wadden Sea. Figure 4 shows an example of the differences between estimated and potential production which pronounce the increased availability of biogenic substances especially near the estuaries. Kroon[19] and Gieskes[20] investigated the productivity off the mouth of the Rhine river and obtained productivity values up to 3 g C m$^{-2}$ day$^{-1}$. Mommaerts[21-23] undertook extensive primary productivity investigations in the South Bight during the Belgium national program "Study on Pollution in the North Sea" and found figures ranging from 100 to 1500 mg C m$^{-2}$ day$^{-1}$. Recent investigations during the Fladen Ground Experiment 1976 (FLEX) give new detailed information on the primary productivity of the Fladen Ground area.[24] Hagmeier and Weigel's[25] findings indicate a much higher production than was obtained by Steele during the first measurements of primary productivity in the North Sea.

Table 2
PHYTOPLANKTON PRODUCTIVITY IN THE TROPICAL
ATLANTIC DURING SPRING AND AUTUMN[36]

| Daily productivity ($gCm^{-2}$) | Spring 1963 | | | |
|---|---|---|---|---|
| | Area ($km^2 \cdot 10^6$) | Percent of total area | Mean productivity ($tC \cdot 10^6 day^{-1}$) | Percent of total productivity |
| 0.05—0.5 | 7.10 | 88.7 | 1.80 | 68.0 |
| 0.—1.0 | 0.78 | 9.8 | 0.59 | 22.3 |
| 1.00—3.0 | 0.10 | 1.2 | 0.20 | 7.4 |
| 3.0 | 0.02 | 0.3 | 0.06 | 2.3 |
| Total | 8.0 | | 2.65 | |
| | Autumn 1963 | | | |
| 0.05—0.5 | 12.6 | 90.0 | 2.62 | 64.7 |
| 0.50—1.0 | 1.1 | 8.0 | 0.78 | 19.2 |
| 1.00—3.0 | 0.3 | 2.0 | 0.65 | 16.1 |
| 3.0 | — | — | — | — |
| Total | 14.0 | | 4.05 | |

# THE ATLANTIC

A number of primary productivity investigations have been made in different areas of the Atlantic since 1956. Most of these were only short period investigations. Detailed research has been made on the Continental shelf off New York.[26] The authors obtained (from chlorophyll data) daily production values of 0.33 g C $m^{-2}$ corresponding to an annual productivity of 120 g C $m^{-2}$. These values can be compared to a coastal water productivity of 160 to 190 g C $m^{-2}$ $year^{-1}$.[26,27] Quite similar values (0.25 to 0.60 g C $m^{-2} day^{-1}$) have been obtained at 30° N on stations in the North Atlantic Current.[28]

Detailed investigations from September 19, 1975 to December 19, 1976 by Cohen and Wright[29] on Georges Bank using the [14]C method lead to productivity values in the order of 400 to 500 g C $m^{-2}$ $year^{-1}$. Primary production was studied for several years on three standard sections: Faeroers Islands — Scotland, Faeroers Island — Greenland, and Cape Farewell — Ireland.[30-32] Production in these areas depends on the interaction of warm waters of Atlantic origin with cold water of the Polar basin. The productivity values obtained were 0.5 to 0.8 g C $m^{-2}$ $day^{-1}$ in the coastal waters off the Faeroers Islands at the boundary between the East Greenland Polar Current and the Irminger Current, and similar values were found in the regions of the Faeroers Islandic ridge. Low values of below 0.1 g C $m^{-2}$ $day^{-1}$ were obtained in waters of the Deivis Strait and the East Greenland Current. Studies near the west coast of Greenland gave values between 0.2 and 1.3 g C $m^{-2}$ $day^{-1}$ in July 1958.[33] Annual productivity rates were calculated to be between 28 and 98 g C $m^{-2}$.[31] Berge[34] undertook numerous productivity measurements from May to July in the Norwegian Sea. While the average productivity was around 0.5 g C $m^{-2}$ $day^{-1}$, the highest value of 1.8 g C $M^{-2}$ $day^{-1}$ was obtained in the area south of Spitzbergen, where mixing processes of Atlantic with Arctic waters occur. Chmyr and Sherstney[35] undertook productivity investigations in the same area for the same period in 1962 and obtained similar values.[36] Data for higher latitudes are available from the Barents Sea where measurements over a period of 6 years were undertaken.[37,38] Annual productivity values of 34 g C $m^{-2}$ in the upper 10-m layer were estimated near the Murman coast. In the western Barent Sea, Corlett[39]

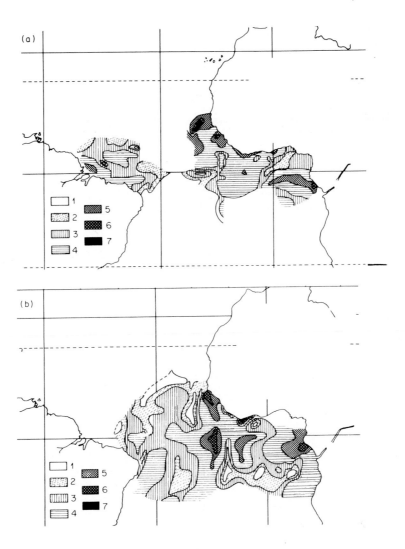

FIGURE 5.    Phytoplankton productivity (g C m$^{-2}$ day$^{-1}$) in the tropical Atlantic Ocean. (a) February to May, 1963: 1, 0.05; 2, 0.05 to 0.10; 3, 0.1 to 0.2; 4, 0.2 to 0.5; 5, 0.5 to 1.0; 6, 1.0 to 3.0; 7, 3.0. (b) August to September 1963: 1, 0.05; 2, 0.05 to 0.10; 3, 0.1 to 0.2; 4, 0.2 to 0.5; 5, 0.5 to 1.0; 6, 1.0 to 2.0; 7, 2.0. (Modified from Finenko.[36])

found that productivity in arctic waters was four times greater than in waters of Atlantic origin.

Detailed investigations on the primary productivity of the tropical Atlantic have been undertaken during an international study of the area referred to as the Egualant Program. The obtained data are summarized by Finenko and Kondratieva.[40] The results show that primary productivities in the tropical Atlantic vary greatly depending on the hydrography of the area and its annual fluctuations (see Table 2). In general, high productivity values of 1 g C m$^{-2}$ day$^{-1}$ were found in the upwelling areas west and south of Dacar and south of Ft. Gentil. The phytoplankton productivity of the western tropical Atlantic is relatively low (0.3 g C m$^{-2}$ day$^{-1}$) except in a high productivity region off the Amazon delta.

Figure 5 gives data for the phytoplankton productivity in the tropical Atlantic during different seasons.[40] These data show that in most areas, daily productivity does not exceed 0.5 g C m⁻² day⁻¹ and that in only 11.3% (10% in autumn) of the whole area production exceeds 0.5 g C m⁻² day⁻¹. Additional data on the productivity of the tropical Atlantic are given by Sorokin and Klyashtorin,[28] Reyssac,[41] Lloyd,[42] and Margelef.[43] Recent ¹⁴C measurements by Gieskes et al[56] showing 5 to 15 times higher production values in the North Equatorial Current are possible due to the method used (4ℓ incubation bottles) and are still under discussion.

Productivity data from the Sargasso Sea exist from the Galethea expedition[11] and from intensive studies in the years 1957 to 1960 by Menzel and Ryther.[44] The highest values obtained in winter and spring are 0.8 to 0.9 g C m⁻² day⁻¹. In summer and autumn, productivity in this area is extremely low with values of 0.15 and 0.05 g C m⁻² day⁻¹. These values agree with data calculated by Riley[45,46] from O₂ bottle experiments, where he obtained a mean net productivity for one year of 0.13 and a gross productivity of 0.27 g C m⁻² day⁻¹.

There are great differences in productivities within the Gulf of Mexico and the Caribbean. At the upwelling areas in the Trinidad-Tobago-Venezuela shelf in the western part of the Gulf of Mexico and the western part of the Gulf of Campeche, highest productivity is in April and June, with values of more than 1 g C m⁻² day⁻¹.[47] Lower productivity values never exceeding 0.2 g C m⁻² day⁻¹ are recorded for the central parts of the Gulfs.[11,48-51] Finenko[36] mentioned 40 g C m⁻² day⁻¹ as an average production value for the open waters of the Caribbean and the Gulf of Mexico. Steven,[52] who studied primary productivity in the north-western tropical areas, showed that considerable changes in phytoplankton productivity can occur, even in the tropical Atlantic where relatively stable hydrographical conditions exist. Finenko[36] subdivided the tropical Atlantic into three regions of different levels of primary productivity:

1. Regions of equatorial divergence with upwelling, including the areas to which these nutrient-rich upwelling waters are carried, with an annual average production of 300 g C m⁻²
2. The equatorial upwelling, extending from 5 to 6° S to 7° to 8° N, with an average productivity of 150 to 220 g C m⁻² year⁻¹
3. The regions of anticyclonic gyrals, in the north and south west, with an annual production of between 40 and 100 g C m⁻²

There is little information on the productivity of the South Atlantic. Some measurements exist from the Patagonian shelf[53,54] and from the upwelling areas along the South African coast.

A large volume of information exists on the productivity of the Baltic, the North Sea, and the North, and tropical Atlantic. For the Atlantic, where extensive investigations do not exist, the most complete information is given on the map of primary productivity distribution by Koblentz-Mishke and coauthors.[55,12] As examples from the North Sea and the Baltic Sea illustrate, repetitive investigations often provide different data. This is partly because of the fluctuations, variability, and heterogeneity of phytoplankton and their productivity within the seas.

## REFERENCES

1. **Niemi, A.,** Ecology of phytoplankton in the Tvarminne area, SW coast of Finland. II. Primary production and environmental conditions in the archipelago and the sea zone, *Acta Bot. Fenn.,* 105, 1, 1975.
2. **Hobro, R.,** *Annual Phytoplankton Successions in a Coastal Area in the Northern Baltic,* Asko Laboratory, Trosa, Sweden, in press.

3. **Smetacek, V.,** Die Sukzession des Phytoplanktons in der Wetlichen Kieler Bucht, Dissertation, Universitat Kiel, 1975.
4. **Hagmeier, E.,** *Dynamische Aspekte der Planktonbestande bei Helgoland,* Biologische Anstalt Helgoland, Jahresberichte, 1966.
5. **Holigan, P. M. and Harbour, D. S.,** The vertical distribution and succession of phytoplankton in the western English Channel in 1975 and 1976, *J. Mar. Biol. Assoc. U.K.,* 57, 1075, 1977.
6. **Gargas, E.,** Phytoplankton Production Chlorophyll *a* and Nutrients in the Open Danish Waters 1955-77, The National Agency of Environmental Protection, Denmark, 1978.
7. **Gargas, E.,** *A Manual for Phytoplankton Primary Production Studies in the Baltic,* Baltic Marine Biologists Publication No. 2, Water Quality Institute, Horsholm, Denmark, 1975.
8. **Steele, J. H.,** Plant production on the Fladen Ground, *J. Mar. Biol. Assoc. U.K.,* 35, 1, 1956.
9. **Steele, J. H.,** A comparison of plant production estimates using C-14 and phosphate data, *J. Mar. Biol. Assoc. U.K.,* 36, 1957.
10. **Cushing, D. H.,** Production of carbon in the sea, *Nature (London),* 179, 876, 1957.
11. **Steemann-Nielsen, E. and Jensen, A.,** Primary oceanic production. The autotrophic production of organic matter in the oceans, *Galathea Rep.,* 1, 49, 1957.
12. **Koblentz-Mishke, O. I., Volkovinski, V. V., and Kabanova, YU.G.,** Primary production of the world ocean (Russian), in *Programma i Metodika Izucheniga Biogecenozov Vodnoj Sredy,* Zenkevitch, L. A., Ed., Nauka, Moscow, 1970, 66.
13. **Hagmeier, E.,** Primarproduktion bei Helgoland, Aus dem Jahresbericht der Biologischen Anstalt Helgoland, 1976.
14. **Glover, R. S.,** The continuous Plankton Recorder survey of the north Atlantic, *Symp. Zool. Soc. London,* 19, 189, 1967.
15. **Cushing, D. H.,** Productivity of the North Sea, in *North Sea Science, Goldberg, E. D., Ed.,* MIT Press, Cambridge, 1973, 249.
16. **Reynolds, N.,** Phytoplankton of the west coast of the British Islands, *Ann. Biol.,* 32, 56, 1975.
17. **Postma, H. and Rommets, J. W.,** Primary production in the Wadden Sea, *Neth. J. Sea Res.,* 4, 470, 1970.
18. **Cadée, G. C. and Hegeman, J.,** Primary production of phytoplankton in the Dutch Wadden Sea, *Neth. J. Sea Res.,* 8, 240, 1974.
19. **Kroon (de), J. C.,** Potentiele productie in het oostelijke deel van de zuiderlijke Noordzee, *Int. Rep. Neth. Inst. Sea Res. (Texel),* 1, 1, 1971.
20. **Gieskes, W. W.,** Primary production, nutrients and size spectra of suspended particles in the Southern North Sea, *Int. Rep. Neth. Inst. Sea Res. (Texel),* 16, 1972.
21. **Mommaerts, J. P.,** L'indice de productivite en Mer du Nord, In Modele mathematique rapport de synthese, Programme national sur l'environnement physique et biologique, Pollution des eaux, projet Mer, Journees D'etude des 24 et 25 Novembre 1971, 1972, 144.
22. **Mommaerts, J. P.,** The relative importance of nannoplankton in the North Sea primary production, *Br. Phycol. J.,* 8, 13, 1973.
23. **Mommaerts, J. P.,** Primary production in the South Bight of the North Sea, *Br. Phycol. J.,* 8, 217, 1973.
24. **Gieskes, W. W. C. and Kraay, G. W.,** Primary Production and Pigment Measurements in the Northern North Sea during FLEX-76, Netherlands Inst. Sea Res. (Texel), in press, 1979.
25. **Hagmeier, E. and Weigel, P.,** Biologische anstalt Helgoland, personal communication, 1978.
26. **Ryther, J. H. and Yentsch, S. C.,** Primary production of continental shelf waters off New York, *Limnol. Oceanogr.,* 3, 327, 1958.
27. **Platt, T.,** The annual production by phytoplankton in St. Margaret's Bay, Nova Scotia, *J. Cons. perm. int. Explor. Mer,* 33, 324, 1971.
28. **Sorokin, YU.I. and Klyashtorin, L. B.,** Primary production in the Atlantic Ocean (Russian), *Tr. Vses. Gidrobiol. Ova.,* 2, 265, 1961.
29. **Cohen, F. B. and Wright, W. R.,** Changes in the Plankton of Georges Bank in Relation to the Physical and Chemical Environment During 1975-76, ICES Paper C.M./L:27, 1978.
30. **Steemann-Nielsen, E.,** Experimental methods for measuring organic production in the sea, *Rapp. P.-V. Reun. Cons. Int. Explor. Mer.,* 144, 38, 1958.
31. **Steemann-Nielsen, E.,** A survey of recent Danish measurement of the organic productivity in the sea, *Rapp. P.-V. Reun. Cons. Int. Explor. Mer,* 144, 92, 1958.
32. **Steemann-Nielsen, E. and Hansen, V. K.,** Light adaptation in marine phytoplankton populations and its interrelations with temperature, *Physiol. Plant.,* 12, 353, 1959.
33. **Steemann-Nielsen, E. and Hanse, V. K.,** The primary production in the waters west of Greenland during July 1958, Rapp. P.-V. Reun. Cons. Int. Explor. Mer, 149, 158, 1959.
34. **Berge, G.,** Measurements of the primary production and recordings of the water transparency in the Norwegian Sea during May-June 1958, *Rapp. P.-V. Reun. Cons. Int. Explor. Mer,* 149, 148, 1958.

35. **Chmyr, V. D. and Sherstney, A. I.,** Investigation of primary production in the Norwegian Sea with radiocarbon (Russian) *Tr. Atl. NIRO,* 10, 28, 1963.

36. **Finenko, Z. Z.,** Production in plant populations, in *Marine Ecology,* John Wiley & Sons, 1978.

37. **Rouchiainen, M. I.,** Primary production of plankton in one of the gulfs of the Barents Sea, *Dokl. Akad. Nauk SSSR,* 141, 205, 1961.

38. **Sokolova, S. A. and Solvyeva, A. A.,** Primary production in the Dalnezelenetskaya Bay (Murman coast) in 1967 (in Russian), *Okeanologiya,* 11, 460, 1971.

39. **Corlett, J.,** Measurement of primary production in the western Barents Sea, *Rapp. P.-V. Reun. Cons. Int. Explor. Mer,* 144, 76, 1958.

40. **Finenko, Z. Z. and Kondratieva, T. M.,** Primary production in tropical part of Atlantic Ocean (Russian), in *Plankton i biologicheskaya productivnost tropicheskoi Atlantiki,* Greze, V. N., Ed., Naukova Dumka, Kiev, 1971, 122.

41. **Reyssac, J.,** Mesures de la production primaire par la methode du $C^{14}$ au large de la Cote d'Ivoire, *Rep.O.R.S.T.O.M., Abidjan,* 35, 1, 1966.

42. **Lloyd, I. J.,** Primary production off the coast of northwest Africa, *J. Cons. permint. Explor. Mer,* 33, 312, 1971.

43. **Margalef, R.,** Fitoplancton de la region de afloramiento del noroeste de Africa, *Res. Exp. Cient B/ O Cornide (Madrid),* 1, 23, 1972.

44. **Menzel, D. W. and Ryther, J. H.,** The annual cycle of primary production in the Sargasso Sea off Bermuda, *Deep-Sea Res.,* 6, 351, 1960.

45. **Riley, G. A.,** Plankton studies. II. The western north Atlantic May-June, 1939, *J. Mar. Res.,* 2, 145, 1939.

46. **Riley, G. A.,** Phytoplankton of the north central Sargasso Sea, 1950-1952, *Limnol. Oceanogr.,* 2, 252, 1957.

47. **Kabanova, YU., G. and Lopes, L. B.,** The primary production in the southern part of the Gulf of Mexiko and off the north western coast of Cuba, (Russian) in *Okeanologicheskije issledovanija,* Vol. 20, Belousov, I. M., Ed., Nauka, Moscow, 1970, 46.

48. **Curl, H.,** Primary production measurements in the north coastal waters of South America, *Deep-Sea Res.,* 7, 183, 1960.

49. **Steele, J. H.,** A study of production in the Gulf of Mexico, *J. Mar. Res.,* 22, 211, 1964.

50. **Margalef, R.,** El ecosistema pelagico del mar Caribe, *Separata de la Memoria de la Sociedad de Ciencias Naturales la Salle,* 29, 5, 1969.

51. **Kabanova, YU.G.,** Dependence of primary production values upon different factors in the northeastern part of the Caribbean Sea (Russian), *Okeanologiya,* 12, 299, 1972.

52. **Steven, D. M.,** Primary productivity of the tropical western Atlantic Ocean near Barbados, *Mar. Biol.,* 10, 261, 1971.

53. **Volkovinskii, V. V.,** Research of primary production in the South Atlantic waters (Russian), Second Int. Oceanography Congr., Abstracts of Papers, Nauka, Moscow, 99, 1966.

54. **El-Sayed, S. Z.,** Phytoplankton production in the Antarctic and subantarctic waters (Atlantic and Pacific sections) (Russian), Second Int. Oceanography Congr., Abstracts of Papers, Nauka, Moscow, 440, 1966.

55. **Koblentz-Mishke, O. I., Volkovinski, V. V., and Kabanova, Yu.G.,** Recent data on the quantity of the world ocean (Russian), *Dokl.Akad. Nauk SSSR,* 183, 1189, 1968.

56. **Gieskes, W. W., Kraay, G. W., and Baars, M. A.,** Current $^{14}C$ methods for measuring primary production: gross underestimates in oceanic waters, *Neth. J. Sea Res,* 13 (1), 58, 1979.

57. **Lindahl, O.,** Studies on the production of phytoplankton and zooplankton in the Baltic in 1975, *Medd. från Havsfiskelab. Lysekil,* nr. 217, 1977.

58. **''Asko'' team, in Gargas, E.,** Phytoplankton Production Chlorophyll *a* and Nutrients in the Open Danish Waters 1955-77, The National Agency of Environmental Protection, Denmark, 1978.

59. **Renk, H.** Primary production and chlorophyll content of the Baltic Sea. Part III. Primary production in the southern part of the Baltic. *Pol. Arch. Hydrobiol.* 21:2, 191, 1974.

60. **Schulz, S. and Kaiser, W.,** Some pecularities of the primary production during the last years in the Baltic and their causes. Presented at the IVth Symposium of the Baltic Marine Biologists, Gdansk, 13-18, October, 1975.

61. **Schulz, S. and Kaiser, W.,** Produktionsbiologische Veranderungen in der Ostsee im Jahre 1973. Fischerei-Forschung. *Wiss. Schriftenreihe* 13, 15, 1975.

62. **Schulz, S. and Kaiser, W.,** Produktionsbiologische Untersuchungen in der Ostsee 1975 und einige spezielle Ergebnisse aus dem Jahre 1974. Fischerei-Forschung. *Wiss. Schriftenreihe* 14, 53, 1976.

# PHYTOPLANKTON PRODUCTIVITY IN THE INDIAN OCEAN

## B. Zeitzschel

As compared to the Atlantic and the Pacific Oceans, the Indian Ocean is unique. Morphologically, the Indian Ocean is landlocked in the north and does not extend into the cold climate regions of the northern hemisphere. According to Wyrtki,[1] this causes an asymmetrical development of its structure and circulation which is most obvious in the huge layers of extremely low oxygen content in the Arabian Sea and the Gulf of Bengal. The land mass of Asia also effects the ocean climatologically, causing the seasonally changing monsoons which in turn reverse the oceanic circulation over its northern part. Connected with this seasonally changing circulation are various upwelling areas which operate only during one season, in contrast to all the other major upwelling areas in the world.

The primary productivity of the Indian Ocean was studied in detail during the International Indian Expedition, 1959 to 1965. Results of this expedition series have been summarized by Ryther et al.,[2] Roblentz-Mishke et al.,[3] Aruga,[4] Cushing,[5], Krey,[6] Bunt,[7] Krey and Babenerd,[8] Finenko,[9] and others. Krey[6] distinguishes eight plankton-geographical regions (see Table 1), indicating the dominant phytoplankton groups in the different physical regimes of the Indian Ocean. Subrahmanyan and Sarma[10] compiled a list of 37 phytoplankton species which they identified as mass forms in neritic waters. These contribute 70 to 80% of the biological activity (see Table 2).

Examples of distributional patterns of primary production are given in Figures 1 to 3. Table 3 lists the relevant data for the primary productivity of the Indian Ocean in comparison with data for other oceans.[11] Summarizing the distribution maps, it is apparent that regions of high productivity are those affected by the monsoon in the Arabian Sea, the Bay of Bengal, and the Indonesian Sea. The period of the SW-monsoon (summer) was in general more productive than the NE-monsoon (winter).

Figures 4 and 5 give two examples of sections for potential primary productivity in milligrams C $m^{-3}$ $hr^{-1}$.[8] They are section 14 from E — W from cruise 1/65 of RV Diamantina (May 16 to June 2, 1965) and section 27 from N — S from cruise 5 of Anton Bruun (April 4 to 30, 1964). It is obvious from these sections, that the primary production in the Indian Ocean is limited to a layer of about 100 m during these time periods. In the oligotroph subtropical area between 20 and 30° S, the potential primary productivity is rather low (between 0.2 and 0.3 mg C $m^{-3}$ $hr^{-1}$), whereas in Section 27 higher values (more than 0.5 mg C $m^{-3}$ $hr^{-1}$) have been measured at and near the equator as well as between 35 to 40° S.

There are few studies of seasonal variations of the primary productivity in the open Indian Ocean. Intensive studies have been carried out by Australian scientists[12,13] along 110° E (see Figure 6). Mean primary productivity integrated over 150 m increased from 50 mg C $m^{-2}$ $hr^{-1}$ in August to a maximum (62 mg C $m^{-2}$ $hr^{-1}$) in October. Then it decreased to a minimum (4 mg C $m^{-2}$ $hr^{-1}$) in January, after which it increased slowly to 25 mg C $m^{-2}$ $hr^{-1}$ in April to May. It sharply increased to 45 mg C $m^{-2}$ $hr^{-1}$ in late May and remained at that level until August. The average yearly production was estimated to be 37 mg C $m^{-2}$ $hr^{-1}$.

Qasim[14] studied the productivity of backwaters and estuaries in South India and came to the conclusion that 90% of the total production is confined to the topmost layer (see Figure 7). Maximum production occurs either at the surface or slightly below. This is caused by the highly turbid conditions in the backwaters, especially during the monsoon season. Monthly values of gross and net production integrated over the

Table 1

PHYTOPLANKTON PREDOMINANCE IN EIGHT
PLANKTON−GEOGRAPHICAL REGIONS OF THE INDIAN
OCEAN[6]

| Area | Predominance | Secondary dominance |
|---|---|---|
| Coastal upwelling areas Southern Arabia, Western Australia, Indonesia | Diatoms | Dinoflagellates, partly blue-green algae |
| Central Arabian Sea and Bay of Bengal | Dinoflagellates, blue-green algae | Diatoms, coccolithophores |
| Somali Current region NE section SW section | Diatoms | Dinoflagellates, partly blue-green algae |
| Mozambique Current region | Diatoms | Dinoflagellates, coccolithophores |
| Equatorial Current region | Dinoflagellates, coccolithophores | Diatoms, blue-green algae |
| Southern subtropical gyre between southern subtropical convergence and the southern tropical front | Dinoflagellates | Coccolithophores, diatoms |
| West—wind drift region | Diatoms | Dinoflagellates, coccolithophores |
| Antarctic gyre up to the subantarctic convergence | Diatoms | Dinoflagellates, coccolithophores |

Table 2

MASS FORMS OF PHYTOPLANKTON IN THE INDIAN OCEAN[6,10]

Bacillariophyceae:
  *Asterionella japonica*
  *Bacteriastrum hyalinum* var. *princeps*
  *Biddulphia heteroceros*
  *Biddulphia mobiliensis*
  *Chaetoceros affinis*
  *Chaetoceros brevis*
  *Chaetoceros compressus*
  Chaetoceros contortum
  *Chaetoceros curvisetus*
  Chaetoceros lasciniosus
  *Chaetoceros lauderii*
  Chaetoceros lorenzianus
  *Chaetoceros pelagicus*
  *Chaetoceros socialis*
  *Coscinodiscus asteromphalus*
  Coscinodiscus oculus-iridis
  *Fragilaria oceanica*
  *Guinardia flaccida*
  *Lauderia annulata*
  *Leptocylindrus danicus*

*Nitzschia seriata*
*Nitzschia sigma* var. *indica*
*Rhizosolenia alata*
*Rhizosolenia robusta*
*Rhizosolenia stolterfothii*
*Sceletonema costatum*
*Schroderella delicatula*
*Thalassiothrix frauenfeldii*
*Thalassiothrix longissima*

Dinophyceae
  *Ceratium fusus*
  *Ceratium macroceros*
  *Ceratium tripos*
  *Dinophysis caudata*
  *Glenodinium lenticula* f. *asymmetrica*
  *Noctiluca miliaris*
  *Peridinium depressum*

Myxophyceae
  *Trichodesmium erythraeum*

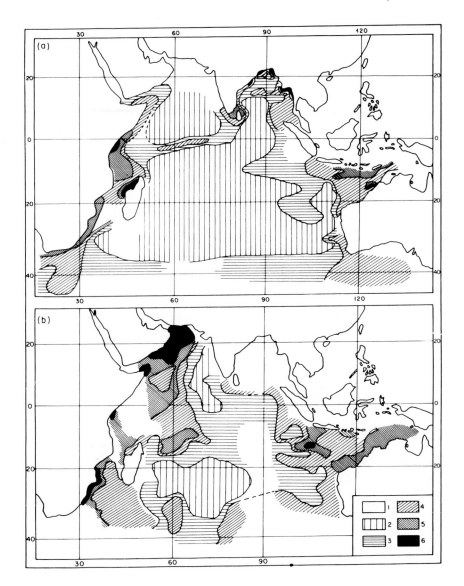

FIGURE 1.   Primary productivity in the Indian Ocean in g C m$^{-2}$ day$^{-1}$. (a) In the NE-monsoon; (b) in the SW-monsoon 1. no data; 2. <0.1; 3. 0.11 to 0.25; 4, 0.26 to 0.50; 5, 0.51 to 1.0; 6, >1.1.[9,11] (Based on information provided by Kabanova, 1968, from Finenko, Z. Z., in *Marine Ecology*, Vol. 4, Kinne, O., Ed., John Wiley & Sons, 1978, chap. 2. With permission.)

euphotic zone are given in Figure 8. The column production shows three peaks: one in April, the second in July, and the third in October. These peaks were of a short duration and amounted to a three- to fourfold increase.

In conclusion, the mean level of primary productivity in the Indian Ocean amounts to 81 g C m$^{-2}$ hr$^{-1}$ as compared to 69.4 and 46.4 for the Atlantic and Pacific Oceans.[3]

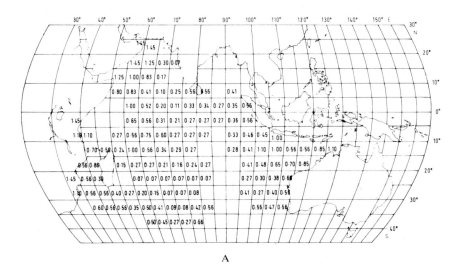

A

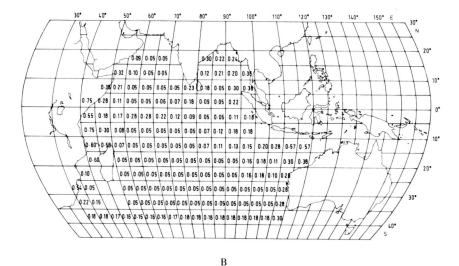

B

FIGURE 2.   Primary productivity in the Indian Ocean in g C m⁻² day⁻¹. (A) In the SW-
monsoon; (B) in the NE-monsoon; (C) the distribution of stations for primary productivity
during the International Indian Ocean Expedition, symbols indicate the different ships. (From
Cushing, D. H., in *The Biology of the Indian Ocean, Ecological Studies,* Vol. 3, Zeitzschel,
B., Ed., Springer-Verlag, New York, 1973, 475. With permission.)

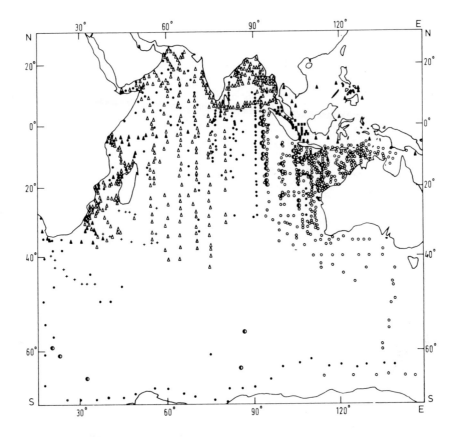

Figure 2C

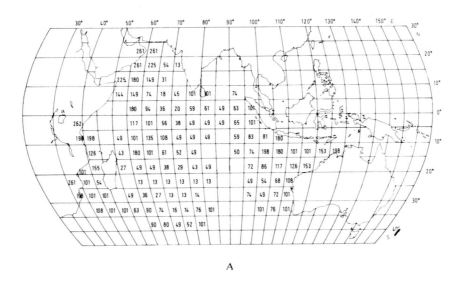

A

FIGURE 3. Primary productivity in the Indian Ocean in g C m⁻² 180 day⁻¹. (A) in the SW-monsoon; (B) in the NE-monsoon.[5] (From Cushing, D. M., in *The Biology of the Indian Ocean, Ecological Studies*, Vol. 3, Zeitzschel, B., Ed., Springer-Verlag, New York, 1973, 475. With permission.)

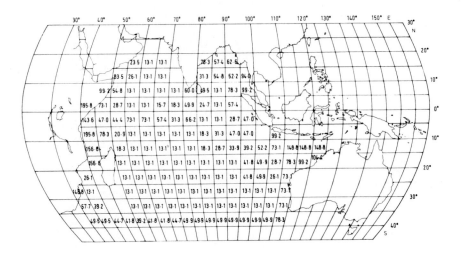

Figure 3B

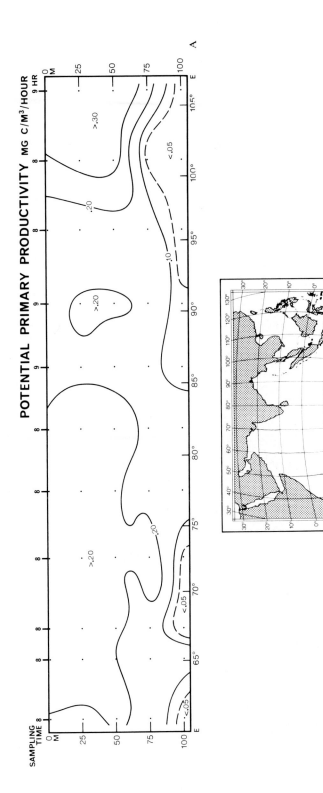

FIGURE 4. (A) Potential primary production in mg C m$^{-3}$ hr$^{-1}$ along Section 14 of RV Diamantina cruise 1/65. (B) Cruise track (see map.)[8]

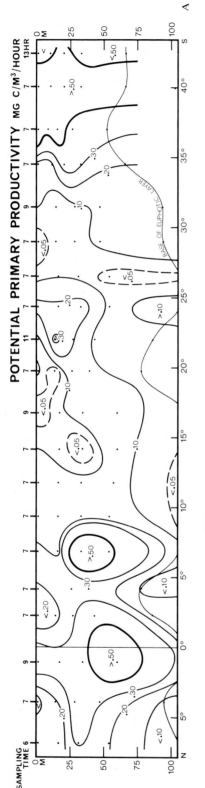

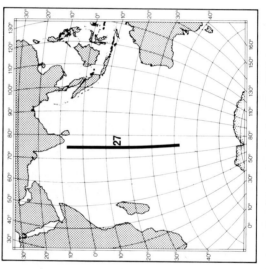

FIGURE 5.    (A)  Potential primary productivity in mg C m$^{-3}$ hr$^{-1}$  along Section 27 of RV Anton Bruun cruise 5.  (B)  Cruise track (see map).[5]

Table 3

GLOBAL PLANKTON PRIMARY PRODUCTION FOR VARIOUS TYPES OF OCEAN WATER[3]

| Type of water | Primary productivity level (mg C/m²/day) Mean value | Limits of fluctuation | Areas of each type of water in different oceans Area[a] | ×10³ km² | World ocean (%) | Summary of annual production of the world ocean for each type of water (10⁹ t C/year) |
|---|---|---|---|---|---|---|
| Oligotrophic waters of the central parts of subtropical halistatic areas | 70 | <100 | PO | 90,105 | 24.6 | 3.79 |
| | | | IO | 19,599 | 5.3 | |
| | | | AO; | 30,624 | 8.3 | |
| | | | OW | 8,000 | 2.2 | |
| | | | WO | 148,329 | 40.4 | |
| Transitional waters between subtropical and subpolar zones; extremity of the area of equatorial divergences | 140 | 100—150 | PO | 33,357 | 9.1 | 4.22 |
| | | | IO | 23,750 | 6.5 | |
| | | | AO | 22,688 | 6.2 | |
| | | | OW | 3,051 | 0.8 | |
| | | | WO | 82,847 | 22.6 | |
| Waters of equatorial divergence and oceanic regions of subpolar zones | 200 | 150—250 | PO | 31,319 | 8.5 | 6.31 |
| | | | IO | 18,886 | 5.2 | |
| | | | AO | 32,650 | 8.9 | |
| | | | OW | 3,642 | 1.0 | |
| | | | WO | 86,498 | 23.6 | |
| Inshore waters | 340 | 250—500 | PO | 10,422 | 2.8 | 4.80 |
| | | | IO | 7,944 | 2.2 | |
| | | | AO | 14,183 | 3.9 | |
| | | | OW | 6,184 | 1.7 | |
| | | | WO | 38,735 | 10.6 | |
| Neritic waters | 1,000 | >500 | PO | 243 | 0.07 | 3.90 |
| | | | IO | 5,289 | 1.4 | |
| | | | AO | 2,717 | 0.74 | |
| | | | OW | 2,433 | 0.66 | |
| | | | WO | 10,683 | 2.9 | |
| Total all waters | | | WO | 367,092 | 100 | 23 × 10⁹ t C/year |

[a] PO, Pacific Ocean; IO, Indian Ocean; AO, Atlantic Ocean; OW, other waters (North Polar ocean, Indonesian seas, Mediterranean, Black, Asov, White, Okhotsk, Bering, Japan, China, Yellow seas); WO, summary value for the whole World Ocean.

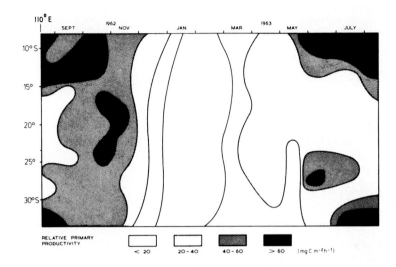

FIGURE 6.    Primary productivity (light-saturated) with respect to latitude and month. Column values to 150 m in mg Cm$^{-2}$ h$^{-1}$. (From Tranter, D. J., in *The Biology of the Indian Ocean, Ecological Studies,* Vol. 3, Zeitszchel, B., Ed., Springer-Verlag, New York, 1973, 487. With permission.)

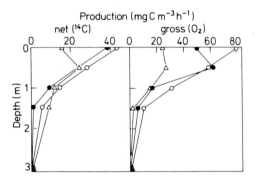

FIGURE 7.    Primary production (gross and net) in the Cochin back-water in relation to depths. The data have been pooled into three seasons. Monsoon season, closed circles; post-monsoon season, triangles; premonsoon months, open circles.[14] (From Qasim, S. Z., in *The Biology of the Indian Ocean, Ecological Studies,* Vol. 3, Zeitzschel, B., Ed., Springer-Verlag, New York, 1973, 143. With permission.)

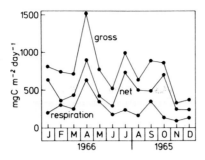

FIGURE 8. Seasonal changes in gross and net primary production in the back-water. Community respiration has also been included in the figure. The integrated values refer to the euphotic zone. (From Qasim, S. Z., in *The Biology of the Indian Ocean, Ecological Studies,* Vol. 3, Zeitzschel, B., Ed., Springer-Verlag, New York, 1973, 143. With permission.)

# REFERENCES

1. **Wyrtki, K.,** Physical Oceanography of the Indian Ocean, in *The Biology of the Indian Ocean, Ecological Studies,* Vol. 3, Zeitzschel, B., Ed., Springer-Verlag, New York, 18, 1973.
2. **Ryther, J. H., Hall, J. R., Pease, A. K., Bakun, A., and Jones, M. M.,** Primary production in relation to the chemistry and hydrography of the western Indian Ocean, *Limnol. Oceanogr.,* 11, 371, 1966.
3. **Koblentz-Mishke, O. J., Volkovinsky, V. V., and Kabanova, J. G.,** Plankton primary production of the world ocean, in *Scientific Exploration of the South Pacific,* Wooster, W. S. Ed., National Academy of Sciences, Washington, D.C., 1970, 183.
4. **Aruga, Y.,** Primary production in the Indian Ocean II, in *The Biology of the Indian Ocean, Ecological Studies,* Vol. 3, Zeitzschel, B., Ed., Springer-Verlag, New York, 1973, 127.
5. **Cushing, D. H.,** Production in the Indian Ocean and the transfer from the primary to the secondary level, in *The Biology of the Indian Ocean, Ecological Studies,* Vol. 3, Zeitzschel, B., Ed., Springer-Verlag, New York, 1973, 475.
6. **Krey, J.,** Primary production in the Indian Ocean I, in *The Biology of the Indian Ocean, Ecological Studies,* Vol. 3, Zeitzschel, B., Ed., Springer-Verlag, New York, 1973, 115.
7. **Bunt, J. S.,** Primary productivity of marine ecosystems, in *Primary Productivity of the Biosphere, Ecological Studies,* Vol. 11, Lieth, H. and Whittaker, R. H., Eds., Springer-Verlag, New York, 1975, chap. 8.
8. **Krey, J. and Babenerd, B.,** Phytoplankton production, in *Atlas of the International Indian Ocean Expedition,* Institut für Meereskunde an der Universität Kiel, 1976.
9. **Finenko, Z. Z.,** Production in plant population, in *Marine Ecology,* Vol. 4, Kinne, O., Ed., John Wiley & Sons, Chichester, 1978, chap. 2.
10. **Subrahmanyan, R. and Sarma, A. H. V.,** Studies on the phytoplankton of the west coast of India. IV. Magnitude of the standing crop for 1955—62 with observations on nanoplankton and its significance to fisheries, *J. Mar. Biol. Assoc. India,* 7, 406, 1965.
11. **Kabanova, J. G.,** Primary production of the northern part of the Indian Ocean, *Oceanology,* 8, 214, 1968.
12. **Jitts, H. R.,** Seasonal variations in the Indian Ocean along 110°E. IV. Primary production, *Aust. J. Mar. Freshwater Res.,* 20, 65, 1969.
13. **Tranter, D. J.,** Seasonal studies of the pelagic ecosystem, in *The Biology of the Indian Ocean, Ecological Studies,* Vol. 3, Zeitszchel, B., Ed., Springer-Verlag, New York, 1973, 487.
14. **Qasim, S. Z.,** Productivity of backwaters and estuaries, in *The Biology of the Indian Ocean, Ecological Studies,* Vol. 3, Zeitzschel, B., Ed., Springer-Verlag, New York, 1973, 143.

# PHYTOPLANKTON PRODUCTIVITY IN THE AUSTRALIAN SEA

## J. S. Bunt

## INTRODUCTION

This section summarizes information currently available and attempts an assessment of likely yields from planktonic photosynthesis for the territorial sea.

## THE PHYSICAL ENVIRONMENT

The continent of Australia occupies an area of $7.6 \times 10^6$ km$^2$ and has a coastline the length of which, using 1:250,000 charts and 0.5-km intercepts, has been estimated to be 45,000 km.[13] Roughly half of it is within the tropics; the other half is temperate. Australia's bounding seas have diverse origins. The region to the northeast, the Coral Sea, is under the influence of the warm south equatorial current. Further south, the western Tasman Sea receives the warm waters of the east Australian current. Southern and south western coasts are directly affected by the west wind drift of cold subantarctic seas. To the north and northwest, the shallow Arafura and Timor Seas represent an extension of the tropical Indian Ocean, although connections to the Coral Sea through Torres Strait also bring links with the equatorial Pacific.

Figure 1 indicates the approximate boundaries of coastal waters to the 200-m depth contour, as well as a 200-mi (320 km) territorial limit, proposed but not yet fixed. Arbitrary zones have been applied in this account for use in considering regional primary production. Only the most prominent coastal features have been identified. It should be noted, however, that a wide range of estuarine environments exists for which information on planktonic productivity is lacking.

## ESTIMATION OF PRIMARY PRODUCTION

The problems of obtaining credible measures of primary production in the sea will not be considered here. For background, the reader may wish to refer to accounts by Strickland and Parsons,[1] Steemann-Nielsen,[2] and others. Suffice to remark that current techniques normally depend on incubation of water samples spiked with $^{14}$C bicarbonate, either *in situ* or under controlled conditions aboard ship. Arriving at values of photosynthetic carbon fixation per unit sea surface necessarily involves well-recognized extrapolation and uncertainties. Spatial and temporal variations in planktonic activity create particular problems for arriving at annual production figures. Unfortunately, regularly repeated observations over extended periods at specified locations are uncommon, so that it is frequently necessary to base annual mean predictions on flimsy information. Limitations such as these apply generally for Australian waters.

## AVAILABLE DATA

Most reviewers in this field turn first to the account by the Russian authors, Kolblentz-Mishke et al.,[3] who considered data from more than 7000 stations collected by the investigators of a number of nations throughout the global sea. The data base covers a period of 15 years since the pioneering expeditions of Steemann-Nielsen.[4] Using information on $^{14}$C uptake as well as figures on phytoplankton standing stocks and other indexes, estimated production rates were assigned in 5 categories from less than 100 to more than 500 mg C·m$^{-2}$ d$^{-1}$. Distributions of primary production were

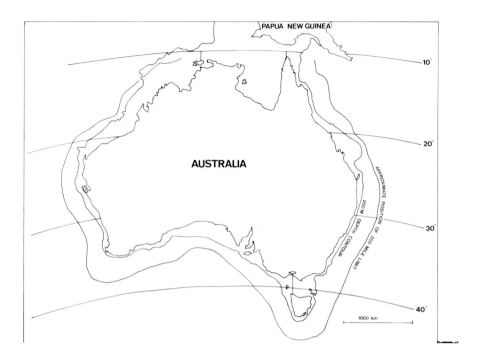

FIGURE 1.    The continent of Australia with approximate limits of the marine environment
to the 200-m depth contour and to a possible 200-mi territorial limit.

mapped. Table 1 gives the total annual primary production to be expected on this basis
in each of the arbitrary regions indicated in Figure 1. Daily rates of production recorded by several individual investigations and those from Koblentz-Mishke et al.[3] also
have been listed for purposes of comparison.

Within the 200-mi territorial sea, as defined in Figure 1, and using the data compiled
by Koblentz-Mishke et al.,[3] the total annual primary production would represent 365
to 662 × 10⁶ tons C. To the 200-m depth contour, the figure would be reduced to 180
to 345 × 10⁶ tons C. Although daily production figures from individual investigators
agree broadly with the Russian compilations, in many cases they suggest substantially
wider ranges. Most, however, were isolated in time. Only the data from Jitts[5] incor-
porate seasonal measurements. Since a scanning of the literature suggests that the data
base available to Koblentz-Mishke et al.[3] was comparably limited, the projections for
Australian waters must be considered speculative. There are suggestions, nonetheless,
that the most productive waters occur to the north and northwest of the continent and
perhaps also in the Great Australian Bight, although summer values from Motoda et
al.[6] were not high. While the same authors reported isolated remarkably high rates of
production during summer in the Gulf of Carpentaria, the mean figure in this area
was 133 mg $C \cdot m^{-2} \cdot hr^{-1}$, equivalent to almost 1.6 g $C \cdot m^{-2} \cdot d^{-1}$ if the rate were maintained for 12 hr. Yearly production at such levels would exceed rates in the highly
fertile Peru upwelling. Clearly, longer-term observations are required, not only here,
but throughout the region, including the apparently less productive waters to the east
and southwest.

There is a particular need for attention to waters within and close to estuaries where
nutrient limitation might be expected to be reduced. Bunt[7] has calculated potential
levels of production between 10° S and 40° S in the range 12 to 15 g $C \cdot m^{-2} \cdot d^{-1}$. It is
unlikely that environmental constraints ever allow such activities to be long sustained

Table 1
PREDICTIONS OF ANNUAL PRIMARY PRODUCTION BY ZONE BASED
ON KOBLENTZ-MISHKE ET AL.[3] AND DAILY RATES OF PRODUCTION
FROM INDIVIDUAL INVESTIGATIONS; ZONAL AREAS FOR
REFERENCE[a]

| Zone | Daily rate mg $C \cdot m^{-2} \cdot d^{-1}$ | Annual rate per zone ($10^6$ tons C) | | Zonal area (km²) | | Ref. |
|---|---|---|---|---|---|---|
| | | 1 | 2 | 1 | 2 | |
| (10—20° S) NE | 100—150 | 4.7—7.1 | 15.9—24.3 | 129,800 | 441,200 | 3 |
| | 180—300 | | | | | 10 |
| | 270—290 | | | | | 11 |
| N | 250—500 | 53.9—108.0 | 53.9—108.0 | 598,600 | 598,600 | 3 |
| | 790—3360 | | | | | 6 |
| | 390 | | | | | 12 |
| NW | 250—500 | 52.0—104.0 | 64.0—128 | 577,900 | 711,100 | 3 |
| | 284 | | | | | 12 |
| (20—30° S) E | 100—150 | 5.8—8.8 | 14.9—22.8 | 160,900 | 415,250 | 3 |
| | 130 | | | | | 10 |
| W | 250—500 | 16.0—32.0 | 41.6—83.2 | 178,200 | 462,000 | 3 |
| | 50—700 | | | | | 5 |
| (30—40° S) E | 150—250 | 6.2—10.1 | 25.5—41.7 | 112,500 | 463,700 | 3 |
| | 324 | | | | | 10 |
| SW | 250—500 | 33.5—67.0 | 105.0—210.0 | 372,000 | 1,166,200 | 3 |
| | 130—270 | | | | | 6 |
| (40—50° S) E | 250 | 7.6 | 44.4 | 84,800 | 493,100 | 3 |
| | 588 | | | | | 10 |
| Totals | | 180—345 | 365—662 | 2,214,700 | 4,751,150 | |

[a]   1, coast to 200-m contour; 2, coast to 200-mi territorial limit as shown in Figure 1.

in marine situations. Nonetheless, annual means of production by estuarine phyto-plankton between 1 and 3 g $C \cdot m^{-2} \cdot d^{-1}$ are not uncommon and could be expected to occur over parts of the Australian coastline. For example, Scott[8] documents levels of *in situ* production up to 1.0 g $C \cdot m^{-2} \cdot d^{-1}$ in a temperate east coast estuary. Contributions to nearshore ecosystems could be substantial.

Using the best global estimates of primary production available, Wassink[9] concluded that field photosynthetic efficiency of marine microphytes overall lies close to 0.2%. On this basis, phytoplankton production within the Australian territorial sea would be expected to deliver approximately 408 × $10^6$ tons $C \cdot year^{-1}$. In fact, this figure lies squarely within the limits 365 to 662 × $10^6$ tons $C \cdot year^{-1}$ projected by Koblentz-Mishke et al.[3] Allowing for the influence on water quality of run-off from a substantial coastline, however, some credence might be placed in the upper range of the estimate. The uncertainty is sufficiently large to justify more careful inquiry through direct expansion of field observations.

# REFERENCES

1. Strickland, J. D. H. and Parsons, T. R., A manual of sea water analysis, *Fish. Res. Board Can.,* No. 125, 1965.
2. Steemann-Nielsen, E., *Marine Photosynthesis with Special Emphasis on the Ecological Aspects*, Elsevier, Amsterdam, 1975.
3. Koblentz-Mishke, O. J., Volkovinsky, V. V., and Kabanova, J. G., Plankton primary production of the world ocean, in Scientific Exploration of the South Pacific, National Academy of Sciences, Washington, D.C., 1970, 183.
4. Steemann-Nielsen, E., The use of radioactive carbon ($^{14}$C) for measuring organic production in the sea, *J. Cons. Perm. Int. Explor. Mer,* 18, 117, 1952.
5. Jitts, H. R., Seasonal variations in the Indian Ocean along 110° E. IV. Primary production, *Aust. J. Mar. Freshwater Res.,* 20, 65, 1969.
6. Motoda, S., Kawamura, T., and Taniguchi, A., Differences in productivities between the Great Australian Bight, *Mar. Biol.,* 46, 93, 1978.
7. Bunt, J. S., An Assessment of Marine Primary Production, *CRC Handbook of Biosolar Resources,* Vol. 1, Mitsui, A. and Black, C. C., Eds., CRC Press, Boca Raton, in press.
8. Scott, B. D., Division of Fisheries and Oceanography, Research Report, 1974-77, Commonwealth Scientific and Industrial Research Organization, Cronulla, N. S. W., Autralia, 1978.
9. Wassink, E. C., Photosynthesis and productivity in different environments, in *Photosynthesis and Productivity in Different Environments*, Cooper, J. P., Ed., Cambridge University Press, Cambridge, 1975, 675.
10. Jitts, H. R., The summer characteristics of primary productivity in the Tasman and Coral Seas, *Aust. J. Mar. Freshwater Res.,* 16, 151, 1965.
11. Scott, B. D. and Jitts, H. R., Photosynthesis of phytoplankton and zooxanthellae on a coral reef, *Mar. Biol.,* 41, 307, 1977.
12. Cushing, D. H., A comparison of production in temperate seas and the upwelling areas, *Trans. R. Soc. S. Afri.* 40, 17, 1971.
13. Galloway, R., personal communication.

# GENERAL DESCRIPTION OF PRIMARY PRODUCTIVITY OF MACROFLORA

## I. T. Show, Jr.

Figure 1 shows the high-production areas of the world for both macroflora and phytoplankton and, not surprisingly, these zones overlap. For the most part, high-production areas for both occur in the temperate zones. With the exception of the Mediterranean Sea, the southern coast of India, and the northwest African coast, all high-production areas are found where summer surface water temperature does not exceed 20°C. Particularly impoverished areas include Pacific Central America, tropical west Africa, and the Red Sea.

The large Laminariceae (large brown kelps) tend to show the highest productivity on a world-wide basis. *Laminaria* has been shown to produce as much as 66 g dry weight/m²/day in the Mediterranean and up to 1500 g dry weight/m²/year under cultivation in Japan. *Macrocystis* produces 400 to 820 g C/m²/year off California and up to 2000 g C/m²/year off southern India. Smaller brown algae, such as *Fucus, Pelvetia*, and *Sargassum*, also show relatively high production. *Fucus* produces as much as 640 to 840 g C/m²/year off Nova Scotia; however, its production may be considerably higher than this in the Mediterranean (6.9 kg dry weight/m²/year). *Pelvetia* is particularly productive in the Mediterranean (35 g dry weight/m²/day), while *Sargassum* produces up to 3.1 kg dry weight/m²/year off Hawaii.

In general, the red algae are next in productivity. Some of the most productive genera are *Chondrus, Gracilaria, Porphyra, Eucheuma, Gigartina, Hidea, Neoagardiella*, and *Hypnea*. Production values range from 16.4 kg dry weight/m²/year for *Chondrus* in the Canary Islands to 4.4 kg dry weight/m²/year for *Hypnea* cultured in Florida. Very little reliable data is available on green algae production; however, they are probably comparable to the red algae.

Several marine spermatophytes (flowering plants) have high productivity; in particular, *Spartina, Zostera*, and *Thalassia*, Zostera production ranges from 340 g C/m²/year in Denmark to 1500 g C/m²/year on the U.S. east coast and in the Bering Sea. *Spartina* produces up to 897 g C/m²/year on the U.S. east coast. *Ruppia* and most mangroves produce much less. *Ruppia* produces only 6.4 mg C/m²/day in North Carolina, while mangrove production does not generally exceed a modest 400 g C/m²/year anywhere in the world.

Table 1 gives a large sample of usable data available on macroflora production in the original units used by the authors. Table 2 gives some harvest values for various genera.

As is evident from Table 1, usable data are widely scattered in the literature and are often very sparse. In addition much production data are not directly comparable. The first factor in lack of comparability is obvious from Table 1, the units vary considerably. Units used include various forms of dry weight per area per time and weight carbon per area per time. Dry weight and weight carbon are difficult to compare because of the great variability of any conceivable conversion factor and the dependence of the conversion factor on rarely known physical, chemical, and biological variables. The problem of comparing time in units of day$^{-1}$ and year$^{-1}$ is also usually very difficult. Production year$^{-1}$ is the integral of production day$^{-1}$. Therefore, a knowledge of the production curve over the entire year is required, but rarely known. Here, the problem of comparing perennials and annuals is encountered.

A second and more serious problem involves methods of defining and measuring production. Production is expressed as gross production, net production, and yield.

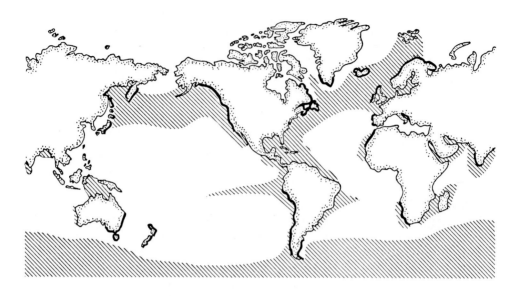

FIGURE 1. High productivity areas of the world. Highlighted dotted areas of the coasts indicate areas of high macrophyte production. Highlighted lines in the oceanic areas indicate the same for phytoplankton. The figure is drawn from reference data given in Table 1.

## Table 1
## MARINE PLANT PRODUCTION

| Species | Phylum | Production | Location[a] | Ref. |
|---|---|---|---|---|
| *Alaria* sp. | Brown | 14 g dry wt/m²/day | Mediterranean (NP) | 1 |
| *Alaria* sp. | Brown | 5.11 kg dry wt/m²/year | Nova Scotia, Scotland, Norway (NP) | 1 |
| *Ascophyllum nodosum* | Brown | 1 kg dry wt/m²/year | Nova Scotia (NP) | 2 |
| *Asparagopsis taxiformis* | Red | 6.71 g C/m²/day | Canary Islands (NP) | 3 |
| *Avicenna* sp. | Mangrove | 400 g C/m²/year | Caribbean (NP) | 2 |
| *Caulerpa mexicana* | Brown | 1.51 g C/m²/day | Canary Islands (NP) | 3 |
| *Caulerpa prolifera* | Brown | 4.52 g C/m²/day | Canary Islands (NP) | 3 |
| *Chaetoceros curvisetus* | Dinoflagellate | 0.9 g N/m²/day 5.3g C/m²/day | Virgin Islands (culture) | 4 |
| *Chlorella* sp. | Green | 10 kg dry wt/m²/year | Thailand (culture) | 5 |
| *Chlorella* sp. | Green | 3.21 kg dry wt/m²/year | Japan (commercial) | 6 |
| *Chondeas crispus* | Red | 45 g dry wt/m²/day | — | 7 |
| *C. crispus* | Red | 1.72 g C/m²/day | Canary Islands (NP) | 3 |
| *Cystosiera abies-marina* | Brown | 6.41 g C/m²/day | Canary Islands (NP) | 3 |
| *Cystosiera fimbriata* | Brown | 5.53 g C/m²/day | Canary Islands (NP) | 3 |
| *Dictyota dichotoma* | Brown | 21.36 g C/m²/day | Canary Islands (NP) | 3 |
| *Egregia* sp. | Brown | 9.13 kg C/m²/year | Australia (NP) | 1 |
| *Eucheuma spinosum* | Brown | 7.41 kg dry wt/m²/year | Philippines (commercial) | 8 |
| *Eucheuma striatum* | Brown | 7.41 kg dry wt/m²/year | Sacol I. Philippines (commercial) | 8 |
| *E. striatum* | Brown | 3.21 kg dry wt/m²/year | Tapean I. Philippines (commercial) | 8 |
| *Eucheuma* sp. | Brown | 1.53 kg dry wt/m²/year | (NP) | 7 |
| *Fucus* sp. | Brown | 19—42 g dry wt/m²/day | Mediterranean (NP) | 1 |
| *Fucus* sp. | Brown | 6.9 kg dry wt/m²/year | Mediterranean (NP) | 2 |
| *Fucus* sp. | Brown | 640—840 g C/m²/year | Nova Scotia (NP) | 2 |
| *Fucus* sp. | Brown | 0.84 kg dry wt/m²/year | Nova Scotia (NP) | |
| *Galaxaura Squalida* | Brown | 3.89 g C/m²/day | Canary Islands (NP) | 3 |
| *Gelidium* sp. | Red | 0.6 kg dry wt/m²/year | Japan (commercial) | 7 |
| *Gelidium* spp. | Red | 1.0 kg dry wt/m²/year | Japan (commercial) | 9 |

## Table 1 (continued)
## MARINE PLANT PRODUCTION

| Species | Phylum | Production | Location[a] | Ref. |
|---|---|---|---|---|
| *Gigartina exasperata* | Red | 7.6 kg dry wt/m²/year | Washington (state) (NP) | 10 |
| *Gigartina* sp. | Red | 54 g dry wt/m²/day | Mediterranean (NP) | 1 |
| *Gloeopeltis* spp. | Red | 0.05 kg dry wt/m²/year | Japan (commercial) | 9 |
| -*Gloeopeltis* spp. | Red | 0.03 kg dry wt/m²/year | Japan (commercial) | |
| *Gracilaria edulis* | Red | 0.23 kg dry wt/m²/year | India (commercial harvest) | 11 |
| *G. foliifera* | Red | 3.3 kg dry wt/m²/year | Florida (waste recycling polyculture) | 12 |
| *Gracilaria* spp. | Red | 0.2 kg dry wt/m²/year | Taiwan (commercial) | 9 |
| *Gracilaria* sp. | Red | 2.8 kg dry wt/m²/year | — | 7 |
| *Gracilaria* sp. | Red | 15.7 kg dry wt/m²/year | Japan | 13 |
| *Gracilaria* sp. | Red | 6.21 kg dry wt/m²/year | Florida (culture) | 14 |
| *Halopteris scoparia* | Red | 8.28 g C/m²/day | Canary Islands (NP) | 3 |
| *Hypnea musciformis* | Red | 4.38 kg dry wt/m²/year | Florida (culture) | 14 |
| *Iridea cordata* | Red | 1.95 kg ash free dry wt/ m²/year | U.S. Pacific (culture) | 10 |
| *I. cordata* | Red | 7.6 kg dry wt/m²/year | Washington (state) (NP) | 10 |
| *I. cordata* | Red | 4.6 kg dry wt/m²/year | British Columbia, (NP) | 10 |
| *Iridophycus* sp. | Red | 19 g dry wt/m²/day | Mediterranean (NP) | 1 |
| *Laminaria hyperborea* | Brown | 2.13 kg dry wt/m²/year | Scotland (NP) | 15 |
| *Laminaria laponica* | Brown | 0.2 kg dry wt/m²/year | Japan and China (commercial) | 16 |
| *L. iaponica* | Brown | 1.5 kg dry wt/m²/year | Japan and China (commercial) | 7 |
| *Laminaria* sp. | Brown | 66 g dry wt/m²/day | Mediterranean (NP) | 1 |
| *Laminaria* sp. | Brown | 1.2—2.0 kg C/m²/year | Nova Scotia (NP) | 2 |
| *Laminaria* spp. | Brown | 7.5 mg C/m²/day | California (NP) | 17 |
| *Macrocystis pyrifera* | Brown | 400—800 g C/m²/year | California (NP) | 18 |
| *M. pyrifera* | Brown | 1.1—2.2 kg C/m²/year | Indian Ocean (NP) | 2 |
| *M. pyrifera* | Brown | 14.6 g C/m²/day | California (NP) | 7 |
| *Neoagardhiella baileyi* | Red | 5.5 kg dry wt/m²/year | Florida (waste recycling polyculture) | 12 |
| *N. baileyi* | Red | 4.6 kg dry wt/m²/year | — | |
| *Padina pavonia* | Brown | 7.56 g C/m²/day | Canary Islands (NP) | 3 |
| *Pelvetia* sp. | Brown | 35 g dry wt/m²/day | Mediterranean (NP) | 1 |
| *Pocuckiella variegata* | Brown | 7.28 g C/m²/day | Canary Islands (NP) | 3 |
| *Porphyra tenera* | Red | 3.2 kg dry wt/m²/year | Japan (commercial) | 9 |
| *P. tenera* | Red | 1.82 kg dry wt/m²/year | Japan (commercial) | 18 |
| *Porphyra* spp. | Red | 0.5 kg dry wt/m²/year | Japan (NP) | 7 |
| *Porphyra* sp. | Red | 11—21 g dry wt/m²/day | Mediterranean (NP) | 1 |
| *Rhizophoea* sp. | Mangrove | 400 g C/m²/year | Florida (NP) | 2 |
| *Ruppia maritima* | Flowering | 6.4 mg C/m²/day | North Carolina (NP) | 17 |
| *Sargassum* spp. | Brown | 3.13 kg dry wt/m²/year | Hawaii (NP) | 19 |
| *Sargassum* spp. | Brown | 6.4 mg C/m²/day | North and South Carolina (NP) | 17 |
| *Spartina alterniflora* | Flowering | 200 g C/m²/year | Delaware and New Jersey (NP) | 2 |
| *S. alterniflora* | Flowering | 897 g C/m²/year | Georgia (NP) | 2 |
| *S. alterniflora* | Flowering | 1.4 g C/m²/day | North Carolina (NP) | 17 |
| *Thalassia testudinum* | Flowering | 2.17 g dry wt/m²/day | Jamaica (Harvest) | 20 |
| *T. testudinum* | Flowering | 7.85 kg dry wt/m²/year | Florida (NP) | 21 |
| *T. restudinum* | Flowering | 5.15 kg dry wt/m²/year | Bermuda (NP) | 20 |
| *T. testudinum* | Flowering | 3.8 kg dry wt/m²/year | Card Sound, Florida (NP) | 20 |
| *T. testudinum* | Flowering | 1.8 kg dry wt/m²/year | Biscayne Bay, Florida (NP) | 20 |

## Table 1 (continued)
## MARINE PLANT PRODUCTION

| Species | Phylum | Production | Location[a] | Ref. |
|---|---|---|---|---|
| *T. testudinum* | Flowering | 1.8 kg dry wt/m²/year | Playa Habana, Cuba (NP) | 20 |
| *T. testudinum* | Flowering | 1.7 kg dry wt/m²/year | Jamaica (NP) | 20 |
| *Thalassia* sp. | Flowering | 0.6—1.0 kg C/m²/year | Indian Ocean (NP) | 2 |
| *Thalassia* sp. | Flowering | 1.0 kg C/m²/year | Caribbean (NP) | 2 |
| *Ulva lactuca* | Green | 2.63 kg dry wt/m²/year | Mediterranean (NP) | 22 |
| *Ulva* sp. | Green | 3—7.2 g dry wt/m²/day | Mediterranean (NP) | 1 |
| *Undaria pinnatifida* | Brown | 0.1 kg dry wt/m²/year | Japan (commercial) | 16 |
| *Undaria* sp. | Brown | 0.43 kg dry wt/m²/year | Japan (commercial) | 6 |
| *Zostera marina* | Flowering | 0.35 g C/m²/day | Canary Islands (NP) | 3 |
| *Z. marina* | Flowering | 1.0—1.5 kg C/m²/year | Bering Sea (NP) | 23 |
| *Z. marina* | Flowering | 11.0 mg C/m²/day | North Carolina (NP) | 17 |
| *Z. marina* | Flowering | 340 g C/m²/year | Denmark (NP) | 2 |
| *Z. marina* | Flowering | 1.5 kg C/m²/year | U.S. East Coast (NP) | 2 |
| Mixed phytoplankton | — | 5 g C/m²/day 1.0 g N/m²/day | Virgin Island (culture) | 4 |
| Mixed phytoplankton | — | 8.7 g dry wt/m²/day 2.6 g C/m²/day | Woods Hole (culture) | — |
| Mixed phytoplankton | — | 4—645 mg C/m²/day | French Polynesia (NP) | — |
| Mixed phytoplankton | — | 50—830 mg C/m²/day | Sargasso Sea (NP) | 17 |
| Mixed phytoplankton | — | 213—229 g C/m²/year | Nova Scotia | 2 |

[a]  NP denotes natural populations.

Gross production is defined as the amount of energy bound or carbon fixed. Net production is defined as gross production minus metabolic utilization of energy or fixed carbon. Finally, yield is the amount of plant material actually harvested by man. Production often is measured by different methods which are not directly comparable. The most common methods are the measurement of $O_2$ concentrations around the plants or the change in total biomass over time. Very often, the above definitions and methods are used without a clear explanation of what was done.

## Table 2
## YEARLY HARVESTS[25]

| Genus | Yield (wet tons) | Country | Year |
|---|---|---|---|
| *Ascophyllum* | 40,000 | Ireland | 1965 |
| *Ascophyllum* | 1,827 | France | 1967 |
| *Chordrus* and *Iridea* | 8,000 | Japan | 1960—1970 |
| *Eisenia* and *Ecklonia* | 118,000 | Japan | 1941—1945 |
| *Gelidium* | 18,000 | Japan | 1966—1970 |
| *Gloiopeltis* | 2,000 | Japan | 1966—1970 |
| *Gracilaria* | 12,000 | Japan | 1956—1960 |
| *Laminaria* | 37,521 | China | 1958 |
| | 5,800 | France | 1965 |
| | 167,000 | Japan | 1966—1970 |
| | 60,000 | Norway and Great Britain | 1971 |
| *Macrocystis* | 135,129 | U.S. | 1965 |
| *Monostroma* | 15,000 | Japan | 1966—1970 |
| *Porphyra* | 176,000 | Japan | 1966—1970 |
| *Undaria* | 123,000 | Japan | 1966—1970 |
| Chondrus | 22,000 | Canada | 1970 |
| | 1,200 | France | 1965 |
| | 3,600 | Ireland | 1965 |
| | 400 | U.S. | 1970 |

## REFERENCES

1. **Gessner, F. and Hammer, L.,** Die primar production in Mediterranean *Caulerpa — Cymodocea — Wiesen,* Bot. Mar., 2, 157, 1960.
2. **Mann, K. H.,** Seaweeds: their productivity and strategy for growth, *Science,* 182 (4116), 975, 1973.
3. **Johnston, C. S.,** The ecological distribution and primary production of marophytic marine algae in the Eastern Canaries, *Int. Rev. Ges. Hydrobiol.,* 54, 473, 1969.
4. **Roels, D. A., Haines, K. C., and Sunderlin, J. B.,** *Proc. 10th Eur. Marine Biology Symp.,* Ostend, Belgium, 1976.
5. **McGarry, M. G.,** *Proc. 7th Int. Seaweed Symp.,* Sapporo, Japan, 1972, 401.
6. **Ryther, J. H.,** The Use of Flowing Biological Systems in Aquaculture Sewage Treatment, Assay, and Food Chain Studies, Progress report, Woods Hole, Mass., 1971.
7. **Jackson, G. A.,** Biological constraints on seaweed culture, in Mitsui, A. Eds., *Biological Solar Energy Conversion,* Academic Press, New York, 1977.
8. **Parker, H. S.,** The culture of the red algae, genus *Eucheuma,* in the Philippines, *Aquaculture,* 3, 425, 1971.
9. **Tamura, T.,** Marine Aquaculture (Translation), NTIS PB, 1940, 51, 1966.
10. **Waaland, J. R.,** Growth of the red algae *Iridea cordata, J. Exp. Mar. Biol. Ecol.,* 23, 4, 1976.
11. **Raja, P. V. and Thomas, P. S.,** Experimental field cultivation of *Gracilaria edulis* (Mel.) Silva, *Bot. Mar.,* 14, 71, 1971.
12. **De boer, J. A., Ryther, J. H., and La Pointe, B. E.,** Yields of Seaweeds *Neoagardhiella bailey* and *Gracilaria foliifera* in a Waste Recycling Polyculture System, Woods Hole Oceanographie Institute Tech. Rep. No. 76—92, Woods Hole, Mass., 1976.
13. **Jackson, G. A.,** Marine Biomass production through seaweed aquaculture, in *Biochemical and Photosynthetic Aspects of Energy Production,* San Pietro, A., Ed., Academic Press, New York, 1976.
14. **La Pointe, B. E., Williams, L. D., Goldman, J. C., and Ryther, J. H.,** The mass outdoor culture of macroscopic marine algae, 8(1), 9, 1976.
15. **Jupp, B. P. and Drew, D. S.,** Studies on the growth of *Laminara hyperborea* (Gann.) Fosl. I. Biomass and productivity, *J. Exp. Mar. Biol. Ecol.,* 15, 185, 1974.
16. **Bardach, J. E., Ryther, J. H., and McLarney, W. O.,** *Aquaculture: The Farming and Husbandry of Freshwater and Marine Organisms,* John Wiley & Sons, New York, 1972.

17. Howard, K. L. and Menzies, R. J., Distribution and production of *Sargassum* in the waters off the Carolina coast, *Bot. Mar.*, 12, 244, 1969.
18. Glendenning, K. A., *Proc. 4th Int. Seaweed Symp.*, 40, 55, 1963.
19. Minata, D. and Keller, M., Marine plantations, in *Biomass Energy for Hawaii*, Vol. 4, Stanford Institute for Energy Studies, Stanford University, Stanford, Calif., 1977.
20. Greenway, M., The effects of cropping on the growth of *Thalassia testudinum* (Konig) in Jamaica, *Aquaculture*, 4, 199, 1974.
21. Bauersfeld, P., Dunant, N. W., and Sykes, J. E., *Proc. 6th Int. Seaweed Symp.*, Santiago, Chile, 1969, 637.
22. Haq, Q. N. and Percival, E., *Proc. 5th Int. Seaweed Symp.*, Halifax, N. S., 1966, 261.
23. McRoy, P. C. and Barksdale, R. J., Phosphate absorption in eelgrass, *Limnol. Oceanogr.*, 15, 6, 1970.
24. Platt, T. and Irwin, B., Caloric content of phytoplankton, *Limnol. Oceanogr.*, 18, 306, 1973.
25. Jackson, G. A. and North, W. J., Concerning the Selection of Seaweeds Suitable for Mass Cultivation in a Number of Large, Open-Ocean Solar Energy Facilities ("Marine Farms") in Order to Provide a Source of Organic Matter for Conversion to Food, Synthetic Fuels, and Electrical Energy, U.S. Naval Weapons Center, China Lake, 1973.

# PRIMARY PRODUCTIVITY OF MACROFLORA IN THE ATLANTIC, NORTH SEA, AND MEDITERRANEAN

## I. T. Show Jr.

The rocky coasts of the North Atlantic all show the same basic form of zonation and productivity.[1,2] Overall, brown algae are the most conspicuous, both above low water and in the subtidal zone. Typically, there is also a carpet of moss-like red algae near low water. *Pelvetia canaliculata* is usually found growing near high water, with *Fucus spiralis* or *Fucus platycarpus* growing just below. *Ascophyllum nodosum* or *Fucus vesiculosus*, is usually found at mid water, and *Fucus serratus, Chondrus crispus, Gigartina stellata, Porphyra umbilicus*, and *Rhodymenia palmata* are found at low water. *Laminaria digitata, Laminaria saccharina, Saccorhiza bulbosa*, and *Alaria esculenta* are found below low water.

The greatest single contribution to macroflora production in the North Atlantic is the large brown algae, *Laminaria*. Wherever this plant is found, its production tends to be high: 595 g $C/m^2$/year in Scotland,[3] 1200 to 2000 g $C/m^2$/year in Nova Scotia,[4] and as high as 6 kg $C/m^2$/year in some areas of the Mediterranean.[5] *L. digitata* and *L. saccharina* are the most conspicuous members of the subtidal community. *Fucus* is another major contributor, ranging from 640 to 840 g $C/m^2$/year in Nova Scotia,[4] 2400 g $C/m^2$/day in the Canary Islands,[3] to 13 g $C/m^2$/day in Scotland.

It is relatively difficult to draw any generalities on broad patterns from the data available for rocky coast algae communities. However, it would appear that the coast of North America north of Cape Cod shows somewhat higher production than northern Europe. The North American side seems to consistently approach a production of 1800 to 2000 g $C/m^2$/year, while the European side tends to consistently approach about 1000 g $C/m^2$/year. The Mediterranean Sea seems to have high but variable production, with a maximum production range of 2000 to 6000 g $C/m^2$/year.

Soft bottom shores are not conducive to the growth of marine algae. These areas are dominated in the North Atlantic by spermatophytes which are only secondarily marine. *Spartina* forms salt marshes at or just below high water. *Zostera* forms dense beds at or below low water. *Ruppia* and *Thalassia* are found in subtidal zones in the tropics. Mangroves occur at high tide throughout the tropics.

*Spartina* production data are available for the western North Atlantic and show production to vary from 200 g $C/m^2$/ year in Delaware and New Jersey to 897 g $C/m^2$/year in Georgia.[4] About 500 g $C/m^2$/year was also recorded for North Carolina.[7] There appears, therefore, to be a steady increase in the western North Atlantic from north to south. *Zostera*, on the other hand, tends to increase from south to north and tends to be higher on the North American side. Annual production ranges from 340 g $C/m^2$ in Denmark to 1500 g $C/m^2$ north of Cape Cod.[4] Daily production figures range from 0.11 g $C/m^2$ in North Carolina[7] to 0.35 g $C/m^2$ in the Canary Islands.[6]

Only one value was found for *Ruppia*: 6.4 mg $C/m^2$/day in North Carolina.[7] However, *Thalassia* has been studied extensively. This form grows in dense subtidal beds in the tropics. Production values range from 1.7 kg dry wt/$m^2$/year in Jamaica[8] to 5.15 kg dry wt/$m^2$/year in Florida[9] and 7.85 kg dry wt/$m^2$/year in Bermuda.[8] In general, production in the Caribbean might be as high as 1000 g $C/m^2$/year.[4]

# REFERENCES

1. Dawson, E. Y., *Marine Botany: An Introduction,* Holt, Reinhart & Winston, New York, 1966.
2. Chapman, V. I., *Seaweeds and Their Uses,* 2nd ed., Methven, London, 1970.
3. Jupp, B. P. and Drew, D. S., Studies on the growth of *Laminaria hyperborea* (Gann.) Fosl. I. Biomass and productivity, *J. Exp. Mar. Biol. Ecol.,* 15, 185, 1974.
4. Mann, K. H., Seaweeds: their productivity and strategy for growth, *Science,* 182 (4116), 975, 1973.
5. Gessner, F. and Hammer, L., Die primar production in Mediterranean *Canlerpa-cymodocea-Wiesen,* *Mar. Bot.,* 2, 157, 1960.
6. Johnston, C. S., The ecological distribution and primary production of macrophytic marine algae in the eastern Canaries, *Int. Rev. Ges. Hydrobiol.,* 54, 473, 1969.
7. Howard, K. L. and Meszies, R. J., Distribution and production of *Sargassum* in the waters off the Carolina coast, *Bot. Mar.,* 12, 244, 1969.
8. Greenway, M., The effects of cropping on the growth of *Thalassia testudinum* (Konig) in Jamaica, *Aquaculture,* 4, 199, 1974.
9. Bauersfeld, P., Dunant, N. W., and Sykes, J. E., Proc. 6th Int., Seaweed Symp., Santiago, Chile, 1969, 637.

# PRIMARY PRODUCTIVITY OF MACROALGAE IN FLORIDA, THE CARIBBEAN, AND THE SOUTH ATLANTIC

## Clinton J. Dawes

## INTRODUCTION

Studies on the productivity of macroalgae are increasingly available, especially for temperate and cold water floras of the North Atlantic coasts. However, few studies are available for macroalgae of the South Atlantic (Florida, Caribbean, Cuba, and South American coasts). This is in sharp contrast to the increasing number of investigations dealing with seagrass productivity, especially *Thalassia* (Turtle grass) of the Caribbean.[1]

Most studies are concerned with a single genus (e.g., *Laurencia, Hypnea,* and *Euchema*) and not with the entire flora of a site. The studies center in the Caribbean with no evident publication for the eastern or western coasts of the southern hemispheric Atlantic (see Table 1). As might be expected, because of the limited number of studies and variability of techniques, there is some conflict in productivity estimated for macroalgae.

## PHYTOGEOGRAPHIC CONSIDERATIONS

Van den Hoek[2] has described the tropical western Atlantic (Florida, Caribbean, and southern Antilles) as containing the richest macroalgal flora in the northern hemispheric Atlantic (760 species according to Taylor).[3] In his phytogeographic study, van den Hoek[2] ranked the warm temperate eastern Atlantic (northwestern Africa), including the Canary Islands, second since this algal flora is almost as rich, but still poorly known, and apparently quite distinct from the western flora. These two tropical to warm water floras of the Atlantic are the richest floras of the North Atlantic because of impoverishment of the coastal floras of the North Atlantic by glacial damage since the Pleistocene.[2]

Joly[4] presented the first floristic account of marine algae of the Atlantic coast of Latin America in which he listed 274 genera. This work now needs to be expanded to species and a more detailed ecological treatment be included for the dominant algal habitats. It is apparent that the Latin American algae grade from tropical to temperate forms in the southwestern Atlantic, but abundance or defined phytogeographic zones are not yet established. No such major floristic studies are available for the eastern Atlantic or African coast. Simons[5] lists 255 genera for South Africa, but he points out that some of these are not certain and probably many more species exist. Lawson[6] described intertidal zonation for the gold coast of Africa, and John et al.[7] published a quantitative study of subtidal algae off the Ghana coast. Recently, Lawson[8] has demonstrated through quantitative studies that there are two clearly defined groupings of marine algae representing the eastern Atlantic and the western Atlantic. On the western side, the floras are relatively closely knit. However, it is evident that more detailed floristic information is required and productivity data are limited.

## PRODUCTIVITY

In terms of biomass, macroalgae vary widely as to the amount present and rates of production (see Table 1). Calculations based on field studies for the red algae *Laurencia* from Florida show a range from 8.1 to 270 g dry weight/m²/year. Photosynthetic

## Table 1
## PRIMARY PRODUCTIVITY OF MACROALGAE OF THE SOUTH ATLANTIC, FLORIDA, AND THE CARIBBEAN

| Site | Species | Biomass data | Net productivity | | Ref. |
| --- | --- | --- | --- | --- | --- |
| | | | Oxygen production | Carbon fixation | |
| Florida (Card Sound) | *Laurencia* | 8.1 g dry wt/m²/year | — | — | 18[a] |
| Florida (Card Sound) | *Laurencia* | 2190 g dry wt/m²/year | — | — | 19[b] |
| | *Penicillus* and *Halimeda* | 720 g dry wt/m²/year | | | |
| Florida (Keys, west coast) | *Eucheuma* (3 populations) | 10—50 g dry wt/m²/year | 3—8 mL O₂/g dry wt/hr | — | 9[c] |
| Jamaica (coral reef) | *Halimeda* | — | 6—9 mg O₂/L/hr | 839 g C/m²/year | 15[d] |
| Florida (west coast) mangrove vs. open coast | *Hypnea* (2 populations) | — | 4—6 mL O₂/g dry wt/hr | — | 10[e] |
| Curaco (coral reef) | Encrusting Corallines | — | 0.11 g O₂/m²/hr | 715 g C/m²/year | 16[f] |
| | Fleshy/filamentous forms | | 0.14 g O₂/m²/hr | 1204 g C/m²/year | |
| Curaco (coral reef) | *Sargassum* (attached other browns) | 229 g/m²/year 194 g/m²/year | 1.32 g/m²/hr 1.06 g O₂/m²/hr | 2550 g C/m²/year (total) | 17[g] |
| Florida (Card Sound) | *Laurencia* | 21 g—270 g dry wt/m²/year | — | — | 20[h] |
| Florida (west coast) mangrove and salt marsh | *Bostrychia, Catenella, Gracilaria, Spyridia* | — | 4—12 m O₂/g dry wt/hr | — | 11[i] |
| Cuba | 85 species: 39 Chlorophyta 23 Phaeophyta 23 Rhodophyta | — | 5.3 mg O₂/g dry wt/hr 7.6 mg O₂/g dry wt/hr 6.6 mg O₂/g dry wt/hr | — | 12[j] |

| | | | | | |
|---|---|---|---|---|---|
| Florida (west coast) | *Bostrychia* | 18 g dry wt/m²/year | 7.5 mL O₂/g dry wt/hr | 58.4 g C/m²/year | 13[k] |
| | *Gracilaria* | 24 g dry wt/m²/year | 1.1 mL O₂/g dry wt/hr | 65.7 g C/m²/year | 14[l] |
| Florida (east and west coast) | *Hypnea* (2 populations) | — | 3—7 mL O₂/g dry wt/hr | — | |

[a] Data based on laboratory growth studies and square meters estimated from these growth studies; agrees well with (c).

[b] The data presented are based only on estimated field production. Note the high levels of biomass.

[c] Oxygen data obtained from Mathieson and Dawes[21] from manometric techniques.

[d] Oxygen meter was used in studies and a photosynthetic rhythm was noted over continuous monitoring of 24 hours.

[e] A manometric study of two populations of *Hypnea* from an open coast and a mangrove swamp.

[f] Winker titration used to determine oxygen levels and grams C calculated from the values. A series of runs was made to establish average daylight productivity. Encrusting corallines included genera of a blue-green, 9 genera of green, 3 genera of brown, and 17 genera of red algae were included under the fleshy/filamentous forms.

[g] Winkler technique was used, and daily rates were based on a morning and afternoon sampling period. One species of *Sargassum* and three other brown algae were studied.

[h] The study was carried out over a period of 1 year in Card Sound.

[i] Manometric procedures were used in a study of summer plants from a mangrove swamp and a salt marsh.

[j] Winkler titration procedures carried out during a single period of the day; the study was nonseasonal.

[k] A comparative study of two populations from a salt marsh and a mangrove swamp using field procedures (Winkler titration) and laboratory tests (manometric, ¹⁴CO₂). Daily photosynthetic rhythms were found in the field and in the laboratory, thus stressing the need for a number of samplings over the photoperiod to avoid errors in estimation of productivity.

[l] Manometric procedures were used in a seasonal study on an east and west coast population.

studies, measuring oxygen output, are difficult to compare, since both manometric and Winkler titration methods were employed. Furthermore, the oxygen production was based either on a g dry weight of plant material used[9-14] or per square meter. Carbon fixation per square meter per year was the only measurement that could be compared among studies.[13,15-17] However, the grams of C fixed varied widely, since the studies had different goals ranging from the total productivity of a reef flat to the contribution of a single red algal component of a mangrove swamp or a salt marsh. The investigation of Wanders[16,17] at Curacao is probably the most useful for purposes of comparison of macroalgal production with other plant communities. The net production is 25 to 65 times higher than that of the phytoplankton of the open Caribbean Sea, and it appears to be about 2 times higher than that of a rice field (1460 g C/m²/year) or a pine forest (912 g C/m²/year).[16,17]

In summary, productivity data on macroalgae are sparse, localized mostly in the Caribbean, and difficult to compare between the sites. It is evident that investigations are needed, especially in the south Atlantic. These new studies should present grams C per square meter per year data for entire populations of macroalgae and also include seasonal information, as well as determining whether rhythmicity of carbon fixation may be occurring during the day. Rhythmicity is especially critical, since Hoffman[13] has demonstrated that a calculation based on midday rates for *Gracilaria* would result in an 84% error in estimation of productivity for the 24-hr period.

## REFERENCES

1. **Dawes, C. J., Bird, K., Durako, M., Goddard, R., Hoffman, W., and McIntosh, R.,** Chemical fluctuations due to seasonal and cropping effects on an algal-seagrass community, *Aquat. Bot.,* 6, 79, 1979.
2. **van den Hoek, C.,** Phytogeographic provinces along the coasts f the northern Atlantic Ocean, *Phycology,* 14, 317, 1975.
3. **Taylor, W. R.,** *Marine Algae of the Eastern Tropical and Subtropical Coasts of the Americas,* University of Michigan Press, Ann Arbor, 1960.
4. **Joly, A. B.,** *Generos de Algas Marinhas de Coasta Atlantica Latino-Americana,* Editoria da Universidade de Sao Paulo, Sao Paulo, 1967.
5. **Simons, R. H.,** Seaweeds of Southern Africa: Guide-Lines for Their Study and Identifications, Report of the South African Department of Industrial Fisheries, Bull. No. 7, 1976.
6. **Lawson, G. W.,** Rocky shore zonation on the Gold Coast, *Ecology,* 44, 153, 1956.
7. **John, D. M., Lieberman, D., and Liberman, M.** A quantitative study of the structure and dynamics of benthic subtidal algal vegetation in Ghana (tropical west Africa), *J. Ecol.,* 65, 497, 1977.
8. **Lawson, G. W.,** The distribution of seaweed floras in the tropical and subtropical Atlantic Ocean: a quantitative approach, *Bot. J. Linn. Soc.,* 76, 177, 1978.
9. **Dawes, C. J., Mathieson, A. C., and Cheney, D. P.,** Ecological studies of floridian *Eucheuma* (Rhodophyta, Gigartinales). I. Seasonal growth and reproduction, *Bull. Mar. Sci.,* 24, 235, 1974.
10. **Dawes, C. J., Moon, R., and LaClaire, J.,** Photosynthetic responses of the red alga, *Hypnea musciformis* (Wulfen) Lamouroux (Gigartinales), *Bull. Mar. Sci.,* 26, 467, 1976.
11. **Dawes, C. J., Moon, R. E., and Davis, M. A.,** The photosynthetic and respiratory rates and tolerances of benthic algae from a mangrove and salt marsh estuary: a comparative study, *Estuarine Coastal Mar. Sci.,* 5, 175, 1977.
12. **Buesa, R. J.,** Photosynthesis and respiration of some tropical marine plants, *Aquat. Bot.,* 3, 203, 1977.
13. **Hoffman, W. E.,** Photosynthetic rhythms and primary production of two Florida benthic red algal species, *Bostrychia binderi* Harvey (Ceramiales) and *Gracilaria verrucosa* (Hudson) Papenfuss (Gigartinales), Masters thesis, University of South Florida, Tampa, 1978.
14. **Durako, M. J.,** A Comparative Study of the Physiological Ecology of Two Populations of *Hypnea musciformis* (Wulfen) Lamouroux, Masters thesis, University of South Florida, Tampa, 1978.

15. **Hillis-Colinvaux, L.,** Productivity of the coral reef alga *Halimeda* (Order Siphonales), in Proc. Sec. Int. Coral Reef Symp., Brisbane, 1974, 35.
16. **Wanders, J. B. W.,** The role of benthic algae in the shallow reef of Curacao (Netherlands Antilles). I. Primary productivity in the coral reef, *Aquat. Bot.* 2, 235, 1976.
17. **Wanders, J. B. W.,** The role of benthic algae in the shallow reef of Curacao (Netherlands Antilles). II. Primary productivity of the Sargassum beds on the north-east coast submarine plateau, *Aquat. Bot.*, 2, 327, 1976.
18. **Thorhaug, A. and Garcia-Gomez, J.,** Preliminary laboratory and field growth studies of *Laurencia* complex, *J. Phycol.,* 8(Abstr.), 10, 1972.
19. **Thorhaug, A., Segar, D., and Roessler, M.,** Impact of power plant on a subtropical estuarine environment, *Mar. Pollut. Bull.,* 11, 71, 1973.
20. **Josselyn, M. N.,** Seasonal changes in the distribution and growth of *Laurencia poitei* (Rhodophyceae, Ceramiales) in a subtropical lagoon, *Aquat. Bot.,* 3, 217, 1977.
21. **Mathieson, A. C. and Daves, C. J.,** Ecological studies of Floridian *Eucheuma* (Rhodophyta, Gigartinales). II. Photosynthesis and respiration, *Bull. Mar. Sci.,* 24, 274, 1974.

# PRIMARY PRODUCTIVITY OF MACROALGAE IN NORTH PACIFIC AMERICA*

## David Coon

## INTRODUCTION

The marine macroalgal communities of the west coast of North America are unique in several respects. Within these communites is the only occurrence of the largest kelp species, *Macrocystis*, in the northern hemisphere and the occurrence of other large kelp species, such as *Nereocystis* and *Pelagophycus*. In addition to large and productive kelp forests of various species, there are diverse subtidal and intertidal assemblages of smaller algae. These communities are exposed to strong seasonal upwelling along this western continental margin and occur in a wide variety of coastal exposures. As a result of their size, accessability, and production, the kelp forests of this coastline have been used as a source of commercial raw material on a large scale for many years. Dense populations of *Iridaea, Gelidium*, and *Gigartina* have also been examined for commercial utilization at various locations.

Because of the availability of information on species of commercial interest, they will provide the basis for this summary. Production values for natural populations and for populations in laboratory and field culture are included to give measures of potential as well as actual productivities. The production of *Macrocystis* will be emphasized because of the extensive information available from the long history of utilization of this species.

The large kelps form extensive subtidal "beds" along the shore, with almost continuous distribution from Alaska to Mexico.[1] *Nereocystis* is often the dominant plant in beds off British Columbia,[2] but does not dominate over most of the range, which is occupied by *Macrocystis*. *Nereocystis* does not extend below Point Conception, California and the commercially harvested beds of California are essentially all *Macrocystis*. Druehl[3] has described the general distribution pattern of the order Laminarales, containing the larger kelps, for the northeast Pacific region. *Iridaea* occurs from Alaska to Mexico and is often abundant on exposed coasts.[1] *Gelidium, Gigartina*, and *Gracilaria* species of commercial interest are found from British Columbia to Mexico, with *Gelidium* species occurring in the largest quantities in the more southern part of this range.[1]

Productivity of macroalgae can be considered in terms of its components, standing crop and growth rate (percent/day increase) of the plants. This approach is useful, since it includes measures which are more often available than total productivity measurements, and it allows projections and comparisons not otherwise possible. Productivity estimations, which can be based on short-term physiological measurements as well as actual field data, do not always consider factors related to actual sustainable yield, such as grazing and sloughing losses, seasonal changes in plant standing crops, and harvesting efferroduuion. However, in the absence of data on actual long-term yields from macroalgae under various conditions, such estimations are important especially when the plants are of interest for utilization as a resource.

* This research was sponsored in part by NOAA, National Sea Grant Program, Department of Commerce, under Grant # 04-8-MO1-189,Project # R/CZ — 46 through the California Sea Grant Program. The U.S. Government is authorized to produce and distribute reprints for governmental purposes, notwithstanding any copyright notation that may appear hereon.

Table 1
STANDING CROP VALUES OF KELP FORESTS

| Location | Species | Wet kg/m² | Wet ton/acre | Ref. |
|---|---|---|---|---|
| La Jolla, Calif. | *Macrocystis* | 6.0—10.0 | 26.7—44.5 | 4 |
| Southern Calif. and Pacific Mexico | *Macrocystis* | 3.0—22.0 | 13.4—98.2 | 5 |
| Paradise Cove, Calif. | *Macrocystis* | 4.4—5.8 | 19.6—25.8 | 6 |
| Goleta, Calif. | *Macrocystis* | 7.0—9.1 | 31.2—40.6 | 7 |
| Monterey, Calif. | *Macrocystis* | 5.9 | 26.3 | 8 |
| Monterey, Calif., | *Macrocystis* | 0.5—6.0 | 2.2—26.8 | 9 |
| Dundas Archipelago, B.C. | *Macrocystis* and *Nereocystis* | 4.2—4.7 | 18.7—21.0 | 2 |

Productivity will be discussed here in terms of grams carbon per square meter per day, dry grams per square meter per day, kilograms per square meter per year, and tons per acre per year. Standing crop will be discussed in terms of kilograms per square meter and tons per acre.

## STANDING CROP AND GROWTH RATE

Standing crops reported for kelp beds at various locations are listed in Table 1. Most of these values are single measurements, made at one time of the year, and may not be representative of an annual average. Coon[7] and Wheeler[10] have observed two ways in which the standing crop of *Macrocystis* varies in southern California. First, the total number of plants and fronds per unit area changes with seasonal storms and general growth conditions. Second, the weight of individual plant fronds changes per unit length with season. Spring frond weights per unit length may be twice those on the same plant during fall. With occasional severe losses of plant material because of storms, the mean value for a 2.5-year period of observation in northern California kelp beds[9] was 3.5 kg/m², which can be compared to the range shown in Table 1. Based on the values in Table 1, maximal values observed in natural kelp forests may reach 10 to 20 kg/m², and often reach 4 to 6 kg/m² wet weight. Typically, cultivated systems are more dense than wild stands of the same plant. However, *Macrocystis* already occurs in very dense concentrations relative to available substrate — the higher values in the table. Thus, while it is not unreasonable to project standing crops at some time in the future (with genetic changes, etc.) which exceed natural maxima, an initial projection for early cultivation efforts could be equal to maximal natural values.

Table 2 contains standing crop data reported for various red algal species in natural and cultivated populations. Overall, the range of values for standing crops of these plants are similar to the larger kelp species. Maximum values for both are around 10 to 20 kg/m² and populations may undergo large changes, often seasonally.[16]

The growth rates of the giant kelp, *Macrocystis*, have been determined by a number of botanists using various techniques to deal with the difficulties posed by the dimensions of the plant. The usual method is to randomly select fronds and to measure their length at intervals over a period of time. Since growth of individual fronds is not representative of the overall growth of the entire plant, only whole-plant measurements and estimations are considered here. Growth rates reported for macroalgae are shown in Tables 3 and 4.

The recurring values for the growth of *Macrocystis* are about 3%/day, or less, with the exception of the work by North[23] on juvenile plants in culture, where higher values were observed. Similarly, the high rates for certain red algae in Table 4 are those re-

## Table 2
### STANDING CROP VALUES FOR CERTAIN RED ALGAE

| Location | Species | kg/m² | Comments | Ref. |
|---|---|---|---|---|
| Georgia Strait, B.C. | *Iridaea cordata* | 2.0—4.0 | Fresh weight | 11 |
| San Juan Island, Wash. | *I. cordata* | 16.0—20.0 | Fresh weight, cultivated on nets | 12 |
| Seattle, Wash. | *I. cordata* | 4.0 | Fresh weight | 13 |
| San Juan Island, Wash. | *I. cordata* | 1.4—4.9 | Dry weight, cultivated | 14 |
| Central California | *I. cordata* | 1.2 | Dry weight | 15 |
| Central California | *I. cordata* | 0.2—1.4 | Dry weight | 16 |
| Southern California | *Gelidium robustum* | 0.5 | Wet weight | 17 |
| Baja, Calif. | *G. robustum* | 1.1—2.1 | Wet weight, over large areas | 18 |
| Southern California | *G. robustum* | 0.6—20.0 | Wet weight | 19 |
| Vancouver Island, B.C. | *Gracilaria* sp. | 4.6 / 0.1 | Intertidal, sheltered subtidal, exposed | 20 |
| Seattle, Wash. | *Gigartina exasperata* | 4.8 | In culture, fresh weight | 13 |

## Table 3
### GROWTH RATES OF *MACROCYSTIS*

| Location | Growth/day (%) | Comments | Ref. |
|---|---|---|---|
| Los Angeles, Calif. | 3.0 | | 21 |
| La Jolla, Calif. | 2.9 | Juvenile plants | 22 |
| Southern California | 4.8—10.9 | Juvenile plants | 23 |
| Goleta, Calif. | 0.9—1.4 | | 7 |
| Goleta, Calif. | 5.7 | Theoretical rate, based on photosynthetic studies | 10 |

## Table 4
### GROWTH RATES OF CERTAIN RED ALGAE

| Location | Species | Growth/day (%) | Comments | Ref. |
|---|---|---|---|---|
| Clam Bay and Seattle, Wash. | *Iridaea cordata* | 9.5 | Fresh weight | 13 |
| | *Iridaea heterocarpus* | 8.1 | in culture, all | |
| | *Iridaea cornucopia* | 8.7 | maximum | |
| | *Gigartina exasperata* | 8.3 | values | |
| Seattle, Wash. | *G. exasperata* | 3.1 | Fresh weight, average value | 13 |
| Vancouver Island, B.C. | *Gracilaria* sp. | 4.0 | Fresh weight, *in situ* culture | 20 |

### Table 5
### ESTIMATION OF *MACROCYSTIS* PRODUCTIVITY IN CALIFORNIA

| Location | g C/m²/day | Dry g/m²/day[a] | Wet kg/m²/day | Wet ton/acre/year | Ref. |
|---|---|---|---|---|---|
| **Estimates Based Primarily on Field Measurements of Plants** | | | | | |
| Santa Barbara | | 7.1 | 21.5[b] | 95.8 | 7 |
| Monterey | | 7.6 | 23.0 | 102.6 | 9 |
| Palos Verdes | | 9.9 | 30.0 | 133.8 | 24 |
| Southern California | | 6.6 | 20.1 | 89.5[c] | 25 |
| **Estimates Based Primarily on Physiological Data** | | | | | |
| Paradise Cove | 4.7—5.8[d] | 24.7—30.5 | 75.0—92.6 | 335—414 | 6 |
| Santa Barbara | 6.1[d] | 32.0 | 97.4 | 435 | 10 |
| San Diego | 9.5 | 49.9 | 151.7 | 676 | 26 |
| Monterey | 6.8 | 35.7 | 108.6 | 485 | 8 |
| **Estimates Based on Actual Harvest Data** | | | | | |
| Southern California | | | 1.5 | 6.5 | 25 |
| Santa Barbara | | | 2.7 | 11.9 | 27 |

[a]   Based on dry = 0.12 wet.[28]
[b]   Based on 7 kg/m² standing crop.
[c]   Converted from dry ash-free, using values in Lindner, Dooley, and Wade.[28]
[d]   Oxygen values converted to carbon.[10]

ported as the maximum observed in controlled cultures. Overall growth rates in *Macrocystis* in southern California do not appear to vary significantly with season.[5,7,10] Definite periods of high and low growth rate have been observed in a kelp bed in northern California.[9]

## PRODUCTIVITY

Various approaches have been used in estimating the production of giant kelp. Estimates have been made using field measures of plant density, standing crop, turnover rates, and elongation of fronds.[7,9,24,25] Other estimates have been based primarily on physiological measurements made in the lab or field,[6,10,26,8] and estimates can be made using extrapolations from harvest data for large areas.[25,28] This variety of approaches used for *Macrocystis* reflects the difficulty in measuring actual production with marine plants of this size. Strictly for comparative purposes, estimates made from the above methods have been assembled, converted where necessary, and projected up to annual values in Table 5. The accuracy of the annual values is of course limited by the short time periods over which most productivity measurements are made. The measures of red algal productivity listed in Table 6 are based on actual weight changes observed in natural and cultivated populations.

The field-based values shown in Table 5 might be considered an estimate of the optimal net production for giant kelp in natural beds. The much higher values based on physiological measurements may be more indicative of the maximal values of production that might be obtained over short periods of time. In comparison with the other estimations, the harvest values are very low. This reflects the fact that the estimates are based on the total areas leased for harvest and thus include areas which may

## Table 6
## PRODUCTIVITY OF CERTAIN RED ALGAE

| Location | Species | Productivity | Comments | Ref. |
|---|---|---|---|---|
| Vancouver Island, B.C. | *Gracilaria (verrucosa)* | 2.0 g dry/m²/day | Subtidal plants | 20 |
| | | 4.7 g dry/m²/day | Intertidal plants | |
| Seattle, Wash. | *Gigartina exasperata* | 5.2 g dry/m²/day | Semiclosed culture | 13 |
| | | 9.0 g dry/m²/day | Strain M-11, semi-closed culture | |
| | *Iridaea cordata* | 10.8 g dry/m²/day | 1974, semiclosed culture | |
| | | 11.9 g dry/m²/day | 1976, semiclosed culture | |
| | | 1.7 g dry/m²/day | Natural populations, Pacific Northwest | |
| Georgia Strait, B.C. | *I. cordata* | 30.0 g fresh/m²/day | | 11 |
| San Juan Island, Wash. | *I. cordata* | 10.2—15.4 g dry/m²/day | On nets | 12 |
| | *I. cordata* | 20.0 g dry/m²/day | | |
| Central California | *I. cordata* | 19.3 g dry/m²/day | Maximum Values | 15 |

be sparse, inaccessible to harvesting equipment, or infrequently harvested. Based on the field and harvest values of Table 5, it appears that harvesting represents about 10% of the production of the commercially utilized kelp beds.

A more detailed look at long-term yields is provided in Table 7. Data provided by the Marine Resources Branch of the California Department of Fish and Game[29] allows the harvest information of Neushul[30] to be extended from 1970 to 1977. Kelp beds designated as open beds are subject to harvesting by any licensed harvesting firm, whereas leased beds are used only by the leaseholder. As Smith[29] points out, average production from unleased, open beds has dropped 33% below the 15-year average for all beds. On the other hand, the average production from leased beds has increased by 12% during the same period, perhaps reflecting better management. Commercial companies in California have been experiencing brief shortage in harvestable kelp during the last several years. Harvesting of kelp is also carried out in Mexico, and annual production totals are reported from 15 to 30 thousand wet tons.[31] If the U.S. kelp harvesting totals represent about 10% of the total production in commercially utilized beds, then the total annual production would be about 1.5 million wet tons per year in these areas. Using a different basis, Michanek[32] estimated the production of this region at 3 million tons per annum. In the northern kelp beds off British Columbia, Scagel[33] has estimated an available production of 0.75 to 1 million tons annually, of *Macrocystis* and *Nereocystis*. All these estimates have been derived from natural populations of wild plants, and the total production from the kelp beds along the coast of North America may be improved substantially in future years by increased management and by actual cultivation of these resources.

Table 7
## CALIFORNIA KELP HARVEST (IN WET TONS) 1916 TO 1977

| Year | Open beds | Leased beds | Total wet tons |
|------|-----------|-------------|----------------|
| 1916 | 134,537.00 | — | 134,537.00 |
| 1917 | 394,974.00 | — | 394,537.00 |
| 1918 | 395,098.00 | — | 395,098.00 |
| 1919 | 16,673.00 | — | 16,673.00 |
| 1920 | 25,464.00 | — | 25,464.00 |
| 1931 | — | 260.00 | 260.00 |
| 1932 | 301.61 | 10,013.08 | 10,314.69 |
| 1933 | 52.75 | 21,568.65 | 21,621.40 |
| 1934 | 1,827.21 | 14,052.57 | 15,879.78 |
| 1935 | — | 30,601.83 | 30,601.83 |
| 1936 | 14,336.70 | 34,980.34 | 49,317.04 |
| 1937 | 9,612.70 | 34,340.65 | 43,953.35 |
| 1938 | 18,284.06 | 29,413.25 | 47,697.31 |
| 1939 | 25,546.30 | 31,189.70 | 56,736.00 |
| 1940 | 33,322.20 | 25,682.20 | 59,004.40 |
| 1941 | 36,103.00 | 19,614.00 | 55,717.00 |
| 1942 | 44,880.11 | 17,017.65 | 61,897.76 |
| 1943 | 19,547.44 | 28,410.86 | 47,958.30 |
| 1944 | 22,710.50 | 30,319.40 | 53,029.90 |
| 1945 | 37,541.30 | 21,639.90 | 59,181.20 |
| 1946 | 60,384.90 | 30,683.60 | 91,068.50 |
| 1947 | 46,028.80 | 28,207.80 | 74,236.60 |
| 1948 | 50,966.10 | 27,675.00 | 78,641.10 |
| 1949 | 56,076.00 | 27,270.00 | 83,346.00 |
| 1950 | 49,955.00 | 50,647.00 | 100,602.00 |
| 1951 | 30,318.00 | 84,442.00 | 1,760.00 |
| 1952 | 37,906.00 | 72,252.50 | 110,158.50 |
| 1953 | 37,172.50 | 89,476.50 | 126,649.00 |
| 1954 | 40,269.00 | 65,946.50 | 106,215.50 |
| 1955 | 38,992.00 | 85,071.00 | 124,053.00 |
| 1956 | 35,476.50 | 82,339.00 | 117,815.50 |
| 1957 | 32,810.75 | 61,396.50 | 94,207.25 |
| 1958 | 41,105.75 | 72,955.25 | 114,061.50 |
| 1959 | 42,289.65 | 47,309.45 | 89,599.10 |
| 1960 | 61,914.95 | 58,384.85 | 120,299.80 |
| 1961 | 71,952.70 | 57,303.30 | 129,256.00 |
| 1962 | 86,227.70 | 54,005.10 | 140,232.80 |
| 1963 | 57,517.00 | 63,514.80 | 121,031.80 |
| 1964 | 35,592.90 | 91,660.85 | 127,253.75 |
| 1965 | 33,464.10 | 101,664.90 | 135,129.00 |
| 1966 | 11,100.70 | 108,363.20 | 119,463.90 |
| 1967 | 9,331.00 | 122,164.00 | 131,495.00 |
| 1968 | 20,388.20 | 114,465.15 | 134,853.35 |
| 1969 | 10,023.70 | 121,210.50 | 131,239.20 |
| 1970 | 8,543.00 | 118,496.00 | 127,039.00 |
| 1971 | 33,959.00 | 121,600.00 | 155,559.00 |
| 1972 | 13,270.00 | 149,241.00 | 162,511.00 |
| 1973 | 24,539.00 | 128,541.00 | 153,080.00 |
| 1974 | 37,994.00 | 132,187.00 | 170,181.00 |
| 1975 | 30,213.60 | 141,384.00 | 171,597.60 |
| 1976 | 27,297.48 | 131,092.00 | 158,371.48 |
| 1977 | 21,898.55 | 108,698.00 | 130,596.55 |

*Note:* No data is available of 1921 to 1930.

# REFERENCES

1. **Abbott, I. A. and Hollenberg, G. J.**, *Marine-Algae of California*, Standford University Press, Stanford, 1976.
2. **Field, E. J. and Clark, E. A. C.**, Kelp Inventory, 1976, Part 2, The Dundas Group, Fisheries Development Report No. 11, British Columbia Ministry of Recreation and Conservation, Marine Resources Branch, 1978.
3. **Druehl, L. D.**, The pattern of Laminariales distribution in the northeast Pacific, *Phycologia*, 9, 237, 1970.
4. **Aleem, A. A.**, The ecology of a kelp-bed in southern California, *Bot. Mar.*, 16, 38, 1973.
5. **North, W. J.**, Introduction and background, in *Biology of Giant Kelp Beds (Macrocystis)* in California, North, W. J., Ed., J. Cramer, Lehre, 1971, chap. 1.
6. **McFarland, W. N. and Prescott, J.**, Standing crop, chlorophyll content and *in situ* metabolism of a giant kelp community in southern California, *Publ. Inst. Mar. Sci. Univ. Tex.*, 6, 109, 1959.
7. **Coon, D. A.**, Studies of whole plant growth in *Macrocystis angustifolia*, *Bot. Mar.* 24, 19, 1981.
8. **Towle, D. W. and Pearse, J. S.**, Production of the giant kelp, *Macrocystis*, estimated by the *in situ* incorporation of $^{14}$C in polyethelene bags, *Limnol. Oceanogr.*, 18, 155, 1973.
9. **Gerard, V. A.**, Some Aspects of Material Dynamics and Energy Flow in a Kelp Forest in Monterey Bay, California, Ph.D. thesis, University of California, Santa Cruz, 1976.
10. **Wheeler, W. N.**, Ecophysiological Studies on the Giant Kelp, *Macrocystis*, Ph.D. thesis, University of California, Santa Barbara, 1978.
11. **Adams, R. W. and Austin, A.**, Potential yields of *Iridaea cordata* (Florideophyceae) in natural and artificial populations in the Pacific Northwest, in *Proc. 9th Int. Seaweed Symp.*, Jensen, A. and Stein, J. R., Eds., Science Press, Princeton, 1978, 499.
12. **Mumford, T. F.**, Field and laboratory experiments with *Iridaea cordata* (Florideophyceae) grown on nylon netting, in *Proc. 9th Int. Seaweed Symp.*, Jensen, A. and Stein, J. R., Eds., Science Press, Princeton, 1978, 515.
13. **Waaland, J. R.**, Growth of Pacific Northwest marine algae in semi-closed culture, in *Marine Plant Biomass of the Pacific Northwest Coast*, Krauss, R. W., Ed., Oregon State University Press, Corvallis, 1977, chap. 7.
14. **Mumford, T. F.**, Growth of Pacific Northwest marine algae on artificial substrates — Potential and practice, in *Marine Plant Biomass of the Pacific Northwest Coast*, Krauss, R. W., Ed., Oregon State University Press, Corvallis, 1977, chap. 8.
15. **Hansen, J. E.**, Studies on the population dynamics of *Iridaea cordata* (Gigartinaceae, Rhodophyta), in *Proc. 8th Int. Seaweed Symp.*, 1974, in press.
16. **Hansen, J. E. and Doyle, W. T.**, Ecology and natural history of *Iridaea cordata* (Rhodophyta; Gigartinacea): population structure, *J. Phycol.*, 12, 273, 1976.
17. **Silverthorne, W.**, Optimal production from a seaweed resource, *Bot. Mar.*, 20, 75, 1977.
18. **Guzman del Proo, S. A. and de la Campa de Guzman, S.**, *Gelidium robustum* (Florideophyceae), an agarophyte of Baja, California, Mexico, in *Proc. 9th Int. Seaweed Symp.*, Jensen, A. and Stein, J. R., Eds., Science Press, Princeton, 1978, 303.
19. **Woessner, J. W.**, unpublished data, 1979.
20. **Saunders, R. G. and Lindsay, J. G.**, Growth and enhancement of the agarophyte *Gracilaria* (Florideophyceae), in *Proc. 9th Int. Seaweed Symp.*, Jensen A. and Stein, J. R., Eds., Science Press, Princeton, 1978, 249.
21. **Sargent, M. C. and Lantrip, L. W.**, Photosynthesis, growth and translocation in giant kelp, *Am. J. Bot.*, 39, 99, 1952.
22. **Neushul, M. and Haxo, F. T.**, Studies on the giant kelp, *Macrocystis*, I. Growth of young plants, *Am. J. Bot.*, 50, 349, 1963.
23. **North, W. J.**, unpublished data, 1977.
24. **Kirkwood, P. D.**, Seasonal Patterns in the Growth of the Giant Kelp, *Macrocystis pyrifera*, Ph.D. thesis, California Institute of Technology, Pasadena, 1977.
25. **Clendenning, K. A.**, Organic productivity in kelp areas, in *Biology of Giant Kelp Beds (Macrocystis)* in California, North, W. J., Ed., J. Cramer, Lehre, 1971, chap. 12.
26. **Jackson, G. A.**, Nutrients and production of giant kelp, *Macrocystis pyrifera*, off southern California, *Limnol Oceanogr.*, 22, 979, 1977.
27. **Coon, D. A.**, unpublished data, 1978.
28. **Lindner, E., Dooley, C. A., and Wade, R. H.**, Chemical variation of chemical constituents in *Macrocystis pyrifera*, Naval Undersea Center Report, San Diego, 1977.
29. **Smith, E. J.**, personal communication, 1978.

30. **Neushul, M.,** The domestication of the giant kelp, *Macrocystis,* as a marine plant biomass producer, in Krauss, R. W., Ed., *The Marine Plant Biomass of the Pacific Northwest Coast,* Oregon State University Press, Corvallis, 1978, chap. 9.
31. **Guzman del Proo, S. A., de la Campa de Guzman, S. and Pineda Barrera, J.,** La cosecha de algas comerciales en Baja California, *Ser. Divulg. Inst. Nac. Pesca, Mex.,* 6, 1974, 15.
32. **Michanek, G.,** Seaweed resources of the ocean, *FAO Fish. Tech. Pap.,* 138, 1, 1975.
33. **Scagel, R. F.,** Marine plant resources of British Columbia, *Fish Res. Board Canada,* 127, 1, 1961.

# PRIMARY PRODUCTIVITY OF MACROALGAE IN JAPANESE REGIONS

## Y. Aruga

Macroalgae (seaweeds) and phanerogams, as well as phytoplankton, are essential primary producers in the coastal regions of the sea. Primary productivity of these plants is determined principally by the standing crop, production structure, metabolic activities (i.e., photosynthesis and respiration), and environmental conditions.

In Asian countries, especially Japan, various kinds of seaweeds are used extensively, principally directly or indirectly as foodstuffs for man. The most representative is Nori, red algae *Porphyra* spp., the most commonly cultivated edible seaweed. A number of reports on the production of useful seaweeds are available. However, there are very few analytical studies of primary productivity of macroalgae, including the useful seaweeds in the Asian areas.

The standing crop of macroalgae usually is reported in terms of fresh weight (wet weight), very few in dry weight, though not ash-free dry weight. Thus, comparison of standing crop is difficult, especially with the inclusion of calcareous algae. Often the results of photosynthesis and respiration measurements are reported in milligrams or milliliters $O_2$ on a dry weight basis, surface area basis, or chlorophyll *a* basis which also makes comparison difficult.

The data in Tables 1 to 11 are restricted to those obtained in Japanese regions.

## Table 1
## SEASONAL AND VERTICAL VARIATIONS IN THE STANDING CROP OF SEAWEEDS MEASURED IN OYASHIO CURRENT AREA, KUROSHIO CURRENT AREA, AND SETO INLAND SEA, JAPAN

| | Locality | Depth[b] | Dominant species | Standing crop (kg wet weight/m²) | | | | | | | | | | | | |
|---|---|---|---|---|---|---|---|---|---|---|---|---|---|---|---|---|
| | | | | Jan. | Feb. | Mar. | Apr. | May | June | July | Aug. | Sept. | Oct. | Nov. | Dec. | Jan. |
| Oyashio Current Area | Muroran | U | Gloiopeltis furcata | | | | | | 0.48 | | | | | 0.17 | | |
| | | M | Sargassum thunbergii | | | | | | 4.02 | | | | | 1.33 | | |
| | | L | Laminaria japonica | | | | | | 15.69 | | | | | 3.69 | | |
| | | L | S. thunbergii[a] | | | 4.80 | 10.56 | 9.90 | 9.60 | 13.86 | 10.20 | 5.94 | 4.92 | 4.02 | 4.08 | 4.74 |
| | | L | L. japonica[a] | | | 2.52 | 6.60 | 5.58 | 5.52 | 12.12 | 19.08 | 17.76 | | | | |
| | | L | Alaria crassifolia[a] | | | | | 13.44 | 13.56 | | 9.60 | | | | | |
| | | L | Chondrus yendoi[a] | | | 2.37 | 2.89 | 2.44 | 2.96 | 2.00 | 1.22 | 1.41 | 1.33 | 3.81 | 3.52 | 3.96 |
| Kuroshio Current Area | Manazuru | M | Hizikia fusiformis[a] | | | | | | | | 1.90 | | | | | |
| | | L-1 | Grateloupia livida | | | | | | | | 2.20 | | | | | |
| | | L-2 | | | | | | | | | 2.15 | | | | | |
| | | L-3 | | | | | | | | | 1.50 | | | | | |
| | | L-4 | | | | | | | | | 1.95 | | | | | |
| | | 1 m | Pachimeniopsis elliptica | | | | | | | | 1.35 | | | | | |
| | | 5 m | Gelidium amansii | | | | | | | | 2.00 | | | | | |
| | | 10 m | Dictyopteris undulata | | | | | | | | 1.05 | | | | | |
| | Inamuragasaki | L-1 | Laurencia sp.[a] | | | | | | | | 0.40 | | | | | |
| | | L-2 | Hypnea charoides | | | | | | | | 2.35 | | | | | |
| | | L-3 | Sargassum patens | | | | | | | | 2.20 | | | | | |
| | | 0 m | Eisenia bicyclis | | | | | | | | 5.75 | | | | | |
| | | 1 m | E. bicyclis[a] | | | | | | | | 4.80 | | | | | |
| | | 5 m | Ecklonia cava[a] | | | | | | | | 9.00 | | | | | |
| | | 10 m | E. cava[a] | | | | | | | | 3.25 | | | | | |
| | Arasaki | M | Ishige okamurai[a] | | | | | | | | 0.40 | | | | | |
| | | L-1 | H. fusiformis[a] | | | | | | | | 1.05 | | | | | |
| | | L-2 | H. fusiformis | | | | | | | | 0.45 | | | | | |
| | | 0 m | E. bicyclis | | | | | | | | 3.45 | | | | | |
| | | 1 m | E. bicyclis | | | | | | | | 6.55 | | | | | |
| | | 5 m | E. cava[a] | | | | | | | | 6.95 | | | | | |
| | | 7 m | E. cava[a] | | | | | | | | 5.50 | | | | | |

| Location | Station[b] | Species | | | | | | |
|---|---|---|---|---|---|---|---|---|
| Bishamon | M-1 | I. okamurai[a] | | | 0.30 | | | |
| | M-2 | H. fusiformis[a] | | | 3.05 | | | |
| | L-1 | Chondrus verrucosus | | | 2.85 | | | |
| | L-2 | G. amansii | | | 7.50 | | | |
| | L-3 | Chondrus elatus | | | 2.40 | | | |
| | 1 m | E. bicyclis | | | 3.65 | | | |
| | 4 m | E. cava[a] | | | 7.80 | | | |
| Shimoda | 0 m | Sargassum sagamianum | 4.28 | 3.99 | 3.44 | | | |
| | 2 m | Sargassum patens | | 6.41 | 5.37 | | | |
| | 3 m | E. bicyclis | | | 6.77 | | | |
| | 5 m | E. cava[a] | 7.31 | 9.88 | | 4.60 | | |
| | 10 m | E. cava[a] | 6.00 | 9.10 | 17.03 | | | 4.12 |
| (Shitaru) | 10 m | E. cava[a] | | 15.60 | 15.40 | 13.30 | 11.69 | 7.50 |
| (Shirahama) | 10 m | G. amansii | 1.00 | 1.20 | 1.27 | 1.00 | 0.64 | 0.40 |
| Mochimune | 2 m | Cystophyllum sisymbrioides | 7.26 | 5.57 | 2.99 | 2.24 | 0.83 | 0.87 |
| | 3—4 m | E. bicyclis[a] | 1.61 | 1.41 | 2.69 | | | |
| | 3—4 m | Sargassum ringgoldianum | 11.93 | 10.26 | | | | |
| Seto Inland Sea | | | | | | | | |
| Yura | M | I. okamurai[a] | | | 6.73 | 6.79 | | |
| | M | Gymnogongrus flabelliforme | 2.00 | 5.04 | | | | |
| | L | Grateloupia elliptica | 6.34 | 15.50 | | | | |
| Iwaya | 0 m | E. cava[a] | 2.74 | | | | | |
| | 1.5 m | E. cava | | 6.78 | | | | |
| | 3.5 m | E. cava[a] | | | 7.02 | | | |
| | 4.5 m | E. cava[a] | | | | | | |
| Sumoto | 0 m | H. fusiformis | 5.50 | | | | | |
| | 0 m | I. okamurai | 1.96 | | | | | |
| | 1.5 m | Hypnea charoides | 1.24 | | | | | |
| | 2.5 m | H. charoides | 1.34 | | | | | |
| | 3.5 m | G. amansii | 1.24 | | | | | |
| Myojin | M | G. flabelliforme | 2.84 | | | | | |
| | M | Ulva pertusa | 1.50 | | | | | |
| | L | U. pertusa | 1.36 | | | | | |
| | 0 m | U. pertusa | 1.38 | | | | | |
| | 1 m | U. lactuca[a] | 1.82 | 2.31 | 1.78 | | | |
| | 2 m | U. lactuca[a] | 1.25 | 1.93 | 1.66 | | | |
| | 3 m | U. lactuca[a] | 1.68 | 1.45 | 0.93 | | | |

[a]  Single species community.

[b]  U, upper intertidal zone; M, middle intertidal zone; L, lower intertidal zone.

From Yokohama, Y., in *Productivity of Biocenoses in Coastal Regions of Japan*, Hogetsu, K., Hatanaka, M., Hanaoka, T., and Kawamura, T., Eds., University of Tokyo Press, Tokyo, 1977, 48. With permission.

Table 2
STANDING CROP OF SEAWEEDS IN JAPAN

| Species | Standing crop | | Month | Water depth (m) | Locality | Ref. |
|---|---|---|---|---|---|---|
| | Wet weight[a] (kg/m²) | Dry weight (g/m²) | | | | |
| Green Algae | | | | | | |
| *Enteromorpha* sp. | 0.42 | 63 | April | (S) | Kominato | 2 |
| *Monostroma nitidum* | (1.35) | 270[b] | March | (L) | Uranouchi Bay | 3 |
| *Ulva pertusa* | (0.16) | 29[b] | May | | | 4 |
| *U. pertusa* | 0.25 | | June | | Otaru | 5 |
| *U. pertusa* | 1.33 | | December | (L) | Nakanoumi | 6 |
| *U. lactuca* | 2.31 | | August | 1 | Myojin | 1 |
| Brown Algae | | | | | | |
| *Alaria crassifolia* | 13.56 | | June | (L) | Muroran | 1 |
| *Dictyopteris undulata* | 1.05 | | August | 10 | Manazuru | 1 |
| *Ecklonia cava* | 4.00 | | | | Kominato | 2 |
| *E. cava* | 9.00[b] | | August | 5 | Inamuragasaki | 1 |
| *E. cava* | 6.95[b] | | August | 5 | Arasaki | 1 |
| *E. cava* | 7.50 | | August | 4 | Bishamon | 1 |
| *E. cava* | 17.03[b] | | August | 5 | Shimoda | 1 |
| *E. cava* | 20.70[b] | | July | 10 | Shitaru | 1 |
| *E. cava* | 15.50 | | June | 0 | Yura | 1 |
| *E. cava* | 7.02[b] | | August | 4.5 | Iwaya | 1 |
| *Ecklonia kurome* | 5.76 | | August | | Tanabe Bay | 6 |
| *Eisenia bicyclis* | 11.70 | | June—August | 5—6 | Kominato | 7 |
| *E. bicyclis* | 20.00[b] | | June—July | 2—4 | Matsushima Bay | 8 |
| *E. bicyclis* | 5.75[b] | | August | 0 | Inamuragasaki | 1 |
| *E. bicyclis* | 6.55[b] | | August | 1 | Arasaki | 1 |
| *E. bicyclis* | 3.65 | | August | 1 | Bishamon | 1 |
| *E. bicyclis* | 6.77 | | July | 3 | Shimoda | 1 |
| *E. bicyclis* | 11.93[b] | | August | 3—4 | Mochimune | 1 |
| *Endarachne binghamiae* | 0.88 | | February | (L) | Tanabe Bay | 6 |
| *Hizikia fusiformis* | 19.80[b] | | March | (L) | Kominato | 9 |
| *H. fusiformis* | 16.80[b] | | April | (L) | Tanabe Bay | 6 |
| *H. fusiformis* | 1.90 | | August | (L) | Manazuru | 1 |
| *H. fusiformis* | 1.05[b] | | August | (L) | Arasaki | 1 |
| *H. fusiformis* | 3.05 | | August | (L) | Bishamon | 1 |
| *H. fusiformis* | 5.50 | | October | 0 | Sumoto | 1 |
| *Hydroclathrus clathratus* | 1.47 | | February | (L) | Tanabe Bay | 6 |
| *Ishige okamurai* | 0.40 | | August | (L) | Arasaki | 1 |
| *I. okamurai* | 0.30 | | August | (L) | Bishamon | 1 |
| *I. okamurai* | 2.00 | | June | (L) | Yura | 1 |
| *I. okamurai* | 1.96 | | October | 0 | Sumoto | 1 |
| *Laminaria japonica* | 15.69[b] | | June | (L) | Muroran | 1 |
| *L. japonica* | 19.08[b] | | August | (L) | Muroran | 1 |
| *L. angustata* | (2.50) | 500 | | | | 4 |
| *L. angustata* | (1.51) | 301[b] | July | | | 4 |
| *L. ochotensis* | 6.00 | | July | 8 | Rishiri | 10 |
| *Laminaria religiosa* | 5.10[b] | | June | 1 | Kominato | 5 |
| *L. religiosa* | 11.41[b] | | July | 0.6 | Oshoro Bay | 11 |
| *Myelophycus simplex* | 0.10 | | February | (L) | Tanabe Bay | 6 |
| *Myagropsis myagroides* | 10.01 | | May | 0.5 | Fukuura | 23 |
| *M. myagroides* | 7.26[b] | | April | 2 | Mochimune | 1 |
| *Sargassum confusum* | 3.00 | | May | 3.5 | Fukuura | 23 |
| *Sargassum fulvellum* | 3.00 | | May | 3 | Fukuura | 23 |
| *S. fulvellum* | 10.94 | 3660 | April—May | | Kominato | 2 |
| *Sargassum hemiphyllum* | 4.80 | | May | | Nakanoumi | 6 |

## Table 2 (continued)
## STANDING CROP OF SEAWEEDS IN JAPAN

| Species | Standing crop Wet weight[a] (kg/m²) | Standing crop Dry weight (g/m²) | Month | Water depth (m) | Locality | Ref. |
|---|---|---|---|---|---|---|
| *Sargassum horneri* | 0.70 | | May | 2 | Fukuura | 23 |
| *Sargassum micracanthum* | 3.92 | | May | 1—2 | Fukuura | 23 |
| *Sargassum patens* | 6.61 | | September | 2 | Fukuura | 23 |
| *S. patens* | 19.20[b] | | May | | Tanabe Bay | 6 |
| *S. patens* | 8.24 | 1594 | April—May | | Kominato | 2 |
| *S. patens* | 2.20 | | August | (L) | Inamuragasaki | 1 |
| *S. patens* | 6.41[b] | | June | 2 | Shimoda | 1 |
| *S. patens* | (20.12) | 4023 | April | 4—6 | Iida Bay | 12 |
| *Sargassum piluliferum* | 4.50 | | September | 4 | Fukuura | 23 |
| *S. piluliferum* | 13.27[b] | 3660[b] | April—May | | Kominato | 2 |
| *Sargassum ringgoldianum* | 8.32 | | September | 2 | Fukuura | 23 |
| *S. ringgoldianum* | 10.41[b] | 2668[b] | April—May | | Kominato | 2 |
| *S. ringgoldianum* | 6.79[b] | | October | 3—4 | Mochimune | 1 |
| *Sargassum sagamianum* | 4.60[b] | | September | 0 | Shimoda | 1 |
| *Sargassum serratifolium* | 16.60[b] | | April | 3—7 | Kasaoka | 13 |
| *S. serratifolium* | 5.30 | | May | 0.5 | Fukuura | 23 |
| *S. serratifolium* | 4.93[b] | | May | | Mukaishima | 14 |
| *S. serratifolium* | 28.80[b] | | March—January | | Kasaoka Bay | 6 |
| *S. serratifolium* | (35.36) | 7075 | April | 4—6 | Iida Bay | 12 |
| *Sargassum thunbergii* | 11.70[b] | | February | (L) | Tanabe Bay | 6 |
| *S. thunbergii* | 10.30[b] | | April | (L) | Miho Bay | 6 |
| *S. thunbergii* | 0.93[b] | | December—April | (L) | Nakanoumi | 6 |
| *S. thunbergii* | 1.33 | 244 | April | (L) | Kominato | 2 |
| *S. thunbergii* | 13.86[b] | | July | (L) | Muroran | 1 |
| *Sargassum tortile* | 13.95 | 1594 | April—May | | Kominato | 2 |
| *Undaria pinnatifida* | 1.15 | | May | 2 | Fukuura | 23 |
| **Red Algae** | | | | | | |
| *Chondrus yendoi* | 3.96[b] | | January | (L) | Muroran | 1 |
| *Chondrus elatus* | 2.40 | | August | (L) | Bishamon | 1 |
| *Gelidium amansii* | (6.97) | 2300[b] | | | | 15 |
| *G. amansii* | 9.50 | | May | 2 | Fukuura | 23 |
| *G. amansii* | 2.00 | | August | 10 | Manazuru | 1 |
| *G. amansii* | 7.50 | | August | (L) | Bishamon | 1 |
| *G. amansii* | 1.61[b] | | April | 10 | Izu-Shirahama | 1 |
| *G. amansii* | 1.24 | | October | 3.5 | Sumoto | 1 |
| *Gigartina mikamii* | 2.85 | | August | (L) | Bishamon | 1 |
| *Gloiopeltis furcata* | 0.48[b] | | June | (S) | Muroran | 1 |
| *Gracilaria verrucosa* | 1.17[b] | | December | | Nakanoumi | 6 |
| *Grateloupia divaricata* | 0.82[b] | | December | | Nakanoumi | 6 |
| *Grateloupia livida* | 2.20 | | August | (L) | Manazuru | 1 |
| *Gymnogongrus flabelli-forme* | 5.04 | | June | (L) | Yura | 1 |
| *G. flabelliforme* | 2.84 | | August | (L) | Myojin | 1 |
| *Hypnea charoides* | 2.35 | | August | (L) | Inamuragasaki | 1 |
| *H. charoides* | 1.34[b] | | October | 2.5 | Sumoto | 1 |
| *Laurencia* sp. | 0.40 | | August | (L) | Inamuragasaki | 1 |
| *Pachymeniopsis elliptica* | 1.35 | | August | 1 | Manazuru | 1 |
| *P. elliptica* | 6.34 | | June | (L) | Yura | 1 |
| **Seagrasses** | | | | | | |
| *Zostera marina* | 1.84[b] | | December—January | | Kasaoka Bay | 6 |

Table 2 (continued)
## STANDING CROP OF SEAWEEDS IN JAPAN

| Species | Standing crop | | Month | Water depth (m) | Locality | Ref. |
|---|---|---|---|---|---|---|
| | Wet weight[a] (kg/m²) | Dry weight (g/m²) | | | | |
| *Z. marina* | 5.50 | | May | | Nakanoumi | 6 |
| *Zostera japonica* | 1.75 | | May | | Nakanoumi | 6 |
| *Z. marina* | | 242[b] | May | | Odawa Bay | 24 |

*Note:* L, littoral zone; S, supra-littoral zone.

[a]  Figures in parentheses were estimated from dry weight data.
[b]  Figures are the maximum among several measurements at the same locality during different months.


Table 3
## STANDING CROP OF MARINE ALGAE IN KABIRA BAY AND ITS VICINITY, ISHIGAKI-JIMA, JAPAN[25]

| Species | Date | Coverage (%) | Dry weight (g/m²) |
|---|---|---|---|
| *Calothrix pilosa* | July 27, 1977 | 100 | 785.6 |
| *Monostroma latissimum* | April 4, 1977 | 70 | 434.3 |
| *Ulva conglobata* | July 26, 1976 | 100 | 424 |
| *Ulva* sp. (1) | April 4, 1977 | 100 | 384 |
| *Ulva* sp. (2) | April 4, 1977 | 100 | 211.2 |
| *Enteromorpha* sp. | April 4, 1977 | 80 | 445 |
| *Halimeda opuntia* f. *opuntia* | July 26, 1977 | 100 | 1883.2[a] |
| *Codium repens* | April 23, 1977 | 100 | 399.2 |
| *Padina australis* | April 7, 1977 | 80 | 258 |
| *Chnoospora implexa* | April 7, 1977 | 100 | 396 |
| *Hydroclathrus clathratus* | April 3, 1977 | 100 | 593.6 |
| *Turbinaria ornata* | April 27, 1977 | 65 | 607.5 |
| *Hormophysa triquetra* | July 27, 1977 | 100 | 811.2 |
| *Galaxaura fasciculata* | July 27, 1977 | 100 | 956.8[a] |
| *Halymenia floresia* | April 5, 1977 | 100 | 385.6 |
| *Gracilaria eucheumoides* | April 21, 1977 | 80 | 869 |
| *Ceratodictyon spongiosum* | April 27, 1977 | 100 | 556 |
| | July 27, 1977 | 100 | 718.4 |
| *Centroceros clavulatum* | October 26, 1976 | 100 | 766 |

[a]  Data including calcium carbonate.

Table 4
## STANDING CROP OF SEAGRASSES IN KABIRA BAY, ISHIGAKI-JIMA, JAPAN[25]

| Species | Date | Dry weight (g/m²) | | | | Density (Number/m²) |
|---|---|---|---|---|---|---|
| | | Leaf | Rhizome | Root | Total | |
| *Thalassia hemprichii* | July 26, 1976 | 354.4 | 504.0[a] | | 858.4 | 1008 |
| | October 23, 1976 | 336 | 216 | 252.8 | 804.8 | 1376 |
| | October 26, 1976 | 345.6 | 274.4 | 105.6 | 725.6 | 704 |
| | April 4, 1977 | 124 | 140 | 241.6 | 505.6 | 672 |
| | April 7, 1977 | 199.2 | 119.2 | 123.2 | 441.6 | 704 |
| | April 8, 1977 | 388.8 | 220 | 321.6 | 930.4 | 1648 |
| | July 26, 1977 | 176 | 171.2 | 153.6 | 500.8 | 1504 |
| | July 27, 1977 | 508.8 | 409.6 | 292.8 | 1211.2 | 1040 |
| *Cymodocea rotundata* | October 23, 1976 | 168.8 | 75.2 | 144 | 388 | 2080 |
| | April 8, 1977 | 192 | 93.6 | 99.2 | 384.8 | 2848 |
| | July 26, 1977 | 203.2 | 107.2 | 81.6 | 392 | 2160 |
| *Halodule uninervis* | April 7, 1977 | 33.6 | 32.8 | 64 | 130.4 | 1952 |
| | July 27, 1977 | 56 | 76.8 | 73.6 | 206.4 | 2000 |
| *Halodule pinifolia* | July 27, 1976 | — | — | — | 37.6 | — |
| | October 23, 1976 | — | — | — | 6 | — |
| | April 8, 1977 | — | — | — | 9.9 | — |
| | July 26, 1977 | — | — | — | 21.6 | — |
| *Halophila ovalis* | October 23, 1976 | — | — | — | 21.8 | — |
| | October 26, 1976 | — | — | — | 89.6 | — |
| | April 8, 1977 | — | — | — | 15 | — |
| | July 26, 1977 | — | — | — | 64 | — |
| | July 26, 1977 | — | — | — | 42.4 | — |
| | July 26, 1977 | — | — | — | 56 | — |

[a]   Value of rhizomes plus roots.

Table 5

STANDING CROP OF *PORPHYRA* AND *MONOSTROMA* CULTIVATED IN
JAPAN

| Species | Cultivation facilities | Month | Standing Crop[a] | | Locality | Ref. |
|---|---|---|---|---|---|---|
| | | | g(d.w.)/10 cm (net yarn) | g(d.w.)/m² (net area) | | |
| *Porphyra yezoensis* | Pole system, fixed type | November | 1.11 | 155 | Tokyo Bay | |
| *Monostroma latissimum* | Pole system, fixed type | December | 1.27 | 178 | Ise Bay | 16 |
| | | January | 3.15 | 449 | | |
| | | March | 1.13 | 158 | | |
| | | March | 0.95 | 133 | | |
| *M. latissimum* | Pole system, fixed type | March | 0.72 | 100 | Matoya Bay | 16 |
| | | | 1.08 | 151 | | |
| | | | 1.18 | 165 | | |
| | | | 1.70 | 238 | | |
| | | | 2.56 | 358 | | |
| *Monostroma nitidum* | Pole system, fixed type | February | 3.87 | 469 | Uranouchi Bay | 3 |
| | Floating system | March | 4.60 | 593 | | |

[a]  d.w., dry weight.

Table 6

HARVESTS OF CULTIVATED BROWN ALGAE IN JAPAN

| Species | Harvest [kg(wet weight)/ m(rope)] | Cultivation period | Water depth (m) | Locality | Ref. |
|---|---|---|---|---|---|
| *Eisenia bicyclis* | 3.6 | February—September | 0—5 | Sendai Bay | 17 |
| *Kjellmaniella gyrata* | 12 | February—June | 0—5 | Sendai Bay | 17 |
| *Laminaria angustata f. longissima* | 31.5 | February—June | 0—5 | Sendai Bay | 17 |
| *Laminaria diabolica* | 20.7 | February—June | 0—5 | Sendai Bay | 17 |
| *L. diabolica* | 78 | February—June | 0—5 | Sendai Bay | 17 |
| *Laminaria japonica* | 30 | December—June | 2—3 | Akashi | 18 |
| *L. japonica* | 166.2 | December—June | 0—5 | Sendai Bay | 17 |
| *L. japonica* | 12 | February—June | 0—5 | Sendai Bay | 17 |
| *L. japonica* | 31.2 | February—June | 0—5 | Sendai Bay | 17 |
| *L. japonica* | 46.8 | February—June | 0—5 | Sendai Bay | 17 |
| *L. japonica f. membranacea* | 49.5 | February—June | 0—5 | Sendai Bay | 17 |
| *Laminaria religiosa* | 31.5 | February—June | 0—5 | Sendai Bay | 17 |
| *L. religiosa* | 47.1 | February—June | 0—5 | Sendai Bay | 17 |
| *Undaria pinnatifida* | 10 | | | Akashi | 26 |
| *U. pinnatifida* | 33.3 | December—June | 0—5 | Sendai Bay | 17 |

## Table 7
## CHARACTERISTICS OF PHOTOSYNTHESIS-LIGHT CURVES OF SEAWEEDS IN KABIRA BAY[25]

| Class | Number of species | Net photosynthesis[a] ($P_n$) | Dark respiration[a] (R) | $P_n$/R ratio | Compensation light intensity[b] | $I_k$[b] | Initial slope (mgO$_2$/klx) |
|---|---|---|---|---|---|---|---|
| Chlorophyceae | 8 | 16.3 ( 2.3—80.0) | 1.9 (0.5—4.0) | 9.6 (2.5—21.7) | 0.79 (0.15—2.10) | 4.9 (2.5—10.6) | 3.4 (0.5—15.6) |
| Phaeophyceae | 5 | 16.6 ( 6.5—34.5) | 2.6 (0.3—9.5) | 13.2 (3.6—26.0) | 0.58 (0.20—1.68) | 6.9 (3.0—16.1) | 3.9 (0.4— 6.8) |
| Rhodophyceae | 6 | 7.6 ( 2.0—15.3) | 1.5 (0.2—4.5) | 8.7 (2.8—30.0) | 0.76 (0.23—2.00) | 4.7 (2.4—10.8) | 2.2 (0.5— 4.1) |
| Monocotyledoneae | 4 | 18.5 (11.7—30.0) | 3.0 (0.3—8.0) | 13.7 (3.5—39.0) | 0.81 (0.15—2.60) | 4.6 (3.0— 6.3) | 4.8 (2.3— 7.3) |

a  in mgO$_2$/g(dry weight)/hr.
b  in × 10³ lx.

## Table 8
## LIGHT-SATURATED PHOTOSYNTHETIC RATES ($P_n^{max}$) AND RESPIRATORY RATES (R) OF CULTIVATED MONOSTROMA AND PORPHYRA

| Species | Stages of growth | Water temp (°C) | $P_n^{max}$[a] (mg O$_2$/g(d.w.)/hr) | R[a] (mg O$_2$/g(d.w.)/hr) |
|---|---|---|---|---|
| Monostroma latissimum | Early (November—December) | 15 | 30—40 | 2—5 |
| | Later March—April) | 15 | 12—15 | 2—5 |
| Porphyra yezoensis | Early (November—December) | 15 | 35—50 | 3—7 |
| | Later (January—February) | 15 | 18—26 | 2—4 |

a  d.w., dry weight.

Table 9

PRODUCTIVITY OF *ECKLONIA CAVA* COMMUNITY AT 5-m DEPTH AT
SHIMODA, JAPAN[19]

| Period | Days | Bladelets formed (kg(wet weight)/m²) | Daily net production (g(wet weight)/m²) |
|---|---|---|---|
| December 15—February 1 | 48 | 3.81 | 79.4 |
| February 1—February 15 | 14 | 1.33 | 95.2 |
| February 15—April 14 | 58 | 6.43 | 110.9 |
| April 14—June 6 | 53 | 6.58 | 124.2 |
| June 6—June 21 | 15 | 1.27 | 84.9 |
| June 21—August 9 | 49 | 2.23 | 45.5 |
| August 9—September 1 | 23 | 1.00 | 43.4 |
| September 1—October 26 | 55 | 2.27 | 41.2 |
| October 26—December 12 | 47 | 1.92 | 40.2 |
| December 12—December 15 | 3 | 0.18 | 59.8 |
| | | | |
| Annual net production | | 27.02 (approx. 6 kg(d.w.)/m²)[a] | Average 74.0 (approx. 16.4 g(d.w.)/m², range 9—25 g(d.w.)/m²)[a] |

[a]   d.w., dry weight.

Table 10

DAILY PRODUCTIVITY, g(d.w.)/m²/day, OF CULTIVATED *PORPHYRA*
COMMUNITIES IN JAPAN[a]

| Species | Periods | Mean | | Maximum[b] | | |
|---|---|---|---|---|---|---|
| | | Net production | Respiration | Net production | Respiration | Ref. |
| *Porphyra yezoensis* | October—December (2 months) | 7.5 | 3.7 | 12 | 4.2 | 20 |
| *P. yezoensis* | October—December (66 days) | 3.9 | 0.9 | — | — | 2 |
| | October—December (86 days) | 4 | 1.1 | — | — | |
| | November (7 days) | 11.6 | 2.1 | (15) | (4) | |
| | November (7 days) | 14.6 | 3.5 | (12) | (3) | |
| *P. yezoensis* | | | | 18 | 3 | 21 |

[a]   d.w., dry weight.
[b]   Figures in parentheses are approximate.

## Table 11
## ANNUAL MINIMUM NET PRODUCTION OF CULTIVATED *PORPHYRA* COMMUNITY IN JAPAN

|  |  | Annual net production [g(d.w.)/m²(net area)][a] |
|---|---|---|
| *Porphyra* spp. | Best farm | ca. 3,000 |
|  | Good farm | 500—800 |
|  | National average | 300 |

*Note:* Estimated from the harvest data for the growing season of approximately 6 months (October to March).[22]

[a]  d.w., dry weight.

## REFERENCES

1. **Yokohama, Y.**, Biomasses of seaweeds, in *Productivity of Biocenoses in Coastal Regions of Japan*, Hogetsu, K., Hatanaka, M., Hanaoka, T., and Kawamura, T., Eds., University of Tokyo Press, Tokyo, 1977, 48.
2. **Katada, M. and Satomi, M.**, Ecology of marine algae, in *Advance of Phycology in Japan*, Tokida, J. and Hirose, H., Eds., Gustav Fischer Verlag, 1975, 211.
3. **Maeda, M. and Ohno, M.**, On the productivity of *Monostroma* (green algae) community in the natural and cultivated ground, *Rep. Usa Mar. Biol. Sta. (Usa Rinkai Jikkenjo Kenkyu Hokoku)*, 19, 1, 1972.
4. **Fuji, A. and Kawamura, K.**, Studies on the biology of the sea urchin. VII. Bio-economics of the population of *Strongylocentrotus intermedius* on a rocky shore of southern Hokkaido, *Bull. Jpn. Soc. Sci. Fish.*, 36, 763, 1970.
5. **Sakai, Y.**, Vegetation structure and standing crop of the marine algae in the *Laminaria*-bed of Otaru City, Hokkaido, Japan, *Jpn. J. Ecol. (Nippon Seitai Gakkaishi)*, 27, 45, 1977.
6. **Tokioka, T., Harada, E., and Nishimura, S.**, *Marine Ecology*, Tsukiji Shokan, Tokyo, 1972.
7. **Ueda, S. and Ino, T.**, Biomass of seaweeds for potassium, *Aquiculture*, 3(3), 51, 1956.
8. **Yoshida, T.**, On the productivity of *Eisenia bicyclis* community, *Bull. Tohoku Fish. Res. Lab.*, 30, 107, 1970.
9. **Katada, M.**, Ecological studies on the propagation of *Hizikia fusiformis*, *J. Fish. Sci.*, 35, 320, 1940.
10. **Kaneko, T. and Niihara, Y.**, Ecology of *Laminaria japonica* var. *ochotensis*, in *Recent Researches of the Propagation and Cultivation of Laminaria in Hokkaido and its Vicinity*, Japanese Society of Phycology, Sapporo, 1977, 21.
11. **Abe, E. and Funano, T.**, Grazing on *Laminaria religiosa* by abalones and sea urchins in Oshoro Bay, *Monthly Rep. Hokkaido Fish. Exp. Sta.*, 30, 25, 1973.
12. **Taniguchi, K. and Yamada, E.**, Ecological study on *Sargassum patens* C. Agardh and *S. serratifolium* C. Agardh in the sublittoral zone at Iida Bay of Noto Peninsula in the Japan Sea, *Bull. Jpn. Sea Reg. Fish. Res. Lab. (Nihonkai-Ku Suisan Kenkyusho Kenkyu Hokoku)*, 29, 239, 1978.
13. **Fuse, S.**, The animal community in the *Sargassum* belt., *Physiol. Ecol.*, 11, 23, 1962.
14. **Mukai, H.**, The phytal animals on the thalli of *Sargassum serratifolium* in the *Sargassum* region, with reference to their seasonal fluctuations, *Mar. Biol.*, 8, 170, 1971.
15. **Nonaka, T., Osuga, H., and Sasaki, T.**, Studies on the propagation of Gelidiaceous algae. VII. On the recovery of stock after harvest, *Rep. Izu Br. Shizuoka Pref. Fish Exp. Sta.*, 18, 1, 1962.
16. **Maegawa, M. and Argua, Y.**, Studies on the growth and the variation of photosynthetic activity of cultivated *Monostroma latissimum*, *La Mer*, 12, 197, 1974.
17. **Kikuchi, S., Uki, N., Sakurai, Y., Akiyama, K., and Kito, H.**, Studies on the development of seaweed forestation for abalone food., *Tohoku Reg. Fish. Res. Lab. 1971 Data Rep.*, 1972, 10.
18. **Ii, A.**, Success of the cultivation of *Laminaria* in a warm sea area, *Yoshoku*, 4(11), 12, 1967.

19. **Yokohama, Y.,** Productivity of seaweeds, in *Productivity of Biocenoses in Coastal Regions of Japan*, Hogetsu, K., Hatanaka, M., Hanaoka, T., and Kawamura, T., Eds., University of Tokyo Press, Tokyo, 1977, 119.
20. **Satomi, M., Matsui, S., and Katada, M.,** Net production and the increment in stock of the *Porphyra* community in the culture ground, *Bull. Jpn. Soc. Sci. Fish.*, 33, 167, 1967.
21. **Argua, Y.,** Recherches physio-écologiques sur la production organique par les algues, *La Mer.*, 15, 163, 1977.
22. **Aruga, Y.,** Seaweeds as marine resources, *Iden (Heredity)*, 28(9), 49, 1974.
23. **Ishikawa-ken,** unpublished.
24. **Mukai, H.,** personal communication.
25. **Ohba, H. and Aruga, Y.,** unpublished.
26. **Ii, A.,** personal communication.

# PRIMARY PRODUCTIVITY OF MARINE BENTHIC ALGAE AND MACROPHYTES IN AUSTRALIA

## J. S. Bunt

## INTRODUCTION

A fully integrated estimate of nonplanktonic primary production around Australia must recognize the contributions of mangroves, salt marshes, seagrasses, and algae. The latter group embraces all forms from benthic free-living and epiphytic microalgae to the larger seaweeds and the algal symbionts of corals and other reef organisms. No one has yet attempted such an undertaking. Very few measurements of production in individual communities have been reported. This brief essay will consider the likely first order magnitude of the production.

## THE COASTAL ENVIRONMENT

Galloway[12] of the Commonwealth Scientific and Industrial Research Organization Division of Land Use Research has estimated that the length of the Australian coastline, using steps of 0.5 km and 1:250,000 maps, amounts to 45,000 km. This figure incorporates Tasmania and all islands greater in area than 1 km$^2$. Approximately 50% of the coastline is occupied by beaches, 20% is occupied by cliffs, and 10% is occupied by mangroves. The sea bed to a depth of 200 m extends over approximately $2.2 \times 10^6$ km$^2$. The Great Barrier Reef, a major feature, extends for 1600 km along the eastern coast of Queensland. Smaller reef formations occur off the coast of western Australia.

## PRIMARY PRODUCTION CONTRIBUTORS

### Benthic Microalgae

Measurement of photosynthetic activity by this group of organisms is technically and logistically awkward. To this author's knowledge, data for Australia are lacking. World data reported by Bunt[1] range from <30 to >400 g C·m$^{-2}$·year$^{-1}$. Rates of production around 250 g C·m$^{-2}$·year$^{-1}$ are not uncommon, especially in lower latitudes. Comparable activities might be expected in Australian conditions and could well match or exceed plankton production over the same area.

### Coral Reef Communities

Rates of photosynthesis by corals and coral reef communities are known to be high. Values obtained by independent investigators at various locations[1] range between 4 and 12 g C·m$^{-2}$·day$^{-1}$, approaching the theoretical maxima attainable under field conditions. Studies by the LIMER expeditions on the reefs near Lizard Island in the Great Barrier Reef[13] and also by Borowitzka and Larkum[14] indicate that levels of production are comparably high under Australian conditions. Lacking reliable estimates for the areal extent of active coral cover on the Great Barrier Reef, it is difficult to suggest a regional estimate of primary production. As a guide, however, production over a conservative 1000 km$^2$ at 5 g C·m$^{-2}$·d$^{-1}$ would represent an annual total equivalent to 1.8 × 10$^6$ metric tonnes C. With increasing research activity, more reliable estimates may be anticipated.

## Macroalgae

Mann and Chapman[2], reviewing the primary production of marine macrophytes, emphasize discoveries that the larger fucales of temperate and subarctic waters are capable of yielding 500 to 2000 g $C \cdot m^{-2} \cdot year^{-1}$. Data for other groups are rather limited. Representative statistics have been summarized by Bunt[1] and range from <10 to >12 g $C \cdot m^{-2} \cdot d^{-1}$ on an annual basis. Unfortunately, none of the available information was gatherd on the coastline of Australia. As far this author is aware, productivity studies have not been undertaken in this region. This is unfortunate, since large fucoids occur along the coastline of southeastern Australia and Tasmania, and the Australian macroalgal flora is generally diverse. Production along an estimated 9000 km of rugged coastline alone must be substantial. For example, a zone averaging only 10 m in width over such a distance could yield 45,000 tons $C \cdot year^{-1}$ at a mean rate of only 500 g $C \cdot m^{-2} \cdot year^{-1}$.

## Seagrasses

The seagrasses are well represented and widely distributed around Australia.[3,4] Studies in other countries attest to the high productivity of some species. Annual means of production have been reported to range from <1 to >4 g $C \cdot m^{-2} \cdot d^{-1}$.[1]

Once again, none of these data have originated from work in Australian seagrass communities, and their areal extent has not been assessed. A report by West and Larkum[5] documents rates of production in the range 0.7 to 5.5 g dry weight $m^{-2} \cdot d^{-1}$ by *Posidonia australis* in Botany Bay on the east coast. Larkum[15] also reports that *Zostera capricorni* in the same location was found to yield 1700 g dry weight $m^{-2}$ $year^{-1}$. More extensive data are needed, especially for the communities of the tropical coast.

## Mangroves

Mangrove forests are well developed, floristically rich, and extensive on sheltered coasts, within estuaries and bordering tidal rivers in northern Australia. Some species extend their distribution well to the south. The genus *Avicennia* persists in several estuaries in Victoria, almost at 40° S. Although there has been a significant history of interest by botanists, zoologists, and ecologists, attention to mangrove primary production in this region has been quite recent and not widely based. Studies by the author have concentrated on sites in northeastern Queensland. Systematic observations over 3 years in a wide range of mangrove communities reveal rates of litter fall ranging from <1 to almost 4 g $C \cdot m^{-2} \cdot d^{-1}$ as annual means. Bunt et al.[6] consider that litter fall probably represents around 50 to 70% of the total net primary production. Measurements of photosynthesis by $CO_2$ exchange in northern Queensland support this view. Rates of production between 3 and 4 g $\cdot C \cdot m^{-2} \cdot d^{-1}$ are probably extensive. Comparable figures have been reported by workers in the Caribbean.[7] More recently, similar results have been obtained in Thailand.[8]

According to Galloway,[12] mangroves in Australia probably occupy a total of 9000 $km^2$, largely in subtropical and tropical environments. On the basis of current information, a total primary production attributable to these communities might lie in the range <10 to >13 × 10[6] tons $C \cdot year^{-1}$.

## Salt Marshes

Information on the salt marshes of Australasia has been reviewed by Saenger et al.[9] Their productivity, however, was not considered, clearly for lack of data, although King[17] now has studies in progress near Sydney on the east coast.

The temperate salt marshes of the U.S. and Europe are highly productive.[10] This is likely to be so in Australia. At the same time, topographic and other conditions along the southern coastline seem not to be conducive for extensive marsh development. In

Table 1
PRIMARY PRODUCTIVITY OF
AUSTRALIAN MARINE BENTHIC
MICROALGAE AND MACROPHYTES

| Organism | Primary productivity ($10^6$ t C·year$^{-1}$) |
|---|---|
| Benthic microalgae | 180 |
| Mangroves | 10 |
| Coral reefs | 2 |
| Macroalgae (not reef-associated) | 0.045 |
| Seagrasses | ? |
| Salt marshes | ? |
| Total | 190 |

*Note:* The figure for the benthic microalgae has been taken as possibly equal to a conservative estimate of planktonic production within the area between the coast and the 200-m depth contour. Even if the estimate were high by an order of magnitude, it would substantially exceed the mangroves estimate. Accepting the total as it stands would represent a mean value for benthic production of approximately 90 g C·m$^{-2}$·year$^{-1}$ between the coast and the 200-m depth contour. This seems reasonably conservative and acceptable as a first estimate.

lower and even intermediate latitudes, the marsh niche is either occupied by mangroves or sparse samphires. Under harsher conditions further into the tropics, the upper intertidal is commonly extensive, but largley bare of plant life. At present, it is not possible to suggest a salt marsh contribution to coastal primary production.

## CONCLUSION

According to Karo,[11] 4.4% of the global coastline is Australian. At the present time, we can do little better than guess at the magnitude of primary production attributable to the several distinctive and individually diverse plant communities which exist between upper tidal limits and the sea bed to the depth of the effective photic zone around this continent. Possible values are given as in Table 1.

## REFERENCES

1. **Bunt, J. S.,** An Assessment of Marine Primary Production, *CRC Handbook of Biosolar Resources,* Vol. 1, Mitsui, A. and Black, C. C., Eds., CRC Press, Boca Raton, Fla., in press.
2. **Mann, K. H. and Chapman, A. R. O.,** Primary production of marine macrophytes, *Photosynthesis and Productivity in Different Environments,* Cooper, J. P., Ed., Cambridge University Press, Cambridge, 1975, 207.
3. **Den Hartog, C.,** *The Seagrasses of the World,* North-Holland, Amsterdam, 1970.
4. **Larkum, A. W. D.,** Recent research on seagrass communities in Australia, in *Seagrass Ecosystems; a Scientific Perspective,* McRoy, P. and McHerrick, C., Eds., Marcel Dekker, New York, 1977, 247.
5. **West, R. J. and Larkum, A. W. D.,** Leaf productivity of the seagrass *Posidonia australis* in eastern Australian waters, *Aquat. Bot.,* in press.

6. **Bunt, J. S., Boto, K., and Boto, G.,** A survey method for estimating potential levels of mangrove forest primary production, *Mar. Biol.,* in press.

7. **Lugo, A. E. and Snedaker, S. C.,** The ecology of mangroves, *Annu. Rev. Ecol. Syst.,* 5, 39, 1974.

8. **Christensen, B.,** Biomass and primary production of *Rhizophora apiculata* B1 in a mangrove in Southern Thailand, *Aquat. Bot.,* 4, 43, 1977.

9. **Saenger, P., Specht, M. M., Specht, R. L., and Chapman, V. J.,** Mangal and coastal salt-marsh communities in Australasia, in *Wet Coastal Ecosystems,* Chapman, V. J., Ed., Elsevier, Amsterdam, 1977, 393.

10. **Turner, A. E.,** Geographic variations in salt marsh macrophyte production, *Mar. Sci.,* 20, 47, 1976.

11. **Karo, H. A.,** World coastline measurements, *Int. Hydrog. Res.,* 33, 131, 1956.

12. **Galloway, R.,** personal communication.

13. **Barnes, D.,** personal communication.

14. **Borowitzka, M. and Larkum, A.,** personal communication.

15. **Larkum, A.,** personal communication.

16. **Andrews, A.,** personal communication.

17. **King, R.J.,** personal communication.

# PRIMARY PRODUCTIVITY OF SEAGRASSES

## Anitra Thorhaug

## INTRODUCTION

Seagrasses utilize solar energy in the shallow water estuarine and nearshore oceanic regions in the photosynthetically active region (300- to 700-nm wavelength). In general seagrasses are among the highest primary producers in the ocean and can, in the tropics, produce up to 2000 g m$^{-2}$ year$^{-1}$ compared to upwellings (500 g m$^{-2}$ year$^{-1}$). In general, a great deal less is known about seagrass than phytoplankton primary production. The seagrasses are the only group of higher plants which have evolved the strategy of complete marine submersion for extended times, but they are remarkably successful, usually dominating the shallow water communities they inhabit in terms of primary production, standing biomass, and abundance.

The distribution of seagrasses in the shallow portions of the oceans of the world occurs from the subarctic to the tropics. A main difference from the other major absorber of solar energy in the oceans, the phytoplankton, is that seagrasses need an appropriate benthic substrate. Thus, one does not find them in open ocean conditions with plankton, but rather in estuarine and coastal areas. The depth to which seagrasses are able to colonize is strongly dependent on the light penetration and level. Therefore, in turbid estuarine waters, one finds them limited to shallower waters, while in clear tropical coastal waters, one may find them to a depth of 40 m in Jamaica[1] and 30 m in Cuba.[2] Den Hartog[3] recognizes 12 genera with 49 species of seagrasses found in the oceans and estuaries. Generally there are more species found in the Pacific than in the Atlantic or Mediterranean.[3] The Atlantic communities are often made up of a single dominant species or a dominant species and one successional stage species, whereas Pacific communities often have codominants of several species. In the north Atlantic, *Zostera marina*, and in the tropical Atlantic, *Thalassia testudinum*, are important dominant species.[3]

Thus, in terms of distribution patterns of primary productivity, shallow tropical areas, such as the Bahamas, Florida Bay, and certain Pacific Island chains, have thousands of square miles of productive seagrasses. The primary productivity of seagrasses is, in general, extremely high and rivals that of plants in tropical rain forests in carbon-fixed m$^{-2}$ year$^{-1}$. It exceeds that of phytoplankton upwellings on a yearly basis.[4] Conditions which allow high production include

1.  Plants are continuously bathed in seawater, often exposed to optimal light, since the long blades are swayed back and forth in the tides and waves, so that their surfaces, although somewhat prostrated, are continually exposed to sunlight.
2.  Trace minerals are usually in abundance.
3.  Frequently phosophate as well as other nutrients are abundant in the sediment.

Both the blades and the roots are active sites of absorption of materials, so that the plants can take up nutrients either through the blades as other marine plants do or through the sediment as marsh plants do.[5] Particularly, in the tropical waters the growing season is 365 days/year.[6] In the subarctic, there is some indication that plants are green and do not die off appreciably, even in the cold arctic waters under the ice.[7]

Before discussing details of production of fixed carbon, one must discuss the general fate of the plant material. A major portion of the blade material which seagrasses produce goes into the food web, making the nearshore fisheries, especially in the nurs-

Table 1
## SPATIAL DISTRIBUTION OF SEAGRASS PRODUCTIVITY, ABUNDANCE, AND BIOMASS

| Location | Species | Density (blades m⁻²) | Biomass (g dry weight m⁻²) | Productivity (g C m⁻² d⁻¹) | Ref. |
|---|---|---|---|---|---|
| **Pacific** | | | | | |
| Alaska | *Zostera marina* | 710—2101 | 186—324 | 8 | 25 |
| Alaska | *Z. marina* | 599—4576 | 62—1840 | 3.3—3.8 | 26, 27 |
| Washington | *Z. marina* | — | 94—539 | 0.6 | 16 |
| **Atlantic** | | | | | |
| Denmark | *Z. marina* | — | 272—960 | 2—7.3 | 28 |
| Massachusetts | *Z. marina* | — | 15—29 | 0.04 | 29 |
| Rhode Island | *Z. marina* | — | 100 | 2.6 | 30 |
| | | | 225 | 0.4—2.9 | |
| North Carolina | *Z. marina* | — | 225 | 0.2—1.2 | 31 |
| North Carolina | *Halodule beaudettei* | — | 200 | 0.48—2.0 | 9 |
| | | | (105—200) | | |
| Canary Islands | *H. beaudettei* | — | 89 | 0.35 | 32 |
| | | | (70—89) | | |
| Biscayne Bay, FL | *Thalassia testudinum* | 679—4456 | — | 0.68—0.18 | 6 |
| Card Sound, FL | *T. testudinum* | 235—5469 | — | 0.45 | 6 |
| Biscayne Bay, FL | *T. testudinum* | — | 835 | 0.9—2.5 | 33 |
| | | | (300—1800) | | |
| Biscayne Bay, FL | *T. testudinum* | — | — | 1.7—2.3 | 34 |
| **Gulf of Mexico** | | | | | |
| Texas | *T. testudinum* | — | — | 0.9—9.0 | 35 |
| Anclote, FL | *T. testudinum* | 110—475 | — | 0.3—1.24 | 10 |
| Gulf of Mexico | *T. testudinum* | — | 81 | 0.35—1.14 | 36 |
| | | | (74—89) | | |
| **Caribbean** | | | | | |
| North coast of Cuba | *T. testudinum* | | 340 | 9.3 | 8, 2 |
| | | | (200—800) | | |
| Jamaica | *T. testudinum* | | 224 | 3.1 | 11 |
| Puerto Rico | *T. testudinum* | | — | 2.4—4.5 | 37 |
| **Mediterranean** | | | | | |
| Malta | *Posidonia oceanica* | | 742 | 2—5 | 38 |
| | | | (543—1072) | | |
| **Indian Ocean** | | | | | |
| Laccadives | *Syringodium isoetifolia* | | | 5.81 | 39 |

ery areas, direct recipients of seagrasses. Some material is dissolved into seawater to be used by phytoplankton or other plants as their nutrient source. Some is translocated to growing blades. Still other material is exported directly by storms to the more open ocean areas or indirectly to the open ocean by juvenile fishes and invertebrates migrating in their adulthood. Species such as shrimp, crab, lobster, and juvenile fishes are in the direct food chain from seagrasses.

## PRODUCTION OF PLANT MATERIAL

Table 1 presents the few primary productivity measurements made in the Pacific, Atlantic, Gulf of Mexico, Caribbean, Mediterranean, and Indian Ocean ares of several seagrasses. In particular, *Z. marina* and *T. testudinum* have been studied more extensively in terms of productivity. This is partially because of their dominance and importance in the food web. It should be noted that the production of material runs very high in comparison to other primary producers, both in the ocean and on land, and compares favorably to productivities of tropical rain forests and productivity of fertilized crops (although comparison with a crop having man-made energy input such as fertilizer is not really a just comparison.) Reported productivities range from 9.3 g C m$^{-2}$ day$^{-1}$ in Cuba for *T. testudinum*[8] to 0.2 g C m$^{-2}$ day$^{-1}$ in North Carolina for *Z. marina.*[9]

In general, the author's hypothesis is that as one approaches the centers of distribution of a species, the productivity increases and productivity decreases at the limits of distribution. However, since the variability within those measurements subjected to statistics in Table 1 are so large and since the various methods of measuring productivity have provided different results and different amounts of extrapolation* and in addition, since the seasonal (even in the tropics) and other variations (such as between stations at a location) appear so great, no exact statment can be made on spatial distribution patterns of production of seagrasses from Table 1. From our work and that of others, ranging from near the northern limits of *Thalassia* distribution in the Gulf of Mexico (Anclote, 28° N)[10] to near the center of its distribution (Jamaica, 18° N),[11] some spatial patterns emerge. Strong seasonality within productivity of *Thalassia* was found near the northernmost limits.[12] The middle portion, which was studied near Miami, Fla. (done at biweekly intervals at 32 stations over 5 years), grew vigorously all year long. About 50% of the abundance of blades in the winter decreased from the summer values,[6] making the total productivity m$^{-2}$ much less than in Jamaica, where a large standing stock was present all year long (but some seasonality occurred). In the northernmost range near Anclote, the abundance of blades was drastically reduced in winter, and the productivity slowed appreciably. Evidently this was in response to the colder winter waters, since the productivity was higher in the artificially heated thermal effluents during the winter months in the same estuary.[10] To the author's knowledge, this is the only wide-ranging set of seagrass productivity studies done with coordination and by the same method over a sufficiently long period of time and with sufficient number of readings to establish the basis for comparative results.

Table 1 shows productivity values taken by a series of various methods which are thus not directly comparable as are phytoplankton values (which for some time have been taken chiefly by a standardized $^{14}$C method).** An extensive discussion of the variability of the methods is found in McRoy and McMillan[13] and in Bittaker and Iverson.[14] It appears that the tropical species of seagrass do have, in specific instances, higher productivity per day than the same species in subtropical or species in temperate regions. Tropical plants produce material 365 days of the year, whereas in temperate areas the growing season may be limited to only 120 days.[15,16] Not enough data have accumulated to fully corroborate whether temperate species fix less carbon material per day than subarctic, subtropical or tropical. One should carefully note that most of Table 1 has not been taken with sufficient replicates at station-dates, at enough sites within a body of water, or over more than a 1-year period (see Lieth[17] about phenology measurements) to establish their variability or reliability. Care should be taken in extrapolating these numbers for use along with those of phytoplankton or other producers where much more replication over longer time periods has been used within and between stations representing an area as well as at more frequent intervals.

---

\*    For instance, the direct harvest is a measurement of net production of blade carbon over 14 days, whereas $^{14}$C and O$_2$ are a few hours' measurement at midday and include material eventually to be transported to subsand portions.

\*\*  In general, there are several orders of magnitude and fewer reports on seagrass production than on phytoplankton.

## C₃ OR C₄ METABOLISM

The photosynthetic carbon metabolism of marine seagrasses is a point of some controversy. Benedict and Scott,[18] studying carbon isotope fractionation as well as electron microscopy, came to the conclusion that *Thalassia* is a $C_4$ plant. If $CO_2$ fixation is catalyzed by phosphoenolpyruvate carboxylase, the predicted $\delta$ [13]C value for *Thalassia* would be $-9.7$ to $-5.7$ % which compares favorably with measured values near $-6$. The early products of fixation of $HCO_3^-$ in the leaf sections are malic acid and aspartic acid which are similar to early products of $CO_2$ fixation in $C_4$ terrestrial plants. On the contrary, Wetzel and Hough* feel, mainly on morphological evidences, that the plant is a $C_3$ plant. They have not been able to demonstrate $C_4$ metabolism in six seagrasses which leads them to conclude that seagrasses are $C_3$ plants. Parker[19,20] has examined 40 of the almost 50 species of seagrass for 12C/13C ratios and finds most of them in the range of $C_4$ plants. Smith, working with McMillan,[21] has found that the 12C/13C ratios in plants change when they are grown in cultures and under various laboratory conditions. They feel that the seagrasses could be "oddballs" in terms of the $C_3$-$C_4$ problem, i.e., intermediate in metabolism. However recent work shows unequivacally that seagrasses are $C_3$ plants.[22]

## NUTRIENT SUPPLY

Productivity in the marine waters, especially in shallow waters which have a plentiful light supply, is often determined by the nutrient supply. Until recently, little was known about the relationship of seagrass productivity to nutrient supply. In the past few years, the author's laboratory has determined that the seagrass *Thalassia* takes up trace metals, either through the roots as rooted marsh plants do or through the blades as marine algae do, depending on the concentration of trace metals in the compartment.[23] In addition to this, the seagrasses are able to transport trace metals from one compartment to another and accumulate metals over a long period of time.[23] McRoy and Barsdate[24] showed that phosphorus and nitrogen could be absorbed across the surface of the leaves or of the roots in *Z. marina*, a temperate seagrass.

* Quoted in Reference 13.

## REFERENCES

1. **Thorhaug, A.,** unpublished data, 1978.
2. **Buesa, R. J.,** Population and biological data on turtle grass (*Thalassia testudinum* König, 1805) on the northwestern Cuban shelf, *Aquaculture,* 4, 207, 1974.
3. **den Hartog, C.,** *The Sea-grasses of the World,* North-Holland, Amsterdam, 1970, chap. 1.
4. **Ryther, J. H.,** Photosynthesis and fish production in the sea, *Science,* 166, 72, 1969.
5. **Thorhaug, A. and Schroeder, P.,** A model of heavy metal cycling through subtropical and tropical estuarine systems with energy related industry, in Proc. Waste Heat Symp. II, 1978.
6. **Thorhaug, A. and Roessler, M. A.,** Seagrass community dynamics in a subtropical estuarine lagoon, *Aquaculture,* 12, 253, 1977.
7. **McRoy, C. P.,** Eelgrass under arctic winter ice, *Nature (London),* 224, 818, 1969.
8. **Buesa, R. J.,** Produccion primaria de las praderas de *Thalassia testudinum* de la plataforma noroccidental de Cuba, INP, Cuba Cent. Inv. Pesqueras, *Reva. Bal. Trab. CIP,* 3, 101, 1972.
9. **Williams, R. B.,** Annual Production of Marsh Grass and Eelgrass at Beaufort, N.C., Annual Report, USBCF Biological Laboratory, Beaufort, N.C., 1966.
10. **Ford, E., Moore, S., and Humm, H. J.,** *A Study of the Seagrass Beds in the Anclote Estuary During the First Year of Operation of the Anclote Power Plant,* Florida Power Corp., St. Petersburg, Fla., 1975.

11. **Greenway, M.**, The effects of cropping on the growth of *Thalassia testudinum* (König) in Jamaica, *Aquaculture,* 4, 199, 1974.

12. **Thorhaug, A., Roessler, M. A., and McLaughlin, P.**, Benthic biology, in *Post-Operational Ecological Monitoring Program, 1976. Final Report, Anclote Unit 1,* Part 4, Florida Power Corp. St. Petersburg, Fla., 1978.

13. **McRoy, C. P. and McMillan, C.**, Production ecology and physiology of seagrasses, in *Seagrass Ecosystems,* McRoy, C. P. and Helfferich, C., Eds., Marcel Dekker, New York, 1977, chap. 3.

14. **Bittaker, H. F. and Iverson, R. L.**, *Thalassia testudinum* productivity: a field comparison of measurement methods, *Mar. Biol.,* 37, 39, 1976.

15. **Dillon, C. R.**, A Comparative Study of the Primary Productivity of Estuarine Phytoplankton and Macrobenthic Plants, Ph.D. thesis, University of North Carolina, Chapel Hill, 1971.

16. **Phillips, R. C.**, Temperate grass flats, in *Coastal Ecological Systems of the United States,* Odum, H. T., Copeland, B. J., and McMahan, E. A., Eds., Institute of Marine Sciences, University of North Carolina, Chapel Hill, 1969, 737.

17. **Lieth, H.**, Primary productivity in ecosystems: comparative analysis of global patterns, in *Unifying Concepts in Ecology,* van Dobben, W. H. and Lowe-McConnell, R. H., Eds., W. Junk, Wageningen, 1975, 67.

18. **Benedict, C. R. and Scott, J. R.**, Photosynthetic carbon metabolism of a marine grass, *Plant Physiol.,* 57, 876, 1976.

19. **Parker, P. L.**, The biogeochemistry of the stable isotopes of carbon in a marine bay, *Geochim. Cosmochim. Acta,* 28, 1155, 1964.

20. **Parker, P. L.**, unpublished data, 1978.

21. **Smith**, personal communication.

22. **Andrews, J. T. and Abel, K. M,**. Photosynthetic carbon metabolism in seagrasses, *Plant Physiol.,* 63, 650, 1979.

23. **Schroeder, P. and Thorhaug, A.**, Trace metal cycling in seagrass ecosystems, *Am. J. Bot.,* in press,

24. **McRoy, C. P. and Barsdate, R. J.**, Phosphate absorption in eelgrass, *Limnol. Oceanogr.,* 15, 6, 1970.

25. **McRoy, C. P.**, Standing Stock and Ecology of Eelgrass (*Zostera marina* L.) in Izembek Lagoon, Alaska, M. S. thesis, University of Washington, Seattle, 1966.

26. **McRoy, C. P.**, On the Biology of Eelgrass in Alaska, Ph.D. thesis, University of Alaska, Fairbanks, 1970.

27. **McRoy, C. P.**, Standing stocks and related features of eelgrass populations in Alaska, *J. Fish. Res. Board Can.,* 27, 1181, 1970.

28. **Peterson, C. G. J.**, Om Baendeltangens (*Zostera marina*) Aarsproduktion i de danske Farvande, in *Mindeskrift i Anledning af Hundredaaret for Japetus Steenstrups,* Jungersen, F. E. and Warming, J. E. B., Eds., Fodsel. B. Lunos Bogtrykkeri, Copenhagen, 1914.

29. **Conover, J. T.**, Seasonal growth of benthic marine plants as related to environmental factors in an estuary, *Publ. Inst. Mar. Sci. Univ. Tex.,* 5, 97, 1958.

30. **Nixon, S. W. and Oviatt, C. A.**, Preliminary measurements of midsummer metabolism in beds of eelgrass, *Zostera marina, Ecology,* 53, 150, 1972.

31. **Conover, J. T.**, Importance of natural diffusion gradients and transport of substances related to benthic marine plant metabolism, *Bot. Mar.,* 9, 1, 1968.

32. **Johnston, C. S.**, The ecological distribution and primary production of macrophytic marine algae in the Eastern Canaries, *Int. Rev. Ges. Hydrobiol.,* 54, 473, 1969.

33. **Jones, J. A.**, Primary Productivity by the Tropical Marine Turtle Grass, *Thalassia testudinum* König, and Its Epiphytes, Ph.D. thesis, University of Miami, Coral Gables, 1968.

34. **Zieman, J. C.**, A Study of the Growth and Decomposition of the Seagrass, *Thalassia testudinum,* M. S. thesis, University of Miami, Coral Gables, 1968.

35. **Odum, H. T. and Hoskin, C. M.**, Comparative studies of the metabolism of marine waters, *Publ. Inst. Mar. Sci. Univ. Tex.,* 5, 16, 1958.

36. **Pomeroy, L. R.**, Primary productivity of Boca Ciega Bay, Florida, *Bull. Mar. Sci. Gulf Caribb.,* 10, 1, 1960.

37. **Odum, H. T., Burkholder, P. R., and Rivero, J.**, Measurements of productivity of turtle-grass flats, reefs, and the bahia fosforesante of southern Puerto Rico, *Publ. Inst. Mar. Sci. Univ. Tex.,* 6, 159, 1959.

38. **Drew, E. A.**, Botany, in *Underwater Science,* Woods, J. D. and Lythgoe, J. N. Eds., Oxford University Press, London, 1971, chap. 6.

39. **Qasim, S. Z. and Bhattathiri, P. M. A.**, Primary production of a seagrass bed on Kavaratti Atoll (Laccadives), *Hydrobiologia,* 38, 29, 1971.

# PRIMARY PRODUCTIVITY OF MANGROVES

## Samuel C. Snedaker and Melvin S. Brown

Mangroves are a collection of woody, vascular plants that utilize a coastal saline environment. Mangroves are widely distributed in subtropical and pantropical areas. Their geographical distribution is shown in Figure 1. Table 1 lists the principal mangrove genera and gives their distribution.

Since the first field investigation of the primary productivity of mangroves by Golley et al.[1] in 1962, further research has been conducted primarily to determine the natural variation and the controlling environmental factors.[2,3] This field research has been dominated by measurements of gas ($CO_2$) exchange[3] where the uptake of $CO_2$ (carboxylation) is assumed equal to net photosynthesis and the release of $CO_2$ is assumed equal to respiration. For subsequent analyses and interpretation of the results, the terms gross primary productivity (GPP), net primary productivity (NPP), and respiration (R) are used in accordance with the recommendation of the International Biological Program[4] meaning that GPP − R = NPP. Parallel long-term monitoring studies of the rate of litter production have been made because such data are indicative of relative NPP and therefore useful for comparative studies.[5,6] Although Teas[7] has used litter production as a direct estimate of "productivity", litter production underestimates true NPP by not taking into account herbivory and the net accumulations of biomass in wood and roots. Four tables of data have been assembled from the literature to show the range and magnitudes of values for rates of carbon flux and for compartmental and ecosystem GPP, NPP, and R. Litter production data are also presented, with the caveat cited above, because of their comparative value.

Table 2 presents a selection of diurnal records for daytime photosynthesis and nighttime respiration taken from the work of Lugo and Snedaker.[8] The empirically determined rates of uptake and release of carbon (as $CO_2$) have been intergrated over the designated time period of measurement (day vs. night) and normalized to 1 m[2] of surface exchange area. Negative values during the daytime for net photosynthesis, in fact, estimate daytime respiration for nonphotosynthesizing tissues. The range in values reflects differing climatic conditions as well as short-term changes taking place in the substrate as a function of the frequency and duration of tidal inundation. The work of Carter et al.[3] is of special importance because it focuses attention on the significance of correlating substrate conditions with synoptic measurements of gas exchange.[2,12]

In Table 3, data similar to that presented in Table 2 have been summarized by species and by compartment for each of the sites in Florida and Puerto Rico where gas exchange studies have been made.[1-3,9,10] A set of measurements by Miller,[11] who used the eddy flux method, are also included in Table 3. These data show that mangroves, individually by species and compartment, can attain relatively high rates of GPP comparable to other species and communities cited as having unusually high productivity. The variation shown in Table 3 reflects both seasonal and site differences. Carter et al.[3] have proposed that site differences are dominated by differences in soil salinity, and Lugo et al.[12] have used this concept to explain species zonation on the basis of metabolism. This is based on the differences in R (where R is interpreted as the energetic cost of survival) measured for different mangrove species exposed to similar substrate conditions.

The most complete integrated summary of GPP, NPP, and R was prepared by Lugo and Snedaker[13] in 1975 and appears as Table 4. The data[1,3,10-12] therein represent site-specific summaries integrating all of the compartments in a mangrove ecosystem, but

FIGURE 1.   General world distribution of mangroves. (From Chapman, V. J., in *Proc. Int. Symp. Biology and Management of Mangroves,* Vol. 1, Walsh, G. E., Snedaker, S. C., and Teas, H. J., Eds., University of Florida, Gainesville, 1975, 3.

Table 1
### DISTRIBUTION OF PRINCIPAL MANGROVE GENERA AND SPECIES

| Genus | Total species | Indo-Pacific East Africa | Pacific U.S. | Atlantic U.S. | West Africa |
|---|---|---|---|---|---|
| *Rhizophora* | 7 | 5 | 2 | 3 | 3 |
| *Bruguiera* | 6 | 6 | — | — | — |
| *Ceriops* | 2 | 2 | — | — | — |
| *Kandelia* | 1 | 1 | — | — | — |
| *Avicennia* | 11 | 8 | 3 | 3 | 1 |
| *Xylocarpus* | 10? | 8? | ? | 2 | 1 |
| *Laguncularia* | 1 | — | 1 | 1 | 1 |
| *Conocarpus* | 1 | — | 1 | 1 | 1 |
| *Lumnitzera* | 2 | 2 | | | |
| *Camptostemon* | 2 | 2 | | | |
| *Aegialitis* | 2 | 2 | | | |
| *Sonneratia* | 5 | 5 | | | |
| *Scyphiphora* | 1 | 1 | | | |
| *Nypa* | 1 | 1 | | | |
| *Osbornia* | 1 | 1 | | | |
| Total | 53? | 44? | 7? | 10 | 7 |
| Other genera | 15 | 18 | 1 | — | — |
| Grand total | 68? | 62? | 8? | 10 | 7 |

From Chapman, V. J., in *Proc. Int. Symp. Biology and Management of Mangroves,* Vol. 1, Walsh, G. E., Snedaker, S. C., and Teas, H. J., Eds., University of Florida, Gainesville, 1975, 3.

does not include respiration values of mobile consumers (i.e., animals). Thus, they represent estimates of community metabolism. It is apparent from the ranking of GPP, high to low, relative to the associated NPP values, that these parameters are uncorrelated. Furthermore, neither parameter is necessarily correlated with standing stock biomass.[13] From these data, it is implied that the dominant characteristics of the structure and functioning of a mangrove ecosystem are controlled by the level of site-specific respiration. It is also pertinent to note that GPP estimates span an order of

## Table 2
## SUMMARY OF GAS EXCHANGE RATES OF
## MANGROVE SPECIES AND COMPARTMENTS[8]

| Species and compartment | Date | Net daytime photosynthesi (g C/ m²·day) | Nighttime respiration (g C/m²·night) |
|---|---|---|---|
| *Rhizophora mangle* | | | |
| Canopy sun leaves | 8-12-71 | 0.851 | 0.201 |
| | 8-13-71 | 2.870 | 0.432 |
| | 1-28-72 | 0.283 | 0.134 |
| | 1-29-72 | 0.297 | 0.108 |
| | 1-21-73 | 0.760 | 0.410 |
| | 1-21-73 | 1.070 | 1.520 |
| | 1-26-73 | 1.704 | 0.456 |
| | 1-26-73 | 1.164 | 0.248 |
| Mean (SE) | | 1.125 (0.298) | 0.439 (0.162) |
| Canopy shade leaves | 8-10-72 | 0.545 | 0.619 |
| | 1-26-73 | 0.920 | 0.496 |
| | 1-26-73 | 0.800 | 0.268 |
| Mean (SE) | | 0.755 (0.111) | 0.461 (0.103) |
| Trunks | 8-16-71 | −4.25 | 1.87 |
| | 1-28-72 | −1.96 | 0.85 |
| | 1-29-72 | −1.29 | 0.65 |
| Mean (SE) | | −2.50 (0.896) | 1.123 (0.378) |
| Branches | 1-28-73 | 0.001 | 0.007 |
| | 1-28-73 | 0.0004 | 0.006 |
| Mean (SE) | | 0.001 (0) | 0.007 (0.001) |
| Prop roots | 8-11-72 | −0.04 | 0.02 |
| | 8-17-72 | −0.01 | 0.05 |
| | 2-01-73 | 0.013 | 0.172 |
| Mean (SE) | | −0.042 (0.33) | 0.081 (0.046) |
| Seedlings | 8-06-71 | 1.06 | 0.367 |
| | 2-13-72 | −0.657 | 0.300 |
| | 2-14-72 | 0.464 | 1.950 |
| | 2-15-72 | 0.364 | 3.370 |
| | 1-28-73 | 3.54 | 1.0 |
| Mean (SE) | | 0.954 (0.703) | 1.397 (0.575) |
| *Avicennia germinans* | | | |
| Canopy sun leaves | 7-31-71 | 0.445 | 0.189 |
| | 8-02-71 | 0.645 | 1.58 |
| | 1-27-72 | 0.358 | 0.06 |
| | 1-28-72 | 0.106 | 0.04 |
| | 2-06-72 | 5.340 | 1.01 |
| | 2-14-72 | 0.308 | 0.135 |
| | 2-15-72 | 0.724 | 0.398 |
| | 1-21-73 | 1.100 | 0.54 |
| | 1-26-73 | 1.276 | 0.260 |
| Mean (SE) | | 1.145 (0.539) | 0.468 (0.172) |
| Canopy shade leaves | 8-03-71 | 0.723 | 2.31 |
| Trunks | 8-05-71 | −1.30 | 0.65 |
| | 1-26-72 | −0.66 | 0.33 |
| | 2-05-72 | −2.43 | 1.20 |
| Mean (SE) | | −1.463 (0.517) | 0.727 (0.254) |
| Pneumatophores | 8-04-71 | −0.2 | 0.10 |
| | 8-07-71 | −0.132 | 0.68 |
| | 2-03-72 | −0.01 | 0.009 |
| | 2-10-72 | −0.01 | 0.090 |
| | 2-01-73 | 0.04 | 0.39 |
| | 2-01-73 | 0.009 | 0.29 |
| Mean (SE) | | 0.051 (0.038) | 0.26 (0.102) |

Table 2 (continued)
## SUMMARY OF GAS EXCHANGE RATES OF
## MANGROVE SPECIES AND COMPARTMENTS[8]

| Species and compartment | Date | Net daytime photosynthesi (g C/ m²·day) | Nighttime respiration (g C/m²·night) |
|---|---|---|---|
| *Laguncularia racemosa* | | | |
| Canopy sun leaves | 8-14-71 | 1.40 | 0.30 |
| | 1-26-72 | 0.04 | 0.13 |
| | 2-10-72 | 0.265 | 0.08 |
| | 1-21-73 | 0.260 | 0.11 |
| | 1-26-73 | 0.592 | 0.076 |
| Mean (SE) | | 0.511 (0.239) | 0.139 (0.041) |
| Canopy shade leaves | 1-26-73 | 0.524 | 0.348 |
| Trunk | 2-10-72 | −0.831 | 0.40 |
| | | | |
| *Conocarpus erectus* | | | |
| Canopy sun leaves | 2-12-72 | 0.239 | 0.208 |
| | | | |
| *Salicornia* sp. | 2-12-72 | 1.83 | 0.04 |
| | | | |
| *Sesuvium portulacastrum* | 2-13-72 | 2.49 | — |
| Soil | 2-01-73 | 0.04 | 0.49 |

*Note:* All values expressed in grams carbon for unit period normalized
to a gas exchange surface area of 1 m².

Table 3
## SUMMARY OF PRIMARY PRODUCTIVITY AND RESPIRATION OF
## MANGROVE SPECIES AND COMPARTMENTS FOR SEVERAL LOCATIONS
## IN FLORIDA AND PUERTO RICO

| Geographic location, compartment | Date | Number diurnals | Gross primary productivity | Net primary productivity | Total 24-hr respiration | Ref. |
|---|---|---|---|---|---|---|
| Upper Fahka Union River, Florida | | | | | | 3 |
| Red mangrove | December 1972 | 10 | | | | |
| Shade leaves | | | 3.43 | 2.56 | 0.873 | |
| Sun leaves | | | 4.80 | 3.82 | 0.972 | |
| Prop roots | | | | | 0.592 | |
| Trunk | | | | | 1.05 | |
| Black mangrove | December 1972 | 10 | | | | |
| Shade leaves | | | 0.890 | 0.461 | 0.429 | |
| Sun leaves | | | 1.21 | 0.643 | 0.570 | |
| Trunk | | | | | 1.92 | |
| Buttonwood | December 1972 | 10 | | | | |
| Shade leaves | | | 3.17 | 2.00 | 1.17 | |
| Sun leaves | | | 3.90 | 2.66 | 1.24 | |

**Table 3 (continued)**
## SUMMARY OF PRIMARY PRODUCTIVITY AND RESPIRATION OF MANGROVE SPECIES AND COMPARTMENTS FOR SEVERAL LOCATIONS IN FLORIDA AND PUERTO RICO

| Geographic location, compartment | Date | Number diurnals | g C/m²·day | | | Ref. |
| --- | --- | --- | --- | --- | --- | --- |
| | | | Gross primary productivity | Net primary productivity | Total 24-hr respiration | |
| Lower Fahka Union River, Florida | | | | | | 3 |
| Red mangrove | December 1972 | 7 | | | | |
| Sun leaves | | | 2.81 | 2.54 | 0.265 | |
| Prop roots | | | | | 0.406 | |
| Trunk | | | | | 0.747 | |
| Black mangrove | December 1972 | 7 | | | | |
| Sun leaves | | | 2.20 | 1.96 | 0.240 | |
| Trunk | | | | | 0.323 | |
| White mangrove | December 1972 | 7 | | | | |
| Sun leaves | | | 1.66 | 1.43 | 0.225 | |
| Trunk | | | | | 0.126 | |
| Fahkahatchee Bay, Florida (tidal stream) | | | | | | 3 |
| Red mangrove | December 1972 | 10 | | | | |
| Sun leaves | | | 1.14 | 0.661 | 0.477 | |
| Prop roots | | | | | 0.559 | |
| Trunk | | | | | 0.876 | |
| Black mangrove | December 1972 | 10 | | | | |
| Sun leaves | | | 2.80 | 1.60 | 1.20 | |
| Trunk | | | | | 1.12 | |
| White mangrove | December 1972 | 10 | | | | |
| Sun leaves | | | 3.43 | 1.80 | 1.63 | |
| Trunk | | | | | 2.07 | |
| Key Largo, Florida | | | | | | 11 |
| Red mangrove | January 1970 | 6 | | | | |
| Leaves | | | 5.6 | 1.60 | 4.0 | |
| Rookery Bay, Florida | | | | | | 12 |
| Red mangrove (fringe) | August 1971, January /February 1972 | 15 | | | | |
| Leaves | | | 5.63 | 4.02 | 1.61 | |
| Stems | | | | | 0.23 | |
| Surface roots | | | | | 0.06 | |
| Periphyton | | | | | 1.1 | |
| Black mangrove | August 1971, January/ February 1972 | 17 | | | | |
| Leaves | | | 9.02 | 3.62 | 5.40 | |
| Stems | | | | | 0.14 | |
| Surface roots | | | | | 0.54 | |
| Soil | | | | | 0.20 | |
| Turkey Point, Florida | 1974 | | | | | 10 |
| Red mangrove (dwarf) | | 6 | | | | |
| Leaves, trunk, prop roots | | | 7.82 | 0.10 | 7.72 | |
| Soil | | | | | 2.23 | |

Table 3 (continued)
## SUMMARY OF PRIMARY PRODUCTIVITY AND RESPIRATION OF MANGROVE SPECIES AND COMPARTMENTS FOR SEVERAL LOCATIONS IN FLORIDA AND PUERTO RICO

| Geographic location, compartment | Date | Number diurnals | g C/m²·day | | | Ref. |
|---|---|---|---|---|---|---|
| | | | Gross primary productivity | Net primary productivity | Total 24-hr respiration | |
| Turkey Point, Fla. | | 2 | | | | |
| Black mangrove | | | | | | |
| (dwarf) | | | | | | |
| Leaves, trunk, | | | 1.76 | 0.06 | 1.70 | |
| surface roots | | | | | | |
| Soil | | | | | 2.23 | |
| Guayanilla Bay, | — | ᵃ | | | | 9 |
| Puerto Rico | | | | | | |
| Red mangrove | | | | | | |
| forest | | | | | | |
| — stressed | | | | | | |
| Leaves | | | 13.3 | 8.3 | 5.0 | |
| Woody structures | | | | | 0.4 | |
| Root community | | | 9.9 | 5.2 | 4.7 | |
| Jobos Bay, Puerto | — | ᵃ | | | | 9 |
| Rico | | | | | | |
| Red mangrove forest | | | | | | |
| — unstressed | | | | | | |
| Leaves | | | 10.8 | 5.2 | 5.6 | |
| Woody structures | | | | | 0.6 | |
| Root community | | | 9.5 | 3.8 | 5.7 | |
| La Parguera, Puerto | January 1958, | ᵃ | | | | 1 |
| Rico | May 1959, 1960 | | | | | |
| Red mangrove forest | | | | | | |
| Leaves | | | 8.11 | 3.08 | 5.03 | |
| Seedlings | | | 0.12 | | 0.36 | |
| Surface roots | | | | | 2.03 | |
| Soil | | | | | 0.37 | |

*Note:* All values expressed in grams carbon per square meter of ground area for a 24-hr period.

ᵃ    Sporadic hourly measurements.

magnitude and thus demonstrate that at the community, or ecosystem, level, mangroves are not uniform in observed primary productivity.

Largely because of the ease with which litter production can be monitored, many such studies have been and are being made. To the extent that they serve as estimates, albeit conservatively low, they extend the geographic coverage of the relative NPP for mangroves. Litter production studies are also made and reported because of the importance of this source of organic debris in detrital food webs; litterfall in this regard is equatable to net yield. The data summarized in Table 5 are particularly noteworthy because they tend to be less variable than the data presented in the preceding tables. This is evident in the data for the dwarf-forest forms of mangrove communities which

Table 4

SUMMARY OF PRIMARY PRODUCTIVITY AND RESPIRATION FOR
MANGROVE ECOSYSTEMS IN SEVERAL LOCATIONS IN FLORIDA AND
PUERTO RICO

| Location | Date | Number diurnals | g C/m²·day | | | Ref. |
|---|---|---|---|---|---|---|
| | | | Gross primary productivity | Net primary productivity | Total 24-hr respiration | |
| Fahkahatchee Bay, Florida (small tidal stream), red, black, and white mangroves | December 1972 | 10 | 13.9 | 4.8 | 9.1 | 3 |
| Lower Fahka Union River Basin, Florida (red, black, and white mangroves) | December 1972 | 7 | 11.8 | 7.5 | 4.3 | 3 |
| Upper Fahka Union River, Florida (red, black, and button-wood mangroves) | December 1972 | 10 | 10.3 | 6.6 | 3.7 | 3 |
| Rookery Bay, Florida (black mangrove forest) | August 1971 January/February 1972 | 17 | 9 | 2.8 | 6.2 | 12 |
| La Parguera, Puerto Rico (red mangrove) | January 1958 May 1959, 1960 | [a] | 8.2 | 0 | 9.1 | 1 |
| Rookery Bay, Florida (red mangrove) | August 1971 January/February 1972 | 15 | 6.3 | 4.4 | 1.9 | 12 |
| Key Largo, Florida (red mangrove) | June 1968 January 1970 | 6 | 5.3 | 0 | 6.0 | 11 |
| Hammock Forest, Dade County, Florida (red mangrove) | October 1973 | 3 | 1.9 | 1.3 | 0.6 | 10 |
| Dwarf Forest, Dade County, Florida (red mangrove) | October 1973 | 4 | 1.4 | 0 | 2.0 | 10 |

*Note:* All values expressed in grams carbon per square meter of ground area for a 24-hr period.

[a]   Sporadic hourly measurements.

are noted for both a depauperate standing stock biomass and low productivity; the
litter production rates are disproportionately high.

In all of the information available to the author on primary productivity of mangroves *in situ*, it can be seen that most of these data are from the Caribbean area.
Considering the small geographical area these data represent, the order of magnitude

Table 5

## LITTER PRODUCTION IN MANGROVE FORESTS (GRAMS DRY WEIGHT ORGANIC MATTER PER SQUARE METER PER YEAR)

| Geographic region, specific location | Mangrove forest type or dominant species | Sampling duration (year) | Sample number | Litter production (g/m²·year) | Ref. |
|---|---|---|---|---|---|
| Southwestern Florida | | | | | |
| Naples | Riverine | 1.0 | 70 | 1175 | 14 |
| Rookery Bay | Basin | 4.5 | 1600 | 741 | 6 |
| Ten Thousand Islands | Fringe | 3.1 | 1240 | 981 | 6 |
| Ten Thousand Islands | Riverine | 3.1 | 1240 | 1120 | 6 |
| Ten Thousand Islands | Overwash | 3.1 | 620 | 1024 | 6 |
| | | | | | |
| Everglades National Park | | | | | |
| Whitewater Bay | Riverine | 0.33 | — | 876[a] | 15 |
| Everglades City | Riverine | 1.0 | 70 | 1180 | 14 |
| | | | | | |
| Southeastern Florida | | | | | |
| Turkey Point | Fringe | 2.3 | 940 | 1082 | 6 |
| Turkey Point | Hammock | 2.3 | 940 | 750 | 6 |
| Turkey Point | Dwarf | 2.2 | 704 | 220 | 6 |
| Biscayne Bay | Fringe | 1.0 | 480 | 878 | 7 |
| Biscayne Bay | Dwarf | 1.0 | 180 | 449 | 7 |
| | | | | | |
| Puerto Rico | | | | | |
| Vacia Talega | Riverine | 0.67 | 100 | 1445 | 5 |
| Pinones | Basin | 0.67 | 195 | 971 | 5 |
| Ceiba | Fringe | 0.67 | 250 | 664 | 5 |
| La Parguera | Fringe | — | — | 475 | 1 |
| Bahia Sucia | Fringe | 0.75 | 45 | 1580 | 16 |
| | | | | | |
| Panama | | | | | |
| Darien | Fringe | — | — | 710[a] | 14, 19 |
| | | | | | |
| Thailand | | | | | |
| Phuket Island | *Rhizophora apiculata* | 1.3 | 741 | 704 | 17 |
| | | | | | |
| Okinawa | | | | | |
| Oura | *Bruguiera gymnorrhiza* | 1.0 | — | 590 | 18 |
| Oura | *B. gymnorrhiza* | 1.0 | — | 445 | 18 |
| Oura | *Kandelia candel* | 1.0 | — | 281 | 18 |
| Oura | *Bruguiera-Kandelia* | 1.0 | — | 296 | 18 |

[a]  Leaf material only.

that the estimates span for different communities, and the lack of information on surface area covered by mangroves, no extrapolation to a global scale is attempted.

## REFERENCES

1. **Golley, F. B., Odum, H. T., and Wilson, R. F.,** The structure and metabolism of a Puerto Rican red mangrove forest in May, *Ecology,* 43, 9, 1962.
2. **Snedaker, S. C. and Lugo, A. E.,** The Role of Mangrove Ecosystems in the Maintenance of Environmental Quality and a High Productivity of Desirable Fisheries, Final report, Contract No. 14-16-008-606, Bureau of Sports Fisheries and Wildlife, NTIS, Department of Commerce, Springfield, Va., 1973.

3. **Carter, M. R., Burns, L. A., Cavinder, T. R., Dugger, K. R., Fore, P. L., Hicks, D. B., Revels, H. L., and Schmidt, T. W.,** Ecosystems Analysis of the Big Cypress Swamp and Estuaries, EPA 904/0-74-002, U.S. Environmental Protection Agency, Region IV, Atlanta, Ga., 1973.

4. **Newbould, P. J.,** *Methods for Estimating the Primary Productivity of Forests,* IBP Handbook No. 2, Blackwell Scientific, Oxford, 1967.

5. **Pool, D. J., Lugo, A. E., and Snedaker, S. C.,** Litter production in mangrove forests of southern Florida and Puerto Rico, in *Proc. Int. Symp. Biology and Management of Mangroves,* Vol. 1, Walsh, G. E., Snedaker, S. C., and Teas, H. J., Eds., University of Florida, Gainesville, 1975, 213.

6. **Snedaker, S. C. and Brown, M. S.,** Water Quality and Mangrove Ecosystem Dynamics, U.S.E.P.A. Office of Pesticides and Toxic Substances, EPA-600/4-81-022, 1981, 79.

7. **Teas, H. J.,** Mangroves of Biscayne Bay, Report to Metropolitan Dade County Commission, Miami, Florida, 1974.

8. **Lugo, A. E. and Snedaker, S. C.,** Properties of a mangrove forest in southern Florida, in *Proc. Int. Symp. Biology and Management of Mangroves,* Vol. 1, Walsh, G. E., Snedaker, S. C., and Teas, H. J., Eds., University of Florida, Gainesville, 1975, 170.

9. **Canoy, M. J.,** Diversity and stability in a Puerto Rican *Rhizophora mangle* L. forest, in *Proc. Int. Symp. Biology and Management of Mangroves,* Vol. 1, Walsh, G. E., Snedaker, S. C., and Teas, H. J., Eds., University of Florida, Gainesville, 1975, 344.

10. **Snedaker, S. C. and Stanford, R. L.,** Ecological Studies of a Subtropical Terrestrial Biome, Final report, Florida Power and Light Company, Miami, Florida, 1976.

11. **Miller, P. C.,** Bioclimate, leaf temperature, and primary production in red mangrove canopies in south Florida, *Ecology,* 53, 22, 1972.

12. **Lugo, A. E., Evink, G., Brinson, M. M., Broce, A., and Snedaker, S. C.,** Diurnal rates of photosynthesis, respiration, and transpiration in mangrove forests of south Florida, in *Tropical Ecological Systems,* Golley, F. B. and Medina, E., Eds., Springer-Verlag, New York, 1975, 335.

13. **Lugo, A. E. and Snedaker, S. C.,** The ecology of mangroves, *Annu. Rev. Ecol. Syst.,* 5, 39, 1974.

14. **Sell, Maurice G., Jr.,** Modeling the Response of Mangrove Ecosystems to Herbicide Spraying, Hurricanes, Nutrient Enrichment and Economic Development, Ph.D. thesis, University of Florida, Gainesville, 1977.

15. **Heald, E. J.,** The production of organic detritus in a south Florida estuary, *Sea Grant Tech. Bull. No. 6,* University of Miami, Miami, 1971.

16. **Lugo, Ariel E., Cintron, G., and Goenaga, C.,** Mangrove ecosystems under stress, in *Stress Effects on Natural Ecosystems,* Barret, G. and Rosenberg, R, Eds., John Wiley & Sons, New York, in press.

17. **Christensen, B.,** Biomass and primary production of *Rhizophora apiculata* Bl. in a mangrove in southern Thailand, *Aquat. Bot.,* 4, 43, 1978.

18. **Nishira, Moritaka,** Aspects of Japanese mangroves, in *Proc. UNESCO Regional Seminar on Human Uses of Mangrove Environment and Its Management Implications,* BANSDOC, Dacca, Bangladesh, in press.

19. **Golley, F. B., McGinnis, J. T., Clements, R. G., Child, G. I., and Duever, M. J.,** *Mineral Cycling in a Tropical Moist Forest Ecosystem,* University of Georgia Press, Athens, 1975.

20. **Chapman, V. J.,** Mangrove biogeography, in *Proc. Int. Symp. Biology and Management of Mangroves,* Vol. 1, Walsh, G. E., Snedaker, S. C., and Teas, H. J., Eds., University of Florida, Gainesville, 1975, 3.

# PRIMARY PRODUCTIVITY OF TERRESTRIAL PLANTS

## Clanton C. Black

Numerous measurements of terrestrial plant productivity are available, such as biomass or a specific product such as grain or timber, per unit of land. Despite mankind's long-standing interest in measuring plant productivity, it is not easy to obtain accurate values. The simplest and probably still the best first estimation is to obtain the aboveground plant dry weight per unit of land surface per unit of growth time. This chapter reports most productivity data in units of tons of oven-dry matter produced per hectare per year, and terrestrial plant growth rates are in units of grams of dry matter accumulated per square meter of land surface per day.

Though these units are almost self-explanatory, conducting properly designed experiments to obtain accurate values is subject to many flaws. A detailed discussion of flaws to be avoided in conducting plant productivity studies is beyond the scope of this chapter. However, interested readers are referred to the incisive works of Monteith,[1-3] Loomis et al.,[4] and Loomis and Gerakis[5] for such discussions. Some common flaws to be avoided are tall plants grown in plots such that light is intercepted laterally, small plots, border effects, uncertain light interceptions throughout the canopy, and sampling or other numerical errors. These and other flaws are present in much of the literature data on plant productivity, particularly the data on maximum growth rates by specific species.

Before presenting productivity data on individual species, the total net primary productivity of the land of the world will be considered. Probably the best estimation of primary productivity was made by Whittaker and Likens for 1950 (see Table 1).[6] Others have made similar estimations and obtained higher values,[7] so readers should note that each value in Table 1 is an estimation! Even so, forests clearly are the major producers of plant biomass, and cultivated lands produce less than 10% on a world land basis (see Table 1).

Data on productivity of tree species are given in Table 2. Note many of these values span several decades of growth. Though short-term of "maximum" growth rates for tree species are not readily available, *Pinus radiata* (see Table 2) has a maximum growth rate comparable to other $C_3$ plants (most tree species are $C_3$ photosynthesis plants).

Dry matter production data are available for many $C_3$ plant in agricultural systems (see Table 3). Table 3 presents representative values, but not an exhaustive survey, from the literature on $C_3$ plants. Also shown in Table 3 are some short-term growth rates for $C_3$ plants obtained at various locations. Similar information on dry matter production and short-term growth rates is presented for $C_4$ photosynthesis plants in Table 4.

Little information is available on the productivity of Crassulacean acid metabolism (CAM) plants. Table 5 gives field weights of CAM plants. Data on dry matter are not available, and the moisture percentages for the plants in Table 5 are uncertain. The two values available on short-term growth rates (see Table 5) indicate that CAM plants may have growth rates comparable to $C_3$ plants (Table 3) under proper growing conditions. However, in dry areas where CAM plants often grow,[17] their sustained growth rates are likely to be much lower.

The usefulness of short-term growth rates is in comparing the maximum growth capabilities of various species and in predicting yields. From a growth-rate value for a species under specific environmental conditions, one can, in theory, predict yields, i.e., if a given growth rate can be maintained, a plant should yield accordingly. Thus,

Table 1
## ESTIMATION OF THE NET PRIMARY PRODUCTION OF LAND BIOTA[a]

| Type | Total net production ($10^9$ ton/year) |
|---|---|
| Tropical rain forest | 15.3 |
| Tropical seasonal forest | 5.1 |
| Temperate forest, evergreen | 2.9 |
| Temperate forest, deciduous | 3.8 |
| Boreal forest | 4.3 |
| Woodland and shrubland, savanna | 6.9 |
| Temperate grassland | 2.0 |
| Tundra and alpine | 0.5 |
| Desert and semidesert | 0.6 |
| Extreme desert, rock, ice, sand | 0.04 |
| Cultivated land | 4.1 |
| Swamp and marsh | 2.2 |
| Lake and stream | 0.6 |
| Total | 48.3 |

[a]    Estimations by Whittaker and Likens.[6]

Table 2
## ABOVE GROUND BIOMASS PRODUCTION BY TREE SPECIES

| Species | Average standing crop[a] (tons/acre/year) | Age years | Location and Comments |
|---|---|---|---|
| Sycamore | 8.1 | 4 | Schull Shoals, Ga., plantation grown |
| *Abies grandis* | 8.6 | 21 | U.K. |
| *Abies balsama* | 2.3 | 40 | Canada, Natural stand |
| *Picea abies* | 5.4 | 20 | U.K. |
| *Picea abies* | 2.8 | 47 | U.K. |
| *Picea abies* | 2.2 | 39 | Japan |
| *Picea abies* | 1.3 | 52 | Sweden |
| *Larix decidua* | 2.1 | 46 | U.K. |
| *Pinus densiflora* | 3.2 | 16 | Japan |
| *Pinus elliottii* | 5.6 | 8 | North Carolina |
| *Pinus nigra* | 4.8 | 18 | U.K. |
| *Pinus radiata*[b] | 8.4,20 | 18 | New Zealand |
| *Pinus radiata* | 5.0 | 12 | Australia |
| *Pinus strobus* | 2.5 | 41 | Japan |
| *Pinus sylvestris* | 2.0 | 31 | U.K. |
| *Pinus thunbergii* | 2.8 | 10 | Japan |
| *Pinus taeda* | 8.4 | 12 | North Carolina |
| *Pseudotsuga menziesii* | 2.8 | 36 | U.S. |
| *Pseudotsuga menziesii* | 3.8 | 22 | U.K. |
| *Tsuga heterophylla* | 5.4 | 23 | U.K. |

[a]    Data collated by Nemeth[8] or the University of Georgia School of Forestry.
[b]    Maximum growth rate of 41 g/m²/day.

Table 3

TYPICAL RANGES OF DRY MATTER PRODUCTION AND SHORT-TERM
GROWTH RATES OF C$_3$ PHOTOSYNTHESIS PLANTS

| Species | Dry matter (tons/ha/year) | Short-term growth rate (g/m²/day) | Location and comments | Ref. |
|---|---|---|---|---|
| *Festuca arundinacea* | — | 43 | U.K. | 10 |
| *Dactylis glomerata* | 22 | 19,40 | U.K., New Zealand | 10,11 |
| *Lolium perenne* | 16,25 | 21,28 | U.K. | 10,11 |
| | — | 20 | Netherlands | 10 |
| | — | 19 | New Zealand | 10 |
| *Brassica oleracea* | 21 | 21 | U.K. | 10,11 |
| | | 16 | New Zealand | 10,11 |
| Alfalfa | 6.2—10.2 | 23,8(mean) | U.S. | 5,9 |
| *Manihot esculenta* Crantz | 10,41 | 11(mean) | Roots included, Java | 9,12 |
| Soybean | 8.9 | 7(mean) | Japan | 5,9 |
| Sugarbeet | 33,22 | 31,14(mean) | California, Netherlands | 5,9 |
| Barley | 6.5 | 23 | U.K. | 10,13 |
| Wheat | 7—11 | 18 | Netherlands, Sudan, Aust. | 10,13 |
| Red Clover | — | 23 | New Zealand | 13 |
| Potato | 11 to 18 | 37 | California, U.K., Netherlands | 10,13 |
| Cotton | 1.37 | 27 | Southeast U.S., lint plus seed | 13,14 |

Table 4

TYPICAL RANGES IN ANNUAL DRY MATTER PRODUCTION VALUES AND
SHORT TERM GROWTH RATES OF C$_4$ PHOTOSYNTHESIS TERRESTRIAL
PLANTS

| Species | Dry matter (ton/ha/year) | Short-term growth rate (g/m²/day) | Location and comments | Ref.[a] |
|---|---|---|---|---|
| *Cynodon dactylon* | 27 | — | California | 10,11 |
| | 20,27 | 21 | Georgia | 10,11 |
| | 22,26 | 13 | Alabama | 10,11 |
| | 30 | — | Texas | 10 |
| *Paspalum notatum* | 22 | — | Alabama | 10,11 |
| *Paspalum plicatulum* | 31 | — | Australia 10 | |
| *Pennisetum clandestinum* | 29,30 | — | Australia | 10 |
| *Pennisetum purpureum* | 42,43,66,84,85 | 22,39 | Puerto Rico, El Salvador, Trinidad, Hawaii | 10,11 |
| *Pennisetum typhoides* | 22 | 54 | Australia | |
| *Digitaria decumbens* | 24,33,49,50,39 | 19,21,26 | Australia, Puerto Rico, Columbia, Cuba, Trinidad | 10,11 |
| *Chloris gayana* | 23 | — | Australia | 10 |
| *Panicum maximum* | 48,30 | 16 | Puerto Rico, El Salvador | 10,11 |
| *Panicum purpurasens* | 40 | — | Puerto Rico | 10 |
| *Panicum barbinode* | 43 | — | Trinidad | 10 |
| *Hyparrhenia rufa* | 31 | — | El Salvador | 10 |
| *Tripsacum laxum* | 36 | — | Trinidad | 10 |
| *Saccharum officinale* | 64 | 38 | Hawaii | 10 |
| *Zea mays* | 26,22,16,40,29 | 29,27,24,52,52, 38,52,40,31 | New Zealand, New Hampshire, Nebraska, New York, California, Japan, Kentucky, Thailand, Egypt, Peru | 10,11 |
| *Sorghum vulgare* | 30,47 | 51 | California | 11 |
| *Sorgham* sp. | 16 | — | Illinois, 140 days of growth | 10 |

[a]  Cooper[10,11] has collated most of these data and cites the original references.

Table 5

## FIELD WEIGHTS AND GROWTH RATES OF CRASSULACEAN ACID METABOLISM SPECIES

| Species | Field weight (tons/ha/year) | Growth rate (g/m²/day) | Location and comments | Ref. |
|---|---|---|---|---|
| *Opuntia lindheimeri* | 31.5 | — | Texas, 5-year average without cultivation | 15 |
| *Opuntia Fiscus-Indica* | 30.8 | — | North Africa | 16 |
| *Opuntia gommei* | 147 | — | Texas, with cultivation, 3-year average | 15 |
| *Ananas comosus* | 44 | — | Malaysia, fruits only | 17 |
| *Ananas comosus* | 17.1[a] | 15 | Hawaii, sustained for 250 days | 18 |
| *Mesembryanthemum crystallinum* | — | 1.3[b] | San Francisco, Calif. | 19 |

[a]  Tons of sugar plus starch.[18]

[b]  Growth rate unit is nanomoles of $CO_2$ fixed per square centimeter per sec. A comparable value 1.6 was obtained for the $C_3$ plant *Bromus mollis*.[19]

Table 6

## MAXIMUM SHORT-TERM GROWTH RATES FOR CROP PLANTS

|  | (g/m²/day) |
|---|---|
| $C_4$ species | |
| *Pennisetum typhoides* | 54[a] |
| *Zea mays* | 52,52 |
| *Sorghum vulgare* | 51 |
|  | |
| $C_3$ species | |
| *Agrostemma githago* | 39 |
| *Solanum tuberosum* | 37 |
| *Oryza sativa* | 36 |
| *Typha latifolia* | 34 |

[a]  Monteith collated these data and cites the original references.[3]

in studying the values such as those in Tables 3 and 4, it seems that, in general, $C_4$ plants have growth rates higher than $C_3$ plants and subsequently higher yields.

However, these conclusions have been strongly questioned by Gifford[20] and by Evans.[21] They claim $C_3$ and $C_4$ growth rates are indistinguishable, and crop yields show no advantage for $C_4$ photosynthesis. Their ideas do not appear to be acceptable from the detailed analysis of Monteith.[3] Monteith studied the methods and experiments used to collect the growth rate data cited by Gifford and Evans[20,21] and found only the data in Table 6 to be reliable.[3] Clearly the range of maximum values in Table 6 for $C_3$ and $C_4$ crop plants show that the maximum values for $C_4$ plants are higher than those for $C_3$ plants.

These maximum short-term growth rates (see Table 6) should be reflected in yields, even though optimum environmental conditions cannot be maintained throughout the lifetime of a crop nor do plants have similar lifetimes. Monteith circumvented some of these differences by plotting standing dry weight at harvest vs. the length of the growing season (collated in References 5 and 10) and obtained the results for $C_3$ and

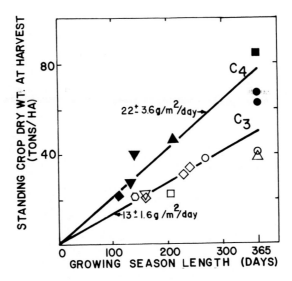

FIGURE 1.    Standing crop dry weight at harvest vs. length of growing season.[3] The closed symbols are $C_4$ plants, and open symbols are $C_3$ plants. The symbols refer to the following crops: bulrush millet, ◆; maize ▼; sorghum, ▲; sugar cane, ●; napier grass, ■; kale, ◯; potatoes, ▽; sugar beet, ◇; rice, □; cassava, ◯; and oil palm, △.

$C_4$ crops as shown in Figure 1.[3] Clearly the dry weight of $C_4$ crops is consistently higher than $C_3$ crops independent of the growing season. Furthermore, he obtained a seasonal mean growth rate for each crop. The mean seasonal growth rates (noted on the two lines on Figure 1) are $22 \pm 3.6$ g/m²/day and $13 \pm 1.6$ g/m²/day for $C_4$ and $C_3$ plants, respectively. Therefore, either from maximal or seasonal growth rates or yields, $C_4$ crop plants are different from $C_3$ crops plants (see Tables 3, 4, 6, and Figure 1). However, each plant species is better adapted to certain environments, and the environmental constraints of productivity may be much more important for yields than the pathway of photosynthetic carbon assimilation employed by a species. The final decisions about specific plants to grow for maximum productivity at a given site will rest on a combination of the environmental constraints plus the genetic capability of the species.

## REFERENCES

1. **Monteith, J. L., Ed.,** *Vegetation and the Atmosphere*, Vol. 1, Academic Press, London, 1975.
2. **Monteith, J. L., Ed.,** *Vegetation and the Atmosphere*, Vol. 2, Academic Press, London, 1976.
3. **Monteith, J. L.,** Reassessment of maximum growth rates for $C_3$ and $C_4$ crops, *Exp. Agric.*, 14, 1, 1978.
4. **Loomis, R. S., Williams, W. A., and Hall, A. E.,** Agricultural productivity, *Annu. Rev. Plant Physiol.*, 22, 431, 1971.
5. **Loomis, R. S. and Gerakis, P. A.,** in *Photosynthesis and Productivity in Different Environments*, Cooper, J. P., Ed., Cambridge University Press, Cambridge, Great Bitain, 1975, 145.
6. **Whittaker, R. and Likens, G.,** *Primary Productivity of the Biosphere*, Lieth, H. and Whittaker, R., Eds., Springer-Verlag, New York, 1973.
7. **Bazilevich, N. I., Rodin, L. Ye., and Rozov, N. N.,** Geographical aspects of biological productivity, *Sov. Geogr.*, 12, 293, 1971.
8. **Nemeth, J. C.,** Dry matter production in young loblolly (*Pinus taeda* L.) and slash pine (*Prinus elliottii* Eng elm.) plantations, *Ecol. Monogr.*, 43, 21, 1973.

9. **Boardman, N. K.,** The energy budget in solar energy conversion in ecological and agricultural systems, in *Living Systems as Energy Converters,* Buvet, R., Allen, M. J., and Massue, J. P., Eds., Elsevier, Amsterdam, 1977, 307.

10. **Cooper, J. P.,** Control of photosynthetic production in terrestrial systems, in *Photosynthesis and Productivity in Different Environments,* Cooper, J. P., Ed., Cambridge University Press, Cambridge, Great Britain, 1975, 593.

11. **Cooper, J. P.,** Potential production and energy conversion in temperatue and tropical grasses, *Herb. Abstr.,* 40, 1, 1970.

12. **Mahon, J. D., Lowe, S. B., and Hunt, L. A.,** Photosynthesis and assimilate distribution in relation to yield of cassava grown in controlled environments, *Can. J. Bot.,* 54, 1322, 1976.

13. **Hall, D. O.,** Solar energy use through biology — past, present and future, *Sol. Energy,* 22, 307, 1979.

14. **Stanhill, G.,** Cotton, in *Vegetation and the Atmosphere,* Vol. 2, Monteith, J. L., Ed., Academic Press, London, 1976, 121.

15. **Griffiths, D.,** U.S. Department of Agriculture Bull. 208, 1915.

16. **LeHouerou, H. N.,** *Arid Lands in Transition,* Dregne, H. E., Ed. American Association Advances in Science Publ. No. 90. Washington, D.C., 1970, 227.

17. **Ting, I. P., Johnson, H. B., and Szarek, S. R.,** Net $CO_2$ fixation in Crassulacean acid metabolism plants, in *Net Carbon Dioxide Assimilation in Higher Plants,* Black, C. C., Ed., Southern Section American Society of Plant Physiology, Mobile, Ala., 1972, 26.

18. **Marzola, D. L. and Bartholomew, D. P.,** Photosynthetic pathway and biomass energy production, *Science,* 205, 555, 1979.

19. **Bloom, A. J. and Troughton, J. H.,** High productivity and photosynthetic flexibility in a CAM plant, *Oceologia,* 38, 35, 1979.

20. **Gifford, R. M.,** A comparison of potential photosynthesis, productivity and yield of plant species with differing photosynthetic metabolism, *Aust. J. Plant Physiol.,* 1, 107, 1974.

21. **Evans, L. T.,** The physiological basis of crop yield, in *Crop Physiology,* Evans, L. T., Ed., Cambridge University Press, Cambridge, Great Britain, 1975, 327.

*Section 8*
*Physical Resources and Inputs*

# SPECTRAL CHARACTERISTICS AND GLOBAL DISTRIBUTION OF SOLAR RADIATION

## Roland L. Hulstrom

The amount and spectral characteristics of solar radiation incident at the surface of the earth are determined by the following major factors:

1. The extraterrestrial solar radiation or "solar constant"
2. The angle of incidence of the direct solar radiation to the surface of the earth
3. The transmission properties, absorption, and scattering of the intervening atmosphere

The total or global solar radiation on a horizontal surface at the surface of the earth is made up of the direct beam radiation and the diffuse skylight. Mathematically, this can be expressed as

$$G_\lambda = I_\lambda + D_\lambda \tag{1}$$

where $G_\lambda$ is the total global spectral solar radiation on a horizontal surface, $I_\lambda$ is the spectral direct beam solar radiation on the horizontal surface, $D_\lambda$ is the spectral diffuse skylight on the horizontal surface, and $\lambda$ is the wavelength.

For clear (cloud-free) sky conditions, the direct solar beam dominates the total available solar radiation, making up about 85 to 90% of the total. Mathematically, the direct solar radiation on a horizontal surface is given as

$$I_\lambda = I_\lambda (o) \, T_\lambda \, \text{Cos} \, \Theta_o \tag{2}$$

where

$I_\lambda (o)$ is the extraterrestrial radiation, and $T_\lambda$ is the atmospheric fractional transmittance at the solar zenith angle of $\theta_o$, where the solar zenith angle is the angle between the surface zenith (vertical) and the sun.

The cos $\theta_o$ function is a very dominating element of the incident direct solar beam on a horizontal surface having values of 0.866 at $\theta_o = 30°$, 0.5000 at $\theta_o = 60°$, and 0.174 at $\theta_o = 80°$. The solar zenith angle is, therefore, a dominating factor that determines the direct solar beam incident on a horizontal surface. The solar zenith angle is determined by geographical location, site latitude and longitude, and time of day, month, and year. Figure 1 shows the availability of the total extraterrestrial solar radiation as a function of time of year and geographic location. The amount of langleys, Figure 1, represents the total solar radiation over all wavelengths of the solar spectrum from 0.30 to 3.0 $\mu$m. This total amount of solar radiation represents the sum over all wavelengths of the product $I_\lambda(o) \cos \theta_o$.

The atmospheric fractional transmittance, $T_\lambda$, is a very complex function determined by atmospheric molecules, aerosols, water vapor, and clouds. These constitutents result in wavelength selective attenuation caused by scattering (molecular and aerosol) and absorption. For a detailed treatment of these processes, one should consult Paltridge and Platt.[1] For monochromatic radiation, the atmospheric transmittance is given by

$$T_\lambda = e^{-\tau\lambda \, m} \tag{3}$$

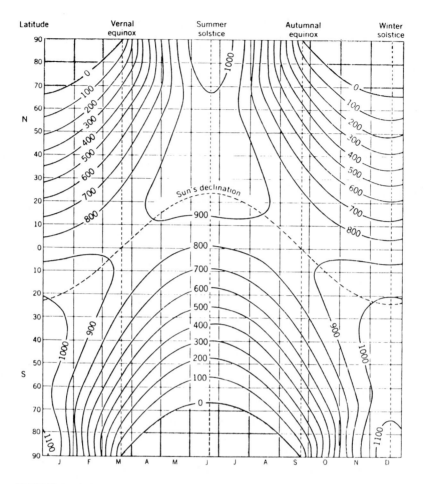

FIGURE 1.   Solar radiation, in langeleys per day, on a horizontal surface outside of the atmosphere of the earth, as a function of latitude and time of year. (After Fritz, S., *Compendium of Meteorology,* American Meteorological Society, Boston, 1951; as cited in Strahler, A. N., *The Earth Sciences,* Harper & Row, New York, 1963.

where $\tau_\lambda$ is the spectral (wavelength dependent) vertical (zenith) attenuation coefficient, and m is the relative slant path through the (to zenith) atmosphere. The relative slant path-airmass can be approximated by the sec $\theta_o$ up to $\theta_o \leqslant 70°$. At greater solar zenith angles, atmospheric refraction and spherical shape has to be taken into account.[1] The atmospheric spectral attenuation coefficient, $\tau_\lambda$, is made up of the sum of various components because of molecular scattering, $\tau(m)$; aerosol scattering, $\tau(p)$; ozone absorption, $\tau(o_3)$; carbon dioxide, $\tau(CO_2)$; oxygen, $\tau(O_2)$; and water vapor, $\tau(H_2O)$. This relationship is given by

$$\tau_\lambda = \tau_\lambda(m) + \tau_\lambda(p) + \tau(CO_2) + \tau(O_2) + \tau(H_2O) \qquad (4)$$

Each component is a complex function of wavelength; a description of this is given in Paltridge and Platt.[1] The overall effects of these processes are shown in Figure 2 which was generated using a computer code developed by the Solar Energy Research Institute and the Air Force Geophysics Laboratory.[2] The atmospheric conditions are defined by the U.S. Standard Atmosphere model considered to be typical/representative conditions of clear continental atmospheric conditions. The marked effect of increasing slant path through the atmosphere (relative air mass) is obvious. As the relative air

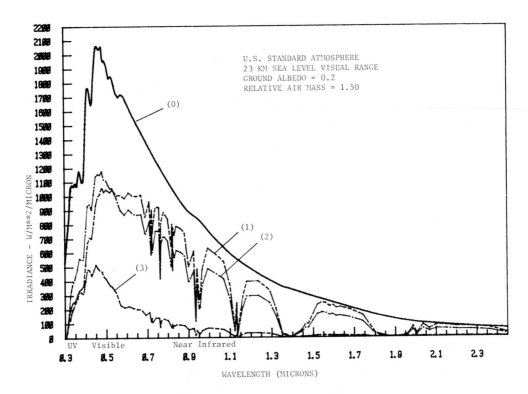

**FIGURE 2.** Spectral distribution of the direct solar beam at sea level and the diffuse skylight; (0), air mass zero — outside of atmosphere; (1) air mass one; (2) air mass two; (3) air mass three; (4) diffuse skylight for air mass one.

mass increases (lower sun angles toward sunrise and sunset), the direct beam is markedly reduced and the wavelength distribution ("color") is shifted toward the red wavelengths. In the visible or light spectral region, about 0.40 to 0.75 μm, the dominant atmospheric attentuation is because of the scattering of the sunlight by atmospheric molecules (molecular scattering) and aerosols. As shown in Figure 2, the amount of attenuation is significant — as evidenced by the difference between the solar irradiation curve outside the atmosphere and solar irradiation curve at sea level. Some of this attenuation, in the visible region, is also because of the absorption of sunlight by atmospheric ozone. The spectral distribution of the visible sunlight is determined by the spectral distribution of the extraterrestrial sunlight and the combination of molecular and aerosol scattering. Molecules scatter the shorter blue wavelengths much more severely than the longer, red wavelengths, thereby creating — on very clear days — a blue sky (as shown by the diffuse component, clear sky, in Figure 2). Aerosols scatter the sunlight in a more "white or gray" manner and create a more grayish or white appearance to the diffuse skylight, as shown in Figure 2 — diffuse component (some haze). This diffuse component was calculated by the use of Monte Carlo radiative transfer techniques/computer program.[3] Such calculations are extremely complex because of the complex manner by which molecules and aerosols scatter the sunlight.

As shown in Equation 1, the total solar radiation is the sum of the direct and diffuse components. As shown in Figure 2, the diffuse skylight component tends to add significant amounts of energy in the 0.30 to 0.60-μm region. It is also extremely variable, depending on the amount of haze in the atmosphere. The Monte Carlo computer program, along with other radiative transfer programs, is also capable[3] of producing detailed spectral distributions of the global radiation incident at the surface of the earth for various atmospheric conditions. The Solar Energy Research Institute also will

(1980) be making actual, high spectral resolution measurements of the direct, diffuse, and global solar radiation. Tables that would adequately represent the solar radiation spectral distributions would be too voluminous (some 575 wavelength points) to include in this publication. The Solar Energy Research Institute, Golden, Colo. Energy Resource Assessment Branch, can provide such extensive tables and/or actual computer codes.

## GLOBAL AND SEASONAL DISTRIBUTION

Fairly extensive measurements of the variations in the spectral distribution of solar radiation-daylight at various geographical locations on the surface of the earth have been made by Goldberg and Klein.[4] Measurements were made at three locations — Barrow, Alaska; Rockville, Md.; and the Pacific end of the Panama Canal. Data were collected from 1968 to 1974. They concluded that

"one cannot compute the daily insolation on a horizontal surface with variations of less than 50 percent from some mean value. The spectral quality of daylight is not the same at all places, nor is the rate of change in the spectral quality the same at all locations. The greatest variations in spectral quality occur in the peak response region of silicon, 0.600 — 0.800 μm. The blue 0.400 — 0.500 μm, also shows large variations. However, like the red band, most of these variations are due to local atmospheric conditions."

Local atmospheric conditions, in terms of the amount of haze and water vapor, largely control the spectral quality of the solar radiation and daylight. However, cloud cover conditions will, to a large extent, control the daily, seasonal, and global distribution of solar radiation and daylight. The global and seasonal distribution of the total broadband (0.3 to 3.0 μm) solar radiation on a horizontal surface is shown in Figures 3, 4, 5, and 6.

## REFERENCES

1. **Paltridge, G. W. and Plaft, C. M. R.,** *Radiative Processes in Meteorology and Climatology,* Elsevier Amsterdam, 1976.
2. **Ireland, P. J., Wagner, S., Kazmerski, L. L., and Hulstrom, R. L.,** A combined irradiance-transmittance solar spectrum and its application to photovoltaic efficiency calculations, *Science,* 204, 611, 1979.
3. **Bird, R. and Hulstrom, R. L.,** Application of Monte Carlo Techniques to Insolation Characterization and Prediction-Status Report June 1, 1979, Solar Energy Research Institute Report No. SERI/RR-36-306, National Technical Information Service, Washington, D.C., October 1, 1978 and June 1, 1979.
4. **Goldburg, B. and Klein, W. H.,** Variations in the spectral distribution of daylight at various geographical locations on the earth's surface, *Sol. Energy,* 19(1), 3, 1977.

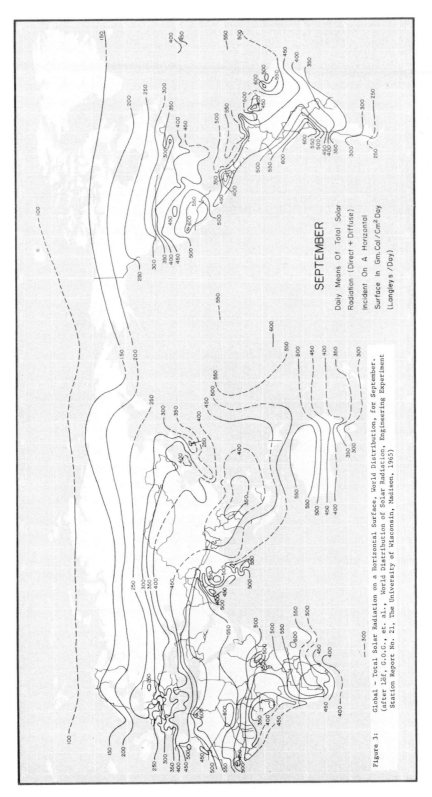

Figure 3:    Global — Total Solar Radiation on a Horizontal Surface, World Distribution, for September.
(after Löf, G.O.G., et. al., World Distribution of Solar Radiation, Engineering Experiment
Station Report No. 21, The University of Wisconsin, Madison, 1965)

SEPTEMBER

Daily Means Of Total Solar
Radiation (Direct + Diffuse)
Incident On A Horizontal
Surface In Gm. Cal./Cm² Day
(Langleys /Day)

FIGURE 3.    Global — total solar radiation on a horizontal surface, world distribution, for September.   (After Löf, G.O.G., et. al., World Distribution of Solar Radiation, Engineering Experiment Station Report No. 21, The University of Wisconsin, Madison, 1965.)

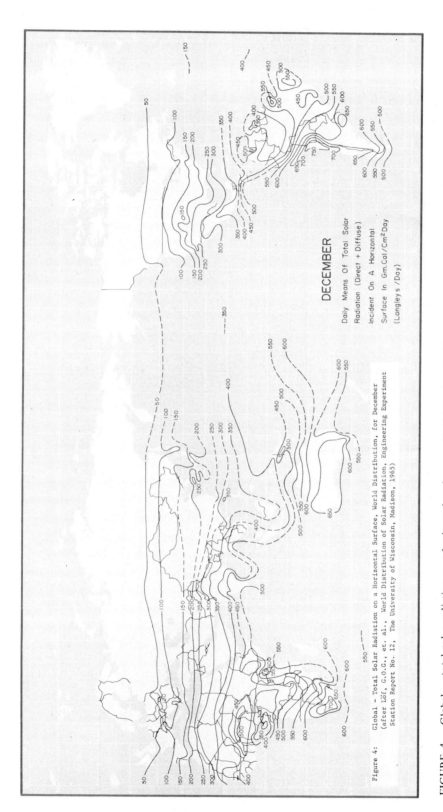

Figure 4:  Global - Total Solar Radiation on a Horizontal Surface, World Distribution, for December (after Löf, G.O.G., et. al., World Distribution of Solar Radiation, Engineering Experiment Station Report No. 12, The University of Wisconsin, Madison, 1965)

FIGURE 4.    Global — total solar radiation on a horizontal surface, world distribution, for December.    (After Löf, G.O.G., et. al., World Distribution of Solar Radiation, Engineering Experiment Station Report No. 12, The University of Wisconsin, Madison, 1965.)

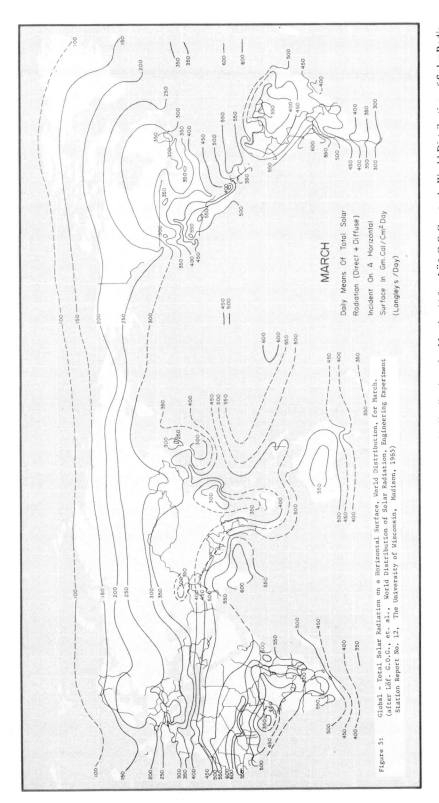

Figure 5: Global – Total Solar Radiation on a Horizontal Surface, World Distribution, for March. (after Löf, G.O.G., et. al., World Distribution of Solar Radiation, Engineering Experiment Station Report No. 12, The University of Wisconsin, Madison, 1965)

FIGURE 5. Global – total solar radiation on a horizontal surface, world distribution, for March. (After Löf, G.O.G., et. al., World Distribution of Solar Radiation, Engineering Experiment Station Report No. 12, The University of Wisconsin, Madison, 1965.)

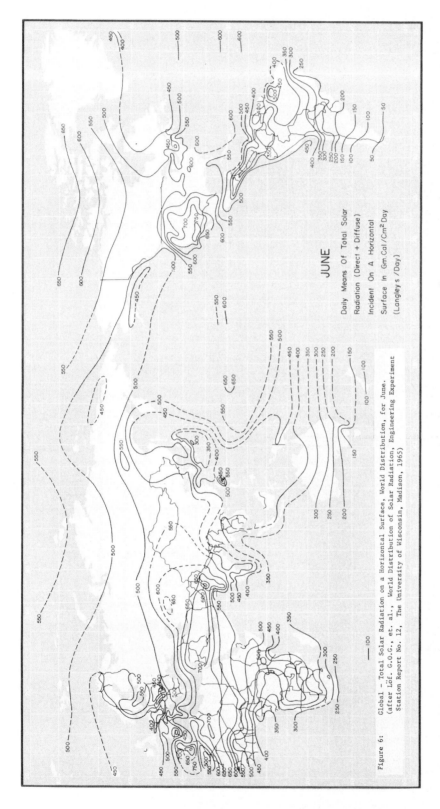

JUNE

Daily Means Of Total Solar
Radiation (Direct + Diffuse)
Incident On A Horizontal
Surface In Gm. Cal./Cm² Day
(Langleys /Day)

Figure 6:  Global – Total Solar Radiation on a Horizontal Surface, World Distribution, for June.
(after Löf. G.O.G., et. al., World Distribution of Solar Radiation, Engineering Experiment
Station Report No. 12, The University of Wisconsin, Madison, 1965)

FIGURE 6.   Global—total solar radiation on a horizontal surface, world distribution, for June.   (After Löf, G.O.G., et al., World Distribution of Solar Radiation, Engineering Experiment Station Report No. 12, The University of Wisconsin, Madison, 1965.)

# PENETRATION OF RADIANT ENERGY INTO THE AQUATIC ENVIRONMENT*

### Howard R. Gordon

## INTRODUCTION

The penetration of light into the sea is usually described in terms of the radiometric quantity irradiance, given the symbol E. This is defined[1] as the amount of radiant power (Watts) with wavelengths (nm) between $\lambda$ and $\lambda + \Delta\lambda$ crossing a flat surface of area $A(m^2)$ divided by the product of A and $\Delta\lambda$. If the radiant power is moving downward and the surface is horizontal, the irradiance is said to be downwelling, and at a depth Z is indicated by $E_d(Z,\lambda)$, while if the radiant power is moving upward, it is said to be upwelling and the irradiance is denoted by $E_u(Z,\lambda)$. Except for very clear blue waters,[2] $E_u(Z,\lambda) / E_d(Z,\lambda)$ is usually less than about 0.05 for $\lambda \geqslant 500$ nm, and $E_u(Z,\lambda)$ can usually be ignored in terms of its contribution to photosynthesis. Techniques for measuring $E_d(Z,\lambda)$ and $E_u(Z,\lambda)$ are given by Jerlov[1] and Tyler and Smith.[2] In biological studies, illuminance, the photometric equivalent of irradiance, is often measured, e.g., with a commercial foot candle meter. Tyler[3,4] has pointed out that since illuminance refers to the spectral response of the human eye, the significance of such measurements in photosynthetic studies is highly questionable.

The irradiance is not a fundamental property of the medium such as the absorption coefficient and volume scattering function,[1] but has been directly related to these properties using radiative transfer theory.[5] The dependence of the spectral composition of irradiance on depth is governed by the optical properties of the water and its constituents. These constituents include suspended organic and inorganic material such as phytoplankton, detritus, and silt, which both scatter and absorb, as well as dissolved detrital material (gelbstoff),[6] which only absorbs light. Natural waters, for which the properties of water itself play a significant role in determining the optical properties of the medium at all wavelengths, appear deep blue when observed from above. Otherwise, the water can appear various shades of green through yellow to even dark brown. Morel and Prieur[7] classified green ocean waters into two broad groups: Case 1 waters (MP-1), for which phytoplankton play a major role in determining the optical properties of the medium, and Case 2 waters (MP-2), for which they play a relatively minor role. This section deals largely with MP-1 waters.

### Spectral Irradiance

Corresponding to the large range of variation in the constituent concentration in natural waters, the spectral irradiance observed at a given depth can vary by several orders of magnitude. An example of this is provided in Figure 1, in which $E_d(Z,\lambda)$ for Crater Lake[2] (CL), well known for its extreme clarity, is compared to that for San Vicente Reservoir[2] (SVR) which is highly turbid because of large concentrations of phytoplankton and detrital material. Note that at Z = 5 m, the CL irradiance at 450 nm is roughly 200 times that in SVR. The strong absorption in the blue for the SVR data is largely because of phytoplankton pigments and gelbstoff.

A convenient way of characterizing the effect of the medium on the spectral composition of subsurface light is through the attenuation coefficient $K_d(\lambda)$ of downwelling irradiance defined by

$$K_d(\lambda) = \frac{-\text{Ln}[E_d(Z_2,\lambda)/E_d(Z_1,\lambda)]}{Z_2 - Z_1} \qquad (1)$$

* This manuscript was completed while the author was on leave of absence at NOAA Pacific Marine Environmental Laboratory, Seattle, Wash.

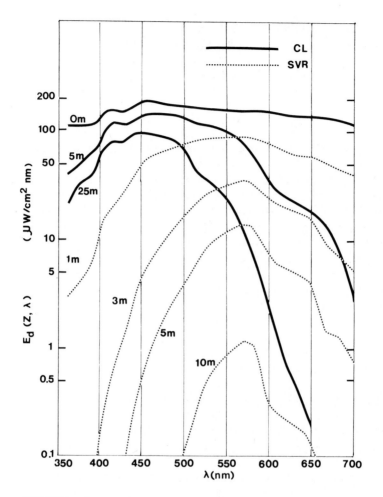

FIGURE 1. Spectra of downwelling irradiance[2] for CL and SVR. In order to facilitate comparison, the SVR data has been normalized to the CL $E_d(0,\lambda)$.

In general $K_d(\lambda)$ depends on $Z_1$ and $Z_2$; however, for $\Delta\lambda$ sufficiently small $K_d(\lambda)$ is nearly constant for homogeneous waters (except for very clear waters at wavelengths less than about 500 nm). For the data presented here, $K_d(\lambda)$ is for the surface layer with $Z_1 = 0$ and $Z_2$ is usually 20 m or less. When $K_d(\lambda)$ is constant,

$$E_d(Z,\lambda) = E_d(O,\lambda)e^{-K_d(\lambda)Z} \qquad (2)$$

and the spectral irradiance can be computed from that just beneath the surface. Jerlov[1] developed an optical classification scheme based on the observation that $K_d(\lambda)$ is largely controlled by the absorption properties of the water and its constituents, and the assumption that since gelbstoff is produced by the decay of suspended material, its absorption is an indicator of the particulate absorption as well. The resulting classification, which should apply only to MP-1 waters, is shown in Figure 2, where I to III refer to ocean types and 1 to 9 refer to coastal types. Note the progressive shift in the minimum from about 465 nm for Type I to 550 nm for Type 9. The essential difference between ocean III and coastal 1 is that the coastal regions are influenced by freshwater runoff from rivers which contain relatively more gelbstoff, resulting in enhanced attenuation at wavelengths less than about 550 nm. The Sargasso Sea has $K_d(\lambda)$ values some-

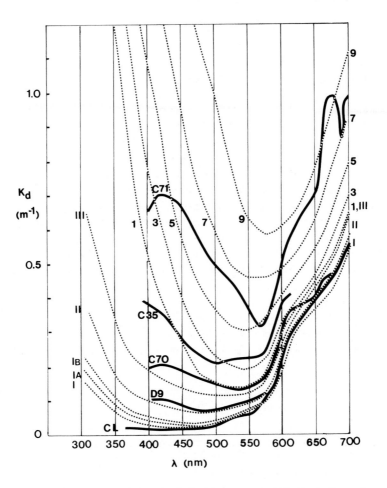

FIGURE 2.    Spectra of $K_d(\lambda)$ for Jerlov's[1] water types (thin lines). Shown for comparison are spectra from the African coast[7,8] (C35, C70, C71), the coast of Peru[9] (D9), and CL[2] (thick lines).

what lower than Type I, while in most of the open ocean the water is IB or clearer except in upwelling areas, for which II and III are common.

Superimposed on Figure 2 are samples of measurements of $K_d(\lambda)$ off Africa[7,8] near 21°N (C70 and C71 are MP-1, and C35 is MP-2) and off the coast of Peru[9] near 3°S (D9, also MP-1), along with those for CL. The CL data are seen to fall near I and the Peru data fall near II, while the African data cannot be uniquely classified with Jerlov's scheme. Morel and Caloumenos[10] presented a similar diagram which suggests that for concentrations of chlorophyll $a$ less than about 1 mg/m³, $K_d(\lambda)$ has a spectrum which falls between I and III.

For MP-1 waters, one expects that $K_d(\lambda)$ can be estimated from the concentration of chlorophyll and other phytoplankton pigments. Smith and Baker[11] have found, from statistical analysis of a rather large data set, that $K_d(\lambda)$ can be approximated by

$$K_d(\lambda) = K_w(\lambda) + k_1(\lambda)C_k, \qquad C_k \leqslant 1 \qquad (3)$$

$$K_d(\lambda) = K_w(\lambda) + K_x(\lambda) + k_2(\lambda)C_k, \qquad C_k \geqslant 1 \qquad (4)$$

where $K_w(\lambda)$ refers to the attenuation coefficient of water (e.g., CL in Figure 2), $K_x(\lambda)$ refers to the contribution to $K_d(\lambda)$ which cannot be directly related to phytoplankton pigments or water, and $k_1(\lambda)$ and $k_2(\lambda)$ are the specific attenuation coefficients ($m^{-1}$/mg $m^{-3}$) for the sum of the pigments ($C_K$) chlorophyll *a* and phaeophytin *a* averaged over one optical depth (defined to be the depth at which the number of quanta (see below) falls to $1/e$ of the number just beneath the surface). The various constants in Equations 3 and 4 are given in Table 1. The column labeled $\Delta K(\lambda)$ refers to the quantity

$$(k_1(\lambda) - k_2(\lambda) - K_x(\lambda))/k_1(\lambda)$$

which should be zero, as can be seen by equating Equations 3 and 4 for $C_K = 1$ mg/$m^3$. The deviation of $\Delta K(\lambda)$ from zero gives an indication of the lack of consistency of the analysis. Figure 3 compares $K_d(\lambda)$ computed using Equation 3 and 4 with the measured values from Figure 2. The agreement is generally good for MP-1 waters except for $\lambda \leqslant 450$ nm with $C_k \geqslant 1$ mg/$m^3$. Note, however, that for C35 (MP-2), Equation 4 will yield a considerably lower $K_d(\lambda)$ than observed, in contrast to the MP-1 waters for which the computed $K_d(\lambda)$ is usually too large, especially in the blue. Several examples of excellent agreement between computed and measured spectra are available.[11]

### Quantum Irradiance

Photosynthesis is essentially a quantum process; hence, it is important to be able to determine the number of quanta available as a function of depth in natural waters. Given an irradiance $E_d(Z,\lambda)$, the total number of quanta $Q(Z,\lambda_1,\lambda_2)$ between $\lambda_1$ and $\lambda_2$ is given by

$$Q(Z,\lambda_1,\lambda_2) = \frac{1}{hc} \int_{\lambda_1}^{\lambda_2} E_d(Z,\lambda)\,\lambda d\lambda \qquad (5)$$

and the total energy is given by

$$W(Z,\lambda_1,\lambda_2) = \int_{\lambda_1}^{\lambda_2} E_d(Z,\lambda)\,d\lambda \qquad (6)$$

where h is Planck's constant ($6.62 \times 10^{-34}$ J sec) and c is the speed of light ($2.99 \times 10^8$ m/sec). The units of Q are quanta/$m^2$/sec while those of W are W/$m^2$. For photosynthetic studies, Working Group 15[12] has recommended $\lambda_1 = 350$ nm and $\lambda_2 = 700$ nm, and henceforth these limits are assumed (unless otherwise specified), and $Q(Z,\lambda_1,\lambda_2)$ will be written $Q(Z)$. It is useful to define an attenuation coefficient for downwelling quanta through

$$K_q = \frac{-Ln[Q(Z_2)/Q(Z_1)]}{(Z_2 - Z_1)} \qquad (7)$$

however, since $E_d(Z,\lambda)$ decreases nearly exponentially with a decay constant $K_d(\lambda)$, which depends strongly on $\lambda$, $K_q$ must depend on depth. This dependence is especially strong near the surface because $K_d(\lambda)$ is large in the red region of the spectrum, while at great depth, $K_q$ will approximately equal the minimum $K_d(\lambda)$. An example of these effects is provided in Figure 4.

Q(Z) can be determined through measurement of $E_d(Z,\lambda)$ and evaluation of the integral in Equation 5. This, however, is a costly and time-consuming process, and investigators have been prompted to search for simpler methods. Equation 5 shows that an irradiance meter equipped with a filter such that the combined detector-filter response is proportional to $\lambda$ between 350 and 700 nm, and zero elsewhere can be used

Table 1
SPECTRAL ATTENUATION
COEFFICIENT $K_w(m^{-1})$ AND $K_x$ $(m^{-1})$ AND
SPECTRAL VALUES OF THE SPECIFIC
ATTENUATION COEFFICIENTS $k_1(\lambda)$ and $k_2(\lambda)$[11]

| $\lambda$ (nm) | $K_w(\lambda)$ | $K_x(\lambda)$ | $k_1(\lambda)$ | $k_2(\lambda)$ | $\Delta K(\lambda)$ |
|---|---|---|---|---|---|
| 350 | 0.059 | 0.177 | 0.249 | 0.066 | 0.024 |
| 355 | 0.055 | 0.177 | 0.249 | 0.066 | 0.024 |
| 360 | 0.051 | 0.177 | 0.249 | 0.666 | 0.024 |
| 365 | 0.045 | 0.178 | 0.248 | 0.063 | 0.028 |
| 370 | 0.044 | 0.179 | 0.245 | 0.061 | 0.020 |
| 375 | 0.043 | 0.179 | 0.240 | 0.058 | 0.013 |
| 380 | 0.040 | 0.179 | 0.237 | 0.055 | 0.014 |
| 385 | 0.036 | 0.179 | 0.232 | 0.053 | 0.000 |
| 390 | 0.031 | 0.177 | 0.227 | 0.051 | −0.009 |
| 395 | 0.029 | 0.175 | 0.223 | 0.050 | −0.009 |
| 400 | 0.027 | 0.172 | 0.216 | 0.049 | −0.025 |
| 405 | 0.026 | 0.167 | 0.210 | 0.048 | −0.027 |
| 410 | 0.025 | 0.162 | 0.205 | 0.047 | −0.024 |
| 415 | 0.024 | 0.156 | 0.200 | 0.046 | −0.013 |
| 420 | 0.024 | 0.150 | 0.194 | 0.045 | −0.005 |
| 425 | 0.023 | 0.145 | 0.187 | 0.044 | −0.010 |
| 430 | 0.022 | 0.137 | 0.181 | 0.042 | 0.006 |
| 435 | 0.022 | 0.132 | 0.175 | 0.041 | 0.007 |
| 440 | 0.022 | 0.125 | 0.168 | 0.039 | 0.021 |
| 445 | 0.023 | 0.121 | 0.163 | 0.038 | 0.022 |
| 450 | 0.023 | 0.116 | 0.158 | 0.037 | 0.030 |
| 455 | 0.023 | 0.112 | 0.150 | 0.036 | 0.013 |
| 460 | 0.023 | 0.110 | 0.146 | 0.034 | 0.011 |
| 465 | 0.023 | 0.104 | 0.141 | 0.033 | 0.029 |
| 470 | 0.023 | 0.100 | 0.135 | 0.031 | 0.027 |
| 475 | 0.022 | 0.095 | 0.130 | 0.030 | 0.038 |
| 480 | 0.022 | 0.091 | 0.125 | 0.029 | 0.034 |
| 485 | 0.024 | 0.087 | 0.120 | 0.027 | 0.042 |
| 490 | 0.025 | 0.084 | 0.115 | 0.026 | 0.043 |
| 495 | 0.027 | 0.080 | 0.110 | 0.025 | 0.045 |
| 500 | 0.029 | 0.077 | 0.105 | 0.024 | 0.035 |
| 505 | 0.033 | 0.074 | 0.102 | 0.022 | 0.056 |
| 510 | 0.037 | 0.071 | 0.096 | 0.021 | 0.039 |
| 515 | 0.043 | 0.069 | 0.093 | 0.020 | 0.045 |
| 520 | 0.048 | 0.066 | 0.088 | 0.019 | 0.033 |
| 525 | 0.050 | 0.064 | 0.085 | 0.017 | 0.047 |
| 530 | 0.050 | 0.061 | 0.084 | 0.016 | 0.085 |
| 535 | 0.052 | 0.060 | 0.080 | 0.015 | 0.062 |
| 540 | 0.055 | 0.059 | 0.076 | 0.014 | 0.042 |
| 545 | 0.059 | 0.056 | 0.073 | 0.013 | 0.049 |
| 550 | 0.063 | 0.055 | 0.070 | 0.012 | 0.044 |
| 555 | 0.067 | 0.054 | 0.070 | 0.011 | 0.070 |
| 560 | 0.071 | 0.053 | 0.070 | 0.011 | 0.087 |
| 565 | 0.074 | 0.052 | 0.071 | 0.010 | 0.120 |
| 570 | 0.077 | 0.053 | 0.072 | 0.009 | 0.133 |
| 575 | 0.082 | 0.054 | 0.074 | 0.009 | 0.154 |
| 580 | 0.088 | 0.056 | 0.077 | 0.008 | 0.160 |
| 585 | 0.099 | 0.059 | 0.085 | 0.008 | 0.213 |
| 590 | 0.107 | 0.066 | 0.095 | 0.007 | 0.223 |
| 595 | 0.121 | 0.091 | 0.110 | 0.007 | 0.105 |
| 600 | 0.131 | 0.131 | 0.125 | 0.007 | −0.106 |
| 605 | 0.146 | 0.150 | 0.148 | 0.007 | −0.060 |

Table 1 (continued)
## SPECTRAL ATTENUATION
### COEFFICIENT $K_w(m^{-1})$ AND $K_x$ $(m^{-1})$ AND SPECTRAL VALUES OF THE SPECIFIC ATTENUATION COEFFICIENTS $k_1(\lambda)$ and $k_2(\lambda)$[11]

| $\lambda$ (nm) | $K_w(\lambda)$ | $K_x(\lambda)$ | $k_1(\lambda)$ | $k_2(\lambda)$ | $\Delta K(\lambda)$ |
|---|---|---|---|---|---|
| 610 | 0.170 | 0.159 | 0.168 | 0.007 | 0.014 |
| 615 | 0.188 | 0.165 | 0.184 | 0.006 | 0.069 |
| 620 | 0.212 | 0.167 | 0.195 | 0.006 | 0.109 |
| 625 | 0.244 | 0.169 | 0.205 | 0.006 | 0.146 |
| 630 | 0.277 | 0.161 | 0.213 | 0.006 | 0.213 |
| 635 | 0.300 | 0.137 | 0.222 | 0.007 | 0.350 |
| 640 | 0.327 | 0.117 | 0.227 | 0.007 | 0.449 |
| 645 | 0.339 | 0.095 | 0.231 | 0.008 | 0.554 |
| 650 | 0.336 | 0.061 | 0.225 | 0.009 | 0.686 |
| 655 | 0.337 | 0.037 | 0.205 | 0.011 | 0.765 |
| 660 | 0.390 | 0.015 | 0.180 | 0.012 | 0.850 |
| 665 | 0.425 | 0.002 | 0.156 | 0.014 | 0.896 |
| 670 | 0.460 | 0.0 | 0.118 | 0.015 | 0.873 |
| 675 | 0.485 | 0.0 | 0.088 | 0.016 | 0.823 |
| 680 | 0.510 | 0.0 | 0.068 | 0.015 | 0.779 |
| 685 | 0.540 | 0.0 | 0.045 | 0.014 | 0.693 |
| 690 | 0.570 | 0.0 | 0.028 | 0.011 | 0.596 |
| 695 | 0.600 | 0.0 | 0.015 | 0.008 | 0.460 |
| 700 | 0.630 | 0.0 | 0.008 | 0.004 | 0.450 |

to measure Q(Z) (in relative units). The design, construction, and calibration of such a quantum meter are described by Jerlov and Nygard.[13] Hojerslev[14] used this meter to show that the depth of the euphotic zone (the depth at which the number of quanta is reduced to 1% of the number just beneath the sea surface) is virtually independent of solar elevation ($h_s$) for elevations greater than $10°$. Hojerslev and Jerlov[15] related the color ratio ($F_E$) defined by

$$F_E \equiv \frac{E_u(1 \text{ m}, 450 \text{ nm})}{E_u(1 \text{ m}, 520 \text{ nm})}$$

to the depth $Z_p$, where Q(Z) falls to p% of its value just beneath the surface, through

$$Z_p = A_p + B_p F_E + C_p F_E^2 \qquad (8)$$

where $A_p$, $B_p$, and $C_p$ are given in Table 2. It has also been shown[16] that for water Types I to III, a simple relationship exists between Q(Z) and E(Z,465 nm):

$$\frac{Q(Z)}{Q(O)} = \frac{E(Z, 465 \text{ nm})}{E(O, 465 \text{ nm})} \left[ 0.23 + \left( \frac{5.6}{7.3 + Z} \right) \right] \qquad (9)$$

(Compare Figure 4 and Equation 9). Using Equation 9, an uncalibrated irradiance meter with a spectral filter peaked at 465 nm can be used to determine Q(Z)/Q(O) (and hence the depth of the euphotic zone) in the open ocean. The value of Q(O) on clear days (cloud cover $<^3/_8$) when the sun is not obscured by clouds can be estimated from[17]

$$Q(O) = (1.2 \times 10^{21} \text{ quanta/m}^2 \text{ sec}) (\text{Sin } h_s)^{1.4}$$

where again $h_s$ is the solar elevation.

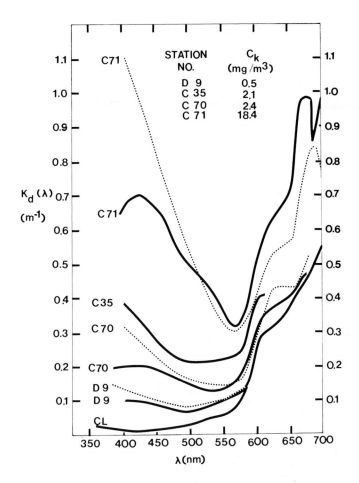

FIGURE 3.   Comparison between $K_d(\lambda)$ observations cited in Figure 2 (solid lines) and $K_d(\lambda)$ values computed from $C_K$ using Equations 3 and 4 (dashed lines). No computation is made for C35 which is an example of MP-2 waters.

As with $K_d(\lambda)$, $Q(Z)$ should depend on the concentration of pigments $C_K$. Smith and Baker[18] have established a relationship between $K_q$ (Equation 7 with $Z_1 = 0$ and $Z_2$ the depth of the euphotic zone) and $C'_k$, the chlorophyll *a* concentration averaged over a depth 0 to $K_q^{-1}$ (called the remote sensing penetration depth).[19] Their statistical analysis indicated that about 65% of the variance in $K_q^{-1}$ can be explained by variations in $C'_k$ (and vice versa) by

$$K_q^{-1} = 8.78 - 7.51 \log_{10} C'_K \tag{10}$$

where $K_q^{-1}$ is in meters and $C'_K$ in mg/m³. Morel and Smith[20] carried out an analysis of spectral irradiance data which demonstrated that between 400 and 700 nm the ratio $Q(Z)/W(Z)$ is essentially constant, depending only on the chlorophyll *a* concentration. For concentrations between 0.02 and 10 mg/m³, Table 3 gives the associated ratio $Q(Z)/W(Z)$. Furthermore, the analysis indicated that for $22° \leq h_s \leq 90°$ and sky conditions from clear to overcast, $Q/W$ just above the sea surface is $2.77 \times 10^{18}$ quanta/ W/sec to within about $\pm 0.5\%$.

It must be emphasized that the only method of accurately determining $E_d(Z,\lambda)$, $K_d(\lambda)$,

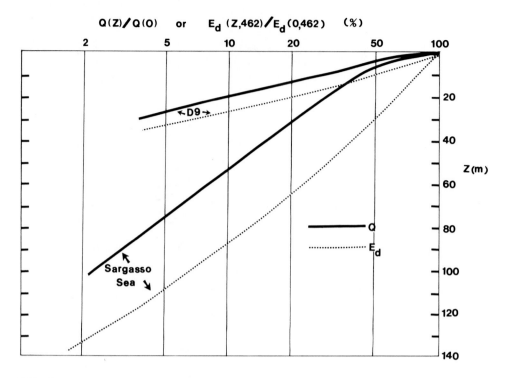

FIGURE 4.  Comparison between profiles of quantum irradiance [Q(Z)] and irradiance at 462 nm [$E_d$(Z,462)] for the Sargasso Sea and D9.[9]

<table>
<tr><td colspan="4">

Table 2

COEFFICIENTS $A_p$, $B_p$, AND $C_p$
FOR USE IN EQUATION 8[15]
</td></tr>
</table>

| p (%) | $A_p$ (m) | $B_p$ (m) | $C_p$ (m) |
|---|---|---|---|
| 30 | 2.0 | 5.9 | −0.35 |
| 10 | 4.0 | 12.2 | 0.00 |
| 3 | 7.0 | 19.5 | 0.00 |
| 1 | 9.8 | 25.2 | 0.00 |

Table 3

Q(Z)/W(Z) AS A FUNCTION OF
THE CHLOROPHYLL A(CNL A)
CONCENTRATION[20]

| Chl a (mg/m³) | Q(Z)/W(Z) ($10^{18}$ quanta/W/sec) |
|---|---|
| <0.10 | 2.35 ± 0.13 |
| 0.10—1.0 | 2.50 ± 0.13 |
| >1.0 | 2.65 ± 0.13 |

or Q(Z) is by actually performing the measurement at the desired location and time. The relationships summarized here (Equations 3, 4, 10, and to a lesser extent 9) are the result of statistical analysis of experimental data and as such are not exact. They can, however, be used for the purpose of designing experiments and obtaining an estimate of irradiance and quanta penetration into MP-1 waters.

## ADDENDUM

A revised version of Table 1 is scheduled to be published by K. S. Baker and R. C. Smith in *Limnology and Oceanography* in 1981.

# REFERENCES

1. Jerlov, N. G., *Marine Optics, Elsevier Oceanography Series*, Vol. 14, Elsevier, Amsterdam, 1976.
2. Tyler, J. E. and Smith, R. C., *Measurements of Spectral Irradiance Under Water. Ocean Series 1*, Gordon and Breach, New York, 1970.
3. Tyler, J. E., Lux vs. quanta, *Limnol. Oceanogr.*, 18, 810, 1973.
4. Tyler, J. E., Applied radiometry, *Oceanogr. Mar. Biol. Annu. Rev.*, 11, 11, 1973.
5. Gordon, H. R., Brown, O. B., and Jacobs, M. M., Computed relationships between the inherent and apparent optical properties of a flat homogeneous ocean, *Appl. Opt.*, 14, 417, 1975.
6. Kalle, K., The problem of the Gelbstoff in the sea, *Oceanogr. Mar. Biol. Annu. Rev.*, 4, 91, 1966.
7. Morel, A. and Prieur, L., Analysis of the variations in ocean color, *Limnol. Oceanogr.*, 22, 709, 1977.
8. Morel, A. and Prieur, L., Analyse Spectrale des Coefficients d'Attenuation Diffuse, de Reflecion Diffuse, d'Absorption et de Retrodiffusion pour Diverse Regions Marines, Rep. No. 17, Laboratoire d'Oceanographie Physique, Villefranche-sur Mer, 1975.
9. Morel, A., Measurement of the Spectral and Total Radiant Flux, in SCOR Data Rep. Discoverer Expedition, Vol. 2, Scripps Inst. Oceanogr. Ref. 73-16, 1973.
10. Morel, A. and Caloumenos, L., Variabilite de la repartition spectrale de l'energie photosynthetique, *Tethys*, 6, 93, 1974.
11. Smith, R. C. and Baker, K. S., Optical classification of natural waters, *Limnol. Oceanogr.*, 23, 260, 1978.
12. Working Group 15, Report of the First Meeting of the Joint Group of Experts in Photosynthetic Radiant Energy, UNESCO Technical Pap. Marine Science 2, 1965.
13. Jerlov, N. G. and Nygard, K., A Quanta and Energy Meter for Photosynthetic Studies, University of Copenhagen, Institute of Phys. Oceanography Rep. No. 10, 1969.
14. Hojerslev, N. K., Daylight Measurements for Photosynthetic Studies in the Western Mediterranean, University of Copenhagen, Institute of Phys. Oceanography Rep. No. 26, 1974.
15. Hojerslev, N. K. and Jerlov, N. G., The Use of the Colour Index for Determining Quanta Irradiance in the Sea, University of Copenhagen, Institute of Phys. Oceanography Rep. No. 35, 1977.
16. Jerlov, N. G., A Simple Method for Measuring Quanta Irradiance in the Ocean, University of Copenhagen, Institute of Phys. Oceanography Rep. No. 24, 1974.
17. Hojerslev, N. K., Water Colour and Its Relation to Primary Productivity, Boundary-Layer Meteorology, 18, 302, 1980.
18. Smith, R. C. and Baker, K. S., The bio-optical state of ocean water and remote sensing, *Limnol. Oceanogr.*, 23, 247, 1978.
19. Gordon, H. R. and McCluney, W. R., Estimation of the depth of sunlight penetration in the sea for remote sensing, *Appl. Opt.*, 14, 413, 1975.
20. Morel, A. and Smith, R. C., Relation between total quanta and total energy for aquatic photosynthesis, *Limnol. Oceanogr.*, 19, 591, 1974.

# CARBON DIOXIDE IN TERRESTRIAL ENVIRONMENTS

## Clanton C. Black

The atmosphere of the earth provides a large reservoir of $CO_2$ which photosynthetic organisms use as the oxidized carbon source for photosynthesis. However, this large reservoir does not insure a constant concentration of $CO_2$. In general, one can state that air near sea level now contains about 330 $\mu L$ of $CO_2$/L of air. However, the $CO_2$ supply available to a specific plant can vary markedly from this value. The $CO_2$ supply can vary each day and with each season of the year. In addition, we face the prediction of drastic changes in atmospheric $CO_2$ levels within the next century.

$CO_2$ levels vary about a plant throughout a day/night cycle, but plants readily cope with these daily changes. Plant photosynthesis responds to $CO_2$ level in the plant's atmosphere in a linear fashion.[1-3] However, it is well established that $C_3$ and $C_4$ photosynthesis respond differently to changes in $CO_2$ level in the atmosphere. Photosynthesis in $C_3$ plants is not saturated with $CO_2$ at current atmospheric $CO_2$ levels, whereas $C_4$ plants are close to saturation in air.[3] Similarly short-term growth (21 days) studies also showed a 200 to 400% increase in the dry matter production of $C_3$ plants, but only small changes ($\pm 10\%$) in the dry matter production of $C_4$ plants occurred, even at $CO_2$ levels eight times air.[3]

Within a stand or canopy of plants, the $CO_2$ levels can drop during a day when photosynthesis is maximal to values near 250 $\mu L$/L and rise to values near 400 $\mu L$/L in the early morning. These daily fluctuations depend upon many factors, such as the intensity of photosynthesis and stand thickness plus the degree and speed of mixing the large atmospheric reservoir of $CO_2$ within the microclimate of each plant.

In addition to daily microclimatic changes in $CO_2$ levels around a plant, seasonal changes occur in atmospheric $CO_2$ levels. However, not only is there a seasonal fluctuation in air $CO_2$, but we now believe that there is a definite trend for the levels of atmospheric $CO_2$ to increase.

Hence, today the responses of plants to atmospheric $CO_2$ levels is a matter of increasing concern. In continuous studies on the $CO_2$ levels in air, several air monitoring stations now have reported seasonal changes in air $CO_2$ levels and reported that the $CO_2$ concentration in air is rising. Data from Hawaii are shown in Figure 1 for nearly the last two decades.[4,5] There is an annual seasonal cycle of a winter high and summer low in $CO_2$ concentration which correlates well with the seasonal photosynthetic activity of plants (see Figure 1).

However, there also is an increase in the air $CO_2$ concentration over the past two decades in which air $CO_2$ has been closely monitored. There is little doubt that air $CO_2$ levels are rising with increases ranging from 0.5 to 2.1 $\mu L$/L each year (see Figure 1).

Since there is a trend for the atmospheric levels of $CO_2$ to increase, several workers have extrapolated these values into the next century.[4-10] Some of the data in Figure 1 are plotted along with other $CO_2$ level values in Figure 2 to reveal a very disturbing pattern. It is true that the values past 1978 in Figure 2 are predicted, but they do seem to be reasonable predictions. The predictions are based on consumption of fossil fuel, deforestation, ocean carbon sinks, and other considerations.[4-10] Of course, time will show the correct values; however, in anticipation of rising air $CO_2$ levels, workers in biosolar energy should consider the long term effects of elevated $CO_2$ levels on plants.

Plants respond to $CO_2$ in a great variety of physiological processes in addition to photosynthesis, such as respiration, allocation of photosynthetic products, growth rates of a variety of organs, as well as the whole plant, transpiration, stomatal conductance, seed germination, nitrogen fixation, and tuber formation. This incomplete

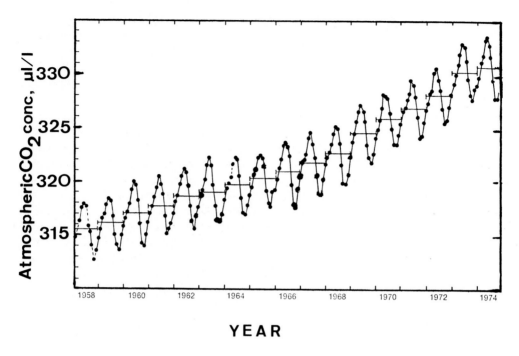

FIGURE 1.   Atmospheric changes in the $CO_2$ concentration at Mauna Loa Observatory, Hawaii. Each point is a monthly average.[4,5]

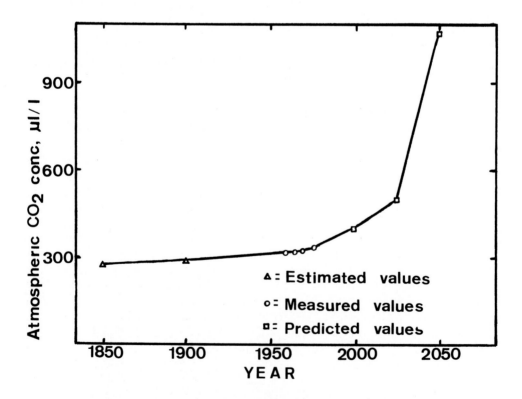

FIGURE 2.   Past, current, and predicted changes of atmospheric $CO_2$ concentrations. The values prior to 1979 were collated from the literature[4-10] and the predicted values were extrapolated.

list illustrates the fact that, if Figure 2 is correct, we do not know what will happen to the growth of plants over the next century!

In brief, a static or constant atmospheric $CO_2$ supply is not available for photosynthesis. Certainly all indications are that the concentration of $CO_2$ in air is increasing. Biosolar research workers should measure and consider the $CO_2$ levels their photosynthetic organisms will encounter when predicting such parameters as yields or products of photosynthesis.

## REFERENCES

1. **Gaastra, P.**, Photosynthesis of crop plants as influenced by light, carbon dioxide, temperature, and stomatal diffusion resistance, *Meded. Landbouwhogesch. Wageningen,* 59, 1, 1959.
2. **Hesketh, J. D.**, Limitations to photosynthesis responsible for differences among species, *Crop Sci.,* 3, 493, 1963.
3. **Akita, S. and Tanaka, I.**, Studies on the merchanism of differences in photosynthesis among species. IV. The differential response in dry matter production between $C_3$ and $C_4$ species to atmospheric carbon dioxide enrichment, *Proc. Crop Sci. Soc. Jpn.,* 42, 288, 1973.
4. **Keeling, C. D., Bacastow, R. B., Bainbridge, A. E., Ekdahl, C. A., Guenther, P. R., Waterman, L. S., and Chin, J. F. S.**, Atmospheric carbon dioxide variations at Mauna Loa Observatory, Hawaii, *Tellus,* 28, 538, 1976.
5. **Baes, C. F., Goeller, H. E., Olson, J. S., and Rotty, R. M.**, Carbon dioxide and climate: the uncontrolled experiment, *Am. Sci.,* 65, 310, 1977.
6. **Bolin, B.**, The carbon cycle, *Sci. Am.,* September, 49, 1970.
7. **Woodwell, G. M., Whittaker, R. M., Reiners, W. A., Likens, G. E., Delwiche, C. C., and Botkin, D. B.**, The biota and the world carbon budget, *Science,* 199, 141, 1978.
8. **Stuiver, M.**, Atmospheric carbon dioxide and carbon reservoir changes, *Science,* 199, 253, 1978.
9. **Siegenthaler, U. and Oeschger, H.**, Predicting future atmospheric carbon dioxide levels, *Science,* 199, 388, 1978.
10. **Wong, C. S.**, Atmospheric input of carbon dioxide from burning wood, *Science,* 200, 197, 1978.

# THE CARBONATE SYSTEM IN SEAWATER

Frank J. Millero and John W. Morse

## INTRODUCTION

The carbonate system in seawater has received increased attention in recent years. This attention has been the result of concern over the rate at which oceans can absorb the $CO_2$ which is being added to the atmosphere from the burning of fossil fuels and the environmental consequences of a significant increase in oceanic $CO_2$. A major sink in the oceans for $CO_2$ is in deep sea sediments where it is deposited as calcium carbonate. The transformation of $CO_2$ to calcium carbonate takes place primarily in surface ocean waters which are supersaturated with respect to the two important calcium carbonate phases, calcite and aragonite. Although the surface ocean waters are supersaturated, the most important mode of calcium carbonate formation is by pelagic organisms, which secrete calcium carbonate, rather than by direct precipitation. As these organisms die, the calcium carbonate which they have secreted sinks into the deep oceans. The saturation state of deep ocean waters decreases with increasing depth because of the combined influences of pressure, temperature, and the production of $CO_2$ from the oxidation of organic matter. This ultimately results in the waters becoming saturated and dissolution of a portion of the calcium carbonate before it can be buried and isolated from the system. This chapter briefly reviews the various components of the carbonate system in seawater.

## THERMODYNAMIC EQUATIONS

The carbonate system can be characterized by the following equations:

$$CO_2 \, (gas) \rightleftarrows CO_2 \, (aq) \tag{1}$$

$$CO_2 \, (aq) + H_2O \rightleftarrows H_2CO_3 \tag{2}$$

$$H_2CO_3 \rightleftarrows H^+ + HCO_3^- \tag{3}$$

$$HCO_3^- \rightleftarrows H^+ + CO_3^{2-} \tag{4}$$

$$CaCO_3 \, (solid) \rightleftarrows Ca^{2+} + CO_3^{2-} \tag{5}$$

The thermodynamic constants for these equations can be defined in terms of the activities of the products and reactants. Since the activity coefficients of the various components are difficult to determine, it is more convenient to use apparent constants.[1] Apparent constants are analogous to thermodynamic constants; however, they are defined in terms of the total concentration of the various species and the apparent activity of the proton. Apparent constants are functions of temperature, pressure, and the composition of the solution. Since seawater has nearly the same relative concentration, the concentration dependence of the apparent constants for open ocean waters can be a function of the salinity (the total solids expressed in parts per thousand).

A number of studies have been made to determine the apparent constants of seawater.[2-10] Differences in the pH scale used to determine the constants have led to a different definition of the constants. In the early work of Buch,[2-4] the Sorensen pH scale was used; in the work of Lyman[5] and Mehrbach et al.,[6] the National Bureau of Standards (NBS) scale[11] was used. In the work of Hansson,[7-10] a new TRIS buffer scale[12] was used. At a given temperature and salinity, the various scales are related to each other; thus, it is possible to derive a relationship between the various constants.[13] We will use the apparent constants based on the apparent activity of the proton $a_H$ obtained by using a glass electrode standardized by using NBS buffers.[11] The value of $a_H$ determined by using the NBS scale is not equal to the true activity of the proton in seawater because of liquid junction potentials.[14] The true activity of the proton at $35^o/_{oo}$ salinity is 1.13 times the measured value.[14] In actual practice, this difference is not important, since the constants used have been determined with the same pH scale.

The apparent constants for the first and second ionization of carbonic acid are given by

$$K_1' = a_H [HCO_3]/[CO_2^*] \tag{6}$$

$$K_2' = a_H [CO_3]/[HCO_3] \tag{7}$$

where [ ] indicates concentration in moles per kilogram of seawater. The first ionization uses the hydration convention, $[CO_2^*] = [H_2CO_3] + [CO_2]$ which is obtained by combining Equations 2 and 3,

$$CO_2 (aq) + H_2O \rightleftarrows H^+ + HCO_3^- \tag{8}$$

The concentration of $CO_2^*$ in seawater in equilibrium with $CO_2$ in the gas phase at a given partial pressure $P_{CO2}$ is determined from

$$[CO_2^*] = P_{CO_2}/\alpha \tag{9}$$

where $\alpha$ is the Henry's law coefficient. The values of $\alpha$ can be determined from the equation of Weiss[15] which is based on the measurements of Murray and Riley:[16]

$$\ln \alpha = -60.2409 + 93.4517 (100/T) + 23.3585 \ln(T/100)$$

$$+ S[0.023517 - 0.023656 (T/100) + 0.0047036 (T/100)^2] \tag{10}$$

where S is the salinity ($^o/_{oo}$), T is the temperature ($^\circ K$), and $\alpha$ is in units of mol kg$^{-1}$ atm ($\alpha = 1.4 \times 10^{-4}$ mol kg$^{-1}$ atm). Values of $\alpha$ from Equation 10 are given in Table 1.

The various components of the carbonate system can be determined by measuring[17] at least two of the following parameters: (1) pH, (2) $A_T$, (3) $\Sigma CO_2$, and (4) $P_{CO2}$. Park[17] has derived the various equations relating these measured parameters to the components of the carbonate system and also discussed the best parameters to use for studying the $CO_2$ system. The classical method used to study the carbonate system is measuring the pH and total alkalinity, $A_T$. The pH is measured with glass and calomel electrodes calibrated with NBS buffers. The total alkalinity, $A_T$, is defined by

$$A_T = [HCO_3^-] + 2[CO_3^{2-}] + [B(OH)_4^-] + [OH^-] - [H^+] + \Sigma[bases] \tag{11}$$

where $[B(OH)_4^-]$ is the concentration of borate ion from the ionization of boric acid

$$B(OH)_3 + H_2O \rightleftarrows H^+ + B(OH)_4^- \tag{12}$$

The value of $[OH^-] - [H^+]$ and other bases able to accept a proton (phosphate ions, silicate ions, $Mg(OH)^+$, etc.) generally do not contribute significantly to $A_T$ and usually

## Table 1
### THE SOLUBILITY OF CARBON DIOXIDE IN WATER AND SEAWATER
$\alpha$(mol kg$^{-1}$ atm)

| S (%$_{oo}$) | 0°C | 5°C | 10°C | 15°C | 20°C | 25°C | 30°C | 35°C | 40°C |
|---|---|---|---|---|---|---|---|---|---|
| 0 | 0.07758 | 0.06407 | 0.05367 | 0.04556 | 0.03916 | 0.03406 | 0.02995 | 0.02662 | 0.02389 |
| 5 | 0.07528 | 0.06221 | 0.05215 | 0.04430 | 0.03812 | 0.03319 | 0.02922 | 0.02600 | 0.02336 |
| 10 | 0.07305 | 0.06040 | 0.05067 | 0.04308 | 0.03710 | 0.03233 | 0.02850 | 0.02539 | 0.02285 |
| 15 | 0.07089 | 0.05865 | 0.04923 | 0.04189 | 0.03611 | 0.03150 | 0.02780 | 0.02480 | 0.02235 |
| 20 | 0.06880 | 0.05695 | 0.04784 | 0.04074 | 0.03916 | 0.03070 | 0.02712 | 0.02422 | 0.02186 |
| 25 | 0.06676 | 0.05529 | 0.04648 | 0.03961 | 0.03421 | 0.02991 | 0.02645 | 0.02366 | 0.02138 |
| 30 | 0.06479 | 0.05369 | 0.04516 | 0.03852 | 0.03330 | 0.02914 | 0.02580 | 0.02311 | 0.02091 |
| 35 | 0.06287 | 0.05213 | 0.04388 | 0.03746 | 0.03241 | 0.02839 | 0.02517 | 0.02257 | 0.02045 |
| 40 | 0.06101 | 0.05062 | 0.04263 | 0.03643 | 0.03154 | 0.02766 | 0.02455 | 0.02204 | 0.02000 |

need not be considered. The $A_T$ is obtained by titrating seawater with HCl to the carbonate end point.[18] The carbonate alkalinity, $A_C$, given by

$$A_C = A_T - [B(OH)_4^-] = [HCO_3^-] + 2[CO_3^{2-}] \tag{13}$$

is determined from $A_T$ at a given pH using the apparent constant of boric acid defined by

$$K_B' = a_H[B(OH)_4^-]/[B(OH)_3] \tag{14}$$

The concentration of $B(OH)_4^-$ is obtained from

$$[B(OH)_4^-] = K_B'[B]_T/(K_B' + a_H) \tag{15}$$

where $a_H$ and $K_B'$ are for the *in situ* conditions and the total concentration of boron, $[B]_T$, is determined from[19]

$$[B]_T = 1.212 \times 10^{-5} S \tag{16}$$

The apparent constants of boric acid in seawater at 1 atm are given by[13]

$$\ell n\, K_B' = 148.0248 - 8966.90/T - 24.4344\, \ell n\, T + (0.0473 + 49.10/T)S^{1/2} \tag{17}$$

which are based on the measurements of Lyman[5] and has a $\sigma = 0.05$ in $\ell n\, K_B'$. Values of $\ell n\, K_B'$ determined from Equation 17 are given in Table 2. The effect of applied pressure (P = depth/10m) on the apparent constants of various acids is given by

$$\ell n(K_i^P/K_i^O) = (-\Delta V_i/RT)P + (0.05\, \Delta K_i/RT)P^2 \tag{18}$$

where R is the gas constant 82.06 cm$^3$ atm K$^{-1}$ mol$^{-1}$, T is the absolute temperature (°K), $\Delta V_i$ (cm$^3$mol$^{-1}$) and $\Delta K_i$(cm$^3$mol$^{-1}$atm$^{-1}$) are the partial molal volume and compressibility changes for the ionization. The volume ($\Delta V_B$) and compressibility change ($\Delta K_B$) for the ionization of boric acid can be determined from[13]

$$-\Delta V_B = 29.69 - 0.300\, (S-35) - 0.1674t + 1.66 \times 10^{-3}\, t^2 \tag{19}$$

$$-10^3\, \Delta K_B = 3.34 - 0.368\, (S-35) - 9.173 \times 10^{-3}\, t$$

$$- 2.16 \times 10^{-3}\, t^2 \tag{20}$$

Table 2
APPARENT CONSTANTS FOR THE IONIZATION OF BORIC ACID IN
SEAWATER ($-\mathit{l}$n $K_B$)

| S (%) | 0°C | 5°C | 10°C | 15°C | 20°C | 25°C | 30°C | 35°C | 40°C |
|---|---|---|---|---|---|---|---|---|---|
| 0 | 21.880 | 21.734 | 21.600 | 21.478 | 21.367 | 21.268 | 21.178 | 21.098 | 21.026 |
| 5 | 21.373 | 21.233 | 21.106 | 20.991 | 20.887 | 20.794 | 20.710 | 20.636 | 20.570 |
| 10 | 21.162 | 21.026 | 20.902 | 20.789 | 20.688 | 20.597 | 20.516 | 20.444 | 20.381 |
| 15 | 21.001 | 20.867 | 20.745 | 20.635 | 20.536 | 20.447 | 20.368 | 20.297 | 20.236 |
| 20 | 20.865 | 20.733 | 20.613 | 20.504 | 20.407 | 20.320 | 20.242 | 20.174 | 20.114 |
| 25 | 20.745 | 20.614 | 20.496 | 20.389 | 20.293 | 20.208 | 20.132 | 20.065 | 20.006 |
| 30 | 20.637 | 20.508 | 20.391 | 20.285 | 20.191 | 20.107 | 20.032 | 19.966 | 19.909 |
| 35 | 20.537 | 20.409 | 20.294 | 20.190 | 20.097 | 20.014 | 19.940 | 19.875 | 19.819 |
| 40 | 20.444 | 20.318 | 20.204 | 20.101 | 20.009 | 19.927 | 19.854 | 19.791 | 19.736 |

Table 3
EFFECT OF PRESSURE ON THE APPARENT
CONSTANTS FOR THE IONIZATION OF CARBONIC
AND BORIC ACIDS AND THE SOLUBILITY OF
CALCIUM CARBONATE AT 2°C AND 34.5% S

| Depth (m) | $K_1^P/K_1^O$ | $K_2^P/K_2^O$ | $K_B^P/K_B^O$ | $K_C^P/K_C^O$ | $K_A^P/K_A^O$ |
|---|---|---|---|---|---|
| 1,000 | 1.117 | 1.074 | 1.139 | 1.232 | 1.217 |
| 1,500 | 1.179 | 1.113 | 1.214 | 1.365 | 1.340 |
| 2,000 | 1.245 | 1.154 | 1.295 | 1.511 | 1.474 |
| 2,500 | 1.315 | 1.196 | 1.380 | 1.670 | 1.619 |
| 3,000 | 1.387 | 1.240 | 1.470 | 1.844 | 1.776 |
| 3,500 | 1.464 | 1.285 | 1.565 | 2.033 | 1.946 |
| 4,000 | 1.544 | 1.332 | 1.666 | 2.239 | 2.130 |
| 4,500 | 1.628 | 1.381 | 1.773 | 2.462 | 2.329 |
| 5,000 | 1.716 | 1.432 | 1.886 | 2.705 | 2.542 |
| 5,500 | 1.809 | 1.485 | 2.005 | 2.968 | 2.772 |
| 6,000 | 1.906 | 1.540 | 2.131 | 3.253 | 3.019 |
| 6,500 | 2.007 | 1.596 | 2.263 | 3.560 | 3.285 |
| 7,000 | 2.114 | 1.655 | 2.404 | 3.892 | 3.569 |
| 7,500 | 2.225 | 1.717 | 2.552 | 4.250 | 3.872 |
| 8,000 | 2.342 | 1.781 | 2.708 | 4.635 | 4.197 |
| 8,500 | 2.464 | 1.847 | 2.872 | 5.048 | 4.543 |
| 9,000 | 2.592 | 1.915 | 3.046 | 5.492 | 4.912 |
| 9,500 | 2.725 | 1.987 | 3.228 | 5.968 | 5.305 |
| 10,000 | 2.865 | 2.061 | 3.420 | 6.477 | 5.721 |

where t is the temperature (°C). These values of $\Delta V_B$ and $\Delta K_B$ were determined from the measurements of Culberson and Pytkowicz.[20] Values of $K_B^P/K_B^O$ calculated from Equation 18 at t = 2°C and S = 34.5, which is typical for deep ocean waters, are given in Table 3. The $\Sigma CO_2$ given by

$$\Sigma CO_2 = [HCO_3^-] + [CO_3^{2-}] + [CO_2^*] \qquad (21)$$

can be determined by direct measurements of $CO_2$ stripped from seawater after the addition of acid[21-23] or by titrating the seawater with HCl as a function of pH.[18] The $P_{CO2}$ of a seawater solution can be determined by analyzing the $CO_2$ in air that has been equilibrated with seawater.[24-28]

In most seawater studies, it is convenient to determine the various components of

## Table 4
### APPARENT CONSTANTS FOR THE IONIZATION OF CARBONIC ACID IN SEAWATER ($-\ln K_1$)

| S (⁰/₀₀) | 0°C | 5°C | 10°C | 15°C | 20°C | 25°C | 30°C | 35°C | 40°C |
|---|---|---|---|---|---|---|---|---|---|
| 0 | 15.147 | 15.006 | 14.885 | 14.782 | 14.695 | 14.625 | 14.569 | 14.527 | 14.498 |
| 5 | 14.819 | 14.683 | 14.567 | 14.468 | 14.386 | 14.320 | 14.269 | 14.231 | 14.206 |
| 10 | 14.683 | 14.550 | 14.435 | 14.339 | 14.259 | 14.194 | 14.144 | 14.108 | 14.085 |
| 15 | 14.579 | 14.447 | 14.334 | 14.239 | 14.160 | 14.097 | 14.049 | 14.014 | 13.992 |
| 20 | 14.491 | 14.360 | 14.249 | 14.155 | 14.078 | 14.016 | 13.968 | 13.935 | 13.914 |
| 25 | 14.413 | 14.284 | 14.174 | 14.081 | 14.005 | 13.944 | 13.897 | 13.865 | 13.845 |
| 30 | 14.343 | 14.215 | 14.106 | 14.014 | 13.939 | 13.879 | 13.833 | 13.801 | 13.782 |
| 35 | 14.279 | 14.152 | 14.043 | 13.953 | 13.878 | 13.819 | 13.774 | 13.743 | 13.725 |
| 40 | 14.219 | 14.093 | 13.985 | 13.895 | 13.822 | 13.763 | 13.720 | 13.689 | 13.671 |

## Table 5
### APPARENT CONSTANTS FOR THE IONIZATION OF BICARBONATE ION IN SEAWATER ($-\ln K_2$)

| S (⁰/₀₀) | 0°C | 5°C | 10°C | 15°C | 20°C | 25°C | 30°C | 35°C | 40°C |
|---|---|---|---|---|---|---|---|---|---|
| 0 | 24.474 | 24.305 | 24.153 | 24.016 | 23.894 | 23.785 | 23.689 | 23.606 | 23.534 |
| 5 | 23.205 | 23.022 | 22.856 | 22.707 | 22.572 | 22.452 | 22.345 | 22.251 | 22.168 |
| 10 | 22.775 | 22.587 | 22.416 | 22.261 | 22.122 | 21.997 | 21.885 | 21.786 | 21.698 |
| 15 | 22.484 | 22.292 | 22.116 | 21.958 | 21.814 | 21.685 | 21.570 | 21.467 | 21.376 |
| 20 | 22.265 | 22.068 | 21.890 | 21.728 | 21.581 | 21.449 | 21.330 | 21.224 | 21.131 |
| 25 | 22.091 | 21.891 | 21.710 | 21.544 | 21.395 | 21.260 | 21.139 | 21.030 | 20.934 |
| 30 | 21.950 | 21.747 | 21.563 | 21.395 | 21.242 | 21.105 | 20.981 | 20.871 | 20.772 |
| 35 | 21.833 | 21.628 | 21.441 | 21.270 | 21.115 | 20.976 | 20.850 | 20.737 | 20.636 |
| 40 | 21.736 | 21.528 | 21.338 | 21.166 | 21.099 | 20.867 | 20.739 | 20.624 | 20.521 |

the carbonate system by using pH and $A_T$, or $A_T$ and $\Sigma CO_2$. If the pH and $A_T$ are determined, the components of the carbonate system are determined from

$$[HCO_3^-] = A_C/[1 + 2K_2'/a_H] \tag{22}$$

$$[CO_3^{2-}] = A_C K_2'/(a_H + 2K_2') \tag{23}$$

$$[CO_2^*] = (A_C a_H/K_1')/(1 + 2K_2'/a_H) \tag{24}$$

where the values of $K_i'$ and $a_H$ are for the *in situ* applied pressure ($P$ = depth/10 m) and temperature. The apparent constants at 1 atm are given

$$\ln K_1' = 290.9097 - 14554.21/T - 45.0575 \ln T + (0.0221 + 34.02/T)S^{1/2} \tag{25}$$

$$\ln K_2' = 207.6548 - 11843.79/T - 33.6485 \ln T + (0.9805 - 92.65/T)S^{1/2} - 3.294 \times 10^{-2}S \tag{26}$$

where $\sigma$ = 0.01 and 0.03, respectively, in $\ln K_1'$ and $\ln K_2'$. The values of $\ln K_1'$ and $\ln K_2'$ determined from Equations 25 and 26 are given in Tables 4 and 5.

The effect of applied pressure on the apparent constants of carbonic acid can be determined from Equation 18. The values of $\Delta V_i$ and $\Delta K_i$ for the ionization are given by[13]

$$-\Delta V_1 = 25.02 + 0.1757 \, (S-35) + 0.0629t - 8.225 \times 10^{-3} t^2 \tag{27}$$

$$-\Delta V_2 = 15.92 - 0.327 \, (S-35) + 0.0120t \tag{28}$$

$$-10^3 \Delta K_1 = 2.17 + 0.2865 \, (S-35) + 0.2789t - 1.59 \times 10^{-2} t^2 \tag{29}$$

$$-10^3 \Delta K_2 = -0.64 + 0.262 \, (S-35) + 0.02098t + 4.355 \times 10^{-3} t^2 \tag{30}$$

Values of $K_1^P/K_1^o$ and $K_2^P/K_2^o$ calculated from Equation 18 at t = 2°C and S = 34.5, which is typical for deep ocean waters, are given in Table 3. The effect of temperature and pressure on $a_H$ needed to calculate the *in situ* value from pH measurements at 1 atm and 25°C can be determined in two ways:

1. An iterative computer technique[29-31]
2. Using explicit functions of pH as a function of T and P[13]

The iterative method requires the solving of the cubic equation [29,32]

$$a_H^3 + a_H^2 \, [K_1'(A-1) + K_B'(A-B)]/A + a_H[K_1'K_B'(A-B-1)$$

$$+ K_1'K_2'(A-2)]/A + K_1'K_2'K_B'(A-B-2)/A = 0 \tag{31}$$

which is defined by combining Equations 6, 7, and 13 to 16. The values of A and B in Equation 31 are given by

$$A = A_T/\Sigma CO_2 \tag{32}$$

$$B = [B]_T/\Sigma CO_2 \tag{33}$$

The value of $\Sigma CO_2$ is first estimated using the 1 atm pH measurements using the equation

$$\Sigma CO_2 = A_C(1 + a_H/K_1' + K_2'/a_H)/(1 + 2K_2'/a_H) \tag{34}$$

Equation 31 is solved for $a_H$ using the solutions for a cubic equation[33] or by iterative solutions.[30,31] The process is repeated until a self-consistent value of $\Sigma CO_2$ is obtained within certain prescribed limits. Once the *in situ* $a_H$ is determined, the various components of the carbonate system can be determined from Equations 21 to 24.

Since many workers do not have computer systems available, we have generated values of pH from Equation 31 as a function of temperature, pressure, and salinity.[13] The effect of temperature on the pH at 1 atm is given by[13]

$$pH_t = pH_{25} + A(t-25) + B(t-25)^2 \tag{35}$$

$$10^3 A = -9.702 - 2.378 \, (pH_{25}-8) \; 3.885 \, (pH_{25}-8)^2 \tag{36}$$

$$10^4 B = 1.123 - 0.003 \, (pH_{25}-8) + 0.933 \, (pH_{25}-8)^2 \tag{37}$$

The equation is valid from t = 0 to 40°C, S = 30 to 40°/$_{oo}$, and $pH_{25}$ = 7.6 to 8.2

and has a $\sigma = 0.002$. The effect of applied pressure on the pH is a linear function of pressure

$$pH^P = pH^o + AP \qquad (38)$$

where $pH_o$ is the value at 1 atm and A is given by

$$10^3 A = 0.424 - 0.0048 (S-35) - 0.00282t - 0.0816 (pH^o-8) \qquad (39)$$

which is valid from S = 32 to 38°/₀₀ and t = 0 to 25°C ($\sigma = 0.003$).

If the $A_T$ and $\Sigma CO_2$ are determined[18] the various components of the carbonate system can be determined from[29]

$$[CO_2*] = \Sigma CO_2 - A_C + \frac{A_C K_R - \Sigma CO_2 K_R - 4A_C + Z}{2(K_R - 4)} \qquad (40)$$

$$[HCO_3^-] = \frac{\Sigma CO_2 K_R - Z}{(K_R - 4)} \qquad (41)$$

$$[CO_3^{2-}] = \frac{A_C K_R - \Sigma CO_2 K_R - 4A_C + Z}{2(K_R - 4)} \qquad (42)$$

where $K_R = K_1'/K_2'$ and Z is given by

$$Z = [(4A_C + \Sigma CO_2 K_R - A_C K_R)^2 + 4(K_R - 4)A_C^2]^{1/2} \qquad (43)$$

The *in situ* value of $a_H$ needed to determine $A_C$ from $A_T$ can be calculated from the methods described earlier.

The saturation state ($\Omega$) of seawater solutions with respect to $CaCO_3$ is determined from

$$\Omega = I.P./K_{sp}' \qquad (44)$$

where the ion product (I.P.) is given by

$$I.P. = [Ca^{2+}] [CO_3^{2-}] \qquad (45)$$

and $K_{sp}'$ is the solubility product at the *in situ* conditions. The concentration (moles per kilogram) of calcium $[Ca^{2+}]$ can be estimated from

$$[Ca^{2+}] = 2.934 \times 10^{-4} S \qquad (46)$$

The value of $[CO_3^{2-}]$ is determined from pH and $A_T$ or $A_T$ and $\Sigma CO_2$ measurements. The solubility product of calcite at 1 atm can be calculated from[13]

$$\ell n\, K_{sp}' \text{ (calcite)} = 303.1308 - 13348.09/T - 48.7537\, \ell n\, T$$

$$+ (1.6233 - 118.64/T)S^{1/2} - 6.999 \times 10^{-2} S \qquad (47)$$

where $\sigma = 0.065$ in $\ell n\, K_{sp}$. This equation is based on the pure water work of Berner[35] and Jacobson and Langmuir[36] and the seawater work of Ingle et al.[37,38] Values of $\ell n$

Table 6
APPARENT SOLUBILITY PRODUCT OF CALCITE IN SEAWATER
$(-\ln K_{sp}')$

| S (⁰/₀₀) | 0°C | 5°C | 10°C | 15°C | 20°C | 25°C | 30°C | 35°C | 40°C |
|---|---|---|---|---|---|---|---|---|---|
| 0 | 19.246 | 19.251 | 19.273 | 19.308 | 19.357 | 19.418 | 19.490 | 19.573 | 19.666 |
| 5 | 16.937 | 16.925 | 16.930 | 16.949 | 16.982 | 17.028 | 17.085 | 17.154 | 17.234 |
| 10 | 16.186 | 16.167 | 16.164 | 16.177 | 16.203 | 16.243 | 16.294 | 16.357 | 16.431 |
| 15 | 15.691 | 15.666 | 15.658 | 15.666 | 15.687 | 15.722 | 15.769 | 15.827 | 15.897 |
| 20 | 15.328 | 15.299 | 15.287 | 15.290 | 15.307 | 15.338 | 15.381 | 15.435 | 15.501 |
| 25 | 15.051 | 15.018 | 15.001 | 15.000 | 15.014 | 15.041 | 15.080 | 15.132 | 15.194 |
| 30 | 14.833 | 14.796 | 14.776 | 14.772 | 14.782 | 14.806 | 14.842 | 14.891 | 14.950 |
| 35 | 14.661 | 14.621 | 14.598 | 14.590 | 14.597 | 14.618 | 14.652 | 14.697 | 14.754 |
| 40 | 14.526 | 14.482 | 14.456 | 14.445 | 14.449 | 14.467 | 14.498 | 14.541 | 14.596 |

Table 7
APPARENT SOLUBILITY PRODUCT OF ARAGONITE IN SEAWATER
$(-\ln K_{sp}')$

| S (⁰/₀₀) | 0°C | 5°C | 10°C | 15°C | 20°C | 25°C | 30°C | 35°C | 40°C |
|---|---|---|---|---|---|---|---|---|---|
| 0 | 18.840 | 18.846 | 18.867 | 18.903 | 18.951 | 19.012 | 19.085 | 19.168 | 19.261 |
| 5 | 16.532 | 16.520 | 16.524 | 16.543 | 16.576 | 16.622 | 16.680 | 16.749 | 16.828 |
| 10 | 15.780 | 15.762 | 15.759 | 15.771 | 15.798 | 15.837 | 15.889 | 15.952 | 16.026 |
| 15 | 15.285 | 15.261 | 15.253 | 15.260 | 15.282 | 15.316 | 15.363 | 15.422 | 15.491 |
| 20 | 14.923 | 14.894 | 14.881 | 14.884 | 14.901 | 14.932 | 14.975 | 15.030 | 15.095 |
| 25 | 14.645 | 14.612 | 14.596 | 14.595 | 14.608 | 14.635 | 14.675 | 14.726 | 14.789 |
| 30 | 14.428 | 14.391 | 14.371 | 14.366 | 14.377 | 14.400 | 14.437 | 14.485 | 14.545 |
| 35 | 14.256 | 14.216 | 14.192 | 14.185 | 14.192 | 14.213 | 14.246 | 14.292 | 14.348 |
| 40 | 14.120 | 14.077 | 14.050 | 14.040 | 14.044 | 14.062 | 14.093 | 14.136 | 14.190 |

$K_{sp}'$ for calcite determined from Equation 47 are given in Table 6. The effect of pressure on the solubility of calcite can be determined from Equation 18, where[13]

$$\Delta V_C = -35.5 - 0.53\,(25\text{-}t) \qquad (48)$$

$$10^3 \Delta K_C = -2.529 - 0.369\,(25\text{-}t) \qquad (49)$$

The $K_{sp}'$ for the solubility of aragonite in seawater at 1 atm can be determined from[13]

$$\ln K_{sp}' \text{ (aragonite)} = 303.5363 - 13348.09/T - 48.7537 \ln T$$

$$+(1.6233 - 118.64/T)S^{1/2} - 6.999 \times 10^{-2}S \qquad (50)$$

This equation is based on the pure water work of Berner[35] and the seawater work of Morse et al.[39] Values of $\ln K_{sp}'$ for aragonite determined from Equation 50 are given in Table 7. The effect of pressure on the solubility of aragonite can be estimated from Equation 18, where[13]

$$\Delta V_A = 32.7 - 0.53\,(25\text{-}t) \qquad (51)$$

and $\Delta K_A$ is assumed to be equal to $\Delta K_C$. Values of $K_C^P/K_C^O$ and $K_A^P/K_A^O$ calculated from Equation 18 at t = 2°C and S = 34.5, which is typical for deep ocean waters, are given in Table 3.

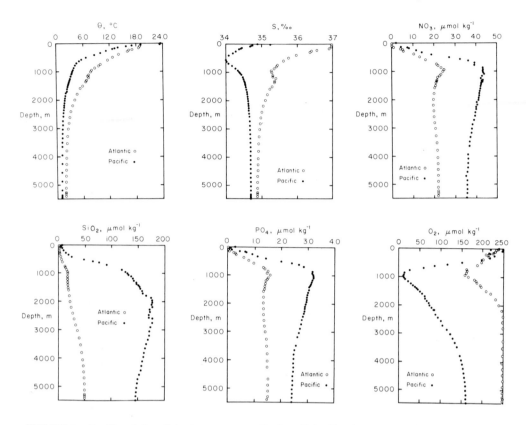

FIGURE 1.   Profiles of the adiabatic temperature ($\theta$), the salinity (S), nitrate ($NO_3$), silica ($SiO_2{}^{aq}$), phosphate ($PO_4$), and oxygen ($O_2$) in two GEOSECS stations in the North Atlantic (#115) and North Pacific (#204) Oceans.

## DISTRIBUTION OF $CO_2$ IN OCEANS

The distribution of the various components of the $CO_2$ system in oceans have been studied by a number of workers. Skirrow[32] has reviewed much of this work. In recent years, extensive measurements of $A_T$ and $\Sigma CO_2$ have been made as part of the Geochemical Ocean Sections (GEOSECS) program. This section examines the distribution of the various parameters in two stations from the GEOSECS work. We have selected stations in the North Atlantic (GEOSECS, #115, 28° N, 26° W) and North Pacific (GEOSECS, #204, 31° N, 150° E) oceans to demonstrate the depth dependence and major differences between the two oceans.

Figure 1 gives profiles of the adiabatic temperature ($\theta$), salinity (S), the concentration of $O_2$ ($\mu$mol kg$^{-1}$), $SiO_2$ ($\mu$mol kg$^{-1}$), $NO_3$ ($\mu$mol kg$^{-1}$), and $PO_4$ ($\mu$mol kg$^{-1}$). The near surface concentrations of nutrients are quite low because of the biological activity in the photic zone. The deep waters have higher concentrations of nutrients because of the oxidation of sinking biological matter. This oxidation causes a decrease in $O_2$. The deep waters have higher concentrations of $O_2$, since they have been formed from the sinking of cold waters at high latitudes saturated with $O_2$. These combined factors cause a minimum in the $O_2$ profile. The maximum concentrations of $SiO_2$ (aq), $NO_3$, and $PO_4$ occur at a depth close to the $O_2$ minimum.

The concentrations of nutrients are higher and the $O_2$ is lower in the deep Pacific Ocean than in the Atlantic Ocean. This is because of the fact that deep Pacific Ocean waters are older and therefore have had a longer time to oxidize biological matter.

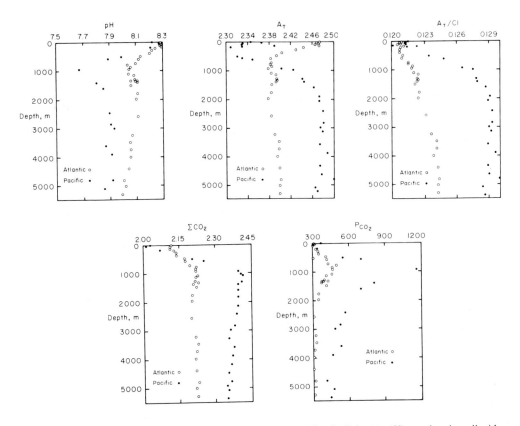

FIGURE 2.    Profiles of the pH, titration alkalinity ($A_T$), specific alkalinity ($A_T/Cl$), total carbon dioxide ($\Sigma CO_2$), and partial pressure of carbon dioxide ($P_{CO2}$) in two GEOSECS stations in the North Atlantic (#115) and North Pacific (#204) Oceans.

The various components of the carbonate system ($P_{CO2}$ and $\Sigma CO_2$) shown in Figure 2 follow a similar pattern to that found for the nutrients. They both go through a maximum near the $O_2$ minimum depth. The pH shows the opposite effect going through a minimum. This is primarily the result of increase in $P_{CO2}$. The total alkalinity of ocean waters is largely due to $[HCO_3^-]$ which is fairly conservative. This can be demonstrated by examining the specific alkalinity (S.A.) defined by

$$S.A. = A_T \times 1.80655/S \qquad (52)$$

Most ocean waters have a S.A. between 0.12 and 0.13. The increase in $A_T$ in the deep ocean is largely due to the dissolution of $CaCO_3$. Depth profiles of these components of the $CO_2$ system are presented in Figure 2.

The surface waters are all supersaturated with respect to calcite ($\Omega = 5.0$) and aragonite ($\Omega = 3.0$). The deep waters become unsaturated largely because of the effect of temperature and pressure. The unsaturation occurs in deeper waters in the Atlantic Ocean than in the Pacific Ocean. This is largely caused by the higher $P_{CO2}$ resulting from the oxidation of biological material found in the Pacific ocean. Saturation profiles for calcite and aragonite are presented in Figure 3.

## III. CALCIUM CARBONATE ACCUMULATION IN DEEP SEA SEDIMENTS

Calcium carbonate, produced by small pelagic organisms living near the surface of

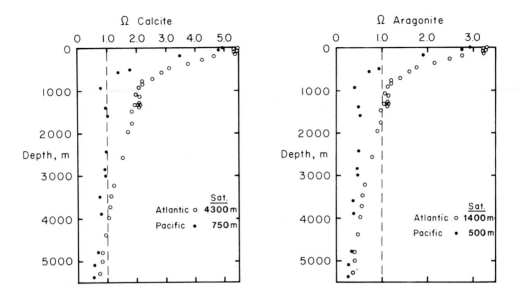

FIGURE 3.    Profiles of the saturation state (Θ) of calcite and aragonite in two GEOSECS stations in the North Atlantic (#115) and North Pacific (#204) Oceans.

the ocean, is a common component of deep sea sediments (see Figure 4).[40,41] Sediments, where the overlying water depth is less than approximately 4 km, are frequently composed of 70 to 90% calcium carbonate by dry weight. In different regions of oceans, different factors are predominantly responsible for the distribution of calcium carbonate. Calcium carbonate is not abundant in sediments from high latitudes because most of the important calcium carbonate secreting organisms do not thrive in the cold surface waters. The concentration of calcium carbonate in sediments accumulating near the edge of continents is generally less than that found in sediments accumulating far from continents. This is primarily the result of two factors. The first is that the accumulation rate of noncalcium carbonate sediment components is generally higher near continents causing a dilution of the calcium carbonate. The second factor is the generally higher content of organic matter in the near-continent sediments. Oxidation of the organic matter within the sediments results in the production of $CO_2$ and organic acids which cause dissolution of calcium carbonate. Throughout much of the world oceans, the concentration of calcium carbonate in sediments is found to be close to constant until a critical water depth is reached. Below this depth, the concentration of calcium carbonate decreases rapidly with increasing water depth.

The relation between the saturation state of seawater, with respect to calcite and aragonite, overlying deep sea sediments, and the concentration of calcium carbonate within the sediments has been of major interest to marine chemists and sedimentologists for over 100 years. This topic has recently been reviewed by Morse and Berner,[41] and only major concepts are briefly presented here. The most important observations is that both calcite and aragonite are accumulating in sediments where the overlying water is unsaturated. Because aragonite is more soluble than calcite, under the conditions found in the deep oceans, it is preserved in sediments only to depths considerably shallower (on the order of 3 km) than calcite. Since both calcium carbonate phases can accumulate in sediments where the overlying water is unsaturated, dissolution kinetics must play an important role in determining under what conditions calcium carbonate will be preserved in sediments. Both laboratory and open ocean experiments indicate that the relationship between saturation state and dissolution rate is complex and highly nonlinear.[41]

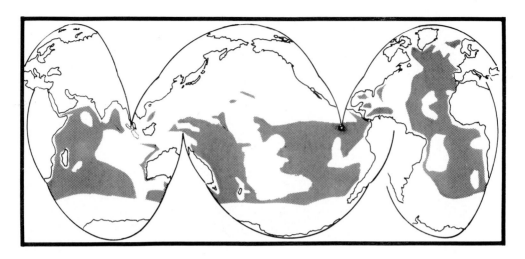

FIGURE 4. The distribution of recent pelagic sediments containing greater than 30 wt% calcium carbonate by dry weight.[40,41]

## ACKNOWLEDGMENT

Frank J. Millero acknowledges the support of the Office of Naval Research (N00014-75-C-0173) and the oceanographic section of the National Science Foundation (OCE73-00351-A01) for this study. John W. Morse acknowledges the support of the National Science Foundation (OCE78-18072), Marine Chemistry Program.

## REFERENCES

1. **Brönsted, J. N.,** Acid and base catalysis, *Chem. Rev.,* 5, 231, 1928.
2. **Buch, K.,** On boric acid in the sea and its influence on the carbonic acid equilibrium, *J. Conserv.,* 8, 309, 1933.
3. **Buch, K.,** New determinations of the second dissociation constant of carbonic acid in sea water, *Acta Acad. Abo. Ser. B,* 11, 18, 1938.
4. **Buch, K., Harvey, H. W., Wattenberg, H., and Gripenberg, S.,** Uber der kohlensauresystem im meerwasser, *Rapp. P.-V. Reun. Cons. Int. Explor. Mer.,* 79, 1, 1932.
5. **Lyman, J.,** Buffer Mechanism of Sea Water, Ph.D. thesis, University of California, Los Angeles, 1957.
6. **Mehrbach, C., Culberson, C. H., Hawley, J. E., and Pytokowicz, R. M.,** Measurement of the apparent dissociation constants of carbonic acid in seawater at atmospheric pressure, *Limnol. Oceanogr.,* 18, 897, 1973.
7. **Hansson, I.,** An A nalytical Approach to the Carbonate System in Seawater, Ph.D. thesis, University of Goteborg, Sweden, 1972.
8. **Hansson, I.,** A new set of acidity constants for carbonic acid and boric acid in sea water, *Deep-Sea Res.,* 20, 461, 1973.
9. **Hansson, I.,** Determination of the acidity constant of boric acid in synthetic sea water media, *Acta Chem. Scand.,* 27, 924, 1973.
10. **Hansson, I.,** The determination of dissociation constants of carbonic acid in synthetic sea water in the salinity range of 20-40°/oo and temperature range of 5 − 30°C, *Acta Chem. Scand.,* 27, 931, 1973.
11. **Bates, R. G.,** *Determination of pH, Theory and Practice,* John Wiley & Sons, New York, 1964.
12. **Hansson, I.,** A new set of pH-scales and standard buffers for sea water, *Deep-Sea Res.,* 20, 479, 1973.
13. **Millero, F. J.,** The thermodynamics of the carbonate system in seawater, *Geochim. Cosmochim. Acta,* 43, 1651, 1979.

14. **Hawley, J. E. and Pytkowicz, R. M.,** Interpretation of pH measurements in concentrated electrolyte solutions, *Mar. Chem.,* 1, 245, 1973.

15. **Weiss, R. F.,** Carbon dioxide in water and seawater: the solubility of a non-ideal gas, *Mar. Chem.,* 2, 203, 1974.

16. **Murray, C. N. and Riley, J. P.,** The solubility of gases in distilled water and sea water. IV. Carbon dioxide, *Deep-Sea Res.,* 18, 533, 1971.

17. **Park, P. K.,** Oceanic $CO_2$ system: an evaluation of ten methods of investigation, *Limnol. Oceanogr.,* 14, 179, 1969.

18. **Edmond, J. M.,** High precision determination of titration alkalinity and total carbon dioxide content of sea water by potentiometric titration, *Deep-Sea Res.,* 17, 737, 1970.

19. **Culkin, F.,** The major constituents of seawater, in *Chemical Oceanography,* Vol. 1, Riley, J. P. and Skirrow, G., Eds., Academic Press, New York, 1965, 121.

20. **Culberson, C. and Pytkowicz, R. M.,** Effect of pressure on carbonic acid, boric acid and the pH in sea water, *Limnol. Oceanogr.,* 13, 403, 1968.

21. **Saruhashi, K.,** On the total carbonaceous matter and hydrogen ion concentration in sea water, *Meteorol. Geophys. (Tokyo),* 3, 202, 1953.

22. **Park, K., Kennedy, G. H., and Dobson, H. H.,** Comparison of gas chromatographic method and pH-alkalinity method for the determination of total carbon dioxide in sea water, *Anal. Chem.,* 36, 1686, 1964.

23. **Li, Y.-H.,** The Degree of Saturation of $CaCO_3$ in the Oceans, Ph.D. thesis, Columbia University, New York, 1967.

24. **Keeling, C. D., Rakestraw, N. W., and Waterman, L. S.,** Carbon dioxide in surface waters of the Pacific Ocean. I. Measurements of the distribution, *J. Geophys. Res.,* 70, 6087, 1965.

25. **Hood, D. W., Berkshire, D., Supernaw, I., and Adams, R.,** Calcium Carbonate Saturation Level of the Ocean from Latitudes of North America to Antartica and Other Chemical Oceanographic Studies During Cruise III of the USNS *Eltanin,* Data report, National Science Foundation, Texas A & M University, College Station, Texas, 1963.

26. **Kanwisher, J.,** $pCO_2$ in seawater and its effect on the movement of $CO_2$ in nature, *Tellus,* 12, 209, 1960.

27. **Teal, J. M. and Kanwisher, J.,** The use of $pCO_2$ for the calculation of biological production, with examples from waters off Massachusetts, *J. Mar. Res.,* 24, 4, 1966.

28. **Takahashi, T.,** Carbon dioxide in the atmosphere and in the Atlantic Ocean water, *J. Geophys. Res.,* 66, 477, 1961.

29. **Edmond, J. M. and Gieskes, J. M. T. M.,** On the calculation of the degree of saturation of seawater with respect to calcium carbonate under *in situ* conditions, *Geochim. Cosmochim. Acta,* 34, 1261, 1970.

30. **Ben-Yaakov, S.,** A method for calculating the *in situ* pH of seawater, *Limnol. Oceanogr.,* 15, 326, 1970.

31. **Almgren, T., Dyrssen, D., and Strandberg, M.,** Determination of pH on the moles per kg seawater scale ($M_w$), *Deep-Sea Res.,* 22, 635, 1975.

32. **Skirrow, G.,** The dissolved gases-carbon dioxide, in *Chemical Oceanography,* Vol. 2, 2nd ed., Riley, J. P. and Skirrow, G., Eds., Academic Press, London, 1975.

33. **Weast, R. C., Ed.,** *The Handbook of Chemistry and Physics,* 55th ed., CRC Press, Cleveland, Ohio, 1975.

34. **Millero, F. J.,** The thermodynamics of seawater, in *Oceans Handbook,* Horne, R. A., Ed., Marcel Dekker, New York, in press.

35. **Berner, R. A.,** The solubility of calcite and aragonite in seawater at atmospheric pressure and 34.5°/$_{oo}$ salinity, *Am. J. Sci.,* 276, 713, 1976.

36. **Jacobson, R. L. and Langmuir, D.,** Dissociation constants of calcite and $CaHCO_3^+$ from 0 to 50°C, *Geochim. Cosmochim. Acta,* 38, 301, 1974.

37. **Ingle, S. E., Culberson, C. H., Hawley, J. E., and Pytkowicz, R. M.,** The solubility of calcite in seawater at atmospheric pressure and 35°/oo salinity, *Mar. Chem.,* 1, 295, 1973.

38. **Ingle, S. E.,** Solubility of calcite in the ocean, *Mar. Chem.,* 3, 301, 1975.

39. **Morse, J. W., Mucci, A., and Millero, F. J.,** The solubility of calcite and aragonite in seawater of 35°/$_{oo}$ salinity at 25°C and atmospheric pressure, *Geochim. Cosmochim. Acta,* 44, 85, 1980.

40. **Sverdrup, H. U., Johnson, M. W., and Fleming, R. H.,** *The Oceans,* Prentice-Hall, Englewood Cliffs, N.J., 1942.

41. **Morse, J. W. and Berner, R. A.,** The chemistry of calcium carbonate in the deep oceans in, *Chemical Modelling — Speciation, Sorption, Solubility and Kinetics in Aqueous Systems,* Jenne, E. A., Ed., American Chemical Society Symposium Series, Washington, D.C., 1978.

# OXYGEN IN AQUATIC ENVIRONMENTS

## James H. Carpenter

Oxygen in the environment appears to have been released from water by photosynthetic reactions over geologic time to produce an inventory in which most of the free $O_2$ occurs in the atmosphere as the diatomic molecule, $O_2$. The composition of dry air shows an $O_2$ content of 20.95%. Free $O_2$ also occurs by simple solution of the gas in natural waters, primarily in oceans.

The inventory amounts to $0.230$ kg/cm$^2$ of earth surface in the atmosphere and $0.002$ kg/cm$^2$ of ocean surface. Thus, the predominance of $O_2$ in the atmosphere results in only small variations in atmospheric abundance in response to perturbations in the water-air equilibrium.

Oxygen dissolves in water from the atmosphere to an extent that is dependent on atmospheric pressure, water temperature, and salinity. The dependence of solubility on pressure is a direct proportionality, in accordance with Henry's law (see Equation 1):

$$X = K \cdot P \tag{1}$$

The dependency of $O_2$ solubility on temperature and salinity is more complex. The currently accepted values are based on measurements by Carpenter[1] and Murray and Riley[2] that were fitted with Equation 2 by Weiss.[3]

$$\ln C = -173.4292 + 249.6339 \left(\frac{100}{T}\right) + 143.3483 \ln \left(\frac{T}{100}\right) - 21.8492 \left(\frac{T}{100}\right)$$

$$+ S\,°/oo = [-0.033096 + 0.014259 \left(\frac{T}{100}\right) - 0.0017 \left(\frac{T}{100}\right)^2] \tag{2}$$

where T is the absolute temperature (K) and S $_o$/oo is the salinity in parts per thousand.

Representative values for the $O_2$ solubilities as a function of temperature and salinity computed using Equation 2 are shown in Table 1. These "air solubility" values are for the normal atmosphere at a total pressure of 760 mmHg and 100% relative humidity. Most workers consider that the air in immediate contact with the water will have 100% relative humidity, even though the local atmosphere may be drier. The values may be corrected for deviations from the standard pressure to account for local atmospheric fluctuations or for work with lakes at elevated altitudes by Equation 3.

$$\text{Pressure correction factor} = \frac{P_a - P_w}{760 - P_w} \tag{3}$$

where $P_a$ is the ambient pressure, and $P_w$ is the vapor pressure of water at the prevailing temperature.

The concentration units for the dissolved $O_2$ in Table 1 are milliliters per liter which have been used in oceanography since the turn of the century and reflect the fact that the earlier measurements were made with gasometric methods. Workers on fresh water systems more commonly use the unit of milligrams per liter, or approximately parts per million. The molar volume of $O_2$ at STP is 22.385 L/mol, according to Weast,[4] so that 1 mL/L is equal to 1.4295 mg/L.

If the water is not at equilibrium with the atmosphere, the gas transfer processes

Table 1
SOLUBILITY OF OXYGEN (mL/L) IN NATURAL
WATERS IN EQUILIBRIUM WITH AN ATMOSPHERE
OF 20.95% OXYGEN WITH 100% RELATIVE
HUMIDITY AND PRESSURE 760 mmHg

| Temperature | Salinity (%) | | | | | | | |
|---|---|---|---|---|---|---|---|---|
| | 0 | 5 | 10 | 15 | 20 | 25 | 30 | 35 |
| 0 | 10.22 | 9.87 | 9.54 | 9.22 | 8.91 | 8.61 | 8.32 | 8.05 |
| 1 | 9.94 | 9.60 | 9.28 | 8.97 | 8.68 | 8.39 | 8.11 | 7.84 |
| 2 | 9.67 | 9.35 | 9.04 | 8.74 | 8.45 | 8.17 | 7.90 | 7.64 |
| 3 | 9.41 | 9.10 | 8.80 | 8.51 | 8.23 | 7.96 | 7.70 | 7.45 |
| 4 | 9.16 | 8.86 | 8.57 | 8.29 | 8.02 | 7.76 | 7.51 | 7.26 |
| 5 | 8.93 | 8.64 | 8.36 | 8.09 | 7.83 | 7.57 | 7.33 | 7.09 |
| 6 | 8.70 | 8.42 | 8.15 | 7.89 | 7.64 | 7.39 | 7.15 | 6.92 |
| 7 | 8.49 | 8.22 | 7.95 | 7.70 | 7.45 | 7.22 | 6.98 | 6.76 |
| 8 | 8.28 | 8.02 | 7.76 | 7.52 | 7.28 | 7.05 | 6.82 | 6.61 |
| 9 | 8.08 | 7.83 | 7.58 | 7.34 | 7.11 | 6.89 | 6.67 | 6.46 |
| 10 | 7.89 | 7.64 | 7.41 | 7.17 | 6.95 | 6.73 | 6.52 | 6.32 |
| 11 | 7.71 | 7.47 | 7.24 | 7.01 | 6.80 | 6.58 | 6.38 | 6.18 |
| 12 | 7.53 | 7.30 | 7.08 | 6.86 | 6.65 | 6.44 | 6.24 | 6.05 |
| 13 | 7.37 | 7.14 | 6.92 | 6.71 | 6.50 | 6.31 | 6.11 | 5.93 |
| 14 | 7.20 | 6.98 | 6.77 | 6.57 | 6.37 | 6.17 | 5.99 | 5.80 |
| 15 | 7.05 | 6.84 | 6.63 | 6.43 | 6.24 | 6.05 | 5.87 | 5.69 |
| 16 | 6.90 | 6.69 | 6.49 | 6.30 | 6.11 | 5.93 | 5.75 | 5.58 |
| 17 | 6.75 | 6.55 | 6.36 | 6.17 | 5.99 | 5.81 | 5.64 | 5.47 |
| 18 | 6.61 | 6.42 | 6.23 | 6.05 | 5.87 | 5.69 | 5.53 | 5.36 |
| 19 | 6.48 | 6.29 | 6.11 | 5.93 | 5.75 | 5.59 | 5.42 | 5.26 |
| 20 | 6.35 | 6.17 | 5.99 | 5.81 | 5.64 | 5.48 | 5.32 | 5.17 |
| 21 | 6.23 | 6.05 | 5.87 | 5.70 | 5.54 | 5.38 | 5.22 | 5.07 |
| 22 | 6.11 | 5.93 | 5.76 | 5.60 | 5.44 | 5.28 | 5.13 | 4.98 |
| 23 | 5.99 | 5.82 | 5.65 | 5.49 | 5.34 | 5.18 | 5.04 | 4.89 |
| 24 | 5.88 | 5.71 | 5.55 | 5.39 | 5.24 | 5.09 | 4.95 | 4.81 |
| 25 | 5.77 | 5.61 | 5.45 | 5.30 | 5.15 | 5.00 | 4.86 | 4.73 |
| 26 | 5.66 | 5.51 | 5.35 | 5.20 | 5.06 | 4.92 | 4.78 | 4.65 |
| 27 | 5.56 | 5.41 | 5.26 | 5.11 | 4.97 | 4.83 | 4.70 | 4.57 |
| 28 | 5.46 | 5.31 | 5.17 | 5.03 | 4.89 | 4.75 | 4.62 | 4.50 |
| 29 | 5.37 | 5.22 | 5.08 | 4.94 | 4.81 | 4.67 | 4.55 | 4.42 |
| 30 | 5.28 | 5.13 | 4.99 | 4.86 | 4.73 | 4.60 | 4.47 | 4.35 |

(invasion or evasion) will lead to changes in the dissolved $O_2$ concentration in accordance with the Adeney and Becker[5] relationship

$$\frac{dc}{ct} = k \left(\frac{A}{v}\right) (S - C) \tag{4}$$

or

$$\ln \frac{S - C_t}{S - C_o} = k \left(\frac{A}{v}\right) t \tag{5}$$

where C is the $O_2$ concentration, k is the transfer coefficient, A is the surface area of the water, v is the volume of the water, and t is time. If a unit area is considered, A/v is essentially the depth. The information on the gas transfer coefficients in the oceans and lakes has been reviewed by Liss[6] and Broecher and Peng,[7] and a typical value is 11 cm/hr.

As may be seen in Table 1, cooler water tends to contain more dissolved $O_2$ than warmer water, and fresher waters contain more dissolved $O_2$ than saline waters. These basic chemical properties are reflected in the observed distributions of dissolved $O_2$ in the oceans and lakes or rivers, but the distributions also reflect the intensity of the biochemical processes of photosynthesis that releases $O_2$ from water and respiration by plants and animals that combines the $O_2$ back into $CO_2$. The magnitude of the formation of organic matter and the associated $O_2$ production appears to be correlated with the fixed nitrogen (nitrate or ammonia) and the phosphate content of the water as described by Redfield et al.[8]

$$106CO_2 + 16HNO_3 + H_3PO_4 + 122H_2O \xrightleftharpoons[\text{respiration}]{\text{photosynthesis}}$$

$$\underset{\text{organic matter}}{(CH_2O)_{106}\,(NH_3)_{16}\,H_3PO_4} + 138O_2 \qquad\qquad (6)$$

Open ocean surface waters contain only small amounts of fixed nitrogen and phosphate, and the dissolved $O_2$ content deviates from equilibrium with the atmosphere by only a few percent, as do nutrient-poor (oligotrophic) lakes. In oceanic areas where deeper water containing higher nutrient concentrations is brought to the surface (upwelling) by winds or in nutrient-rich (eutrophic) lakes, substantial supersaturation of dissolved $O_2$ can be observed, and there is significant transfer of $O_2$ from the water to the atmosphere. As the organic matter that is produced in the sunlight-illuminated surface (euphotic) zone moves to the dark depths, either by flow and mixing of the waters or the sinking of organic particles, the reverse respiratory or decomposition reaction predominates, and decreases in the dissolved $O_2$ content are observed. The decrease in dissolved $O_2$ in the deeper water of stratified lakes has been shown by Hutchinson[9] to be related to the trophic status of various lakes, and in extreme cases of high nutrient levels, the dissolved $O_2$ in the deep water (hypolimion) may be completely depleted.

These processes produce the intricate patterns of dissolved $O_2$ distributions in the oceans that are shown in Figures 1, 2, and 3 that are based on the recent GEOSECS expeditions, for which the data were reported by Bainbridge.[10,11] Oceanic-dissolved $O_2$ distributions and processes are discussed in detail by Kester[12] and Richards.[13] However, the salient features may be seen in these figures. The most obvious feature is the $O_2$ minimum at intermediate depths which is more intense on the eastern side of the Atlantic (see Figure 1) than on the western side (see Figure 2), probably as a reflection of the fact that highest production of organic matter occurs in the equatorial upwelling areas off the coast of Africa. A low $O_2$ concentration zone, centered at 1000 m at 50°S, may be seen in Figure 2 which may be related to the high productivity of Antarctic surface waters that are nutrient-rich because of vertical mixing. The $O_2$ values in the minimum are lower in the Pacific (see Figure 3) than the Atlantic which correlates with the higher (nearly double) nutrient concentrations of the Pacific waters.

The deep waters of the Atlantic have a renewal time of approximately 500 years based $^{-14}C$ dating, and since the deeper waters of the Atlantic show only small losses of $O_2$, compared to the concentrations that existed when these waters were previously at the surface, $O_2$ consumption in the deep waters is quite weak. The deep waters in the North Pacific, as shown in Figure 3, have lower dissolved $O_2$ contents than the North Atlantic, at least partly as a manifestation of the slower renewal rate for the North Pacific of approximately 1000 years.

As a generalization, one can see that the dissolved $O_2$ content in aquatic environments is always the result of the dynamic balance between the physical processes of exchange with the atmosphere, internal circulation and mixing, and the biochemical reactions of photosynthesis and respiration. Since the intensity of these processes varies

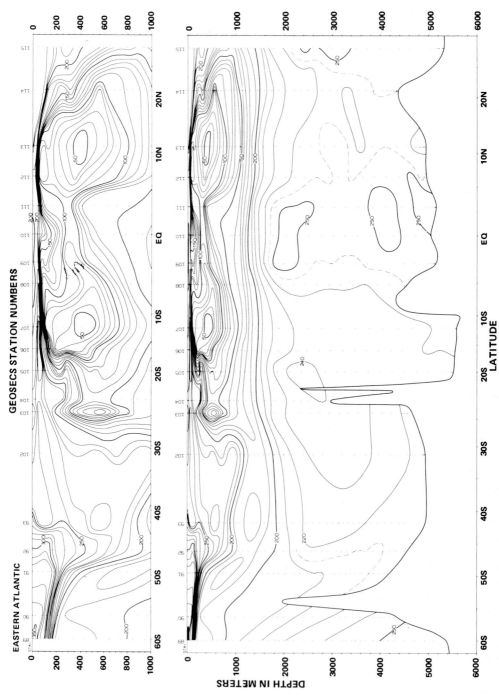

FIGURE 1.    Dissolved oxygen distribution in a north-south transect of the eastern Atlantic from GEOSECS data.[10] Oxygen concentrations in units of micromoles per liter.

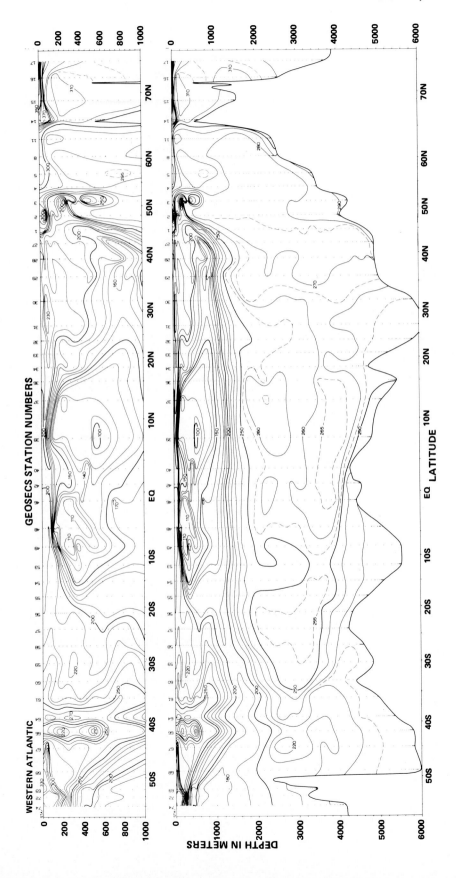

FIGURE 2. Dissolved oxygen distribution in a north-south transect of the western Atlantic from GEOSECS data.[10] Oxygen concentrations in units of micromoles per liter.

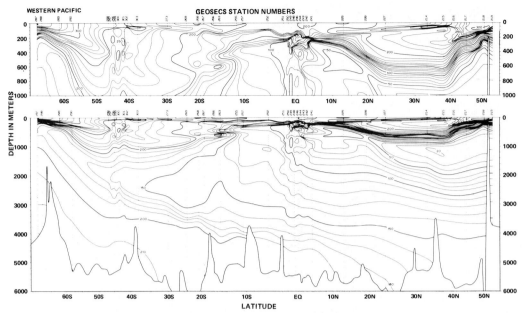

FIGURE 3.    Dissolved oxygen distribution in a north-south transect of the western Pacific from GEOSECS data.[11] Oxygen concentrations in units of micromoles per liter.

from environment to environment, no *a priori* statements about the dissolved $O_2$ can be made, but each particular case can be described in terms of a time-varying dynamic balance.

# REFERENCES

1.  **Carpenter, J. H.,** New measurements of oxygen solubility in pure and natural water, *Limnol. Oceanogr.,* 11, 264, 1966.
2.  **Murray, C. N. and Riley, J. P.,** The solubility of gases in distilled water and sea water. II. Oxygen, *Deep-Sea Res.,* 16, 311, 1969.
3.  **Weiss, R. F.,** The solubility of nitrogen, oxygen and argon in water and seawater, *Deep-Sea Res.,* 17, 721, 1970.
4.  **Weast, R. C., Ed.,** *Handbook of Chemistry and Physics,* 50th ed., CRC Press, Cleveland, Ohio, 1969.
5.  **Adeney, W. E. and Becker, H. G.,** The determination of the rate of solution of atmospheric nitrogen and oxygen by water, *Philos. Mag.,* 39, 385, 1920.
6.  **Liss, P. S.,** Processes of Gas Exchange across an air-water interface, *Deep-Sea Res.,* 20, 221, 1973.
7.  **Broecher, W. S. and Peng, T. H.,** Gas exchange rates between air and sea, *Tellus,* 26, 21, 1974.
8.  **Redfield, A. C., Ketchum, B. H., and Richards, F. A.,** The influence of organisms on the composition of seawater, in *The Sea: Ideas and Observations on Progress in the Study of the Seas,* Vol. 2, Mill, N. N., Ed., Interscience, New York, 1963, chap. 2.
9.  **Hutchinson, C. E.,** Oxygen in lake waters, in *A Treatise on Limnology,* Vol. 1 (Part 2) John Wiley & Sons, New York, 1957, chap. 2.
10. **Bainbridge, A. E.,** GEOSECS Atlantic Expedition, Vol. 2, Sections and Profiles, National Science Foundation, Washington, D.C., 1978.
11. **Bainbridge, A. E.,** GEOSECS Pacific Expedition, Vol. 5, Sections and Profiles, National Science Foundation, Washington, D.C., 1978.
12. **Kester, D. R.,** Dissolved gases other than $CO_2$, in *Chemical Oceanography,* Riley, J. P. and Skirrow, G., Eds., Academic Press, New York, 1975, chap. 8.
13. **Richards, F. A.** Oxygen in the ocean, in *Treatise on Marine Ecology and Paleontology,* Vol. 1., Hedgpeth, J., Ed., Geol. Society of America Mem., 1957, 67.

# THE MARINE MICRONUTRIENTS OF NITROGEN, PHOSPHORUS, AND SILICON

Rod G. Zika and Rana A. Fine

## INTRODUCTION

Seawater contains low concentrations of certain elements which are essential to the production and maintenance of living organisms. Some of these micronutrients are often utilized faster than they can be replenished, and in regions of high biological demand, their low concentration can become a limiting factor in the level of biological activity. This is particularly important in controlling the distribution and the yield of primary productivity in the ocean. Various transition metals, certain other inorganic anions and cations, and some organic compounds may be included in the micronutrient category, but traditionally the emphasis has been on the biologically available form of nitrogen, phosphorus, and silicon (which is essential to the growth of siliceous organisms like diatoms and silicoflagellates). The observed spatial and temporal distribution of the micronutrient chemical species of each of these elements is determined by a complex myriad of physical, chemical and biological properties of the environment which are illustrated in the marine cycles for the three elements (see Figures 1, 2, and 3). It is not possible in this brief article to go into a detailed discussion of these properties; however, comprehensive treatments of the subject are available.[1,2] Instead, information is provided on aspects of the chemical characteristics of each of these elements in the marine environment and on major aspects of their temporal and spatial large-scale distributions.

## ASPECTS OF MICRONUTRIENT CHEMISTRY IN SEAWATER

### Nitrogen

Only 0.5% of the nitrogen budget of the earth is present in the hydrosphere, the bulk of this being in the form of dissolved $N_2$. Nitrogen also occurs in seawater in various oxidation states between $-3$ and $+5$, with $NO_3^-$, which is the most highly oxidized form, having the only thermodynamically stable oxidation state at the commonly accepted p$\varepsilon$ of 12.5 and pH of 8.1. At equilibrium the partial pressure of $N_2$ should be only $1 \times 10^{-4}$ of its present value. Yet, in the hydrosphere, the total amount of major dissolved nitrogen constituents consists of $2.3 \times 10^{13}$ tons of $N_2$, $1 \times 10^{12}$ tons of $NO_3^-$, $1 \times 10^{10}$ tons of $NH_3$ and $NO_2^-$, and $3.7 \times 10^{11}$ tons of organic nitrogen. The fact that the amount of $NO_3^-$ is far less than anticipated on thermodynamic grounds suggests that the reactions leading to the oxidation of $N_2$ are very slow or that there are significant unknown competing denitrification reactions occurring. The high mobility of the reduction-oxidation processes in the nitrogen cycle are demonstrated by the simultaneous presence of significant concentrations of the lower oxidation state members, e.g., $NH_3$ and $NO_2^-$.

Nitrite is the intermediate oxidation level between $NO_3^-$ and $NH_3$ and is formed as the result of the oxidation of ammonia or the reduction, either microbially or photochemically, of $NO_3^-$. It is also an excretion product of phytoplankton where large surpluses of nitrate and phosphate exist. The highest concentrations of nitrate are generally found in transition zones between oxic and anoxic regions, but in general its concentrations tend to be very low and often not detectable by current analytical methods. Nitrite ion is the conjugate base of the weak acid, $HNO_2$, but since the ionization constant for this acid is only $4.5 \times 10^{-5}$, little of the un-ionized form will be present in seawater.

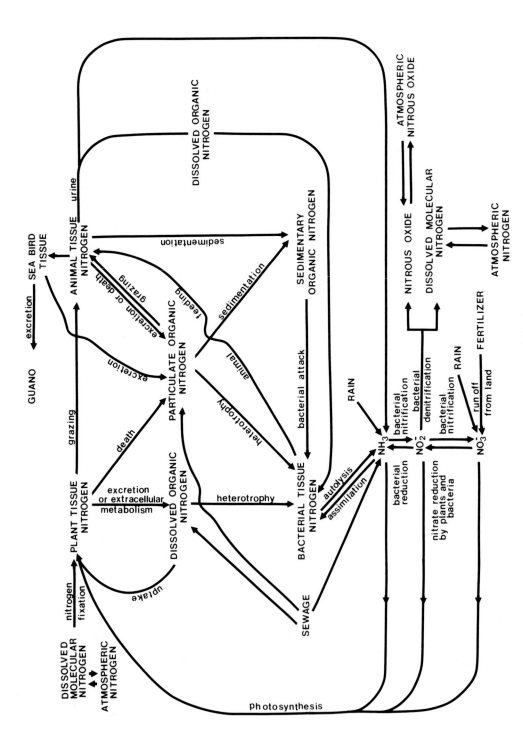

FIGURE 1.    The nitrogen cycle for the ocean.

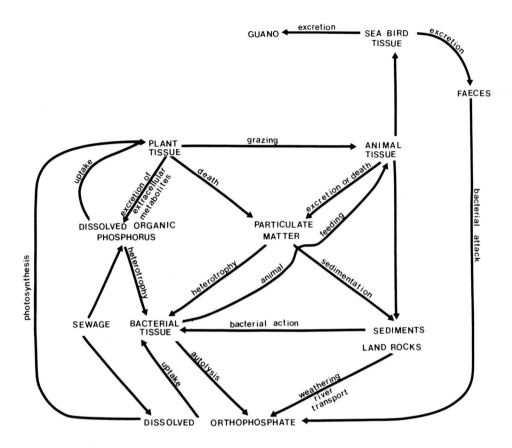

FIGURE 2.    The phosphorus cycle for the ocean.[1]

Ammonia contains nitrogen in the most highly reduced form, and, like the unstable nitrite ion, it is generally found in the oxic regions of the oceans in low concentrations, seldom in excess of 3 $\mu M$/kg. However, because it is the immediate product of the microbial turnover of organic nitrogen, it can rapidly be generated, and it can become the predominant form present, relative to $NO_3^-$ and $NO_2^-$. As well as its formation by the action of proteolytic bacteria on organic compounds, it also is a major excretion product of animals. The highest concentrations of $NH_3$ are found under near-anoxic to anoxic conditions, where bacterial denitrification processes utilizing $NO_3^-$ generate $N_2$ and $NH_3$. In such zones, the concentration of $NH_3$ can reach 100 $\mu M$/kg. In the normal oceanic pH range, the ammonium ion will be the dominant form present; the $pK_a$ for the dissociation of the ionic form is 9.3.

## Phosphorus

Phosphorus exists in seawater in both dissolved and particulate forms; the nature of the chemical species is characterized primarily on the basis of their separability and their reactivity in various analytical schemes. They are generally divided into the categories of dissolved inorganic phosphorus, polyphosphates, and total phosphorus. The dissolved inorganic phosphorus is reactive in the standard molybdate analytical method and is probably primarily ortho-phosphate. The other two categories are also determined as ortho-phosphate, but only after preliminary vigorous hydrolysis and hydrolysis-oxidation steps are employed. The occurrence of significant concentrations of polyphosphates is usually only connected with estuarine and coastal waters, where a principal source is often the polyphosphate-based detergents. Where polyphosphates

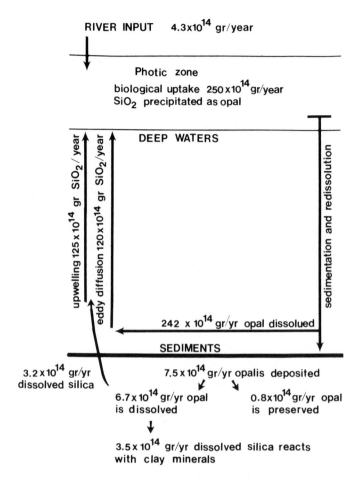

FIGURE 3.   A tentative silica cycle for the ocean.[7]

are not likely to be part of the total, the difference between total phosphorus (measured after acidic oxidative digestion) and free ortho-phosphate has been assumed to be organic phosphorus. Little is known about the composition of this fraction, but the relative inertness to hydrolysis might be explained by the presence of compounds with stable carbon-phosphorus covalent bonds. The measurement of organic and other unreactive forms of phosphorus probably does not give an accurate value for the nutrient availability, for some of these forms may be biologically refractory or only slowly assimilated. In general, it is reactive ortho-phosphate which is more abundant and is an immediate source of the nutrient element. It is present in seawater as various anionic species of phosphoric acid. The apparent ionization constants for phosphoric acid at 20°C and 33⁰/₀₀ salinity are $K_1' = 2.53 \times 10^{-3}$, $K_2' = 8.8 \times 10^{-7}$, and $K_3 = 1.37 \times 10^{-10}$.[3] At pH 8.0, ortho-phosphate is present in seawater at about 1% $H_2PO_4^-$, 87% $HPO_4^-$, and 12% $PO_4^{3-}$. Much of the $PO_4^{3-}$ and $HPO_4^{2-}$ exists in the form of ion pairs with $Ca^{2+}$ and $Mg^{2+}$.

## Silicon

Silicon is present in seawater in both particulate and soluble forms. The particulate form is a combination of the remains of siliceous organisms and various minerals (i.e., quartz, feldspars, and clays) derived primarily from fluvial or eolian sources. The soluble form is probably silicic acid, $Si(OH)_4$, which is a very weak acid ($pK_a = 9.41$) and is essentially nondisassociated in seawater. It is this soluble form which is detected with the standard methods of seawater analysis for silicate. Seawater is highly unsaturated, with respect to dissolved silica, for the concentrations found in the oceans are only a fraction of the reported 1600 to 1800 $\mu M/kg$ solubility.[4] The sedimentation of siliceous plankton is the principal biological mechanism for the removal of dissolved silica, but appears to account for only a fraction of the 5 to $6 \times 10^{14}$g of silicon supplied to the oceans yearly. The apparent imbalance has been explained by invoking various removal mechanisms which involve reactions of the dissolved silica with suspended matter.[5,6] Such proposals have, however, been criticized,[7] and the solution to the problem of accurately describing the silica cycle in the oceans is still forthcoming.

## SOURCES OF ANALYTICAL CHEMISTRY PROCEDURES

The literature dealing with the analytical chemistry of nitrogen, phosphorus, and silicon micronutrients has recently been reviewed.[8] Detailed analytical procedures and information on sampling and storage are provided in books on marine analytical chemical procedures by Strickland and Parsons[9] and Grasshoff.[10] The procedures given in those two sources are probably the most commonly used.

## MICRONUTRIENT DISTRIBUTIONS

### Typical Profiles

The nutrient distributions in the oceans are a consequence of biological, chemical, and physical processes. Table 1 contains some of the major references dating from 1964 to 1977 of studies of nutrient distributions in the world oceans. The Geochemical Ocean Sections[11] expeditions in the Atlantic (1972 to 1973), Pacific (1973 to 1974), and Indian Oceans (1978) have provided, perhaps, the most comprehensive data base. To examine the general vertical distribution of the nutrient species in the open ocean, a typical station (30° N, 50° W) from the GEOSECS Atlantic expedition has been chosen. Figure 4 contains profiles of nitrate, phosphate, and silicate vs. depth. The concentration of nutrients in the surface mixed layer and/or photic zone is the most variable both spatially and temporally, and it is generally the lowest in the water column because of biological productivity. Throughout the pycnocline nutrient concentrations increase rapidly to a maximum; this is because of intensified biological regenerative processes in this zone (see Figures 1, 2, and 3). The maximum occurs at about the base of the pycnocline and just below the oxygen minimum (500 to 2000 m); it is most marked in the Atlantic. Below the maximum, the nutrient concentrations remain fairly constant to the bottom, mainly as a result of a slower deep water circulation.

### Spatial Similarities

To examine the spatial similarities of the nutrients in the Atlantic, Pacific, and Indian oceans, three vertical sections (see Figures 5, 6, and 7) from the western Atlantic GEOSECS expedition[11] have been chosen. In all three oceans, there are large-scale latitudinal variations in the nutrients of surface and near-surface waters that are a consequence of the amount of exchange with the deeper waters. The ocean circulation is such that the nutrient-rich waters rise to the photic zone in the equatorial regions and in the cold subarctic cyclonic gyres, whereas they lie below the photic zone within the large subtropical anticyclonic gyres, and the shallow waters of the tropical regions also have very low nutrient concentrations. Nutrients in the gyre boundary currents to some extent reflect the gyres adjacent to them; for example, the California Current (of the subtropical gyre) may be considered an extension of the richer subarctic gyre.[12]

In all three oceans, there are small-scale features which affect the nutrients, namely, fronts and eddies or rings. These act to both contain the nutrients (in the center of

Table 1
## NUTRIENTS IN THE OCEANS

| Location | Nutrients | Ref. |
|---|---|---|
| South Pacific | $NO_3$, $SiO_2(aq)$ | 15 |
| Western Bay of Bengal | $NO_3$, $PO_4$ | 16 |
| Western Bay of Bengal | $NO_3$, $PO_4$ | 17 |
| Arabian Sea | $NO_3$, $PO_4$ | 18 |
| Aleutian Islands | $NO_3$, $PO_4$, $SiO_2(aq)$ | 19 |
| North Pacific | $NO_3$, $NO_2$ | 20 |
| Central Atlantic | $NO_3$, $PO_4$ | 21 |
| Eastern tropical North Pacific | $NO_3$ | 22 |
| Mid-Atlantic Coast | $NO_3$, $NO_2$, $PO_4$, $SiO_2(aq)$ | 23 |
| Arabian Sea | $PO_4$ | 24 |
| Black Sea | $PO_4$ | 25 |
| Eastern Pacific | $NO_3$ | 26 |
| Peru Coast | $NO_3$, $PO_4$ | 27 |
| Review | $PO_4$ | 28 |
| Eastern tropical North Pacific | $NO_3$ | 29 |
| Baltic Sea | $NO_3$, $PO_4$ | 30 |
| Black Sea | $NO_3$, $PO_4$ | 31 |
| Indian Ocean | $NO_3$, $PO_4$ | 32 |
| Western Atlantic | $SiO_2(aq)$ | 33 |
| North Atlantic | $NO_3$, $PO_4$, $SiO_2(aq)$ | 34 |
| Japan Coast | $NO_3$ | 35 |
| Southeastern region of the Bering Sea | $NO_3$, $PO_4$, $SiO_2(aq)$ | 36 |
| Black Sea | $NO_3$ | 37 |
| Equatorial Pacific | $NO_3$ | 38 |
| Western North Atlantic | $PO_4$ | 39 |
| North Pacific | $NO_3$ | 40 |
| Weddell Sea | $NO_3$, $PO_4$, $SiO_2(aq)$ | 41 |
| Cariaco Trench | $NO_3$ | 42 |
| Mediterranean Sea | $NO_3$ | 43 |
| Labrador and Baffin Island coast | $NO_3$, $PO_4$, $SiO_2(aq)$ | 44 |
| North Indian | $PO_4$ | 45 |
| Bay of Bengal | $PO_4$ | 46 |
| Arabian Sea | $NO_3$, $PO_4$ | 47 |
| West coast of South America | $NO_3$, $NO_2$ | 48 |
| Central Western North Indian | $PO_4$, $SiO_2(aq)$ | 49 |
| Labrador Sea | $PO_4$, $SiO_2(aq)$ | 50 |
| Japan coast | $PO_4$ | 51 |
| Sea of Japan | $PO_4$ | 52 |
| Pacific Ocean east | $PO_4$, $SiO_2(aq)$ | 53 |
| Kuroshio-Taiwan Island | $SiO_2(aq)$ | 54 |
| Australia | $NO_3$, $PO_4$ | 55 |
| Southeast Pacific | $NO_3$, $PO_4$ | 56 |
| North Indian | $PO_4$, $SiO_2(aq)$ | 57 |
| North Indian | $NO_3$ | 58 |
| Western North Pacific, Indian, and Antarctic | $PO_4$, $SiO_2(aq)$ | 59 |
| Northern equatorial Pacific | $NO_3$ | 60 |
| Southeast Pacific | $NO_3$, $PO_4$ | 61 |
| Pacific Ocean, subarctic boundary near 170° W | $NO_3$, $PO_4$, $SiO_2(aq)$ | 62 |
| Equatorial Atlantic | $NO_3$ | 63 |
| Japan Sea | $PO_4$ | 64 |
| Southern California coast | $NO_3$, $SiO_2(aq)$ | 65 |
| Western Indian | $NO_3$ | 66 |
| Southwest Indian | $PO_4$ | 67 |
| Southwest Indian | $PO_4$ | 68 |
| Eastern equatorial Atlantic | $NO_3$, $PO_4$ | 69 |
| North Sea | $NO_3$, $SiO_2(aq)$ | 70 |

**Table 1 (continued)**
**NUTRIENTS IN THE OCEANS**

| Location | Nutrients | Ref. |
|---|---|---|
| North Sea | $NO_3$, $PO_4$, $SiO_2$(aq) | 71 |
| Barents Sea | $NO_3$, $PO_4$, $SiO_2$(aq) | 72 |
| Peru Current | $NO_3$ | 73 |
| Subarctic Pacific region | $PO_4$ | 74 |
| Labrador Sea | $NO_3$, $PO_4$ | 75 |
| Atlantic Ocean | $PO_4$ | 76 |
| Oyashio and the Northern part of Kuroshio regions | $PO_4$ | 77 |
| Agulhas Current | $PO_4$ | 78 |

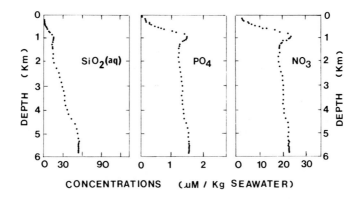

FIGURE 4.   Depth profiles of $NO_3^-$, $PO_4^{3-}$, and $SiO_2$(aq), at a GEO-SECS station near 30° N, 50° W.[11]

these features) and to disperse them (enhanced mixing along the boundaries). En-hanced vertical mixing also occurs in upwelling regions where nutrient-rich waters from below mix up into the photic zone. The major upwelling areas are the equatorial re-gions of all three oceans and coastal regions, such as off the west coast of the U.S. and Africa. In addition, nutrient concentrations of coastal waters, near-shore waters, and estuaries are also increased by land runoff of fertilizer, input from sewage outfalls, and rivers.

Because of a balance between the circulatory supply of oxygen and biological con-sumption, some oceanographic regions and basins can be classified as oxygen deficient, intermittently anoxic, or permanently anoxic. As a consequence, in these regions, there is a large build-up of nutrients. Some examples are, respectively, the eastern tropical North Pacific where denitrification appears to be occurring and nitrate levels are high, the Norwegian fjords, and the Black Sea.[13]

**Spatial Variations**

Generally, the concentration of all nutrients in the deep waters increases from the Atlantic to the Indian to the Pacific oceans because of the respective increase in the age of the waters. For the same reason, nutrient concentrations increase going north in the deep waters of the Indian and Pacific oceans. The opposite situation exists in the Atlantic because of the formation of deep water in the North. Particularly high concentrations of silicate are found in the eastern South Atlantic, Pacific, and Antarc-tic because of continental weathering and the presence of siliceous organisms in the shallow water.

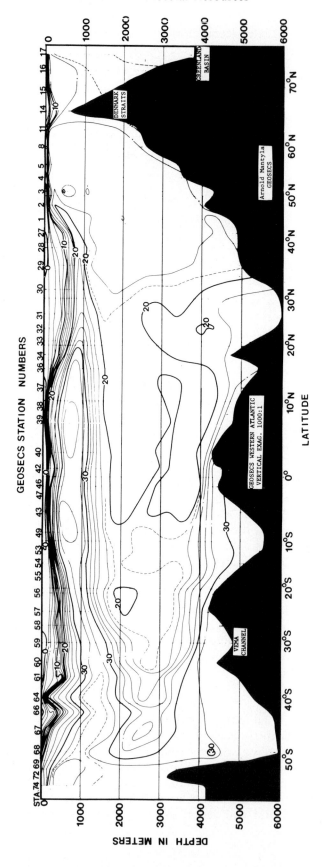

FIGURE 5.   Vertical section of NO$_3^-$ ($\mu M$/kg) vs. depth (m) from the western Atlantic GEOSECS expedition.[11]

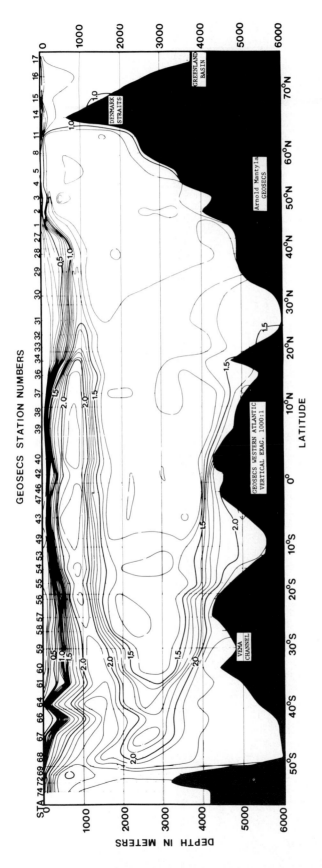

FIGURE 6.  Vertical section of $PO_4^{3-}$ ($\mu M/kg$) vs. depth (m) from the western Atlantic GEOSECS expedition.[11]

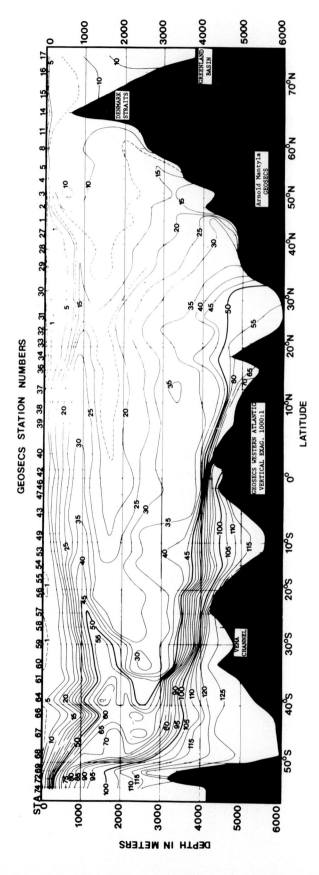

FIGURE 7.   Vertical section of SiO₃ ($\mu M$/kg) vs. depth (m) from the western Atlantic GEOSECS expedition.[11]

Higher concentrations of all nutrients are found in the cold waters of the Arctic and Antarctic oceans; for example, Figures 5, 6, and 7 show nutrient maxima in Antarctic Intermediate waters (near 1000 m, originating at 40° S). Those oceans also show large seasonal variability because of the great variation in light intensity near the poles.

## Temporal Variations

Largest seasonal variations occur in the photic zones of temperate latitude coastal waters and estuaries;[14] smaller changes occur in the tropics. Seasonal variations also occur in the open ocean, especially where there is a seasonal pycnocline and in the Indian Ocean where there are monosoonal effects. Maximum nutrient concentrations occur in November through March, when there is a lack of daylight and no seasonal pycnocline to inhibit mixing; then as the seasonal pycnocline forms, concentrations begin to decrease to the lowest values which are found in June and September. In general, nutrients are regenerated in the winter and consumed in the summer.

## REFERENCES

1. **Riley, J. P. and Chester, R.,** *Introduction to Marine Chemistry,* Academic Press, New York, 1971, chap. 7.
2. **Spencer, C. P.,** The micronutrient elements, in *Chemical Oceanography,* Vol. 2, 2nd ed., Riley, J. P. and Skirrow, G., Eds., Academic Press, New York, 1975, chap. 11.
3. **Kester, D. R. and Pytkowicz, R. M.,** Determination of the apparent dissociation constants of phosphoric acid in seawater, *Limnol. Oceanogr.,* 12, 243, 1967.
4. **Krauskopf, K. B.,** Dissolution and precipitation of silica at low temperatures, *Geochim. Cosmochim. Acta,* 10, 1, 1956.
5. **MacKenzie, F. T. and Garrels, R. M.,** Chemical mass balance between rivers and oceans, *Am. J. Sci.,* 264, 507, 1966.
6. **Burton, J. D. and Liss, P. S.,** Oceanic budget of dissolved silicon, *Nature (London),* 220, 905, 1968.
7. **Wollast, R.,** The silica problem, in *The Sea,* Vol. 5, Goldberg, E. D., Ed., John Wiley & Sons, New York, 1974, chap. 11.
8. **Riley, J. P.,** Analytical chemistry of sea water (determination of micronutrient elements), in *Chemical Oceanography,* Vol. 3, 2nd ed., Riley, J. P. and Skirrow, G., Eds., Academic Press, London, 1975, chap. 19.
9. **Strickland, J. D. H. and Parsons, T. R.,** *A Practical Handbook of Seawater Analysis,* 2nd ed., Fisheries Research Board of Canada, Ottawa, 1972, part 2.
10. **Grasshoff, K.,** *Methods of Seawater Analysis,* Verlag Chemie, Weinheim, 1976, chap. 9.
11. **Bainbridge, A . E.,** GEOSECS Atlantic Expedition, Vol. 2, U.S. Government Printing Office, Washington, D.C., 1979.
12. **Reid, J. L., Brinton, E., Fleminger, A., Venrick, E. L., and McGowan, J. A.,** Ocean circulation and marine life, in *Advances in Oceanography,* Chamock, H. and Deacon, G., Eds., Plenum Press, New York, 1978.
13. **Richards, F. A.,** Anoxic basins and fjords, in *Chemical Oceanography,* Vol. 1, Riley, J. P. and Skirrow, G., Eds., Academic Press, New York, 1965, chap. 13.
14. **Aston, S. R.,** Estuarine chemistry, in *Chemical Oceanography,* Vol. 7, 2nd ed., Riley, J. P. and Chester, R., Eds., Academic Press, New York, 1978, chap. 41.
15. **Friederich, G. E., Codispoti, L. A., Friebertshauser, M. A., and Bishop, D. D.,** Coastal Upwelling Ecosystems Analysis, Technical Rep. 33, PB-279173, National Technical Information Service, Springfield, Va., 1977.
16. **Naqvi, S. W. A., De Sousa, S. N., and Reddy, C. V. G.,** Relationship between nutrients and dissolved oxygen with special reference to water masses in western Bay of Bengal, *Indian J. Mar. Sci.,* 7(1), 15, 1978.
17. **Sen Gupta, R., De Sousa, S. N., and Joseph, T.,** On nitrogen and phosphorus in the western Bay of Bengal, *Indian J. Mar. Sci.,* 6(2), 107, 1977.
18. **Sen Gupta, R., Sankaranarayanan, V. N., De Sousa, S. N., and Fondekar, S. P.,** Chemical oceanography of the Arabian Sea. III. Studies on nutrient fraction and stoichiometric relations in the northern and the eastern basins, *Indian J. Mar. Sci.,* 5(1), 58, 1976.

19. **Hood, D. W. and Kelley, J. J.,** Evaluation of mean vertical transports in an upwelling system by carbon dioxide measurements, *Mar. Sci. Commun.,* 2(6), 387, 1976.
20. **Kiefer, D. A., Olson, R. J., and Holm-Hansen, O.,** Another look at the nitrite and chlorophyll maxima in the central North Pacific, *Deep-Sea Res.,* 23(12), 1199, 1976.
21. **Broecker, W. S., Takahashi, T., and Li, Y. H.,** Hydrography of the central Atlantic. I. The two-degree discontinuity, *Deep-Sea Res.,* 23(12), 1083, 1976.
22. **Richards, E. A. and Codispoti, L. A.,** An analysis of the horizontal regime of denitrification in the eastern tropical North Pacific, *Limnol. Oceanogr.,* 21(3), 379, 1976.
23. **Szekielda, K. H.,** General chemistry of the coastal waters of the Mid-Atlantic coast, in Proc. Estuarine Research Federation, Outer Continental Shelf Conference Workshop, Estuarine Research Federation, Wachapreague, Va., 1975, 153.
24. **D'Souza, R. S. and Sastry, J. S.,** Oceanography of the Arabian Sea during the south-west monsoon season. V. Distribution of inorganic phosphate, *Indian J. of Mar. Sci.,* 4(2), 145, 1975.
25. **Fonselius, S. H.,** Phosphorus in the Black Sea, *Mem. Am. Assoc. Pet. Geol.,* 20, 144, 1974.
26. **Pak, H., Zaneveld, J., and Ronald, V.,** Equatorial front in the eastern Pacific Ocean, *J. Phys. Oceanogr.,* 4(4), 570, 1974.
27. **Guillen, O. and Ezaguirre de Rondan,** Nutrients in the Peru coastal current, in Proc. Int. Symp. Oceanography of the South Pacific, New Zealand National Committee, UNESCO, Wellington, N.Z., 1973, 397.
28. **Gulbrandsen, R. A. and Roberson, C. E.,** Inorganic phosphorus in sea water, in *Environmental Phosphorus Handbook,* John Wiley & Sons, New York, 117, 1973.
29. **Codispoti, L. A.,** Chemical and Physical Properties of the Eastern Tropical North Pacific with Emphasis on the Oxygen Minimum Layer, 774117/6GA, Government Rep. Announce., U.S. National Technical Information Service, Washington, D.C., 1973.
30. **Sen Gupta, R.,** Nitrogen and phosphorus budgets in the Baltic Sea, *Mar. Chem.,* 1(4), 267, 1973.
31. **Brewer, P. G. and Murray, J. W.,** Carbon, nitrogen, and phosphorus in the Black Sea, *Deep-Sea Res.,* 20(9), 803, 1973.
32. **McGill, D. A.,** Light and nutrients in the Indian Ocean, *Ecol. Stud.,* 3, 53, 1973.
33. **Garner, D. M., Coote, A. R., and Mann, C. R.,** The meridional distribution of silicate in the western Atlantic Ocean, *Deep-Sea Res.,* 20(9), 791, 1973.
34. **Spencer, D. W.,** GEOSECS II, The 1970 north Atlantic station: hydrographic features, oxygen, and nutrients, *Earth Planet. Sci. Lett.,* 16, 91, 1972.
35. **Wada, E. and Hattori, A.,** Nitrite distribution and nitrate reduction in deep sea waters, *Deep-Sea Res.,* 19(2), 123, 1972.
36. **Alvarez-Borrego, S., Gordon, L. I., Jones, L. B., Park, P. K., and Pytkowicz, R. M.,** Oxygen-carbon dioxide-nutrients relationships in the southeastern region of the Bering Sea, *J. Oceanogr. Soc. Jpn.,* 28(2), 71, 1972.
37. **Sen Gupta, R.,** Oceanography of the Black Sea: inorganic nitrogen compounds, *Deep-Sea Res.,* 18(5), 457, 1971.
38. **Hattori, A. and Wada, E.,** Nitrite distribution and its regulating processes in the equatorial Pacific Ocean, *Deep-Sea Res.,* 18(6), 557, 1971.
39. **Stefansson, U. and Atkinson, L. P.,** Nutrient-density relationships in the western North Atlantic between Cape Lookout and Bermuda, *Limnol. Oceanogr.,* 16(1), 51, 1971.
40. **Sagi, T.,** Distribution of nitrate nitrogen in the western North Pacific Ocean, *Oceanogr. Mag.,* 22(2), 63, 1970.
41. **Hufford, G. L. and Tennyson, E. J., Jr.,** Distribution of nutrients in the Weddell Sea: Feb.-Mar. 1968—Feb.-Mar. 1969, Oceanography Rep. No. 33, U.S. Coast Guard, Washington, D.C., 1970.
42. **Okuda, T., Benitez, J., and Fernandez, A. E.,** Vertical distribution of inorganic and organic nitrogen in the Cariaco Trench, *Inst. Oceanogr. Bull.,* 8(1 and 2), 28, 1969.
43. **Yegorova, V. A.,** Nitrites in the waters of the Mediterranean Sea, *Oceanology,* 8(4), 502, 1969.
44. **McGill, D. A. and Corwin, N.,** Nutrient Distribution Along the Labrador and Baffin Island Coast, 1965, WHOI-Ref-69-37, WHOI-Contrib-1838, U.S. Coast Oceanography Rep. No. 12, CG-373-12, 1969.
45. **Viswanathan, R. and Ganguly, A. K.,** Distribution of phosphorus in Northern Indian Ocean, 1962-63, *Bull. Natl. Inst. Sci. India,* 38(1), 350, 1968.
46. **Rao, V. C. and Rao, T. S. S.,** Distribution of total phosphorus in the Bay of Bengal, *Bull. Natl. Inst. Sci. India,* 38(1), 93, 1968.
47. **Reddy, C. V. G. and Sankaranarayanan, V. N.,** Distribution of nutrients in the shelf waters of the Arabian Sea along the West Coast of India, *Bull. Natl. Inst. Sci. India,* 38(1), 206, 1968.
48. **Fiadeiro, M. and Strickland, J. D. H.,** Nitrate reduction and the occurrence of a deep nitrite maximum in the ocean off the west coast of South America, *J. Mar. Res.,* 26(3), 187, 1968.

49. Reddy, C. V. G. and Sankaranarayanan, V. N., Distribution of phosphates and silicates in the central western North Indian Ocean, in relation to some hydrographical factors, *Bull. Natl. Inst. Sci. India,* 38, 103, 1968.

50. Jones, P. G. W. and Folkard, A. R., The Distribution of Phosphate, Silicate and Dissolved Oxygen in the 0- to 100-m Layer During Norwestlant 1-3, Special Publication No. 7, International Commission for the Northwest Atlantic Fisheries, Dartmouth, N.S., 57, 1968.

51. Akiyama, T., The distribution of dissolved oxygen and phosphate phosphorus in the adjacent seas of Japan, *Oceanogr. Mag.,* 20(2), 147, 1968.

52. Yamamoto, K., The total and organic phosphorous in the Japan Sea, *Oceanogr. Mag.,* 20(1), 39, 1968.

53. Park, P. K., Hager, S. W., Pirson, J. E., and Ball, D. S., Surface phosphate and slicate distributions in the northeastern Pacific Ocean, *J. Fish. Res. Board Can.,* 25(12), 2739, 1968.

54. Hung, T.-C. and Lee, C.-W., Chemical oceanography in the Kuroshio around Taiwan Island, *Bull. Inst. Chem. Acad. Sin.,* 14, 80, 1967.

55. Kirkwood, L. F., Inorganic Phosphate, Organic Phosphorus, and Nitrate in Australian Waters, Division of Fish. Oceanography Technical Paper, Commonwealth Scientific and Industrial Research Organization, Australia, 25, 1967.

56. Goering, J. J. and Wallen, D., Vertical distribution of phosphate and nitrite in the upper one-half meter of the southeast Pacific Ocean, *Deep-Sea Res.,* 14(1), 29, 1967.

57. Rozanov, A. G., Distribution of Phosphates and Silicic Acid in the Water of the Northern Part of the Indian Ocean, Translation No. 324, U.S. Naval Oceanographic Office, Washington, D.C., 1967.

58. Rozanov, A . G. and Bykova, V. S., Distribution of Nitrates and Nitrites in the Water of North Indian Ocean, Translation No. 323, U.S. Naval Oceanographic Office, Washington, D.C., 1967.

59. Toyota, Y. and Okabe, S., Vertical distribution of iron, aluminum, silicon, and phosphorus and particulate matter collected in the western North Pacific, Indian, and Antarctic oceans, *J. Oceanogr. Soc. Jpn.,* 23(1), 1, 1967.

60. Wooster, W. S., Further observations on the secondary nitrite maximum in the northern equatorial Pacific, *J. Mar. Res.,* 25(2), 154, 1967.

61. Goering, J. J. and Wallen, D., The vertical distribution of phosphate and nitrite in the upper one-half meter of the southeast Pacific Ocean, *Deep-Sea Res.,* 14(1), 29, 1967.

62. Park, K., Chemical features of the subarctic boundary near 170 degrees W, *J. Fish. Res. Board Can.,* 24(5), 899, 1967.

63. Okuda, T., Vertical distribution of inorganic nitrogen in the equatorial Atlantic Ocean, *Inst. Oceanogr. Bol.,* 5(1-2), 67, 1966.

64. Ohwada, M. and Yamamoto, K., Some chemical elements in the Japan Sea, *Oceanogr. Mag.,* 18(1-2), 31, 1966.

65. Armstrong, F. A. J. and Lafond, E. C., Chemical nutrient concentrations and their relationship to internal waves and turbidity off Southern California, *Limnol. Oceanogr.,* 11(4), 538, 1966.

66. Fraga, F., Distribution of particulate and dissolved nitrogen in the western Indian Ocean, *Deep-Sea Res.,* (13(3), 413, 1966.

67. Orren, M. J., Hydrology of the South West Indian Ocean, Investigational Report No. 55, Division of Sea Fisheries, Department of Commerce and Industries, South Africa, 1966.

68. Mostert, S. A., Distribution of Inorganic Phosphate and Dissolved Oxygen in the South West Indian Ocean, Investigational Report No. 54, Division of Sea Fisheries, Department of Commerce and Industries, South Africa, 1966.

69. Jones, P. G. W., The nitrate and phosphate content of the eastern equatorial Atlantic Ocean, *Sci. Naturelles,* 28(2), 444, 1966.

70. Olsen, O. V., Hydrographic investigations in the Skagerak and northern North Sea, *Ann. Biol.,* 22, 40, 1965.

71. Burns, R. B., Chemical observations in the North Sea in 1965, *Ann. Biol.,* 22, 29, 1965.

72. Norina, A. M., Hydrochemical Characteristics of the Northern Part of the Barents Sea, Gosudarstvennyi Okeanograficheskii Institut, Moscow, No. 83, 1965.

73. Wooster, W. S., Chow, T. J., and Barrett, I., Nitrite distribution in Peru current waters, *J. Mar. Res.,* 23(3), 210, 1965.

74. Sugiura, Y., Distribution of reserved, preformed, phosphate in the subarctic Pacific region, *Pap. Meterol. Geophys.,* 15(3-4), 208, 1965.

75. McGill, D. A. and Corwin, N., The Distribution of Nutrients in the Labrador Sea, Summer 1964, U.S. Coast Guard Oceanographic Report No. 10, CG 373, 25, 1965.

76. McGill, D. A. and Sears, M., The distribution of phosphorus and oxygen in the Atlantic Ocean, as observed during the I.G.Y., 1957-1958, in *Progress in Oceanography,* Vol. 2, Sears, M., Ed., Pergamon Press, New York, 1964, 127.

77. **Sugiura, Y. and Yoshimura, H.,** Distribution and mutual relation of dissolved oxygen and phosphate in the Oyashio and the northern part of Kuroshio regions, *J. Oceanogr. Soc. Jpn.*, 20(1), 14, 1964.
78. **Darbyshire, J.,** A hydrological investigation of the Agulhas current area, *Deep-Sea Res.*, 11(5), 781, 1964.

# LAND RESOURCES

## Jean François Henry and Oskar R. Zaborsky

### Table 1
### WORLD LAND RESOURCES (1,000 ha)

| Region[a] | Total area[b] | Land area[c] | Arable land[d] | Permanent crops[e] | Permanent pasture[f] | Forest and woodland[g] | Other land[h] | Inland water |
|---|---|---|---|---|---|---|---|---|
| Africa | 3,031,168 | 2,964,613 | 194,910 | 14,465 | 800,437 | 639,602 | 1,315,199 | 66,555 |
| North and Central America | 2,246,443 | 2,140,488 | 264,974 | 6,568 | 346,735 | 718,311 | 803,900 | 105,955 |
| South America | 1,781,980 | 1,753,691 | 81,892 | 22,176 | 441,834 | 924,263 | 283,526 | 28,289 |
| Asia | 2,757,442 | 2,676,621 | 454,908 | 26,347 | 538,310 | 603,645 | 1,053,411 | 80,821 |
| Europe | 487,032 | 472,809 | 127,438 | 14,942 | 87,606 | 153,444 | 89,379 | 14,223 |
| Oceania | 850,956 | 842,906 | 46,212 | 912 | 469,761 | 185,950 | 140,071 | 8,050 |
| U.S.S.R. | 2,240,220 | 2,227,200 | 227,400 | 4,906 | 373,400 | 920,000 | 701,494 | 13,020 |
| World | 13,395,241 | 13,078,328 | 1,397,734 | 90,316 | 3,058,083 | 4,145,215 | 4,386,980 | 316,913 |

*Note:* Specific country notes pertaining to land-use categories are as follows: Total Area — Greenland: data refer to area free from ice. Mauritius: data exclude dependencies. Namibia: data include the territory of Walvis Bay. New Caledonia: data include dependencies. South Africa: data exclude the territory of Walvis Bay. U.S.S.R.: data include the White Sea (9,000,000 ha) and the Azov Sea (3,730,000 ha) Arable Land and Land Under Permanent Crops — Australia: data on arable land include about 27,000,000, ha of cultivated grassland. Cuba: data refer to the State sector only. Portugal: data include about 800,000 ha of temporary crops grown in association with permanent crops and forests. Permanent Meadows and Pastures — Australia: data refer to balance of area of rural holdings. Egypt: rough grazing land is included under Other land. U.S.S.R.: data exclude pastures for reindeer.

[a] Countries included in the regions: Africa — Algeria, Angola, Benin, Botswana, British Indian Ocean Territory, Burundi, Cameroon, Cape Verde, Central African Empire, Chad, Comoros, Congo, Djibouti, Egypt, Equatorial Guinea, Ethiopia, Gabon, Gambia, Ghana, Guinea, Guinea-Bissau, Ivory Coast, Kenya, Lesotho, Liberia, Libya, Madagascar, Malawi, Mali, Mauritania, Mauritius, Morocco, Mozambique, Namibia, Niger, Nigeria, Reunion, Rhodesia, Rwanda, St. Helena, Sao Tome and Principe, Senegal, Seychelles, Sierra Leone, Somalia, South Africa, Spanish North Africa, Sudan, Swaziland, Tanzania, Togo, Tunisia, Uganda, Upper Volta, Western Sahara, Zaire, Zambia; North and Central America — Antigua, Bahamas, Barbados, Belize, Bermuda, Canada, Cayman Islands, Costa Rica, Cuba, Dominica, Dominican Republic, El Salvador, Greenland, Grenada, Guadeloupe, Guatemala, Haiti, Honduras, Jamaica, Martinique, Mexico, Montserrat, Netherlands Antilles, Nicaragua,

Table 1 (continued)
## WORLD LAND RESOURCES (1,000 ha)

Panama, Panama Canal Zone, Puerto Rico, St. Kitts-Nevis-Anguilla, St. Lucia, St. Pierre and Miquelon, St. Vincent, Trinidad and Tobago, Turks and Caicos Islands, United States, Virgin Islands (U.K.), Virgin Islands (U.S.); South America — Argentina, Bolivia, Brazil, Chile, Colombia, Ecuador, Falkland Islands (Malvinas), French Guiana, Guyana, Paraguay, Peru, Surinam, Uruguay, Venezuela; Asia — Afghanistan, Bahrain, Bangladesh, Bhutan, Brunei, Burma, China, Cyprus, East Timor, Gaza Strip (Palestine), Hong Kong, India, Indonesia, Iran, Iraq, Israel, Japan, Jordan, Kampuchea, Democratic Korea, Democratic People's Republic of Kuwait, Lao, Lebanon, Macau, Malaysia: peninsula Malaysia, Malaysia: Sabah, Malaysia: Sarawak, Maldives, Mongolia, Nepal, Oman, Pakistan, Philippines, Qatar, Saudi Arabia, Singapore, Sri Lanka, Syria, Thailand, Turkey, United Arab Emirates, Viet Nam, Yemen Arab Republic, Yemen, Democratic; Europe — Albania, Andorra, Austria, Belgium-Luxembourg, Bulgaria, Czechoslovakia, Denmark, Faeroe Islands, Finland, France, German Democratic Republic, Germany, Federal Republic of, Gibraltar, Greece, Holy See, Hungary, Iceland, Ireland, Italy, Liechtenstein, Malta, Monaco, Netherlands, Norway, Poland, Portugal, Romania, San Marino, Spain, Sweden, Switzerland, United Kingdom, Yugoslavia; Oceania — American Samoa, Australia, Canton and Enderbury Islands, Christmas Island (Australia), Cocos (Keeling) Islands, Fiji, French Polynesia, Gilbert Islands, Guam, Johnston Island, Midway Islands, Nauru, New Caledonia, New Hebrides, New Zealand, Niue Island, Norfolk Island, Pacific Islands (Trust Territory), Papua New Guinea, Pitcairn Island, Samoa, Solomon Islands, Tokelau, Tonga, Tuvalu, Wake Island, Wallis and Futuna Islands; U.S.S.R.

b    Total area refers to the total area of the country, including area under inland water bodies.

c    Land area refers to total area, excluding area under inland water bodies. The definition of inland water bodies generally includes major rivers and lakes.

d    Arable land refers to land under temporary crops (double-cropped areas are counted only once), temporary meadows for mowing or pasture, land under market and kitchen gardens (including cultivation under glass), and land temporarily fallow or lying idle.

e    Land under permanent crops refers to land cultivated with crops which occupy the land for long periods and need not be replanted after each harvest, such as cocoa, coffee, and rubber; it includes land under shrubs, fruit trees, nut trees, and vines, but excludes land under trees grown for wood or timber.

f    Permanent meadows and pastures refers to land used permanently (5 years or more) for herbaceous forage crops, either cultivated or growing wild (wild prairie or grazing land).

g    Forests and woodland refers to land under natural or planted stands of trees, whether or not productive, and includes land from which forests have been cleared, but which will be reforested in the foreseeable future.

h    Other land includes unused but potentially productive land, built-on areas, wasteland, parks, ornamental gardens, roads, lanes, barren land, and any other land not specifically listed under items (c) through (g).

Adapted from FAO Production Yearbook 1977, volume 31, FAO Statistics Series No. 15, Food and Agriculture Organization of the United Nations, Rome, Italy, 1978.

## Table 2
## WORLD IRRIGATED LAND

| Region[a] | Irrigated area[b] (1,000 ha) | Percent of arable land |
|---|---|---|
| Africa | 7,697 | 3.95 |
| North and Central America | 23,174 | 8.75 |
| South America | 6,730 | 8.22 |
| Asia | 163,637 | 35.97 |
| Europe | 12,393 | 9.72 |
| Oceania | 1,625 | 3.72 |
| U.S.S.R. | 15,300 | 6.73 |
| World | 230,556 | 16.49 |

*Note:* Specific country notes pertaining to irrigation are as follows: France: data exclude market and kitchen gardens. Hungary: data exclude complementary farm plots and individual farms. U.K.: data exclude Scotland and Northern Ireland. For the following countries, data refer to land provided with irrigation facilities: Bulgaria, Norway, Romania, and Surinam. For the following countries, data refer to irrigated rice only: Japan, Republic of Korea, and Sri Lanka.

[a]   See footnote to Table 1.
[b]   Data on irrigation relate to areas purposely provided with water, including land flooded by river water for crop production or pasture improvement, whether this area is irrigated several times or only once during the year stated.

Adapted from FAO Production Yearbook 1977, Volume 31, FAO Statistics Series No. 15, Food and Agriculture Organization of the United Nations, Rome, Italy, 1978.

## Table 3
## FOREST LAND IN THE WORLD (1,000 ha)

| Region | Forest land in use | Unproductive | Unstocked | Total | Percent of total land |
|---|---|---|---|---|---|
| Africa | NA[a] | NA | 10,000 | 710,000 | 24 |
| North and Central America | NA | 287,103 | 42,160 | 826,219 | 38 |
| South America | NA | NA | 50,000 | 890,000 | 47 |
| Asia | NA | NA | 50,000 | 550,000 | 19 |
| Europe | 123,000 | 11,700 | 6.000 | 144,000 | 29 |
| U.S.S.R. | 317,945 | 27,273 | 171,892 | 910,009 | 34 |
| World | (440,945) | (326,076) | (330,052) | (4,030,228) | 29 |

*Note:* Numbers are only approximate and may not match exactly data of Table 1 because of time difference.

[a]   NA, not available.

Adapted from Food and Agriculture Organization of the United Nations,   World Forest Inventory 1963, FAO, Rome, Italy.

Table 4
LAND UTILIZATION IN THE U.S., 1974 (1,000 acres)

| State | Cropland Used for crops[a] | Idle | Used only for pasture | Grassland pasture[b] | Forest land[c] | Special use areas[d] | Other land | Total land area |
|---|---|---|---|---|---|---|---|---|
| Alabama | 3,243 | 419 | 2,135 | 2,917 | 21,333 | 1,768 | 637 | 32,452 |
| Alaska | 17 | 6 | 2 | 1,625 | 118,076 | 32,902 | 209,888 | 362,516 |
| Arizona | 1,310 | 218 | 117 | 40,941 | 17,420 | 8,709 | 3,872 | 72,587 |
| Arkansas | 7,296 | 521 | 2,409 | 2,559 | 18,236 | 1,566 | 598 | 33,245 |
| California | 9,276 | 463 | 1,404 | 23,910 | 39,826 | 16,292 | 8,900 | 100,071 |
| Colorado | 9,159 | 508 | 1,292 | 29,274 | 19,387 | 3,238 | 3,552 | 66,410 |
| Connecticut | 163 | 15 | 56 | 46 | 1,846 | 661 | 325 | 3,112 |
| Delaware | 513 | 12 | 21 | 8 | 390 | 186 | 138 | 1,268 |
| District of Columbia | | | | | | 39 | | 39 |
| Florida | 2,732 | 380 | 1,086 | 6,026 | 17,652 | 5,647 | 1,005 | 34,618 |
| Georgia | 4,764 | 556 | 1,828 | 1,731 | 24,869 | 2,766 | 653 | 37,167 |
| Hawaii | 151 | 169 | 37 | 1,018 | 1,626 | 842 | 269 | 4,112 |
| Idaho | 5,272 | 249 | 874 | 20,840 | 18,030 | 4,078 | 3,570 | 52,913 |
| Illinois | 22,502 | 652 | 1,856 | 1,834 | 3,745 | 3,211 | 1,879 | 35,679 |
| Indiana | 11,978 | 678 | 1,423 | 1,487 | 3,870 | 2,041 | 1,625 | 23,102 |
| Iowa | 24,134 | 425 | 3,630 | 2,152 | 2,430 | 1,968 | 1,063 | 35,802 |
| Kansas | 26,821 | 909 | 3,895 | 15,950 | 1,363 | 2,210 | 1,196 | 52,344 |
| Kentucky | 4,339 | 571 | 4,487 | 2,013 | 11,888 | 1,519 | 559 | 25,376 |
| Louisiana | 4,037 | 595 | 1,427 | 2,270 | 15,342 | 1,731 | 3,353 | 28,755 |
| Maine | 456 | 96 | 101 | 142 | 17,505 | 779 | 710 | 19,789 |
| Maryland | 1,525 | 110 | 258 | 209 | 2,925 | 1,028 | 275 | 6,330 |
| Massachusetts | 193 | 16 | 55 | 52 | 2,848 | 1,208 | 637 | 5,009 |
| Michigan | 6,567 | 869 | 909 | 1,241 | 19,000 | 3,867 | 3,910 | 36,363 |
| Minnesota | 20,519 | 1,394 | 1,993 | 1,954 | 18,415 | 4,004 | 2,466 | 50,745 |
| Mississippi | 5,467 | 764 | 2,478 | 2,620 | 16,892 | 1,281 | 767 | 30,269 |
| Missouri | 12,842 | 947 | 6,692 | 6,610 | 12,661 | 2,321 | 2,084 | 44,157 |
| Montana | 14,504 | 372 | 1,145 | 49,465 | 19,899 | 4,633 | 3,158 | 93,176 |
| Nebraska | 19,724 | 412 | 3,274 | 22,137 | 1,032 | 1,618 | 752 | 48,949 |
| Nevada | 564 | 36 | 153 | 46,673 | 7,255 | 7,235 | 8,412 | 70,328 |
| New Hampshire | 120 | 11 | 43 | 34 | 5,046 | 304 | 219 | 5,777 |
| New Jersey | 569 | 62 | 83 | 54 | 1,856 | 1,598 | 591 | 4,813 |
| New Mexico | 1,553 | 207 | 542 | 50,525 | 17,256 | 5,325 | 2,295 | 77,703 |
| New York | 4,396 | 461 | 1,228 | 1,580 | 14,897 | 5,502 | 2,548 | 30,612 |
| North Carolina | 4,819 | 628 | 1,099 | 1,050 | 20,223 | 2,731 | 651 | 21,231 |
| North Dakota | 27,275 | 753 | 2,270 | 10,528 | 419 | 1,532 | 1,562 | 44,339 |
| Ohio | 10,457 | 733 | 1,515 | 1,610 | 6,422 | 2,983 | 2,504 | 26,224 |
| Oklahoma | 10,731 | 449 | 4,651 | 16,235 | 9,296 | 2,029 | 629 | 44,020 |
| Oregon | 4,242 | 219 | 815 | 23,172 | 29,387 | 2,655 | 1,667 | 61,557 |
| Pennsylvania | 4,437 | 470 | 1,023 | 1,026 | 17,638 | 3,714 | 470 | 28,778 |
| Rhode Island | 22 | 2 | 5 | 5 | 395 | 190 | 52 | 671 |
| South Carolina | 2,610 | 251 | 704 | 667 | 12,402 | 1,599 | 1,111 | 19,344 |
| South Dakota | 16,655 | 255 | 2,770 | 24,670 | 1,700 | 1,590 | 971 | 48,611 |
| Tennessee | 4,439 | 436 | 3,501 | 1,899 | 12,820 | 2,193 | 1,162 | 26,450 |
| Texas | 24,383 | 2,275 | 11,280 | 95,803 | 24,043 | 6,852 | 3,130 | 167,766 |
| Utah | 1,426 | 132 | 438 | 23,711 | 14,720 | 5,185 | 6,929 | 52,541 |
| Vermont | 543 | 36 | 233 | 234 | 4,384 | 263 | 238 | 5,931 |
| Virginia | 2,685 | 315 | 1,690 | 1,819 | 16,076 | 2,127 | 747 | 25,459 |
| Washington | 7,470 | 209 | 688 | 6,679 | 20,534 | 5,337 | 1,688 | 42,605 |
| West Virginia | 771 | 107 | 739 | 717 | 12,126 | 568 | 377 | 15,405 |
| Wisconsin | 9,538 | 566 | 1,762 | 2,095 | 14,891 | 2,664 | 3,341 | 34,857 |
| Wyoming | 2,101 | 69 | 560 | 46,016 | 5,885 | 5,375 | 2,204 | 62,210 |

### Table 4 (continued)
### LAND UTILIZATION IN THE U.S., 1974 (1,000 acres)

Cropland

| State | Used for crops[a] | Idle | Used only for pasture | Grassland pasture[b] | Forest land[c] | Special use areas[d] | Other land | Total land area |
|-------|-------|------|------|------|------|------|------|------|
| U.S. | 361,340 | 21,008 | 82,736 | 597,833 | 718,177 | 181,664 | 300,829 | 2,263,587 |

[a]  Cropland harvested, crop failure, and cultivated summer fallow.
[b]  Grassland and other nonforest pasture and range.
[c]  Excludes reserved forest land in parks and other special uses of land. Includes forested grazing land.
[d]  Includes urban and transportation areas. Federal and State areas used primarily for recreation and wildlife purposes, military areas, farmsteads, farm roads and lanes, and miscellaneous other uses.

From Agricultural Statistics, 1978, U.S. Department of Agriculture, Washington, D.C., 1978.

### Table 5
### LAND UTILIZATION, U.S., SELECTED YEARS, 1910 TO 1974 (IN MILLION ACRES)

| Major land uses | 1910 | 1920 | 1930 | 1940 | 1950 | 1959 | 1969 | 1974 |
|-----------------|------|------|------|------|------|------|------|------|
| Cropland used for crops[a] | 330 | 368 | 382 | 368 | 377 | 359 | 333 | 361 |
| Idle cropland | 23 | 34 | 31 | 31 | 32 | 33 | 51 | 21 |
| Cropland used only for pasture | 84 | 78 | 67 | 68 | 69 | 66 | 88 | 83 |
| Grassland pasture[b] | 693 | 652 | 652 | 650 | 631 | 633 | 604 | 598 |
| Forest land[c] | 600 | 602 | 601 | 608 | 601 | 728 | 723 | 718 |
| Special uses[d] | — | — | — | — | — | 147 | 174 | 182 |
| Other land | 174 | 170 | 171 | 179 | 194 | 305 | 291 | 301 |
| Total land area[e] | 1,904 | 1,904 | 1,904 | 1,904 | 1,904 | 2,271 | 2,264 | 2,264 |

[a]  Cropland harvested, crop failure, and cultivated summer fallow.
[b]  Grassland and other nonforest pasture and range.
[c]  Excludes reserved forest land in parks and other special uses of land. Includes forested grazing land.
[d]  Includes urban and transportation areas. Federal and State areas used primarily for recreation and wildlife purposes, military areas, farmsteads, farm roads and lanes, and miscellaneous other uses.
[e]  Remeasurement and increases in reservoirs account for changes in total land area except for the major increase in 1959 when data for Alaska and Hawaii were added.

From Agricultural Statistics, 1978, U.S. Department of Agriculture, Washington, D.C., 1978.

Table 6

LAND AREAS IN THE U.S., BY MAJOR CLASS OF LAND, SECTION, REGION, AND STATE, JANUARY 1, 1977[a] (THOUSAND ACRES)

| Section, region, and state | Total land area[b] | Forest land | | | | | Range land | Other land[c] |
|---|---|---|---|---|---|---|---|---|
| | | Total | Commercial | Productive reserved | Productive deferred | Other forest | | |
| **New England** | | | | | | | | |
| Connecticut | 3,111.7 | 1,860.8 | 1,805.6 | 30.5 | 0.0 | 24.7 | 0.0 | 1,250.9 |
| Maine | 19,788.8 | 17,748.6 | 16,894.3 | 220.7 | 0.0 | 633.6 | 0.4 | 2,039.8 |
| Massachusetts | 5,008.6 | 2,952.3 | 2,797.7 | 104.5 | 0.0 | 50.1 | 0.1 | 2,056.2 |
| New Hampshire | 5,777.3 | 5,013.5 | 4,692.0 | 58.7 | 25.0 | 237.8 | 0.0 | 763.8 |
| Rhode Island | 671.4 | 404.2 | 395.3 | 8.9 | 0.0 | 0.0 | 0.0 | 267.2 |
| Vermont | 5,930.9 | 4,511.7 | 4,429.9 | 61.5 | 0.0 | 20.3 | 0.0 | 1,419.2 |
| Total | 40,288.7 | 32,491.1 | 31,014.8 | 484.8 | 25.0 | 966.5 | 0.5 | 7,797.1 |
| **Middle Atlantic** | | | | | | | | |
| Delaware | 1,268.5 | 391.8 | 384.4 | 1.8 | 0.0 | 5.6 | 0.0 | 876.7 |
| Maryland | 6,330.2 | 2,653.2 | 2,522.7 | 108.9 | 0.0 | 21.6 | 83.7 | 3,593.3 |
| New Jersey | 4,813.4 | 1,928.4 | 1,856.8 | 34.0 | 0.0 | 37.6 | 60.5 | 2,824.5 |
| New York | 30,611.8 | 17,377.7 | 14,489.0 | 2,480.9 | 0.0 | 407.8 | 1.8 | 13,232.3 |
| Pennsylvania | 28,778.2 | 17,832.0 | 17,478.0 | 194.0 | 0.0 | 160.0 | 0.0 | 10,946.2 |
| West Virginia | 15,404.8 | 11,668.6 | 11,483.7 | 124.4 | 36.0 | 24.5 | 0.0 | 3,736.2 |
| Total | 87,206.9 | 51,851.7 | 48,214.6 | 2,944.0 | 36.0 | 657.1 | 146.0 | 35,209.2 |
| **Lake states** | | | | | | | | |
| Michigan | 36,363.5 | 19,270.4 | 18,778.2 | 268.2 | 19.0 | 205.0 | 0.4 | 17,092.7 |
| Minnesota | 50,745.0 | 18,335.3 | 16,100.0 | 685.0 | 3.0 | 1,547.3 | 68.2 | 32,341.5 |
| North Dakota | 44,334.7 | 421.8 | 405.0 | 3.2 | 0.0 | 13.6 | 12,296.0 | 31,616.9 |
| South Dakota (East) | 41,732.4 | 334.7 | 223.0 | 0.0 | 0.0 | 111.7 | 18,740.4 | 22,657.3 |
| Wisconsin | 34,857.0 | 14,907.7 | 14,478.0 | 34.2 | 14.0 | 381.5 | 7.0 | 19,942.3 |
| Total | 208,032.6 | 53,269.9 | 49,984.2 | 990.6 | 36.0 | 2,259.1 | 31,112.0 | 123,650.7 |
| **Central** | | | | | | | | |
| Illinois | 35,678.7 | 3,810.4 | 3,692.3 | 44.9 | 8.2 | 65.0 | 0.4 | 31,867.9 |

| | | | | | | | |
|---|---|---|---|---|---|---|---|
| Indiana | 23,102.1 | 3,942.9 | 3,815.0 | 38.5 | 9.0 | 80.4 | 3.2 | 19,156.0 |
| Iowa | 35,802.2 | 1,561.3 | 1,460.2 | 75.9 | 0.0 | 25.2 | 38.3 | 34,202.6 |
| Kansas | 52,343.7 | 1,344.4 | 1,187.0 | 0.0 | 0.0 | 157.4 | 16,278.3 | 34,721.0 |
| Kentucky | 25,376.0 | 12,160.8 | 11,901.9 | 212.8 | 0.0 | 46.1 | 0.0 | 13,215.2 |
| Missouri | 44,156.8 | 12,876.0 | 12,288.6 | 256.1 | 33.0 | 298.3 | 1,447.6 | 29,833.2 |
| Nebraska | 48,949.1 | 1,029.1 | 788.8 | 13.8 | 0.0 | 226.5 | 24,274.4 | 23,645.6 |
| Ohio | 26,224.0 | 6,504.1 | 6,422.0 | 76.1 | 6.0 | 0.0 | 0.0 | 19,719.9 |
| Total | 291,632.6 | 43,229.0 | 41,555.8 | 718.1 | 56.2 | 898.9 | 42,042.2 | 206,361.4 |
| | | | | | | | | |
| Total, North | 627,160.8 | 180,841.7 | 170,769.4 | 5,137.5 | 153.2 | 4,781.6 | 73,300.7 | 373,018.4 |
| | | | | | | | | |
| South Atlantic | | | | | | | | |
| North Carolina | 31,230.7 | 20,043.3 | 19,562.2 | 433.8 | 1.1 | 46.2 | 0.0 | 11,187.4 |
| South Carolina | 19,344.0 | 12,249.4 | 12,176.1 | 59.2 | 1.5 | 12.6 | 20.0 | 7,074.6 |
| Virginia | 25,459.2 | 16,417.4 | 15,938.8 | 374.6 | 34.0 | 70.0 | 17.2 | 9,024.6 |
| Total | 76,033.9 | 48,710.1 | 47,677.1 | 867.6 | 36.6 | 128.8 | 37.2 | 27,286.6 |
| | | | | | | | | |
| East Gulf | | | | | | | | |
| Florida | 34,617.6 | 17,039.7 | 15,330.0 | 114.5 | 1.1 | 1594.1 | 2,189.0 | 15,388.9 |
| Georgia | 37,166.7 | 25,256.0 | 24,812.3 | 413.5 | 0.0 | 30.2 | 0.0 | 11,910.7 |
| Total | 71,784.3 | 42,295.7 | 40,142.3 | 528.0 | 1.1 | 1,624.3 | 2,189.0 | 27,299.6 |
| | | | | | | | | |
| Central Gulf | | | | | | | | |
| Alabama | 32,453.1 | 21,361.1 | 21,333.1 | 28.0 | 0.0 | 0.0 | 54.0 | 11,038.0 |
| Mississippi | 30,269.4 | 16,912.4 | 16,891.9 | 20.5 | 0.0 | 0.0 | 19.7 | 13,337.3 |
| Tennessee | 26,449.9 | 13,160.5 | 12,819.8 | 322.2 | 18.5 | 0.0 | 400.4 | 12,889.0 |
| Total | 89,172.4 | 51,434.0 | 51,044.8 | 370.7 | 18.5 | 0.0 | 474.1 | 37,264.3 |
| | | | | | | | | |
| West Gulf | | | | | | | | |
| Arkansas | 33,244.8 | 18,281.5 | 18,206.7 | 39.1 | 13.3 | 22.4 | 0.4 | 14,962.9 |
| Louisiana | 28,755.2 | 14,558.1 | 14,526.6 | 18.3 | 13.2 | 0.0 | 516.6 | 13,680.5 |
| Oklahoma | 44,020.8 | 8,513.3 | 4,323.4 | 32.4 | 0.0 | 4,157.5 | 9,301.0 | 26,206.5 |
| Texas | 167,765.8 | 23,279.3 | 12,512.5 | 14.9 | 18.1 | 10,733.8 | 91,603.5 | 52,883.0 |
| Total | 273,786.6 | 64,632.2 | 49,569.2 | 104.7 | 44.6 | 14,913.7 | 101,421.5 | 107,732.9 |
| | | | | | | | | |
| Total, South | 510,777.2 | 207,072.0 | 188,433.4 | 1,871.0 | 100.8 | 16,666.8 | 104,121.8 | 199,583.4 |

Table 6 (continued)
## LAND AREAS IN THE U.S., BY MAJOR CLASS OF LAND, SECTION, REGION, AND STATE, JANUARY 1, 1977ᵃ (THOUSAND ACRES)

| Section, region, and state | Total land areaᵇ | Forest land | | | | | Range land | Other landᶜ |
|---|---|---|---|---|---|---|---|---|
| | | Total | Commercial | Productive reserved | Productive deferred | Other forest | | |
| **Pacific Northwest** | | | | | | | | |
| Alaska | | | | | | | | |
| Coastal | 32,926.0 | 13,340.9 | 7,040.2 | 193.3 | 318.8 | 5,788.6 | 16,550.5 | 3,034.6 |
| Interior | 329,590.0 | 105,804.0 | 4,109.9 | 0.0 | 0.0 | 101,694.1 | 214,921.1 | 8,864.9 |
| Summary | 362,516.0 | 119,144.9 | 11,150.1 | 193.3 | 318.8 | ᵈ107,482.7 | 231,471.6 | 11,899.5 |
| Oregon | 19,167.0 | 15,367.0 | 13,875.0 | 303.0 | 109.0 | 1,080.0 | 2,009.0 | 1,791.0 |
| Western | | | | | | | | |
| Eastern | 42,391.0 | 14,656.0 | 10,560.0 | 413.0 | 175.0 | 3,508.0 | 20,313.7 | 7,421.3 |
| Summary | 61,558.0 | 30,023.0 | 24,435.0 | 716.0 | 284.0 | 4,588.0 | 22,322.7 | 9,212.3 |
| Washington | 15,843.0 | 12,607.0 | 9,788.0 | 1,024.0 | 150.0 | 1,645.0 | 894.4 | 2,341.6 |
| Western | | | | | | | | |
| Eastern | 26,762.0 | 10,574.0 | 8,134.0 | 720.0 | 168.0 | 1,552.0 | 7,236.2 | 8,951.8 |
| Summary | 42,605.0 | 23,181.0 | 17,922.0 | 1,744.0 | 318.0 | 3,197.0 | 8,130.6 | 11,293.4 |
| Total | 466,679.0 | 172,348.9 | 53,507.1 | 2,653.3 | 920.8 | 115,267.7 | 261,924.9 | 32,405.2 |
| **Pacific Southwest** | | | | | | | | |
| California | 100,071.0 | 40,152.0 | 16,303.0 | 1,365.0 | 268.0 | 22,216.0 | 43,039.7 | 16,879.3 |
| Hawaii | 4,112.0 | 1,986.0 | 948.0 | 114.0 | 0.0 | 924.0 | 968.0 | 1,158.0 |
| Total | 104,183.0 | 42,138.0 | 17,251.0 | 1,479.0 | 268.0 | 23,140.0 | 44,007.7 | 18,037.3 |
| Total, Pacific Coast | 570,862.0 | 214,486.9 | 70,758.1 | 4,132.3 | 1,188.8 | 138,407.7 | 305,932.6 | 50,442.5 |
| **Northern Rocky Mountain** | | | | | | | | |
| Idaho | 52,913.1 | 21,726.6 | 13,540.6 | 1,913.0 | 935.3 | 5,337.7 | 23,598.0 | 7,588.5 |
| Montana | 93,175.2 | 22,559.3 | 14,359.4 | 2001.8 | 708.7 | 5,489.4 | 53,334.1 | 17,281.8 |
| South Dakota (West) | 6,878.8 | 1,367.3 | 1,244.1 | 11.1 | 0.0 | 112.1 | 4,652.4 | 859.1 |
| Wyoming | 62,209.9 | 10,028.3 | 4,334.2 | 2,688.6 | 331.3 | 2,674.2 | 46,896.3 | 5,285.3 |
| Total | 215,177.0 | 55,681.5 | 33,478.3 | 6,614.5 | 1,975.3 | 13,613.4 | 128,480.8 | 31,014.7 |

Southern Rocky Mountain

| | | | | | | | |
|---|---|---|---|---|---|---|---|
| Arizona | 72,587.3 | 18,493.9 | 3,895.6 | 382.8 | 19.1 | 14,196.4 | 45,167.8 | 8,925.6 |
| Colorado | 66,410.2 | 22,271.0 | 11,314.7 | 684.0 | 752.2 | 9,520.1 | 27,821.7 | 16,317.5 |
| Nevada | 70,329.0 | 7,683.3 | 134.3 | 5.9 | 0.0 | 7,543.1 | 56,101.6 | 6,544.1 |
| New Mexico | 77,703.7 | 18,059.8 | 5,537.5 | 550.5 | 279.5 | 11,692.3 | 48,705.5 | 10,938.4 |
| Utah | 52,540.7 | 15,557.4 | 3,404.6 | 152.6 | 157.3 | 11,842.9 | 27,526.3 | 9,457.0 |
| Total | 339,570.9 | 82,065.4 | 24,286.7 | 1,775.8 | 1,208.1 | 54,794.8 | 205,322.9 | 52,182.6 |
| Total, Rocky Mountain | 554,747.9 | 137,746.9 | 57,765.0 | 8,390.3 | 3,183.4 | 68,408.2 | 333,803.7 | 83,197.3 |
| Total, All Regions | 2,263,547.7 | 740,147.4 | 487,725.9 | 19,531.1 | 4,626.2 | 228,264.3 | 817,158.8 | 706,241.5 |

a   Zeros indicate no data or negligible amounts.

b   U.S. Department of Commerce, Bureau of Census, Area Measurement Reports, GE-20 No. 1, 1970.

c   Includes pasture, swampland, industrial and urban areas, and other nonforest land.

d   Some parts of this area in Interior Alaska meet standards for commercial forest land, but the detailed survey of the Interior is not complete.

From Forest Service, Forest Statistics of the U.S., 1977, Review Draft, U.S. Department of Agriculture, Washington, D.C., 1978.

FIGURE 1.    Land resource regions and major land resource areas of the U.S.  (From Soil Conservation Service, Land Resource Regions and Major Land Resource Areas of the United States, Agriculture Handbook 296, U.S. Department of Agriculture, Washington, D.C., 1965.

**A. Northwestern Forest, Forage, and Specialty Crop Region**
1 Northern Pacific Coast Range and Valleys
2 Willamette and Puget Sound Valleys
3 Olympic and Cascade Mountains (Western Slope)
4 California Coastal Redwood Belt
5 Siskiyou - Trinity Area

**B. Northwestern Wheat and Range Region**
6 Cascade Mountains (Eastern Slope)
7 Columbia Basin
8 Columbia Plateau
9 Palouse and Nez Perce Prairies
10 Upper Snake River Lava Plains and Hills
11 Snake River Plains
12 Lost River Valleys and Mountains
13 Eastern Idaho Plateaus

**C. California Subtropical Fruit, Truck, and Specialty Crop Region**
14 Central California Valleys
15 Central California Coast Range
16 California Delta
17 Sacramento and San Joaquin Valleys
18 Sierra Nevada Foothills
19 Southern California Coastal Plain
20 Southern California Mountains

**D. Western Range and Irrigated Region**
21 Klamath and Shasta Valleys and Basins
22 Sierra Nevada Range
23 Malheur High Plateau
24 Humboldt Area
25 Owyhee High Plateau
26 Carson Basin and Mountains
27 Fallon - Lovelock Area
28 Great Salt Lake Area
29 Southern Nevada Basin and Range
30 Sonoran Basin and Range
31 Imperial Valley
32 Northern Intermountain Desertic Basins

33 Semiarid Rocky Mountains
34 Central Desertic Basins, Mountains, and Plateaus
49 (See E)
35 Colorado and Green Rivers Plateaus
36 New Mexico and Arizona Plateaus and Mesas
37 San Juan River Valley Mesas and Plateaus
38 Black, Hualpai, and Cerbat Mountains
39 Arizona and New Mexico Mountains
40 Central Arizona Basin and Range
41 Southeastern Arizona Basin and Range
42 Southern Desertic Basins, Plains, and Mountains

**E. Rocky Mountain Range and Forest Region**
43 Northern Rocky Mountains
44 Northern Rocky Mountain Valleys
45 Alpine Meadows and Rockland
46 Northern Rocky Mountain Foothills
47 Wasatch and Uinta Mountains
48 Southern Rocky Mountains
49 Southern Rocky Mountain Foothills
50 San Luis Valley
51 High Intermountain Valleys

**F. Northern Great Plains Spring Wheat Region**
52 Brown Glaciated Plain
53 Dark Brown Glaciated Plain
54 Rolling Soft Shale Plain
55 Black Glaciated Plains
56 Red River Valley of the North
57 Western Minnesota Forest - Prairie Transition

**G. Western Great Plains Range and Irrigated Region**
58 Northern Rolling High Plains
59 Northern Smooth High Plains
60 Pierre Shale Plains and Badlands
61 Black Hills Foot Slopes
62 Black Hills
63 Rolling Pierre Shale Plains
64 Mixed Sandy and Silty Tableland
65 Nebraska Sand Hills

66 Dakota - Nebraska Eroded Tableland
67 Central High Plains
68 Irrigated Upper Platte River Valley
69 Upper Arkansas Valley Rolling Plains
70 Pecos - Canadian Plains and Valleys

**H. Central Great Plains Winter Wheat and Range Region**
71 Central Nebraska Loess Hills
72 Central High Tableland
73 Rolling Plains and Breaks
74 Central Kansas Sandstone Hills
75 Central Loess Plains
76 Bluestem Hills
77 Southern High Plains
78 Central Rolling Red Plains
79 Great Bend Sand Plains
80 Central Rolling Red Prairies

**I. Southwestern Plateaus and Plains Range and Cotton Region**
81 Edwards Plateau
82 Texas Central Basin
83 Rio Grande Plain

**J. Southwestern Prairies Cotton and Forage Region**
84 Cross Timbers
85 Grand Prairie
86 Texas Blackland Prairie
87 Texas Claypan Area

**K. Northern Lake States Forest and Forage Region**
88 Northern Minnesota Swamps and Lakes
89 Minnesota Rockland Hills
90 Central Wisconsin and Minnesota Thin Loess and Till
91 Wisconsin and Minnesota Sandy Outwash
92 Superior Lake Plain
93 Northern Michigan and Wisconsin Stony, Sandy, and Rocky Plains and Hills

Figure 1 continued

94 Northern Michigan Sandy Drift

**L. Lake States Fruit, Truck, and Dairy Region**
95 Southeastern Wisconsin Drift Plain
96 Western Michigan Fruit Belt
97 Southwestern Michigan Fruit and Truck Belt
98 Southern Michigan Drift Plain
99 Erie - Huron Lake Plain
100 Erie Fruit and Truck Area
101 Ontario - Mohawk Plain

**M. Central Feed Grains and Livestock Region**
102 Loess, Till, and Sandy Prairies
103 Central Iowa and Minnesota Till Prairies
104 Eastern Iowa and Minnesota Till Prairies
105 Northern Mississippi Valley Loess Hills
106 Nebraska and Kansas Loess - Drift Hills
107 Iowa and Missouri Deep Loess Hills
108 Illinois and Iowa Deep Loess and Drift
109 Iowa and Missouri Heavy Till Plain
110 Northern Illinois and Indiana Heavy Till Plain
111 Indiana and Ohio Till Plain
112 Cherokee Prairies
113 Central Claypan Areas
114 Southern Illinois and Indiana Thin Loess and Till Plain
115 Central Mississippi Valley Wooded Slopes

**N. East and Central General Farming and Forest Region**
112 (See M)
116 Ozark Highland

117 Boston Mountains
118 Arkansas Valley and Ridges
119 Ouachita Mountains
120 Kentucky and Indiana Sandstone and Shale Hills and Valleys
121 Kentucky Bluegrass
122 Highland Rim and Pennyroyal
123 Nashville Basin
124 Western Allegheny Plateau
125 Cumberland Plateau and Mountains
126 Central Allegheny Plateau
127 Eastern Allegheny Plateau and Mountains
128 Southern Appalachian Ridges and Valleys
129 Sand Mountain
130 Blue Ridge

**O. Mississippi Delta Cotton and Feed Grains Region**
131 Southern Mississippi Valley Alluvium
132 Eastern Arkansas Prairies
134 (See P)

**P. South Atlantic and Gulf Slope Cash Crop, Forest, and Livestock Region**
86 (See J)
133 Southern Coastal Plain
134 Southern Mississippi Valley Silty Uplands
135 Alabama and Mississippi Blackland Prairies
136 Southern Piedmont
137 Carolina and Georgia Sand Hills
138 North - Central Florida Ridge

**R. Northeastern Forage and Forest Region**
139 Eastern Ohio Till Plain
140 Glaciated Allegheny Plateau and Catskill Mountains
141 Tughill Plateau
142 St. Lawrence - Champlain Plain
143 Northeastern Mountains
144 New England and Eastern New York Upland
145 Connecticut Valley
146 Aroostook Area

**S. Northern Atlantic Slope Truck, Fruit, and Poultry Region**
147 Northern Appalachian Ridges and Valleys
148 Northern Piedmont
149 Northern Coastal Plain

**T. Atlantic and Gulf Coast Lowland Forest and Truck Crop Region**
150 Gulf Coast Prairies
151 Gulf Coast Marsh
152 Gulf Coast Flatwoods
153 Atlantic Coast Flatwoods

**U. Florida Subtropical Fruit, Truck Crop, and Range Region**
154 South - Central Florida Ridge
155 Southern Florida Flatwoods
156 Florida Everglades and Associated Areas

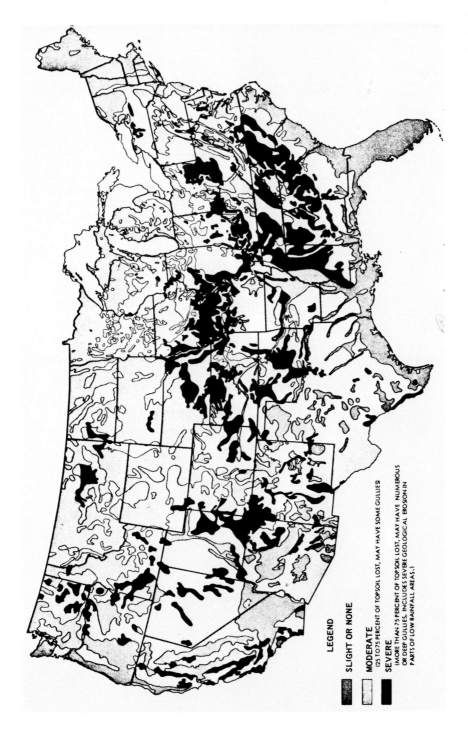

FIGURE 2.   Extent of erosion damage in the U.S. (From U.S. Department of Agriculture, Climate and Food: Climatic Fluctuation and U.S. Agricultural Production, U.S. National Academy of Sciences, Washington, D.C., 1976.

LEGEND

SLIGHT OR NONE

MODERATE
(25 TO 75 PERCENT OF TOPSOIL LOST, MAY HAVE SOME GULLIES)

SEVERE
(MORE THAN 75 PERCENT OF TOPSOIL LOST, MAY HAVE NUMEROUS OR DEEP GULLIES, INCLUDES SEVERE GEOLOGICAL EROSION IN PARTS OF LOW RAINFALL AREAS.)

# WATER RESOURCES

## Jean François Henry and Oskar R. Zaborsky

### Table 1
### VOLUME OF WATER ON EARTH

| Source | Volume (10⁶ Km³) | Percentage |
|---|---|---|
| Ocean water | 1351 | 97.3 |
| Fresh water | 38 | 2.7 |
| Total | 1389 | 100.0 |

Adapted from Skinner, B. J. and Turekian, K. K., *Man and the Ocean,* Prentice Hall, Englewood Cliffs, N.J., 1973 and Baumgartner, A. and Reichel, E., *The World Water Balance,* Elsevier, New York, 1975.

### Table 2
### DISTRIBUTION OF FRESH WATER

| Source | Volume (10⁶ km³) | Percentage |
|---|---|---|
| Polar ice and glaciers | 29.3 | 77.2 |
| Ground water | 8.5$^a$ | 22.3 |
| Lakes | 0.12 | 0.3 |
| Rivers | 0.012 | 0.03 |
| Soil moisture | 0.024 | 0.06 |
| Atmosphere | 0.013 | 0.03 |
| Total | 38 | 100.0 |

$^a$ Approximately two thirds lies deeper than 750 m below surface.

Adapted from Skinner, B. J. and Turekian, K. K., *Man and the Ocean,* Prentice Hall, Englewood Cliffs, N.J., 1973 and Baumgartner, A. and Reichel, E., *The World Water Balance,* Elsevier, New York, 1975.

## Table 3
## INLAND WATER AREAS OF THE U.S., 1970 (million acres)

| State | Total land and inland water area | Inland water area | State | Total land and inland water area | Inland water area |
|---|---|---|---|---|---|
| Alabama | 33.0 | 0.5 | Nebraska | 49.4 | 0.5 |
| Alaska | 375.3 | 12.8 | Nevada | 70.7 | 0.4 |
| Arizona | 72.9 | 0.2 | New Hampshire | 6.0 | 0.2 |
| Arkansas | 34.0 | 0.6 | New Jersey | 5.0 | 0.2 |
| California | 101.6 | 1.4 | New Mexico | 77.9 | 0.1 |
| Colorado | 66.7 | 0.3 | New York | 31.7 | 1.1 |
| Connecticut | 3.2 | 0.1 | N. Carolina | 33.7 | 2.4 |
| Delaware | 1.3 | 0.1 | N. Dakota | 45.2 | 0.9 |
| Florida | 37.5 | 2.8 | Ohio | 26.4 | 0.1 |
| Georgia | 37.7 | 0.4 | Oklahoma | 44.7 | 0.6 |
| Hawaii | 4.1 | —b | Oregon | 62.1 | 0.5 |
| Idaho | 53.5 | 0.6 | Pennsylvania | 29.0 | 0.2 |
| Illinois | 36.1 | 0.3 | Rhode Island | 0.8 | 0.1 |
| Indiana | 23.2 | 0.1 | S. Carolina | 19.9 | 0.5 |
| Iowa | 36.0 | 0.2 | S. Dakota | 49.3 | 0.7 |
| Kansas | 52.6 | 0.1 | Tennessee | 27.0 | 0.6 |
| Kentucky | 25.9 | 0.3 | Texas | 171.1 | 2.8 |
| Louisiana | 31.1 | 2.2 | Utah | 54.3 | 1.6 |
| Maine | 21.3 | 1.5 | Vermont | 6.1 | 0.2 |
| Maryland | 6.8 | 0.4 | Virginia | 26.1 | 0.6 |
| Massachusetts | 5.3 | 0.3 | Washington | 43.6 | 1.0 |
| Michigan | 37.3 | 0.9 | W. Virginia | 15.5 | 0.1 |
| Minnesota | 53.8 | 3.1 | Wisconsin | 35.9 | 1.1 |
| Mississippi | 30.5 | 0.2 | Wyoming | 62.7 | 0.4 |
| Missouri | 44.6 | 0.4 | All states | 2,313.7 | 47.5 |
| Montana | 94.2 | 1.0 | | | |

ᵃ   Columns may not add to totals because of rounding.
ᵇ   Less than 50,000 acres

From Forest Service, The Nation's Renewable Resources — An Assessment, 1975, Forest Resource Report No. 21, U.S. Department of Agriculture, Washington, D.C., 1977.

## Table 4
## EXPECTED WATER SUPPLIES IN THE U.S., BY WATER RESOURCE REGION[a] (BILLION GALLONS A DAY)

| Region | Confidence level[b] | |
|---|---|---|
| | Mean | 95% |
| New England | 74.7 | 46.0 |
| Middle Atlantic | 91.0 | 56.2 |
| South Atlantic-Gulf | 227.8 | 121.6 |
| Great Lakes | 74.8 | 47.0 |
| Ohio | 240.0 | 138.6 |
| Tennessee | 68.9 | 52.0 |
| Upper Mississippi | 266.1 | 140.4 |
| Lower Mississippi | 1,196.7 | 749.6 |
| Souris-Red-Rainy | 6.0 | 18.3 |
| Missouri | 156.8 | 97.3 |
| Arkansas-White-Red | 778.0 | 30.7 |
| Texas-Gulf | 33.9 | 12.2 |
| Rio Grande | 58.1 | 40.0 |
| Upper Colorado | 24.6 | 13.7 |
| Lower Colorado | 10.6 | 10.2 |
| Great Basin | 72.4 | 13.1 |
| Columbia-North Pacific | 611.4 | 323.7 |
| California-South Pacific | 74.2 | 34.3 |
| Alaska | 905.0 | 704.6 |
| Hawaii | 6.2 | 3.3 |

[a] Estimates are based primarily on adjusted natural runoff, which is the annual flow of water that would appear in surface streams, adjusted to account for upstream water development. In areas of surface water/ground water continuity, the adjusted natural runoff include the perennial recharge or yield of ground water aquifers.

The mean annual runoff in the contiguous U.S. is about 1,200 billion gal/day. The summary data in the table cannot be totaled to estimate total water supply for the U.S. because flows from upstream areas are included in downstream estimates. Therefore, while these data represent an estimate of water available in each region, to total them would be double counting in many cases.

[b] These data represent the average daily supply of water that would be exceeded on the average and in 95% of the years. For example, the New England region can expect an average daily water supply of 46.0 billion gal/day or more in 95 out of 100 years.

From Forest Service, The Nation's Renewable Resources — An Assessment, 1975, Forest Resource Report No. 21, U.S. Department of Agriculture, Washington, D.C., 1977.

Table 5

**GROUND WATER WITHDRAWALS AND PERCENTAGE OF
OVERDRAFT — "1975"**[a,b]

| Water resources region and number | Total withdrawal (mgd)[c] | Overdraft Total (mgd) | Percent | Number in region | Number with overdraft | Range in overdraft (%) |
|---|---|---|---|---|---|---|
| New England (1) | 635 | 0 | 0 | 6 | 0 | — |
| Mid-Atlantic (2) | 2,661 | 32 | 1.2 | 6 | 3 | 1—9 |
| South Atlantic-Gulf (3) | 5,449 | 339 | 6.2 | 9 | 8 | 2—13 |
| Great Lakes (4) | 1,215 | 27 | 2.2 | 8 | 1 | 30 |
| Ohio (5) | 1,843 | 0 | 0 | 7 | 0 | — |
| Tennessee (6) | 271 | 0 | 0 | 2 | 0 | — |
| Upper Mississippi (7) | 2,366 | 0 | 0 | 5 | 0 | — |
| Lower Mississippi (8) | 4,838 | 412 | 8.5 | 3 | 3 | 7—13 |
| Souris-Red-Rainy (9) | 86 | 0 | 0 | 1 | 0 | — |
| Missouri (10) | 10,407 | 2,557 | 24.6 | 11 | 10 | 4—36 |
| Arkansas-White-Red (11) | 8,846 | 5,457 | 61.7 | 7 | 7 | 2—76 |
| Texas-Gulf (12) | 7,222 | 5,578 | 77.2 | 5 | 5 | 24—95 |
| Rio Grande (13) | 2,335 | 657 | 28.1 | 5 | 4 | 22—43 |
| Upper Colorado (14) | 126 | 0 | 0 | 3 | 0 | — |
| Lower Colorado (15) | 5,008 | 2,415 | 48.2 | 3 | 3 | 7—53 |
| Great Basin (16) | 1,424 | 591 | 41.5 | 4 | 4 | 7—75 |
| Pacific Northwest (17) | 7,348 | 627 | 8.5 | 7 | 6 | 4—45 |
| California (18) | 19,160 | 2,197 | 11.5 | 7 | 5 | 7—31 |
| Regions 1—18 | 81,240 | 20,889 | 25.7 | 99 | 59 | 1—95 |
| Alaska (19) | 44 | 0 | 0 | 1 | 0 | — |
| Hawaii (20) | 790 | 0 | 0 | 4 | 0 | — |
| Caribbean (21) | 254 | 13 | 5.1 | 2 | 1 | 5 |
| Regions 1—21 | 82,328 | 20,902 | 25.4 | 106 | 60 | 1—95 |

[a]  "1975" — base year for the Second National Water Assessment data; it represents assumed average conditions and quotation marks are used throughout to indicate that the 1975 data are not actual data.

[b]  Definition of terms: Region — water resources region as designated by the U.S. Water Resources Council; there are 21 regions — 18 in the conterminous U.S. and one each for Alaska, Hawaii, and the Caribbean; subregion — subdivision of a region; there are 106 subregions used exclusively in the Second National Water Assessment; ground water overdraft — that part of the ground water withdrawals which exceeds recharge; sometimes referred to as ground water mining; withdrawal — water taken from a surface- or ground water source for offstream use.

[c]  mgd = millions of gallons per day.

From U.S. Water Resources Council, The Nation's Water Resources, 1975—2000, Vol. 1, Summary, Second National Water Assessment, Washington, D.C., 1979.

Table 6
## TOTAL WITHDRAWALS AND CONSUMPTION, BY FUNCTIONAL USE, FOR THE 21 WATER RESOURCES REGIONS — "1975", 1985, 2000 (MILLION GALLONS PER DAY)

| Functional use | Total withdrawals | | | Total consumption | | |
|---|---|---|---|---|---|---|
| | "1975" | 1985 | 2000 | "1975" | 1985 | 2000 |
| Fresh water | | | | | | |
| Domestic | 21,164 | 23,983 | 27,918 | 4,976 | 5,665 | 6,638 |
| Central (municipal) | | | | | | |
| Noncentral (rural) | 2,092 | 2,320 | 2,400 | 1,292 | 1,408 | 1,436 |
| Commercial | 5,530 | 6,048 | 6,732 | 1,109 | 1,216 | 1,369 |
| Manufacturing | 51,222 | 23,687 | 19,669 | 6,059 | 8,903 | 14,699 |
| Agriculture | 158,743 | 166,252 | 153,846 | 86,391 | 92,820 | 92,506 |
| Irrigation | | | | | | |
| Livestock | 1,912 | 2,233 | 2,551 | 1,912 | 2,233 | 2,551 |
| Steam electric | 88,916 | 94,858 | 79,492 | 1,419 | 4,062 | 10,541 |
| generation | | | | | | |
| Minerals industry | 7,055 | 8,832 | 11,328 | 2,196 | 2,777 | 3,609 |
| Public lands and others[a] | 1,866 | 2,162 | 2,461 | 1,236 | 1,461 | 1,731 |
| Total fresh water | 338,500 | 330,375 | 306,397 | 106,590 | 120,545 | 135,080 |
| Saline water,[b] total | 59,737 | 91,236 | 118,815 | | | |
| Total withdrawals | 398,237 | 421,611 | 425,212 | | | |

*Note:* Consumption is the portion of water withdrawn for offstream uses and not returned to a surface- or groundwater source.

[a]  Includes water for fish hatcheries and miscellaneous uses.
[b]  Saline water is used mainly in manufacturing and steam electric generation.

From U.S. Water Resources Council, The Nation's Water Resources, 1975—2000, Vol. 1, Summary, Second National Water Assessment, Washington, D.C., 1979.

Table 7
TOTAL FRESH- AND SALINE-WATER WITHDRAWALS —
"1975"

| Water resources region and number | Withdrawals, in million gallons per day | | | | |
|---|---|---|---|---|---|
| | Freshwater | | | Saline water | Total |
| | Surface | Ground | Total | | |
| New England (1) | 4,463 | 635 | 5,098 | 5,216 | 10,314 |
| Mid-Atlantic (2) | 15,639 | 2,661 | 18,300 | 19,625 | 37,925 |
| South Atlantic-Gulf (3) | 19,061 | 5,449 | 24,510 | 7,460 | 31,970 |
| Great Lakes (4) | 41,598 | 1,215 | 42,813 | 0 | 42,813 |
| Ohio (5) | 33,091 | 1,843 | 34,934 | 0 | 34,934 |
| Tennessee (6) | 7,141 | 271 | 7,412 | 0 | 7,412 |
| Upper Mississippi (7) | 10,035 | 2,366 | 12,401 | 0 | 12,401 |
| Lower Mississippi (8) | 9,729 | 4,838 | 14,567 | 1,253 | 15,820 |
| Souris-Red-Rainy (9) | 250 | 86 | 336 | 0 | 336 |
| Missouri (10) | 27,609 | 10,407 | 38,016 | 0 | 38,016 |
| Arkansas-White-Red (11) | 4,022 | 8,846 | 12,868 | 0 | 12,868 |
| Texas-Gulf (12) | 9,703 | 7,222 | 16,925 | 9,163 | 26,088 |
| Rio Grande (13) | 3,986 | 2,335 | 6,321 | 0 | 6,321 |
| Upper Colorado (14) | 6,743 | 126 | 6,869 | 0 | 6,869 |
| Lower Colorado (15) | 3,909 | 5,008 | 8,917 | 0 | 8,917 |
| Great Basin (16) | 6,567 | 1,424 | 7,991 | 0 | 7,991 |
| Pacific Northwest (17) | 30,147 | 7,348 | 37,495 | 131 | 37,626 |
| California (18) | 20,476 | 19,160 | 39,636 | 14,569 | 54,205 |
| Total regions 1—18 | 254,169 | 81,240 | 335,409 | 57,417 | 392,826 |
| Alaska (19) | 261 | 44 | 305 | 57 | 362 |
| Hawaii (20) | 1,089 | 790 | 1,879 | 1,139 | 3,018 |
| Caribbean (21) | 653 | 254 | 907 | 1,124 | 2,031 |
| Total, Regions 1—21 | 256,172 | 82,328 | 338,500 | 59,737 | 398,237 |

From U.S. Water Resources Council, The Nation's Water Resources, 1975-2000, Vol. 1, Summary, Second National Water Assessment, Washington, D.C., 1979.

Table 8
COMPOSITION OF STREAMS AND THE OCEAN

| Element | Sea water ($\mu g/\ell$) | Streams ($\mu g/\ell$) |
|---|---|---|
| Hydrogen | $1.10 \times 10^8$ | $1.10 \times 10^8$ |
| Lithium | 170 | 3 |
| Boron | 4,450 | 10 |
| Carbon (inorganic) | 28,000 | 11,500 |
| Carbon (dissolved organic) | 500 | —[a] |
| Nitrogen (dissolved $N_2$) | 15,000 | —[a] |
| Nitrogen (as $NO_3^1$, $NO_2^1$, $NH_4^{+1}$ and dissolved organic) | 670 | 226 |
| Oxygen (dissolved $O_2$) | 6,000 | —[a] |
| Oxygen (as $H_2O$) | $8.83 \times 10^8$ | $8.83 \times 10^8$ |
| Fluorine | 1,300 | 100 |
| Sodium | $1.08 \times 10^7$ | 6,300 |
| Magnesium | $1.29 \times 10^6$ | 4,100 |
| Aluminum | 1 | 400 |
| Silicon | 0—2,900 | 6,100 |
| Phosphorus | 0—88 | 20 |
| Sulfur | $9.04 \times 10^5$ | 5,600 |

Table 8 (continued)
## COMPOSITION OF STREAMS AND THE OCEAN

| Element | Sea water ($\mu g/L$) | Streams ($\mu g/L$) |
|---|---|---|
| Chlorine | $1.94 \times 10^7$ | 7,800 |
| Potassium | $3.92 \times 10^5$ | 2,300 |
| Calcium | $4.11 \times 10^5$ | 15,000 |
| Vanadium | 1.9 | 0.9 |
| Chromium | 0.2 | 1 |
| Manganese | 1.9 | 7 |
| Iron | 3.4 | 670 |
| Cobalt | 0.05 | 0.1 |
| Nickel | 2 | 0.3 |
| Copper | 2 | 7 |
| Zinc | 2 | 20 |
| Arsenic | 2.6 | 2 |
| Bromine | 67,300 | 20 |
| Molybdenum | 10 | 0.6 |
| Silver | 0.28 | 0.3 |
| Cadmium | 0.11 | —[a] |
| Antimony | 0.33 | 2 |
| Iodine | 64 | 7 |
| Cesium | 0.30 | 0.02 |
| Barium | 20 | 20 |
| Gold | 0.011 | 0.002 |
| Mercury | 0.15 | 0.07 |
| Lead | 0.03 | 3 |
| Bismuth | 0.02 | —[a] |

[a]   No data or reasonable estimates available.

Adapted from Skinner, B. J. and Turekian, K. K., *Man and the Ocean*, Prentice Hall, Englewood Cliffs, N.J., 1973.

## Table 9
### AVERAGE SURFACE OCEAN TEMPERATURES (WORLD LOCATIONS, IN °C)

| | Annual | Coldest month |
|---|---|---|
| Red Sea | 4.3 | 0.3 |
| Mediterranean | 8.3 | 2.6 |
| Guam | 24.5 | 23.5 |
| Brazil (South Atlantic) | 22.1 | 21.2 |
| Equator, East Pacific | 20.5 | 19.5 |
| Central Pacific | 23.0 | 22.5 |
| West Pacific | 25.0 | 24.6 |
| Equator, Indian Ocean | 22.5 | 22.0 |
| Equator, West Africa | 21.2 | 18.6 |
| Saudi Arabia | 17.1 | 15.2 |
| Persian Gulf | 19.0 | 13.5 |
| Iran, Gulf of Oman | 18.5 | 14.2 |
| India | 21.5 | 20.0 |

From Wolff, P. M., Temperature difference resource, in *Proc. 5th Ocean Thermal Energy Conversion Conf.*, February 20 to 22, 1978, Lavi, A. and Veziroglu, T. N., Eds., Miami Beach, 1978.

## Table 10
### AVERAGE SURFACE OCEAN TEMPERATURE (U.S. LOCATIONS, IN °C)

| | Annual | Coldest month | Range |
|---|---|---|---|
| Hawaii | 21.4 | 20.0 | 20—23 |
| Puerto Rico | 22.3 | 20.9 | 20—26 |
| West Gulf | 20.6 | 16.3 | 17—26 |
| East Gulf | 22.6 | 20.8 | 18—25 |
| North Gulf | 21.0 | 16.0 | 16—26 |
| Key West | 20.1 | 17.4 | 15—23 |
| East Florida | 20.6 | 18.7 | 15—23 |

From Wolff, P. M., Temperature difference resource, in Proc. 5th Ocean Thermal Energy Conversion Conf., February 20 to 22, 1978, Miami Beach, 1978.

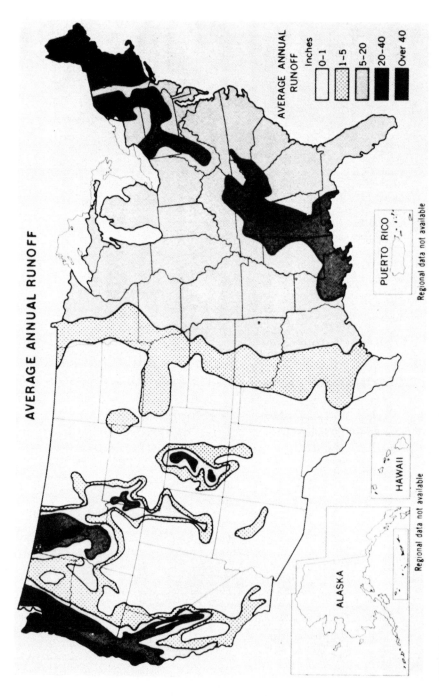

FIGURE 1.   Average annual runoff, 1921 to 1960. Lines show points of equal runoff at its place of origin where it first collects in the stream channel. Yield shown on map does not show losses from evapotranspiration. (From Forest Service, The Nation's Renewable Resources — An Assessment, 1975, Forest Resource Report No. 21, U.S. Department of Agriculture, Washington D.C., 1977.)

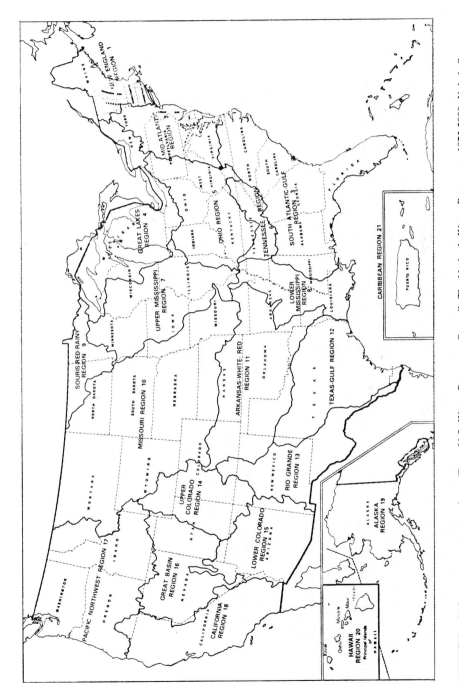

FIGURE 2.   U.S. water resources regions. (From U.S. Water Resources Council, The Nation's Water Resources, 1975-2000, Vol. 1, Summary, Second National Water Assessment, Washington, D.C., 1979.)

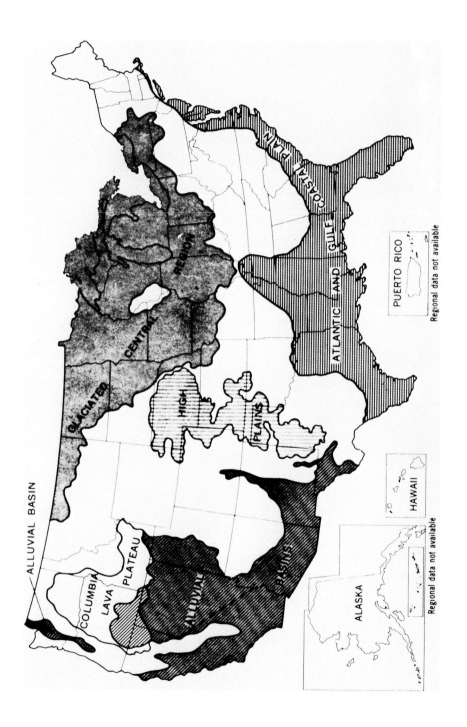

FIGURE 3.    Major areas of potential groundwater development. Ground water aquifers presently supply more than 20% of the U.S. withdrawal use of water. Ground water also provides the base flow of streams, and in some regions, ground water flows provide streams with a continuity of flow that they would not otherwise possess. The water supply information presented in the surface water section includes considerable water that enters from ground water aquifers. Part of this ground resource does not naturally get into the surface water supply and can be developed only by drilling. Ground water is very difficult to inventory because of its limited access. It has been estimated that the total storage greatly exceeds the volume at all five Great Lakes. About one half of the country is underlain by rock material that could yield at least 50 gal/min to wells.

The Atlantic and Gulf Coastal Plains contain the largest reserve of ground water in the nation. Present pumpage is but a small fraction of the supplies that could be developed. However, saltwater encroachment along the Gulf and Atlantic coasts is a limiting factor in ground water development.

Perhaps the next most significant area for ground water potential is the series of alluvial basins in the Far West. These are alluvium-filled valleys surrounded by mountains from which they receive run-off recharge. The surface is very dry, but the alluvial deposits are usually very thick, and they now store equivalent to centuries of recharge. In this area, conjunctive development of streamflow and ground water in storage is becoming a necessity because of heavy water use for irrigation and domestic needs in large cities, such as Los Angeles, Phoenix, and Albur-querque.

Still another area with important ground water potential is the area of glacial deposits in the Great Lakes area, extending from central Montana to eastern New York. The glacial deposits contain beds of watersorted permeable sand and gravel that constitute an important source of water.

The High Plains area has large quantities of ground water in storage. In the southern part of the Plains, irrigation pumpage from wells accounts for more than 10% of the total ground water pumpage of the nation. This area is a classic example of ground water mining, where withdrawals greatly exceed recharge. Estimates on the life of the recoverable storage in the High Plains range from 10 to 50 years.(From Forest Service, The Nation's Renewable Resources — An Assessment, 1975, Forest Resource Report No. 21, U.S. Department of Agriculture, Washington, D.C., 1977.)

# CLIMATIC INPUT

## Jean François Henry and Oskar R. Zaborsky

### Table 1
### MEAN DATE OF LAST 32°F TEMPERATURE IN SPRING, FIRST 32°F IN AUTUMN, AND MEAN LENGTH OF FREEZE-FREE PERIOD (DAYS)

| State and station[a] | Mean date last 32°F in spring[b] | Mean date first 32°F in fall[b] | Mean freeze-free period (number of days)[b] | State and station[a] | Mean date last 32°F in spring[b] | Mean date first 32°F in fall[b] | Mean freeze-free period (number of days)[b] |
|---|---|---|---|---|---|---|---|
| Alabama | | | | Arkansas | | | |
| Birmingham | March 19 | November 14 | 241 | Fort Smith | March 23 | November 9 | 231 |
| Mobile (U) | February 17 | December 12 | 298 | Little Rock | March 16 | November 15 | 244 |
| Montgomery (U) | February 27 | December 3 | 279 | | | | |
| | | | | California | | | |
| Alaska | | | | Bakersfield | February 14 | November 28 | 287 |
| Anchorage | May 18 | September 13 | 118 | Eureka (U) | January 24 | December 25 | 335 |
| Barrow | June 27 | July 5 | 8 | Fresno | February 3 | December 3 | 303 |
| Cordova | May 10 | October 2 | 145 | Los Angeles (U) | * | * | * |
| Fairbanks | May 24 | August 29 | 97 | Red Bluff | February 25 | November 29 | 277 |
| Juneau | April 27 | October 19 | 176 | Sacramento | January 24 | December 11 | 321 |
| Nome | June 12 | August 24 | 73 | San Diego | * | * | * |
| | | | | San Francisco (U) | * | * | * |
| Arizona | | | | Colorado | | | |
| Flagstaff | June 8 | October 2 | 116 | Denver (U) | May 2 | October 14 | 165 |
| Phoenix | January 27 | December 11 | 317 | Palisades | April 22 | October 17 | 178 |
| Tucson | March 6 | November 23 | 261 | Pueblo | April 28 | October 12 | 167 |
| Winslow | April 28 | October 21 | 176 | | | | |
| Yuma (U) | January 11 | December 27 | 350 | | | | |

[a]   (U) indicates urban.
[b]   * occurs in less than 1 year in 10. No freeze of record in Key West, Fla.

## Table 1 (continued)
## MEAN DATE OF LAST 32°F TEMPERATURE IN SPRING, FIRST 32°F IN AUTUMN, AND MEAN LENGTH OF FREEZE-FREE PERIOD (DAYS)

| State and station[a] | Mean date last 32°F in spring[b] | Mean date first 32°F in fall[b] | Mean freeze-free period (number of days)[b] |
|---|---|---|---|
| Connecticut | | | |
| Hartford | April 22 | October 19 | 180 |
| New Haven | April 15 | October 27 | 195 |
| D.C. | | | |
| Washington (U) | April 10 | October 28 | 200 |
| Florida | | | |
| Apalachicola (U) | February 2 | December 21 | 322 |
| Fort Myers | * | * | * |
| Jacksonville (U) | February 6 | December 16 | 313 |
| Key West | * | * | * |
| Lakeland | January 10 | December 25 | 349 |
| Miami | * | * | * |
| Orlando | January 31 | December 17 | 319 |
| Pensacola (U) | February 18 | December 15 | 300 |
| Tallahassee | February 26 | December 3 | 280 |
| Tampa | January 10 | December 26 | 349 |
| Georgia | | | |
| Atlanta (U) | March 20 | November 19 | 244 |
| Augusta | March 7 | November 22 | 260 |
| Macon | March 12 | November 19 | 252 |
| Savannah | February 21 | December 9 | 291 |
| Idaho | | | |
| Boise | April 29 | October 16 | 171 |
| Pocatello | May 8 | September 30 | 145 |
| Salmon | June 4 | September 6 | 94 |
| Illinois | | | |
| Cairo (U) | March 23 | November 11 | 233 |
| Chicago (U) | April 19 | October 28 | 192 |
| Freeport | March 8 | October 4 | 149 |
| Peoria | April 22 | October 16 | 177 |
| Springfield (U) | April 8 | October 30 | 205 |
| Indiana | | | |
| Evansville | April 2 | November 4 | 216 |
| Fort Wayne | April 24 | October 20 | 179 |
| Indianapolis (U) | April 17 | October 27 | 193 |
| South Bend | May 3 | October 15 | 165 |
| Iowa | | | |
| Des Moines (U) | April 20 | October 19 | 183 |
| Dubuque (U) | April 19 | October 19 | 184 |
| Koekuk | April 12 | October 26 | 197 |
| Sioux City | April 28 | October 12 | 167 |
| Kansas | | | |
| Concordia (U) | April 16 | October 24 | 191 |
| Dodge City | April 22 | October 24 | 184 |
| Goodland | May 5 | October 9 | 157 |
| Topeka (U) | April 9 | October 26 | 200 |
| Wichita | April 5 | November 1 | 210 |
| Kentucky | | | |
| Lexington | April 13 | October 18 | 198 |
| Louisville (U) | April 1 | November 7 | 220 |

| Station | | | |
|---|---|---|---|
| **Louisiana** | | | |
| Lake Charles | February 18 | December 6 | 291 |
| New Orleans | February 13 | December 12 | 302 |
| Shreveport | March 1 | November 27 | 272 |
| **Maine** | | | |
| Greenville | May 27 | September 20 | 116 |
| Portland | April 29 | October 15 | 169 |
| **Maryland** | | | |
| Annapolis | March 4 | November 15 | 225 |
| Baltimore (U) | March 28 | November 17 | 234 |
| Frederick | March 24 | October 17 | 176 |
| **Massachusetts** | | | |
| Boston | April 16 | October 25 | 192 |
| Nantucket | April 12 | November 16 | 219 |
| **Michigan** | | | |
| Alpena (U) | May 6 | October 9 | 156 |
| Detroit | April 25 | October 23 | 181 |
| Escanaba (U) | May 14 | October 6 | 145 |
| Grand Rapids (U) | April 25 | October 27 | 185 |
| Marquette (U) | May 14 | October 17 | 156 |
| Sault Ste. Marie | May 18 | October 3 | 138 |
| **Minnesota** | | | |
| Albert Lee | May 3 | October 6 | 156 |
| Big Falls R.S. | June 4 | September 7 | 95 |
| Brainerd | May 16 | September 24 | 131 |
| Duluth | May 22 | September 24 | 125 |
| Minneapolis | April 30 | October 13 | 166 |
| St. Cloud | May 9 | September 29 | 144 |
| **Mississippi** | | | |
| Jackson | March 10 | November 13 | 248 |
| Meridian | March 13 | November 14 | 246 |

| Station | | | |
|---|---|---|---|
| Vicksburg (U) | March 8 | November 15 | 252 |
| **Missouri** | | | |
| Columbia | April 9 | October 24 | 198 |
| Kansas City | April 5 | October 31 | 210 |
| St. Louis (U) | April 2 | November 8 | 220 |
| Springfield | April 10 | October 31 | 203 |
| **Montana** | | | |
| Billings | May 15 | September 24 | 132 |
| Glasgow (U) | May 19 | September 20 | 124 |
| Great Falls | May 14 | September 26 | 135 |
| Havre (U) | May 9 | September 23 | 138 |
| Helena | May 12 | September 23 | 134 |
| Kalispell | May 12 | September 23 | 135 |
| Miles City | May 5 | October 3 | 150 |
| Superior | June 5 | August 30 | 35 |
| **Nebraska** | | | |
| Grand Island | April 29 | October 6 | 160 |
| Lincoln | April 20 | October 17 | 180 |
| Norfolk | May 4 | October 3 | 152 |
| North Platta | April 30 | October 7 | 160 |
| Omaha | April 14 | October 20 | 189 |
| Valentine Lakes | May 7 | September 30 | 146 |
| **Nevada** | | | |
| Elko | June 6 | September 3 | 89 |
| Las Vegas | March 13 | November 13 | 245 |
| Reno | May 14 | October 2 | 141 |
| Winnemucca | May 18 | September 21 | 125 |
| **New Hampshire** | | | |
| Concord | May 11 | September 30 | 142 |
| **New Jersey** | | | |
| Cape May | April 4 | November 15 | 225 |
| Trenton (U) | April 8 | November 5 | 211 |

Table 1 (continued)
MEAN DATE OF LAST 32°F TEMPERATURE IN SPRING, FIRST 32°F IN AUTUMN, AND MEAN
LENGTH OF FREEZE-FREE PERIOD (DAYS)

| State and station[a] | Mean date last 32°F in spring[b] | Mean date first 32°F in fall[b] | Mean freeze-free period (number of days)[b] |
|---|---|---|---|
| New Mexico | | | |
| Albuquerque | April 16 | October 29 | 196 |
| Roswell | April 9 | November 2 | 208 |
| New York | | | |
| Albany | April 27 | October 13 | 169 |
| Binghamton (U) | May 4 | October 6 | 154 |
| Buffalo | April 30 | October 25 | 179 |
| New York (U) | April 7 | November 12 | 219 |
| Rochester | April 28 | October 21 | 176 |
| Syracuse | April 30 | October 15 | 168 |
| North Carolina | | | |
| Asheville (U) | April 12 | October 24 | 195 |
| Charlotte (U) | March 21 | November 15 | 239 |
| Greenville | March 28 | November 5 | 222 |
| Hatteras | February 25 | December 18 | 296 |
| Raleigh (U) | March 24 | November 16 | 237 |
| Wilmington (U) | March 8 | November 24 | 262 |
| North Dakota | | | |
| Bismarck | May 11 | September 24 | 136 |
| Devils Lake (U) | May 18 | September 22 | 127 |
| Fargo | May 13 | September 27 | 137 |
| Williston (U) | May 14 | September 23 | 132 |
| Ohio | | | |
| Akron-Canton | April 29 | October 20 | 173 |
| Cincinnati (Abbe) | April 15 | October 25 | 192 |
| Cleveland | April 21 | November 2 | 195 |
| Columbus (U) | April 17 | October 30 | 196 |
| Dayton | April 20 | October 21 | 184 |
| Toledo | April 24 | October 25 | 184 |
| Oklahoma | | | |
| Oklahoma City (U) | March 28 | November 7 | 223 |
| Tulsa | March 31 | November 2 | 216 |
| Oregon | | | |
| Astoria | March 18 | November 24 | 251 |
| Bend | June 17 | August 17 | 62 |
| Medford | April 25 | October 20 | 178 |
| Pendleton | April 27 | October 8 | 163 |
| Portland (U) | February 25 | December 1 | 279 |
| Salem | April 14 | October 27 | 197 |
| Pennsylvania | | | |
| Allentown | April 20 | October 16 | 180 |
| Harrisburg | April 10 | October 28 | 201 |
| Philadelphia (U) | March 30 | November 17 | 232 |
| Pittsburgh | April 20 | October 23 | 187 |
| Scranton (U) | April 24 | October 14 | 174 |
| Rhode Island | | | |
| Providence (U) | April 13 | October 27 | 197 |

| Station | Spring | Fall | Days |
|---|---|---|---|
| **South Carolina** | | | |
| Charleston (U) | February 19 | December 10 | 294 |
| Columbia (U) | March 14 | November 21 | 252 |
| Greenville | March 23 | November 17 | 239 |
| **South Dakota** | | | |
| Huron (U) | May 4 | September 30 | 149 |
| Rapid City (U) | May 7 | October 4 | 150 |
| Sioux Falls (U) | May 5 | October 3 | 152 |
| **Tennessee** | | | |
| Chattanooga (U) | March 26 | November 10 | 229 |
| Knoxville (U) | March 31 | November 6 | 220 |
| Memphis (U) | March 20 | November 12 | 237 |
| Nashville (U) | March 28 | November 7 | 224 |
| **Texas** | | | |
| Albany | March 30 | November 9 | 224 |
| Balmorhea | April 1 | November 12 | 226 |
| Beeville | February 21 | December 6 | 288 |
| College Station | March 1 | December 1 | 275 |
| Corsicana | March 13 | November 27 | 259 |
| Dalhart Exp. Sta. | April 23 | October 18 | 178 |
| Dallas | March 18 | November 22 | 249 |
| Del Rio | February 12 | December 9 | 300 |
| Encinal | February 15 | December 12 | 301 |
| Houston | February 5 | December 11 | 309 |
| Lampasas | April 1 | November 10 | 223 |
| Matagorda | February 12 | December 17 | 308 |
| Midland | April 3 | November 6 | 218 |
| Mission | January 30 | December 21 | 325 |
| Mount Pleasant | March 23 | November 12 | 233 |
| Nocogdoches | March 15 | November 13 | 243 |
| Plainview | April 10 | November 6 | 211 |
| Presidio | March 20 | November 13 | 238 |
| Quanah | March 31 | November 7 | 221 |
| San Angelo | March 25 | November 15 | 235 |
| Ysleta | April 5 | October 30 | 207 |
| **Utah** | | | |
| Blanding | May 18 | October 14 | 148 |
| Salt Lake City | April 12 | November 1 | 202 |
| **Vermont** | | | |
| Burlington | May 8 | October 3 | 148 |
| **Virginia** | | | |
| Lynchburg (U) | April 6 | October 27 | 205 |
| Norfolk (U) | March 18 | November 27 | 254 |
| Richmond (U) | April 2 | November 8 | 220 |
| Roanoke | April 20 | October 24 | 187 |
| **Washington** | | | |
| Bumping Lake | June 17 | August 16 | 60 |
| Seattle (U) | February 23 | December 1 | 281 |
| Spokane | April 20 | October 12 | 175 |
| Tatoost Island | January 25 | December 20 | 329 |
| Walla Walla (U) | March 28 | November 1 | 218 |
| Yakima | April 19 | October 15 | 179 |
| **West Virginia** | | | |
| Charleston | April 18 | October 28 | 193 |
| Parkersburg | April 16 | October 21 | 189 |
| **Wisconsin** | | | |
| Green Bay | May 6 | October 13 | 161 |
| La Crosse (U) | May 1 | October 8 | 161 |
| Madison (U) | April 26 | October 19 | 177 |
| Milwaukee (U) | April 20 | October 25 | 188 |
| **Wyoming** | | | |
| Casper | May 18 | September 25 | 130 |
| Cheyenne | May 20 | September 27 | 130 |
| Lander | May 15 | September 20 | 128 |
| Sheridan | May 21 | September 21 | 123 |

Charts and tabulation were derived from the Freeze Data tabulation in *Climatography of the United States No. 60 — Climates of the States.*

Table 2

MEAN NUMBER OF DAYS MAXIMUM TEMPERATURE 90°F AND ABOVE, EXCEPT 70°F AND ABOVE IN ALASKA[a]

| States and stations | Years | January | February | March | April | May | June | July | August | September | October | November | December | Annual |
|---|---|---|---|---|---|---|---|---|---|---|---|---|---|---|
| **Alabama** | | | | | | | | | | | | | | |
| Birmingham | 17 | 0 | 0 | 0 | * | 4 | 15 | 19 | 20 | 8 | 1 | 0 | 0 | 67 |
| Mobile | 19 | 0 | 0 | * | * | 5 | 17 | 18 | 20 | 9 | 1 | 0 | 0 | 70 |
| Montgomery | 16 | 0 | 0 | 0 | * | 8 | 19 | 22 | 24 | 13 | 1 | 0 | 0 | 87 |
| **Arizona** | | | | | | | | | | | | | | |
| Flagstaff (U) | 43 | 0 | 0 | 0 | 0 | 0 | 1 | 1 | * | * | 0 | 0 | 0 | 2 |
| Phoenix | 21 | 0 | 0 | * | 8 | 22 | 29 | 30 | 31 | 28 | 14 | * | 0 | 163 |
| Prescott | 18 | 0 | 0 | 0 | 0 | 1 | 10 | 17 | 11 | 8 | * | 0 | 0 | 18 |
| Tucson | 20 | 0 | * | * | 5 | 18 | 28 | 29 | 29 | 26 | 10 | * | 0 | 146 |
| Winslow | 29 | 0 | 0 | 0 | * | 2 | 18 | 25 | 18 | 9 | * | 0 | 0 | 72 |
| Yuma | 10 | 0 | 1 | 3 | 14 | 24 | 30 | 31 | 31 | 29 | 22 | 2 | 0 | 187 |
| **Arkansas** | | | | | | | | | | | | | | |
| Fort Smith | 15 | 0 | 0 | * | * | 4 | 17 | 24 | 24 | 14 | 1 | 0 | 0 | 84 |
| Little Rock | 19 | 0 | 0 | 0 | * | 4 | 16 | 22 | 21 | 10 | 1 | 0 | 0 | 74 |
| Texarkana | 18 | 0 | 0 | * | * | 3 | 10 | 25 | 24 | 13 | 1 | 0 | 0 | 85 |
| **California** | | | | | | | | | | | | | | |
| Bakersfield | 23 | 0 | 0 | 0 | 2 | 9 | 19 | 30 | 27 | 18 | 5 | * | 0 | 110 |
| Bishop | 13 | 0 | 0 | 0 | 1 | 5 | 19 | 30 | 28 | 17 | 1 | 0 | 0 | 101 |
| Blue Canyon | 17 | 0 | 0 | 0 | 0 | 0 | * | * | * | * | 0 | 0 | 0 | 1 |
| Burbank | 18 | 0 | * | 0 | 1 | 2 | 3 | 12 | 12 | 13 | 4 | 1 | * | 47 |
| Eureka | 50 | 0 | 0 | 0 | 0 | 0 | 0 | 0 | 0 | 0 | 0 | 0 | 0 | 0 |
| Fresno | 21 | 0 | 0 | 0 | 1 | 7 | 16 | 29 | 26 | 17 | 4 | 0 | 0 | 100 |
| Long Beach | 17 | 0 | 0 | 0 | * | 1 | * | 2 | 2 | 3 | 2 | 1 | * | 11 |
| Los Angeles (U) | 20 | 0 | * | 0 | 1 | 1 | 1 | 4 | 3 | 6 | 2 | 1 | 0 | 19 |
| Mt. Shasta | 18 | 0 | 0 | 0 | 0 | * | 1 | 8 | 6 | 4 | * | 0 | 0 | 20 |
| Oakland | 30 | 0 | 0 | 0 | 0 | * | * | * | * | 2 | * | 0 | 0 | 4 |
| Red Bluff | 18 | 0 | 0 | * | 2 | 7 | 15 | 28 | 25 | 17 | 4 | * | 0 | 98 |
| Sacramento | 27 | 0 | 0 | 0 | * | 4 | 11 | 20 | 18 | 12 | 2 | 0 | 0 | 67 |

| | | | | | | | | | | | | |
|---|---|---|---|---|---|---|---|---|---|---|---|---|
| Sandberg | 28 | 0 | 0 | 0 | * | 2 | 8 | 7 | 4 | 0 | 0 | 0 | 21 |
| San Diego | 20 | 0 | 0 | * | * | * | * | * | 1 | 1 | * | * | 3 |
| San Francisco | 24 | 0 | 0 | 0 | * | * | * | * | 1 | * | 0 | 0 | 1 |
| Santa Maria | 18 | 0 | 0 | * | * | * | * | * | 1 | 1 | 1 | * | 4 |
| **Colorado** | | | | | | | | | | | | | |
| Alamosa | 15 | 0 | 0 | 0 | 0 | * | 1 | 0 | 0 | 0 | 0 | 0 | 1 |
| Colorado Springs | 12 | 0 | 0 | 0 | 0 | 4 | 8 | 5 | 2 | 0 | 0 | 0 | 19 |
| Denver | 26 | 0 | 0 | * | * | 7 | 14 | 11 | 3 | 0 | 0 | 0 | 35 |
| Grand Junction (U) | 64 | 0 | 0 | 1 | 1 | 11 | 22 | 15 | 2 | 0 | 0 | 0 | 51 |
| Pueblo | 20 | 0 | 0 | 1 | 1 | 12 | 19 | 17 | 7 | * | * | 0 | 57 |
| **Connecticut** | | | | | | | | | | | | | |
| Bridgeport | 12 | 0 | 0 | 0 | 0 | 2 | 4 | 2 | * | 0 | 0 | 0 | 8 |
| Hartford | 51 | 0 | 0 | * | * | 2 | 4 | 3 | 1 | * | * | 0 | 10 |
| New Haven | 17 | 0 | 0 | 0 | 0 | 1 | 1 | 1 | * | 0 | 0 | 0 | 3 |
| **Delaware** | | | | | | | | | | | | | |
| Wilmington | 13 | 0 | 0 | 1 | 1 | 5 | 9 | 5 | 1 | * | 0 | 0 | 21 |
| **D.C.** | | | | | | | | | | | | | |
| Washington (U) | 88 | 0 | * | 2 | 2 | 6 | 10 | 7 | 3 | * | 0 | 0 | 28 |
| **Florida** | | | | | | | | | | | | | |
| Apalachicola | 31 | 0 | 0 | 1 | 1 | 4 | 5 | 5 | 2 | * | 0 | 0 | 17 |
| Daytona Beach | 17 | 0 | * | 6 | 6 | 13 | 17 | 16 | 8 | 1 | 00 | 62 | |
| Everglades | 26 | 0 | 1 | 15 | 15 | 25 | 27 | 28 | 22 | 9 | 1 | 0 | 133 |
| Fort Myers | 20 | 0 | 1 | 16 | 16 | 22 | 25 | 27 | 21 | 5 | * | 0 | 121 |
| Jacksonville | 19 | 0 | * | 10 | 10 | 19 | 24 | 22 | 11 | 1 | 0 | 0 | 88 |
| Key West | 12 | 0 | 1 | 1 | 1 | 10 | 19 | 20 | 10 | 1 | 0 | 0 | 81 |
| Key West (U) | 87 | 0 | 0 | 1 | 1 | 5 | 10 | 12 | 5 | * | * | 0 | 33 |
| Lakeland (U) | 20 | 0 | * | 9 | 9 | 18 | 21 | 22 | 12 | 2 | 0 | 0 | 85 |
| Miami | 18 | 0 | 1 | 3 | 3 | 12 | 17 | 22 | 11 | 1 | 0 | 0 | 67 |
| Miami Beach | 17 | 0 | * | 1 | 1 | 2 | 2 | 6 | 2 | * | 0 | 0 | 13 |
| Orlando | 18 | 0 | 3 | 13 | 13 | 21 | 24 | 25 | 17 | 3 | 0 | 0 | 107 |
| Pensacola | 21 | 0 | * | 2 | 2 | 7 | 12 | 14 | 4 | * | 0 | 0 | 39 |
| Tallahassee | 21 | 0 | 1 | 9 | 9 | 18 | 20 | 21 | 11 | 1 | 0 | 0 | 81 |
| Tampa | 14 | 0 | 1 | 9 | 9 | 17 | 20 | 20 | 14 | 3 | 0 | 0 | 84 |
| West Palm Beach | 20 | 0 | 2 | 5 | 5 | 16 | 23 | 24 | 15 | 3 | * | * | 89 |

Table 2 (continued)

MEAN NUMBER OF DAYS MAXIMUM TEMPERATURE 90°F AND ABOVE, EXCEPT 70°F AND ABOVE IN ALASKA[a]

| States and stations | Years | January | February | March | April | May | June | July | August | September | October | November | December | Annual |
|---|---|---|---|---|---|---|---|---|---|---|---|---|---|---|
| Georgia | | | | | | | | | | | | | | |
| Athens | 17 | 0 | 0 | 0 | * | 4 | 14 | 16 | 17 | 6 | * | 0 | 0 | 57 |
| Atlanta | 12 | 0 | 0 | 0 | 0 | 3 | 12 | 14 | 15 | 4 | * | 0 | 0 | 48 |
| Augusta | 10 | 0 | 0 | 0 | 1 | 9 | 19 | 24 | 25 | 9 | 1 | 0 | 0 | 88 |
| Columbus | 15 | 0 | 0 | 0 | * | 8 | 18 | 21 | 22 | 10 | 1 | 0 | 0 | 80 |
| Macon | 12 | 0 | 0 | 0 | 1 | 12 | 21 | 24 | 25 | 12 | 2 | 0 | 0 | 97 |
| Rome | 15 | 0 | 0 | 0 | * | 4 | 15 | 20 | 21 | 8 | 1 | 0 | 0 | 69 |
| Savannah | 10 | 0 | 0 | * | 1 | 7 | 15 | 20 | 21 | 8 | 1 | 0 | 0 | 71 |
| Thomasville | 36 | 0 | 0 | * | * | 7 | 17 | 19 | 20 | 12 | 1 | 0 | 0 | 76 |
| Hawaii | | | | | | | | | | | | | | |
| Hilo | 15 | 0 | 0 | 0 | 0 | 0 | 0 | 0 | * | * | * | 0 | 0 | * |
| Honolulu | 38 | 0 | 0 | 0 | 0 | 0 | 0 | 0 | 0 | 0 | 0 | 0 | 0 | 0 |
| Lihue | 11 | 0 | 0 | 0 | 0 | 0 | 0 | 0 | 0 | 0 | * | 0 | 0 | * |
| Idaho | | | | | | | | | | | | | | |
| Boise | 21 | 0 | 0 | 0 | * | 1 | 4 | 19 | 14 | 3 | 0 | 0 | 0 | 41 |
| Idaho Falls | 11 | 0 | 0 | 0 | 0 | * | 2 | 13 | 10 | 2 | 0 | 0 | 0 | 27 |
| Lewiston | 14 | 0 | 0 | 0 | 0 | 1 | 3 | 16 | 12 | 4 | 0 | 0 | 0 | 36 |
| Pocatello | 22 | 0 | 0 | 0 | 0 | * | 2 | 16 | 12 | 2 | 0 | 0 | 0 | 32 |
| Illinois | | | | | | | | | | | | | | |
| Cairo (U) | 18 | 0 | 0 | 0 | 0 | 2 | 13 | 18 | 17 | 5 | * | 0 | 0 | 55 |
| Chicago | 18 | 0 | 0 | 0 | 0 | 1 | 6 | 8 | 8 | 3 | * | 0 | 0 | 26 |
| Chicago (U) | 85 | 0 | 0 | 0 | * | 1 | 3 | 5 | 3 | 1 | * | 0 | 0 | 13 |
| Moline | 26 | 0 | 0 | 0 | * | 1 | 6 | 10 | 8 | 3 | * | 0 | 0 | 28 |
| Peoria | 21 | 0 | 0 | 0 | 0 | 1 | 6 | 8 | 8 | 3 | * | 0 | 0 | 26 |
| Rockford | 10 | 0 | 0 | 0 | 0 | 1 | 5 | 7 | 6 | 3 | 0 | 0 | 0 | 22 |
| Springfield | 13 | 0 | 0 | 0 | 0 | 1 | 7 | 10 | 8 | 4 | * | 0 | 0 | 30 |

| | | | | | | | | | | | | | | |
|---|---|---|---|---|---|---|---|---|---|---|---|---|---|---|
| **Indiana** | | | | | | | | | | | | | | |
| Evansville | 20 | 0 | 0 | 0 | 0 | 1 | 10 | 14 | 13 | 5 | * | 0 | 0 | 43 |
| Fort Wayne | 14 | 0 | 0 | 0 | 0 | * | 4 | 6 | 6 | 2 | * | 0 | 0 | 18 |
| Indianapolis | 20 | 0 | 0 | 0 | 0 | * | 5 | 7 | 7 | 2 | * | 0 | 0 | 21 |
| South Bend | 20 | 0 | 0 | 0 | * | * | 4 | 6 | 6 | 2 | 0 | 0 | 0 | 18 |
| **Iowa** | | | | | | | | | | | | | | |
| Burlington (U) | 62 | 0 | 0 | 0 | * | 1 | 7 | 12 | 10 | 4 | * | 0 | 0 | 34 |
| Des Moines | 27 | 0 | 0 | 0 | 0 | 1 | 5 | 10 | 9 | 3 | * | 0 | 0 | 28 |
| Dubuque | 10 | 0 | 0 | 0 | 0 | * | 2 | 4 | 3 | 2 | 0 | 0 | 0 | 11 |
| Sioux City | 20 | 0 | 0 | 0 | * | 1 | 7 | 12 | 10 | 4 | * | 0 | 0 | 34 |
| Waterloo | 12 | 0 | 0 | 0 | * | * | 3 | 6 | 8 | 2 | * | 0 | 0 | 17 |
| **Kansas** | | | | | | | | | | | | | | |
| Concordia (U) | 76 | 0 | 0 | * | * | 1 | 9 | 16 | 15 | 7 | * | 0 | 0 | 46 |
| Dodge City | 18 | 0 | 0 | * | 1 | 3 | 13 | 18 | 19 | 8 | 1 | 0 | 0 | 63 |
| Goodland (U) | 40 | 0 | 0 | 0 | * | 2 | 11 | 20 | 18 | 8 | 1 | 0 | 0 | 60 |
| Topeka | 14 | 0 | 0 | 0 | 1 | 1 | 10 | 15 | 17 | 8 | 1 | 0 | 0 | 53 |
| Wichita | 7 | 0 | 0 | 0 | 1 | 2 | 12 | 21 | 23 | 10 | 1 | 0 | 0 | 70 |
| **Kentucky** | | | | | | | | | | | | | | |
| Lexington | 16 | 0 | 0 | 0 | 0 | 1 | 6 | 10 | 9 | 4 | * | 0 | 0 | 30 |
| Louisville | 13 | 0 | 0 | 0 | * | 2 | 10 | 16 | 14 | 6 | 1 | 0 | 0 | 49 |
| **Louisiana** | | | | | | | | | | | | | | |
| Baton Rouge | 9 | 0 | 0 | 0 | 1 | 7 | 22 | 26 | 24 | 15 | 2 | 0 | 0 | 97 |
| Lake Charles | 22 | 0 | 0 | * | * | 3 | 19 | 23 | 24 | 15 | 2 | 0 | 0 | 86 |
| New Orleans | 45 | 0 | 0 | * | * | 3 | 16 | 20 | 19 | 10 | 1 | 0 | 0 | 69 |
| Shreveport (U) | 80 | 0 | 0 | * | * | 4 | 18 | 24 | 23 | 14 | 2 | 0 | 0 | 85 |
| **Maine** | | | | | | | | | | | | | | |
| Caribou | 21 | 0 | 0 | 0 | 0 | * | 1 | 1 | 1 | * | 0 | 0 | 0 | 3 |
| Portland | 20 | 0 | 0 | 0 | 0 | * | 1 | 2 | 2 | * | 0 | 0 | 0 | 5 |
| **Maryland** | | | | | | | | | | | | | | |
| Baltimore (U) | 88 | 0 | 0 | * | * | 1 | 5 | 10 | 6 | 2 | * | 0 | 0 | 24 |
| Frederick | 11 | 0 | 0 | 0 | * | 1 | 7 | 10 | 8 | 2 | * | 0 | 0 | 28 |

Table 2 (continued)
## MEAN NUMBER OF DAYS MAXIMUM TEMPERATURE 90°F AND ABOVE, EXCEPT 70°F AND ABOVE IN ALASKA[a]

| States and stations | Years | January | February | March | April | May | June | July | August | September | October | November | December | Annual |
|---|---|---|---|---|---|---|---|---|---|---|---|---|---|---|
| **Massachusetts** | | | | | | | | | | | | | | |
| Blue Hill OBS. | 75 | 0 | 0 | 0 | 0 | * | 1 | 2 | 1 | * | 0 | 0 | 0 | 4 |
| Boston | 9 | 0 | 0 | 0 | 0 | * | 3 | 5 | 4 | 1 | * | 0 | 0 | 13 |
| Nantucket | 14 | 0 | 0 | 0 | 0 | 0 | 0 | * | * | 0 | 0 | 0 | 0 | * |
| Pittsfield | 22 | 0 | 0 | 0 | 0 | 0 | * | 1 | 1 | * | 0 | 0 | 0 | 2 |
| Worcester (U) | 10 | 0 | 0 | 0 | 0 | 0 | 2 | 2 | 2 | * | 0 | 0 | 0 | 6 |
| **Michigan** | | | | | | | | | | | | | | |
| Alpena (U) | 45 | 0 | 0 | 0 | 0 | * | 1 | 1 | 1 | * | 0 | 0 | 0 | 3 |
| Detroit | 27 | 0 | 0 | 0 | 0 | * | 3 | 6 | 4 | 1 | 0 | 0 | 0 | 14 |
| Escanaba (U) | 51 | 0 | 0 | 0 | 0 | * | * | * | * | * | 0 | 0 | 0 | * |
| Flint | 19 | 0 | 0 | 0 | 0 | * | 3 | 4 | 4 | 1 | 0 | 0 | 0 | 12 |
| Grand Rapids | 21 | 0 | 0 | 0 | 0 | * | 2 | 5 | 5 | 1 | 0 | 0 | 0 | 13 |
| Lansing | 43 | 0 | 0 | 0 | 0 | * | 2 | 5 | 2 | 1 | 0 | 0 | 0 | 10 |
| Marquette | 23 | 0 | 0 | 0 | 0 | * | 1 | 2 | 2 | 1 | 0 | 0 | 0 | 6 |
| Wuskegon | 21 | 0 | 0 | 0 | 0 | 0 | 1 | 2 | 2 | 1 | 0 | 0 | 0 | 6 |
| Sault Ste. Marie | 19 | 0 | 0 | 0 | 0 | 0 | * | * | 1 | * | 0 | 0 | 0 | 1 |
| **Minnesota** | | | | | | | | | | | | | | |
| Duluth | 19 | 0 | 0 | 0 | 0 | 0 | * | 1 | 1 | 0 | 0 | 0 | 0 | 2 |
| International Falls | 21 | 0 | 0 | 0 | * | * | 1 | 2 | 1 | * | 0 | 0 | 0 | 4 |
| Minneapolis | 22 | 0 | 0 | 0 | * | * | 3 | 7 | 6 | 2 | 0 | 0 | 0 | 18 |
| Rochester | 10 | 0 | 0 | 0 | 0 | * | 3 | 5 | 5 | 1 | 0 | 0 | 0 | 14 |
| St. Cloud | 21 | 0 | 0 | 0 | 0 | * | 2 | 4 | 4 | 1 | * | 0 | 0 | 11 |
| **Mississippi** | | | | | | | | | | | | | | |
| Jackson (U) | 58 | 0 | 0 | * | 1 | 7 | 21 | 24 | 25 | 17 | 3 | 0 | 0 | 98 |
| Meridian | 15 | 0 | 0 | 0 | 1 | 8 | 21 | 24 | 24 | 12 | 2 | 0 | 0 | 82 |
| Vicksburg | 23 | 0 | 0 | 0 | * | 2 | 15 | 20 | 21 | 9 | 1 | 0 | 0 | 68 |

| | | | | | | | | | | | | |
|---|---|---|---|---|---|---|---|---|---|---|---|---|
| **Missouri** | | | | | | | | | | | | |
| Columbia | 21 | 0 | 0 | * | 1 | 8 | 14 | 14 | 6 | * | 0 | 0 | 43 |
| Kansas City | 27 | 0 | 0 | * | 2 | 10 | 18 | 17 | 7 | 1 | 0 | 0 | 55 |
| St. Joseph (U) | 43 | 0 | 0 | * | 1 | 8 | 16 | 13 | 6 | * | 0 | 0 | 43 |
| St. Louis (U) | 23 | 0 | 0 | * | 2 | 9 | 16 | 13 | 6 | * | 0 | 0 | 46 |
| Springfield | 15 | 0 | 0 | * | * | 8 | 13 | 14 | 6 | * | 0 | 0 | 41 |
| **Montana** | | | | | | | | | | | | |
| Billings | 26 | 0 | 0 | * | 1 | 3 | 13 | 11 | 3 | 0 | 0 | 0 | 31 |
| Butte | 29 | 0 | 0 | 0 | 0 | * | 3 | 1 | * | 0 | 0 | 0 | 4 |
| Glasgow | 17 | 0 | 0 | * | 1 | 2 | 13 | 9 | 2 | * | 0 | 0 | 28 |
| Great Falls | 23 | 0 | 0 | 0 | * | 1 | 8 | 5 | 1 | * | 0 | 0 | 15 |
| Havre (U) | 57 | 0 | 0 | 0 | 1 | 2 | 10 | 7 | 1 | * | 0 | 0 | 21 |
| Helena | 20 | 0 | 0 | 0 | * | 1 | 8 | 5 | 1 | 0 | 0 | 0 | 15 |
| Kalispell | 11 | 0 | 0 | 0 | 0 | * | 5 | 3 | 1 | 0 | 0 | 0 | 9 |
| Miles City | 16 | 0 | 0 | * | 1 | 3 | 16 | 14 | 4 | * | 0 | 0 | 38 |
| Missoula | 16 | 0 | 0 | 0 | * | 1 | 11 | 6 | 1 | 0 | 0 | 0 | 19 |
| **Nebraska** | | | | | | | | | | | | |
| Grand Island | 22 | 0 | 0 | * | 1 | 8 | 16 | 14 | 6 | 0 | 0 | 0 | 45 |
| Lincoln (U) | 66 | 0 | * | * | 2 | 7 | 14 | 12 | 5 | * | 0 | 0 | 40 |
| Norfolk | 15 | 0 | 0 | * | 1 | 7 | 11 | 11 | 5 | * | 0 | 0 | 35 |
| North Platte | 9 | 0 | 0 | 0 | * | 8 | 15 | 14 | 6 | * | 0 | 0 | 43 |
| Omaha | 25 | 0 | 0 | * | 1 | 8 | 14 | 12 | 5 | * | 0 | 0 | 40 |
| Scottsbluff | 17 | 0 | 0 | 0 | 1 | 6 | 15 | 13 | 5 | * | 0 | 0 | 40 |
| Valentine (U) | 64 | 0 | 0 | * | 1 | 5 | 12 | 10 | 4 | * | 0 | 0 | 32 |
| **Nevada** | | | | | | | | | | | | |
| Elko | 30 | 0 | 0 | 0 | * | 4 | 20 | 16 | 3 | 0 | 0 | 0 | 44 |
| Ely | 22 | 0 | 0 | 0 | 0 | 1 | 9 | 5 | 1 | 0 | 0 | 0 | 16 |
| Las Vegas | 12 | 0 | 0 | 4 | 15 | 27 | 31 | 30 | 25 | 8 | 0 | 0 | 140 |
| Reno | 18 | 0 | 0 | 0 | 1 | 5 | 20 | 15 | 6 | 0 | 0 | 0 | 47 |
| Winnemucca | 11 | 0 | 0 | 0 | 1 | 7 | 22 | 17 | 6 | 0 | 0 | 0 | 53 |
| **New Hampshire** | | | | | | | | | | | | |
| Concord | 19 | 0 | 0 | 0 | 1 | 4 | 5 | 3 | 1 | 0 | 0 | 0 | 14 |
| Mt. Washington | 28 | 0 | 0 | 0 | 0 | 0 | 0 | 0 | 0 | 0 | 0 | 0 | 0 |

Table 2 (continued)

MEAN NUMBER OF DAYS MAXIMUM TEMPERATURE 90°F AND ABOVE, EXCEPT 70°F AND ABOVE IN ALASKA[a]

| States and stations | Years | January | February | March | April | May | June | July | August | September | October | November | December | Annual |
|---|---|---|---|---|---|---|---|---|---|---|---|---|---|---|
| New Jersey | | | | | | | | | | | | | | |
| Atlantic City (U) | 79 | 0 | 0 | 0 | 0 | * | 1 | 1 | 1 | * | * | 0 | 0 | 3 |
| Newark | 19 | 0 | 0 | 0 | * | 1 | 6 | 9 | 7 | 2 | * | 0 | 0 | 25 |
| Trenton | 28 | 0 | 0 | 0 | * | 1 | 4 | 7 | 5 | 1 | * | 0 | 0 | 18 |
| New Mexico | | | | | | | | | | | | | | |
| Albuquerque | 21 | 0 | 0 | 0 | 0 | 2 | 18 | 23 | 18 | 6 | 0 | 0 | 0 | 67 |
| Clayton | 15 | 0 | 0 | 0 | * | 1 | 11 | 14 | 11 | 4 | * | 0 | 0 | 41 |
| Raton | 15 | 0 | 0 | 0 | 0 | 0 | 5 | 7 | 4 | 1 | 0 | 0 | 0 | 17 |
| Roswell | 13 | 0 | 0 | 0 | 3 | 11 | 25 | 26 | 26 | 15 | 2 | 0 | 0 | 106 |
| New York | | | | | | | | | | | | | | |
| Albany | 14 | 0 | 0 | 0 | 0 | * | 3 | 6 | 4 | 1 | 0 | 0 | 0 | 14 |
| Binghamton (U) | 64 | 0 | 0 | 0 | * | * | 2 | 4 | 3 | 1 | * | 0 | 0 | 10 |
| Buffalo | 17 | 0 | 0 | 0 | 0 | * | 1 | 3 | 3 | 1 | 0 | 0 | 0 | 8 |
| New York (U) | 44 | 0 | 0 | 0 | * | 1 | 3 | 6 | 4 | 1 | * | 0 | 0 | 15 |
| New York | 20 | 0 | 0 | 0 | 0 | 1 | 4 | 7 | 5 | 1 | * | 0 | 0 | 18 |
| Rochester | 20 | 0 | 0 | 0 | 0 | * | 3 | 5 | 4 | 2 | * | 0 | 0 | 14 |
| Syracuse | 11 | 0 | 0 | 0 | 0 | 0 | 2 | 4 | 3 | 1 | 0 | 0 | 0 | 10 |
| North Carolina | | | | | | | | | | | | | | |
| Asheville | 30 | 0 | 0 | 0 | 0 | * | 4 | 6 | 3 | 1 | * | 0 | 0 | 14 |
| Cape Hateras (U) | 78 | 0 | 0 | 0 | 0 | 0 | * | 1 | * | * | * | 0 | 0 | 1 |
| Charlotte | 21 | 0 | 0 | * | * | 4 | 14 | 16 | 15 | 6 | 1 | 0 | 0 | 56 |
| Greensboro | 32 | 0 | 0 | * | * | 2 | 10 | 12 | 9 | 4 | * | 0 | 0 | 37 |
| Raleigh | 16 | 0 | 0 | * | 1 | 2 | 12 | 15 | 12 | 5 | 1 | 0 | 0 | 48 |
| Wilmington | 9 | 0 | 0 | 0 | * | 3 | 10 | 15 | 14 | 4 | * | 0 | 0 | 46 |
| Winston-Salem (U) | 57 | 0 | 0 | * | * | 3 | 10 | 13 | 10 | 5 | * | 0 | 0 | 41 |

|  | | | | | | | | | | | | | |
|---|---|---|---|---|---|---|---|---|---|---|---|---|---|
| **North Dakota** | | | | | | | | | | | | | |
| Bismarck | 21 | 0 | 0 | * | 1 | 2 | 9 | 8 | 3 | * | 0 | 0 | 23 |
| Devils Lake (U) | 56 | 0 | 0 | * | * | 1 | 4 | 4 | 1 | 0 | 0 | 0 | 10 |
| Fargo | 19 | 0 | 0 | * | 1 | 2 | 5 | 5 | 1 | * | 0 | 0 | 14 |
| Williston (U) | 44 | 0 | 0 | * | 1 | 2 | 8 | 6 | 1 | * | 0 | 0 | 18 |
| **Ohio** | | | | | | | | | | | | | |
| Akron-Canton | 12 | 0 | 0 | 0 | 0 | 2 | 3 | 3 | 2 | 0 | 0 | 0 | 10 |
| Cincinnati (ABBE.) | 45 | 0 | 0 | * | 1 | 5 | 11 | 8 | 4 | * | 0 | 0 | 29 |
| Cleveland | 19 | 0 | 0 | 0 | * | 5 | 6 | 5 | 2 | * | 0 | 0 | 18 |
| Columbus | 21 | 0 | 0 | 0 | 1 | 6 | 9 | 7 | 3 | * | 0 | 0 | 26 |
| Dayton | 17 | 0 | 0 | 0 | * | 4 | 6 | 6 | 2 | 0 | 0 | 0 | 18 |
| Sandusky | 33 | 0 | 0 | * | 1 | 4 | 7 | 5 | 2 | * | 0 | 0 | 19 |
| Toledo (U) | 81 | 0 | 0 | 0 | * | 3 | 5 | 3 | 1 | 0 | 0 | 0 | 12 |
| Youngstown | 17 | 0 | 0 | 0 | * | 2 | 4 | 4 | 1 | 0 | 0 | 0 | 11 |
| **Oklahoma** | | | | | | | | | | | | | |
| Oklahoma City (U) | 62 | * | * | * | 2 | 11 | 20 | 20 | 10 | 1 | 0 | 0 | 64 |
| Tulsa | 22 | 0 | * | * | 2 | 13 | 21 | 22 | 11 | 2 | 0 | 0 | 71 |
| **Oregon** | | | | | | | | | | | | | |
| Astoria | 7 | 0 | 0 | 0 | 0 | * | * | 0 | * | 0 | 0 | 0 | 1 |
| Burns | 10 | 0 | 0 | 0 | * | 1 | 11 | 7 | 2 | 0 | 0 | 0 | 21 |
| Eugene | 18 | 0 | 0 | 0 | * | 1 | 6 | 4 | 3 | 0 | 0 | 0 | 14 |
| Meacham | 16 | 0 | 0 | 0 | 0 | 0 | 2 | 1 | * | 0 | 0 | 0 | 3 |
| Medford | 31 | 0 | 0 | * | 2 | 5 | 16 | 15 | 8 | 0 | 0 | 0 | 47 |
| Pendleton | 25 | 0 | 0 | 0 | 1 | 4 | 14 | 9 | 3 | 1 | 0 | 0 | 31 |
| Portland (U) | 58 | 0 | 0 | * | * | 1 | 3 | 2 | 1 | 0 | 0 | 0 | 6 |
| Roseburg | 8 | 0 | 0 | * | 1 | 2 | 7 | 6 | 3 | * | 0 | 0 | 20 |
| Salem | 23 | 0 | 0 | 0 | * | 2 | 6 | 5 | 3 | * | 0 | 0 | 16 |
| Sexton Summit | 16 | 0 | 0 | 0 | 0 | * | 1 | 1 | * | 0 | 0 | 0 | 2 |

## Table 2 (continued)
## MEAN NUMBER OF DAYS MAXIMUM TEMPERATURE 90°F AND ABOVE, EXCEPT 70°F AND ABOVE IN ALASKA[a]

| States and stations | Years | January | February | March | April | May | June | July | August | September | October | November | December | Annual |
|---|---|---|---|---|---|---|---|---|---|---|---|---|---|---|
| Pennsylvania | | | | | | | | | | | | | | |
| Allentown | 17 | 0 | 0 | 0 | 0 | * | 5 | 8 | 4 | 1 | * | 0 | 0 | 18 |
| Erie (U) | 77 | 0 | 0 | 0 | 0 | * | 1 | 2 | 1 | * | 0 | 0 | 0 | 4 |
| Harrisburg | 22 | 0 | 0 | 0 | * | 1 | 5 | 9 | 7 | 2 | * | 0 | 0 | 24 |
| Philadelphia | 20 | 0 | 0 | 0 | * | 1 | 6 | 10 | 6 | 2 | * | 0 | 0 | 25 |
| Pittsburgh | 8 | 0 | 0 | 0 | 0 | 0 | 1 | 3 | 3 | 2 | 0 | 0 | 0 | 9 |
| Reading | 20 | 0 | 0 | 0 | * | 1 | 6 | 9 | 6 | 2 | * | 0 | 0 | 24 |
| Williamsport | 16 | 0 | 0 | 0 | * | * | 5 | 8 | 4 | 1 | * | 0 | 0 | 18 |
| Rhode Island | | | | | | | | | | | | | | |
| Block Island | 10 | 0 | 0 | 0 | 0 | 0 | * | * | 0 | 0 | 0 | 0 | 0 | * |
| Providence | 7 | 0 | 0 | 0 | 0 | * | 2 | 2 | 2 | * | 0 | 0 | 0 | 6 |
| South Carolina | | | | | | | | | | | | | | |
| Charleston | 18 | 0 | 0 | 0 | * | 4 | 13 | 13 | 14 | 5 | * | 0 | 0 | 49 |
| Columbia | 13 | 0 | 0 | 0 | 1 | 9 | 19 | 24 | 23 | 9 | 1 | 0 | 0 | 86 |
| Florence | 12 | 0 | 0 | 0 | 1 | 7 | 16 | 21 | 21 | 7 | 1 | 0 | 0 | 74 |
| Greenville | 7 | 0 | 0 | 0 | * | 3 | 11 | 17 | 18 | 6 | 0 | 0 | 0 | 55 |
| Spartanburg | 17 | 0 | 0 | 0 | * | 3 | 13 | 15 | 13 | 4 | 1 | 0 | 0 | 49 |
| South Dakota | | | | | | | | | | | | | | |
| Huron | 21 | 0 | 0 | 0 | * | 1 | 5 | 11 | 10 | 3 | * | 0 | 0 | 30 |
| Rapid City | 18 | 0 | 0 | 0 | 0 | 1 | 3 | 12 | 12 | 4 | * | 0 | 0 | 32 |
| Sioux Falls | 15 | 0 | 0 | 0 | * | 1 | 4 | 10 | 9 | 3 | * | 0 | 0 | 27 |
| Tennessee | | | | | | | | | | | | | | |
| Bristol | 15 | 0 | 0 | 0 | 0 | * | 5 | 5 | 7 | 2 | * | 0 | 0 | 19 |
| Chattanooga | 20 | 0 | 0 | 0 | * | 4 | 14 | 18 | 17 | 6 | 1 | 0 | 0 | 60 |
| Knoxville | 18 | 0 | 0 | 0 | * | 2 | 11 | 15 | 14 | 5 | * | 0 | 0 | 47 |
| Memphis | 19 | 0 | 0 | 0 | * | 3 | 16 | 22 | 21 | 9 | 1 | 0 | 0 | 72 |
| Nashville | 19 | 0 | 0 | 0 | * | 2 | 14 | 19 | 17 | 8 | * | 0 | 0 | 60 |
| Oak Ridge | 15 | 0 | 0 | 0 | 0 | 2 | 9 | 12 | 11 | 5 | * | 0 | 0 | 39 |

| | | | | | | | | | | | | | | |
|---|---|---|---|---|---|---|---|---|---|---|---|---|---|---|
| **Texas** | | | | | | | | | | | | | | |
| Abilene | 21 | 0 | * | 1 | 3 | 10 | 22 | 27 | 27 | 16 | 3 | * | 0 | 109 |
| Amarillo | 20 | 0 | 0 | * | 1 | 4 | 10 | 20 | 20 | 9 | 1 | 0 | 0 | 71 |
| Austin | 19 | 0 | * | 1 | 1 | 9 | 23 | 29 | 28 | 18 | 5 | * | * | 114 |
| Brownsville | 18 | 0 | * | 1 | 2 | 9 | 23 | 29 | 28 | 18 | 5 | * | 0 | 115 |
| Corpus Christi | 22 | 0 | * | * | 1 | 4 | 19 | 29 | 28 | 16 | 3 | * | 0 | 100 |
| Dallas | 20 | 0 | 0 | * | 1 | 7 | 21 | 27 | 27 | 17 | 2 | 0 | 0 | 102 |
| El Paso | 21 | 0 | 0 | 0 | 1 | 10 | 25 | 26 | 24 | 14 | 1 | 0 | 0 | 101 |
| Ft. Worth | 13 | 0 | 0 | * | 1 | 6 | 20 | 26 | 27 | 15 | 2 | 0 | * | 98 |
| Galveston | 10 | 0 | 0 | 0 | * | 1 | 4 | 13 | 14 | 3 | * | 0 | 0 | 35 |
| Houston | 22 | * | * | * | * | 4 | 19 | 25 | 25 | 15 | 2 | 1 | * | 90 |
| Laredo | 17 | 0 | 2 | 6 | 13 | 23 | 28 | 31 | 29 | 25 | 12 | 1 | 0 | 171 |
| Lubbock | 14 | 0 | 0 | 0 | 2 | 8 | 21 | 23 | 22 | 11 | 1 | 0 | 0 | 88 |
| Midland | 13 | 0 | 0 | * | 4 | 12 | 25 | 27 | 27 | 15 | 3 | 0 | 0 | 113 |
| Port Arthur | 7 | 0 | 0 | 0 | * | 3 | 19 | 24 | 22 | 13 | 2 | 0 | 0 | 83 |
| San Angelo | 13 | 0 | * | 1 | 4 | 12 | 23 | 27 | 27 | 16 | 3 | * | * | 113 |
| San Antonio | 18 | 0 | * | 1 | 3 | 10 | 24 | 30 | 28 | 18 | 5 | * | * | 119 |
| Victoria | 14 | 0 | * | * | 1 | 7 | 25 | 30 | 28 | 19 | 5 | * | 0 | 115 |
| Waco | 18 | 0 | * | * | 1 | 7 | 23 | 28 | 28 | 18 | 4 | * | * | 109 |
| Wichita Falls | 17 | 0 | 0 | 1 | 3 | 8 | 21 | 26 | 27 | 17 | 3 | 0 | 0 | 106 |
| **Utah** | | | | | | | | | | | | | | |
| Melford | 12 | 0 | 0 | 0 | 0 | 1 | 11 | 18 | 18 | 7 | 0 | 0 | 0 | 61 |
| Salt Lake City | 32 | 0 | 0 | 0 | 0 | 1 | 8 | 18 | 18 | 4 | 0 | 0 | 0 | 53 |
| Wendover | 11 | 0 | 0 | 0 | 0 | 1 | 8 | 19 | 19 | 4 | 0 | 0 | 0 | 54 |
| **Vermont** | | | | | | | | | | | | | | |
| Burlington | 17 | 0 | 0 | 0 | 0 | * | 2 | 3 | 3 | 1 | 0 | 0 | 0 | 8 |
| **Virginia** | | | | | | | | | | | | | | |
| Lynchburg | 16 | 0 | 0 | 0 | * | 1 | 7 | 10 | 8 | 3 | * | 0 | 0 | 29 |
| Norfolk | 12 | 0 | 0 | 0 | 1 | 2 | 9 | 13 | 9 | 3 | * | 0 | 0 | 37 |
| Richmond | 31 | 0 | 0 | * | 1 | 3 | 10 | 14 | 11 | 5 | 1 | 0 | 0 | 45 |
| Roanoke | 13 | 0 | 0 | 0 | 2 | 2 | 9 | 14 | 11 | 4 | * | 0 | 0 | 42 |

Table 2 (continued)

## MEAN NUMBER OF DAYS MAXIMUM TEMPERATURE 90°F AND ABOVE, EXCEPT 70°F AND ABOVE IN ALASKA[a]

| States and stations | Years | January | February | March | April | May | June | July | August | September | October | November | December | Annual |
|---|---|---|---|---|---|---|---|---|---|---|---|---|---|---|
| **Washington** | | | | | | | | | | | | | | |
| Olympia | 19 | 0 | 0 | 0 | 0 | * | 1 | 3 | 2 | * | 0 | 0 | 0 | 6 |
| Seattle (U) | 27 | 0 | 0 | 0 | 0 | * | * | 1 | * | * | 0 | 0 | 0 | 2 |
| Spokane | 13 | 0 | 0 | 0 | 0 | * | 1 | 9 | 5 | 1 | 0 | 0 | 0 | 16 |
| Stampede Pass | 17 | 0 | 0 | 0 | 0 | 0 | 0 | 0 | * | 0 | 0 | 0 | 0 | * |
| Tatoosh Island (U) | 58 | 0 | 0 | 0 | 0 | 0 | 0 | 0 | 0 | 0 | 0 | 0 | 0 | 0 |
| Walla Walla (U) | 46 | 0 | 0 | 0 | * | 1 | 4 | 16 | 13 | 3 | 0 | 0 | 0 | 37 |
| Yakima | 14 | 0 | 0 | 0 | 0 | 2 | 4 | 15 | 10 | 4 | 0 | 0 | 0 | 35 |
| **Virgin Islands** | | | | | | | | | | | | | | |
| San Juan P. R. | 62 | * | * | * | 1 | 2 | 1 | * | 1 | 2 | 2 | * | * | 9 |
| **West Virginia** | | | | | | | | | | | | | | |
| Charleston | 13 | 0 | 0 | 0 | * | 1 | 5 | 9 | 7 | 3 | * | 0 | 0 | 25 |
| Huntington | 11 | 0 | 0 | 0 | 1 | 3 | 10 | 14 | 12 | 6 | 1 | 0 | 0 | 47 |
| Parkersburg | 72 | 0 | 0 | 0 | * | 1 | 5 | 8 | 6 | 3 | * | 0 | 0 | 23 |
| **Wisconsin** | | | | | | | | | | | | | | |
| Green Bay | 11 | 0 | 0 | 0 | 0 | * | 1 | 1 | 2 | 2 | 0 | 0 | 0 | 6 |
| La Crosse | 10 | 0 | 0 | 0 | * | * | 3 | 5 | 4 | 2 | 0 | 0 | 0 | 14 |
| Madison | 21 | 0 | 0 | 0 | * | * | 4 | 7 | 6 | 2 | 0 | 0 | 0 | 19 |
| Milwaukee | 20 | 0 | 0 | 0 | 0 | * | 2 | 3 | 4 | 1 | 0 | 0 | 0 | 10 |
| **Wyoming** | | | | | | | | | | | | | | |
| Casper | 21 | 0 | 0 | 0 | 0 | 0 | 3 | 12 | 8 | 2 | 0 | 0 | 0 | 25 |
| Cheyenne | 25 | 0 | 0 | 0 | 0 | * | 1 | 5 | 4 | 1 | 0 | 0 | 0 | 11 |
| Lander | 14 | 0 | 0 | 0 | 0 | * | 2 | 10 | 7 | 2 | 0 | 0 | 0 | 21 |
| Sheridan | 20 | 0 | 0 | 0 | 0 | * | 2 | 12 | 11 | 3 | 0 | 0 | 0 | 28 |
| Yellowstone | 30 | 0 | 0 | 0 | 0 | 0 | 0 | 1 | * | * | 0 | 0 | 0 | 1 |

**Mean Number of Days Maximum Temperature 70°F. and Above**

| Alaska | Yrs[a] | Jan | Feb | Mar | Apr | May | Jun | Jul | Aug | Sep | Oct | Nov | Dec | Ann. |
|---|---|---|---|---|---|---|---|---|---|---|---|---|---|---|
| Anchorage | 35 | 0 | 0 | 0 | 0 | * | 4 | 6 | 5 | * | 0 | 0 | 0 | 14 |
| Annette | 13 | 0 | 0 | 0 | * | 1 | 3 | 4 | 5 | 2 | 0 | 0 | 0 | 15 |
| Barrow | 40 | 0 | 0 | 0 | 0 | 0 | * | * | * | 0 | 0 | 0 | 0 | * |
| Barter Island | 13 | 0 | 0 | 0 | 0 | 0 | 0 | * | * | 0 | 0 | 0 | 0 | * |
| Bothel | 16 | 0 | 0 | 0 | 0 | * | 3 | 4 | 1 | * | 0 | 0 | 0 | 9 |
| Cold Bay | 16 | 0 | 0 | 0 | 0 | 0 | 0 | * | * | * | 0 | 0 | 0 | * |
| Cordova | 15 | 0 | 0 | 0 | * | * | 2 | 3 | 3 | * | * | 0 | 0 | 8 |
| Fairbanks | 31 | 0 | 0 | 0 | * | 3 | 17 | 19 | 9 | 1 | 0 | 0 | 0 | 49 |
| Juneau | 17 | 0 | 0 | 0 | 0 | 2 | 7 | 7 | 5 | * | 0 | 0 | 0 | 21 |
| King Salmon | 15 | 0 | 0 | 0 | 0 | * | 3 | 6 | 3 | * | 0 | 0 | 0 | 13 |
| Kotzebue | 17 | 0 | 0 | 0 | 0 | * | 1 | 2 | 1 | 0 | 0 | 0 | 0 | 4 |
| McGrath | 18 | 0 | 0 | 0 | 0 | 1 | 11 | 13 | 4 | * | 0 | 0 | 0 | 29 |
| Nome | 14 | 0 | 0 | 0 | 0 | 0 | 1 | 1 | 1 | 0 | 0 | 0 | 0 | 3 |
| St. Paul Island | 43 | 0 | 0 | 0 | 0 | 0 | 0 | 0 | 0 | 0 | 0 | 0 | 0 | 0 |
| Shemya | 15 | 0 | 0 | 0 | 0 | 0 | 0 | 0 | 0 | 0 | 0 | 0 | 0 | 0 |
| Yakutat | 14 | 0 | 0 | 0 | 0 | 1 | 1 | 2 | 1 | * | 0 | 0 | 0 | 5 |

*Note:*  Data from airport, except those marked with U for urban.

[a]  * Less than once in 2 years.

## Table 3
## SUMMARY OF SURFACE WATER $p_{CO_2}$ VALUES OBTAINED DURING THE 1957 TO 1958 IGY AND 1972 GEOSECS EXPEDITIONS IN THE WESTERN AND SOUTHERN ATLANTIC OCEAN

| Latitude range | 1957—1958 | | 1972 | | Temperature change, $\Delta T$ (°C) | $\Delta p_{CO_2}$ ($10^{-6}$ atm) |
|---|---|---|---|---|---|---|
| | Temperature (°C) | $p_{CO_2}$ ($10^{-6}$ atm) | Temperature (°C) | $p_{CO_2}$ ($10^{-6}$ atm) | | |
| 60—50° N | 8.3 | 254 ± 10[a] | 9.2 | 296 ± 13[a] | 0.9 | 32 |
| 30—20° N | 25.5 | 302 ± 2 | 27.4 | 331 ± 3 | 1.0 | 3 |
| 30—20° N | 25.5 | 302 ± 2 | 27.4 | 331 ± 3 | 1.0 | 3 |
| 20—10° N | 26.6 | 299 ± 4 | 27.8 | 318 ± 6 | 1.2 | 3 |
| 10—0° N | 27.3 | 300 ± 6 | 27.5 | 339 ± 17 | 0.2 | 36 |
| 0—10° S | 27.2 | 326 ± 8 | 26.6 | 350 ± 8 | −0.6 | 33 |
| 10—20° S | 27.5 | 327 ± 3 | 26.0 | 335 ± 4 | −1.5 | 30 |
| 20—30° S | 25.3 | 345 ± 6 | 23.4 | 312 ± 6 | −1.9 | −3 |
| 30—40° S | 23.2 | 315 ± 7 | 18.0 | 291 ± 5 | −4.8 | 44 |
| 40—50° S | 16.6 | 249 ± 10 | 13.0 | 284 ± 5 | −2.4 | 62 |
| 50—60° S | 4.1 | 294 ± 5 | 2.3 | 317 ± 18 | −1.8 | 47 |
| 30—40° S(E) | 17.2 | 256 ± 5 | 22.3 | 328 ± 9 | 5.1 | 13 |
| 40—50° S(E) | 8.3 | 283 ± 8 | 8.5 | 294 ± 2 | 0.2 | 8 |
| 50—60° S(E) | 3.5 | 289 ± 21 | 1.1 | 300 ± 5 | −2.4 | 42 |

*Note:* Mean increase in surface water $p_{CO_2}$ = 27(±6) × $10^{-6}$ atm, where (±6) is the SD of the mean. Mean annual rate of increase in the 15-year period = 1.8(±0.4) × $10^{-6}$ atm/year, where (±0.4) is the SD of the mean.

[a]   One SD.

From Carbon Dioxide Effects — Research and Assessment Program, Workshop on the Global Effects of Carbon Dioxide from Fossil Fuels, U.S. Department of Energy, Washington, D.C., 1979.

*Figures 1 to 15*

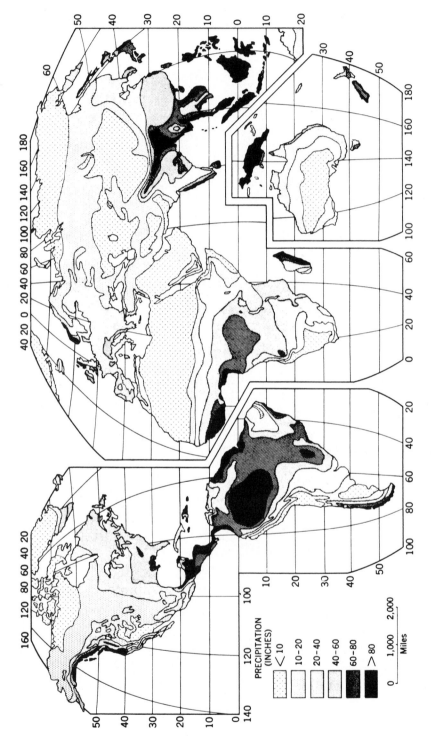

FIGURE 1.    Average annual precipitation over the globe. (From Longwell, G. R., Flint, R. F., and Sanders, J. E., *Physical Geology*, John Wiley & Sons, New York, 1969. With permission.)

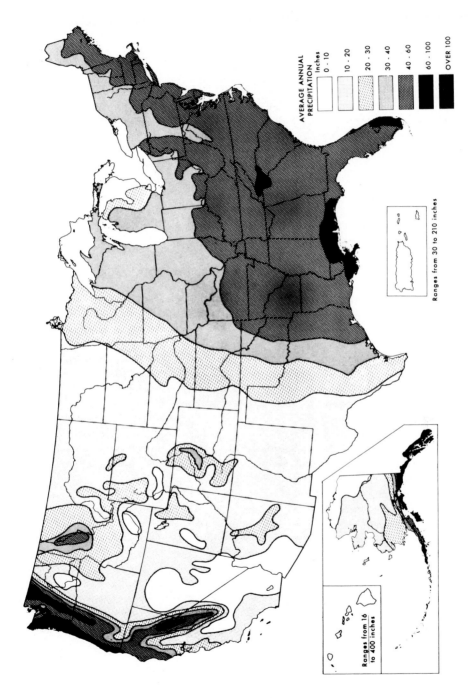

FIGURE 2. Average annual precipitation. (From Forest Service, The Nation's Renewable Resources — An Assessment, 1975, Forest Resource Report No. 21, U.S. Department of Agriculture, Washington, D.C., 1977.)

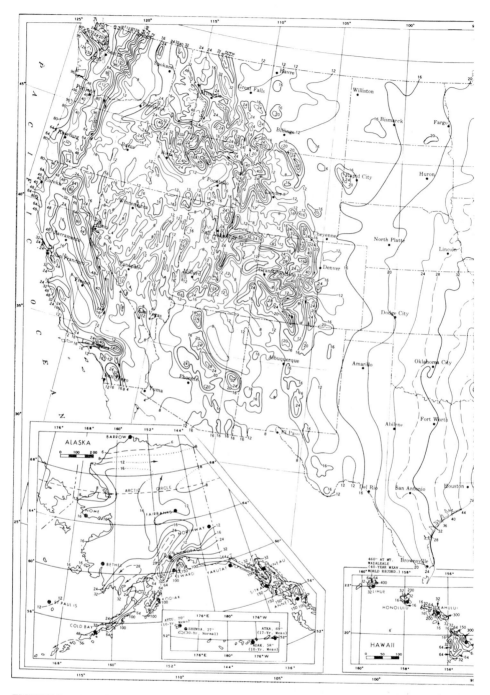

FIGURE 3.  Normal annual total precipitation (inches). (From Weather Atlas of the United States, Environmental Data Service, Environmental Science Services Administration, U.S. Department of Commerce, National Climatic Center, Asheville, N.C., 1977.)

Figure 3 (continued)

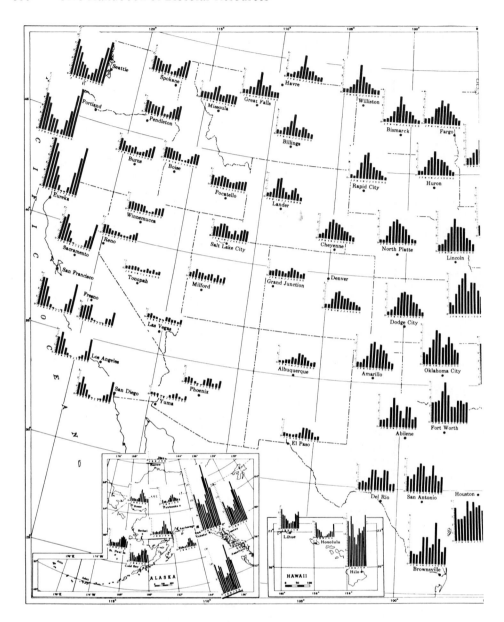

FIGURE 4.    Normal monthly total precipitation (inches). (From Weather Atlas of the United States, Environmental Data Service, Environmental Science Service Administration, U.S. Department of Commerce, National Climatic Center, Asheville, N.C., 1977.)

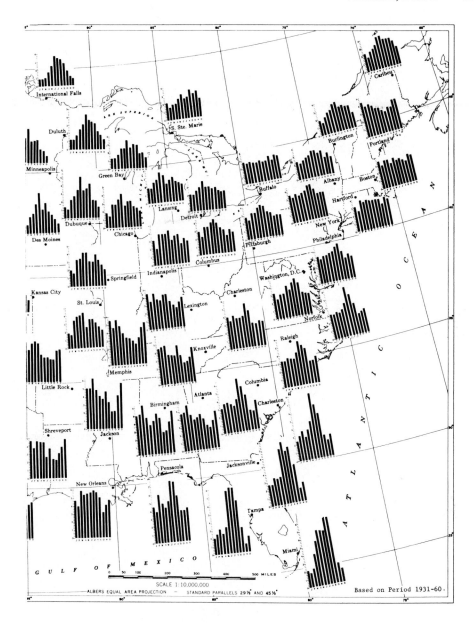

Figure 4 (continued)

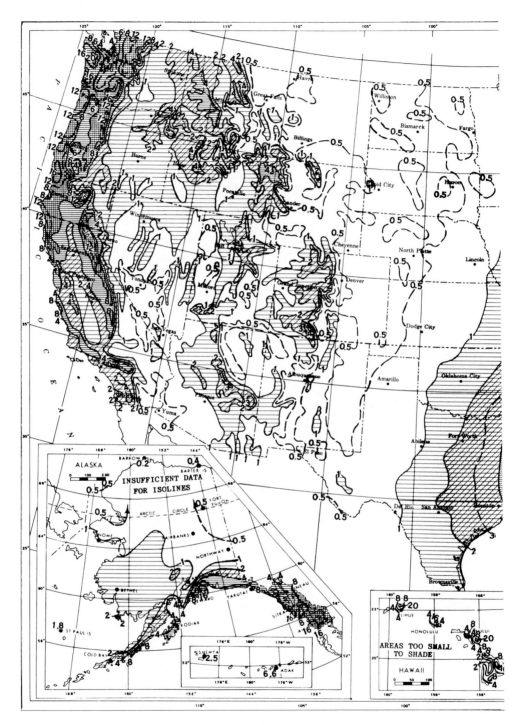

FIGURE 5.    Normal total precipitation (inches), January. (From Weather Atlas of the United States, Environmental Data Service, Environmental Science Services Administration, U.S. Department of Commerce, National Climatic Center, Asheville, N.C., 1977.)

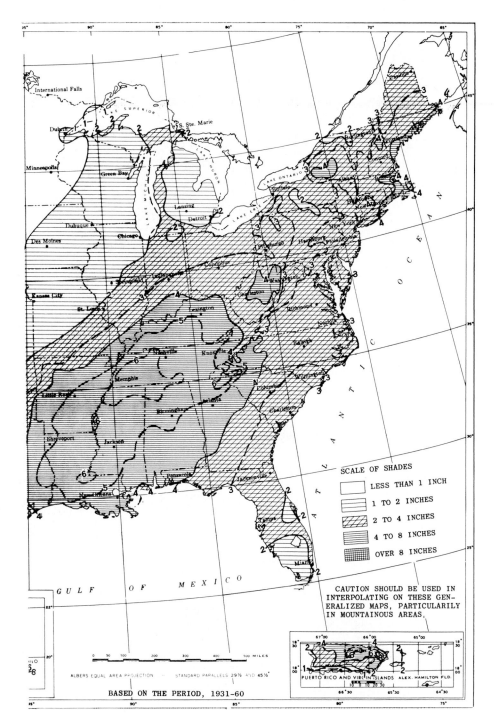

Figure 5 (continued)

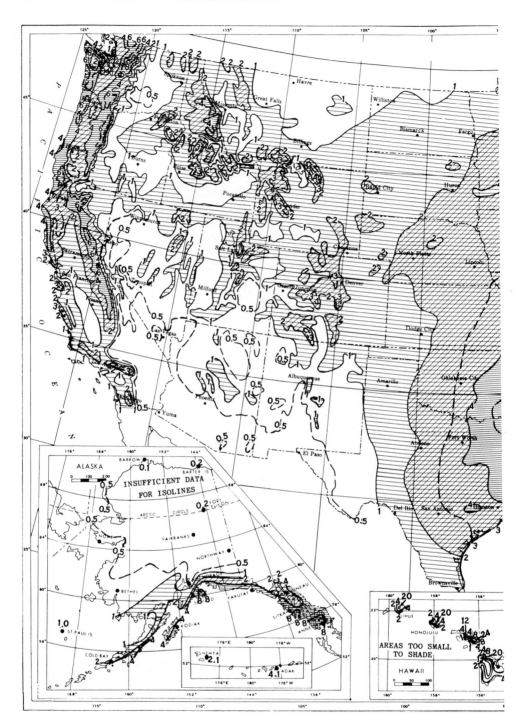

FIGURE 6.    Normal total precipitation (inches), April. (From Weather Atlas of the United States, Environmental Data Service, Environmental Science Services Administration, U.S. Department of Commerce, National Climatic Center, Asheville, N.C., 1977.)

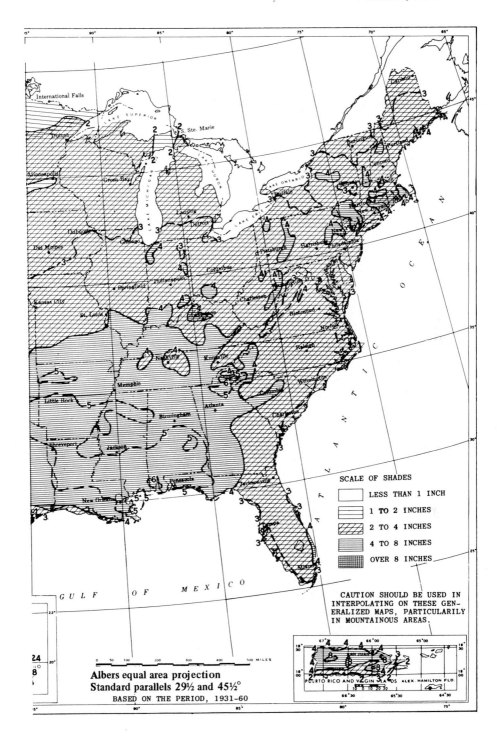

Figure 6 (continued

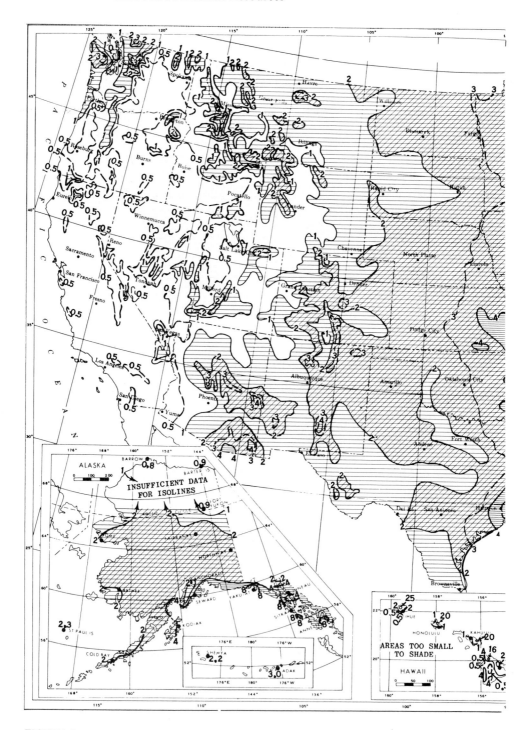

FIGURE 7.  Normal total precipitation (inches), July. (From Weather Atlas of the United States, Environmental Data Service, Environmental Science Services Administration, U.S. Department of Commerce, National Climatic Center, Asheville, N.C., 1977.)

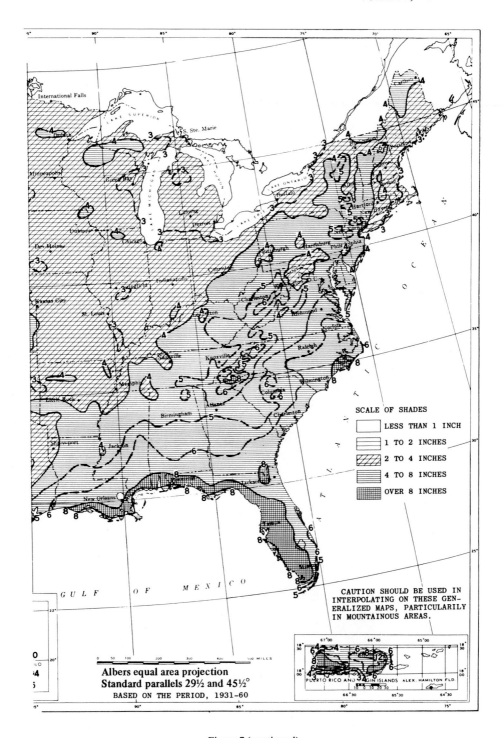

Figure 7 (continued)

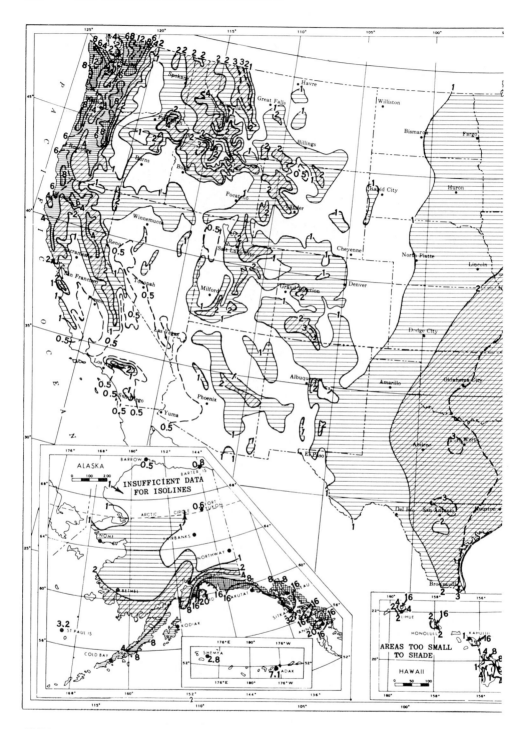

FIGURE 8.    Normal total precipitation (inches), October. (From *Weather Atlas of the United States*, Environmental Data Service, Environmental Science Services Administration, U.S. Department of Commerce, National Climatic Center, Asheville, N.C., 1977.)

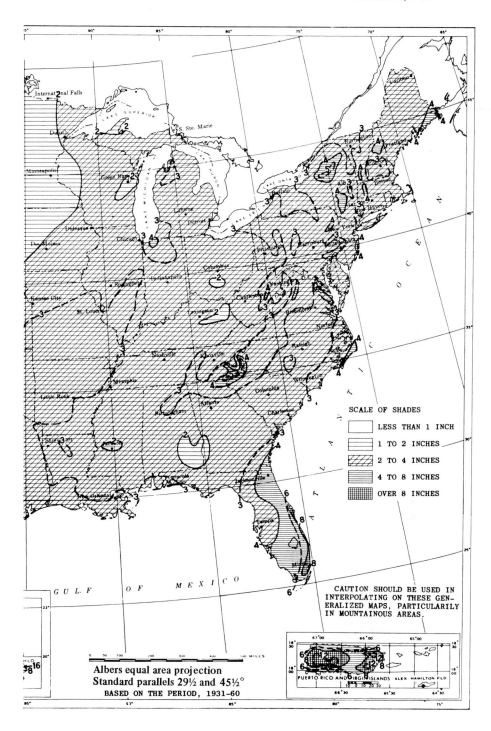

SCALE OF SHADES

| | |
|---|---|
| | LESS THAN 1 INCH |
| | 1 TO 2 INCHES |
| | 2 TO 4 INCHES |
| | 4 TO 8 INCHES |
| | OVER 8 INCHES |

CAUTION SHOULD BE USED IN
INTERPOLATING ON THESE GEN-
ERALIZED MAPS, PARTICULARILY
IN MOUNTAINOUS AREAS.

Albers equal area projection
Standard parallels 29½ and 45½°
BASED ON THE PERIOD, 1931-60

PUERTO RICO AND VIRGIN ISLANDS   ALEX. HAMILTON FLD.

Figure 8 (continued)

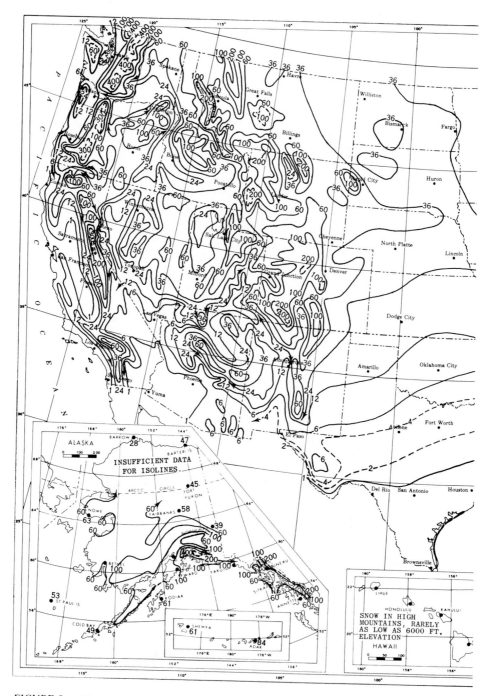

FIGURE 9.    Mean annual total snowfall (inches). (From Weather Atlas of the United States, Environmental Data Service, Environmental Science Services Administration, U.S. Department of Commerce, National Climatic Center, Asheville, N.C., 1977.)

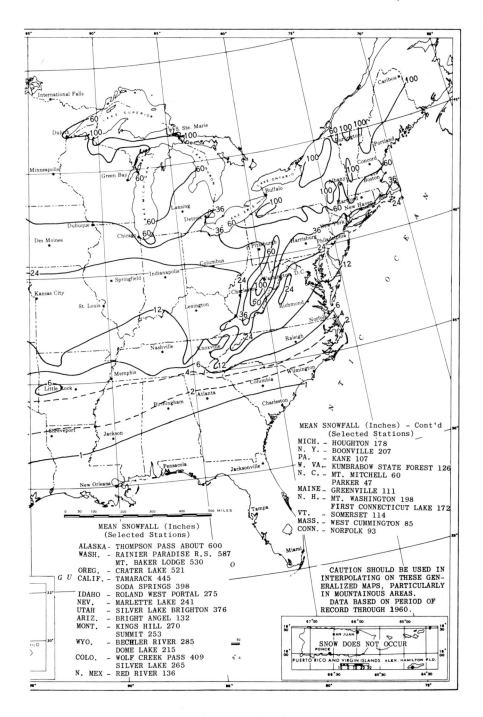

MEAN SNOWFALL (Inches) - Cont'd
(Selected Stations)

MICH. – HOUGHTON 178
N. Y. – BOONVILLE 207
PA. – KANE 107
W. VA.– KUMBRABOW STATE FOREST 126
N. C. – MT. MITCHELL 60
         PARKER 47
MAINE – GREENVILLE 111
N. H.– MT. WASHINGTON 198
         FIRST CONNECTICUT LAKE 172
VT. – SOMERSET 114
MASS.– WEST CUMMINGTON 85
CONN. – NORFOLK 93

MEAN SNOWFALL (Inches)
(Selected Stations)

ALASKA– THOMPSON PASS ABOUT 600
WASH. – RAINIER PARADISE R.S. 587
         MT. BAKER LODGE 530
OREG. – CRATER LAKE 521
CALIF. – TAMARACK 445
         SODA SPRINGS 398
IDAHO – ROLAND WEST PORTAL 275
NEV. – MARLETTE LAKE 241
UTAH – SILVER LAKE BRIGHTON 376
ARIZ. – BRIGHT ANGEL 132
MONT. – KINGS HILL 270
         SUMMIT 253
WYO. – BECHLER RIVER 285
         DOME LAKE 215
COLO. – WOLF CREEK PASS 409
         SILVER LAKE 265
N. MEX – RED RIVER 136

CAUTION SHOULD BE USED IN
INTERPOLATING ON THESE GEN-
ERALIZED MAPS, PARTICULARLY
IN MOUNTAINOUS AREAS.
   DATA BASED ON PERIOD OF
RECORD THROUGH 1960.

SNOW DOES NOT OCCUR

PUERTO RICO AND VIRGIN ISLANDS   ALEX. HAMILTON FLD.

Figure 9 (continued)

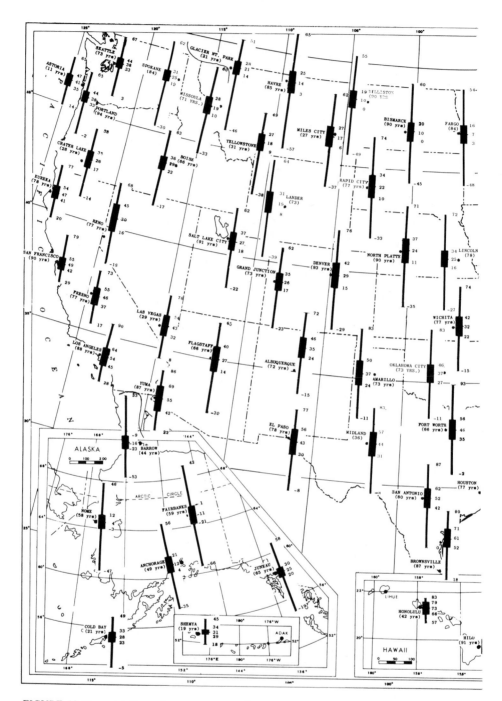

FIGURE 10. Normal daily maximum, average, and extreme temperatures (°F), January. (From Weather Atlas of the United States, Environmental Data Service, Environmental Science Services Administration, U.S. Department of Commerce, National Climatic Center, Asheville, N.C., 1977.)

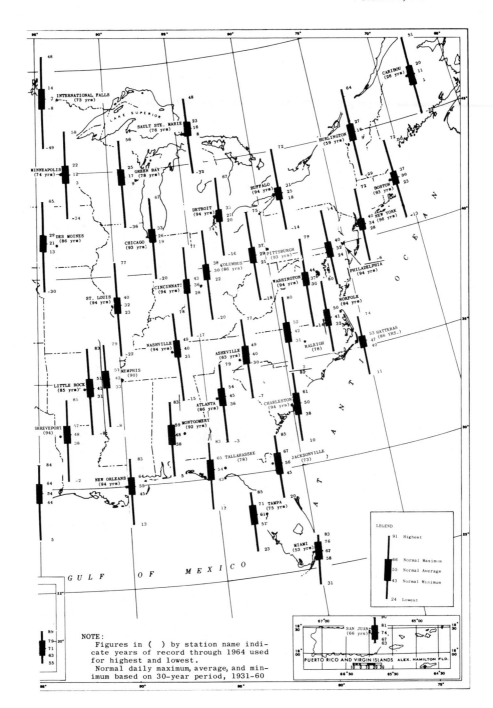

Figure 10 (continued)

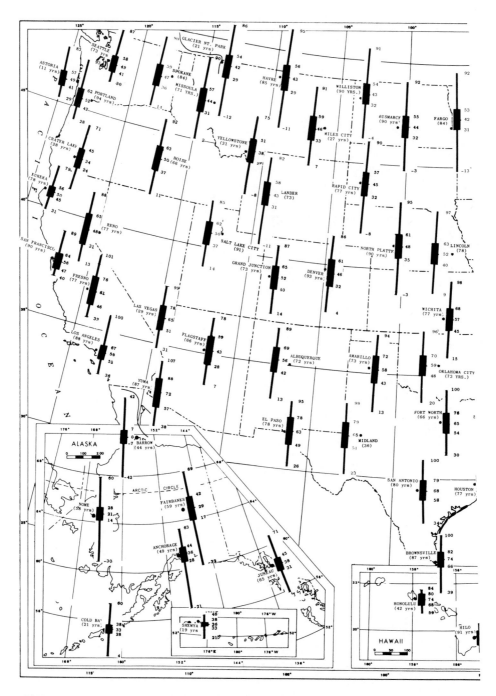

FIGURE 11. Normal daily maximum, average, and extreme temperatures (°F), April. (From Weather Atlas of the United States, Environmental Data Service, Environmental Science Services Administration, U.S. Department of Commerce, National Climatic Center, Asheville, N.C., 1977.)

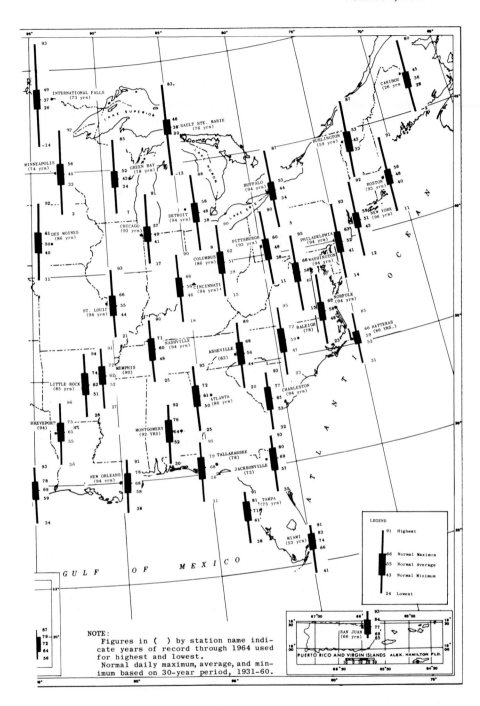

Figure 11 (continued)

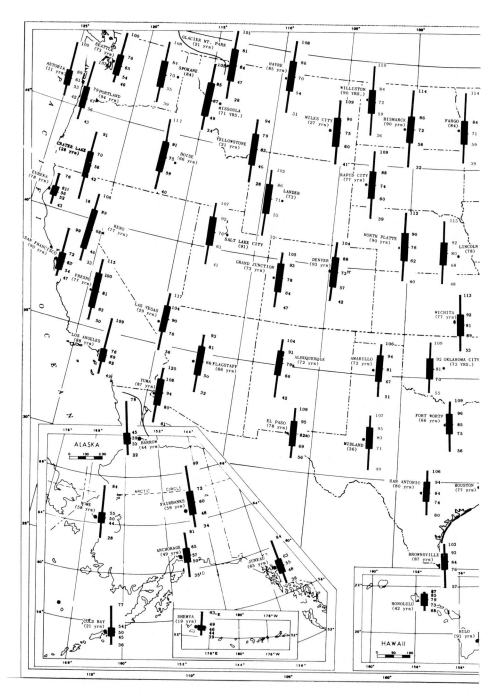

FIGURE 12. Normal daily maximum, average, and extreme temperatures (°F), July. (From Weather Atlas of the United States, Environmental Data Service, Environmental Science Services Administration, U.S. Department of Commerce, National Climatic Center, Asheville, N.C., 1977.)

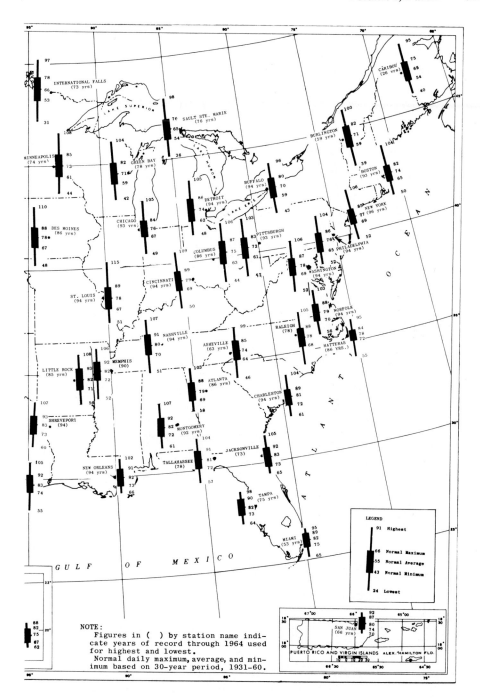

Figure 12 (continued)

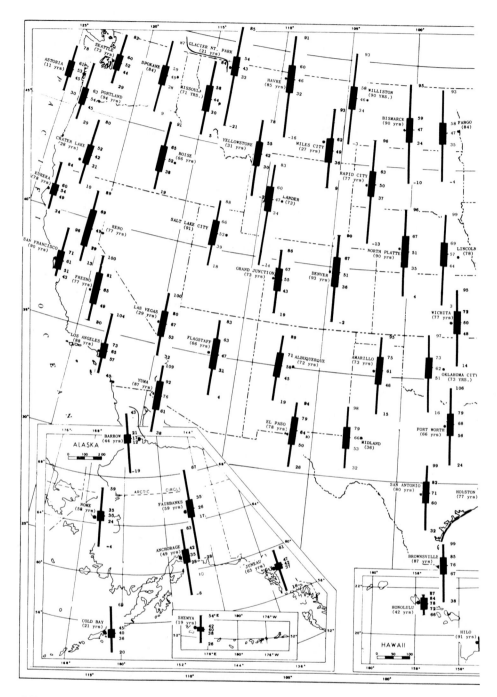

FIGURE 13. Normal daily maximum, average, and extreme temperatures (°F), October. (From Weather Atlas of the United States, Environmental Data Service, Environmental Science Services Administration, U.S. Department of Commerce, National Climatic Center, Asheville, N.C., 1977.)

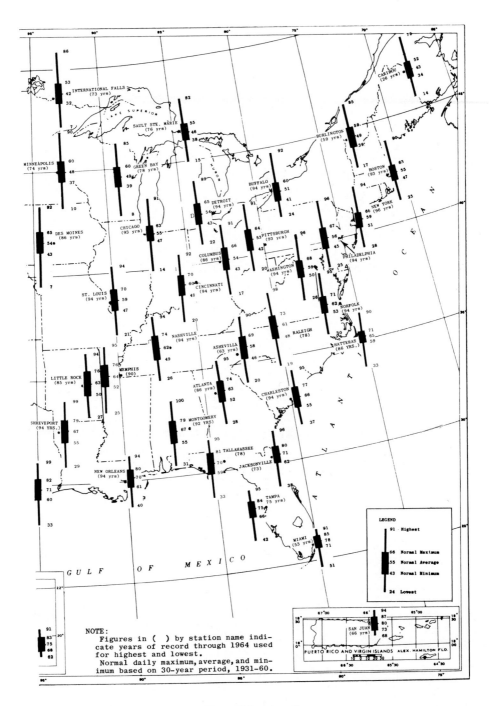

Figure 13 (continued)

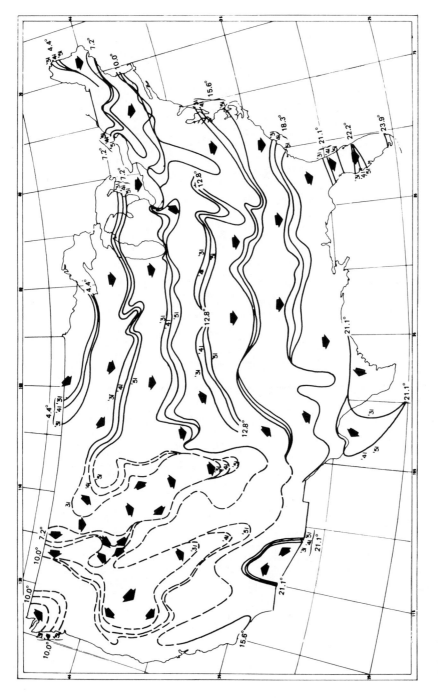

FIGURE 14. Average annual temperature progression (°C) by 20-year periods ('31 is 1931-1950; '41 is 1941-1960; '51 is 1951-1970; dashed lines are mountain regions). (From Climate and Food, Climatic Fluctuation and U.S. Agricultural Production, U.S. National Academy of Sciences, Washington, D.C., 1976.)

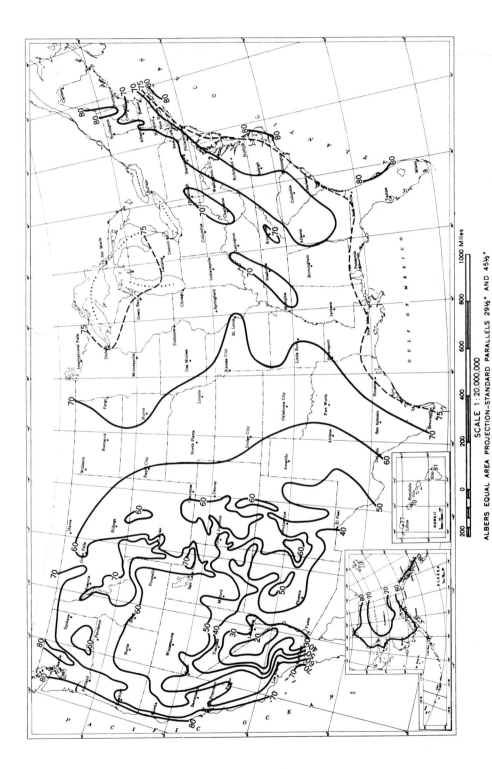

FIGURE 15.   Mean relative humidity (%), annual. (From Weather Atlas of the United States, Environmental Science Services Administration, U.S. Department of Commerce, National Climatic Center, Asheville, N.C., 1977.)

*Index*

# INDEX

# G

# O

# P

## S

(1): 46

Sargasso Sea, productivity data from, (2): 414

*Sargassum ringgoldianum*, light intensity vs. photosynthesis in, (2): 317

*Scenedesmus*
cytochromes in, (1): 150
elemental composition of, (2): 25, 28
plastocyanin in, (1): 163
QR by, (1): 56
temperature vs. photosynthesis in, (2): 339

*Scenedesmus quadricauda*
temperature vs. growth in, (2): 345
emperature ves. photosynthesis in, (2): 339

Schizophyceae, taxonomic characteristics of, (1): 529, 530, 536—537

Scytosiphonales, taxonomic characteristics of, (1): 551

*Scytosiphon lomentaria*, light intensity vs. photosynthesis in, (2): 317

Seagrasses
Australian, (2): 468
chemical composition of, (2): 92, 94
distribution of, (2): 91—92, 472
ecological adaptation of, (2): 91
evolution of, (2): 91
Japanese standing crop of, (2): 359, 461
nutrient supply and, (2): 474
primary productivity of, (2): 471—474
productivity of, (2): 92
trace elements in, (2): 94, 96

Seawater, see also Oceans
carbonate system in, (2): 517—528
composition of, (2): 580—581
ionization of boric acid in, (2): 520
micronutrient chemistry in, (2): 537—541
movement of, (2): 397
optical properties of, (2): 157

Seaweeds, see also Macroalgae
Japanese standing crop of, (2): 456—460
photosynthesis-light curves of, (2): 463

Sediment, deep sea, calcium carbonate in, (2): 526—528

Sedimentation, (2): 397

*Sedum praetum*, PEP carboxylases from, (1): 207

Seedlings, sources of, (1): 597

Seed plants, classification system for, (1): 569—571

Seeds, sources of, (1): 597

Serine, (1): 501, 508

Sesquiterpenoids
from brown algae, (1): 468
from red algae, (1): 471

Shade adaptation, see also Light intensity
aquatic photosynthetic microorganisms, (2): 257—259
geographical distribution of, (2): 300

Shikimate pathway, (1): 479

Shikimic acid pathway, (1): 406, 408

Shoot tip cultures, (1): 601

Silicoflagellatophyceae, taxonomic characteristics of, (1): 533, 547

Silicon

diatom requirement for, (2): 33
distribution in oceans, (2): 546
in seawater, (2): 540—541

Sinapate, formed from ferulate, (1): 480, 481

*Sinapis alba*, QR for photosynthesis by, (1): 48

Siphonaxanthin, (1): 77; (2): 257

Siphonein, (1): 77

Siphonocladales, taxonomic characteristics of, (1): 543

Siphonocladophycidae, taxonomic characteristics of, (1): 543

Sitosterol, (1): 445

*Skeletonema*, (2): 34
temperature vs. growth in, (2): 343
temperature vs. photosynthesis in, (2): 338

Snowfall, mean annual total in U. S., (2): 620—621

Sodium
in photosynthetic bacteria, (2): 140—141
response of photosynthetic bacteria to, (2): 140—141

Solaniflorae, taxonomic characteristics of, (1): 579, 584—585

Solar radiation
global and seasonal distribution of, (2): 498—502
in Pacific Ocean, (2): 403
spectral characteristics of, (2): 495

Sonneratiaceae, (2): 252

Sorensen pH scale, (2): 517

*Sorghum almum*, effect of $CO_2$ concentration on AP of, (2): 198

*Sorghum bicolor*, drought resistance of, (2): 380, 381

*Sorghum halepense*, response of AP to light intensity in, (2): 188

South America
forests in, (2): 553
irrigated land in, (2): 552
land resources of, (2): 551—552

South Carolina
freeze-free periods in, (2): 591
high temperatures reported for, (2): 600

South Dakota
freeze-free periods in, (2): 591
high temperatures reported for, (2): 600

Southern United States
land resource regions of, (2); 572
land utilizaton in, (2): 557

Soybean
$CO_2$ concentrations in, (2): 203
response to $CO_2$ and $O_2$ concentrations, (2): 205, 206
zinc concentrations in, (2): 219

Spanish moss, adaptation of, (1): 191

*Spartina*, primary production of, (2): 439

Spectral attenuation coefficients, (2): 507—508

Spectral irradiance, (2): 503—506

Spectroscopy EPR, (1): 137

Sphacelarales, taxonomic characteristics of, (1): 551

Sphaeropleales, taxonomic characteristics of, (1):

# U

# V